Asphaltenes, Heavy Oils, and Petroleomics

Asphaltenes, Heavy Oils, and Petroleomics

Edited by

OLIVER C. MULLINS
Scientific Advisor
Schlumberger-Doll Research

ERIC Y. SHEU
Chief Scientist
Vanton Research Laboratory, Inc.

AHMED HAMMAMI
New Venture Project Manager
Schlumberger Oilfield Services

and

ALAN G. MARSHALL
Robert O. Lawton
Professor of Chemistry & Biochemistry
Florida State University

Springer

Library of Congress Control Number: 2005939171

ISBN 10: 0-387-31734-1 Printed on acid-free paper.
ISBN 13: 978-0387-31734-2

9 8 7 6 5 4 3 2 1

springer.com

This book is dedicated to all those scientists and technologists who have and will become enthralled and enchanted by the wiles of the asphaltenes and heavy oils, and to the families and friends of our fold who at least feign enthusiasm when subjected to renderings of the mysterious objects of our study.

—OCM

Preface

This book represents an amalgam of objectives related to the study of petroleum at many, diverse levels. The most important attribute any thriving technical field must have is an injection and infusion of dedicated, expert, young scientists who have absorbed from their elders the fascination of scientific mystery coupled with the fundamental satisfaction of revelation and providing contribution. And, of course, these youthful practitioners must also learn to challenge the authority of their elders. From experiences with my own students, this seems not to be a problem. Many chapters in this book are coauthored by young scientists yielding the prognosis of continued health of our scientific field. Indeed, I am quite proud that several of my own chapters in this book are coauthored with students and young engineers of enormous capability. It is a humbling honor to help delineate direction of this formidable talent. It is incumbent upon my generation of scientists to provide a vision of the future. In this book, we connect the scientific excellence of the past with a vision for petroleum science, Petroleomics. Medical science of the past has been of singular societal focus with scientific discoveries of enormous import. Nevertheless, Genomics is revolutionary in that causal relations in medical science are being established with scientific exactitude and fundamental understanding. Genomics is creating a predictive medical science that was but a dream for previous generations. In a similar way, scientific advances described in this book are laying the foundations for Petroleomics—the challenge and framework to agitate our youthful contributors. Petroleomics embodies the establishment of structure—function relations in petroleum science with particular focus on asphaltenes, the most enigmatic of petroleum components. Correlative phenomenology is giving way to proper predictive science based in detailed petroleum chemical composition. This book describes the nascent development of the Petroleome, the complete listing of all components in a crude oil. As is shown herein, causal scientific relations in petroleum and asphaltene science are now being established that were merely plausible conjectures in the recent past.

This book also serves the purpose to reinforce the seemless continuity in petroleum science of basic scientific discovery with application of technology in a major and growing economic sphere. Longer standing concerns such as flow assurance are treated herein within a much more rigorous setting. In addition, very recent advances in the use of Downhole Fluid Analysis to address the most important issues in deepwater production of oil motivate renewed vigor in detailed chemical investigations in petroleum science. Oil operating companies and oil services companies are at the forefront of many of these technologic developments of enormous import. The economic impact of these new directions mandates development of exacting scientific underpinnings from leading universities and national facilities. Research dollars are too scarce and the technological challenges too great to employ research models of redundant effort in different institutions or of moving directionless unaware of impact. The new model promulgated in this book is to have cohesive collegial, international teams across corporate and university boundaries, across scientific and technological disciplines with research portfolios consisting of basic science and applied technology with a mix of near term and long term objectives. Certainly, internecine scientific battles will rage, and proprietary knowledge must be managed. (This book attempts to settle several of the most fierce, long-standing battles.) Nevertheless, this new research model delivers efficient use of expert human capital to address concerns of major scientific and economic impact. Life's experiences are greatly broadened by participation in such endeavors. As Chief Editor of this book, I have tried to reflect in this book the spirit of my own experiences of visiting six continents recently to grow our new business segment which I had the good fortune to initiate, to visit universities around the world, to interact with our field engineers, reservoir engineers, university professors and their students, male and female, of so many interesting cultures and nationalities. Science and technology are truly enriching for those lucky enough to participate.

Oliver C. Mullins

Contents

3. Petroleomics: Advanced Characterization of Petroleum-Derived Materials by Fourier Transform Ion Cyclotron Resonance Mass Spectrometry (FT-ICR MS)

Ryan P. Rodgers and Alan G. Marshall

4. Molecular Orbital Calculations and Optical Transitions of PAHs and Asphaltenes

Yosadara Ruiz-Morales

5. Carbon X-ray Raman Spectroscopy of PAHs and Asphaltenes

Uwe Bergmann and Oliver C. Mullins

6. Sulfur Chemical Moieties in Carbonaceous Materials

Sudipa Mitra-Kirtley and Oliver C. Mullins

7. Micellization

Stig E. Friberg

8. Insights into Molecular and Aggregate Structures of Asphaltenes Using HRTEM

Atul Sharma and Oliver C. Mullins

11. Molecular Composition and Dynamics of Oils from Diffusion Measurements

Denise E. Freed, Natalia V. Lisitza, Pabitra N. Sen, and Yi-Qiao Song

12. Application of the PC-SAFT Equation of State to Asphaltene Phase Behavior

P. David Ting, Doris L. Gonzalez, George J. Hirasaki, and Walter G. Chapman

13. Application of Isothermal Titration Calorimetry in the Investigation of Asphaltene Association

Daniel Merino-Garcia and Simon Ivar Andersen

14. Petroleomics and Characterization of Asphaltene Aggregates Using Small Angle Scattering

Eric Y. Sheu

15. Self-Assembly of Asphaltene Aggregates: Synchrotron, Simulation and Chemical Modeling Techniques Applied to Problems in the Structure and Reactivity of Asphaltenes

Russell R. Chianelli, Mohammed Siadati, Apurva Mehta, John Pople, Lante Carbognani Ortega, and Long Y. Chiang

16. Solubility of the Least-Soluble Asphaltenes
Jill S. Buckley, Jianxin Wang, and Jefferson L. Creek

17. Dynamic Light Scattering Monitoring of Asphaltene Aggregation in Crude Oils and Hydrocarbon Solutions
Igor K. Yudin and Mikhail A. Anisimov

18. Near Infrared Spectroscopy to Study Asphaltene Aggregation in Solvents
Kyeongseok Oh and Milind D. Deo

23. Precipitation and Deposition of Asphaltenes in Production Systems: A Flow Assurance Overview

Ahmed Hammami and John Ratulowski

Contributors

Simon Ivar Andersen
Professor of Chemical Engineering
Center for Phase Equilibria and Separation Processes
Department of Chemical Engineering, Building 229
Technical University of Denmark
DK-2800 Kgs. Lyngby
Denmark

Gaelle Andreatta
Schlumberger Doll Research
36 Old Quarry Road
Ridgefield, Connecticut 06877
United States

Mikhail A. Anisimov
Professor of Chemical Engineering and Institute for Physical Science and Technology
University of Maryland, College Park
Maryland 20742
United States

Uwe Bergmann
Stanford Synchrotron Radiation Laboratory
PO Box 20450, Stanford
California 94309
USA

Neil Bostrom
Schlumberger Doll Research
36 Old Quarry Road, Ridgefield
Connecticut 06877
United States

Jill S. Buckley
Petroleum Recovery Research Center
New Mexico Tech, Socorro,
New Mexico 87801
United States

Lante Carbognani Ortega
Consultant, Caracas, Venezuela;
Present address:
Department of Chemical and Petroleum Engineering
University of Calgary, 2500
University Drive NW, Calgary, AB, T2N 1N4
Canada

Walter G. Chapman
William W. Akers Chair in Chemical Engineering
Department of Chemical Engineering
Rice University, Houston, Texas-77005
United States

Russell R. Chianelli
Professor of Chemistry, Materials and Environmental Science and Engineering
Director of the Materials Research and Technology Institute
University of Texas, El Paso, Burges 300, El Paso, Texas, 79968
United States

Long Y. Chiang
Professor of Chemistry
University of Massachusetts
Lowell, Massachusetts 01850
United States

Jefferson L. Creek
Chevron Energy Technology Company
Flow Assurance Team, 1500 Louisiana St.
Houston, Texas 77002
United States

Tadeusz Dabros
CANMET Energy Technology Centre
Natural Resources Canada
1 Oil Patch Drive, Devon, Alberta T9G 1A8
Canada

Milind D. Deo
Professor of Chemical Engineering and Director of Petroleum Research Center
University of Utah, 50 S Central Campus Drive
Salt Lake City, Utah 84112
United States

Denise E. Freed
Schlumberger Doll Research
36 Old Quarry Road
Ridgefield, Connecticut 06877
United States

Stig E. Friberg
Visiting Scientist
Chemistry Department

University of Virginia
Charlottesville, Virginia 22903
United States

Go Fujisawa
Schlumberger K.K.
2-2-1 Fuchinobe, Sagamihara-shi
Kanagawa-ken, 229-0006
Japan

Doris L. Gonzalez
Department of Chemical Engineering
Rice University, Houston, Texas-77005
United States

Henning Groenzin
Schlumberger-Doll Research
36 Old Quarry Road
Ridgefield, Connecticut 06877
United States

Ahmed Hammami
Schlumberger Oilfield Services
Edmonton, Alberta, T6N 1M9
Canada

Hassan Hamza
CANMET Energy Technology Center
Natural Resources Canada
1 Oil Patch Drive, Devon, Alberta T9G 1A8
Canada

Pål V. Hemmingsen
Norwegian University of Science and Technology (NTNU)
Ugelstad Laboratory, Department of Chemical Engineering
Trondheim N-7491
Norway

George J. Hirasaki
A. J. Hartsook Professor in Chemical Engineering
Rice University
Houston, Texas-77005
United States

Harald Kallevik
Statoil R&D Center, Rotvoll
Trondheim N-7005
Norway

Natalia V. Lisitza
Schlumberger-Doll Research
36 Old Quarry Road
Ridgefield, Connecticut 06877
United States

Yicheng Long
CANMET Energy Technology Centre
Natural Resources Canada,
1 Oil Patch Drive
Devon, Alberta T9G 1A8
Canada

Alan G. Marshall
Robert O. Lawton Professor of Chemistry and Biochemistry
Director, Ion Cyclotron Resonance Program
National High Magnetic Field Laboratory
Florida State University
1800 East Paul Dirac Drive
Tallahassee, FL 32310-4005
United States

Apurva Mehta
Stanford Synchrotron Radiation Laboratory SSRL/SLAC
2575 Sand Hill Road, MS 69, Menlo Park
California, 94025

Daniel Merino-Garcia
Consultant, Pedro Barruecos 2 4C
47002 Valladolid
Spain

Sudipa Mitra-Kirtley
Professor, Physics and Optical Engineering
Rose-Hulman Institute of Technology
Terre Haute, Indiana 47803
United States

Oliver C. Mullins
Scientific Advisor
Schlumberger-Doll Research
36 Old Quarry Road
Ridgefield, Connecticut 06877
United States

Kyeongseok Oh
Department of Chemical Engineering
University of Utah
50 S Central Campus Drive
Salt Lake City, Utah 84112
United States

John Pople
Stanford Synchrotron Radiation Laboratory SSRL/SLAC
2575 Sand Hill Road, MS 69, Menlo Park
Calfifornia, 94025

John Ratulowski
Schlumberger Well Completion and Productivity
Subsea-Flow Assurance
14910 Airline Rd. Bldg. 20
Rosharon, Texas, 77583
United States

Ryan P. Rodgers
Director of Environmental and Petrochemical Applications
FT-1CR Mass Spectrometry Facility
National High Magnetic Field Laboratory
Florida State University
1800 East Paul Dirac Drive
Tallahassee, FL 32310-4005
United States

Yosadara Ruiz-Morales
Programa de Ingeniería Molecular
Instituto Mexicano del Petróleo
Eje Central Lázaro Cárdenas 152
México, DF 07730
México

Pabitra N. Sen
Scientific Advisor
Schlumberger-Doll Research
36 Old Quarry Road
Ridgefield, Connecticut 06877
United States

Atul Sharma
Advanced Fuel Group
Energy Technology Research Institute
National Institute of Advanced Industrial Science and Technology
16-1 Onogawa, Tuskuba 305 8569, Ibaraki
Japan

John M. Shaw
Professor and NSERC Industrial Research Chair in Petroleum Thermodynamics
Department of Chemical and Materials Engineering
Chemical Materials Engineering Building
University of Alberta
Edmonton, Alberta T6G 2G6
Canada

Eric Y. Sheu
Vanton Research Laboratory, Inc.
7 Old Creek Place
Lafayette, California 94549
United States

Mohammed Siadati
Materials Research and Technology Institute
University of Texas
El Paso, Texas
United States

Johan Sjöblom
Professor in Chemical Engineering and Head of the Ugelstad Laboratory
Norwegian University of Science and Technology (NTNU)
Ugelstad Laboratory
N-7491 Trondheim
Norway

Yi-Qiao Song
Schlumberger-Doll Research
36 Old Quarry Road
Ridgefield, Connecticut 06877
United States

P. David Ting
Shell Global Solutions (US)
Westhollow Technology Center
Houston, Texas 77082
United States

Jianxin Wang
Petroleum Recovery Research Center
New Mexico Tech, Socorro
New Mexico 87801
United States

Igor K. Yudin
Oil and Gas Research Institute
Russian Academy of Sciences
Moscow 117971
Russia

Xiangyang Zou
Oilphase-DBR, Schlumberger, 9419-20th Avenue
Edmonton, Alberta T6N 1E5
Canada

1

Petroleomics and Structure–Function Relations of Crude Oils and Asphaltenes

Oliver C. Mullins

1. Introduction

Petroleum science and technology are advancing at a rapid pace due to a myriad of considerations. The efficient generation and utilization of energy are increasingly being recognized as a societal necessity from economic and environmental vantages. Increasing concerns regarding physical limits of total hydrocarbon resources are colliding with rapidly expanding economies in heavily populated regions of the world, that require plentiful, affordable transportation fuels to realize expectations of impatient populaces. Geopolitical instabilities are magnified by disparate distributions of hydrocarbons attracting attention of powerful hydrocarbon consuming nations commensurate with the perceived value of these resources. Exploitation of hydrocarbon resources in many cases is the best hope for lifting nations out of grinding poverty. However, in large measure, the "easy" hydrocarbon resources have already been drained, increasing the technical demand for exploitation of the remainder. Heavy oils and bitumens that were bypassed in favor of their lighter bedfellows constitute an increasing fraction of remaining hydrocarbon resources. Deepwater production of hydrocarbon resources involves tremendous costs, thereby mandating efficiencies that can be achieved only with proper understanding of petroleum chemistry. Exploitation of marginal reserves in mature markets rich in infrastructure, such as the North Sea, hinges on accurate prediction of production. The insightful characterization of reservoir architecture and of reservoir dynamics, very challenging tasks, rests in large part on the detailed understanding of the contained fluids.

The confluence of these diverse considerations has created a welcome challenge amongst those scientists and technologists who find crude oils and asphaltenes worthy subjects of study. At the same time, investigative methods are inexorably improving; new technology, greater sensitivity, higher resolution coupled with improved theoretical modeling and simplifying formalisms more clearly

Oliver C. Mullins • Scientific Advisor, Schlumberger-Doll Research, Ridgefield, CT 06877

rooted in physical foundation are providing the scientist sharper, more powerful tools to prod, probe, inspect, and interrogate the carbonaceous materials of our concern. The petroleum technical community has been galvanized applying sophisticated new techniques and advanced application of mature methods; this focus is bearing fruit in all areas of petroleum science and technology. The most enigmatic component of crude oil, the asphaltenes are finally revealing their secrets; in particular, basic asphaltene molecular structure is now understood, an absolute necessity for development of predictive petroleum science. Simplifying governing principles of asphaltenes are being uncovered enabling development of structure–function relationships, one of the pillars of Petroleomics. Connection of molecular scale knowledge of asphaltenes is helping to provide the basis of the phase behavior of asphaltenes at the different length scales, thus vertically integrating diverse studies. Petroleomics, the establishment of structure–function relations for asphaltenes and crude oils, is being implemented. New mass spectral and other analytic techniques are of sufficient resolution that generation of the petroleome is in sight, the complete listing of every component even for heavy crude oil. For the first time, asphaltene science and petroleum science are poised to join the pantheon of scientific disciplines sufficiently developed that new phenomena can be treated within a framework of first principles. It is an exciting time to be involved in the study of asphaltenes and crude oils.

"If you want to understand function, study structure" advises Francis Crick.[1] To perform proper predictive science, the structure of the system under study must be known. This necessary step allows structure–function relations to be established. Further study then reveals detailed mechanistic processes and identifies broad, underlying governing principles. In a perfect scientific world, structure can be determined and these investigative precepts are followed without interruption. Results are questioned, but not the process. Consider the evolution of the understanding of a rather important liquid other than petroleum(!). Water has played a central role in all aspects of life since life started on the planet. It is certainly true that the use of water by sentient beings greatly preceded the understanding of this life enabling substance. Nevertheless, the concept of understanding and explaining properties of water is unimaginable without knowing its molecular structure and its intermolecular interactions. The water molecule is a bent triatomic with D2h symmetry. The oxygen in water is sp^3 hybridized and has two lone electron pairs; as such the H-O-H bond angle is close to that expected for a tetrahedron, 109.5° but due to the increased repulsion of the unshared nonbonding electrons, the bond angle of water is 105.5°. The large electronegativity contrast of constituent water elements creates a large dipole moment and large dielectric constant of the bulk enabling water to dissolve a large number of ionic compounds. The lone pairs of electrons can engage in hydrogen bonding giving water an unusually high boiling point for a molecule of 18 amu, contrasted by methane and ethane for example. The very directional hydrogen bond structure in the solid (ice I) causes the lattice to open up, thereby creating a lower density of the solid than the liquid. Knowing the structure does not imply that the understanding all properties of water follows immediately. In fact, recent results are changing the understanding of the extent of H-bonding per molecule in liquid water.[2] Petroleum chemists are forgiven for not "solving" the multicomponent, complex object of their study since pure liquid water still retains controversy. It is important to recognize that asphaltene-rich materials, such as bitumen, are perhaps best described as composites. Composites such as bone, steel, and wood possess properties that are defined by the integration of their constituents.[3] Certain crude oils share this trait. Nevertheless, in the case of water, and every other substance, pure or otherwise, it is of paramount importance to realize function follows structure.

System complexity generally retards predictive science and of course the platitude "necessity is the mother of invention" continues to prevail. Advances in materials that portend the greatest distinctions from previous human eras identify archeological ages. The stone age, the bronze age, and the iron age all corresponded to fundamental advances in the mastery of the natural world, and always preceded detailed structural understanding. While samurai sword makers followed a ritualistic process to create the world's best blades; the explanation of this process and of the metallurgy of steel followed much later.[3] Rubber was utilized long before polymer science matriculated to an academic discipline. Superconductivity was discovered long before it was understood at a fundamental level. Many advances proceed with an intriguing mix of some predictive conceptualization coupled with indefatigable Edisonian searches. In such cases, structure is not known a priori. History has taught that alert, perceptive minds can recognize patterns that yield valuable advances, even without knowing basic structure. There may even be a natural human aversion to alter processes known to yield phenomenological successes; we may all have a little of the samurai sword makers in us. Nevertheless, to *understand* function, structure simply must be known.

The endeavor of human medicine is exquisitely enshrouded in phenomenology. The subject is too important and the complexity too great to wait for scientific validation. Shamans embodied some of the earliest approaches to medicine mixing mysticism with natural curative agents perceptively discovered. Of course, medical science has made tremendous advances through the ages. Still much of the methodology has remained unchanged. The small pox vaccine developed by Edward Jenner rested upon the astute observation by that milk maidens (thus exposed to cow pox) did not develop small pox. Countless serendipitous advances in medical science have similarly occurred. Nevertheless, in many ways medicine is practiced by responding to symptoms. We collectively are individually in the wait-and-see mode regarding our health. It is true that diagnostic medical science continues to improve and will continue to be exploited in ever expanding ways. However, this approach is fundamentally flawed; the disease must develop to be detected. It is greatly preferred to predict and treat disease *prior* to the development of symptoms. Early detection of symptoms requires repeated, sensitive, thus costly testing; without prediction, the diagnostic search is not directed. But repeated Edisonian searches cannot be sensitive and cost effective. The deficiency of predictive medical science is not due to the lack of focus. Any physical scientist trying to acquire funding is well aware of the behemoth engine of medical research which must be sated first. And as a scientist who studies asphaltenes, it is hard for this author to argue against this priority. Beepers are not the norm for asphaltene emergencies. Of course, asphaltene science does directly impact the oil business, which is not inconsiderable. The biggest impediment to predictive medical science has been the lack of understanding structure, known to Crick when he expounded the guiding principle cited above.

Millennia after humans initiated medical science, Watson, Crick, Franklin, and Wilkins discovered the structure of the alphabet of human life in 1953. It took 50 years, but/and in 2003 the book of human life, the human genome has been read. This event is a turning point in human history—but there was some disappointment accompanying this great achievement. It was known that the *C. elegans* roundworm (a popular subject of study) has ~19,000 gene. Naturally, speculation was rife that we humans, so much better than the roundworm, must have perhaps 100,000 genes or more. (Some limits of human DNA were known at that time, or undoubtedly the estimates would have been much higher.) Well, humans only have about 30,000 genes. Now we are using this modest excess of our genes versus the roundworm in an exponent or as a factorial where it would clearly show our superiority again. Tautology notwithstanding, reading the book of human life is a monumental achievement.

Now that the structure of the human genome is known, structure–function relations can finally be established in medicine. Deleterious genes are being uncovered that relate to a variety of medical problems; major public health issues are being addressed. For instance, an article in the *New England Journal of Medicine*[4] (and on the front page of the *New York Times*) that a particular variant of a gene is associated with a factor of five increased risk of congestive heart failure. In the United States there are more people hospitalized with congestive heart failure than all cancers combined, thus is of enormous public policy concern. The initial application of genomics may be screening for particular deleterious genes for congestive heart failure, for stroke, for specific cancers. For those with the offending genes, specific sensitive diagnostic analyses can be performed searching for the corresponding symptoms, controlling costs while being sensitive.

In the longer term, genomics promises to change the way medical science is practiced. By knowing the deleterious genes, the hope and expectation is that one will know the proteins encoded by the normal and defective genes; one will know the biomedical pathways involving these proteins. One will know precisely the impact of the deleterious gene. Effective treatments can then be developed for those who possess the deleterious genes. In the future, the medical community will read *your* genome. (But the reader may have to live a considerable while for this to come to fruition.) A bar chart will be generated for the probability of your developing specific maladies. If the probability of a specific ailment is high, the treatment for this problem can be launched. One can treat the disease prior to the development of symptoms. In this way, genomics will revolutionize medicine. The absolute foundation and requirement for genomics are knowing the structure of DNA and reading the human genome. Without this structural foundation, we would revert back to phenomenology, the analysis of symptoms, as the predictive approach would be precluded.

In addition to improving the direct application of medical science, genomics has enormous public policy implications as well. It is known that black Americans have a congestive heart failure rate a factor of five greater than white Americans. Had one been asked to identify likely causality for this observation prior to the discovery of the deleterious gene for congestive heart failure, factors including socioeconomic differences, access to health care, and a myriad of other plausible origins would be listed. Solutions to problems of congestive heart failure in the black American community would then be based on these "likely" candidates. These solutions, ignoring the importance of genetics, would have little or no impact on the rate of congestive heart in the black American community. Understanding the importance of the genetics is critical to understanding the origins of congestive heart failure and developing the proper remedies. The origins of congestive heart failure in black and white Americans are linked in large measure to our genes.[4] Expenditure of public funds in the United States to address these genetic origins and corresponding curative measures is in fact unifying and effective for the population at large. One may also wish to address racial imbalances regarding access to and exploitation of societal resources; however, inaccurate identification of causality leads to ineffective and wasteful "solutions", engendering division and reduced allocation of resources.

There is always concern that application of first principles to complex systems may fail; the less adventurous path is to default to phenomenology when the complexity is perceived too formidable. One does not need an acute acoustic sense to hear such foreboding expressed about petroleum. One might choose a bold path. It is known that a broad array of factors have helped shaped human development including the shapes of continents and variations in natural flora and fuana.[5] Nevertheless, E.O. Wilson makes a strong case that various elements of human behavior, with its extreme complexity, can be understood from a genomics vantage.[6] A forceful point is that social scientists neglect genetics to their considerable detriment. For instance, Wilson describes in detail the Westermarck effect, named after a Finish anthropologist. The effect is simply that inbreeding amongst human siblings and between parents and children is very uncommon. Indeed, human societies envelop close kin mating in taboo. The Westermarck effect has been observed not only in most human societies but all primates studied.[6] A plausible cause for this effect is the documented destructive concentration of double recessive, deleterious genes with inbreeding. The suggestion is that the Westermarck effect is controlled in part by genetic impulse. However, note that major components of Freud's Oedipal complex run counter to the Westermark effect. At the least, plausible genetic influences on human behavior should be understood by social scientists in their endeavors. It behooves all scientists to understand the foundations to locate and decipher phenomenology.

2. Evolution of the Oil Patch

As currently practiced, petroleum science shares many traits with medical science. The analysis of crude oil for issues of economic concern is often rooted in phenomenology. For instance, in the upstream side of the petroleum business, crude oil phase transitions can be quite problematic. Figure 1.1 shows several

Figure 1.1. Various solids that obstruct oil pipelines.

solid phases that can form during the production of crude oil; all but one directly involve hydrocarbons. These phase transitions of crude oil include the formation of solid deposits of asphaltene, wax, gas hydrate, organic scale, and diamondoids, possibly in combination. The appearance of organic scale accurately reflects what production engineers think of it. For completeness, an inorganic scale is also shown.

The crude oil chemistry involving the formation of a solid precipitant or flocculant is complex. The factors that determine whether a newly precipitated solid phase actually forms a deposit which then grows and occludes tubulars, pipelines and production facilities involve not only the oil chemistry but are compounded by interfacial interactions of the organics with oil, water, gas, mineral, and metal surfaces, altered by natural corrosive and erosive interactions. As with biological systems, the complexities are significant, but not preclusive. As with medical science, the petroleum industry has had to develop operational solutions to the problems displayed in Figure 1.1 prior to development of proper scientific description of the problems; the approach has largely been phenomenological. "Does a crude oil have a wax problem?" stick it in the refrigerator and see if wax forms. "Does the live oil have an asphaltene deposition problem?" drop the pressure on the live oil and see if asphaltenes precipitate. Flocculation or asphaltene destabilization is a necessary but not sufficient condition for the formation of deposits. It is much harder to determine if deposits form under high shear and realistic conditions (cf. Chapter 23). Thus fairly basic and phenomenological methods have been employed to uncover problems associated with oil chemistry.

Petroleum science mandates establishing the first principles that govern the behavior of crude oil in all of its sundry manifestations. Utilizing a complete chemical description of crude oil to predict all properties is the ultimate objective of

Petroleomics. The Petroelome, the complete listing of all chemical constituents in a crude oil thus enables Petroleomics. Phase behavior (cf. Fig. 1.1), interfacial activity, viscoelasticity, and solubility, which is the defining characteristic of asphaltenes, are subsumed within this overarching agenda. Molecular structure of crude oil components and especially of their enigmatic constituents asphaltenes must be understood as the root source of all that follows. In addition, crude oils and asphaltenes exhibit hierarchical aggregation behavior in different physical length scales; for corresponding accurate characterization, petroleum science mandates establishment of causal relations between different hierarchical regimes. In the broadest sense, structure–function relations must be developed providing vertical integration of this hierarchy. Ultimately, petroleum science rests upon developing the complete listing of every component in a crude oil. Analogous to the genome, the complete representation of petroleum provides a clear and only path toward establishment of all structure–function relations in crude oil. In practice, it might be sufficient to determine the elemental composition of each component in a crude oil concatenated with bulk structural determination for the whole crude or important bulk fractions. Nevertheless, the objectives remain—full resolution of crude oil chemical constituents and full determination of structure-function relations in all crude oil hierarchies.

3. Phenomological Petroleum Analysis

The phenomenological approach to the analysis of oil chemistry issues has served the petroleum industry reasonably well for many years, but the efficacy of this approach has deteriorated substantially in recent years due to the dramatic changes in the petroleum market. According to the Minerals Management Service, the arm of the United States Government, which oversees oil production offshore, many experts believed as late as 1990 that formations in deepwater environments would contain no oil of economic value. Since that time, intrepid oil operating companies moved off the continental shelf and continued to find oil in deeper water. Either we have had very recent reservoir charging, or many experts were in error! The understanding of turbidity currents resulting in turbidites in river-fed marine basins has helped explain large discoveries in deepwater. Deepwater is now recognized as a global play and includes deepwater basins corresponding the Mississippi River, the Niger River, the Congo River, the Nile River, the Paraiba River, the Mahakam River. Other high cost markets such as the North Sea and offshore eastern Canada have also contributed substantially to the changing the oil market. Some estimates conclude that 50% of the world's undiscovered oil is offshore. A sea change has taken place with regard to the location of new oil.

In addition to Flow Assurance issues, the efficient production of oil is now known to depend critically on petroleum analysis, but within an entirely new context (cf. Chapter 22), thereby providing new opportunities for scientific and technological contributions. The oil industry operating practices have routinely incorporated two large physics errors in reservoir exploitation. In spite of the concerns from knowledgeable technologists, the operations side of the oil industry has often been forced, not unreluctantly, to presume the most optimistic scenario for the production of crude oil. The erstwhile default scenario is that, unless proven otherwise, oil fields were considered to consist of giant tanks of homogeneous hydrocarbons. Of course, gas caps, oil columns and the occasional tar mat were recognized, as was gross compartmentalization. Nevertheless, the industry defaulted to an overly optimistic scenario for several reasons. First, there had been no cost effective means of acquiring accurate information on fluid compositional variation, and on compartmentalization prior to production. (A compartment is defined as a single flow unit that must be penetrated by a well to be drained.) Second, the identification of either fluid compositional variation or compartmentalization is "bad news", decreasing reserves and increasing costs. It is difficult to justify inclusion of costly complexity without the existence of corresponding established procedures for data acquisition and analysis.

The use of these reservoir descriptions, optimistic to a fault, has led to the commonplace occurrence that the prediction of production and the actual production are rarely in agreement, often with regard to both the quantity and type of fluids produced. In a low cost environment, one can tolerate large initial errors in prediction by updating prediction as more wells are drilled and put into production. It is illustrative to consider that the cost structure in the land production of crude oil is commensurate with the existence of many small oil companies. A relatively small amount of capital is needed to explore and, with luck produce oil. But beware, as the principal owner of the Harvard oil company told this author, "the oil business is not for the weak hearted".

However, in high cost markets such as deepwater, prediction of production is of paramount importance. Entire production projects must be forward modeled to justify requisite billion dollar sea floor installations. In this environment, errors in prediction have cost operating companies billions of dollars in individual fields. It is no longer tolerable nor economically viable in the oil industry to sustain enormous errors in prediction built on frequently invalidated optimism. The relatively recent arrival of deepwater has altered the landscape; proper technical solutions are now mandated. In fact, this represents a new, huge opportunity to hydrocarbon fluid experts around the world.

There is a dramatic revision in thinking taking place regarding the understanding of the distribution of hydrocarbons in subsurface formations. This revision is in fact for operating units. The technologists have been aware of the following issues; however, previously there had been no cost effective method to acquire requisite data prior to development of production facilities and strategies. There are two components to this dramatic revision in thinking; (1) hydrocarbon compositional grading and (2) compartmentalization. In the past, the normal presumption was that the hydrocarbons are present in the subsurface formations as a homogeneous fluid. That is, it was presumed that there was no spatial variation in hydrocarbon properties. Ironically, in the oil business, the formation rocks have been given due respect. It is recognized that rock mineralogy and petrophysical properties can easily change, laterally and vertically, on a centimeter length scale or less. Rock variations could include a change in mineralogy such as going from shale to sandstone, a change in cementation, grain size and/or shape, changes in clay content etc. But the liquid oil columns were presumed to be invariant unless otherwise proven. It turns out that the hydrocarbons are frequently highly graded compositionally in the subsurface formations. The new view is that "hydrocarbons in the formation are considered compositionally graded unless otherwise proven."[7] Contributing factors include gravity, thermal gradients, multiple reservoir charging, current reservoir charging, leaky seals possibly pressure dependent, biodegradation, water washing, and reservoir alteration during charging. All but the first two factors move the hydrocarbon column away from equilibrium.

A second component in understanding complexities of hydrocarbon fluids in the formation relate to compartmentalization. In the deepwater arena, it is very difficult to determine compartment size. Traditional methods of finding compartment size such as well testing (essentially a production test) are often precluded due to cost. A well test can cost nearly what a new well would cost in deepwater. Consequently, this expensive solution is not performed on a routine basis. For many years, the primary method used to find compartment size had been to determine hydraulic (pressure) communication. In a well, pressure communication is established by obtaining a single pressure gradient at different points in the fluid column. Pressure communication was then presumed to imply flow communication. However, pressure communication in geologic time is a necessary but insufficient condition to establish flow communication in a production time frame. Geologic to production time differs by 6 orders of magnitude; requisite permeabilities for flow versus pressure communication differ by several orders of magnitude. Thus, the standard industry method for identification of compartments is in error by up to 9 orders of magnitude. Given this gross technical failure to identify compartments, it is no wonder that compartmentalization is generally viewed as public enemy number one in the oil industry today, at least for deepwater production. For a technologist, discovery of such a gargantuan disconnect in the application of technology is fertile ground for revolutionary innovation.

Downhole Fluid Analysis (DFA) is a new technology that is enabling cost effective identification of fluid compositional variation and of compartmentalization. DFA (Chapter 22) enables important and different fluids to be identified at the point of sample acquisition in the subsurface. Thus, DFA is aiding the laboratories to get a proper representative sampling of the variation of fluids in the formation. Without DFA, requisite random sample acquisition and analysis had been too expensive to employ on a routine basis. In addition, DFA is identifying compartmentalization by virtue of identifying fluid density inversions in the hydrocarbon column.[7] That is, DFA is routinely identifying higher density fluids higher in the column. In general, the most likely explanation for such an occurrence is compartmentalization. This new technical solution to some of the industry's most important problems directly involves fluid complexities and places a new focus on understanding petroleum. It is important for the academic community that has a strong focus on fluids (e.g., all academic contributing authors in this book!) to understand this new use of fluid analysis to address the largest problems in the oil business.

4. Petroleomics

Again, we consider Francis Crick's axiom, "If you want to understand function, study structure." For the first time, the basic structural issues of asphaltene science are sufficiently well developed that Crick's axiom has become an achievable goal. It behooves the asphaltene scientist to place his/her own results within the context of structural information at adjacent length scales. In the past, the asphaltene literature had been rather contradictory. Consequently, structure–function relations had been largely precluded since the foundations were so uncertain. Often, measurements at a particular length scale were extrapolated to other length scales without regard to direct measurements from other laboratories at that length scale. A cynical characterization of this approach might be "if I didn't measure it, it doesn't exist." However, asphaltene science is too complex for a single laboratory to measure everything there is to know. This difficulty has been exacerbated by the existence of simple, low cost measurements that consistently generate the wrong answer. Improper asphaltene molecular weight determination via vapor pressure osmometry comes to mind. As this book demonstrates, there is now considerable consistency regarding the resolution of fundamental issues in asphaltene and petroleum science.

5. Building Up Petroleum Science—A Brief Outline

Low molecular weight components are treated within a proper chemical framework. For instance, if a subsurface hydrocarbon reservoir contains H_2S, all aspects of resource utilization will incorporate treatment of this pernicious chemical component. However, the fundamental chemical description of the most enigmatic components of crude oil, asphaltenes, has been the subject of debate for decades. The most fundamental question of any chemical compound, its elemental constituents, is easily determined for asphaltenes and agreement prevails here. Within this agreement, one never hears that the polydispersity of asphaltenes precludes determination of their elemental composition. The second most basic property of a chemical compound, its molecular weight, has been the subject of dispute by one or more orders of magnitude in asphaltene science for decades. It turns out that for molecules, size counts. This is also true for quantum mechanics, and bank accounts so the importance of size for asphaltene molecules should not be a surprise. In large measure, the debate regarding asphaltene molecular weight reduces to the question whether asphaltenes are monomeric or polymeric. Clearly, asphaltenes are polydisperse so there will be a molecular weight distribution with its various moments. It is important to understand not only the mean asphaltene molecular weight, but also the width of the distribution, and the (asymmetric) tails on the small and large mass sides. Nevertheless, the debate on asphaltene molecular weight has been one to several orders of magnitude, so resolving the mean is the first important task. More specifically, the asphaltenes are known to be interfacially active. Any question involving interfacial science of crude oils is likely to have a component, potentially critical, involving asphaltenes. Issues such as emulsion stability, deposition, and wettability all involve interfaces. Prediction of asphaltene phase behavior clearly necessitates proper understanding of asphaltenes at the molecular level. We believe chapters herein (Chapters 2 and 3) present compelling evidence that this longstanding controversy is resolved, asphaltenes are small molecules.

After molecular weight, the next question is to understand asphaltene molecular structure. There has been some convergence on this topic. Here it is important to acknowledge polydispersity at the outset. The chemistry of interest for a particular observable might be dominated by a component of the asphaltenes that is present in small mass fraction. While it is unlikely that this would prevail in the formation of asphaltene nanoaggregates, this situation plausibly applies in at least some cases of interfacial interactions. Nevertheless, in high concentrations, the highest energy asphaltene sites might be tightly complexed and thus unavailable for facile interfacial access. Regarding molecular structure, the asphaltene molecular weights are not high. This fortuitous circumstance limits possible candidate structures. A polymeric structure consisting of covalent linkages with many large fused aromatic ring systems is incompatible with measured asphaltene molecular weights. An issue of primary concern is the size of the average aromatic fused ring system in asphaltenes. There is convergence from several lines of investigation. Asphaltenes are deeply colored in the visible and extending into the near infrared spectral range. Small aromatic ring systems, even those containing heteroatoms, are nonabsorptive or of very low absorptivity in the visible (e.g. benzene, naphthalene, anthracene, dibenzothiophene, dibenzopyrrole, pyrene, phenanthrene, etc.). The smallest fused ring systems that are optically absortive such as pentacene are catacondensed while x-ray raman spectroscopy (Chapter 5) as well as energetic considerations (Chapter 4) clearly show that asphaltenes are pericondensed. Consequently, what one sees visually is evidence that asphaltene ring systems contain more than a few rings. Detailed molecular orbital calculations (Chapter 4) coupled with detailed optical studies confirm intuition. Direct molecular imaging studies of asphaltenes indicate the asphaltene ring systems contain on order 7 fused rings (Chapter 8). Measurement of rotational diffusion of asphaltene molecules is consistent with this mean number with a width of roughly 4 to 10 rings (Chapter 2). 13C NMR studies also indicate a ratio of interior to exterior carbon that is consistent with this assessment. Known asphaltene molecular weights coupled with these determinations of fused ring systems leads to the conclusion that generally asphaltene molecules are shaped "like your hand" with the palm representing the single aromatic fused ring system in the molecule (with possible alicyclic substituents) and the fingers, alkane substituents. This description is consistent with the very definition of asphaltenes. Aspahltenes are defined by a solubility classification. The intermolecular attraction of the polarizable π-bond ring systems is counterbalanced by steric repulsions of alkane substituents. Thus, asphaltenes exhibit a strong correlation between the size of their fused ring systems and the extent of alkyl substitution. Asphaltene sulfur and nitrogen chemistry have been elucidated by x-ray spectroscopy methods (Chapter 6).

Asphaltene molecules aggregate at low concentrations, for instance at ~150 mg per liter in toluene, to form nanoaggregates (Chapters 9, 10, and 11).

Plausibly the governing physics is that the nanoaggregates grow until steric hindrance from the alkane and alicyclic substituents impedes further close approach of fused aromatic portions of molecules in the aggregate. At this point growth of this aggregate terminates and new nanoaggregates grow upon increasing concentration. The relation of the aggregates to standard micelles is explored by careful consideration of the respective governing physics (Chapter 7). Small angle neutron scattering and small angle x-ray scattering clearly show a fundamental length scale is observed in asphaltenes, the radius of gyration is a few nanometers (Chapter 14). Most importantly, x-ray scattering data shows that these results apply to crude oils, not just to isolated asphaltenes (Chapter 15). These rather tightly-bound but perhaps somewhat open aggregates then undergo higher order clustering at longer length scales. Neutron and x-ray scattering exhibit a variety of higher length scales (Chapter 14). The energetics involved in aggregation and clustering have been directly measured by microcalorimetry (Chapter 13). In addition, these studies point out that water may play an important role in asphaltene aggregation. Water is always present in the natural crude oil systems; this provides insight into the relation of asphaltene in toluene versus asphaltenes in crude oil. The fundamental importance of van der Waals interactions has been established by experiment and applied theory in the formation of asphaltene flocs (Chapter 16). Remarkably, this result fits within the framework of the governing chemical principles of asphaltenes identified at the molecular length scale (Chapter 2). Master equations are found to treat enormous volumes of dynamic light scattering data thereby identifying the underlying physics (Chapter 17). In particular, the important change in aggregation kinetics indicates that the fundamental nature of flocculation changes at the concentration of several grams asphaltene per liter implying clustering of nanoaggregates at this concentration (Chapter 17). Near-infrared studies of asphaltene flocculation corroborate this concentration dependent transition (Chapter 18). In addition, the applicability of SAFT modeling for measured asphaltene phase behavior also is consistent within this picture (Chapter 12). The predictive success of the SAFT modeling regarding properties of asphaltene phase behavior encourages yet broader approaches (Chapter 12). The overall phase behavior of carbonaceous systems can be very complex, with up to four thermodynamically stable phases. X-ray transmission measurements are best suited for these measurements (Chapter 19). Understanding the possible phase behavior complexities of hydrocarbons is vital and has been underappreciated in the past (Chapter 19). Many of these complexities are now being observed in subsurface formations and have an inordinate impact on production. Control of the phase behavior of bitumen can lead to substantial increases in efficiencies in resource utilization (Chapter 20). The increase in heavy oil and bitumen utilization mandates progressive thinking identifying new, cost effective processing methods. The oil–water emulsion characteristics of asphaltenic oils is an especially important topic which involves emulsion stabilization by a variety of complex interfacial interactions (Chapters 20 and 21).

Treatment of proper live crude oil samples starts first with the acquisition of proper representative samples (Chapters 22 and 23). In addition, the recent development of DFA has shown that fluid analysis can be used in an efficient manner to address some of the most important difficulties in the production of crude oil (Chapter 22). Deposition of asphaltene from these live crude oils in realistic flow conditions has been most problematic to recreate in the laboratory. Chapter 23 describes the latest solution to this problem. Chapter 23 also delineates the many Flow Assurance issues that impact oil production.

6. Asphaltenes: An Update of the Yen Model

Asphaltenes are the most enigmatic component of crude oil and as such are of special concern when attempting to characterize the chemistry of crude oils. Professor Teh Fu Yen proposed a hierarchy of structures within heavy crude oil, asphalt and asphaltene.[8] He employed the term micelle to describe the small stacks of fused aromatic ring systems of asphaltene molecules. These micelles were able to grow to a small limiting size. He then proposed that these asphaltene micelles can cluster into aggregates when the concentration is sufficiently high. Various types of structures were suggested for the aggregates. This hierarchical structure of asphaltenes has been termed the Yen model.

This book presents considerable evidence that the hierarchical structures for asphaltenes are indeed correct. The concentration for primary aggregation of asphaltenes in toluene is now known to be rather low. Furthermore, this book essentially resolves that asphaltenes are monomeric species not polymeric and that for the most part asphaltenes contain one binding site per molecule. These concepts have been developed subsequent to the Yen model and place restrictions on the Yen model and yet expand the applicability of this model. In particular, the dynamics of asphaltene solutions at low concentrations are explained by the additional constraints of small molecular size for asphaltenes. As established herein, dilute toluene solutions of asphaltenes exhibit nanoaggregate formation at $\sim$150 mg/liter. If asphaltene molecules were large with many binding sites, then single molecules would participate in multiple nanoaggregates. In other words, the nanoaggregates would be covalently linked to each other. Thus, upon nanoaggregate formation, the asphaltenes would form a gel. This is counter to observation, for example, as presented herein. Instead, asphaltene nanoaggregates form at low concentration. Upon increasing the concentration more than 10 times, clustering commences. Each asphaltene molecule participates in a single nanoaggregate. The binding is somewhat high with favorable van der Waals interactions of geometrically positioned ring systems. After several molecules are in the nanoaggregate, steric hindrance precludes further molecular addition. At much higher concentrations the nanoaggreates cluster—but with much weaker binding (thus necessitating higher concentrations) due to excessive steric hindrance. By understanding the molecular structure as well as the predominant intermolecular interactions as developed within, the Yen model can be extended to dilute solutions of asphaltenes and can be understood based on fundamental principles of molecular structure–function relations. The additional restriction of small molecular size with a (predominantly) single binding site separates the structures which form at different concentrations.

7. Future Outlook in Petroleum Science

One standard way of treating dead crude oils (gases already liberated) is to represent their components within the SARA classification,—saturates, aromatics, resins, and asphaltenes (cf. Chapter 23). These designations are focused on operational procedures associated with solubility and adhesion in column chromatography. (The designation SARA remains fixed but the corresponding operational separation procedures vary widely.) There has not been a clear chemical designation for crude oil components that readily captures important chemical classes. In fact, as discussed above, there had been no agreement regarding asphaltene molecular weight, which essentially precludes chemical definition. The SARA scheme is useful for providing a rough description of crude oils and the procedures can be followed in a routine manner. Consequently, the SARA classification has been widely utilized. Nevertheless, the SARA scheme is seriously flawed for utilization as a predictive tool first because it utilizes only four pseudo components for a dead crude oils and second because it is based on cursory chemical properties and does not differentiate the different chemical moieties in the heavy ends.

To enable petroleomics it is a necessary but not sufficient condition to have the basics of asphaltene molecular structure worked out; subsequent chapters indicate this is largely accomplished for the bulk of asphaltenes. This knowledge has given us the ability to understand structure–function relations in asphaltene science. We note the caveat that interfacial asphaltene science could be strongly dependent on components present in small mass fraction. Petroleomics extends these concepts beyond a generalized understanding of structure–function relationships. Petroleomics holds the promise of looking at constituents of a given crude oil and from its constituents predicting specific properties. Thus, what is needed is the petroleome—the analogue of the genome. For instance, the presence or absence of heavy, hydrogen deficient, hetroatom containing aromatic hydrocarbons could be the harbinger of asphaltene deposition problems. To fully engage the concepts of petroleomics, it is necessary to obtain the complete listing of all components in a crude oil. Of course, there are pragmatic issues associated with detection thresholds vs. mass fraction that deleterious chemical components require to display their undesirable traits. Another pragmatic component is deciphering which are the pernicious chemical constituents that may be hiding amongst a forest of benign components. But one can easily imagine lumping together closely related chemical species to form a chemical family thereby reducing the number of parameters involved. For instance, one could lump together all chemical constituents in the molecular weight range of 750–850 amu, with a carbon aromaticity in a specified range, with no heteroatoms except sulfur. By such a process, one could develop say 60 chemical families to characterize a crude oil.

With such a petroleome, the process of petroleomics progresses much as genomics. One would generate the petroleome for a series of crude oils—the structure. One would also generate the analyses of relevant crude oil properties—the function. Relevant properties could include phase behavior, interfacial properties including the related multiphase stability (emulsion stability, foaming heavy oil), corrosive tendencies, acid and base numbers and perhaps even commingling phase stability. Matricies would then be developed that relate structure and function. Chemical intuition would be utilized to define the chemical families, while the mathematical machinery of standard chemometric methods would be utilized to generate the structure-function matricies. Petroleomics proceeds in the same way that analysis of the genome can identify likely health problems. With a new oil sample, one could obtain the petroleome and predict likely "health" problems of this oil. The health of the crude oil includes all aspects in the production, transportation, refining, and sale of the end products. Petroleomics allows molecular mapping in this entire process. For instance, the likelihood of organic deposits during production and transportation of crude oil would be predicted. Petroleomics continues to proceed as genomics; with identification of likely health problems, high-end laboratories are directed to provide detailed and specific information relevant to the sample of interest. Petroleomics enables a much more accurate assessment the econometrics for each project by removing uncertainties associated with unanticipated problems. Production of marginal reserves is much more likely to proceed if the efficiency can be monitored accurately and if the value of the crude oil is determined precisely.

A rather important question arises, "where do we obtain the petroleome?" If we actually need to have the molecular structure of each of the tens of thousands of components, then we will all be waiting a while for the technology to develop. But this ultimate solution is not necessary to extract a great deal of the value of the process embodied in petroleomics. The development of fourier-transform, ion-cyclotron-resonance mass spectroscopy (FT-ICR-MS; see Chapter 3) by Professor Alan Marshall and coworkers has pushed the resolution and mass accuracy of mass spectroscopy to new heights. With large, homogeneous magnetic fields coupled with FT-ICR-MS methods, the achievable resolving power is in excess of one million. The mass defect of individual nuclides is on order of 1 to 10 millidaltons. Consequently, unique elemental listings for each peak in the mass spectrum of a crude oil can be obtained because the heaviest crude oil components are on order one kilodalton.

Chemical structural information has long been obtained on crude oils and crude oil components by a variety of techniques. This structural information could be concatenated to the mass spectral information to obtain an effective petroleome. If needed new separation procedures could be devised if petroleomics directs that specific chemical families are inordinately important; those families would be subject to close scrutiny. Furthermore, crude oils consist of so many components that idiosyncracies of particular compounds tend to be averaged out. For instance, in a given crude oil, the population of the largest fused ring systems in that crude oil have been shown to obey the Urbach tail description, which is a thermally induced statistical relationship between the different photoabsorbers in the system. This finding from solid-state physics applies to all crude oils and asphaltenes, illustrating the overriding simplicity of a statistical ensemble vs a small collection of a few chromophores. Of course, there are certainly technical hurdles remaining with the development of the petroleome. Obtaining properly normalized mass spectra across a broad mass range is a requirement. Nevertheless, the least tractable components for mass spectroscopy, the saturates, can be treated utilizing high temperature gas chromatography (HTGC) and two-dimensional gas chromatography (2D-GC). It is plausible that the first petroleome will be a concatenation of FT-ICR-MS with advanced GC methods.

Petroleomics is the future vision for petroleum science; yet, many components are already in place. The concept of performing predictive science based on the petroleome will come to fruition in time. The establishment of structure–function relations in petroleum science is well developed and progressing. Debate will continue about specifics of these relations but hopefully not about the process. In petroleum science, technical success is increasingly enabling commercial success, representing much need exploitation by society at large. The petroleum scientist must achieve in this setting; this rewarding challenge is a gift.

References

[1] Crick, F. (1988). *What Mad Pursuit, a Personal View of Scientific Discovery*, Basic Books, New York.

[2] Wernet, Ph., D. Nordlund, U. Bergmann, M. Cavalleri, M. Odelius, H. Ogawasara, L.A. Naslund, T.K. Hirsch, L. Ojarnae, P. Glatzel, L.G.M. Pettersson, A. Nilsson (2004). The structure of the first coordination sphere in liquid water, *Science* **304**, 995.

[3] Sass, S.L. (1998). *The Substance of Civilization*, Arcade Publishing, New York.

[4] Small, K.M., L.E. Wagoner, A.M. Levin, S.L.R. Kardia, S.B. Liggett (2002). *N. Engl. J. Med.*, **347**, 1135.

[5] Diamond, J. (1997). *Guns, Germs and Steel*, W.W. Norton & Co., New York.

[6] Wilson, E.O. (1998). *Consilience, The Unity of Knowledge*, Vintage Books, New York.

[7] Mullins, O.C., G. Fujisawa, M.N. Hashem, H. Elshahawi (2005). Determination of coarse and ultra-fine scale compartmentalization by downhole fluid analysis coupled with other logs, *Intl. Petrol. Tech. Conf. Paper*, 10036.

[8] Yen, T.F. (1990). *ACS Div. Pet. Chem. Preprint*, 35, 314.

2

Asphaltene Molecular Size and Weight by Time-Resolved Fluorescence Depolarization

Henning Groenzin and Oliver C. Mullins

1. Introduction

1.1. Overview

The most important attribute of any chemical compound is its elemental constituents. There is, fortunately, no uncertainty about the elemental composition of asphaltenes. The second most important attribute of any chemical compound is its molecular structure and, as a prerequisite to that information, molecular weight. Although the set of structures of individual chemical units constituting asphaltene, such as the number of fused aromatic rings, length of aliphatic chains, and common functional groups is mostly agreed upon, the asphaltene molecular weight has been the subject of a large and long-standing controversy. For the most part, literature reports differ by a factor of 10, but some reports differ by many orders of magnitude. The question is essentially if and how the chemical units are linked. These uncertainties are exacerbated by the corresponding possibilities that different asphaltenes are variable, thus prohibiting facile comparison of results across different laboratories on different asphaltenes. This controversy has retarded the development of asphaltene science in that knowledge of structure–function relations is precluded if the structure is unknown. Consequently, a phenomenological approach has been routine in asphaltene science.

We employ time-resolved fluorescence depolarization (TRFD) to measure the molecular rotational correlation time of a large variety of asphaltenes. TRFD methods naturally allow interrogation of different chromophore classes in the asphaltenes enabling stringent predictions to be tested regarding molecular weight and molecular structure. *n*-Heptane asphaltenes from virgin crude oils are found to have a molecular weight distribution with a mean at ~750 g/mol, and a FWHM at 500 g/mol and 1000 g/mol, with a rapidly diminishing tail at higher molecular weight. There is little variation of molecular weight among virgin crude

Henning Groenzin and Oliver C. Mullins • Schlumberger-Doll Research, Ridgefield, CT 06877.

oil (petroleum) asphaltenes. Coal asphaltenes are significantly smaller, with a mean ~500 g/mol (or perhaps smaller). A variety of other asphaltene samples are investigated as well. Furthermore, all TRFD results are consistent with a molecular structure that has a single fused ring system of 4 to 10 rings per (petroleum) asphaltene molecule including a small number of aliphatic chains. These results are exploited to develop structure-function relations for asphaltenes; implications are discussed in terms of asphaltene nanoaggregate formation. Finally, we note that asphaltenes are polydisperse, other molecular structures and likely present but only in small mass fraction.

1.2. Chemical Bonding of Functional Groups in Asphaltenes

Molecular weight is one of the most fundamental attributes of any chemical compound. Although it appears as a byproduct of a structure, it becomes a critical parameter of unique structures that consist of small reoccurring units, e.g., polymers and proteins. In such structures molecular weight has a profound impact on the physical properties of the compound such as solubility, density, phase behavior, rheology, and intermolecular interaction. As in so many other important arenas, size counts. For instance, in quantum mechanics, size appears explicitly in equations governing nonclassical behavior of particles. In chemistry, molecular size is inextricably tied to properties. Monomers differ fundamentally from corresponding polymeric systems. A chemist would never permit ethylene and polyethylene to be considered as equivalent substances. Styrene and polystyrene are completely different from any rheological and phase behavior perspective and would never be considered as equivalent. For a system such as asphaltenes, which are defined by a solubility classification, molecular weight is a crucial attribute. However, the issue of molecular weight of asphaltenes has often been treated cavalierly, with the perspective that as long as one understands the constituent groups more or less, the issue of whether these fundamental units are covalently linked or simply aggregated in solution is secondary. This perspective is reinforced when limitations are recognized within laboratories for measuring asphaltene molecular weight. Rather than acknowledging limitations, workers have been known to be unrealistically optimistic in the assessment of fundamentally flawed techniques. This pernicious and irreverent treatment of such a fundamental molecular property has impeded advances in asphaltene science, essentially limiting discovery to be phenomenological rather than causal. It is inconceivable to imagine the tremendous advances currently taking place in the field of genetics if the prevailing view were that DNA base linkages whether covalent or merely associative are essentially equivalent. The field of asphaltene science deserves proper treatment and respect for first principles. It is thus essential to resolve the debate over asphaltene molecular weight.

1.3. Techniques Employed to Study the Size of Asphaltenes

Ironically, the central focus of asphaltene molecular weight has helped maintain the controversy on this issue. There is no standard set of asphaltene samples that would allow calibration of results from different laboratories around the world. In addition, the degree of heterogeneity among different asphaltenes is uncertain, creating concern that results from different laboratories are not universal. Consequently, the different asphaltene samples of interest in various laboratories interrogated by divergent techniques lack any standard of comparison. In essence, this situation seems to mandate that each laboratory determine key attributes of the asphaltene sample under study. Thus, many different laboratories "measure" asphaltene molecular weight for routine sample characterization, and then embark on specific studies unique to that laboratory. The problem is that the molecular weight determination of asphaltenes is not a trivial task. The literature is filled with reports utilizing demonstrably inappropriate techniques to determine asphaltene molecular weight. Incorrect parameter determination is worse than no parameter determination; this truism has been slow to penetrate the body of asphaltene science.

Colligative techniques such as vapor pressure osmometry (VPO) have been popular for "molecular" weight determination of asphaltenes. The primary difficulty with this technique is that for VPO, requisite concentrations of asphaltenes ($\sim$1%) greatly exceed the critical nanoaggregate concentrations (CNAC) of asphaltenes. For instance, in toluene, the nanoaggregate concentrations are on the order of $\sim$100 mg/L (cf. Chapters 9–11). The requisite VPO concentrations also exceed that of nanoaggregate clustering (cf. Chapter 17, 18). VPO has been used to report "molecular" weights of asphaltenes, but in fact reports aggregate weights of asphaltenes. The aggregate weight is related to both molecular weight and aggregate number. Some VPO studies report the impact of solvent, temperature and concentration on asphaltene molecular weight. The variable of interest here is aggregation tendency, not the molecular weight. Of course, asphaltene molecular weight is not a function of any of these parameters, this just illustrates that VPO is an improper technique for determination of asphaltene molecular weight. Extrapolations of VPO results to low concentrations are also problematic. Asphaltenes in solution are known to exhibit aggregation at different length scales at different concentrations. The concentration range of VPO experiments may extrapolate below that of clustering of nanoaggreagtes but not below that of nanoaggregate formation. Any technique such as VPO that exhibits a rapidly changing molecular weight value with extrapolation to zero concentration cannot be considered robust.

Similarly, gel permeation chromatography (GPC) has been used to characterize asphaltene molecular weights, but the application of this technique to molecular weight determination suffers from major problems. Surprisingly, some GPC results on asphaltenes employ solvents that do not dissolve all of the asphaltene, such as N-methyl pyrrolidone (NMP). Obviously, these reports are fundamentally flawed. GPC requires the use of standards; typically polystyrene. But there are reasons to expect polystyrene and asphaltenes to behave differently in any chromatography setting. In addition, GPC studies often employ concentrations that are not well characterized and may exceed the asphaltene aggregation concentrations. Furthermore, some GPC column materials are incompatible with toluene.

Mass spectroscopy is perhaps the most obvious candidate to determine asphaltene molecular weight. Mietec Boduszynski published results from field-ionization mass spectroscopy (FIMS) on n-heptane asphaltenes reporting a mean asphaltene molecular weight of $\sim$850 g/mol.[1] These results were at odds with conventional wisdom and so were questioned based on two issues, the ability to obtain gas phase of large components and possible fragmentation. Laser desorption mass spectroscopy (LDMS) and matrix-assisted laser desorption ionization (MALDI) were subsequently utilized to study asphaltenes. Both of these techniques are complicated by severe baseline issues. Corresponding reports in the literature vary by more than a factor of 10 on asphaltene molecular weight. Some studies[2,3] obtained values quite close to those of Boduszynski, others much higher. It has been shown both laser power and asphaltene concentration have a significant impact on the mass spectra using laser desorption ionization (including LDMS and MALDI). At low laser power *and* low asphaltene concentration asphaltene molecular weights of ~850 amu are obtained. With either higher laser power or higher asphaltene concentration, then artificially elevated molecular weights are obtained.[3] It is probable that laser desorption studies that report high asphaltene molecular weight suffer from these artifacts.

Other ionization techniques have been employed to determine asphaltene molecular weight. Fortunately, these techniques are in agreement with the original FIMS results and with the TRFD results. Recently, electrospray ionization, ion-cyclotron-resonance mass spectroscopy (ESI-FT-ICR) has been employed to study asphaltenes[4] and heavy Venezuelan crude oils[5]. Results in accord with Boduszynski have been obtained. It is important to note that ESI (for which John Fenn won the Nobel Prize in 2002) does not evaporate the asphaltene. Rather the solvent is evaporated leaving the original solution with no more solvent, thus the solute in a vacuum. Coulomb repulsion prevents aggregation of the ions in the evaporating solvent droplets.[6] In addition, the method of ionization is very soft; there is no fragmentation. Very delicate and heavy systems can be successfully studied with ESI. Asphaltenes are not pushing this technique to the limits. ESI is only applicable to molecules containing at least one heteroatom.

A question arises why ESI investigating heteroatom containing asphaltene molecules should issue a molecular weight that is not very different than for non-heteroatom containing asphaltene molecules; there is expected an inverse relation between polarity and molecular weight.[7] Sulfur is often the heteroatom in greatest abundance in asphaltenes. The sulfur moieties in asphaltenes are predominantly thiophene and sulfide.[8–11] These sulfur species are not very polar and thus have only a small impact on intermoleuclar interactions. An occasional asphaltene contains sulfoxide, which is very polar (~4 Debye). As we will see later in this chapter, this chemical species does influence the average molecular weight of this specific asphaltene. Finally, from a statistical point of view, one expects the bulk of asphaltene molecules to have at least one heteroatom. Thus, ESI interrogates the bulk of asphaltenes and largely derives the same molecular weights as FIMS.

Atmospheric pressure chemical ionization (APCI) has been applied to asphaltenes in different laboratories and is producing consistent results again.[12,13] APCI has been performed on different solubility classes of asphaltenes and the expected results are obtained, less soluble fractions consist of higher molecular weight.[13] The studies report that the bulk of the asphaltene population lies below 1000 Da, but there is certainly a diminishing tail reaching ~1300 Da. Nevertheless, there is some question as to whether the high mass fraction contains some noncovalent dimers. Essentially, all mass spectral studies of asphaltenes are in accord excepting some LDMS and MALDI studies that the mass centroid is roughly 700 g/mol. With the laser ionization techniques there are a many conflicting literature studies with some in accord with the original field ionization measurements. Nevertheless, the mass spectral studies of asphaltenes have not been universally accepted and the application of some of the ionization methods is relatively recent, hence it is important to employ other experimental methods to investigate asphaltene molecular weight.

1.4. Time-Resolved Fluorescence Depolarization (TRFD)

In our laboratory, we have selected to use optical techniques to investigate asphaltene molecular weight, in particular, time-resolved, fluorescence depolarization (TRFD). With this technique, a polarized laser of selected wavelength excites a subset of chromophores in the asphaltene solution. This excitation process creates a net polarization vector for the molecular ensemble. Rotational diffusion causes reorientation of the molecular polarization. This results in a net decrease in the magnitude versus time for the ensemble polarization; the rate of this decrease is directly dependent on the rate of rotational diffusion. Polarized fluorescence emission of a particular wavelength is then detected at some later time after excitation. Detecting the polarization of the fluorescence emission enables a recording of the polarization of the molecular ensemble as a function of time. The selection of the excitation and emmission wavelength selects a subset of the chromophores. By using relatively small wavelength differences between excitation and emission we are observing the HOMO-LUMO transition thereby avoiding depolarization due to non-radiative processes within the excited states of the electronic manifold. (The HOMO-LUMO gap is the energy gap between the highest occupied molecular orbital to lowest unoccupied molecular orbital.) Unlike nuclear polarization techniques such as NMR relaxation, the polarization vector of the excited electronic state of the individual chromophore follows exactly the molecular rotation (for nondegenerate excited states). Since the absorption and emission dipoles are collinear for HOMO-LUMO transitions, the polarization of the photon emitted upon fluorescence relaxation to the electronic ground state is (essentially) the same as the polarization of the adsorbed photon in the now-rotated molecular coordinates. This technique relies on the assumption that the transition dipole moment is uniquely defined in the molecular coordinates. For an ensemble of molecules undergoing rotational random walk, the net effect of molecular rotation is the loss of polarization. At sufficiently long times, all polarization is lost.

Figure 2.1 shows a cartoon of this concept. In the first step, the molecule absorbs a photon that is linearly polarized, say in the Z direction of the laboratory frame. The cartoon depicts a case where the dipole moment of the transition is perpendicular to the fused ring system of the chromophore, the light-absorbing group in the molecule. The molecule undergoes rotational random walk or equivalently, rotational diffusion. The molecule constantly undergoes rotational diffusion but prior to photoabsorption, there is no way to monitor this process. The rotational

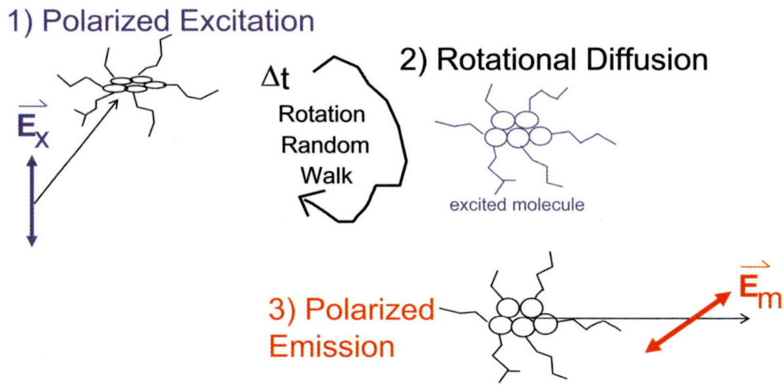

Figure 2.1. Schematic illustrating the process of time-resolved fluorescence depolarization (TRFD). Absorption of a polarized photon E_{ex} polarizes the excited electronic state of the fluorophore. The fluorophore undergoes rotational random walk; the electronic transition dipole moment is fixed in the molecular frame. The emitted fluorescence photon reflects the polarization of the rotated molecule. For an ensemble of molecules, rotational random walk causes an exponential decay of net polarization.

diffusion of the excited state results in a continuous reorientation of the polarization vector of that molecule as the orientation of that vector is fixed within the molecular system. At some point, the excited molecule emits a fluorescence photon thereby de-exciting the molecule from the excited state back down to the ground state. The polarization of the emitted photon is correlated to the new angular coordinates of the molecule which are rotated from the initial angular coordinates at the time of photoabsorption. The cartoon in Figure 2.1 depicts a net 90° rotation of the polarization vector from the initial direction. Upon photoabsorption, the ensemble average of the direction of the transition dipole in the molecules adsorbing polarized light is aligned with the photon electric field at time zero. As time progresses, the net polarization vector diminishes and at long times, the individual dipole moments in the ensemble point equally in all directions. Thus, the emission of the system becomes unpolarized. Molecules with a smaller hydrodynamic volume exhibit a faster rotational diffusion. The exact relation between correlation time τ_r to molecular size will be discussed in Section 2.

1.5. The Optical Range Relevant to Asphaltene Investigations

To perform TRFD on asphaltene molecules, the first issue of concern is to identify the relevant optical spectral range of interest. Asphaltenes are highly absorptive in the visible even into the near-infrared, while standard polycyclic aromatic hydrocarbons with four fused rings or fewer are mostly colorless. Pericyclic rings, which may be predominant in asphaltenes[14], tend to have blue-shifted absorption compared to linear catacyclic ring systems for example. More accurately, asphaltene ring systems are dominated by Sextet Carbon (cf. Ch. 5) which have blue-shifted optical transitions (cf. Ch. 4). This tells us that asphaltenes certainly possess some large chromophores with their corresponding small HOMO-LUMO

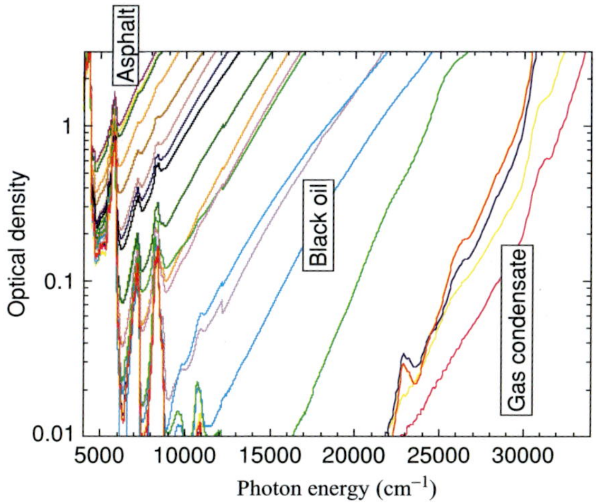

Figure 2.2. The electronic absorption edge of crude oils including very heavy oil. For each sample, the long wavelength edge of the absorption decreases exponentially (versus photon energy) corresponding to the Urbach tail. This decline reflects the exponential decline of the molecular population with low energy transitions, that is, big fused ring systems.

gaps. Optical absorption experiments place limits on the long wavelength end. If there is no optical absorption beyond certain long wavelengths by asphaltenes, then this wavelength is of no concern for TRFD applied to asphaltenes. The asphaltene electronic absorption edge (on the long wavelength side) is characterized by the "Urbach tail" in the Fermi edge, a result familiar from solid state physics. Figure 2.2 shows the electronic absorption spectra of many crude oils, from very heavy to light. The Urbach tail is the exponential decay of optical absorption with decreasing photon energy. The Urbach tail corresponds to the electronic absorption edge of various materials exhibiting thermal excitation so the electronic absorption edge has an exponential decline with slope kT on the long wavelength side. All crude oils[15,16] and all asphaltenes[15,17] have very similar electronic edge slopes characterized by ~10 kT.

We have a good understanding of the long wavelength spectrum of crude oils and asphaltenes. The electronic edge of these carbonaceous materials is not determined by the thermal excitation of individual chromophores: that is, crude oils are not black due to the presence of hot benzene. Instead, the coloration of crude oils and asphaltenes is determined by the thermal production of big chromophores from small chromophores (in the catagenesis of kerogen). Consequently, our Urbach scaling is not restricted to be kT. Bigger chromophores absorb at longer wavelengths in accord with the quantum particle-in-a-box formalism. Just as increasing the distance between nodes on a guitar string produces lower notes, increasing the delocalization area in a π system of an electron in a larger aromatic box increases the wavelength of the electron wavefunction thereby decreasing its transition frequencies. Thus, the population distribution of large aromatic ring systems determines the optical absorption profile in the long wavelength range. Large asphaltene chromophores are produced by chemical reaction from the small chromophores. Consequently, the population of big chromophores exponentially declines in asphaltenes. One can consider the increased colorization of toasting white bread as a related phenomenon. After all, catagenesis of kerogen is colloquially referred to as occurring in the geological "kitchen". Toasting white bread evolves through a sequence of coloration—yellow, tan, brown, and eventually black if the toaster remains on too long. All of these colors are represented by an exponential decrease in absorption at longer wavelength (in accord with the exponential decrease in the population of larger chromophores). The exact color is determined by the size of the chromophores produced. (The green color observed for some crude oils is actually due in part to crude oil fluorescence, not simply from optical absorption.)

For asphaltenes, there is a limit in coloration; all petroleum asphaltenes show an electronic absorption edge at ~650 nm.[15,17] Thus, TRFD studies do not need to employ longer wavelengths than 650 nm for investigating the bulk of asphaltenes. This limit on molecular and chromophore size is related to solubility characteristics of asphaltenes. Since solubility is the requisite property defining asphaltenes, one cannot have arbitrarily large ring systems. If the fused aromatic ring systems get too big, solubility is precluded due to large intermolecular interaction. Essentially, van der Waals interaction scales with the number of fused rings, the greater the contact area, the greater the binding energy. Since binding energy occurs in the exponential argument of the Boltzman factor, fused ring systems of too great a size are simply excluded from any solubility class. Note that a single large fused ring system is much "stickier" than two smaller fused ring systems of equal ring number, where the smaller ring systems are tethered by an alkane chain. For a single large ring system, there is a single entropy reduction upon binding to a surface. For two ring systems, there is a large reduction in entropy for the binding of each ring to a surface. This extra entropy reduction retards binding.

These concepts explain the physisorption of "decolorizing carbon" familiar to all who have matriculated from undergraduduate organic chemistry laboratories. After performing organic synthesis of small products, the resulting product solutions often assume a brownish coloration even though often none of the reactants or products is colored. This brown color is the result of some degree of aromatization perhaps accompanied by some polymerization. P.J. Flory, a Nobel laureate in polymer chemistry, once remarked that the "brown stuff on the bottom of the reactant vessel" is what interested him. In any event, to remove this colored material, one adds to the solution then filters out decolorizing carbon. This insoluble material provides ample surface with a high degree of aromatic carbon content. Colored reaction byproducts containing many fused aromatic rings will stick to the aromatic surface of the decolorizing carbon thereby being removed from solution. Asphaltene solubility considerations will be seen to relate to decolorizing carbon and these freshman chemistry principles. The asphaltene solubility classification captures the largest ring systems that can remain stable as a (micro)colloidal suspension in crude oil for geologic time. The solubility classification is fundamental to the nature of asphaltenes and needs to be understood from the point of view of chemical structure.

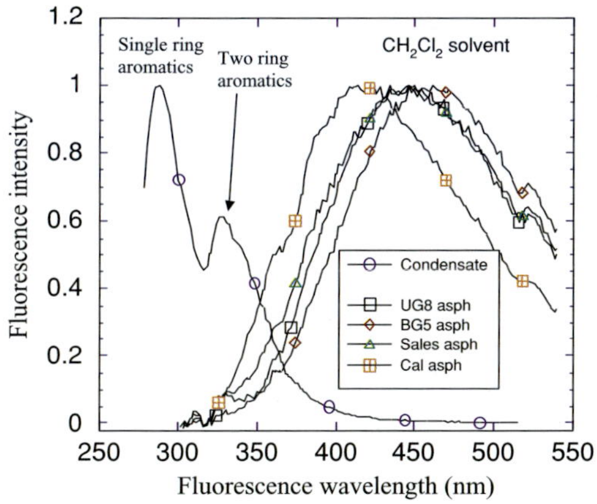

Figure 2.3. The fluorescence emission spectra of petroleum asphaltenes and of a condensate for excitation at 265 nm. The spectra of the asphaltenes lack emission from small aromatics that are evident in the condensate. Asphaltenes lack substantial populations of these small ring systems.

For optical fluorescence interrogation of asphaltenes, we need to establish the short wavelength spectral limits. We cannot look for optical absorption to define the smallest chromophores that must be investigated because big chromophores also absorb at short wavelengths due to excitation of higher lying electronic states, but we can look for fluorescence emission because only small chromophores emit short wavelength light. All fluorescence spectra of asphaltenes lack much emission from aromatics with one and two rings, at 290 nm and 320 nm respectively.[15] Figure 2.3 shows the fluorescence emission spectra for several typical asphaltenes and for a gas condensate where fluorescence emission from one- and two-ring aromatics is evident. Asphaltenes also lack much fluorescence emission from three-ring aromatics, but there is more emission here than for one- and two-ring aromatics. This is widely known and repeated in all laboratories that measure fluorescence spectra of asphaltenes. In our laboratory, we have probed the reason for this lack of emission from small fused ring systems. Either one- and two-ring systems are not present in abundance in asphaltenes or they are present but predominantly undergo radiationless transitions in asphaltenes such as fluorescence resonant energy transfer (FRET) to the large ring systems. If the former explanation is correct, the lifetime of the UV fluorescence of asphaltenes should match that of maltenes. If the latter explanation is correct, then the small chromophores have a new decay path (radiationless transition) thereby decreasing their fluorescence lifetimes. We have established the occurrence of collisional energy transfer with concomitant lifetime reduction in high concentrations of crude oils and asphaltenes.[15,18] We found essentially that the UV emission from dilute solutions of asphaltenes and maltenes are comparable, thus we conclude that asphaltenes lack UV fluorescence emission because for the most part they lack one-, two- and largely three- fused ring aromatics.[19] Either way the relevant spectral range to interrogate asphaltenes is thus established to be between 370 nm on the high energy side and 650 nm on the low energy side.

1.6. Structure Predictions from TRFD

TRFD is presented in this chapter; it is our method to obtain asphaltene molecular weight.[20–25] TRFD is employed to determine rotational correlation times of asphaltene fluorophores in very dilute solution. This technique allows us to determine the size of the asphaltene molecules. The measured τ_r's (rotational correlation time) are analyzed using standard theoretical formalisms to obtain molecular size. In addition, fluorophores of known size and structure are run to compare the sizes obtained from the τ_r's for the asphaltenes directly. These known fluorophores have comparable alkyl to aromatic carbon ratios and comparable fused ring systems as asphaltenes. The τ_r's obtained for the asphaltenes are all shown to be small and comparable to standard aromatic dyes that are on order 750 g/mol. The smallest asphaltene molecules are comparable in size to alkyl porphyrins ~500 g/mol. These results are in accord with all mass spectral results that do not use laser desorption, and some mass spectral results that do use laser desorption. A variety of source materials for asphaltenes are used as well as various solubility subfractions of asphaltenes. Related samples are also measured. These sample sets are utilized to test universality and invariants of asphaltene structure. In addition, these sample sets are designed to test directly key predictions from the TRFD results.

The asphaltenes exhibit an order of magnitude monotonic variation of τ_r's as a function of wavelength over the relevant spectral range. Blue-emitting chromophores undergo rotational diffusion ~10 times faster than the red-emitting chromophores. This large variation is an independent and additional check on asphaltene molecular size. The implication is that small chromophores are not attached to big chromophores; if they were, both chromophores would exhibit large correlation times of somewhat comparable magnitude. Any degree of cross-linking of small to large chromophores would affect their rotational correlation times. For asphaltenes, the short correlation times, and the factor of 10 variation of correlation times, both imply that there is a single fused ring system per molecule. If one constructs a proposed asphaltene molecule with seven fused rings, with 40% aromatic carbon, 60% saturate carbon, with one or two heteroatoms, one obtains a molecular weight of roughly 750 g/mol. The prediction is that solubility mandates this structure; that if a molecule had several of these fused ring systems per molecule, it would not be soluble in toluene, and correspondingly would not be stable in crude oil for geologic time. We are now in a position to test these concepts. Instead of being mystified by asphaltene properties due to unknown asphaltene structure, we take the approach that understanding the structure allows predicting function, in particular asphaltene solubility.

The results discussed in this chapter lead to the conclusion that asphaltene identity is actually quite easy to understand in terms of widely known chemical principles with regard to solubility. TRFD is shown to lead uniquely to these simplifying and powerful concepts. These underlying principles provide clear predictions for TRFD results on particular carbonaceous materials as well as providing predictions for results from other experimental methods. All predictions are confirmed; many of which are presented in other chapters in this book.

All asphaltenes are shown to share certain key characteristics. Some important differences are noted particularly between coal and petroleum asphaltenes providing a stringent test platform for ideas advanced in this chapter. High temperature hydrotreatment samples are shown to behave as expected based on the coal and petroleum results lending credence to interpretations herein. Certain asphaltenes of unusual heteroatom chemistry also provide an excellent test for the simple arguments advanced in the following. The TRFD results are shown to grade continuously for solubility sub-fractions of asphaltenes as well as for resins and maltenes. Such continuity follows from the fairly simple framework for asphaltene molecular weight and molecular structure.

The solubility behavior of asphaltenes is shown to follow simple freshman chemistry ideas, balancing steric hindrance and van der Waals energy in π-bonding stacking. These simple ideas are shown to be a basis to build structure—function relations, which are the foundation of concepts embodied in petroleomics. The basis of nanoaggregate growth is discussed in this context, especially with regard to the new ultrasonics result on asphaltenes. It is noted however, that the interfacial properties of a polydisperse system are inherently more complicated because a small mass fraction may dominate particular interfacial characteristics. Nevertheless, understanding molecular structure and nanoaggregate formation will aid in understanding asphaltene interfacial chemistry.

2. Theory

In order to develop the equations which describe the anisotropy decay of the fluorescence emission of fluorophores in solution, several different approaches were made.[26–32] The theory, based on the work of Einstein[33,34] and Debye[35] on rotational diffusion by Brownian Motion was extended and presented in the final and complete version as it pertains to fluorescence depolarization caused by rotational diffusion.[26]

In the following we want to present two different approaches. The first treatment approximating molecules as spheres is based on reference 28. This approach is widely used to analyze experimental data;[36–42] the second is the complete description of the fluorescence anisotropy decay for asymmetric rotators based on operator algebra.[32]

2.1. The Spherical Model

In the following we will assume a spherical molecule rotating in a viscous medium subject to a sticking boundary condition.

The following definitions are used:[43]

$$D(t) = I_{\parallel}(t) - I_{\perp}(t), \tag{2.1}$$

$$S(t) = I_{\parallel}(t) + 2I_{\perp}(t), \tag{2.2}$$

and

$$r(t) = \frac{D(t)}{S(t)}. \tag{2.3}$$

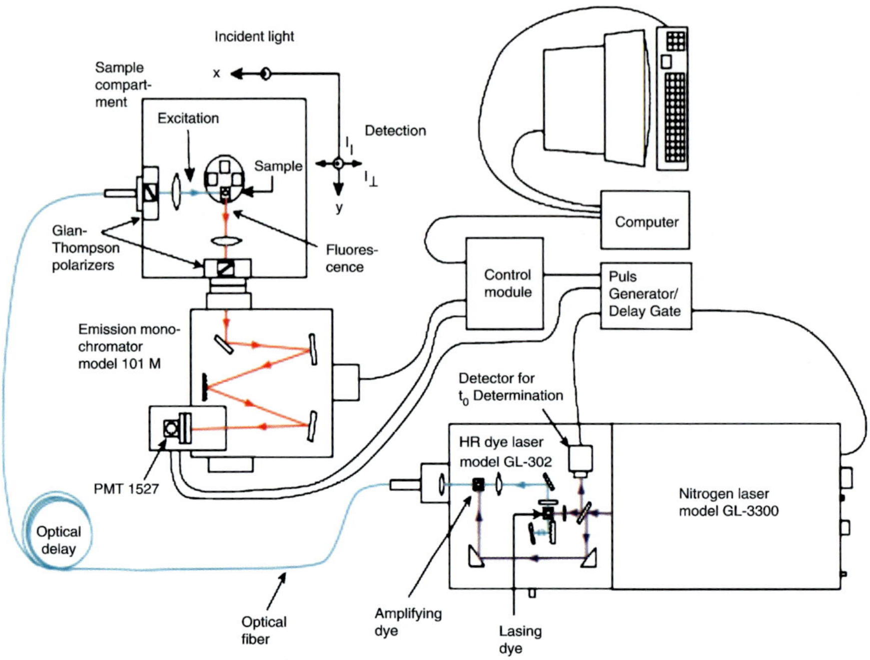

Figure 2.4. Schematic showing the PTI C-72 system used to perform the fluorescence depolarization measurements. A pulsed nitrogen laser pumps a tunable dye laser providing the start time for fluorescence excitation. A high voltage pulse on the PMT provides the stop time for fluorescence emission. Fluorescence intensity is plotted vs delay time for different polarizations.

where $I_{\parallel}(t)$ and $I_{\perp}(t)$ denote the intensity of detected light linearly polarized parallel and perpendicular to the linearly polarized excitation and $r(t)$ represents the anisotropy of the fluorescence emission. $S(t)$ standing for "sum" is equal to the total fluorescence intensity if the two directions perpendicular to the polarization of the excitation source are exchangeable, which is usually the case for fluorescence in an isotropic medium. Experimentally, $r(t)$ is obtained in time-resolved fluorescence experiments, such as described in the experimental section, directly by measuring $I_{\parallel}(t)$ and $I_{\perp}(t)$ independently. Under the assumption that the rotational diffusion is isotropic, $r(t)$ is a single exponential and its two parameters r_0 and τ_r can also be determined in steady-state fluorescence experiments from the Perrin equation through variation of either temperature T or viscosity η. Obviously, the time-dependent approach is far more direct and bound to yield more accurate results.

The following describes how the fluorescence anisotropy $r(t)$ can be related to the hydrodynamic volume of the molecule. We will assume an experimental setup as depicted in Figure 2.4. The exciting light is propagating in the negative x-direction and its polarization is oriented along the z-coordinate of a laboratory-fixed system. The fluorescence will propagate along the positive y-axis and the polarization will be detected either in z- or in x-direction. The origin is placed conveniently at the position of the fluorophore. The transition dipole moment $\vec{\mu}$ shall have an arbitrary orientation with respect to the molecular axes. While the orientation of $\vec{\mu}$ stays constant in the molecule system, it is time dependent in the laboratory system. This implies that the emission dipole moment is assumed to be collinear with the absorption dipole moment, an assumption that is generally justifiable for HOMO-LUMO transitions. $W(\theta, \phi, t)$ shall now denote the probability that the vector $\vec{\mu}$ is oriented (θ, ϕ) at time t. $W(\theta, \phi, t)$ obeys the diffusion equation

$$\frac{\partial W(\theta, \phi, t)}{\partial t} = D\nabla^2 W(\theta, \phi, t), \qquad (2.4)$$

where D is the diffusion constant of a sphere of volume V. The diffusion equation can be solved in terms of its Green's function $G(\theta_0, \phi_0|\theta, \phi, t)$ resulting in

$$W(\theta, \phi, t) = \int_0^{2\pi} d\phi_0 \int_0^{\pi} \sin\theta_0 d\theta_0 W(\theta_0, \phi_0) G(\theta_0, \phi_0|\theta, \phi, t). \qquad (2.5)$$

$G(\theta_0, \varphi_0|\theta, \varphi, t)$ can be interpreted as the time evolution of the probability $W(\theta, \phi, t)$ with $\vec{\mu}$ oriented in (θ_0, ϕ_0) at $t = 0$.

Since the probability for a molecule with a dipole $\vec{\mu}$ to absorb a photon with its electric vector $\hat{\varepsilon}$ polarized in z-direction is $\left|\vec{\mu} \cdot \hat{\varepsilon}\right|^2 = \mu_z^2$, it is seen that the normalized initial distribution is[28]

$$\begin{aligned} W(\theta_0, \phi_0, t) &= \frac{3}{4\pi} \cos^2\theta_0 \\ &= \frac{1}{4\pi}[1 + 2P_2(\cos\theta_0)], \end{aligned} \qquad (2.6)$$

where P_2 is the Legendre polynomial of second order.

It is therefore convenient to expand the Green's function in terms of Legendre functions[28]

$$G(\theta_0, \phi_0|\theta, \phi, t) = \sum_{l=0}^{\infty} \sum_{m=-l}^{l} e^{-l(l+1)Dt} Y_{l,m}^*(\theta_0, \phi_0) Y_{l,m}(\theta, \phi). \qquad (2.7)$$

Applying boundary conditions and the normalization conditions together with the starting condition one derives that only one expansion coefficient $c_{2,0}(t)$ is non zero[28]

$$c_{2,0}(t) = e^{-6Dt}. \qquad (2.8)$$

The probability $W(\theta, \phi, t)$ is thereby calculated to

$$W(\theta, \phi, t) = \frac{1}{4\pi}[1 + 2e^{-6Dt} P_2(\cos\theta)]. \qquad (2.9)$$

The orientation probability can now be used to calculate the intensity for the two different directions of polarization which are given by[28]

$$\begin{aligned} I_{\parallel}(t) &= \iint \sin\theta d\theta d\phi\, \tilde{I}_{\parallel}(\theta, \phi, t) W(\theta, \phi, t) \\ I_{\parallel}(t) &= \iint \sin\theta d\theta d\phi\, \tilde{I}_{\perp}(\theta, \phi, t) W(\theta, \phi, t), \end{aligned} \qquad (2.10\text{ a,b})$$

where $\tilde{I}_{\parallel}(\theta, \phi, t)$ and $\tilde{I}_{\perp}(\theta, \phi, t)$ are proportional to the fluorescence decay and the transition dipole vector component in z- and x-direction respectively. By solving the integrals one obtains[26]

$$I_{\parallel}(t) = \frac{F(t)}{3}\left(1 + \frac{4}{5}e^{-6Dt}\right)$$

$$I_{\perp}(t) = \frac{F(t)}{3}\left(1 - \frac{2}{5}e^{-6Dt}\right), \qquad (2.11\text{ a,b})$$

where $F(t)$ denotes the fluorescence decay. The anisotropy can now be calculated using Equation (2.3)

$$r(t) = \frac{2}{5}e^{-6Dt}. \qquad (2.12)$$

Based on a method used by Einstein[27], an equation[34] was derived for the diffusion tensor $\underline{\underline{D}}$, which can be evaluated with the help of the drag tensor[44] $\underline{\underline{\beta}}$. For a sphere, the tensor reduces to a scalar D

$$6D = \frac{kT}{V\eta}, \qquad (2.13)$$

where η is the viscosity of the solvent, which makes it easy to relate the fluorescence anisotropy to the volume of the sphere. The decay time of the anisotropy $\tau_{r,sph}$, the parameter of our experiment, can now be written as

$$\tau_{r,sph}^{-1} = \frac{kT}{V\eta}. \qquad (2.14)$$

2.2. The Anisotropic Rotator

The more complex model of the anisotropic rotator can be treated completely in the operator formalism.[28,32]

One starts by seeking a solution for a diffusion equation which is given for the complete asymmetric body by

$$\frac{\partial W(\vec{\Omega}, t)}{\partial t} = -\left(\vec{L} \cdot \underline{\underline{D}} \cdot \vec{L}\right) W(\vec{\Omega}, t), \qquad (2.15)$$

where $\vec{L}$ is the quantum mechanical angular momentum operator in units of $\hbar$, $\underline{\underline{D}}$ is the diffusion tensor and $W(\vec{\Omega}, t)$ is again the probability that the vector of a molecular dipole transition $\vec{\mu}$ is oriented in the angle $\vec{\Omega}$ at time t, the angle $\vec{\Omega}$ refers to the orientation of a body fixed coordinate system with respect to a laboratory system.

Again the solution to the differential equation (2.15) is expressed in terms of its Green's function

$$W(\vec{\Omega}, t) = \int_{\vec{\Omega}} W(\vec{\Omega}_0)\, G(\vec{\Omega}_0 \mid \vec{\Omega}, t) d\,\vec{\Omega}_0, \qquad (2.16)$$

where $W(\vec{\Omega}_0)$ is the distribution of the absorption dipoles at time $t = 0$.

By expanding the Green's function in terms of the eigenfunction of the asymmetric rotator which again can be expanded in terms of the eigenfunction of the symmetric rotator for $l \leq 2$, and knowing that[32]

$$W(\vec{\Omega}_{,0}\, t) = P_{\text{abs}}, \tag{2.17}$$

where P_{abs} is the probability for the excitation of a molecule, it is possible to find a closed-form expression for $W(\vec{\Omega}, t)$.[32]

Analogous to Equation (2.10), one can use $W(\vec{\Omega}, t)$ to obtain an expression for $I_{\parallel}$ and $I_{\perp}$

$$\begin{aligned} I_{\parallel}(t) &= \int_{\vec{\Omega}} P_{em}^{\parallel} F(t) W(\vec{\Omega}, t) d\vec{\Omega} \\ I_{\perp}(t) &= \int_{\vec{\Omega}} P_{em}^{\perp} F(t) W(\vec{\Omega}, t) d\vec{\Omega}, \end{aligned} \tag{2.18 a,b}$$

where $P_{em}^{\parallel}$ and $P_{em}^{\perp}$ are the probabilities for the fluorescence emission polarized parallel and perpendicular to the incident light respectively. The solutions to Equation (2.18) when inserted into Equations (2.1)–(2.3), yield in the most general case a five-exponential decay for the anisotropy $r(t)$[32]

$$r(t) = \sum_{i=1}^{5} \alpha_i\, e^{-\frac{t}{\tau_i}}. \tag{2.19}$$

It is possible to reduce the number of decay times to three under the assumption of a body with a rotational symmetry.[28] In such a case, the diagonalized diffusion tensor will have only two elements which we will label as $D_{\parallel}$ and $D_{\perp}$. These two elements govern the molecular relaxation due to Brownian motion parallel and perpendicular to the symmetry axis respectively. The correlation times of the anisotropy decay can be expressed as[28]

$$\begin{aligned} \tau_1^{-1} &= 6D_{\perp} \\ \tau_2^{-1} &= 5D_{\perp} + D_{\parallel} \\ \tau_3^{-1} &= 2D_{\perp} + 4D_{\parallel}. \end{aligned} \tag{2.20}$$

The diffusion coefficients were evaluated as[28]

$$\begin{aligned} D_{\parallel} &= \frac{3}{2}\frac{\rho(\rho - S)}{\rho^2 - 1} D \\ D_{\perp} &= \frac{3}{2}\frac{\rho[(2\rho^2 - S) - \rho]}{\rho^4 - 1} D, \end{aligned} \tag{2.21 a,b}$$

where D is again

$$D = \frac{kT}{6V\eta},$$

the diffusion coefficient calculated for a sphere under the same conditions of volume, temperature and viscosity as the ellipsoid, ρ is the ratio of the longitudinal

Table 2.1. Fluorescence depolarization decay times for an oblate spheroid with an aspect ratio n with respect to a spherical rotator and practically used correction factor $C^{(n)}$ for the transition from spherical model to oblate spheroid, assuming only one exponential decay for the oblate spheroid. The last column compares the ratio of derived radii for an oblate spheroid (semi-major axis) vs a sphere for the same τ_r.

$n = 1/\rho$	τ_1/τ_r	τ_2/τ_r	τ_3/τ_r	$C^{(n)}$	a/r
2	1.13	1.17	1.30	1.41	1.124
4	1.84	1.90	2.1	2.25	1.211
6	2.65	2.71	2.93	3.09	1.248
8	3.47	3.54	3.77	3.94	1.266
10	4.3	4.38	4.62	4.89	1.269

semiaxis to the equatorial semiaxis of the ellipsoid, and[28]

$$S = (\rho^2 - 1)^{-\frac{1}{2}} \ln\left[\rho + (\rho^2 - 1)^{\frac{1}{2}}\right], \quad \rho > 1$$

$$S = (1 - \rho^2)^{-\frac{1}{2}} \arctan\left[\frac{(1 - \rho^2)^{-\frac{1}{2}}}{\rho}\right], \quad \rho < 1.$$

It is now verified, that for an oblate rotator ($\rho < 1$), both diffusion coefficients are approximately equal, and hence the three τ_i's are approximately equal to each other.[26] The anisotropy decay can therefore be described in case of an oblate spheroid by a single exponential decay with a correlation time of[26]

$$\tau_r^{-1} = \frac{3}{2}\frac{kT}{V\eta}\frac{\rho}{\rho^2 - 1}\left(\rho - \left\{(1 - \rho^2)^{-\frac{1}{2}} \arctan\left[\frac{(1 - \rho^2)^{-\frac{1}{2}}}{\rho}\right]\right\}\right). \qquad (2.22)$$

In the most general case for a completely anisotropic rotatator of unknown shape, analysis is difficult as τ_r breaks down into five exponential components that are independent of each other[28]. The assumption of an oblate spheroid with rotational symmetry reduces these five components to three that are similar to each other in magnitude and collapse into a single exponent for the case of an infinitely thin disk. For the shape of an oblate spheroid it is possible to define a single shape parameter, ρ as the ratio of the short axis to the long axis of the spheroid which therefore is always smaller than one, that determines the influence of the shape on the rotational correlation time. To simplify notation the aspect ratio n shall be defined as the reciprocal of ρ. Table 2.1 compares the three exponential decay times to each other for different shape parameters.

Examination of Table 2.1 shows that even for small shape parameter (large n) it introduces no more than 15% error upon collapsing the three exponential decays into one. A correction factor can be introduced that scales the correlation time of a sphere to obtain the correlation time of an oblate spheroid, provided that both are of volume V. The relation between molecular volume and the rotational correlation time is now given by

$$\tau_{r,obl} = C^{(n)} \frac{V_{obl}\eta}{kT} \tag{2.23}$$

Table 2.1 also shows that for a given measured τ_r the calculated molecular volume decreases for increasing n. In plain words, the flatter the assumed shape of the molecule is, the smaller will its calculated volume be. This is intuitively understood since τ_r is dependent on the viscous drag of the molecule. The drag is a function of the surface area of the molecule, and a sphere has of any shape the largest volume for a given surface area. However, a quick calculation shows that the predicted larger half-axis a of the oblate spheroid is larger than the radius r of a sphere calculated for the same rotational correlation time by the factor

$$\frac{a}{r} = \left(\frac{n}{C^{(n)}}\right)^{\frac{1}{3}} \tag{2.24}$$

and will grow with increasing n, despite the fact that the calculated molecular volume is decreasing. The ratio a/r is listed in Table 2.1 for several values of the shape parameter n. Still the error in radius calculated using a spherical model for an oblate spheroid with $n = 4$ is only 21%.

The magnitude of the anisotropy, which is essentially the pre-exponential factor of the anisotropy decay was found to be between 0.27 and 0.4 for all asphaltene solubility fractions at all excitation/emission wavelength combinations, where 0.4 is theoretically the maximum value the magnitude of the anisotropy can assume[28]. The magnitude of the anisotropy does not contain readily extractable size information[32] and was also found experimentally to have a much larger error bar than the measurement of τ_r. For different asphaltene fractions, the variation of the pre-exponent is less than a factor of two, while the variation of the rotational correlation times is an order of magnitude. We therefore find the pre-exponent ill suited to characterize the anisotropy of the molecular species under investigation. Nevertheless, the large anisotropy implies that most of the asphaltene fluorescence is behaving as expected; no other significant anisotropy decay mechanism is observed.

3. Experimental Section

3.1. Optics Methods

For all solutions used for fluorescence work, we checked optical densities using a CARY 5 UV–visible–NIR spectrometer. For collection of fluorescence spectra, we employed the PTI C-72 + A-720 fluorescence spectrometer using a 75 watt Xe compact arc lamp source.

Figure 2.4 shows a schematic of the PTI C-72 system used for collection of time-resolved fluorescence depolarization (TRFD) decay curves. This system employs a PTI GL-3300 nitrogen laser source along with a PTI GL-302 high-resolution dye laser with a fiber optic coupling to the measurement cell to excite the fluorescence. The excitation and emission light from the cell are oriented 90°

from each other with vertical polarization defined to be perpendicular to this plane. The wavelength of the PTI model 101 M emission monochromator is fixed at a selected wavelength while two Glan-Thompson polarizers are used to select the polarizations. One polarizer is placed at the output of the fiber optic, immediately before the measurement cell, and the other polarizer is placed at the entrance to the emission monochromator. TRFD curves are collected for four polarizations; vertical on the source side, vertical on the emission side (v-v), vertical-horizontal (v-h), horizontal-vertical (h-v), and horizontal-horizontal (h-h).

The following procedure is used to acquire the time decay spectra; the laser firing triggers a box car delay gate which then triggers a high voltage pulse at known delay to the PMT. The short duration of the high voltage pulse "turns on" the PMT for a short time interval. The integrated current over this time interval from the PMT is detected. The delay time is sequentially scanned over the desired time range producing the fluorescence decay curve. The time resolution of the system is about 80 picoseconds.

A complete data set for one excitation and emission wavelength pair corresponds to acquisition of the four polarization combinations mentioned above. Typically, the total acquisition time for the four curves is 2 hours. Reproducibility of signal levels was checked periodically during the acquisition time to validate the data. In addition, the v-h and h-h curves should overlay again allowing for excellent quality control. For both cases a 90° rotation is required to align the source polarization vector with the horizontal acceptance polarization on the emission monochromater. The output of the fiber optic is randomized so selecting vertical or horizontal polarization for input into the cell provides the same excitation intensity. Duplicate (or more) runs were performed for all wavelength pairs to assure precision.

Figure 2.5 shows typical data for a given wavelength pair, excitation at 530 nm and emission at 570 nm. The time zero for the dye laser pulse is at 66 nanoseconds. Typically, we used a wavelength shift of 40 nm between the excitation and emission to preclude any possibility of direct detection of scattered light. Also, the small energy gap ensured radiative excitation and de-excitation between the same two electronic states. The h-h and v-h curves overlay as expected. The large difference between the v-v and h-v curves at early times clearly shows a large anisotropy. This anisotropy decays to zero at later times due to molecular rotation.

The h-h curve has a higher intensity than the h-v curve. This is due to the fact that horizontal and vertical polarized light have different transmission efficiencies through the emission channel of the instrument. This effect can be compensated by introducing a calibration factor, which is usually denoted with a capital G and is defined as $G = I_{\mathrm{hv}}/I_{\mathrm{hh}}$ where I_{ij} refers to excitation with i polarization and emission with j polarization. All experimental data sets are corrected by multiplication of I_{vh} with G. Then I_{vv} refers to $I_{\parallel}$, and $I_{\mathrm{vh}} \cdot G$ to $I_{\perp}$. From $I_{\parallel}$ and $I_{\perp}$ the sum and difference curves $S(t)$ and $D(t)$ were generated according to equation (2.1) and (2.2) and fitted to a double and single exponential decay respectively. Typically, chi-square values of 1.2 or less were obtained for a good run. Changes in laser power during the run were associated with large values of chi-square. The assumption of a double exponential decay for the fluorescence decay increases the potential accuracy and is justified by the signal to noise ratio of the sum curve.

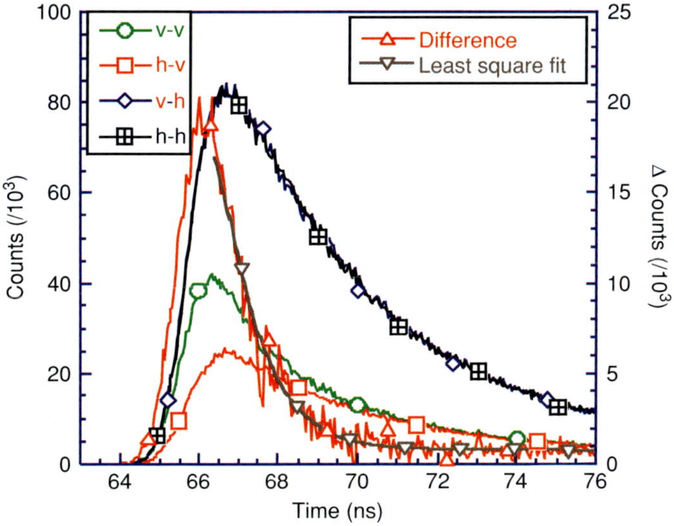

Figure 2.5. Time-resolved fluorescence depolarization. The vertical-horizontal (v-h) and horizontal-horizontal (h-h) curves overlay. The separation between the h-v and v-v curves, which diminishes and vanishes at later times, gives the polarization.

Since the anisotropy decay is much faster than the fluorescence decay for our cases, the difference curve is mainly governed by the anisotropy decay. Consequently, the difference curve was fitted with a single exponential decay. A mean lifetime of the fluorescence intensity decay was calculated from the sum curve fit and the rotational correlation time was obtained by combining the mean fluorescence intensity lifetime of the sum curve with the decay time of the difference curve according to equation (2.3).

3.2. Sample Preparation

The first crude oil sample we used was Kuwait (UG8). We prepared the n-heptane insoluble asphaltenes from this oil using procedures described elsewhere.[15] Briefly, a sample of UG8 crude oil was mixed with 40 times its volume of n-heptane. The resulting solution was stirred in the dark for 24 hours and then filtered using a 1.2 micron pore size nylon Schleicher and Schuell filter. The precipitate was washed with hot n-heptane until the solvent wash was colorless. The resulting powder was air-dried. To check for effects from trapped resin or other oil components, some of this asphaltene was then dissolved in toluene and reprecipitated again using a 40:1 volumetric ratio of n-heptane. After 24 hours of stirring, the resulting precipitate was filtered and washed with hot n-heptane until the heptane wash was colorless. The large heptane volume in this reprecipitation procedure was very light in color indicating that our original asphaltene sample had little if any contamination of materials soluble in n-heptane. Our original and reprecipitated asphaltene exhibit exactly the same rotational correlation times. There is no effect in our data from trapped resins and the like. A crude oil minus its asphaltenes is defined as "the maltenes" or sometimes referred to as de-asphaltened crude oil. The resulting *n*-heptane solution of the maltenes was also used for analysis.

Hydrotreating was performed as described previously[24] in a hydrodesulfurization pilot plant. The feedstock is the resid of atmospheric distillation. Here, metals are first removed followed by deep desulfurization. Product samples were collected at different time intervals from the start of the reactor. During the process the reactor temperature is increased in order to compensate for catalyst deactivation with an aim to keep the sulfur removal on a specified level. Hence samples represent increasing reaction temperatures or lifetime of the reactor bed. Cracking reactions start to dominate around 380°C. The cracking leads to increased product instability seen as asphaltenic sludge formation. Asphaltenes from the feed and the product samples were precipitated by addition of excess heptane (30 cc/g oil) following the IP 143 standard. The oil used was an Arabian heavy atmospheric resid. The asphaltene samples were obtained from the feedstock (TR453-00) and from three process temperatures 359°C (TR453-62), 379°C (TR453-181) and 389°C (TR453-253).

For additional studies, a vacuum residue sample was separated into n-pentane soluble maltenes and n-pentane asphaltenes.[23] The asphaltene fraction (AS) was split into solubility fractions by separating the least soluble fractions first. The asphaltene was dissolved in toluene. After addition of *n*-pentane in a 55/45 pentane/toluene volume-ratio the insolubles were filtered. Both the precipitate and the filtrate were dried/evaporated. The precipitate was labeled the AS6 subfraction of the asphaltene. The remaining asphaltenes of the evaporated filtrate were subject to the same procedure while varying the pentane/toluene ratio to 65/35 (AS5), 75/25 (AS4), 85/15 (AS3), and 95/5 (AS2) respectively. The solvents of the remaining filtrate are evaporated and the residue is classed as AS1. The most soluble asphaltene subfraction AS1 appears more "shiny" than the subfraction AS6 and the other subfractions have a dull black color. In a similar way, n-pentane asphaltenes tend to be shinier than n-heptane asphaltenes. After this fractionation, AS6 had a much lower solubility in toluene. This probably has to do with the significant change in nanoaggregation when lighter asphaltene fractions are removed. In particular, colloidal suspension may be surpressed for more monodisperse samples resulting in flocculation rather than colloidal suspension at the critical nanoaggregate concentration.

Optical densities of all solutions were kept below 0.2 OD to avoid complications from self absorption (although the natural fluorescence red shift coupled with the decreasing absorption at longer wavelength for all asphaltenes mitigates this effect). In addition, at concentrations in the range of 0.06 g/liter and higher, the decay curves exhibited additional anisotropy decay components which may be associated with dimer formation. Consequently, we maintained asphaltene concentrations at or below 0.025 g/L for analysis. Typically, the concentration was 0.006 g/L. For 600 g/mol molecular weight, this concentration is 10 micromolar. All rotational correlation times were determined at room temperature 19°C, and in toluene with a viscosity of 0.59 cP. Two dyes, obtained from Aldrich Chemicals, were also used in this study, octaethyl porphyin (OEP) and a solar dye, *N*,

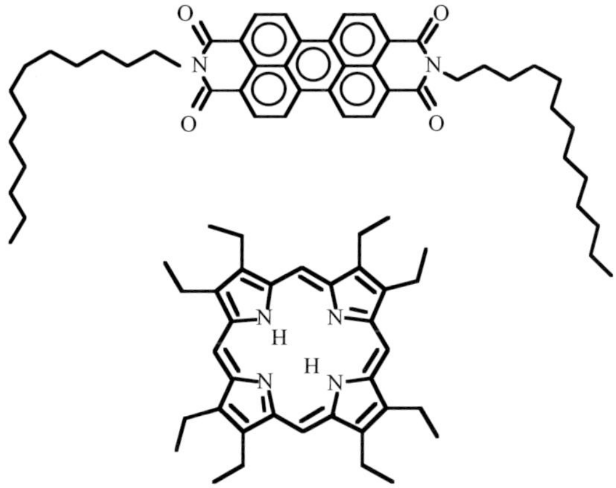

Figure 2.6. The structures of the solar dye and of octaethylporphyrin (OEP). These molecules are comparable in size with asphaltene molecules.

N'-ditridecyl-3,4,9,10-perylenetetracarboxylic diimide; their structures are shown in Figure 2.6.

3.3. Solvent Resonant Quenching of Fluorescence

Any solute or especially solvent that has heavy atoms can quench the fluorescence of organic molecules. The diffusion quenching constant k_q can be related to the diffusion controlled limit,[45] where $k_q = \varepsilon k_D$ and ε is the efficiency of quenching in a collision.

$$k_D = 4\pi N_{av}(D_1 + D_2)R_c, \tag{2.25}$$

where N_{av} is Avogadro's number and D_i are the diffusion constants of the fluorophore and quencher and R_c is the critical quenching radius containing n molecules. For the solvent CS_2, we have observed that the quenching occurs predominantly for short wavelength emission or blue emission. Experiments were performed using various dyes and with CS_2 as a solute in low concentration.

The dependence of efficient quenching on the requirement of proximate energy states implies a resonant interaction. We assume the standard two level mixing scheme where an excited state of the dye resonantly mixes with that of CS_2.

$$\psi = C_d\phi_d + C_{CS_2}\phi_{CS_2} \tag{2.26}$$

where C_i is the coefficient of the pure basis vector ϕ_i for eigenstate ψ. The square of C_{CS_2} gives the efficiency of quenching ε by CS_2 for the dyes. C_{CS_2} is given by equation (2.27) (assuming no diagonal element perturbations)

$$C_{CS_2} = \sin(1/2 \arctan(2W_{12}/\Delta E)), \tag{2.27}$$

where W_{12} is the coupling interaction energy causing wavefunction mixing and ΔE is the energy difference between the two states of interest. For the case treated

Table 2.2. Resonant quenching of fluorophores by CS_2.

Dye	HO-LU energy (cm^{-1})	Lifetime (ns)	$k_q/M_{CS_2}/sec$
Exalite 404	30030	0.61	1.25×10^{10}
Exalite 411	28902	0.62	5.00×10^{9}
Exalite 428	27701	0.53	1.46×10^{9}
Perylene	22779	5.84	3.08×10^{7}
CS_2	31447		

here, ΔE is the energy difference between the lowest electronic excited states of CS_2 and the dye. Table 2.2 shows the dependence of quenching by CS_2 for a series of polycyclic aromatic hydrocarbons, PAH's. The quenching data (determined both by intensity and lifetime measurements) are in accord with wavefunction mixing.[46] The fit is quite good indicating that this resonant interaction model accounts for the fundamentals of the observed energy dependence of k_q. The value of the interaction energy coupling W_{12} is 419 cm^{-1}.

Heavy atom quenching occurs via spin-orbit coupling which is a relativistic effect. This occurs when an electron penetrates the inner electron shell of the heavy atom and is exposed to the high Z unshielded nucleus. The wavefunction mixing shows how the excited electron of the fluorophore can find its way to the core of a solvent molecule. A variety of other halogenated solvents have been shown to exhibit similar energy dependent quenching effects.[47] For all fluorescence experiments, we used toluene as a solvent to avoid any heavy atom quenching effects from the solvent.

Molecular oxygen was found to quench via diffusion with unit efficiency independent of excitation energy.[47] Table 2.3 shows the quenching efficiency of molecular oxygen for fluorophores with fluorescence lifetimes varying three orders of magnitude. Note that all quenching rates are the same within a factor of 2. Thus oxygen had to be excluded from our solutions. The fluorescence cells were fitted with a screw top with a hole in the middle. A silicone/teflon diaphragm sealed the cell. Two GC needles were used to flush the cell with N_2 gas. The N_2 injector was placed deep into the toluene test solution. The vent needle was placed in the gas cap. N_2 was bubbled through the cell for ~15 minutes, care being taken not to evaporate the solvent.

Table 2.3. Quenching of fluorescence by molecular oxygen.

Fluorophore	Lifetime (ns)	S-V slope (Int.)	k_q/M-sec
Exalite 404	0.66	40.4	6.1×10^{10}
Anthracene	3.54	142.7	4.0×10^{10}
Perylene	5.24	201.4	3.8×10^{10}
DiBenzoPent	7.36	237.5	3.2×10^{10}
Ovalene	400	13095	3.3×10^{10}

4. Results and Discussion

4.1. Basic TRFD of Asphaltenes

Figure 2.7A shows the rotational correlation time τ_r of two standard molecules, octaethyl porphyrin (OEP) with a molecular weight of 534 g/mol and solar dye (SD) with a molecular weight of 754 g/mol, a seven ring aromatic with long alkane chains.[20] As discussed later the molecular composition of these model compounds is very similar to the predicted structure of asphaltenes with regard to aromatic/aliphatic carbon ratio as well as the size of the fused ring systems.[22] Figure 2.7A also shows the τ_r's for UG8 asphaltene which will be discussed shortly. Table 2.4 shows the derived parameters for SD and OEP as well as for UG8 asphaltene at the different wavelengths. For OEP, we obtain τ_r equal to the published value from a very different technique, gamma ray-gamma ray perturbed angular correlation.[49] Agreement between very different techniques builds confidence in our results. Figure 2.8 shows the results on TRFD for varying the solvent viscosity on τ_r.[22] The anisotropy decays for a toluene solution ($\eta = 0.59$ cp) and an ethylene glycol solution ($\eta = 16.1$ cp) of SD. The much longer anisotropy decay time predicted by Equation (2.14) is evident in the ethylene glycol solution.

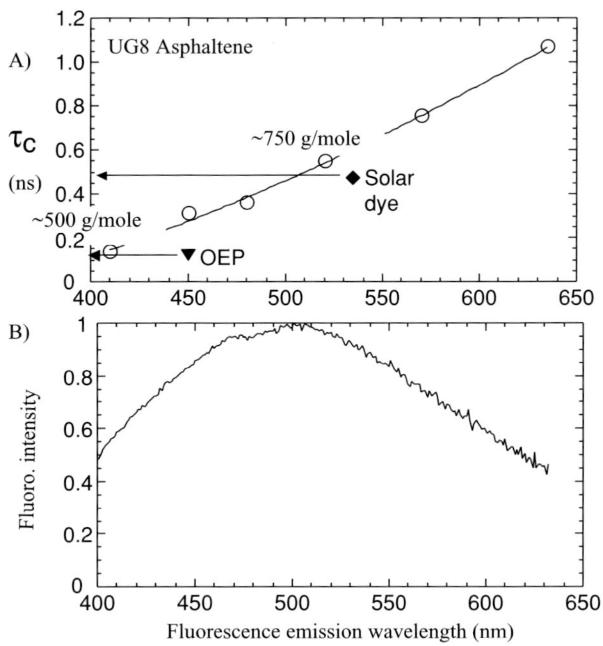

Figure 2.7. (A) Rotational correlation times τ_r for the solar dye, OEP, and UG8 asphaltene for different wavelengths. The τ_r's for OEP and solar dye are comparable to those of asphaltenes. The blue emitting chromophores which are small are in smaller molecules, (B) The fluorescence emission spectrum of UG8 asphaltene showing the optical range of interest; the centroid corresponds to ~750g/mol.

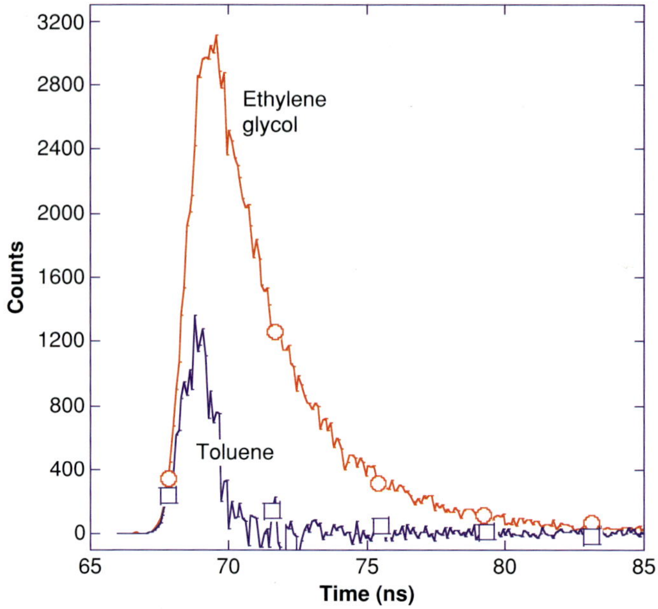

Figure 2.8. Solar dye in ethylene glycol vs in toluene showing the effect of viscosity on fluorescence depolarization in accord with Equation (2.14) Much longer depolarization times are found with high viscosity.

The ratio of the τ_r's is 23.9 and of the ratio of the viscosities is 28.8. Equation (2.14) predicts that these two ratios should be the same. They are close but there is a discrepancy. We may have some error in the measurement of very long decay time for the ethylene glycol solution due to the fact that the anisotropy decay becomes comparable to the fluorescence decay rate. However, the important point is that our measurements are producing expected behavior both with respect to literature values of τ_r's and with respect to viscosity effects.

Table 2.4. The rotational correlation time, anisotropy, and calculated molecular diameters for UG8 asphaltene and two model compounds for various excitation and emission wavelengths.

Sample	λex (nm)	λem (nm)	τ (ns)	Anisotropy	Diameter (Å) sphere	Diameter (Å) oblt. sph. $\rho = 1/2$
asphaltene	365	410	0.1340	0.5907	12.02	13.50
UG8	406	450	0.3115	0.3389	15.92	17.89
	440	480	0.3561	0.3365	16.65	18.70
	480	520	0.5464	0.2623	19.20	21.57
	530	570	0.7518	0.2737	21.35	23.9
	595	635	1.0688	0.2963	24.01	26.98
solar dye	480	535	0.4704	0.3111	18.26	20.52
OEP	406	450	0.1194	0.4248	11.56	12.99

In Figure 2.7A, the τ_r's are presented for several different excitation wavelengths for UG8 asphaltene.[21] Figure 2.7B shows the fluorescence emission spectrum for UG8 asphaltene illustrating the spectral range of interest as discussed above. Two striking features about the τ_r's are immediately apparent. First, the τ_r's of UG8 asphaltene are comparable to those of SD and of OEP. The immediate implication is that the molecular sizes (thus weight) of asphaltenes are comparable to these two model compounds. The maximum fluorescence emission of UG8 asphaltene is approximately 500 nm. The τ_r of UG8 asphaltene matches that of SD at this wavelength (cf. Figure 2.7). Thus we obtain that the asphaltene molecular weight is roughly 750 g/mol. The second striking feature is the large, monotonic variation of τ_r in the relevant spectral range. The factor of 10 variation of rotational rate across the spectral range means that small chromophores (blue-emitting) rotate 10 times faster than large chromophores (red-emitting). (The spectral properties of the PAH and related chromophores are treated in detail experimentally in reference 15 and theoretically in Chapter 4 of this book.) If the large and small chromophores were linked together with any appreciable stiffness, there could not be a factor of 10 difference in τ_r's as they would be unable to rotate independently of each other. The inevitable conclusion is that large and small chromophores do not coexist in the same molecule. That is, there is only one (perhaps two on occasion) chromophore per molecule. Later, in this chapter the relation of TRFD results will be compared with translation diffusion measurements by several other techniques proving that internal rotational relaxation is not a concern for asphaltenes. Figure 2.7 gives the width of the asphaltene distribution. The fluorescence emission curve is roughly half height at $\lambda = 400$ nm as shown in Figure 2.7. The τ_r at this wavelength for UG8 asphaltene is equal to that of OEP. Roughly the width of the asphaltene distribution is 500 g/mol on the short wavelength side and we approximate 1000 g/mol on the long wavelength side. There is a rapidly diminishing population of asphaltene molecules outside this molecular weight range.

The question arises as to whether the interpretation of a single chromophore per molecule is consistent with the observation from Figure 2.7 that the mean molecular weight of asphaltenes is 750 g/mol. We start with coronene for simplicity which consists of seven fused aromatic rings (seven hexagonal rings). The ratio of aromatic sextet carbon to isolated double bond carbon is about right in coronene for asphaltenes.[50] The mean fused ring size of seven for asphaltenes is consistent with direct molecular imaging using STM[51,52] and HRTEM,[53,54] with optical measurements coupled with MO calculations for pericyclic ring systems[55,56] and with 13C NMR[14] results all on asphaltenes. Mass spectral results are consistent with this as well[6] but unlike the other techniques listed, this technique cannot distinguish which rings are fused. Coronene has 24 carbons. Accounting for heteroatomic content, we replace one of the exterior carbons in coronene with nitrogen giving 23 aromatic carbons and one aromatic nitrogen atom in our fused ring core. *n*-Heptane petroleum asphaltenes are approximately 40% aromatic carbon; thus we have approximately 35 saturated carbons in this hypothetical asphaltene molecule. This gives us a total of 83 hydrogen atoms in the molecule. The molecular weight of the hypothetical molecule with $C_{58}H_{83}N$ is 773 g/mol (cf. Figure 2.7)! The

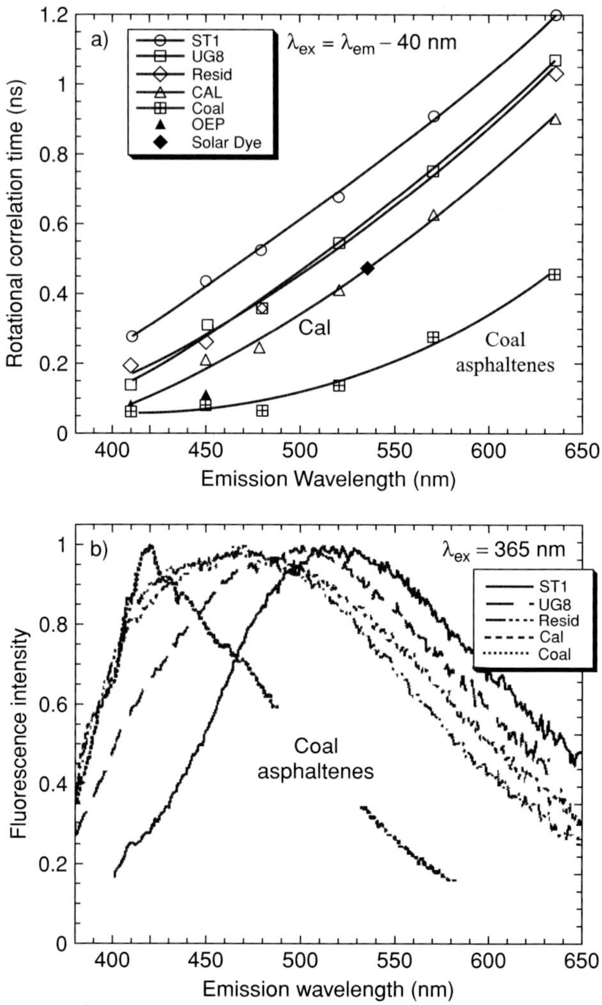

Figure 2.9. (a) τ_r's for a series of asphaltenes. All petroleum asphaltenes are comparable, some differences exist. Coal asphaltenes are much smaller than petroleum asphaltenes. The Cal asphaltene is small for crude oil asphaltenes, it has high sulfoxide content, (b) Fluorescence spectra of the various asphaltenes showing the optical spectral range of interest.

distinct TRFD conclusions regarding the mean molecular weight and the single fused ring system per molecule are absolutely self consistent.

Figure 2.9 compares asphaltenes from a variety of sources. The same trends are observed with all asphaltenes independent of their origin. Figure 2.9a shows that all τ_r's are relatively small indicating that all asphaltenes are small molecules. Figure 2.9b shows that the fluorescence spectra from the crude oil asphaltenes are comparable, the resid spectrum is blue shifted somewhat and the coal asphaltene spectrum is significantly blue shifted. We have never seen evidence that some specific crude oil asphaltenes have much larger molecules. There has been a longstanding uncertainty as to whether different asphaltenes could explain the order(s) of magnitude differences reported for asphaltene molecular weight. Our data show this suggestion is not correct. The TRFD results show that all petroleum asphaltenes are comparable in molecular weight and molecular architecture. That is, the huge increase in τ_r's with increasing wavelength suggests that all asphaltenes have only one fused ring system per molecule. Before going into differences evident in Figure 2.9 for coal versus petroleum asphaltenes, we first explore the more subtle differences among petroleum asphaltenes.

4.2. Many Virgin Crude Oil Asphaltenes—and Sulfoxide

Among the virgin crude oil asphaltenes, Cal asphaltene exhibits the smallest τ_r's and the shortest wavelength of fluorescence emission of all virgin petroleum asphaltenes (cf. Figure 2.9). Thus, Cal has the smallest asphaltene molecules, the centroid of Cal molecules is shifted to shorter wavelength as shown by the spectral shift in the fluorescence emission; at shorter wavelength all τ_r's are smaller. In addition, Cal has the smallest τ_r's of any virgin crude oil asphaltene, thus for both reasons, Cal asphaltene molecules are the smallest we have measured for virgin crude oils. (Virgin crude oil means that no processing or heat treatment has been performed on the corresponding oil or asphaltene sample in contrast to resid asphaltenes.) Cal is one of the heaviest crude oils we have used to generate asphaltenes. It is a bit counterintuitive that a heavy crude oil possesses small asphaltene molecules. Cal has a unique trait amongst all of our virgin crude oil asphaltene samples. Cal asphaltene has several mass percent of sulfur, thus one sulfur atom per every other molecule on average; that is not unusual. However, 44% of the sulfur of Cal is in the form of sulfoxide, while all of our other asphaltenes we have measured have sulfoxide below 5%.[9,10] The sulfoxide group is very polar, ~4 debye. Furthermore, the sulfoxide group is known to be alkyl sulfoxide.[9] Thus, Cal asphaltene is similar to a bidentate ligand. The alkyl sulfoxide is a binding site and the single fused ring system in the molecule is a binding site. Since the sulfoxide represents a tight binding site due to its polarity, then the fused ring system must be smaller than normal to maintain a constant overall binding energy—this being dictated by the asphaltene solubility classification. This makes sense only if there is a single fused ring system per molecule. If there were more than one ring system in a single asphaltene molecule, then the intermolecular binding energy would be determined by both the size and number of fused ring systems per molecule. Reduced binding mandated by the presence of sulfoxide could be achieved by decreasing the number of fused ring systems per molecule. Consequently, the sulfoxide group would not necessarily be associated with smaller fused ring systems.

4.3. Asphaltene Solubility Subfractions

Six subfractions of a virgin crude oil asphaltene where prepared.[23] Figure 2.10 shows the plot of fluorescence spectra for a series of subfractions of a single

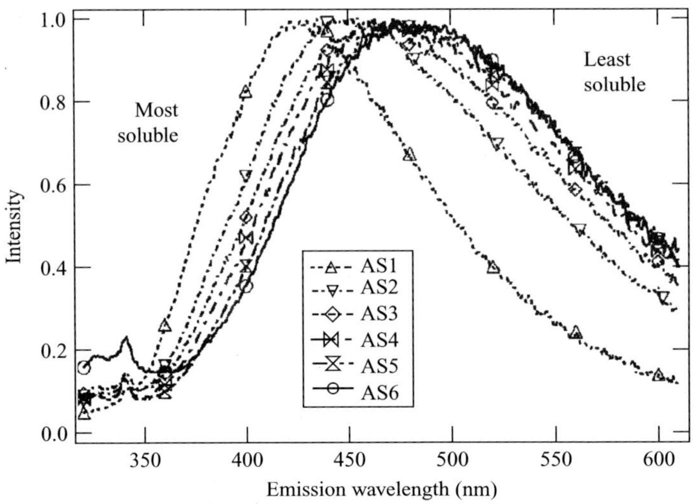

Figure 2.10. Fluorescence spectra of a series of solubility subfractions of an asphaltene. Solubility reduction in n-pentane toluene solutions is associated with a red shift indicating larger aromatic ring systems.

asphaltene. The fractions were obtained by virtue of their solubility in different n-pentane—toluene ratios. The fluorescence spectra exhibit a monotonic variation showing that the most soluble fraction has a population centroid towards smaller fused ring systems while the least soluble fraction has the largest fused ring systems. Figure 2.11 shows the τ_r's for two of the subfractions at many emission wavelengths. Similar behavior of τ_r's is seen for the two subfractions as for all other asphaltenes. Smaller wavelengths correspond to much smaller molecular size. Figure 2.12 shows the behavior of all six subfractions at one emission wavelength. Essentially at a given wavelength all subfractions are close in molecular size, but still the less soluble fractions contains larger molecules for a specific excitation and emission wavelength.

This monotonic and continuous behavior of the asphaltene subfractions is not surprising. First, the solvent system used to obtain the subfractions is toluene and n-pentane. This solvent system interacts primarily via polarizability; toluene is polarizable due to the π-electrons while n-pentane is less polarizable, thus a poor solvent. Asphaltene flocculation is known to depend heavily on van der Waals forces of the π-electron system.[57] Bigger fused ring systems interact more strongly. This is one major reason why the solubility of PAH's in toluene decreases dramatically with increased fused rings. Since the solvent system used to isolate different asphaltene fractions is alkane plus toluene, the primary interaction is van der Waals. Thus, monotonic behavior is obtained for asphaltene molecular size and fused ring size versus solvent quality. This continuous grading for asphaltene subfractions argues against a bimodal distribution for asphaltene molecular weight.

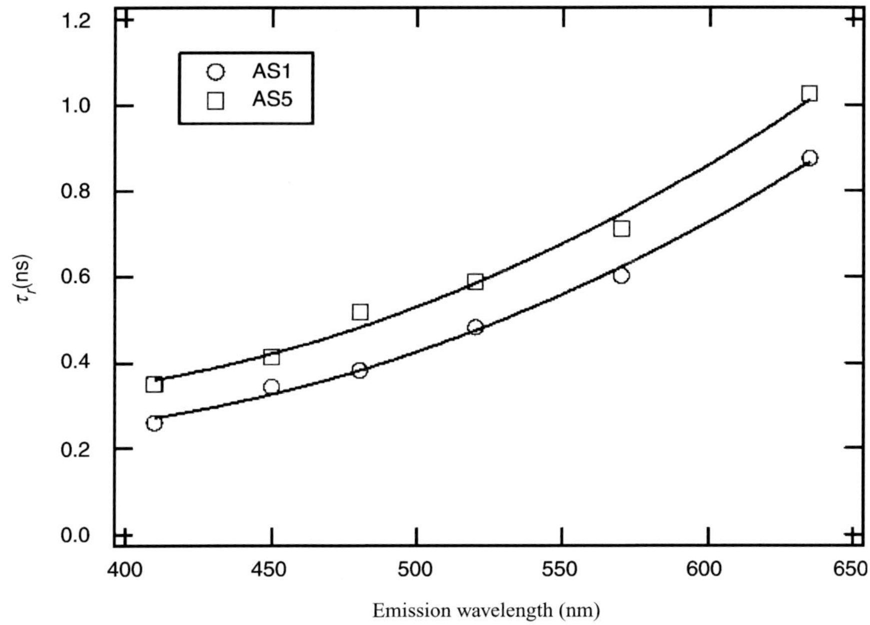

Figure 2.11. Asphaltene solubility fractions. The τ_r's of the most soluble (AS1) and the next-to-least soluble (AS5) samples. The less soluble fractions have some increase in molecular size at a given wavelength but a much bigger variation is seen in molecular size vs. wavelength for both fractions. For the low solubility sample AS5, the molecular population centroid is displaced to larger chromophores (red shifted fluorescence) and larger molecules (larger τ_r's).

4.4. Asphaltenes and Resins

We compare asphaltenes and resins in Figure 2.13.[20] This resin was prepared as being the n-heptane soluble, n-pentane insoluble fraction so it is clearly not so different from asphaltene. Some might refer to this solubility cut as the heaviest fraction of the resins. The fluorescence emission spectrum Figure 2.13b of the resin is blue shifted significantly so the average number of fused rings for resins is less than the asphaltenes. For these resins as defined here, the emission maximum in the fluorescence spectrum has a τ_r corresponding to $\sim$500 g/mol. Nevertheless, at any given spectral range, the resin molecules are nearly as large as those of the asphaltene (cf. Figure 2.13a). This is not surprising; there is only a subtle difference between the asphaltenes and resins in terms of solubility. The continuous grading of asphaltene subfractions extends into the resin fraction.

4.5. Coal Asphaltenes versus Petroleum Asphaltenes

We can use the contrast between coal asphaltenes and petroleum asphaltenes to gain tremendous insight into asphaltene structure and function. It is conventional wisdom, which this time is correct, that coal asphaltenes are in general smaller

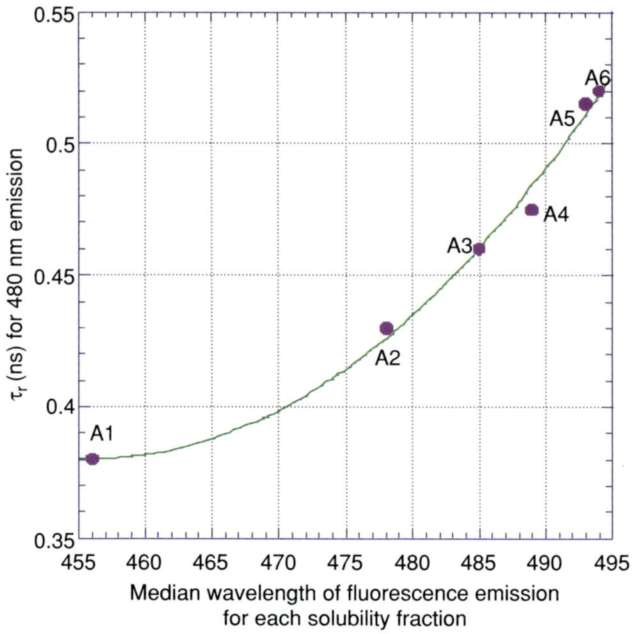

Figure 2.12. The τ_r's for 480 nm emission vs the median wavelength of fluorescence emission ($^1/_2$ above, $^1/_2$ below) for the solubility fractions. The τ_r's are comparable for all solubility fractions but somewhat larger for less soluble fractions. The median molecular wavelengths are obtained from emission spectra shown in Figure 2.10.

than petroleum asphaltenes. Interestingly, laser desorption mass spectral studies indicate this; we will return to this topic shortly. Figure 2.14 shows that the coal asphaltenes are characterized by shorter wavelength emission, thus have smaller chromophores. Figure 2.15 the τ_r's of coal asphaltene molecules at any given wavelength are much smaller than those of petroleum asphaltenes. Figure 2.9 also shows the same trends for a different coal asphaltene sample. Consequently, the centroid for the coal asphaltene molecular population is on order 500 g/mol, while the centroid for petroleum asphaltenes is 750 g/mol. In fact, the coal asphaltenes may be smaller; their τ_r's are nearing our short time limits. The HRTEM results also show by direct imaging that the petroleum asphaltenes have ring systems that are ~1.0 nm on average while for coal, 0.7 nm. Thus, HRTEM is in agreement with fluorescence emission spectroscopy that the coal asphaltene ring systems are significantly smaller than those of the crude oil.[53,54] In fact, the coal asphaltenes are much lighter in color than the petroleum asphaltenes, this is discernable with the unaided eye. But coal is known to possess large ring systems. Coal is after all a solid. It is quite interesting that the toluene soluble portion of coal contains only relatively small molecules while the source coal contains much larger molecules than petroleum.

One might suspect that the very different alkane fractions of the two types of asphaltenes are associated with this large difference in size. Figure 2.16 shows the

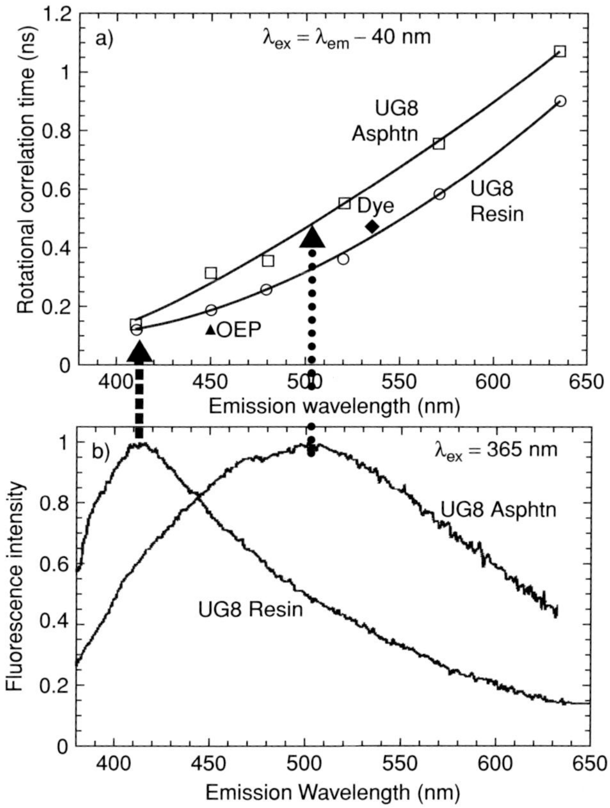

Figure 2.13. (a) The τ_r's for UG8 asphaltene and UG8 resin. The chromophores are comparable at each emission wavelength. (b) The fluorescence emission is shifted substantially to shorter wavelength for the resins; thus, resins molecules are much smaller than asphaltene molecules.

13C NMR comparison of alkane versus aromatic carbon for a petroleum asphaltene and for a coal asphaltene; the aromatic carbon absorption at ~125 ppm is much bigger for the coal asphaltene. The coal asphaltenes lack alkane carbon because the source material lacks much alkyl carbon. The concept that emerges is very simple. The solubility classification of asphaltene mandates a balance between attractive and repulsive intermolecular interactions. The attractive interactions are primarily those found in van der Waals interaction of π-bond systems. Plausibly, the molecules stack like (disordered) pancakes taking advantage of polarizability of the fused ring system along with dipoles found in the rings associated with nitrogen. All the nitrogen in asphaltenes is pyrrolic and pyridinic, thus contained in the ring systems. The attractive forces grow with the number of rings in the fused ring system. The repulsive forces are primarily associated with steric disruption due to the alkane substituents. The petroleum asphaltenes have a large alkane fraction (~60% of the carbon), consequently this large repulsion must be balanced by a large attraction—thus large aromatic ring systems. The coal asphaltenes have

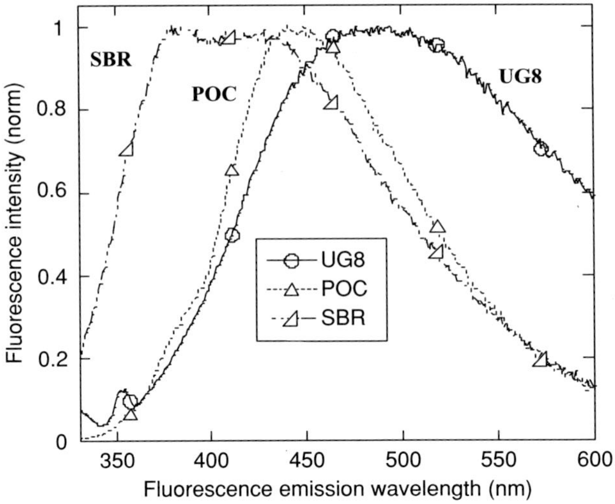

Figure 2.14. The fluorescence spectra for coal asphaltenes POC and SBR versus petroleum asphaltene UG8. The coal asphaltenes have molecular population centroids shifted to much shorter wavelength.

very little disruption from their small fraction of alkane carbon. Consequently, to maintain toluene solubility, the very definition of asphaltene, the fused ring systems of coal asphaltenes must be smaller. These freshman chemistry principles are commonly known; it turns out freshman chemistry helps us understand a great

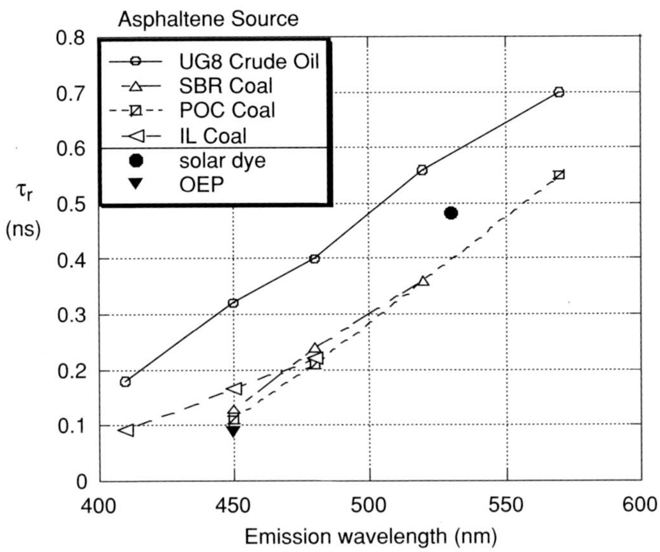

Figure 2.15. The τ_r's of the coal asphaltenes and of the UG8 petroleum asphaltene. The coal asphaltenes have much smaller τ_r's than the petroleum asphaltenes.

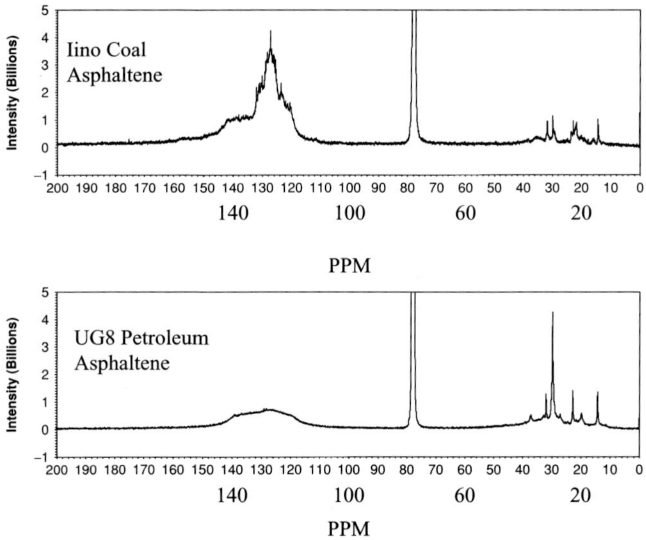

Figure 2.16. The 13C NMR spectra for a coal asphaltene and a petroleum asphaltene. The coal asphaltenes are mostly aromatic carbon whereas the petroleum asphaltenes have substantial saturated carbon in addition to aromatic carbon.

deal about asphaltenes. Figure 2.17 shows molecular structures consistent with coal versus petroleum asphaltenes, illustrating these differences in ring size and alkyl substitution.

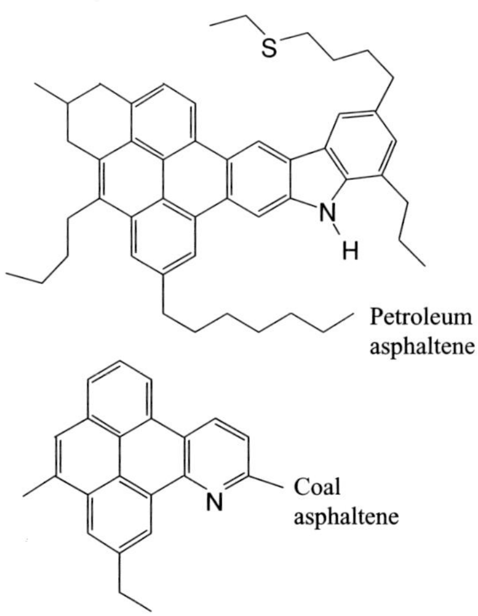

Figure 2.17. Proposed molecular structures for coal and petroleum asphaltenes illustrating the differences in molecular size, ring size, and alkane content.

4.6. Thermally Processed Feed Stock

From these ideas, one develops the prediction that if a hydrocarbon feedstock is hydrocracked, that (1) the least soluble asphaltene fraction minus its alkanes would become insoluble (coke), and (2) molecules previously in the resin fraction minus their alkyl substitution would become smaller asphaltenes similar to the coal asphaltenes. This is in fact observed.[24] Figure 2.18 shows the τ_r's from TRFD for a series of asphaltene samples prepared from a feedstock subjected to increasing thermal cracking. Data is presented for asphaltenes isolated from the initial feedstock and for asphaltenes isolated for the feedstock subjected to increasingly severe conditions. However, the temperature was kept below 400°C. At temperatures above 380°C, there is a substantial increase in severity of reactions.[58] The asphaltene samples are from the conditions: initial feedstock (−00), 359°C (−62), 379°C (−181) and 389°C (−253). The τ_r's become smaller with increasing cracking severity (up to 389°C). Figure 2.18 shows the Iino coal sample as well to illustrate that moderate cracking causes the petroleum asphaltenes to become increasingly similar to coal asphaltenes in terms of molecular size. Figure 2.19 shows that the fluorescence spectra monotonically shift to shorter wavelength with increasing processing time and temperature as well. Figure 2.19 shows the Iino coal sample as well to illustrate that moderate cracking causes the petroleum asphaltenes to become increasingly similar to coal asphaltenes in terms of aromatic ring size and molecular size. The freshman principles, van der Walls attraction of π-bonded ring systems versus steric repulsion explain the variation seen between coal versus oil asphaltenes and between virgin crude versus thermally processed

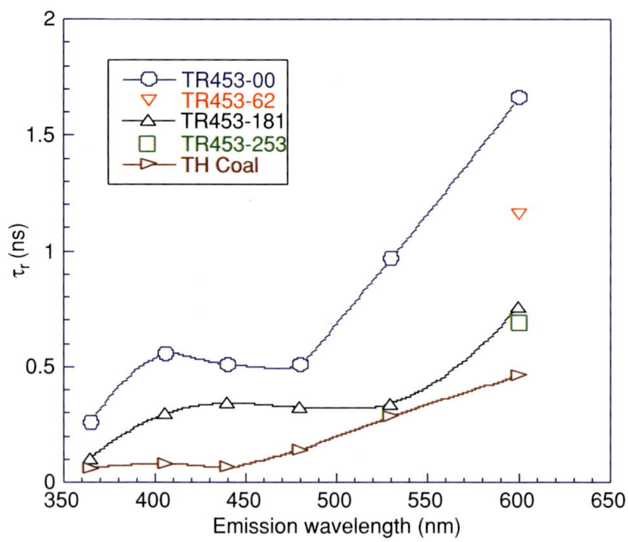

Figure 2.18. The τ_r's of a series of asphaltenes prepared by thermal hydrotreatment. Cracking of feedstock results a reduction of the corresponding asphaltene molecular size.

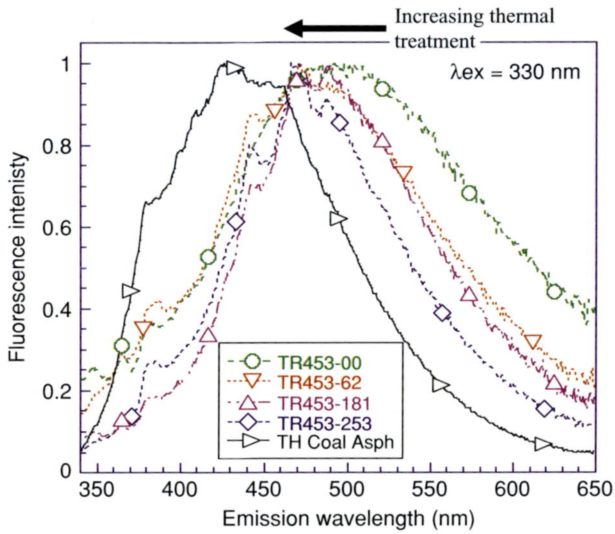

Figure 2.19. The fluorescence spectra of the asphaltenes prepared from thermal hydrocracking of a hydrocarbon feedstock. Cracking of the feedstock shifts the asphaltene fraction to smaller chromophores (blue shifted).

asphaltenes. At higher temperatures more extreme reactions take place that are harder to control.

We can see that the simple chemical principles of steric repulsion and intermolecular π-system attraction are useful to predict observations of a somewhat complex process, thermal hydrotreatment of asphaltene. However, these predictions follow only if asphaltene molecules contain a single fused ring system.

If asphaltene molecules contained multiple fused ring systems, then there is no expectation that smaller fused ring systems must be found in the asphaltene fraction of the treated source material. The intermolecular binding would be determined by both the size *and number* of fused ring systems per molecule. Cracking would reduce the number of fused ring systems per molecule by cleaving alkane linkages between ring systems. Such a cleavage reaction would lead to rather soluble daughters. Only the largest ring systems would remain in the asphaltene fraction. This is contrary to observation. These data support the argument that asphaltene molecules have a single fused ring system per molecule.

As an aside, we note that laser desorption mass spectral studies of asphaltenes have been shown to contain extensive baseline signal. However, laser desorption of polystyrene works rather well. There is the implication that the number of fused rings may impact gas phase (radical?) reactions. The smaller ring systems of coal asphaltenes may yield a lesser baseline issue—this was observed in our study.[23]

One can use different solvent systems to isolate subfractions of asphaltenes. Figure 2.20 shows the τ_r's for a series of subfractions of an asphaltene using acetone and toluene as the solvent system. Acetone with its polar carbonyl function can interact with polar and possible charged groups on asphaltenes. The electron

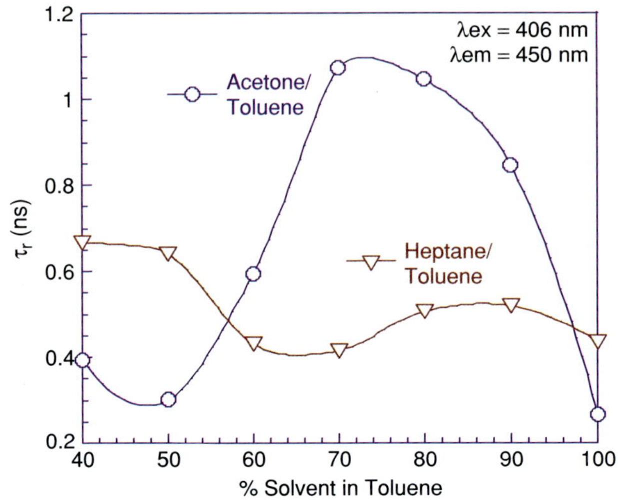

Figure 2.20. The τ_r's for a series of asphaltene solubility subfractions prepared from acetone and toluene. The nonmonotonic behavior shows that polarity of acetone/toluene is important in determining solubility. The τ_r's for the n-heptane/toluene fractions are much more monotonic.

lone pairs of oxygen can participate in hydrogen bonding with corresponding functional groups in asphaltenes. Figure 2.20 shows nonmonotonic behavior of asphaltene molecular size versus acetone fraction in the solvent system. Acetone, with its very different interaction selects for different molecular attributes of asphaltenes rather than the size of the fused ring system. There has been some disagreement as to the most important interactions in asphaltenes with candidates including van der Waals and polar interactions. Part of this disagreement is based on the type of intermolecular interaction being searched for. Here we show that if solvents that select van der Waals interaction are used, one finds asphaltene polarizability paramount; if one uses acetone, interactions other than polarizability are found.

Table 2.5 shows τ_r's for use of N-methyl pyrrolidone (NMP) also called N-methyl pyrrolidinone to isolate subfractions of an asphaltene. Accounting for solvent viscosity, the τ_r's for the NMP soluble and insoluble fractions were measured. Not surprisingly, the derived molecular size for the NMP insoluble fraction

Table 2.5. τ_r values for 410 nm excitation, 450 nm emission for UG8 asphaltene and solubility fractions.

Sample	τ_r (ns)	Solvent viscosity (cp)	Diameter (sphere) (Å)	Diameter (oblate sheroid*) (Å)
Toluene soluble	0.32	0.59	16.1	18.1
NMP soluble	0.65	1.67	14.4	16.2
NMP insoluble	0.47	0.59	18.3	20.6

* Long axis, aspect ratio = 2.

is seen to be bigger than the NMP soluble fraction. These two fractions are seen to straddle the corresponding molecular size for the whole asphaltene.

TRFD measures the rate of chromophore rotational diffusion. For molecules that are roughly half aromatic carbon having a single chromophore, the chromophore size and the molecular size are directly related. The question arises, if two chromophores are tethered by an alkane chain, what is the impact on the rotational diffusion constant and thus on TRFD results. In part, this reduces to a question of the stiffness of the linkage. For petroleum asphaltenes, it is repeatedly reported that the alkane chain length is on the order of four to six carbons.[14] If asphaltene molecules had more than one chromophore, then the alkane tethers connecting different chromophores are expected to be rather short—thus somewhat stiff. A linear arrangement of ring systems and alkane tethers would be rather surprising; more cross-linking would be expected. Cross-linking (two or more alkane linkages between ring systems) would add significant stiffness. Therefore, we believe it is reasonable to expect some intramolecular stiffness if multiple chromophores existed in single asphaltene molecules. Increased stiffness from cross linking chromophores would increase rotational diffusion constants. However, there is no indication from TRFD studies that there is any extra stiffness or larger rotational diffusion beyond those of small model compounds. We conclude there is generally one chromophore per asphaltene compound.

4.7. Alkyl-Aromatic Melting Points

One example of this freshman chemistry principle van der Waals attraction vs steric disruption is shown in Figure 2.21. The melting point of alkyl aromatics is shown to depend dramatically on alkyl substitution and ring number. First, the melting point of benzene is much lower than that of naphthalene which in turn is much lower than anthracene. This shows the increase in van der Waals interaction with increasing numbers of fused rings. Second, these data show the affect of alkyl substitution. For a single ring system (alkyl benzenes), only a single methyl group is needed to interfere with proper crystalline order. Ethyl benzene has a comparable melting point as toluene. As the alkyl group gets sufficiently big, the melting point starts to increase as would normally be expected. For β-alkyl naphthalenes, the methyl group causes some reduction of melting point, but the ethyl group reduces the melting point further. Two ring systems bind more tightly so they require more alkane to disrupt stacking. Finally, for β-alkyl anthracenes, longer chain alkanes are required to disrupt stacking. β-methyl anthracene has nearly the same melting point as anthracene. However, β-ethyl anthracene exhibits a decrease in the melting point. We do not have further data but we suspect that the longer alkanes would continue to reduce the melting point of β-alkyl anthraces until the chain is three or four carbons long. These simple concepts, stacking of π-ring systems vs steric disruption are seen to describe the fundamentals of β-alkyl acenes. The same principles play an important role in defining asphaltene solubility.

One important point is that unsubstituted aromatic ring systems often form T-shaped structures with the edge of one ring T-ing into the middle of another

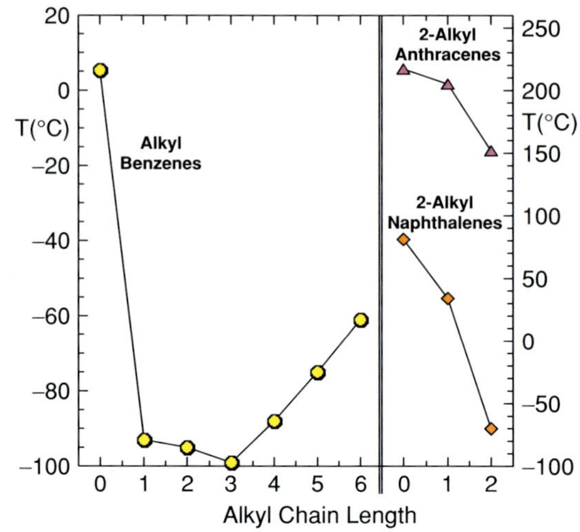

Figure 2.21. Melting point behavior of alkylaromatics showing the effects of steric disruption from alkane substituents versus intermolecular attraction of large π ring systems. Longer alkane chains are needed to disrupt larger fused ring systems.

ring. This structure, which is seen both in crystalline structure and in van der Waals complexes, occurs because the electron density of the (bonding) π-orbitals is interior in the ring leaving an electron deficiency outside the ring system. Thus, the T structure possesses favorable electrostatics. However, increasing peripheral alkane substitution on the ring system increasingly precludes T-shaped binding, thus stacking becomes more favorable.

4.8. Asphaltene Molecular Structure 'Like your Hand' or 'Archipelago'

There are two prevailing descriptions of asphaltene molecular architecture. The description supported by *all* data herein is that asphaltene molecules are "like your hand" where the palm represents the single fused aromatic ring system and the fingers represent alkane substituents. Another description of asphaltene molecular architecture is the archipelago description where each asphaltene molecule contains several fused aromatic ring systems linked together by alkane chains. While asphaltenes may include contributions from both structural classes, here we treat the structure of the bulk of asphaltene molecules.

TRFD. We have shown above that all of the TRFD results are consistent with a single chromophoric group being present in asphaltene molecules. This includes (1) small molecular weights that are incompatible with large polymeric structures (2) small chromophores (fused ring systems) are in small molecules and large chromophores are in large molecules, thus the chromophores are not cross linked, (3) the increasing binding by adding sulfoxide reduces the (single) chromophore size to keep the solubility constant (4) that coal asphaltene chromophores are smaller than petroleum asphaltene chromophores due to the lack of alkane repulsion in coal asphaltenes. This follows only if coal and petroleum asphaltenes have the same *number* of fused ring systems per molecules—thus one ring system (5) thermal degradation leads to a reduction not increase in solubility and coke—breaking up polymers into small subunits (archipelago decomposition) leads to an increase in solubility. Thus TRFD shows that the "hand" model for asphaltene molecular structure wins, hands down. There are other considerations we well.

Electronic Absorption. One immediately notes that the archipelago model is not consistent with molecular weights reported here or in Chapter 3. At 750 g/mol average, one has roughly seven aromatic rings to place. For the "like your hand" description, there would be one seven-membered fused ring system on average per asphaltene molecule. For the archipelago description one would have, say, three ring systems each with two- or three-fused aromatic rings. A molecule with a total of seven rings distributed in isolated 2- and 3-membered ring systems does not possess—COLOR. At last check, asphalt and asphaltene are deeply colored. It would not make sense proposing that asphaltenes are made of colorless ingredients.

Adherence. Small ring systems are not that sticky whereas asphaltenes are notoriously sticky. For example, toluene is a liquid at room temperature and does not even stick to itself that much. The adherence of aromatic ring systems is due in large part to large area with constant binding per unit area. Decolorizing carbon from freshman chemistry works on this principle for removing large aromatic ring systems (which as noted above—are colored) from reaction solutions of smaller molecules. Decolorizing carbon—which is black due to large fused aromatic ring systems—binds large colored compounds. Large aromatic ring systems are adherent not only because of the large area of binding but also because the cost in entropy in binding is paid once. In contrast, for the archipelago model, there is an entropy cost to bind each of the different fused ring systems. Thus, archipelago type structures are much less sticky. Consider the extreme archipelago—polystyrene. Polystyrene which is colorless is a poor model for asphaltenes. In small molecules weights (~800 amu) it is a liquid. The archipelago structure is not a good model for asphaltenes.

Hierarchical Aggregation. There are three chapters in this book that report the formation of asphaltene aggregates at concentrations of ~150 mg/L in toluene, Chapters 9–11. These chapters use direct and indirect methods to conclude that the aggregates are quite small with aggregation numbers on order 10 or even less. Direct molecular imaging from TEM (Chapter 8) as well as SANS results (Chapter 14) are consistent with these nanoaggregates. Universal flocculation behavior is observed that imply clustering of these nanoaggregates at high concentrations (Chapter 17). Near-infrared studies (Chapter 18) and conductivity studies (Chapter 10) corroborate these results. This hierarchy of asphaltene aggregation has implications on asphaltene molecular structure that are consistent with results presented in this Chapter. That is, if the "like-your-hand" model is correct for asphaltene molecular structure, then there is generally one binding site per molecule. This prevents covalent cross-linking across two nanoaggregates. On the other hand, if the archipelago model is correct, then multiple binding sites exist in each molecule. The archipelago model presumes that each asphaltene molecule has multiple islands of fused aromatic rings linked together by alkane chains. Upon aggregate formation, the aggregates would be covalently cross-linked and the entire system would start to gel. That is, it is very unlikely that all binding sites within a single molecule could fold onto themselves to stack. The folding requirements would be too great. The fact that we see a hierarchy in asphaltene aggregation implies that the archipelago model is not correct.

4.9. Considerations of the Fluorescence of Asphaltenes

We have used fluorescence methods to investigate molecular properties of asphaltenes. Here we list a few of the dominant principles of asphaltene and crude oil fluorescence. Our primary focus here is to show that there is no precluding problem associated with the application of fluorescence methods for the investigation of asphaltenes. A thorough review of the photophysics of crude oils and asphaltenes is found elsewhere.[15] Fundamentally, crude oil and asphaltene chromophores are governed by the energy gap law.[59] The well known energy gap law[60] is a consequence of the magnitude of the Frank-Condon factor in optical transitions in molecules. This factor, which depends on energy mismatch, accounts for vibrational state overlap of the initial and final state. For poor overlap in radiationless transitions, this decay is impeded thereby yielding large fluorescence quantum yields. The intensity of fluorescence is reduced by the radiationless relaxation (with rate constant k_{ic}) or the so-called internal conversation when the energy of the transition ΔE (thus photon) is small. Equation (2.28) lists the energy gap law.

$$k_{ic} = A \exp\left(-\frac{\Delta E}{\alpha}\right), \qquad (2.28)$$

Where A is the pre-exponential frequency factor and α is dependent on the decay mechanism. Essentially, the energy gap law accounts for the reduction of quantum yield for large chromophores (with small transition energy). Because asphaltenes have relatively large chromophores, the quantum yield of asphaltenes is somewhat reduced.[15] However, there is no implication that a particular class of asphaltene compounds is excluded from fluorescence interrogation. Figure 2.22 shows the excellent applicability of the energy gap law to crude oils. Similar results are found for asphaltenes. Smaller quantum yields are obtained for asphaltenes due to the photophysics of the energy gap law, but this does not preclude investigation of asphaltenes by fluorescence; one simply uses higher power lasers.

A second issue arises. Is there intramolecular relaxation of electronic excitation in asphaltenes? Potentially this could be problematic. For example, if intramolecular quenching occurred between putative separate fused ring systems in a single molecule, then this energy donor ring system could not be investigated by fluorescence methods. Quenching always increases decay rates; this has been shown conclusively for crude oils.[18] Figure 2.23 shows the effect on lifetime of collisional quenching in asphaltene solutions induced by increasing concentrations. The fluorescence signal exhibits a shorter and shorter lifetime as the concentration increases. Figure 2.24 shows the fluorescence emission from a dilute asphaltene

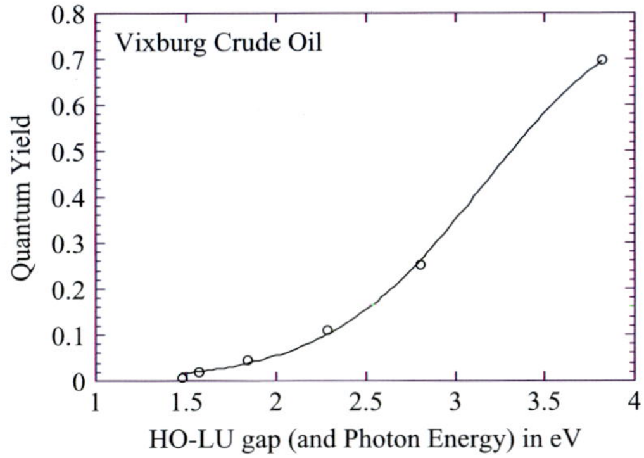

Figure 2.22. The quantum yield of crude oil versus excitation photon energy. The quantum yield varies over a large range depending on photon energy. The energy gap law fits this data.

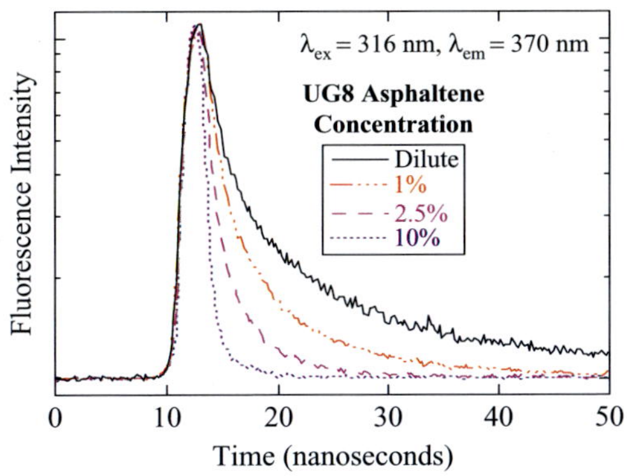

Figure 2.23. Intermolecular interactions of π-systems reduce fluorescence lifetimes. High concentrations create significant intermolecular collisional relaxation of asphaltene molecules and very short fluorescence lifetimes. The fluorescence lifetime of the 10% solution equals the laser time width; the fluorescence lifetime is below 100 picoseconds.

solution, a dilute maltene solution and from a concentrated asphaltene solution. (Maltene is a de-asphaltened oil.) There is no evidence of reduced fluorescence lifetimes in the crude oil solution. Nobody is suggesting that maltene molecules contain multiple chromophores. Yet maltenes and asphaltenes in dilute solutions exhibit similar fluorescence lifetimes. Intramolecular relaxation is not observed in asphaltenes.

4.10. Asphaltene Molecular Diffusion; TRFD vs Other Methods

Several other methods have been employed to measure asphaltene diffusion constants. All these other methods measure translation diffusion constants. These include Taylor dispersion (TD),[61] Fluorescence correlation spectroscopy (FCS)[62]

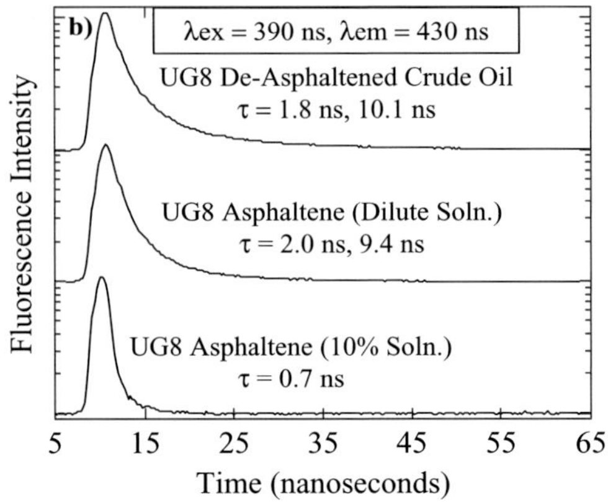

Figure 2.24. The fluorescence lifetimes of dilute solutions of asphaltene, de-asphaltened oil and a concentrated solution of asphaltene. The dilute solutions exhibit long lifetimes; there is no evidence of *intra*molecular energy transfer or quenching effects in asphaltenes, they are similar to other crude oil chromophores. High concentrations allow substantial *inter*molecular interactions which result in greatly reduced lifetimes.

and NMR diffusion methods.[63] TD relies on asphaltene molecules that absorb color, FCS on fluorescent asphaltene molecules and the NMR measurements rely on asphaltene molecules having hydrogen. The only excluded asphaltene molecules would be Type IIA diamonds. Asphaltenes would really become popular if these type of 'molecules' were found therein.

The TD dispersion measurements have been performed on the coal asphaltene (the Tanito Harum or Iino) sample, which is exactly the same sample reported in Figure 2.9. The TD measurements use optical absorption and measure translation diffusion, whereas TRFD uses fluorescence and measures rotational diffusion. The TD results agree exactly with the TRFD results—nearly identical molecular sizes are found.[61] This comparison establishes that there is no appreciable internal rotation in coal asphaltenes and that the TRFD measurements correspond to bulk molecular rotation. Recent FCS results on this same asphaltene sample are also in close agreement with both these measurements.[64] The FCS measurements rely on analyzing the autocorrelation function of the fluorescence signal vs time and monitoring its decay (similar in concept to Dynamic Light Scattering, cf. Ch. 17). These measurements are performed at ultralow concentrations and so avoid any possible aggregation difficulties. FCS measurements on petroleum asphaltenes yield very similar results on molecular size as TRFD, showing that asphaltenes are comparable to porphyrins.[62,64] Thus, FCS results when compared with TRFD results show that there is no internal rotational relaxation in petroleum or coal asphaltene molecules. The large dispersion found in TRFD shows that only one chromophore exists in each asphaltene molecule. The NMR results on UG8 asphaltene yield similar results but with a slightly larger size in the distribution.[63] The NMR results use somewhat higher concentrations; their lowest being roughly 50 mg/l. This is where dimers are thought to form,[65] and so the NMR results might be skewed to slightly larger size. All four diffusion measurements TRFD, TD, FCS and NMR applied to asphaltenes are in close agreement and show that asphaltenes are small molecules, in general with a single chromophore in the molecule. This congruence among all diffusion measurements along with all mass spectral techniques including laser desorption ionization creates a compelling body of work that should largely terminate the debate on asphaltene molecular weight.

5. Conclusions

The TRFD results are rather clear in their interpretation. The rotational diffusion constant of all asphaltene molecules are small, thus, the asphaltene molecules are small. The order of magnitude variation of rotation diffusion constant with chromophore size mandates that the different chromophores are not cross-linked. Consequently, the TRFD results indicate there is one chromophore per asphaltene molecule. The TRFD data indicate that the mean molecular weights of essentially all virgin crude oil asphaltenes is ~750 g/mol with a range of ~500–1000 g/mol. There is a rapidly declining asphaltene molecular population outside this range. All four methods of measuring asphaltene molecular diffusion are in agreement TRFD, TD, FCS, and NMR. These techniques rely on fluorescence, color absoption, or on proton content thereby capturing all asphaltene molecules. The very close agreement with rotation and translation diffusion constants rules out internal rotational relaxation in asphaltene molecules. The large dependence on wavelength of rotation diffusion proves that there is only one chromophore per aspheltene molecule. This molecular weight data is in agreement with all mass spectral studies that do not use laser desorption, and some that do.

The TRFD results and the optical absorption and emission data predict roughly seven fused rings per molecule. We assign a rough range of 4–10 fused rings per chromophore. Here we show the TRFD results are consistent with that assignment. This conclusion is in agreement with 13C NMR analysis, and all direct molecular imaging studies of STM and HRTEM. Furthermore, this conclusion is consistent with recent high resolution mass spectral results that confirm individual asphaltene molecules possess 4–10 rings on average. Finally, a single chromophore of seven fused rings coupled with requisite alkane and heteroatom composition known for petroleum asphaltenes necessarily weighs ~750 g/mol. Consequently, TRFD ties together a large body of data ferreting out both molecular weight and molecular architecture.

A single construct is used to understand studies on many different asphaltene source materials. Coal asphaltenes are found to be smaller; they have much smaller alkyl substitution and they also possess smaller ring systems vs petroleum asphaltenes. This contrast coupled with the systematic trends observed for thermally hydrotreated samples shows that freshman chemistry principles dominate in establishing asphaltene identity. Two freshman chemistry principles are held in balance to determine asphaltene solubility; van der Waals attraction of π-bonded fused ring systems vs steric repulsion dominated by alkane substitution of the ring systems. These competing principles, known from melting point behavior of alkyl aromatics, dominate asphaltene solubility—the defining characteristic of asphaltenes. The asphaltenes *are* alkyl aromatics so this correlation is to be expected.

The objective to establish structure–function relations for asphaltenes has been achieved; asphaltene solubility is shown to relate to asphaltene molecular structure. This is the first step in satisfying petroleomics—relating asphaltene properties to asphaltene structure and composition. Refinement of the picture contained herein of asphaltene molecular structure will likely be mandated by new data. Nevertheless, treatment of extensive results from diverse asphaltene studies are primarily within the bounds of freshman chemical principles. The TRFD studies pave the way towards new simplifying causality in asphaltene science and build expectation for integration of fundamental chemical principles with petroleum science, thereby realizing the vision of petroleomics.

References

[1] Boduszynski, M.M. (1984). Asphaltenes in petroleum asphalt: Composition and formation, Chapter 7 in J.W. Bunger, N.C. Li (Eds.) *Chemistry of Asphaltenes*. American Chemical Society, Washington D.C.

[2] Miller, J.T., Fisher, R.B., Thiyagarajan, P., Winans, R.E., Hunt, J.E. (1998). Subfractionation and characterization of Mayan asphaltene. *Energy & Fuels* **12**, 1290.

[3] Hortal, A.R., Martinez-Haya, B., Lobato, M.D., Pedrosa, J.M., Lago, S. (2006). On the determination of molecular weight distributions of asphaltenes and their aggregates in laser desorption ionization experiments. *J. Mass Spec.* **41**, 960–968.

[4] Rodgers, R.P. (2003). Presentation at *Petroleomics Symp., Astatphys Conference* Puerto Vallarta, Mexico.

[5] Hughey, C.A., Rodgers, R.P., Marshall, A.G. (2002). Resolution of 11,000 compositionally distinct components in a single electrospray ionization Fourier transform ion cyclotron resonance mass spectrum of crude oil. *Anal. Chem.* **74**, 4145.

[6] Rodgers, R.P., Marshall, A.G. (2006). Petroleomics: Advanced characterization of petroleum derived materials by Fourier transform ion cyclotron resonance mass spectrometry (FT-ICR MS). Chapter 3 in this book.

[7] Long, R.B. (1979). *ACS Div. Pet. Chem. Preprints* **24**, 891.

[8] George, G.N., Gorbaty, L.L. (1989). Sulfur K-edge x-ray absorption spectroscopy of petroleum asphaltenes and model compounds. *J. Am. Chem. Soc.* **111**, 3182.

[9] Waldo, G.S., Mullins, O.C., Penner-Hahn, J.E., Cramer, S.P. (1992). Determination of the chemical environment of sulfur in petroleum asphaltenes by X-ray absorption spectroscopy. *Fuel*, **71**, 53.

[10] Mitra-Kirtley, S., Mullins, O.C., Ralston, C.Y., Sellis, D., Pareis, C. (1998). Determination of the chemical environment of sulphur in petroleum asphaltenes by X-ray absorption spectroscopy. *Appl. Spectrosc.* **52**, 1522.

[11] Mitra-Kirtley, S., Mullins, O.C. (2006). Sulfur chemical moieties in carbonaceous materials. Chapter 6 in this book.

[12] Cunico, R.I., Sheu, E.Y., Mullins, O.C. (2004). Molecular weight measurement of UG8 asphaltene by APCI mass spectroscopy. *Petrol. Sci. and Tech.*, **22** (7–8), 787–798. Springer, New York.

[13] Desmazieres, B., Merdrignac, I., Laprevote, O., Terrier, P. (2004). *5th Ann. Conf. Phase Behavior*, Fouling, Banff, Canada.

[14] Scotti, R, Montanari, L. (1998). Molecular structure and intermolecular interaction of asphaltenes by FT-IR, NMR, and EPR, Chapter 3. In O.C. Mullins, E.Y. Sheu (eds.) *Structures and Dynamics of Asphaltenes*. Plenum Press, New York.

[15] Mullins, O.C. (1998). Optical interrogation of aromatic moieties in crude oils and asphaltenes, Chapter 2 in O.C. Mullins, E.Y. Sheu (eds.) *Structures and Dynamics of Asphaltenes*. Plenum Press, New York.
[16] Mullins, O.C., Mitra-Kirtley, S., Zhu, Y. (1992). Electronic absorption edge of petroleum. *Appl. Spectros.* **46**, 1405.
[17] Mullins, O.C., Zhu, Y. (1992). First observation of the Urbach tail in a multicomponent organic system. *Appl. Spectros.* **46**, 354.
[18] Wang, X., Mullins, O.C. (1994). Fluorescence lifetime studies of crude oils. *Appl. Spectrosc.* **48**, 977.
[19] Ralston, C.Y., Mitra-Kirtley, S., Mullins, O.C. (1996). Small population of one to three fused-ring aromatic molecules in asphaltenes. *Energy & Fuels* **10**, 623.
[20] Groenzin, H., Mullins, O.C. (1999). Asphaltene molecular size and structure. *J. Phys. Chem. A* **103**, 11237.
[21] Groenzin, H., Mullins, O.C. (2000). Molecular sizes of asphaltenes from different origin. *Energy & Fuels* **14**, 677.
[22] Buenrostro-Gonzalez, E., Groenzin, H., Lira-Galeana, C., Mullins, O.C. (2001). The overriding chemical principles that define asphaltenes. *Energy & Fuels* **15**, 972.
[23] Groenzin, H., Mullins, O.C., Eser, S., Mathews, J., Yang, M.-G., Jones, D. (2003). Asphaltene molecular size for solubility subfractions obtained by fluorescence depolarization. *Energy & Fuel* **17**, 498.
[24] Buch, L., Groenzin, H., Buenrostro-Gonzalez, E., Andersen, S.I., Lira-Galeana, C., Mullins, O.C. (2003). Molecular size of asphaltene fractions obtained from residuum hydrotreatment. *Fuel* **82**, 1075.
[25] Badre, S., Goncalves, C.C., Norinaga, K., Gustavson, G., Mullins, O.C. (2006). Molecular size and weight of asphaltene and asphaltene solubility fractions from coals, crude oils and bitumen. *Fuel* 85, 1.
[26] Perrin, F. (1926). *J. de Phys. et le Radium* **7**, 390.
[27] Perrin, F. (1936). *J. de Phys. et le Radium* **7**, 1.
[28] Tao, T. (1962). *Biopolymers* **8**, 607.
[29] Weber, G. (1971). *J. Chem. Phys.* **55**, 2399.
[30] Belford, G.G., Belford, R.L., Weber, G. (1972). *Proc. Nat. Acad. Sci. USA* **69**, 1392.
[31] Ehrenberg, M., Rigler, R. (1972). *Chem. Phys. Lett.* **14**, 539.
[32] Chuang, T.J., Eisenthal, K.B. (1972). *J. Chem. Phys.* **57**, 5094.
[33] Einstein, A. (1905). *Ann. d. Phys.* **17**, 549.
[34] Einstein, A. (1906). *Ann. d. Phys.* **19**, 371.
[35] Debye, P., (1929). Chapter 5. *Polar Molecules*, Dover Publications, Inc.
[36] Rice, S.A., Kenney-Wallace, G.A. (1980). *Chem. Phys.* **47**, 161.
[37] Tsunomori, F., Ushiki, H. (1996). *Bull. Chem. Soc. Jpn.* **69**, 1849.
[38] Cross, A.J., Fleming, G.R. (1984). *Biophys. J.* **46**, 45.
[39] Tsunomori, F., Ushiki, H. (1996). *Polym. J.* **28**, 576.
[40] Sasaki, T., Hirota, H., Yamamoto, M., Nishijima, Y. (1986). *Bull. Chem. Soc. Jpn.* **60**, 1165.
[41] Horinaka, J., Ono, K., Yamamoto, M. (1985). *Polym. J.* **14**, 433.
[42] Chang, M.C., Courtney, S.H., Cross, A.J., Gulotty, R.J., Petrich, J.W., Fleming, G.R. (1985). *Anal. Instr.* **14**, 433.
[43] Wahl, P. (1975). Chapter 1 in R.F. Chen, H. Edelhoch (eds.) *Biochemical Fluorescence: Concepts*, Vol. 1–2. Marcel Dekker, Inc., New York.
[44] Wax, N. (1954). (Ed.), *Noise and Stochastic Processes.* Dover Publications, New York.
[45] Berry, R.S., Rice, S.A., Ross, J. (1980). *Physical Chemistry*. John Wiley & Sons, New York.
[46] Groenzin, H., Mullins, O.C., Mullins, W.W. (1999). Energy-dependent quenching of fluorescence by CS_2. *J. Phys. Chem. A* **103**, 1504.
[47] Canuel, C., Badre, S., Groenzin, H., Berheide, M., Mullins, O.C. (2003). Diffusional fluorescence quenching of aromatic hydrocarbons. *Appl. Spectrosc.* **57**, 538.
[48] Andreatta, G., Goncalves, C.C., Buffin, G., Bostrom, N., Quintella, C.M., Arteaga-Larios, F., Perez, E., Mullins, O.C. (2005). Nanoaggregates and structure-function relations in asphaltenes. *Energy & Fuels* **19**, 1282.

[49] Mullins, O.C., Kaplan, M. (1983). Perturbed angular correlation studies of indium metalloporphyrin complexes. *J. Chem. Phys.* **79**, 4475.

[50] Bergmann, U., Groenzin, H., Mullins, O.C., Glatzel, P., Fetzer, J., Cramer, S.P. (2003). Carbon K-edge X-ray Raman spectroscopy supports simple yet powerful description of aromatic hydrocarbons and asphaltenes. *Chem. Phys. Lett.* **369**, 184.

[51] Zajac, G.W., Sethi, N.K., Joseph, J.T. (1994). *Scan. Micros.* **8**, 463.

[52] Battina, N. (2003). STM image analysis of asphaltene molecules, *Astatphys Conference*, Puerto Villarta, Mexico.

[53] Sharma, A., Groenzin, H., Tomita, A., Mullins, O.C. (2002). Probing order in asphaltenes and aromatic ring systems by HRTEM. *Energy & Fuels* **16**, 490.

[54] Sharma, A., Mullins, O.C. (2006). Insights into molecular and aggregate structures of asphaltenes using HRTEM. Chapter 8 in this book.

[55] Ruiz-Morales, Y. (2002). HOMO-LUMO gap as an index of molecular size and structure for polycyclic aromatic hydrocarbons (PHAs) and asphaltenes: a theoretical study. *J. Chem. Phys. A* **106**, 11283.

[56] Ruiz-Morales, Y. (2006). Molecular orbital calculations and optical transitions of PAH's and asphaltenes. Chapter 4 in this book.

[57] Buckley, J., Wang, J., Creek, J.L. (2006). Solubility of the least-soluble asphaltenes. Chapter 16 in this book.

[58] Bartholdy, J., Andersen, S.I. (2000). Changes in asphaltene stability during hydrotreating. *Energy & Fuels* **14**, 52.

[59] Ralston, C.Y., Wu, X., Mullins, O.C. (1992). Quantum yields of crude oils. *Appl. Spectrosc.* **46**, 1405.

[60] Turro, N.J. (1978). *Modern Molecular Photochemistry*. Benjamin/Cummings, Menlo Park, CA.

[61] Wargadalam, V.J., Norinaga, K., Iino, M. (2002). Size and shape of a coal asphaltene studied by viscosity and diffusion coefficient measurements. *Fuel* **81**, 1403.

[62] Andrews, B., Guerra, R., Mullins, O.C., Sen, P.N. Diffusivity of asphaltene molecules by fluorescence correlation spectroscopy. Accepted *J. Phys. Chem. A.*

[63] Freed, D.E., Lisitza, N.V., Sen, P.N., Song, Y.-Q. (2006). Asphaltene molecular composition and dynamics from NMR diffusion measurements. Chapter 11 in this book.

[64] Guerra, R.E., Ladavac, K., Andrews, A.B., Mulins, O.C., Sen, P.N. Submitted Energy & Fuels.

[65] Goncalves, S., Castillo, J., Fernandez, A., Hung, J. (2004). Absorbance and fluorescence spectroscopy on the aggregation behavior of asphaltene–toluene solutions. *Fuel* **83**, 1823.

[66] Mullins, O.C. (2006). Rebuttal to comment by Professors Herod, Kandiyoti, and Bartle on Molecular size and weight of asphaltene and asphaltene solubility fractions from coals, crude oils and bitumen by S. Badre, C.C. Goncalves, K. Norinaga, G. Gustavson and O.C. Mullins. *Fuel* **85**(2006), 1–11. Fuel, in Press.

3

Petroleomics: Advanced Characterization of Petroleum-Derived Materials by Fourier Transform Ion Cyclotron Resonance Mass Spectrometry (FT-ICR MS)

Ryan P. Rodgers and Alan G. Marshall

1. Introduction

The high mass resolving power and mass accuracy of FT-ICR MS allow for the resolution and elemental composition assignment of thousands of species in petroleum-derived materials. Here, we report its application to heavy crude oils, associated production deposits, and isolated asphaltenes to reflect recent advances in the characterization of complex mixtures, as well as low-resolution mass spectrometry experiments aimed at verifying suspected multimer formation. Electrospray ionization (ESI), field desorption/ionization (FD/FI), electron ionization (EI), and atmospheric pressure photoionization (APPI) FT-ICR MS results are discussed. ESI results reveal the compositional complexity of the most polar species in crude oil. Positive-ion electrospray reveals the contributions of the most basic species, many of which contain pyridinic nitrogen. Alternatively, negative-ion electrospray identifies the most acidic species that include pyrrolic nitrogen species and naphthenic acids. At high sample concentration (>1 mg/mL), low-resolution ESI linear ion trap mass analysis of Canadian bitumen shows multimers (up to 2 KDa) that may be isolated and subsequently dissociated to regenerate the parent monomer distribution. Those results support recent claims that asphaltenes and other polar constituents of crude oil and various forms of petroleum-derived materials aggregate extensively at concentrations greater than the critical micelle (nanoaggregation) concentration (CMC). It also suggests that large apparent molecular weights

Ryan P. Rodgers and Alan G. Marshall • National High Magnetic Field Laboratory at Florida State University, Department of Chemistry and Biochemistry.

observed by mass spectrometry (greater than 2 kDa) are due to aggregation. ESI FT-ICR MS results suggest that the molecular weight of polar asphaltenes are between 300 and 1400 Da with the majority of the species between 400 and 800 Da. A high mass tail (at very low signal-to-noise ratio) extending to up to 1400 Da is observed. Isolation and subsequent dissociation (at low internal energy) results in no dissociation, suggesting that these low abundant "high" mass species are a part of the monomer distribution. EI, FD, and APPI mass spectrometry identify nonpolar species that are inaccessible by ESI. Detailed type (i.e., number of rings plus double bonds) and carbon number distributions for the aromatic fraction of crude oil reveal progressive growth of the PAH core as molecular weight increases. Class-based trends in the nonpolar sulfur-containing species are also highlighted. FD results aimed at accessing the higher molecular weight material in a heavy crude oil show stable monomer molecular weight distributions between 1000 and 2000 Da. All results suggest that the molecular weight distributions for the parent oil, SARA-isolated aromatic, resin, and asphaltene fractions are below 2 kDa with the most abundant species between 400 and 800 Da. Consideration of the difficulties in asphaltene molecular weight determination by laser desorption (LD) and matrix-assisted laser desorption (MALDI) mass spectrometry suggest that ESI, FD, and FI currently offer the least discriminatory means for accurate molecular weight determination in polar (ESI) and mixed polar/nonpolar petroleum materials.

The evolution of mass spectrometry's role in petroleum characterization harks back to the birth of commercial mass spectrometry. Briefly, mass spectrometry has long been intimately tied to the petroleum industry and as a result, spawned many of its technological advances. Simply stated, petroleum companies sell molecules and consequently, an oil's composition determines its economic value. Therefore, compositional knowledge equals power: to produce oil reserves more efficiently, to predict production problems and prevent pipe fouling/failures, to reduce refining byproducts and waste, to make money . . . yes, but also to be better stewards of the world's oil reserve. The need to obtain detailed compositional information, on what at the time was considered complex mid- to light distillates, pushed the rapid investment in, and development of mass spectrometry technology. Advances in the 1950s and 1960s led to the development of high-resolution double-focusing sector mass spectrometers and the coupling of gas chromatography to mass spectrometry to form the first hyphenated mass spectrometric technique. Although growth and development continued through the 1990s, mass spectrometry was limited to relatively *low-boiling nonpolar species*, accessed by EI and FD/FI. Fragmentation resulting from traditional 70-eV electron bombardment of a volatilized sample limited its application and eventually led to the development of low-voltage EI to minimize fragmentation. In the analysis of complex petroleum samples, fragmentation is deleterious, because production of more than one signal per analyte complicates an already crowded mass spectrum and hampers parent ion identification. FD/FI minimized the production of fragment ions and accessed a much wider molecular weight range, but was limited by the need to break vacuum between samples.

By the year 2000, the combined efforts of hyphenated mass spectrometric techniques such as GC/MS and LC/MS, high-resolution mass spectrometry, and tandem MS had yielded impressive compositional characterization of many petroleum distillates such as gasoline, diesel, and gas oil. However, comparatively little was known about the less abundant polar species or heavy crude oils and heavy distillates, whose compositional complexity far exceeded the peak capacity of available mass spectrometers. Nevertheless, Boduszynski and others derived a surprisingly inclusive description of heavy petroleum that drew on a variety of analytical techniques.[1–4]

In 2000, Zhan and Fenn[5] pointed out that the most polar species in petroleum distillates could be ionized by ESI (the ionization method for which Fenn received the Nobel Prize) and detected by mass spectrometry. Because the polar constituents of crude oil (those that contain N, S, and O heteroatoms) are believed to be important in petroleum production and processing, Fenn's observation expanded the compositional mass spectral coverage of crude oil and petroleum-derived materials to include polar species, and ultimately led to the development of Petroleomics, namely the goal of determining the relationship between the chemical composition of a fossil fuel and its properties and reactivity. However, Petroleomics is not a new idea. In the early 1990s, Quann and Jaffe pointed out that detailed qualitative and quantitative measurement of compound classes, types, and carbon number distributions of petroleum feed and associated products are the cornerstones of molecular-based management of refinery processes. In view of the limited compositional information available at that time, Quann and Jaffe introduced the idea of structure-oriented lumping to simplify the overwhelming complexity of the petroleum materials.[6–8] APPI emerged later, promising more detailed information on the nonpolar aromatics, as well as accessing the same molecular classes seen by ESI. In summary, prior efforts in the mass spectral characterization of complex petroleum samples laid the groundwork for the ionization and detection of nonpolar species. APPI provided a simple, compact atmospheric pressure ionization method that could easily be coupled to existing mass spectrometers. With the advent of ESI, Fenn expanded the coverage to include the polar species. What was needed next was a mass analyzer that could resolve the mass spectral complexity encountered in the analysis of petroleum-derived materials common to all but the lightest distillates.

2. FT-ICR MS

The development of FT-ICR MS in the early 1970s[9,10] made it possible to obtain ultrahigh resolving power ($m/\Delta m_{50\%} > 100{,}000$, in which $\Delta m_{50\%}$ is the mass spectral peak full width at half-maximum peak height) mass spectra in seconds. However, because many of the figures of merit for ICR performance useful in complex mixture analysis increase linearly or quadratically with magnetic field strength,[11] the development of FT-ICR mass spectrometers capable of analyzing complex mixtures such as petroleum was ultimately tied to the development of high-field, high-homogeneity, temporally stable, solenoidal superconducting magnets. Early commercial FT-ICR mass spectrometers were based on low-field ($\sim$3 T) superconducting magnets. As a result, in order to obtain the high resolution required for individual component identification in complex petroleum samples, only a narrow mass range could be analyzed at a time.[12] Multiple spectral segments were then stitched together to yield the broadband mass spectrum. In a later version of the instrument, that limitation was overcome by simply raising the magnetic field to 5.6 T[13] to enable high resolution, high mass accuracy, broadband mass spectral analysis of petroleum distillates. Recently, the development of temporally stable, high-field (>7 T), high-homogeneity magnets has led to the rapid development of ultrahigh-resolution FT-ICR MS. With a routine mass resolving power of >300,000 and sub-ppm mass accuracy, FT-ICR MS stands poised to shed light on even the most complex materials with little or no dependence on advances in separation science. Its inherent high resolving power and high mass accuracy allow for baseline resolution of closely spaced isobaric species as well as molecular formula assignment through accurate mass determination. For example, Figure 3.1 shows the combined positive-ion (right) and negative-ion (left) ESI FT-ICR mass spectra of a South American crude oil obtained with no chromatographic preseparation. More than 17,000 different compounds are resolved and identified from a single sample.[14] The resulting compositional information may then be conveniently displayed in Kendrick[15,16] or van Krevelen[17–19] plots (see below) for rapid visual comparisons to highlight compositional differences/similarities between samples. Recent advances in FT-ICR MS as well as its role in the growing field of petroleomics have been the subjects of numerous review articles.[20–23]

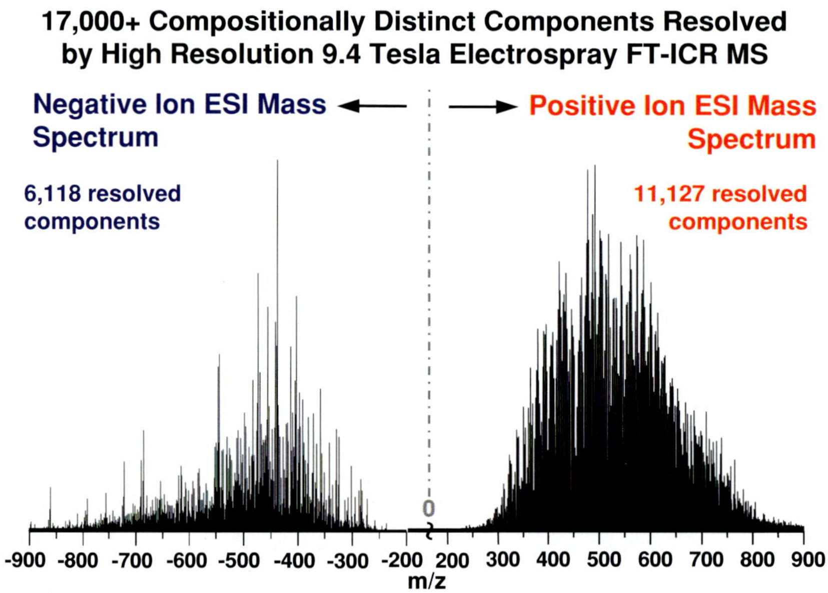

Figure 3.1. Combined positive and negative electrospray ionization 9.4-T Fourier transform ion cyclotron resonance mass spectra of a crude oil. Average mass resolving power, $m/\Delta m_{50\%}$, is ~350,000, allowing for resolution and identification of thousands of basic (right) and acidic (left) species. The 11,127 peaks (right) represent the most complex chemical mixture ever resolved and identified in a single mass spectrum.[14]

2.1. Mass Accuracy and Mass Resolution

The ultrahigh resolution afforded by FT-ICR MS allows for the baseline separation of closely spaced peaks commonly encountered in petroleum-derived materials (e.g., the 3.4-mDa split between isobars differing in elemental composition by C_3 vs. SH_4, both with a nominal mass of 36). Figure 3.2 shows the baseline resolution of the C_3 vs. SH_4 doublet commonly encountered in petroleum samples, in this case, in a South American crude oil. Note that similar information is available at every nominal (nearest-integer) mass in the spectrum ($300 < m/z < 900$). Resolution of this isobar, as well as many others in the mass spectrum, allows for the speciation of heteroatom-containing species that are unobservable by other mass analyzers. Moreover, the mass spacing between $^{12}C_c$ and $^{13}C^{12}C_{c-1}$ forms of otherwise compositionally identical species in turn allows for the determination of ion charge (z), because the two peaks are separated by $1/z$ on the m/z scale.[24] In this way, the charge of virtually all petrochemical species is determined to be either +1 (for +ESI, FD, and EI) or −1 (−ESI). Therefore, from here on, we shall refer to the measured m/z as simply the ion mass in dalton. Finally, resolution of isotopic fine structure[25] (e.g., identification of $^{12}C_c$ and $^{13}C^{12}C_{c-1}$ forms or ^{34}S and ^{32}S forms of otherwise compositionally identical species) provides internal verification of elemental composition assignment.

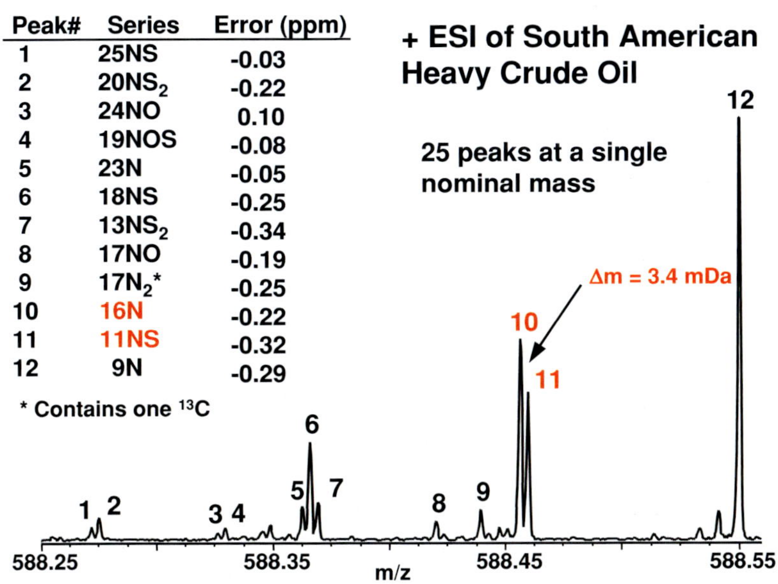

Figure 3.2. Mass scale-expanded segment ($m/z \approx 588$) of the ESI FT-ICR mass spectrum of the South American crude oil shown in Figure 3.1, right, exhibiting baseline resolution of 25 peaks. For brevity, only 12 of the 25 assigned elemental compositions are listed. The two peaks numbered 10 and 11 denote a commonly encountered isobaric mass split (C_3 vs. SH_4, 0.0034 Da) important for petroleomic applications. Elemental compositions are labeled according to their heteroatom content (e.g., NS, NOS), and double-bond equivalence (DBE = rings plus double bonds)–see text.

Ions introduced to the FT-ICR cell, located in the center of a spatially homogenous magnet, rotate about the magnetic field at a "cyclotron" frequency proportional to z/m (in which z is the number of elementary charges per ion and m is the ion mass). The magnetic field confines ions radially, whereas the addition of an axial quadrupolar electrostatic field prevents escape of the ions axially. The addition of the electrostatic "trapping" potential along with a Coulombic contribution from multiple ions in the ICR cell shifts the cyclotron frequency of all ions by an approximately constant amount.[26] Combined, these effects lead to a simple quadratic equation that relates the observed cyclotron frequency to m/z; therefore, ICR frequency-to-m/z calibration for ions of two or more known m/z ratios in the mass spectrum allows for the determination of the m/z ratios of all other ions in the spectrum.[27,28] For petroleum analysis, the calibration equation coefficients obtained by a separate "external" calibration experiment with ions of known masses just prior to analyte analysis provide a mass accuracy ($\sim\pm5$ ppm) sufficient to assign elemental compositions to the universally present ions of a homologous alkylation series of high mass defect (high hydrogen content) class spanning the mass range of interest. Subsequent "internal" calibration of ions of all other m/z values in the analyte spectrum yields substantially higher mass accuracy, because all of the analyte ions are subjected to essentially the same magnetic and electric fields; thus, high mass accuracy (<1 ppm) may be obtained over a broad mass range. Accurate mass measurement allows for the unambiguous assignment of molecular formulas for ions up to $\sim$400 Da. For assignment of those species observed at higher masses, we must rely on helpful spacings in the mass spectrum and Kendrick normalization (see below).

2.2. Kendrick Mass and Kendrick Plots

Close inspection of an FT-ICR mass spectrum of a petroleum-derived material reveals repeated spacing patterns throughout the entire mass range. For example, Figure 3.3 shows two mass scale-expanded segments of South American crude oil obtained by positive-ion ESI FT-ICR MS analysis. At moderate magnification spanning a 100-Da range (Figure 3.3, bottom), a series of 14 Da spacings is readily identified. Each successive 14.01565-Da spacing corresponds to an additional CH_2. Thus, a series of ions differing by multiples of CH_2 comprises the homologous alkylation series (carbon number distribution) for the same *class* (number of heteroatoms N, O, and S) and same *type* (number of rings plus double bonds). At higher magnification (Figure 3.3, top), another series with an exact mass spacing of 2.01565 (the mass of two hydrogens) consists of species with the same *class* and *carbon number*, differing only in their *type* (number of rings plus double bonds).

The Kendrick mass scale exploits the 14.01565 mass spacings to expose members of the same class and type, but varying carbon number. Conversion from IUPAC mass to Kendrick mass simply involves rescaling of the measured mass to the IUPAC mass of a CH_2 unit[15]

$$\text{Kendrick mass} = \text{IUPAC mass} \times (14/14.01565) \qquad (3.1)$$

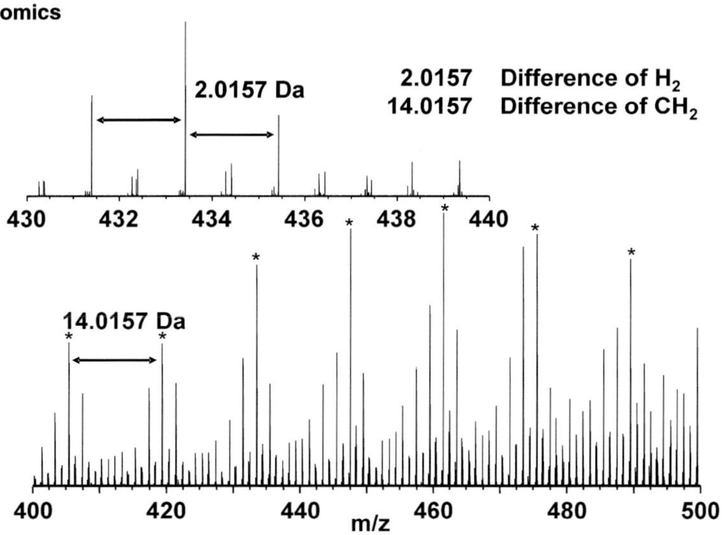

Figure 3.3. Mass spacings commonly encountered in an ESI FT-ICR mass spectrum of a crude oil. Each added ring or double bond lowers the mass by 2.01565 Da (top), and each additional -CH_2 group increases the mass by 14.01565 Da (bottom). Identification of such spacings is essential for the correct assignment of elemental composition.

Thus the IUPAC mass of the CH_2 unit (14.01565) becomes exactly 14. As a result, members of the same class and type have identical Kendrick mass defect (KMD), because the addition of a CH_2 unit increases the parent mass by exactly 14, which is unique to that series.

$$\text{KMD} = (\text{Nominal mass} - \text{Kendrick mass}) \times 1000 \tag{3.2}$$

For example, a series of simple saturated fatty acids (pentanoic, hexanoic, and heptanoic acid) share the same class (O_2) and type (number of rings plus double bonds = 1) and differ only in the number of CH_2 units (4, 5, and 6) attached to the carboxylic acid functional group. Kendrick normalization yields a series, each of whose members has identical KMD and thus appears as a horizontal row of equally spaced points in a plot of KMD vs. nominal (nearest-integer) Kendrick mass. It is important to point out that both the type and the class are commonly abbreviated to identify inclusive members of a carbon number distribution. In the current example, the simple saturated fatty acid series would be abbreviated as $1O_2$, where the 1 is the type (also known as double-bond equivalents (DBE) or the number of rings plus double bonds) and O_2 is the class (heteroatom content). This bookkeeping method allows for the simple abbreviation of the class and type and is used throughout the chapter.

The utility of the Kendrick plot is that once a few members of a class at low mass (<400 Da) have been identified, extension along the horizontal row of data with identical KMD allows for the confident elemental composition assignment of species that would otherwise lie outside the range of unambiguous molecular formula assignment based solely on accurate mass measurement.[16] Thus, Kendrick mass scaling allows for a $\sim 3\times$ increase in the upper mass limit for unique elemental

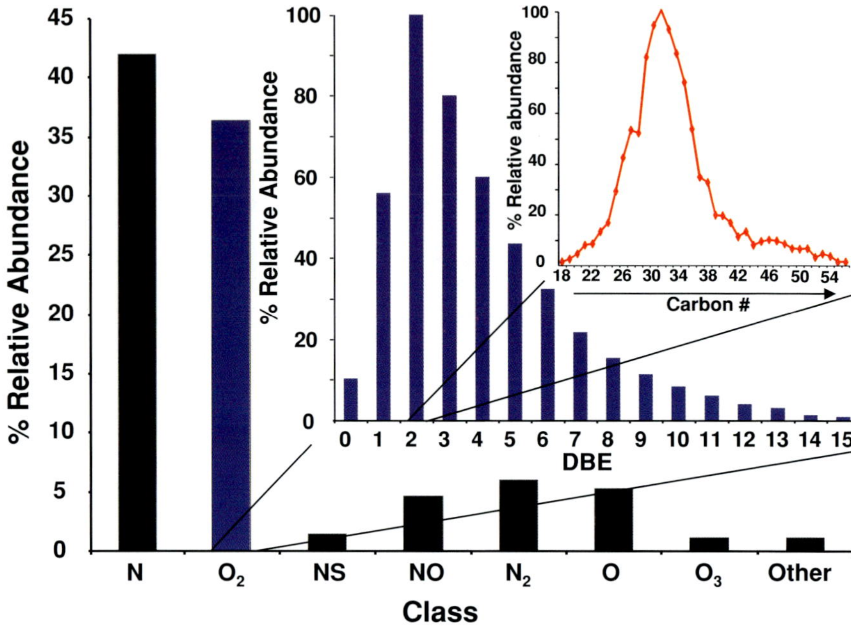

Figure 3.4. Three successive levels of compositional information afforded by Kendrick mass analysis of a negative-ion ESI FT-ICR mass spectrum of South American crude oil. First, the class or heteroatom content (left) may be determined for every species identified in the mass spectrum. For every class, the corresponding DBE (type) distribution may also be determined (middle). Finally, for every type of a given class, the carbon number distribution provides a measure of the extent of alkyl substitution (right). Similar carbon number distributions are provided for every type of a given class identified in the mass spectrum.

composition assignment. Kendrick mass analysis generates elemental compositions that in turn provide three levels of chemical detail, as displayed in Figure 3.4: heteroatom class ($N_nO_oS_s$), type (double-bond equivalents), and carbon distribution. For class comparison, the percent relative abundance is the sum of the class abundance divided by the sum of every identified peak in the mass spectrum. For both type and carbon number, the abundance is scaled to the highest member of the type or carbon number distribution in order to facilitate inner type and carbon number comparisons.

Figure 3.5 shows the Kendrick plot (a plot of KMD vs. nominal Kendrick mass) for the O_2 class of various types for a South American crude oil. Within the O_2 class, ions that contain successively higher numbers of rings plus double bonds are vertically displaced from one another by equal increments in the KMD plot. The equally spaced points on a given horizontal row result from the successive increase in the number of CH_2 groups. Thus, the horizontal range for a given class and type defines the carbon number distribution. In fact, because each heteroatom class has a different Kendrick mass defect, it is possible to display *all* of the assigned elemental compositions in a single Kendrick plot.

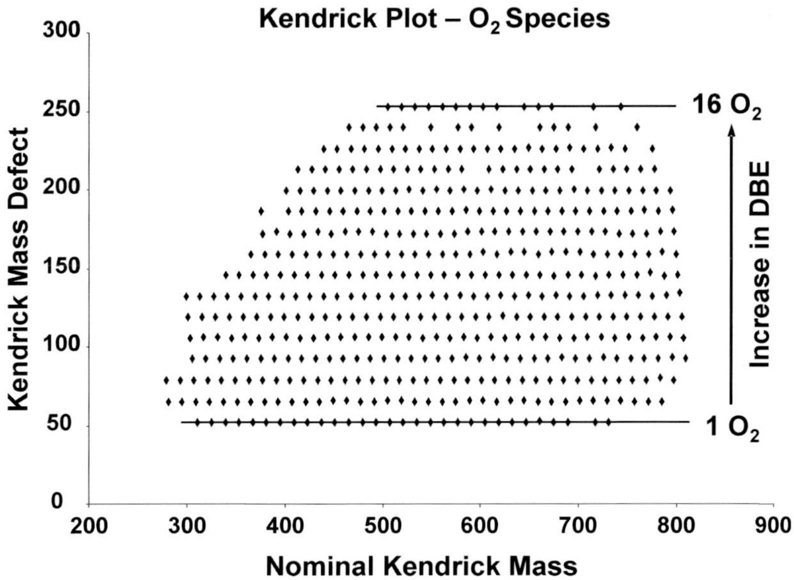

Figure 3.5. Kendrick plot of the O_2 class identified in the negative-ion ESI FT-ICR mass spectrum shown in Figure 3.1, left, exposes both type and carbon number distribution in a single plot. Each increase in DBE displaces the data vertically, whereas each additional CH_2 group displaces the data horizontally.

The Kendrick plot may be rendered more informative by color-coding each elemental composition according to its measured relative abundance in the mass spectrum. The mass spectrum is thereby converted to an image. Figure 3.6 shows the complete three-dimensional (3D) Kendrick plot for *all* classes, types, and carbon numbers identified from North American crude oil. The display enables all compositional information afforded by FT-ICR MS (now also including measured relative abundance) to be visualized in a single plot. Because vertical displacement in the Kendrick plot results from increased aromaticity (rings plus double bonds), changes in the aromatic character of related samples may be quickly visualized. Here, we present an example for North American crude oil (Figure 3.6) and its associated production deposit (Figure 3.7), each obtained from the negative-ion ESI FT-ICR mass spectrum. A clear shift upward in the elemental compositions and their increased relative abundances in Figure 3.7 relative to Figure 3.6 establishes a corresponding increase in aromaticity for the production deposit compared to the whole crude. Thus, the deposit evidently results from the selective deposition of the more aromatic species in the crude oil. Similarly, Figure 3.8 shows a 3D Kendrick plot of a North American crude oil *n*-heptane asphaltene. The large Kendrick mass defect suggests that the majority of the species identified are highly aromatic. Class analysis reveals that most of the species identified are multiheteroatom-containing aromatics. However, the Kendrick plots do not readily expose class-based compositional differences. For that we resort to the van Krevelen diagram.

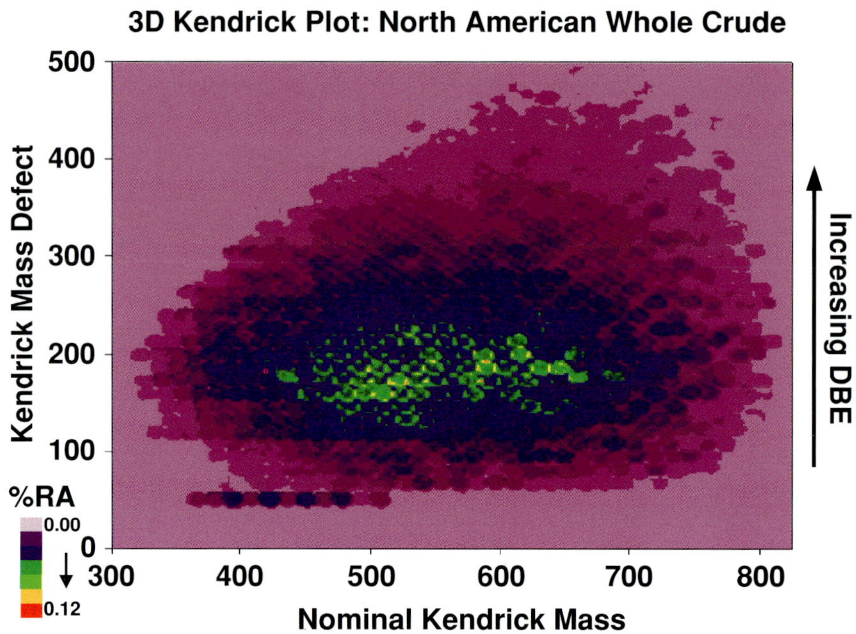

Figure 3.6. 3D Kendrick plot for North American oil generated from an entire broadband negative-ion ESI FTICR mass spectrum, allowing for rapid determination of variation in the carbon number (width in x-direction) and double-bond equivalents (width in y-direction) for all identified species. The relative abundance of each component is shown by color scaling.

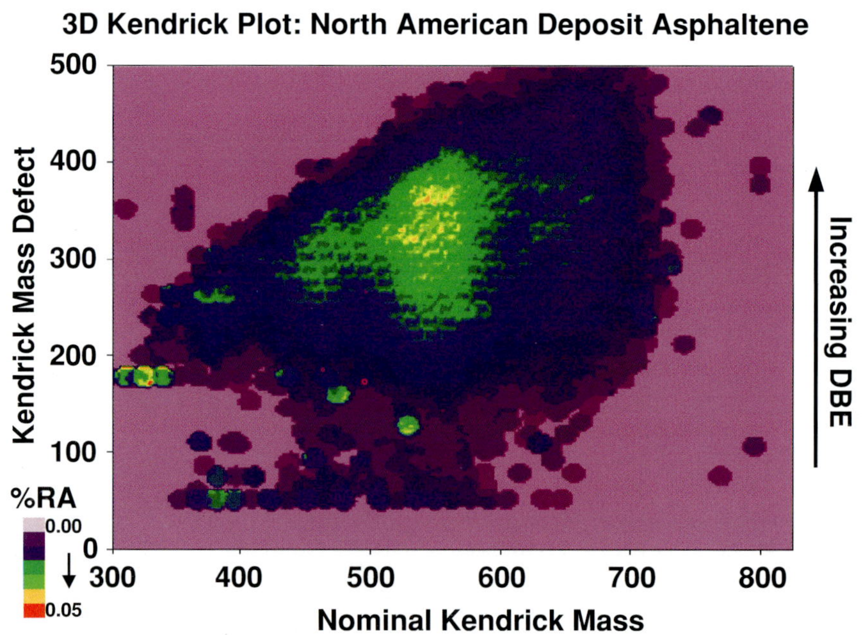

Figure 3.7. 3D Kendrick plot for a North American oil production deposit generated from an entire broadband negative-ion ESI FTICR mass spectrum. Format is as for Figure 3.6. Note the shift to higher Kendrick mass defect (higher DBE), indicating the preferential deposition of more aromatic species.

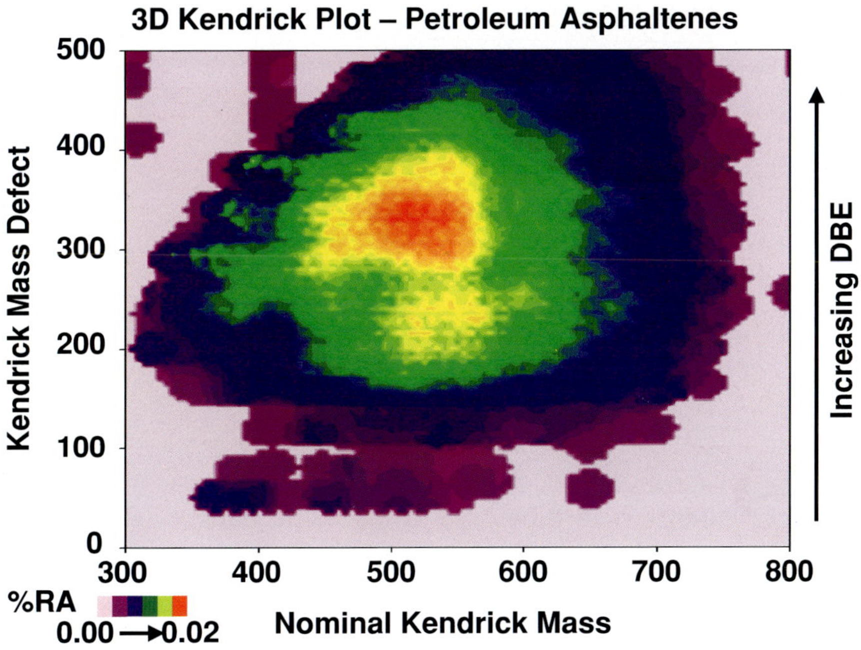

Figure 3.8. 3D Kendrick plot for a North American asphaltene generated from an entire broadband negative-ion ESI FTICR mass spectrum. Format is as for Figure 3.6. Note the shift to higher Kendrick mass defect (higher DBE), indicating that the asphaltenes are highly aromatic species.

2.3. van Krevelen Diagrams

A van Krevelen diagram is a plot based on the ratio of the number of non-carbon atoms (H, N, S, and/or O) to the number of carbon atoms for each elemental composition. Originally proposed for bulk elemental composition (i.e., one data point per sample),[17] the method has been extended to (up to thousands of) individual elemental compositions derived from ultrahigh-resolution FT-ICR MS.[18,19] In the 3D van Krevelen plot, the y-axis is typically hydrogen/carbon ratio and the z-axis is color-coded to reflect the measured relative abundance. The x-axis is X/carbon ratio, in which X can be chosen to highlight the class-specific compositional variation of the X heteroatom (N, S, and O) of interest. Figures 3.9 and 3.10 show van Krevelen diagrams for whole crude and deposit asphaltenes, in which the x-axis is chosen as oxygen/carbon ratio to highlight oxygen-dependent changes in the nitrogen-containing species. Class-specific differences between the oil and the deposit clearly result from selective deposition, namely an increase in the number and relative abundance of more highly oxygenated nitrogen-containing species. Similar information is provided for all identified classes. Although general changes in aromaticity and carbon number are revealed by the 3D *Kendrick* plot, the 3D *van Krevelen* plot readily exposes how those changes are manifested in the class, type and carbon number of the identified species.

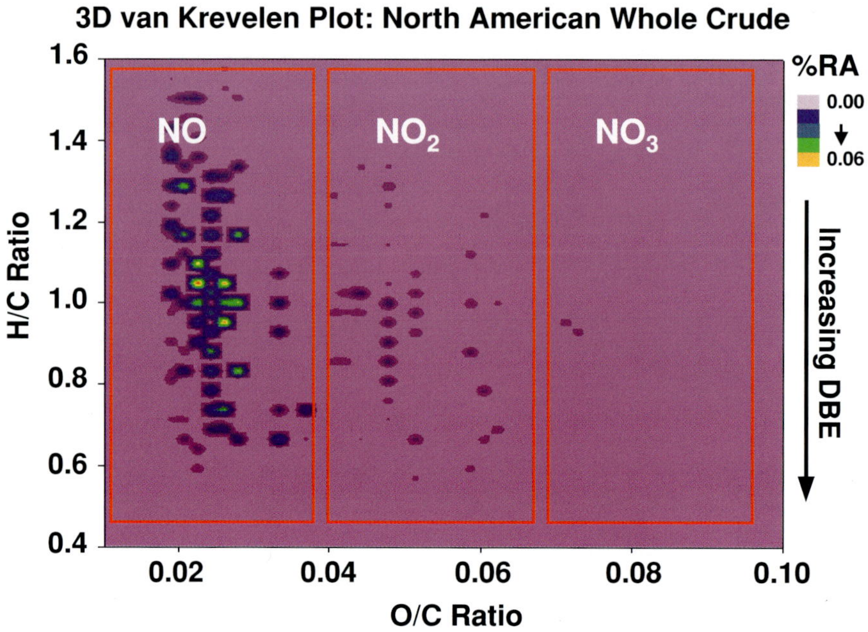

Figure 3.9. 3D van Krevelen diagram for the NO_x species identified from a negative-ion ESI FT-ICR mass spectrum of the North American whole crude oil. The van Krevelen diagram is better suited than the Kendrick plot for exposing class-based variations.

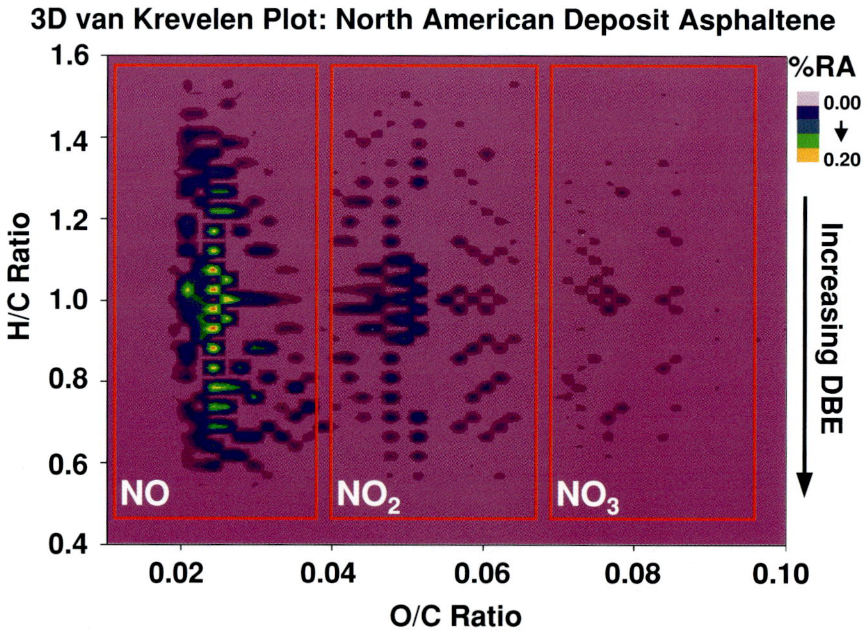

Figure 3.10. 3D van Krevelen diagram as in Figure 3.9, but for the NO_x species identified in a North American production deposit. Comparison to Figure 3.9 provides direct visualization of class-based deposition trends in the NO-containing species to show preferential formation or enrichment of more highly oxygenated species.

2.4. DBE and *Z* Number

Because of historical differences in the manner in which hydrogen deficiency is reported by mass spectrometrists and petrochemists, it is beneficial to clarify both DBE, also known as rings plus double bonds, and *Z* number. Both values should be reported for the *neutral* form of the observed ion and are a direct measure of hydrogen deficiency. The *Z* number is calculated from the general formula, $C_cH_{2c+z}X$, in which X is the heteroatom content (N, O, and S). The *Z* number may simply be calculated from knowledge of the number of hydrogens and carbons in a molecule. DBE is calculated from the general formula $C_cH_hN_nO_oS_s$, by the "nitrogen rule"[29]:

$$\text{DBE} = c - h/2 + n/2 + 1 \qquad (3.3)$$

As for *Z* number, elemental compositions provided by FT-ICR MS easily allow for the determination of DBE. The relationship between DBE and Z number is defined by equation (3.3).

$$Z = -2(\text{DBE}) + n + 2 \qquad (3.4)$$

Equation 3.3 therefore allows for simple conversion between DBE and Z number.

2.5. ESI for Access to Polars

ESI FT-ICR mass spectrometric analysis of petroleum-derived materials has identified thousand of species that comprise tens of different compound classes from a variety of petroleum-derived materials.[14,21,22,30–39] Although mass alone cannot distinguish structural isomers or definitively determine chemical functionality of heteroatom-containing classes, limited chemical speciation of the identified classes may be inferred from the selectivity of the ESI process. Negative-ion ESI favors the ionization of acidic species and thus reflects acidic nitrogen (pyrrolic) and naphthenic acids. Different molecules from the same nitrogen classes (N_1) are generated by positive-ion ESI; however, because positive-ion ESI favors the formation of protonated basic species, the N-containing class now represents pyridinic species. Similar trends may be used to discriminate between O_x, SO_x, NS_x, N_x, and NO_x chemical functionalities. Thus ESI offers limited class speciation without chromatographic preseparation. Class-specific ionization trends have been verified by analysis of a variety of standards, a subset of which included both pyrrolic and pyridinic functionality.[34] Further advances in structural assignments of the species identified by FT-ICR MS will rely on both class-specific chemical derivatization[40] and class-specific chromatographic separation[41–52] or a combination of both.

One benchmark of success of ESI FT-ICR MS analysis of polar species in crude oil appears in Figure 3.1, right. More than 11,000 different elemental compositions could be identified from a single mass spectrum! The ultrahigh resolution provided by FT-ICR MS readily identifies the correct heteroatom class, even for isobaric species differing in mass by as little as 3.4 mDa (C_3 vs. SH_4) (see Figure 3.2). The ability to correctly assign class, type, and carbon number for such closely spaced (but clearly resolved) peaks is paramount in extension to even

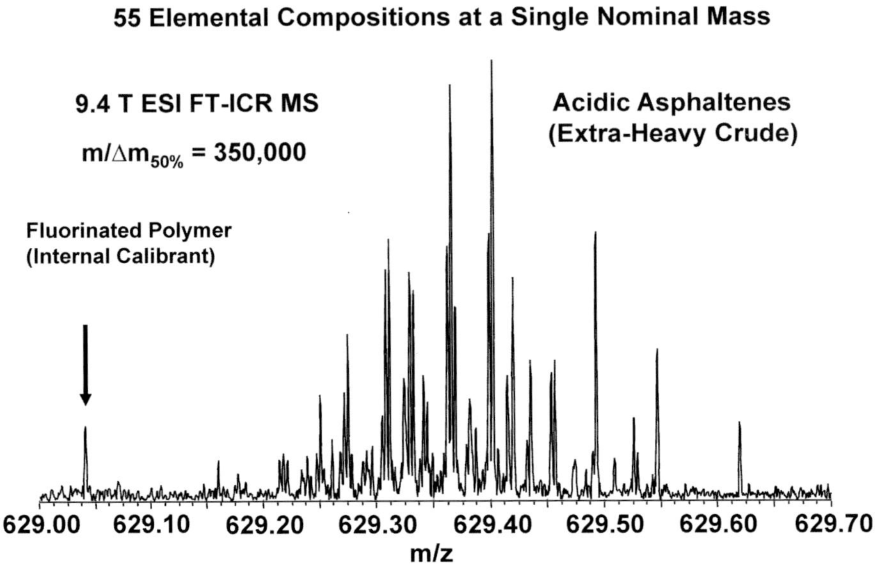

Figure 3.11. Negative-ion ESI selective ion accumulation 9.4-T FT-ICR mass spectrum of acidic asphaltenes. Note the resolution of 55 peaks at a single nominal mass.

more complex petroleum fractions such as asphaltenes. For example, Figure 3.11 shows an ESI FT-ICR mass spectrum of acidic asphaltenes with 55 peaks resolved at a single nominal mass. Currently, detailed chemical composition (class, type, and carbon number) of samples of such complexity is accessible only by FT-ICR MS.

2.6. EI, FD, and APPI for Access to Nonpolars

The success of the first ESI FT-ICR MS analysis of crude oil led to the rapid expansion of the technique to other petroleum-derived materials,[32] coal,[19,35,37] and humic and fulvic acids.[53,54] However, due to the selectivity of ESI for only the most polar species, other ionization methods are necessary to extend the wealth of compositional detail provided by FT-ICR MS to nonpolar species. To that end, we have recently modified our current instruments to accept commercial electron ionization, atmospheric pressure photoionization[55] and field desorption[23,56,57] ion sources. Other researchers have investigated thermal desorption probes coupled with electron ionization[58] or metal complexation[59–61] to gain access to the nonpolars.

EI FT-ICR MS relies on thermal desorption of the sample in an inert heated inlet system prior to ionization. As a result, EI FT-ICR MS is not well suited for analysis of extremely heavy materials such as resids. The operating temperature limit of the oven and thermal stability of the inert inlet coatings prevent operation above 400°C. However, the technique is well suited for the analysis of light to moderately heavy distillates that may be lost to volatilization in FD analysis

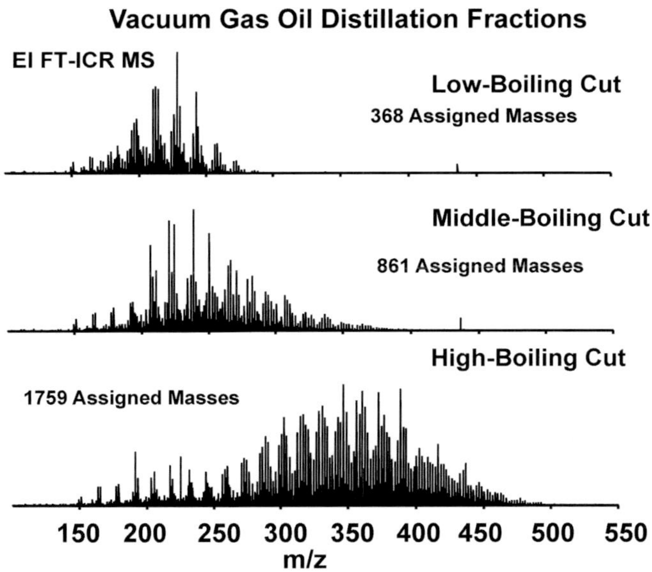

Figure 3.12. Each spectrum represents 100 scan-averaged, 10-eV EI FT-ICR mass spectra of VGO distillation fractions under identical experimental conditions. Low-boiling fraction (top), middle-boiling fraction (middle), high-boiling fraction (bottom).

performed in vacuum. Figure 3.12 shows the EI FT-ICR MS analysis of a series of vacuum gas oil distillates. Access to the nonpolar species identifies hundreds of nonpolar sulfur-containing species (e.g., benzothiophenes) that are unobservable by ESI.

APPI relies on analyte ionization by irradiation from a Krypton lamp (~10 eV). Sample desorption into the gas phase is provided by a heated pneumatically assisted nebulizer just prior to photon exposure. APPI is well suited for the characterization of aromatics (both polar and nonpolar). The ultrahigh resolution of FT-ICR MS is especially important in positive-ion APPI because the process can produce both radical cations and protonated species. Thus, a given neutral analyte can yield isobars differing in composition by ^{13}C vs. ^{12}CH (corresponding to 0.0045-Da mass difference) (Figure 3.13, top). Currently, no other type of mass analyzer can resolve such closely spaced isobars. Even so, other researchers have successfully applied low-resolution APPI MS to estimate molecular weight distributions in heavy petroleum materials.[62]

At this point, it is worth noting that some mass doublets are precluded by the "nitrogen rule" (equation (3.3)), if both ions are even-electron or odd-electron (e.g., ^{13}C vs. ^{12}CH, NH_2 vs. O, etc.).[29] However, such doublets do become possible if one ion is even-electron and the other is odd-electron, as can happen with EI or APPI. Thus, FT-ICR MS becomes even more essential for high-resolution analysis of EI or APPI mass spectra, because it is then necessary to resolve even more possible doublets than for ESI (even-electron), MALDI (even-electron), or FD (odd-electron) ionization.

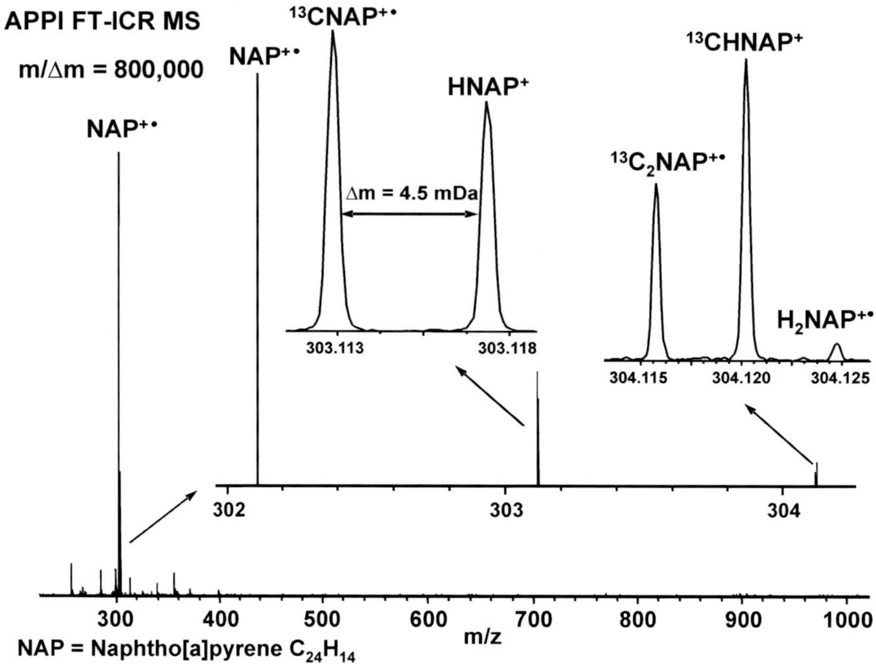

Figure 3.13. APPI FT-ICR mass spectrum of naphtho[a]pyrene highlights the ^{13}C and ^{12}CH isobaric split commonly encountered in APPI mass spectra of petroleum-derived materials due to the tendency of the photoionization process to form both protonated and radical cations.

FD/FI provides thermal desorption and ionization for even the heaviest petroleum fractions. Figure 3.14 shows the FD FT-ICR mass spectrum of a crude oil obtained with no emitter current (i.e., low desorption temperature). Internal calibration from a series of alkyl benzenes produces startlingly good mass accuracy (sub-ppm) throughout the entire mass range. Identified classes include mostly hydrocarbons with lower level contributions from S_1, S_2, S_3, and O_2 classes. FD ionization shows little class discrimination—both polar and nonpolar species ionize with comparable efficiency. Mass resolving power and dynamic range are similar to those obtained for both APPI and ESI, and (as for all FT-ICR MS analyses reported here) the class, type, and carbon number distributions are determined for all species identified in the mass spectrum.

3. Molecular Weight Determination by Mass Spectrometry

As outlined above, advances in mass spectrometric and ionization technology over the past decade have allowed for the characterization of petroleum-derived materials at a level never previously thought possible. However, for all its success in petroleum characterization over the past 50 years, some could argue that mass spectrometry has done little to clarify the true range of molecular weights of

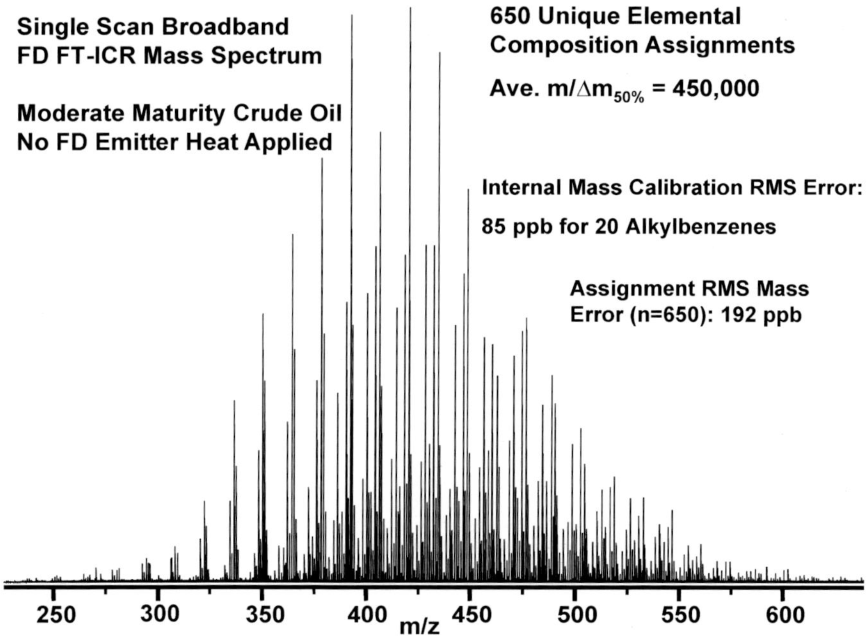

Figure 3.14. Broadband FD FT-ICR mass spectrum of a North American crude oil obtained at low emitter heat. The spectrum is computed from the sum of nine time-domain transients from 10-s external ion accumulations.

petroleum materials such as asphaltenes. Experimentally determined values from various techniques (e.g., vapor phase osmometry, size exclusion chromatography, fluorescence, mass spectrometry, etc.) vary widely in the open literature. Some believe that crude oil and its associated fractions consist mainly of low molecular weight components (100–2000 Da), whereas others believe the molecular weight to be much higher (100–10,000 or even 100–1,000,000 Da). In this abbreviated review, we address only the pertinent mass spectrometric results presented in the open literature during the past 20 years. Problems associated with non-mass-based determination of crude oil and asphaltene molecular weight lie outside the scope of this review and may be found elsewhere.[63–68]

3.1. Low Molecular Weight for Petroleum Components

It is important to recognize that the molecular weight distributions for crude oil and its associated fractions fall well within the range of today's ionization methods and mass analyzers. For example, multiply charged DNA (100,000,000 Da) has been observed by ESI FT-ICR MS,[69] and intact ribosomes (~2,300,000 Da) have been observed by ESI time-of-flight (TOF) mass analysis.[70] Moreover, in the late 1980s, Boduszynski and others proposed a surprisingly inclusive description of complex petroleum-derived materials.[1–4] They relied on a host of analytical techniques, including FI mass spectrometry, and concluded that the majority of the species in the crude oils had masses less than 1.9 kDa. Boduszynski also pointed out the relationship between molecular weight and structure as a function of boiling point, thereby setting practical limits to the size of species in all boiling point-defined fractions.[3,67] Boduszynski and Altgelt summarized these and other relevant findings in a book in the early 1990s.[67] Most notably, they concluded that for Boscan crude oil, one of the heaviest crude oils known, 90 wt% of the material has molecular weight less than 1900 Da. Furthermore, contrary to widely held opinion at the time, the authors further stated that the high boiling point fractions, corresponding to pentane-insolubles (asphaltenes), contain significant amounts of *low* molecular weight (although highly polar) constituents. Thus, in the heaviest oil fractions, progression to higher molecular weight, purely hydrocarbon species could be slowed or even supplanted by an increase in the amount of more polar (lighter) material. The idea is supported by the observation that for a given atmospheric equivalent boiling point (AEBP), heteroatom-containing species exhibit roughly two-thirds the carbon number of their hydrocarbon (PAH) analogs.[3,67] Simply stated, as polarity (N, S, and O content) increases, as for the higher boiling fractions and asphaltenes, the carbon number of the heteroatom-containing species in the same AEBP fraction is approximately two-thirds that of the PAH analog and one-third that of the corresponding *n*-alkane. From their analysis of successively higher boiling point materials, Boscan crude oil was determined to contain only 10 wt% of higher boiling point material. Given the previous observations, it is likely that this 10 wt% contains hydrocarbon material at or slightly above 2 kDa but with a rapidly growing contribution from lighter (roughly two-thirds the molar mass but higher boiling point), more polar heteroatom-containing species. Thus, the molecular weight distribution for the polars (asphaltenes) should be substantially lower than that observed for hydrocarbons and should correspond to two-thirds (from 300 Da up to ~1.3 kDa) that of the highest mass aromatic hydrocarbons.

Our FD FT-ICR MS experiments confirm the upper limit of the *hydrocarbon* molecular weight distribution of heavy oil. Figure 3.15 shows an FD FT-ICR mass spectrum of North American heavy crude oil obtained at high desorption (emitter) temperature. The molecular weight of this material lies between 1000 and 2000 Da, only slightly heavier than previously reported by Boduszynski. However, Boduszynski employed FI rather than FD mass spectrometry for the molecular weight measurements. FD allows for higher desorption temperature than can be achieved with an FI heated probe, and may therefore access higher boiling point material (the unobserved 10 wt%) not thermally desorbed by the FI probe. It is important to note that FT-ICR MS (unlike TOF MS) detection efficiency is independent of molecular weight.

Further support for low molecular weight for petroleum materials is supported by another FI MS study on Canadian bitumen. Like Boduszynski, Del Rio and Philip observed molecular weight distributions below 2 kDa.[71] Recent ESI FT-ICR MS results for the most *polar* species of crude oil also support the low molecular weight theory. With the addition of a mass resolving

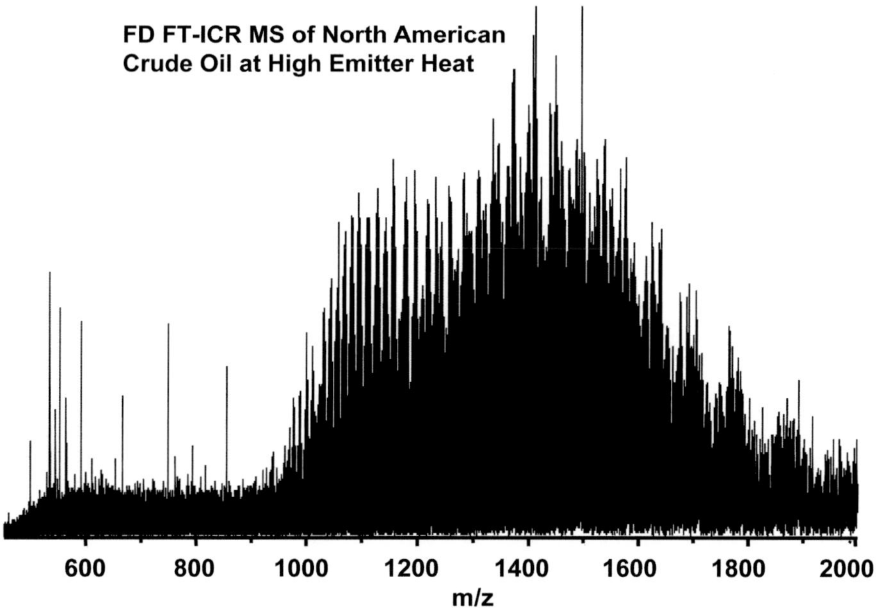

Figure 3.15. Field desorption FT-ICR mass spectrum of North American heavy crude oil obtained at high emitter temperature to access the very high-boiling components of the crude oil. Note that the molecular weight distribution does not extend much past 2 kDa.

quadrupole to the FT-ICR mass spectrometer, preselected mass segments can be preferentially accumulated prior to mass analysis, so that the FT-ICR cell can be filled with ions whose masses span only a limited segment (~40 Da) of the mass spectrum. Selective ion accumulation effectively increases the dynamic range of the analysis and allows for the observation of lower abundance (including higher molecular weight) species unobservable in a single broadband mass spectrum. The broadband mass spectrum may then be reconstituted by "stitching" the individual mass segments together. A recent study of South American crude oil by selected ion accumulation FT-ICR MS revealed abundant polar species from 300 to 900 Da with species extending up to 1.4 kDa (albeit at reduced signal-to-noise ratio), in excellent accord the Boduszynski molecular weight model prediction that the polar aromatics extend to roughly two-thirds (namely, to ~1.3 kDa) of the mass of the previously observed nonpolar species. Sheu recently summarized the use of mass spectrometry as well as many other analytical techniques that support a low molecular weight distribution for asphaltenes.[68] Finally, Mullins and coworkers have published a host of reports in support of the low molecular weight of asphaltenes. Specifically, field depolarization studies[72–75] of petroleum asphaltenes suggest that the average asphaltene molecular weight is approximately 750 Da, almost identical to the FT-ICR MS determined molecular weight of ~650 Da with most species between 350 and 750 Da.

3.2. Mass Spectrometry Caveats

As previously stated, the molecular weight range for crude oil and its associated fractions is well within the current mass analyzer capability. However, the means for introducing the analyte into the gas phase and subsequently ionizing it have associated limitations. EI, FI, FD, LD, APPI, APCI, and MALDI all rely on thermal desorption of the analyte prior to ionization. Therefore, components may not be completely volatilized (e.g., in EI, APPI, or APCI heated inlets) prior to ionization. FI and FD rely on resistively heating the analyte in vacuum and are therefore likely to access the heavy ends. However, because FD employs a carbon microneedle dendrite for sample desorption, it can access much higher temperature than the probe-mounted FI heater. LD and MALDI rely on laser irradiation to heat the sample quickly and subsequently ionize it. Because both depend on focused, short laser pulses of high energy density, they form a plasma of ablated material in vacuum. As a result, several different possible ionization pathways can form unwanted fragments and gas phase multimers. Because the tendency to form unwanted product ions is a function of the analyte class and type, complex mixtures composed of widely varying classes and types of analytes pose significant difficulty. Lowering the laser power may reduce the formation of unwanted product ions, but in turn results in incomplete desorption/ionization. Raising the laser power (in many cases to maximum power) may ensure complete desorption and ionization, but at the expense of increased formation of fragments and multimers that hinder molecular weight determination and complicate an already crowded mass spectrum. Implications of these competing requirements will be discussed in the next section. Although ESI avoids all of the previously discussed problems by producing analyte/solvent microdroplets that are later dried to yield only the analyte in the gas phase, the ionization process depends on acid/base chemistry, thereby restricting its application to the polar materials capable of losing (acids) or acquiring (bases) a proton. Sample volatility is irrelevant—as noted above polymers with mass greater than 100 MDa have been successfully observed by ESI.[69] In fact, broad application to nonvolatiles (mostly biological) formed the basis for the recent Nobel Prize in Chemistry to John Fenn, the inventor of ESI. APPI is performed with a heated pneumatically assisted nebulizer, but its ability to access higher boiling components is currently undetermined. In summary, the current state of the mass analyzer technology does not limit the compositional analysis of high-boiling petroleum-derived materials; rather, it is the inability to representatively introduce to the gas phase and ionize the full complement of analytes prior to mass spectral analysis. Among current ionization methods, ESI, FD, and FI (in that order) offer the most generally representative molecular weight determination with little or no mass (volatility) dependent discrimination. However, because ESI is limited to the analysis of polar species, FD and FI offer the best compromise for determining molecular weights of both polar and nonpolar analytes simultaneously.

Finally, mass measurement alone cannot distinguish structural isomers (either positional or stereioisomers). Because chemical speciation provided by differential ionization yields in positive- and negative-ion ESI is severely limited (typically limited to polar species containing either nitrogen or oxygen), chromatography based, class-specific separations will be needed to chemically speciate compound classes. Furthermore, because ionization efficiency (for any ionization method) for one species can be greatly affected by the presence of other species ("matrix" effect), it is not easy to relate the observed ion relative abundances to the relative abundances of their precursor neutrals in the original sample. Ultimately, it will be necessary to calibrate the class-specific relative ionization efficiencies by spiking the mixture with species (of various chemical functionality, e.g., alcohol, ketone, ester and carboxylic acids, pyridinic, pyrrolic, etc.) of known ionization efficiency.

3.3. High Molecular Weight for Petroleum Components

As noted above, there is convincing evidence for molecular weights lower than ~1500 for the vast majority of species observed by mass analysis of liquid-phase petroleum samples (e.g., ESI, FD). Those results are also highly reproducible. However, prior mass spectra of crude oils and asphaltenes based on LD or MALDI of solid or evaporated samples have been interpreted as evidence for high molecular weight (10,000–100,000 Da) species. Winans and coworkers recently published a fair, systematic, and thorough evaluation of the determination of asphaltene molecular weight by LD and MALDI mass spectrometry with Maya, Khafji, and Iranian Light asphaltenes, conducted under various instrumental conditions and with various analytical techniques.[76] Standards provided a basis to evaluate the formation tendency of parent, fragment, and multimer ions. The authors clearly outlined the difficulties in obtaining reproducible, accurate molecular weight distributions of asphaltenes by LD and MALDI MS. Selective ionization was observed when a matrix (dithranol) was added. It was also noted that the use of the TOF mass spectrometer in reflectron mode (used to obtain higher resolving power vs. linear mode) results in inefficient detection of high molecular weight components. Analyses were therefore carried out in linear TOF mode without the addition of matrix (LD). Standards chosen to represent various compound classes highlighted structural features found in asphaltenes, namely, aromatic rings, alkyl side chains and bridges, as well as presence of heteroatoms. LDMS analysis of the standards revealed class-based ionization and product ion formation tendencies as a function of laser power. Aromatics and alkyl-substituted aromatics ionized at relatively low laser power but produced mostly multimeric species at integer multiples of the parent molecular weight as well as minor fragment ions (alkyl-substituted aromatic). Bridged aromatic standards required moderate laser power for ionization; however, abundant fragment ions resulted from cleavage of the alkyl bridge carbon bonds. Solar dye (N, N'-bis(2,5-di-*tert*-butylphenyl)-3,4,9,10-perylenedicarboximide) was included, because it contains all of the structural features thought to exist in asphaltenes (heteroatoms, aromatic, alkyl and bridged structures). However, high laser power was required for efficient ionization and consequently resulted in the formation of minor fragment ions. Based on LDMS of the standards, the complete desorption/ionization of "heavy" asphlatenic materials requires high laser power. However, high laser power generates polymeric species and fragments for aromatic, alkyl-substituted aromatics, and alkyl-bridged aromatic structures, all thought to be abundant in asphaltenes. Moderate laser power offered the best compromise because it yielded the highest average molecular weights for all crude oil asphaltenes and their associated GPC subfractions. However, based on the results for standards, the laser power was shown to promote polymer and minor fragments in aromatic, alkyl-substituted, and-bridged aromatics as well as minor (alkyl-substituted) and major (bridged) fragment ions. Furthermore, the moderate laser power yielded no observable signal for the "asphaltene like" solar dye.

The above examples illustrate the pitfalls and difficulties in the LDMS and MALDI determination of asphaltene molecular weights. Unfortunately, a complete study would require numerous standards spanning tens of compound classes and types, each with varying degrees of alkylation, and as a result, is not currently feasible due the limited availability of suitable standards. Moreover, to sufficiently understand the trends in ionization efficiencies and matrix effects, mixed standards would be required, because PAH molecules have been reported as effective matrices for the MALDI mass spectrometric analysis of nonpolar species.[77] Therefore, asphaltenes are, in a sense, their own matrix and consequently, LD MS analysis of any asphlatenic material constitutes a form of MALDI with the amount of "matrix" (amount of PAH material present in the sample) being sample dependent. Finally, control "blank" samples are essential, because LD and MALDI can generate a high molecular weight "tail" TOF mass spectrum, even in the absence of sample.

The insight gained from the Winans study sheds new light on previously reported LDMS-determined asphaltene molecular weights. Crude oil and asphaltene molecular weights determined by LD and MALDI MS vary widely. Winans *et al.* note that because none of the LD and MALDI MS reports in the open literature include a measure of laser power density (in Watts per unit area) used in the analysis, it is not possible to compare or repeat those experiments. Winans *et al.* go on to note, "The energy of the laser pulse striking the sample is very sensitive to differences in the optical system and aging of the laser, with a consequent decline in the emitted energy," further complicating comparison of results obtained by different laboratories, on different instruments, or even at different times.

4. Aggregation

Asphaltenes tend to aggregate in solution at concentrations above the so-called "asphaltene critical micelle concentration" (CMC),[78–82] a term taken from colloid science for similar behavior observed for surfactants. Although the aggregate need not be a true micelle, the term is informative, because it sets the maximum concentration at which purely monomer-driven behavior can be expected. Thus, analytical techniques that require analyte concentration above the asphaltene CMC must necessarily report aggregate rather than pure monomer characteristics. Sheu points out that the self-association propensity jeopardizes the relevance of many analytical techniques (vapor phase osmometry (VPO), gel permeation chromatography (GPC), etc.) routinely used for the determination of asphaltene molecular weight.[68] Aggregation of polar components of crude oil and asphaltenes has been confirmed by low-resolution mass spectrometry performed at concentrations well above the widely accepted CMC limit.[68] As noted above, self-aggregation is not a problem for ESI MS if the experiment is conducted at sufficiently low sample concentration (i.e., $\leq$0.1 mg/mL of total sample, so that the concentration of any individual molecular component is less than 1 μM).

Mass analysis of Canadian bitumen confirms the aggregation tendency for the polar fraction and provides an insight into the true monomer molecular weight distribution. Figure 3.16 (top) shows the broadband mass spectrum ($300 < m/z < 1200$) of Canadian Athabasca bitumen obtained (at high analyte concentration, 10 mg/mL) with a low-resolution linear ion trap (LTQ, ThermoFinnigan, Bremen, Germany) mass spectrometer. The high upper mass limit (4 kDa), high ion capacity (ability to trap large ion population), and tandem MS (MS^n) capabilities of the LTQ mass spectrometer are well suited for examining bitumen aggregation. The molecular weight distribution is clearly bimodal below 1.2 kDa, and extends well above 2 kDa (not shown). Figure 3.16 (top right) shows MS^2 isolation of a mass range segment containing suspected multimers. Figure 3.16 (middle) shows the mass spectrum for that isolated segment, confirming the mass selection. Subsequent collision-induced dissociation (at energy too low to break covalent chemical bonds) for ions in the selected mass window regenerates the same monomer distribution (Figure 3.16, bottom) as for the low-mass segment of the original sample. Because dissociation of the high-mass species reproduces the parent monomer distribution and not a collection of broadly distributed mass fragments, these high-mass species may be attributed to multimers.

Further corroboration that the high-mass species are multimers is provided by MS^n ($n = 3, 4$) experiments performed on high-mass species (~1500 and 2000 Da). Figure 3.17 shows that species of molecular weight ~1500 Da in the original mass spectrum dissociate to form ions of ~900 Da, which further dissociate to form ions with the same mass distribution as the monomers in the original broadband mass spectrum. Moreover, ions of ~2000 Da may be isolated and dissociated successively to produce products of 1500 Da, then 900 Da, and finally the monomer distribution. (Not all ions are recovered after each MS/MS stage; thus, the signal-to-noise ratio of the final spectrum is lower than that of the monomers in the original broadband mass spectrum.) Because the dissociation was conducted at ion energy too low to break covalent bonds, it is clear that species of masses 900, 1500, and 2000 represent increasingly aggregated multimers of species whose monomeric molecular weights fall below ~800 Da.

In summary, ESI MS analysis of an Athabasca bitumen at a concentration ~100 times greater than normal exhibits suspected multimers that extend above 2 kDa in mass, with associated dimer, trimer, and other multimer maxima at ~ 900, 1500, and 2000 Da. MS^4 analysis of the 2 kDa, MS^3 analysis of the 1.5 k Da, and MS^2 analysis of the 900 Da segments regenerate the same low molecular weight monomer distribution observed in the original broadband mass spectrum. Clearly, polar species in petroleum-derived materials can aggregate extensively at sufficiently high concentration, and the dissociation of those aggregates may be probed directly by low-energy collisional dissociation tandem mass spectrometry. We demonstrate tetramers up to 2 kDa, and suggest that higher multimers are most

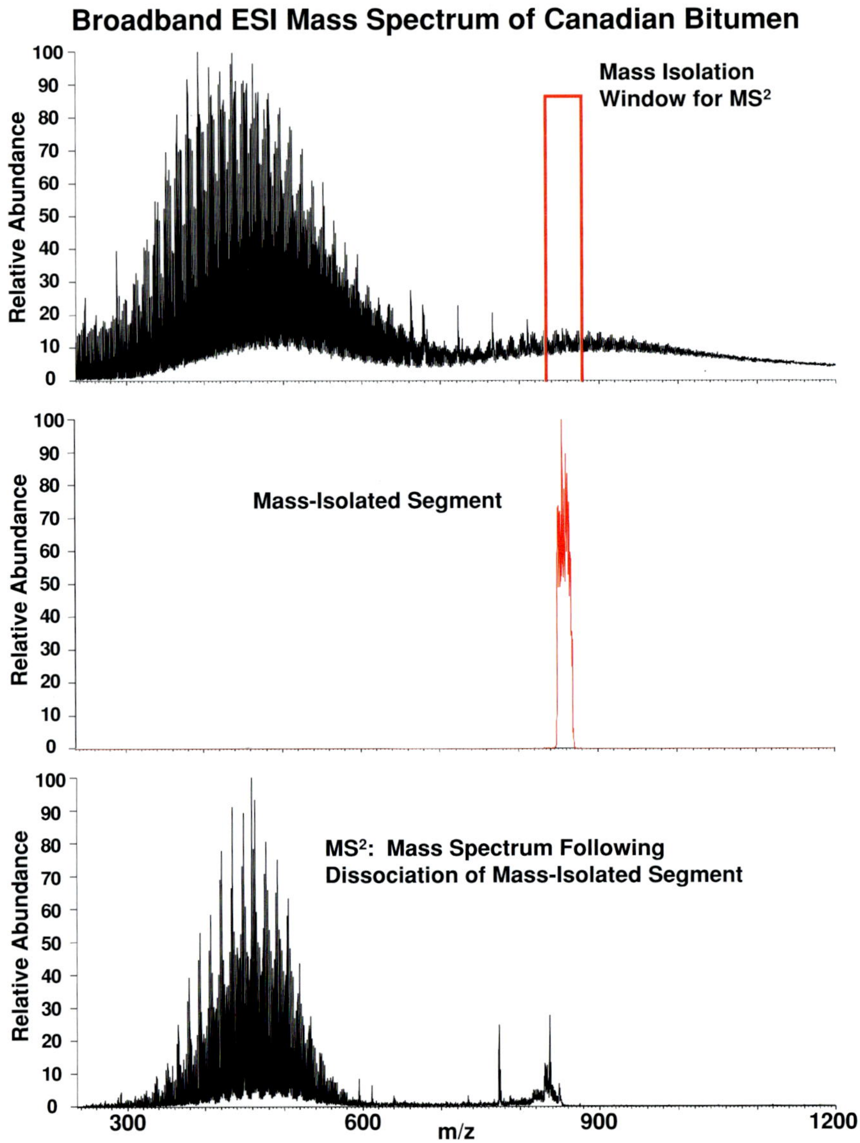

Figure 3.16. Low resolution ESI mass spectrum of a Canadian bitumen (top) obtained at high analyte concentration (10 mg/mL), showing a clear bimodal distribution that suggests multimer formation. Isolation of a mass segment (isolation window shown at upper right) selects the mass segment shown in the middle panel. Subsequent gentle collision-induced dissociation (MS^2) of the isolated species results in the regeneration of the same monomer species (bottom) previously observed in the broadband mass spectrum (top), strongly suggesting that the higher molecular weight species are multimers.

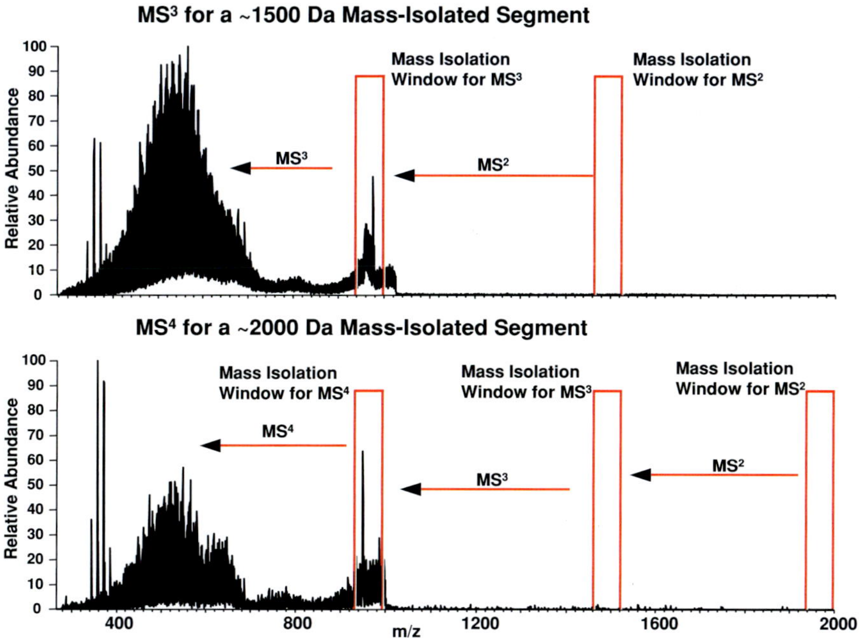

Figure 3.17. Low-resolution ESI MS^n spectra of higher mass material from Canadian bitumen at high analyte concentration. MS^3 (top), starting with isolation of ions in a mass segment near ~1500 Da (right), followed by dissociation and subsequent isolation of the product ions of mass near ~900 Da segment (middle), resulting in the regeneration of the monomer species (left) also present in the original broadband mass spectrum (Figure 3.15, top). MS^4 (bottom), based on successive isolation and dissociation of ions of mass near ~2000 Da, followed by isolation and dissociation of product ions near mass ~1500 Da, and subsequent isolation and dissociation of their product ions near mass ~900 Da, again regenerating monomers found in the original broadband mass spectrum (Figure 3.15, top). These results strongly suggest that the higher molecular weight species are multimers.

likely present. Similar aggregation tendencies in the polar components of other petroleum-derived fractions (e.g., diesel[32]; petroporphyrins[33]) have been observed by ESI FT-ICR mass spectrometry.

5. Petroleomics

It is interesting to note the similarity between the evolution of protein science to proteomics and the current state of asphaltene and crude oil characterization. Proteins, like asphaltenes, were originally classified by differential solubility: albumins (soluble in water and dilute salt solution), globulins (insoluble or sparingly soluble in distilled water; salted out of aqueous solution by half-saturation with ammonium sulfate), prolamins (insoluble in water but soluble in 50–90% aqueous ethanol), glutelins (insoluble in all of the above solvents but dissolve in dilute acid or base solution); scleroproteins (insoluble in most ordinary solvents), etc.[83] Similarly, asphaltenes are currently defined by their solubility in toluene and insolubility in heptane. However, mass spectrometry and many other techniques now provide a much more fundamental compositional (amino acid sequence) and higher order structural basis for sorting and explaining protein functions. Similar advancements in petroleum science are beginning to enable petroleomics. Class, type, and carbon number distributions of monomeric species provide basic insight into the behavior of the polymer (in this case, noncovalently linked multimers) structure, function, and behavior in upstream and downstream processing environments. However, as outlined in a recent review,[22] mass spectrometry cannot be the only analytical tool to enable the field of "petroleomics." Although substantial progress has been made in the characterization of polar and nonpolar species in crude oil, compositional determination of the saturates/olefins remains relatively untouched. The saturates/olefins pose a particularly difficult obstacle to direct mass spectral characterization due to their tendency to fragment and undergo gas phase reactions during the ionization process. For those species, high temperature gas chromatography (HTGC) appears well suited for modeling purposes. Moreover, mass measurement alone does not discern structural isomers. Chromatography-based, class-specific separations will be needed to chemically speciate compound classes, and GC/GC or GC/GC/TOF analyses will be needed to understand the isomer variation and progression in mid to light distillates to guide similar characterization of higher mass (and heavier) materials. Finally, because (as for ESI) ionization efficiency for one species can be greatly affected by the presence of other species (matrix effect), it is not easy to relate the observed *ion* relative abundances to the relative abundances of their precursor *neutrals* in the original sample. For example, negative-ion electrospray ionization favors the most acidic compounds, whose presence can reduce the relative abundance of species of lower acidity. It will be necessary to calibrate ionization efficiencies (at least between different major classes) based on suitable reference compounds (many of which are not currently available synthetically). However, even qualitative and quantitative compositional information derived from mass analyses for each of several ionization methods may not be enough. Ultimately, the compositional data from all available techniques will contribute to form the basis for model-based predictions of oil behavior. Therefore, advances in informatics and predictive modeling will be paramount.

Acknowledgments

We thank the following individuals, who have participated as coauthors in the body of the work underlying this review: Samuel Asomaning, Andrew Yen, Oliver C. Mullins, K.V. Andersen, Erin N. Blumer, Helen J. Cooper, William T. Cooper, Mark R. Emmett, Anne Fievre, Michael A. Freitas, Shenheng Guan, Mark A. Greaney, Larry A. Green, Christopher L. Hendrickson, Christine A. Hughey, Sara Jernström, William M. Landing, Daniel G. McIntosh, Kuangnan Qian, John Quinn, Parviz Rahimi, Winston K. Robbins, Stuart E. Scheppele, Michael V. Senko, Touradj Solouki, Clifford C. Walters, Forest M. White, Sunghwan Kim, Geoffrey C. Klein, Tanner M. Schaub, Lateefah A. Stanford, Jeremiah M. Purcell, and Zhigang Wu. Finally, we thank Carol L. Nilsson for suggesting the term "petroleomics."

The work cited here has been supported by Amoco, ExxonMobil Research and Engineering, The National Science Foundation (currently DMR-00-84173), Florida State University, The Ohio State University, and the National High Magnetic Field Laboratory in Tallahassee, FL.

Glossary

EI – electron ionization
APPI – atmospheric pressure photoIonization
FI – field ionization
FD – field desorption
LD – laser desorption
MALDI – matrix assisted laser desorption/ionization
ESI – electroSpray ionization
FT-ICR MS – fourier transform ion cyclotron resonance mass spectrometry
GC/MS – gas chromatography/mass spectrometry
LC/MS – liquid chromatography/mass spectrometry
SARA – saturates, aromatics, resins, and asphaltenes
CMC – critical micelle concentration
eV – electron Volt
ppm – part per million
DBE – double-bond equivalent or number of rings plus double bonds

References

[1] Boduszynski, M.M. (1987). Composition of heavy petroleums. 1. Molecular weight, hydrogen deficiency, and heteroatom concentration as a function of atmospheric equivalent boiling point up to 1400°F (760°C). *Energy Fuels* **1**(1), 2–11.

[2] Boduszynski, M.M. (1988). Composition of heavy petroleums. 2. Molecular characterization. *Energy Fuels* **2**(5), 597–613.

[3] Boduszynski, M.M. and K.H. Altgelt (1992). Composition of heavy petroleums. 4. Significance of the extended atmospheric equivalent boiling point (AEBP) scale. *Energy Fuels* **6**(1), 72–76.

[4] Altgelt, K.H. and M.M. (1992). Boduszynski, Composition of heavy petroleums. 3. An improved boiling point-molecular weight relation. *Energy Fuels* **6**(1), 68–72.

[5] Zhan, D.L. and J.B. Fenn (2000). Electrospray mass spectrometry of fossil fuels. *Int. J. Mass Spectrom.* **194**(2–3), 197–208.

[6] Jaffe, S.B. (2001). Compositional based modeling. In: *222nd ACS National Meeting*, Chicago, IL.

[7] Quann, R.J. and S.B. Jaffe (1992). Structure-oriented lumping: Describing the chemistry of complex hydrocarbon mixtures. *Ind. Eng. Chem. Res.* **31**(11), 2483–97.

[8] Quann, R.J. and S.B. Jaffe (1996). Building useful models of complex reaction systems in petroleum refining. *Chem. Eng. Sci.* **51**(10), 1615–1635.

[9] Comisarow, M.B. and A.G. Marshall (1974). Fourier transform ion cyclotron resonance spectroscopy. *Chem. Phys. Lett.* **25**, 282–283.

[10] Comisarow, M.B. and A.G. Marshall (1974). Frequency-sweep Fourier transform ion cyclotron resonance spectroscopy. *Chem. Phys. Lett.* **26**, 489–490.

[11] Marshall, A.G. and S. Guan (1996). Advantages of high magnetic field for FT-ICR mass spectrometry. *Rapid Commun. Mass Spectrom.* **10**, 1819–1823.

[12] Guan, S., A.G. Marshall, and S.E. Scheppele (1996). Resolution and chemical formula identification of aromatic hydrocarbons containing sulfur, nitrogen, and/or oxygen in crude oil distillates. *Anal. Chem.* **68**, 46–71.

[13] Rodgers, R.P., F.M. White, D.G. McIntosh, and A.G. Marshall (1998). 5.6 Tesla Fourier transform ion cyclotron resonance mass spectrometer for analysis of volatile complex mixtures. *Rev. Sci. Instrum.* **69**, 2278–2284.

[14] Hughey, C.A., R.P. Rodgers, and A.G. Marshall (2002). Resolution of 11,000 compositionally distinct components in a single electrospray ionization Fourier transform ion cyclotron resonance mass spectrum of crude oil. *Anal. Chem.* **74**, 4145–4149.

[15] Kendrick, E. (1963). A mass scale based on $CH_2 = 14.0000$ for high resolution mass spectrometry of organic compounds. *Anal. Chem.* **35**, 2146–2154.

[16] Hughey, C.A., C.L. Hendrickson, R.P. Rodgers, A.G. Marshall, and K. Qian (2001). Kendrick mass defect spectroscopy: A compact visual analysis for ultrahigh-resolution broadband mass spectra. *Anal. Chem.* **73**, 4676–4681.

[17] van Krevelen, D.W. (1950). Graphical-statistical method for the study of structure and reaction processes of coal. *Fuel* **29**, 269–284.

[18] Kim, S., R.W. Kramer, and P.G. Hatcher (2003). Graphical method for analysis of ultrahigh-resolution broadband mass spectra of natural organic matter, the Van Krevelen diagram. *Anal. Chem.* **75**, 5336–5344.

[19] Wu, Z., R.P. Rodgers, and A.G. Marshall (2004). Two and three dimensional van Krevelen diagrams: A graphical analysis complementary to the Kendrick mass plot for sorting elemental compositions of complex organic mixtures based on ultrahigh-resolution Broadband FT-ICR mass measurements. *Anal. Chem.* **76**, 2511–2516.

[20] Marshall, A.G. (2000). Milestones in Fourier transform ion cyclotron resonance mass spectrometry technique development. *Int. J. Mass Spectrom.* **200**, 331–336.

[21] Marshall, A.G. and R.P. Rodgers (2004). Petroleomics: The next grand challenge for chemical analysis. *Acc. Chem. Res.* **37**, 53–59.

[22] Rodgers, R.P., T.M. Schaub, and A.G. Marshall (2005). Petroleomics: Mass spectrometry returns to its roots. *Anal. Chem.* **77**, 20A–27A.

[23] Schaub, T.M., C.L. Hendrickson, J.P. Quinn, R.P. Rodgers, and A.G. Marshall (2005). Instrumentation and method for ultrahigh resolution field desorption ionization Fourier transform ion cyclotron resonance mass spectrometry of non-polar species. *Anal. Chem.* **77**, 1317–1324.

[24] Senko, M.W., S.C. Beu, and F.W. McLafferty (1995). Automated assignment of charge states from resolved isotopic peaks for multiply charged ions. *J. Am. Soc. Mass Spectrom.* **6**, 52–56.

[25] Shi, S.D.-H., C.L. Hendrickson, and A.G. Marshall (1998). Counting individual sulfur atoms in a protein by ultrahigh-resolution Fourier transform ion cyclotron resonance mass spectrometry: Experimental resolution of isotopic fine structure in proteins. *Proc. Nat. Acad. Sci. U.S.A* **95**(20), 11532–11537.

[26] Marshall, A.G., C.L. Hendrickson, and G.S. Jackson (1998). Fourier transform ion cyclotron resonance mass spectrometry: A primer. *Mass Spectrom. Rev.* **17**, 1–35.

[27] Ledford, E.B., Jr., D.L. Rempel, and M.L. Gross (1984). Space charge effects in Fourier transform mass spectrometry. Mass calibration. *Anal. Chem.* **56**, 2744–2748.

[28] Shi, S.D.-H., J.J. Drader, M.A. Freitas, C.L. Hendrickson, and A.G. Marshall (2000). Comparison and interconversion of the two most common frequency-to-mass calibration functions for Fourier transform ion cyclotron resonance mass spectrometry. *Int. J. Mass Spectrom.* **195/196**, 591–598.

[29] McLafferty, F.W. and F. Turecek (1993). Interpretation of Mass Spectra. University Science Books, Sausalito, CA 371.

[30] Qian, K., R.P. Rodgers, C.L. Hendrickson, M.R. Emmett, and A.G. Marshall (2001). Reading chemical fine print: Resolution and identification of 3000 nitrogen-containing aromatic compounds from a single electrospray ionization Fourier transform ion cyclotron resonance mass spectrum of heavy petroleum crude oil. *Energy Fuels* **15**, 492–498.

[31] Qian, K., W.K. Robbins, C.A. Hughey, H.J. Cooper, R.P. Rodgers, and A.G. Marshall (2001). Resolution and identification of elemental compositions for more than 3000 crude acids in heavy petroleum by negative-ion microelectrospray high field Fourier transform ion cyclotron resonance mass spectrometry. *Energy Fuels* **15**, 1505–1511.

[32] Hughey, C.A., C.L. Hendrickson, R.P. Rodgers, and A.G. Marshall (2001). Elemental composition analysis of processed and unprocessed diesel fuel by electrospray ionization Fourier transform ion cyclotron resonance mass spectrometry. *Energy Fuels* **15**, 1186–1193.

[33] Rodgers, R.P., C.L. Hendrickson, M.R. Emmett, A.G. Marshall, M.A. Greaney, and K. Qian (2001). Molecular characterization of petroporphyrins in crude oil by electrospray ionization Fourier transform ion cyclotron resonance mass spectrometry. *Can. J. Chem.* **79**, 546–551.

[34] Hughey, C.A., R.P. Rodgers, A.G. Marshall, K. Qian, and W.R. Robbins (2002). Identification of acidic NSO compounds in crude oils of different geochemical origins by negative ion electrospray Fourier transform ion cyclotron resonance mass spectrometry. *Org. Geochem.* **33**, 743–759.

[35] Wu, Z., S. Jernström, C.A. Hughey, R.P. Rodgers, and A.G. Marshall (2003). Resolution of 10,000 compositionally distinct components in polar coal extracts by negative-ion electrospray ionization Fourier transform ion cyclotron resonance mass spectrometry. *Energy Fuels* **17**, 946–953.

[36] Hughey, C.A., R.P. Rodgers, A.G. Marshall, K. Qian, C.C. Walters, and P. Mankiewicz (2004). Acidic and neutral polar NSO compounds in smackover oils of different thermal maturity revealed by electrospray high field Fourier transform ion cyclotron resonance mass spectrometry. *Org. Geochem.* **35**, 863–880.

[37] Wu, Z., R.P. Rodgers, and A.G. Marshall (2004). Compositional determination of acidic species in Illinois #6 coal extracts by electrospray ionization Fourier transform ion cyclotron resonance mass spectrometry. *Energy Fuels* **18**, 1424–1428.

[38] Wu, Z., R.P. Rodgers, A.G. Marshall, J.J. Strohm, and C. Song (2005). Comparative compositional analysis of untreated and hydrotreated oil by electrospray ionization Fourier transform ion cyclotron resonance mass spectrometry. *Energy Fuels* **19**(3), 1072–1077.

[39] Wu, Z., R.P. Rodgers, and A.G. Marshall (2005). ESI FT-ICR mass spectral analysis of coal liquefaction products. Fuel **84**(14–15), 1790–1797.

[40] Müller, H., J.T. Andersson, and W. Schrader (2005). Characterization of high-molecular-weight sulfur-containing aromatics in vacuum residues using Fourier transform ion cyclotron resonance mass spectrometry. *Anal. Chem.* **77**, 2536–2543.

[41] Barman, B.N., V.L. Cebolla, and L. Membrado (2000). Chromatographic techniques for petroleum and related products. *Crit. Rev. Anal. Chem.* **30**(2–3), 75–120.

[42] Boduszynski, M.M., R.J. Hurtubise, T.W. Allen, and H.F. Silver (1983). Liquid chromatography/field ionization mass spectrometry in the analysis of high-boiling and nondistillable coal liquids for hydrocarbons. *Anal. Chem.* **55**(2), 225–231.

[43] Conceicao Oliveira, E.V.d.C., M. Cecilia, A.R.V. Sant'Ana Lopes, M. Goreti, and E. Bastos Caramao (2000). Ion-exchange resins in the isolation of nitrogen compounds from petroleum residues. *Crit. Rev. Anal. Chem.* **30**(2–3), 75–120.

[44] Rudzinski, W.E. and T.M. Aminabhavi (2000). A Review on extraction and identification of crude oil and related products using supercritical fluid technology. *Energy Fuels* **14**(2), 464–475.

[45] Roussis, S.G. (1999). Exhaustive determination of hydrocarbon compound type distributions by high resolution mass spectrometry. *Rapid Commun. Mass Spectrom.* **13**(11), 1031–1051.

[46] Lundanes, E. and T. Greibrokk (1994). Separation of fuels, heavy fractions, and crude oils into compound classes: a review. *J. High Resolut. Chromatogr.* **17**(4), 197–202.

[47] Li, M., S.R. Larter, D. Stoddart, and M. Bjoroey (1992). Liquid chromatographic separation schemes for pyrrole and pyridine nitrogen aromatic heterocycle fractions from crude oils suitable for rapid characterization of geochemical samples. *Anal. Chem.* **64**(13), 1337–44.

[48] Carbognani, L. and A. Izquierdo (1989). Preparative and automated compound class separation of Venezuelan vacuum residua by high-performance liquid chromatography. *J. Chromatogr.* **484**, 399–408.

[49] Grizzle, P.L. and D.M. Sablotny (1986). Automated liquid chromatographic compound class group-type separation of crude oils and bitumens using chemically bonded silica-NH2. *Anal. Chem.* **58**(12), 2389–95.

[50] Drushel, H.V. and A.L. Sommers (1966). Isolation and identification of nitrogen compounds in petroleum. *Anal. Chem.* **38**(1), 19–28.

[51] Radke, M., H. Willsch, and D. Welte, H (1984). Class separation of aromatic compounds in rock extracts and fossil fuels by liquid chromatography. *Anal. Chem.* **56**(13), 2538–46.

[52] Selucky, M., T. Ruo, Y. Chu, and O.P. Strausz (1978). Chromatographic studies on oil sand bitumens. Advances in chemistry series, *Anal. Chem. Liq. Fuel Sources* **170**, 117–27.

[53] Stenson, A.C., A.G. Marshall, and W.T. Cooper (2003). Exact masses and chemical formulas of individual suwannee river fulvic acids from ultrahigh resolution electrospray ionization Fourier transform ion cyclotron resonance mass spectra. *Anal. Chem.* **75**, 1275–1284.

[54] Kim, S.W., L.A. Kaplan, R. Benner, and P.G. Hatcher (2004). Hydrogen-deficient molecules in natural riverine water samples–Evidence for the existence of black carbon in DOM. *Marine Chem.* **92**, 224–234.

[55] Purcell, J.M., R.P. Rodgers, C.L. Hendrickson, and A.G. Marshall (2004). Atmospheric pressure photoionization (APPI) Fourier transform ion cyclotron resonance mass spectrometry for analysis of nonpolar hydrocarbons. In: *52nd American Society for Mass Spectrometry Annual conference on Mass spectrometry and Allied Topics*, Nashville, TN.

[56] Schaub, T.M., C.L. Hendrickson, K. Qian, J.P. Quinn, and A.G. Marshall (2003). High-resolution field desorption ionization Fourier transform ion cyclotron resonance high-resolution mass analysis of non-polar molecules. *Anal. Chem.* **75**, 2172–2176.

[57] Schaub, T.M., H.B. Linden, C.L. Hendrickson, and A.G Marshall (2004). Continuous flow sample introduction for field desorption/ionization mass spectrometry. *Rapid Commun. Mass Spectrom.* **18**, 1641–1644.

[58] Miyabayashi, K., Y. Naito, K. Tsujimoto, and M. Miyake (2002). Structure characterization of petroleum vacuum residues by in-beam EI Fourier transform ion cyclotron resonance mass spectrometry. *Int. J. Mass Spectrom.* **221**(2), 93–105.

[59] Roussis, S.G. and R. Proulx (2002). Molecular weight distributions of heavy aromatic petroleum fractions by Ag+ electrospray ionization mass spectrometry. *Anal. Chem.* **74**(6), 1408–1414.

[60] Roussis, S.G. and R. Proulx (2004). Probing the molecular weight distributions of non-boiling petroleum fractions by Ag+ electrospray ionization mass spectrometry. *Rapid Commun. Mass Spectrom.* **18**(15), 1761–75.

[61] Rudzinski, W.E., K. Zhou, and X. Luo (2004). Mass spectrometric characterization of organosulfur compounds using palladium(II) as a sensitivity-enhancing reagent. *Energy Fuels* **18**(1), 16–21.

[62] Cunico, R.L., E.Y. Sheu, and O.C. Mullins (2004). Molecular weight measurement of UG8 asphaltene using APCI mass spectroscopy. *Petrol. Sci. Technol.* **22**(7–8), 787–798.

[63] Nali, M. and A. Manclossi (1995). Size exclusion chromatography and vapor pressure osmometry in the determination of asphaltene molecular weight. *Fuel Sci. Technol. Int.* **13**(10), 1251–1264.

[64] Speight, J.G. (2001). Handbook of petroleum analysis. In: J.D. Winefordner (ed.), *Chemical Analysis: A Series of Monographs on Analytical Chemistry and its Applications*, Vol. 158. John Wiley and Sons, NY, p. 489.

[65] Strausz, O.P., P.A. Peng, and J. Murgich (2002). About the colloidal nature of asphaltenes and the MW of covalent monomeric units. *Energy Fuels* **16**(4), 809–822.

[66] Lang, I. and P. Vavrecka (1981). Standardization of VPO asphaltene molecular weight. *Fuel* **60**(12), 1176–1177.

[67] Algelt, K.H. and M.M. Boduszynski (1994). Compositional Analysis: Dream and Reality. In: K.H. Algelt and M.M. Boduszynski (eds.), Composition and Analysis of Heavy Petroleum Fractions, Dekker, NY, p. 495.

[68] Sheu, E.Y. (2002). Petroleum asphaltene-properties, characterization, and issues. *Energy Fuels* **16**(1), 74–82.

[69] Chen, R., X. Cheng, D.W. Mitchell, S.A. Hofstadler, Q. Wu, A.L. Rockwood, M.G. Sherman, and R.D. Smith (1995). Trapping, detection, and mass determination of coliphage T4 DNA ions of 10^8 Da by electrospray ionization Fourier transform ion cyclotron resonance mass spectrometry. *Anal. Chem.* **67**, 1159–1163.

[70] Videler, H., L.L. Ilag, A.R.C. McKay, C.L. Hanson, and C.V. Robinson (2005). Mass spectrometry of intact ribosomes. *FEBS Lett.* **579**, 943–947.

[71] Del Rio, J.C. and R.P. Philp (1999). Field ionization mass spectrometric study of high molecular weight hydrocarbons in a crude oil and a solid bitumen. *Org. Geochem.* **30**(5), 279–286.

[72] Groenzin, H. and O.C. Mullins (1999). Asphaltene molecular size and structure. *J. Phys. Chem. A* **103**, 11237–11245.

[73] Groenzin, H. and O.C. Mullins (2000). Molecular size and structure of asphaltenes from various sources. *Energy Fuels* **14**, 677–684.

[74] Buch, L., H. Groenzin, E. Buenrostro-Gonzalez, S.I. Andersen, C. Lira-Galeana, and O.C. Mullins (2003). Molecular size of asphaltene fractions obtained from residuum hydrotreatment. *Fuel* **82**, 1075–84.

[75] Buenrostro-Gonzalez, E., H. Groenzin, C. Lira-Galeana, and O.C. Mullins (2001). The overriding chemical principles that define asphaltenes. *Energy Fuels* **15**, 972–978.

[76] Tanaka, R., S. Sato, T. Takanohashi, J.E. Hunt, and R.E. Winans (2004). Analysis of the molecular weight distribution of petroleum asphaltenes using laser desorption-mass spectrometry. *Energy Fuels* **18**(5), 1405–1413.

[77] Robins, C. and P.A. Limbach (2003). The use of nonpolar matrices for matrix-assisted laser desorption/ionization mass spectrometric analysis of high boiling crude oil fractions. *Rapid Commun. Mass Spectrom.* **17**(24), 2839–2845.

[78] Merino-Garcia, D. and S.I. Andersen (2005). Calorimetric evidence about the application of the concept of CMC to asphaltene self-association. *J. Dispersion Sci. Technol.* **26**(2), 217–225.

[79] Andreatta, G., N. Bostrom, and O.C. Mullins (2005). High-Q ultrasonic determination of the critical nanoaggregate concentration of asphaltenes and the critical micelle concentration of standard surfactants. Langmuir **21**(7), 2728–2736.

[80] Andersen, S.I. and K.S. Birdi (1991). Aggregation of asphaltenes as determined by calorimetry. *J. Colloid Interface Sci.* **142**(2), 497–502.

[81] Andersen, S.I. and S.D. Christensen (2000). The critical micelle concentration of asphaltenes as measured by calorimetry. *Energy Fuels* **14**(1), 38–42.

[82] Sheu, E.Y., D.A. Storm, and M.B. Shields (1995). Adsorption kinetics of asphaltenes at toluene/acid solution interface. *Fuel* **74**(10), 1475–9.

[83] Mahler, H.R. and E.H. Cordes (1996). Proteins: Classification, properties and purification. *Biological Chemistry*. Harper & Row Publishers, pp. 18–20.

4

Molecular Orbital Calculations and Optical Transitions of PAHs and Asphaltenes

Yosadara Ruiz-Morales

1. Introduction

In order to understand the chemistry and aggregation of asphaltenes, it is essential to know the size and structure of their fused aromatic ring (FAR) region. It is well established that the FAR region in asphaltenes is similar to polycyclic aromatic hydrocarbons (PAHs); asphaltene molecules additionally may contain heteroatoms and alkyl groups. It is essential to characterize the number of rings in the FAR group as well as their geometry in asphaltenes. Regarding geometry, the terms pericondensed vs. catacondensed are generally used; pericondesation refers to single bridgehead carbons attached to three rings, while catacondensation refers to aromatic carbons shared by at most two rings. ^{13}C NMR and more recently XRRS (X-ray Raman spectroscopy) have been applied to investigate the type of aromatic ring condensation in asphaltenes. A different approach is to employ molecular orbital (MO) calculations especially coupled with the ubiquitous optical absorption and emission data for asphaltenes. This method identifies the types of ring geometries by virtue of understanding their electronic structure. In addition, this method naturally gives the stability of the ring systems thereby enabling one to determine *why* asphaltenes have certain ring geometries. If one considers that asphaltenes are essentially stable for geologic time, then unstable aromatic structures as determined by MO calculations are ruled out. The optical absorption spectra of asphaltenes exhibit an exponential decrease in the neighborhood of 650 nm, showing that large graphitic structures with very low energy electronic absorption do not constitute a significant component of the asphaltene fraction. Likewise, asphaltene fluorescence emission spectra, which exhibit significant intensity in the range of 400–650 nm, reflect the nature and type of FAR structures present in asphaltenes. However, this optical absorption and emission data when considered alone has limited use in particular for determination of aromatic structures. The

Yosadara Ruiz-Morales • Programa de Ingeniería Molecular, Instituto Mexicano del Petróleo, Eje Central Lázaro Cárdenas 152, México D. F. 07730, México. E-mail: yruiz@imp.mx utility of this data is greatly expanded when comparing it with the MO calculations of electronic structure for the many possible candidates, catacondensed and pericondensed PAHs, for asphaltenes. Understanding the implications of different ring geometries and sizes on stability also provides essential information about governing heuristics for the asphaltene aromatics.

Here we show that the characterization of the stability of the FAR systems in terms of kinetic and thermodynamic stability is primary. The defining terms pericondensed and catacondensed are subordinate to the stability considerations. The Clar model which states that the most important representation of a PAH is one having the maximum number of disjoint π-sextets (depicted by inscribed circles) and a minimum number of fixed double bonds captures the essence of the kinetic and thermodynamic stability arguments enabling a simple heuristic for assessing stability. FARs with more sextet carbon are more aromatic and more kinetically and thermodynamically stable; that is, they gain in stability due to delocalization of π-electrons within the resonant sextets. FARs with increased isolated double-bond carbon are much more reactive and unstable; that is, they are kinetically and thermodynamically less stable. This model is readily employed for complex aromatics of the sort to be considered for asphaltenes. We extend these concepts using a methodology with the simple sobriquet "the Y-rule." The Y-rule establishes the most important and most kinetically and thermodynamically stable representation, in terms of π-sextets and double bonds, for any pericondensed PAH. Regarding reactivity, it is well known that a PAH is attacked at the position of the double bonds because it requires less energy to break these bonds than to break a π-sextet. Thus the Clar structures found with the Y-rule also give information about where it is most likely that chemical reactivity would take place. These simple methods are tied to stability arguments and shown to yield asphaltene structures found experimentally. The XRRS results are directly interpretable within this framework validating this direct approach.

In particular, we find that (1) acenes are not allowed as asphaltene FAR region based on stability; (2) fully resonant PAHs are not allowed either based on their high-energy transitions (unless unrealistically large ring systems are assumed). That is, the fully resonant systems are colorless or pale yellow—unlike asphaltenes; (3) almost fully resonant pericondensed structures are stable and are compatible with the large volume of optical absorption and emission data; (4) mostly FARs with 5–10 fused rings and with 2–4 π-sextets satisfy: the requisite of stability, the requisite of optical absorption and emission transitions, and the FAR size constraints imposed by direct molecular imaging and by measurement of the rotational diffusion constants. Finally we identify 56 PAHs, out of thousand of possible isomers, with 5–10 fused rings that fulfill all the experimental constraints and that are most likely structural candidates of the FAR region in asphaltenes.

Asphaltenes are thought to be polycyclic aromatic compounds similar to the PAHs but containing heteroatoms (N, O, S) and alkyl side-chains in their structure.[1] In crude oil, asphaltene micelles are present and they can also aggregate with adsorbed resins and furthermore, can contain metals such as V and Ni which are present in the oil as oil-soluble organometallic compounds.[1–6] There have been a number of experimental studies of asphaltenes to determine their size and structure which point to the conclusion that asphaltene molecules have small absolute sizes. Groenzin and Mullins have measured rotational correlation times of individual petroleum asphaltene molecules using fluorescence depolarization techniques.[7–9] Using simple models and comparisons with known chromophores, they predict a range of asphaltene molecular diameters of 10–20 Å and a molecular weight range of 500–1000 g/mol with a mean value of 750 g/mol. Boduszynski has found molecular weights of asphaltenes in the order of 800 g/mol by using field ionization mass spectroscopy[10,11] and Miller has found asphaltene molecular weights in the range of 200–600 g/mol with a mean value of 400 g/mol by using laser desorption mass spectroscopy.[12] More recently negative-ion electrospray ionization coupled with high-field Fourier transform ion cyclotron resonance mass spectrometry has been used to determine the molecular weight distributions of polar compounds (e.g., asphaltenes and resins) in different oils, finding a mass range of ~250–1000 g/mol.[13]

Scanning tunneling microscopy (STM) has been used to image the aromatic systems in asphaltene molecules. The condensed ring portions in asphaltenes yielded an average dimension of ~11 Å[14] which is consistent with a low molecular weight and 4–10 FARs. High resolution transmission electron microscopy (HRTEM) analysis of asphaltenes has also shown that the length scale of the aromatic ring systems is ~10 Å for petroleum asphaltenes.[15] In addition, ^{13}C NMR analysis predicts a range between 6 and 9 FARs in the condensed aromatic structural units,[14] which also agrees with the STM results. The degree of condensation and/or substitution of aromatic rings has been evaluated by applying solid-state ^{13}C NMR, FT-IR, and EPR. The results of these studies show that asphaltenes with different aromaticity seem to be similar in their aromatic rings condensation and/or substitution degrees and the estimated value of the average number of condensed aromatic rings is nearly 7.[16]

Regarding the geometric distribution of the FARs in the aromatic core of asphaltenes it has been shown experimentally,[17,18] with XRRS,[17] and theoretically,[19] that the FARs region in asphaltenes tend to be pericondensed. Regarding the number of fused rings in each aromatic system in asphaltenes Groenzin and Mullins reported an estimated range of fused rings between 4 and 10, with a mean value of 7, by comparing both rotational correlation times and the experimental optical absorption, and fluorescence emission spectra of known PAH standards and asphaltenes; asphaltene fluorescence emission is significant in the range of 400–650 nm.[7,8] Nevertheless, this fluorescence analysis would benefit greatly by the application of MO calculations to sort out the exact types of PAHs that are consistent with the fluorescence emission, the molecular size data along with other measured molecular parameters. The analysis of spectral data alone is incomplete without the determination of the types of asphaltene ring systems (e.g. pericondensed vs. catacondensed) and without direct investigation of the electronic transitions of PAHs by MO calculations. In fact, relatively few large PAH ring systems have been synthesized; consequently, the MO calculation is the only systematic approach to analyzing asphaltenes. Such a systematic approach can identify trends in PAH chemical stability; asphaltenes are likely to be chemically stable since they are formed in the terrestrial hydrocarbon kitchen and they persist for geologic time. These MO calculations are the main focus of this chapter. The approach of

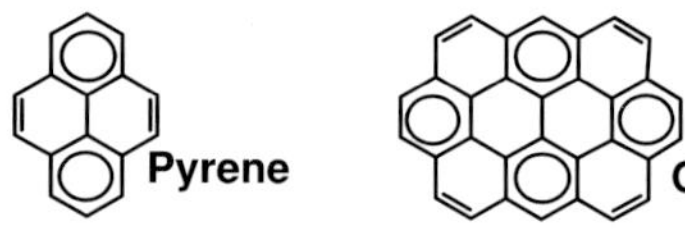

Figure 4.1. Clar structures of pyrene and ovalene.

employing MO calculations coupled with the experimental optical absorption and emission data for asphaltenes identifies the types of ring geometries, by comparing the calculated optical transitions of PAHs that have different geometries, different number of fused rings, and different types of condensation with the observed optical transitions of asphaltenes.

The experimental optical transitions of asphaltenes enclose information on their stability and structure. Stability and structure are two properties that are closely interconnected and in the case of asphaltenes the stability of their FAR region is related to the stability of the associated PAH that composes it. The stability in PAHs is governed by the number of resonant π-sextets present in their structure and in this context the Clar's description[17,19–27] or Clar's model of PAHs is useful. This simple description stipulates that the π-electron density is distributed in electron sextets (π-sextets), indicated by an inscribed circle, assigned to discrete hexagons in benzenoid PAHs and the remaining electrons to double bonds. The components π-sextets and CC double bonds are conjugated, and thus are separated by CC single bonds. In other words, the Clar structures provide a simple way to represent the zeroth order distribution of electron density in the molecule. As such, the Clar representation is far superior to the common representation for PAHs, where all rings are treated as equal with inscribed circles drawn in all hexagons; the electronic structure of the different rings are far from equal. For a particular compound, the Clar structure can be defined as a valence structure which has as many as possible π-sextets and as few as possible CC double bonds. In Figure 4.1 the Clar structure of pyrene and ovalene are presented. These PAHs are characterized by a single Clar structure, which besides π-sextets involves a number of CC double bonds. The distinction between the descriptions of full delocalization vs. isolated sextets and double bonds is key to understanding reactivity and spectroscopy of PAHs, i.e. the disposition and number of the π-electrons in PAHs largely defines their chemistry.

The full resonant structures (FRS),[19] or fully benzenoid,[28–33] have only one Clar structure in which all the π-density is distributed only in localized π-sextets and no double bonds, as it is the case of triphenylene or hexabenzocoronene. In Figure 4.2 the Clar structure of these FARs are presented. Furthermore, there are many PAH compounds that have more than one Clar structure, due to the presence of sextet migration,[20–27] with the highest number of resonant sextets; that is, there is no more a unique way to assign the maximal number of Clar sextets, and not all of these Clar structures represent the real chemical reactivity and real π-electronic distribution. In Figure 4.3 five nonequivalent Clar structures are presented with the highest number of resonant π-sextets and the same number of double bonds that can be drawn for the same PAH.[24] To find these five nonequivalent Clar structures

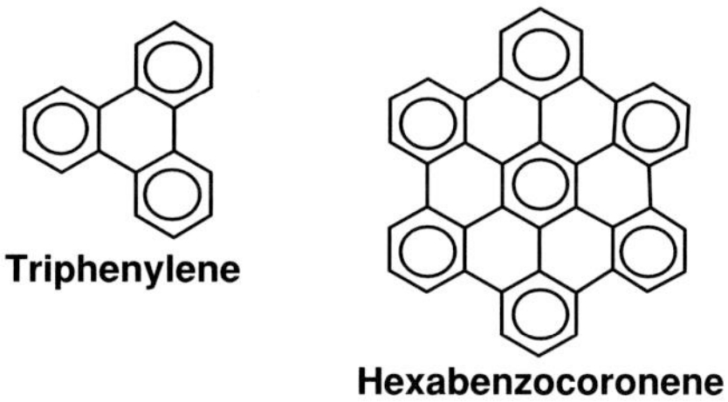

Figure 4.2. Clar structures of two full resonant PAHs: triphenylene and hexabenzocoronene.

is a task in itself[24] and not all of them contribute to the correct representation of the electronic structure and reactivity of this particular PAH. Actually Clar discussed its NMR spectra in terms of the structure 4 shown in Figure 4.3 and its mirror image.[22]

For the case of structural PAH isomers, which have the same π-electron content, it is found that different isomers can have different number of resonant π-sextets due to the different geometrical arrangement of the fused hexagons or rings. The isomers with a larger number of resonant sextets are more kinetically and thermodynamically stable than the isomers with a lower number of resonant sextets. By kinetic stability, we mean stability with respect to the activated complex; that is, in terms of activation energy. The high stability of a molecule reflects its low reactivity toward chemical reactions. It becomes evident the need of figuring

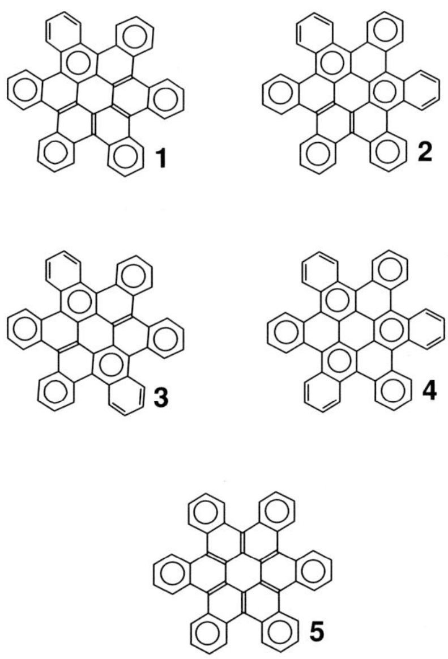

Figure 4.3. Five nonequivalent Clar structures with the highest number of resonant π-sextets, and the same number of double bonds, which can be drawn for the same PAH.

out the number of resonant sextets present in the FAR region in asphaltenes that confers them a high stability.

The establishment of the Clar structure(s) of PAHs that represent the actual chemical stability and reactivity of a given PAH is an open problem that is referred to as the aromaticity in PAHs. For the case of the PAH depicted in Figure 4.3, there are five nonequivalent Clar structures but it turns out that only two of them, which are equivalent, represent the actual electronic distribution. To solve this problem, we have proposed a useful and easy rule called the Y-rule[19,27] that identifies the most *important Clar structures* in pericondensed PAHs of any size, and the superposition of *the most important Clar structures*, if there is more than one, gives as a result the π-electronic distribution. To apply the Y-rule is quite easy as we will show in one of the sections of this chapter. For the particular PAH presented in Figure 4.3, the Y-rule directly establishes that the actual representation of its electronic structure is given by the Clar structure 5 for which the superposition of the Clar structure 4 and its equivalent structure (mirror image) are the major contributors, in agreement with the NMR results. The performance of the Y-rule has been validated theoretically[27] by comparing the π-electronic distribution obtained with it with the π-electronic distribution obtained from nucleus-independent chemical shift (NICS)[25,34–36] calculations.

Only certain PAHs with certain π-electronic distribution, in sextets and double bonds, with certain geometry, and certain type of condensation are most likely to be candidates for the aromatic core of asphaltenes. In this chapter we identify PAH structural candidates of the aromatic core in asphaltenes that fulfill all the asphaltene experimental constraints (optical transitions range, STM, HRTM, molecular weight) and present a π-electron density distribution that agrees with their optical transitions and stability. This is done by coupling experimental data with MO calculations.

2. Computational Details

In section 3.2, we explain why we use the following methodology. The geometry optimization of the PAH systems and the asphaltene structures was done by performing force-field-based minimization using the energy minimization panel in Cerius2 version 4.6 and the COMPASS (Condensed-Phase Optimized Molecular Potentials for Atomistic Simulation Studies)[37,38] consistent force field as it is provided in the Cerius2 package.[39] This type of minimization or geometry optimization is a molecular mechanics simulation where the laws of classical physics are used to predict the structures and properties of molecules. There are many different molecular mechanics methods, each one characterized by its particular force field. A force field has as components a set of equations defining how the potential energy of a molecule varies with the locations of its component atoms, and a series of atom types, defining the characteristics of an element within a specific chemical context. The atom types prescribe different characteristics and behavior for an element depending upon its environment, and one or more parameter sets that fit the equations and atom types to experimental data. The parameter sets define force constants, which are values used in the equations to relate atomic characteristics to energy components and structural data such as bond lengths and angles.

The COMPASS force field (FF) has been tested and validated extensively against experiment for many organic molecules. It is an *ab initio* force field that enables accurate and simultaneous prediction of gas-phase properties (structural, conformational, vibrational, etc.) and condensed-phase properties (equation of state, cohesive energies, etc.) for a broad range of molecules and polymers.[38,39] The molecular systems that are explicitly covered in the COMPASS force field among others include the most common organic molecules, common polymers, and small gas molecules, such as alkanes, alkenes, alkynes, benzenes/aromatics, cycloalkanes, ethers, acetals, alcohols, phenols, amines, ammonia, aldehyde/ketone, acids, esters, carbonates, amides, carbonates/urethanes, siloxanes, silanes, alkyl halides, phosphazenes, nitro groups, nitrils, sulfides, thiols, isocyanides, amineoxides, aromatic halides, cynamide, nitrates, etc. These are well parameterized and rigorously tested. Parameters have been derived using high-level *ab initio* calculations, and optimized to fit experimental data of both gaseous and condensed phases.[39] The COMPASS force field, like all other force fields of this type, is accurate for those molecular classes for which it has been explicitly parameterized. Some examples of COMPASS success include the silico determination of the cohesive properties of the polyether imide, Ultem™ by General Electric-CRD researchers,[40] the understanding of the crystal structure of nickel (II) and cobalt (II) 2,6-naphthalenedicarboxylate tetrahydrate, a new chemical intermediate that leads to high-performance polyesters, by BP Chemicals and Searle researchers;[41] researchers at Rhodia Company have used a combined atomistic and mesoscopic approach to study the binary blend compatibility of polyamide6 and poly (vinyl acetate) with different degrees of hydrolysis.[38,42,43] Using molecular dynamics simulations with the COMPASS force field, they were able to determine cohesive energies (and solubility parameters) with high accuracy.

As we show in section 3.2 the calculated π–π^* electronic transition, using the PAH structures whose geometry optimization was carried out using the COMPASS force field, compares well with the experimental data. The COMPASS geometry optimization of the PAH systems do not require expensive computational resources and can be performed on a personal computer. Usually the geometry optimization of PAHs is done by using semiempirical methods like PM3[44,45] or AM1,[46,47] which use parameters derived from the experimental data. They solve an approximate form of the Schrödinger equation that depends on having appropriate parameters available for the type of chemical system under investigation. However, the calculated π–π^*electronic transition using the PAH structures whose geometry optimization was carried out with the PM3 semiempirical method does not compare that well with the experimental data. The excited electronic states (including the HOMO–LUMO configuration or π–π^* transition) of the PAH compounds and the asphaltenes were calculated using ZINDO/S[48] as it is provided in the Gaussian 98 package,[49] employing the COMPASS force field (FF) geometry optimized structures.

The NICS[25,34–36] calculations were carried out using the GIAO-DFT[50,51] method as implemented in the Gaussian 98[49] package using the COMPASS force field geometry optimized structures. A dummy atom was located on the molecular plane at the geometrical center of each hexagon in the PAH structures to calculate the NICS(0), and at 1Å above the molecular plane to calculate the NICS(1). The Becke's 1988 functional,[52] which includes the Slater exchange along with corrections involving the gradient of the density was used together with the correlation functional of Lee, Yang, and Parr,[53] which includes both local and nonlocal terms; i.e., the B3LYP functional was used. It is well known that the shielding tensor and NICS are very sensitive to the quality and size of the basis set used.[54,55] A basis set is a mathematical representation of the MOs (which in turn combine to approximate the total electronic wavefunction) within a molecule, and the basis set can be interpreted as restricting each electron to a particular region of space. Larger basis sets impose fewer constraints on electrons and more accurately approximate exact MOs but they require correspondingly more computational resources. The minimal basis sets contain the minimum number of basis functions needed for each atom and the minimal basis sets use fixed-size atomic-type orbitals. The first way that a basis set can be made larger is to increase the number of basis functions per atom. Split valence basis sets, such as 3-21G and 6-31G have two (or more) sizes of basis functions for each valence orbital. Split valence basis sets allow orbital to change size, but not to change shape. Polarized basis sets remove this limitation by adding orbitals with angular momentum beyond what is required for the ground state to the description of each atom. For example, polarized basis sets add d functions to carbon atoms and f functions to transition metals, and some of them add p functions to hydrogen atoms. For the calculation of NICS, Schleyer *et al.*[25,34–36] recommend the use of the 6-31G(d) basis which is the 6-31G basis set with d functions added to heavy atoms. However, a sufficiently large basis is needed for an accurate description of the chemical shift. Thus, we used a basis set that is augmented with two sets of polarization functions, i.e., the 6-31G(d, p) basis set. This basis set is the same as recommended by Schleyer *et al.*[25,34–36] but adds p functions to hydrogen atoms in addition to the d functions on heavy atoms.[56]

A single identifying number has been assigned to all the structures in the figures. A combination of Arabic numbers and letters is used. The Arabic number represents the number of FARs in the PAH structure (*n*FAR), i.e., the number of hexagons in the structure, and the letter represents the different structural isomers with the same number of fused aromatic rings (*n*FAR).

3. Results and Discussion

The fused aromatic core region in asphaltenes is similar to PAHs with heteroatoms and alkyl chains. In the process of identifying the most likely PAH structural candidates of the fused aromatic region in asphaltenes, we first analyze the topological characteristics of PAH systems. The topological characteristics give information about the different isomers as well as the effect of the spatial distribution of the fused hexagons on the optical transition of the PAH systems.

The structures of the PAH isomers with optical transitions that fall inside the experimental fluorescence emission of asphaltenes are further more analyzed in terms of size (longest dimension), π-electronic distribution (aromaticity), and percentage of condensation. Those PAH systems that fulfill all the asphaltene experimental constraints (see "Introduction") are identified as most likely structural candidates of the aromatic region in asphaltenes and are presented at the end of this chapter. The size and structure of the FAR region in asphaltenes are comparable to the structural and electronic characteristics of an associated PAH with no heteroatom substitution(s) and no alkyl chains, which we call throughout the text as the associated "free" PAH.

3.1. Topological Characteristics of PAHs

We have only concentrated in studying the neutral even-numbered PAHs with fused six-member rings, i.e., benzenoid-type PAHs. Benzenoid PAHs contain an even number of carbon atoms, and in general, they are restricted to a range of stoichiometries; thus, in order to decide about the particular type of PAH molecules to calculate, we have to find all the possible stoichiometries that define the benzenoid PAHs and their molecular structures. In Figure 4.4 the PAHs stoichiometries are presented in a *n*FAR vs. hydrogen content diagram. All the

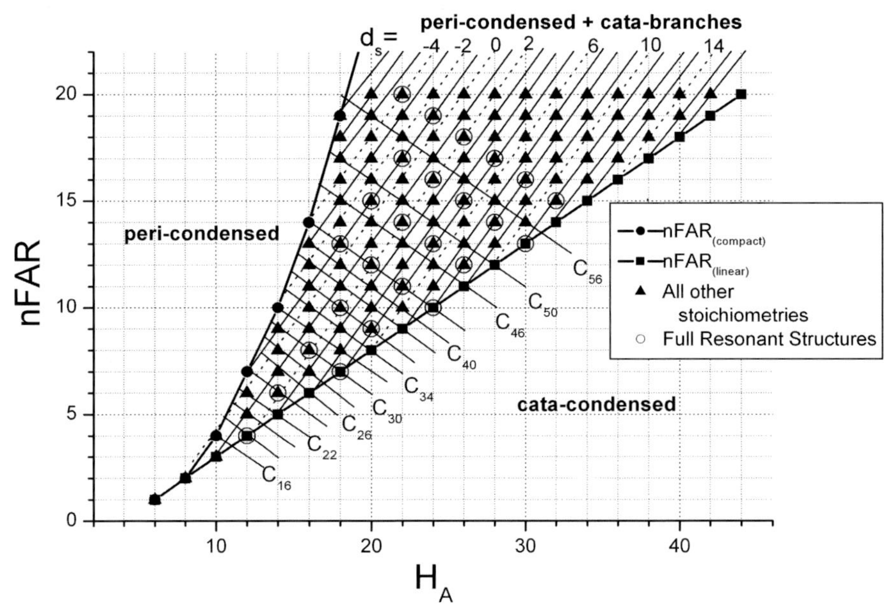

Figure 4.4. Diagram of *n*FAR vs. hydrogen content, H_A, for benzenoid PAHs. In the diagram the diagonal lines mark all the stoichiometries with the same carbon content. Lines that mark all the stoichiometries with the same net number of disconnections among the internal edges (d_s) in the structure are also shown. Reproduced with permission from *J. Phys. Chem. A*, **2002**, 106, 11283–11308. Copyright 2002 Am. Chem. Soc.

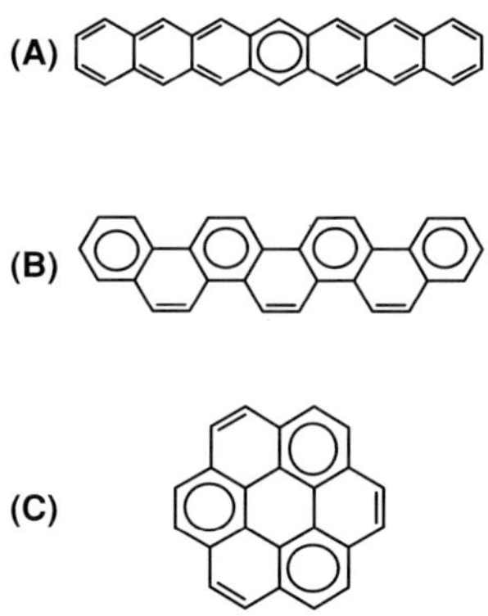

Figure 4.5. (A) Example of a linear PAH structure. (B) Example of a zigzag PAH structure. The linear and zigzag structures are examples of catacondensed systems. (C) Example of a compact circular PAH structure. The circular structures are pericondensed.

benzenoid PAHs stoichiometries are restricted to a region delimited by two borderlines or limit curves. Any other stoichiometry outside the borderlines does not correspond to a benzenoid PAH. The systems on the lower limit curve, represented by diamonds, are pure catacondensed with the least compact structure, like the linear (for example, Figure 4.5A) or zigzag structures (for example, Figure 4.5B). The systems on the upper limit curve, represented by stars, are pure pericondensed systems with the most compact, circular, structure (for example, Figure 4.5C). All the other stoichiometries, represented by solid circles in Figure 4.4, and that lie between the two borderlines, represent systems that have a pericondensed core with catacondensed branches. We found that the best equation[19] that describes the upper border curve is $C_A = -\frac{1}{3} + \frac{H_A^2}{6}$ (where C_A and H_A are the carbon and hydrogen content, respectively). For any benzenoid PAH Dias[28–33] found that the number of FARs or hexagons is given by the equation $n\text{FAR} = \frac{(C_A - H_A) + 2}{2}$.

For each nFAR or each nFAR row, there are a given number of possible stoichiometries. Also, it can be seen in Figure 4.4 that there are many stoichiometries that present the same hydrogen content but different carbon content, which increases by 2. The information obtained from Figure 4.4 is presented in Table 4.1, where all the stoichiometries for benzenoid PAHs from 2FAR to 10FAR are given together with their percentage of compactness (P_C). The percentage of compactness (P_C) is a measure of the degree of condensation of the PAH structure[19,57,58] i.e., it is a measure of the pericondensation. The lowest value of P_C is 0% and the highest value is 100%. The PAHs with 0% of compactness have a structure that is the least compact; this is the case for catacondensed compounds (Figure 4.5). The PAHs with 100% of compactness have a structure that is the most compact; this is the case for the pure pericondensed (circular) systems (Figure 4.5). The percentage of compactness for all the other stoichiometries, between the borderlines in Figure 4.4, is calculated by interpolation.[19,57,58] The systems on the lower borderline (0% P_C) have more hydrogen content than the systems on the upper

Table 4.1. Stoichiometries (Stoic.) associated to each FAR family and the associated structural parameters. Reproduced with permission from *J. Phys. Chem. A*, 2002, **106**, 11283–11308. Copyright 2002 Am. Chem. Soc.

nFAR	Stoic.	C_{Int}^{a}	Number of isomers	P_{C}^{b}	Number of resonant sextets in the structure (N_R)	Number of double bonds in the structure (N_{DB})	d_s^c	C_Y^d	C_{PA3}^e
2FAR	$C_{10}H_8$	2	1	0%	1	2	0	0	2
3FAR	$C_{14}H_{10}$	4	2	0%	2, 1	1, 4	1	0	4
4FAR	$C_{16}H_{10}$	6	1	100%	2	2	0	2	4
	$C_{18}H_{12}$	6	5 (1)	0%	3 (full), 2, 1	0, 3, 6	2	0	6
5FAR	$C_{20}H_{12}$	8	3	50%	3, 2	1, 4	1	2	6
	$C_{22}H_{14}$	8	12	0%	3, 2, 1	2, 5, 8	3	0	8
6FAR	$C_{22}H_{12}$	10	2	82%	3, 2	2, 5	0	4	6
	$C_{24}H_{14}$	10	13 (1)	33%	4 (full), 3, 2	0, 3, 6	2	2	8
	$C_{26}H_{16}$	10	37	0%	4, 3, 2, 1	1, 4, 7, 10	4	0	10
7FAR	$C_{24}H_{12}$	12	1	100%	3	3	−1	6	6
	$C_{26}H_{14}$	12	9	65%	4, 3, 2	2, 4, 7	1	4	8
	$C_{28}H_{16}$	12	62	25%	4, 3, 2	2, 5, 8	3	2	10
	$C_{30}H_{18}$	12	123(1)	0%	5 (full), 4, 3, 2, 1	0, 3, 6, 9, 12	5	0	12
8FAR	$C_{28}H_{14}$	14	8	75%	4, 3, 2	2, 5, 8	0	6	8
	$C_{30}H_{16}$	14	58 (1)	45%	6 (full), 5,4, 3, 2	0, 3, 6, 9, 12	2	4	10
	$C_{32}H_{18}$	14	295	20%	5, 4, 3, 2	1, 4, 7, 10	4	2	12
	$C_{34}H_{20}$	14	446	0%	5, 4, 3, 2, 1	2, 5, 8, 11, 14	6	0	14
9FAR	$C_{30}H_{14}$	16	3	89%	4, 3, 2	3, 6, 9	−1	8	8
	$C_{32}H_{16}$	16	46	60%	5, 4, 3, 2	1, 4, 7, 10	1	6	10
	$C_{34}H_{18}$	16	335	33%	5, 4, 3, 2	2, 5, 8, 11	3	4	12
	$C_{36}H_{20}$	16	1440(1)	16%	6 (full), 5, 4,3 2	0, 3, 6, 9, 12	5	2	14
	$C_{38}H_{22}$	16	1689	0%	6, 5, 4, 3 2	1, 4, 7, 10, 13	7	0	16
10FAR	$C_{32}H_{14}$	18	1	100%	4	4	−2	10	8
	$C_{34}H_{16}$	18	34	66%	5, 4, 3, 2	2, 5, 8, 11	0	8	10
	$C_{36}H_{18}$	18	337 (3)	49%	6 (full), 5,4, 3, 2	0, 3, 6, 9, 11	2	6	12
	$C_{38}H_{20}$	18	1987	28%	6, 5, 4, 3, 2	1, 4, 7, 10, 13	4	4	14
	$C_{40}H_{22}$	18	6973	14%	6, 5, 4, 3,2	2, 5, 8, 11, 14	6	2	16
	$C_{42}H_{24}$	18	6693 (1)	0%	7 (full),6, 5, 4, 3, 2, 1	0, 3, 6, 9, 12, 15, 18	8	0	18

[a] C_{Int} Total internal bridging carbons $C_{\text{Int}} = C_A - H_A = C_Y + C_{PA3} = 2(n\text{FAR}) - 2$, where nFAR is the number of fused aromatic rings. [b] P_C Percentage of compactness. It is a measure of the degree of condensation of the PAH structure. The lowest value of P_C is 0% (like in triphenylene) and the highest value is 100% (like in coronene). [c] d_s Net number of disconnections among the internal edges in the FAR region, $d_s = C_{PA3} - nFAR$. [d] C_Y Number of Y-carbons, internal aromatic carbons with a connectivity of 3. For catacondensed systems, $C_Y = 0$. [e] C_{PA3} Peripheral aromatic carbons in the whole FAR region having a connectivity of 3.

border (100% P_C); thus, the hydrogen content gives the percentage of condensation. In Table 4.1 also the topological elements are given. The topological elements that are common to all PAH isomers having a given stoichiometry are[19,29] (a) the number of σ-bonds and π-electrons; (b) the carbon and hydrogen content (C_A and H_A); (c) the total internal bridging carbons, C_{Int}, $C_{\text{Int}} = C_A - H_A$; (d) the number of fused aromatic rings (nFAR), $n\text{FAR} = \frac{(C_{\text{Int}}+2)}{2}$; (e) the number of internal aromatic carbons with a connectivity of three[28–33] or Y-carbons,[19] because they are in the vertex of what looks like a Y-letter, (C_Y), $C_Y = -2 + n\text{FAR} - d_s$; (f) the number of peripheral aromatic carbons having a connectivity of three (C_{PA3}),

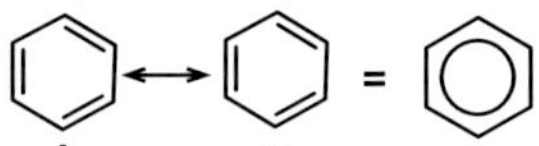

Figure 4.6. Resonant sextet or aromatic sextet. The resonant sextet confers a stability of 36 kcal/mol to benzene and it involves 6π-electrons.

$C_{PA3} = C_{Int} - C_Y = n\text{FAR} + d_s$; and (g) the net number of disconnections among the internal edges, d_s, $d_s = C_{PA3} - n\text{FAR}$. As it can be seen many of these terms are related by mathematical expressions. The d_s, C_Y, C_{PA3}, and C_{Int} structural parameters are the same for all the isomers, which have the same stoichiometry, and correspond to the structural features of the σ-backbone only (see Table 4.1).

For a given stoichiometry (Table 4.1) there are many isomers.[29,31] The number of PAH isomers increases rapidly with the increase of the number of fused rings. It would be impossible to study all the isomers in many cases. But only some of the isomers are more stable and more likely to be good candidates for the fused ring region in asphaltenes. These are the strain-free benzenoid PAHs with a certain number of resonant sextets (see "Introduction"), as we will discuss later. The number of resonant sextets (N_R), or also called aromatic sextets, are depicted by circle notation in all of our figures with chemical structures. Any π-electrons that do not participate in aromatic sextets are depicted as localized double bonds. In Figure 4.6 the resonant sextet of benzene is shown. The circle notation represents the resonance of states I and II and involves six π-electrons. In Table 4.1 the number of resonant sextets that can be reached in a given stoichiometry are presented. There are PAH compounds which have only a single Clar structure (see "Introduction"); in this group the FRS or "fully benzenoid" are included. In the FRS systems the π-electronic density is distributed only in localized resonant sextets and there are no localized double bonds. In Figure 4.4 some of the stoichiometries that present FRS are enclosed by an open circle. In Table 4.1, under the column of number of isomers, the number of FRS isomers is given in parenthesis, while in the N_R column the total number of resonant rings or resonant sextets, that can be reached in the FRS, is given under the label "full".

3.2. The HOMO–LUMO Optical Transition

In this section the theoretical HOMO–LUMO gap (the energetic gap between the highest occupied molecular orbital and the lowest unoccupied molecular orbital) for free PAHs is studied. The HOMO–LUMO gap and structural relationships of several PAHs are analyzed to conclude about the distribution and structure of the fused rings in the aromatic cores of the chromophores in asphaltenes. The fluorescence emission range for asphaltenes has been experimentally measured by Groenzin and Mullins.[7,8] The experimental $\lambda_{0\text{–}0}$ fluorescence emission band, which corresponds to the energy gap between the ground and first excited state (within a given spin manifold) is related to the HOMO–LUMO gap.[7,8,59] The HOMO–LUMO gap is used as a direct indicator of kinetic stability.[19,20–23] The HOMO (highest occupied molecular orbital) and the LUMO (lowest molecular orbital) are the frontier orbitals (FOs) in the MO energy levels diagram of molecules. The difference in energy between the FOs is known as the HOMO–LUMO gap (H–L gap). A large H–L gap implies high kinetic stability and low chemical reactivity since it is energetically unfavorable to add electrons to a high-lying LUMO or to extract electrons from a low-lying HOMO.

3.2.1. Validation of the Calculation Method

Here we carried out the validation of the best combination of theoretical methods (geometry optimization method//excited state calculation method, and geometry optimization method//single point calculation method) that agree better with the experimental fluorescence emission data of PAHs. With this in mind, the excited states (including the HOMO–LUMO transition) of several well-known PAHs compounds were calculated. To find the best method for the geometry optimization of the structures we tested three different methods (1) using COMPASS force-field-based (FF) minimization (for more information about this force field see section 2 named "Computational Details"); (2) using the semiempirical PM3 method, and (3) using density functional theory and the B3LYP functional. The HOMO–LUMO transition was also calculated using different methods: (1) the semiempirical electronic structure method ZINDO/S, (2) the semiempirical PM3 method, (3) using a single point calculation at the Hartree-Fock (HF) self-consistent field level, (4) using a single point calculation with density functional theory (DFT) and the B3LYP functional, and (5) using the time-dependent density functional theory TD-DFT to calculate the transition energy.

In Tables 4.2–4.4 the results from these calculations for three different PAHs (naphthalene **2a**, anthracene **3a**, and pyrene **4b**; see Figures 4.7 and 4.8) are presented. It can be seen in these tables that in general the method combinations B3LYP//B3LYP, PM3//B3LYP, PM3//DFT-TD, PM3//ZINDO, FF//B3LYP, and FF//ZINDO give a HOMO–LUMO gap that compares well with the experimental data, while the other methods (B3LYP//HF or PM3 and PM3//HF or PM3) give bad results. The optimization of structures using the B3LYP functional is expensive as well as the calculation of transition states with TD-DFT. Because of this, we decided to work with the combinations PM3//B3LYP, PM3//ZINDO, FF//B3LYP, and FF//ZINDO to do the validation of the calculation method. In Figures 4.7–4.11 the validation of the calculation method and of the calculated HOMO–LUMO gap (H–L gap) is carried out for several PAH compounds, with 1–7 FARs, and the calculated gap is compared with the reported experimental λ_{0-0} fluorescence emission band. The structures of the different compounds are also provided in Figures 4.7–4.11, and the data plotted in these figures are also provided in Tables 4.5–4.9.

From Figures 4.7 to 4.11, it can be concluded that the HOMO–LUMO gap calculated with the ZINDO/S method and using the structures that were optimized with the COMPASS force field (FF) compares well with the experimental data. As the number of rings increases, there is a slight distortion from the planarity in the PAH structures. From 9FAR onward, this distortion is more significant but

Table 4.2. Calculated HOMO and LUMO energies and HOMO–LUMO gap (ΔE_{H-L}) for naphthalene (2a, for structure see Figure 4.7) using different levels of theory

Calculation method: Optimization of structure // single point calculation	Energy (eV) HOMO	LUMO	ΔE_{H-L} (eV)
B3LYP//B3LYP	−6.0557	−1.2539	4.8018
B3LYP//HF	−7.8465	2.3954	10.2419
B3LYP//PM3	−8.8038	−0.4577	8.3461
PM3//B3LYP	−6.0769	−1.2177	4.8592
PM3//HF	−7.8886	2.0902	9.9788
PM3//PM3	−8.8348	−0.4089	8.4258
PM3//DFT-TD			4.4462
PM3//ZINDO			4.6065
FF//B3LYP			4.8121
FF//ZINDO			4.5671
Experiment			4.1308[a]
			4.6068[b]
			4.5227[c]
			3.8504[d]

[a]Ref. 7. [b]Ref. 60. [c]Ref. 61. [d]Ref. 59.

in general the structures are not concave. Semiempirical methods like PM3 produce a structure that is always planar, despite of the number of fused rings, and the calculated HOMO–LUMO gap, using either the ZINDO/S or the B3LYP methods, do not compare well with the experimental values (see Figures 4.7–4.11). The FF//B3LYP calculation gives a HOMO–LUMO gap that agrees well with the experimental data (Figures 4.7–4.11); however, the best agreement between

Table 4.3. Calculated HOMO and LUMO energies and HOMO–LUMO gap (ΔE_{H-L}) for anthracene (3a, for structure see Figure 4.7) using different levels of theory

Calculation method: Optimization of structure // single point calculation	Energy (eV) HOMO	LUMO	ΔE_{H-L} (eV)
B3LYP//B3LYP	−5.4910	−1.9105	3.5802
B3LYP//HF	−7.0195	1.4966	8.5161
B3LYP//PM3	−8.2119	−1.0280	7.1839
PM3//B3LYP	−5.5082	−1.8618	3.6464
PM3//HF	−7.0614	1.5766	8.6380
PM3//PM3	−8.2173	−0.9908	7.1839
PM3//ZINDO			3.6322
FF//B3LYP			3.3759
FF//ZINDO			3.4450
Experiment			3.4908[a]
			3.6663[b]
			3.2611[c]

[a]Ref. 7. [b]Ref. 62. [c]Ref. 63.

Table 4.4. Calculated HOMO and LUMO energies and HOMO–LUMO gap (ΔE_{H-L}) for pyrene (4b, for structure see Figure 4.8) using different levels of theory

Calculation method: Optimization of structure // single point calculation	Energy (eV)		
	HOMO	LUMO	ΔE_{H-L} (eV)
B3LYP//B3LYP	−5.5887	−1.7502	3.8384
B3LYP//HF	−7.0679	1.6204	8.6883
B3LYP//PM3	−8.1999	−1.0757	7.1242
PM3//B3LYP	−5.6164	−1.7032	3.9132
PM3//HF	−7.1273	1.6976	8.8248
PM3//PM3	−8.2138	−1.0343	7.1795
PM3//ZINDO			3.7936
FF//B3LYP			3.6129
FF//ZINDO			3.6100
Experiment			3.6946[a]
			3.7158[b]
			3.2979[c]
			3.3134[d]
			3.3223[e]
			3.3951[f]

[a]Ref. 62. [b]Ref. 59. [c]Ref. 64. [d]Ref. 63. [e]Ref. 65. [f]Ref. 66.

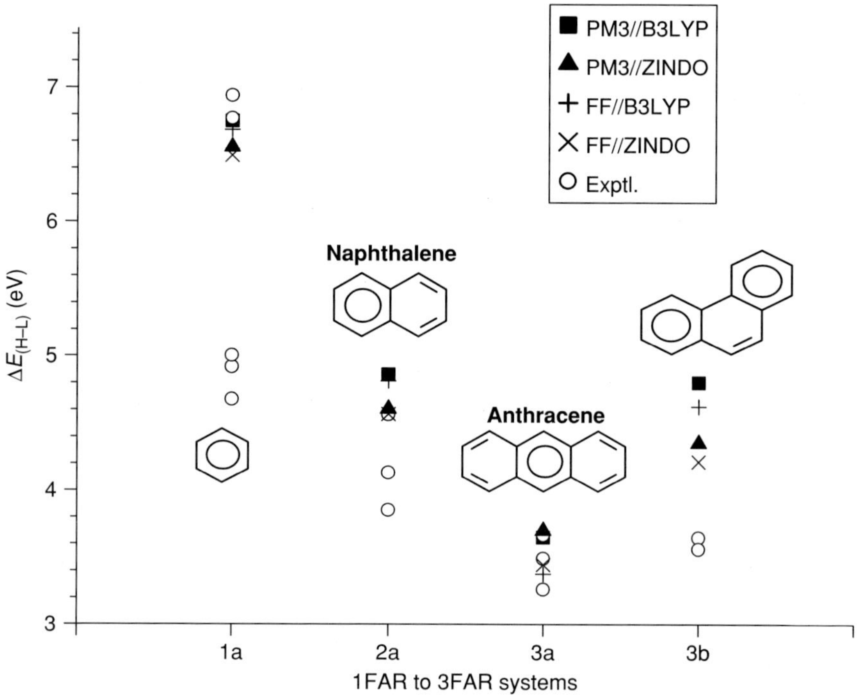

Figure 4.7. Diagram of the experimental and calculated HOMO–LUMO energy gap, at various levels of theory (geometry structure optimization method//energy gap calculation method) for the case of PAH systems with one to three fused aromatic rings (1FAR–3FAR).

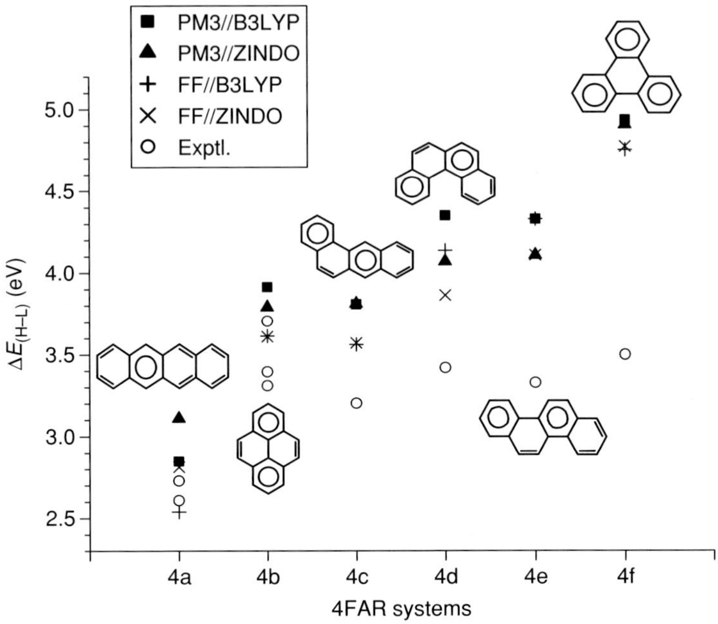

Figure 4.8. Diagram of the experimental and calculated HOMO–LUMO energy gap, at various levels of theory (geometry structure optimization method//energy gap calculation method) for the case of PAH systems with four fused aromatic rings (4FAR).

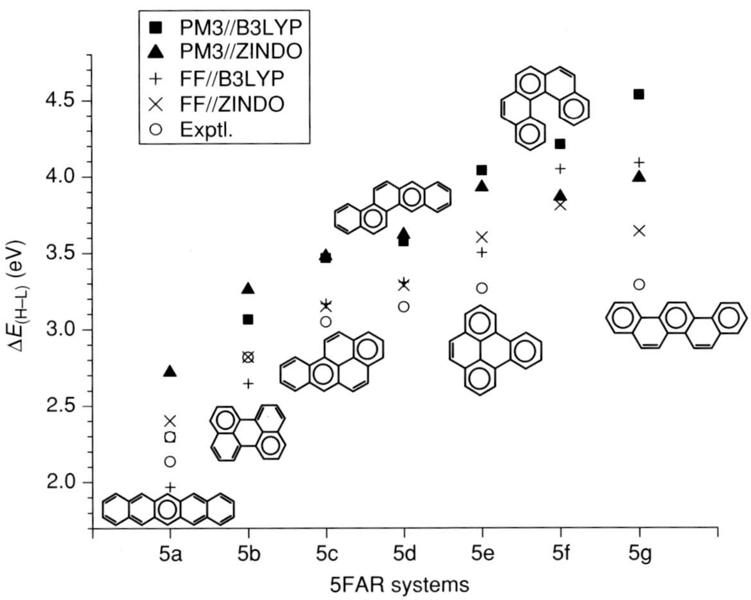

Figure 4.9. Diagram of the experimental and calculated HOMO–LUMO energy gap, at various levels of theory (geometry structure optimization method//energy gap calculation method) for the case of PAH systems with five fused aromatic rings (5FAR).

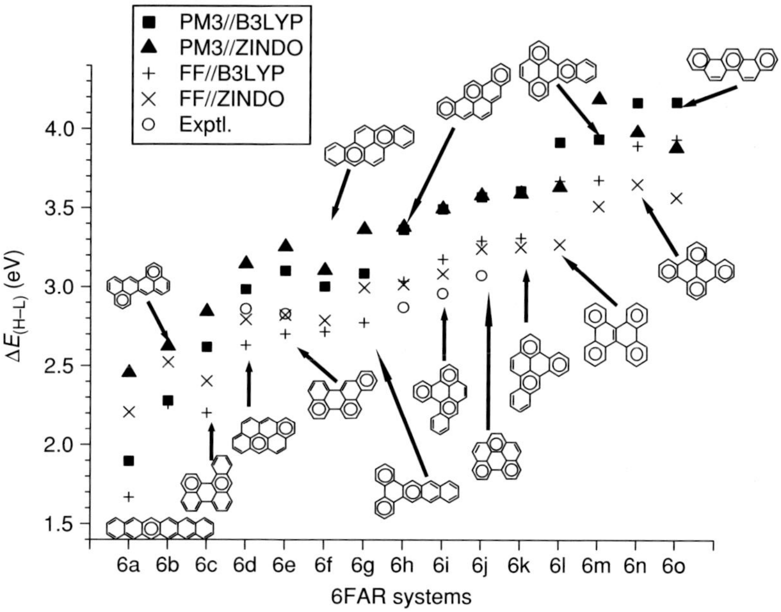

Figure 4.10. Diagram of the experimental and calculated HOMO–LUMO energy gap, at various levels of theory (geometry structure optimization method//energy gap calculation method) for the case of PAH systems with six fused aromatic rings (6FAR).

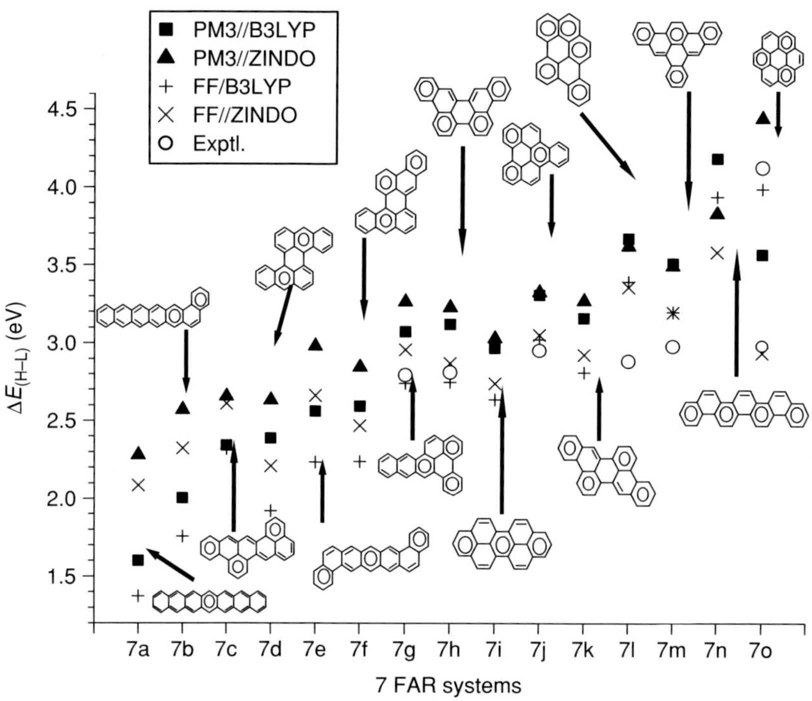

Figure 4.11. Diagram of the experimental and calculated HOMO–LUMO energy gap, at various levels of theory (geometry structure optimization method//energy gap calculation method) for the case of PAH systems with seven fused aromatic rings (7FAR).

Table 4.5. ΔE_{H-L} obtained from single point B3LYP calculations and ZINDO calculations for the 1 to 3 FAR structures optimized with PM3 and COMPASS force field

	ΔE_{H-L} for PM3 structure (eV)		ΔE_{H-L} for FF structure (eV)		
System	B3LYP	ZINDO	B3LYP	ZINDO	Exptl. (eV)
1a	6.7495	6.5568	6.6850	6.4949	6.9400[a,d]
					6.7716[b,d]
					4.6763[c,e]
					4.9176[a,e]
					5.0030[a,e]
2a	4.8592	4.6065	4.8121	4.5671	4.1308[c]
					4.6068[f,i]
					4.5227[g,i]
					3.8504[h]
3a	3.6465	3.6984	3.3759	3.4450	3.4908[c]
					3.6663[j]
					3.2611[k]
3b	4.8015	4.3498	4.6241	4.2118	3.6448[j]
					3.5610[k]

[a]Ref. 67. [b]Ref. 68. [c]Ref. 7. [d]These values, which are similar, were averaged and the averaged value is plotted in Figure 4.7. [e]These values, which are similar, were averaged and the averaged value is plotted in Figure 4.7. [f]Ref. 60. [g]Ref. 61. [h]Ref. 59. [i]These values, which are similar, were averaged and the averaged value is plotted in Figure 4.7. [j]Ref. 62. [k]Ref. 63.

Table 4.6. ΔE_{H-L} obtained from single point B3LYP calculations and ZINDO calculations for the 4FAR structures optimized with PM3 and COMPASS force field

	ΔE_{H-L} for PM3 structure (eV)		ΔE_{H-L} for FF structure (eV)		
System	B3LYP	ZINDO	B3LYP	ZINDO	Exptl. (eV)
4a	2.8485	3.1101	2.5383	2.8132	2.7276[a]
					2.6089[b]
4b	3.9132	3.7920	3.6129	3.6100	3.6946[e,d]
					3.7158[a,d]
					3.2979[c,f]
					3.3134[g,f]
					3.3223[h,f]
					3.3951[k]
4c	3.8082	3.8114	3.5617	3.5708	3.2021[g]
4d	4.3499	4.0692	4.1364	3.8632	3.4249[a,i]
					3.4138[g,i]
4e	4.3278	4.1075	4.3277	4.1078	3.3268[a]
4f	4.9340	4.9049	4.7536	4.7729	3.4963[a,j]
					3.5006[g,j]

[a]Ref. 59 [b]Ref. 7 [c]Ref. 64 [d]These values, which are similar, were averaged and the averaged value is presented in Figure 4.8. [e]Ref. 62. [f]These values which are similar, were averaged and the averaged value is presented in Figure 4.8 [g]Ref. 63 [h]Ref. 65 [i]These values, which are similar, were averaged and the averaged value is presented in Figure 4.8. [j]These values, which are similar, were averaged and the averaged value is presented in Figure 4.8. [k]Ref. 66.

Table 4.7. ΔE_{H-L} obtained from single point B3LYP calculations and ZINDO calculations for the 5FAR structures optimized with PM3 and COMPASS force field

	ΔE_{H-L} for PM3 structure (eV)		ΔE_{H-L} for FF structure (eV)		
System	B3LYP	ZINDO	B3LYP	ZINDO	Exptl. (eV)
5a	2.2956	2.7200	1.9693	2.4028	2.1366[a] 2.3111[b,c] 2.2834[c,d]
5b	3.0654	3.2604	2.6458	2.8193	2.8178[e,f] 2.8164[f,g]
5c	3.4668	3.4812	3.1669	3.1500	3.0598–3.0225[h,i,j] 3.0598[g]
5d	3.5765	3.6195	3.3065	3.2865	3.1452[g]
5e	4.0388	3.9286	3.5029	3.6020	3.2871–3.2440[h,i,k]
5f	4.2080	3.8669	4.0477	3.8128	
5g	4.5333	3.9874	4.0861	3.6423	328.71[g]

[a]Ref. 7 [b]Ref. 69 [c]These similar values were averaged and the averaged value is plotted in Figure 4.9 [d]Ref. 70. [e]Ref. 59. [f]These similar values were averaged and the averaged value is plotted in Figure 4.9. [g]Ref. 63. [h]Depending on the solvent. [i]Ref. 71. [j]These similar values were averaged and the averaged value is plotted in Figure 4.9. [k]These similar values were averaged and the averaged value is plotted in Figure 4.9.

Table 4.8. ΔE_{H-L} obtained from single point B3LYP calculations and ZINDO calculations for the 6FAR structures optimized with PM3 and COMPASS force field

	ΔE_{H-L} for PM3 structure (eV)		ΔE_{H-L} for FF structure (eV)		
System	B3LYP	ZINDO	B3LYP	ZINDO	Exptl. (eV)
6a	1.8961	2.4547	1.6673	2.2080	
6b	2.2781	2.6218	2.2561	2.5232	
6c	2.6175	2.8414	2.2020	2.4027	
6d	2.9831	3.1440	2.6289	2.7916	2.8516[a,e] 2.8392[b,e] 2.8752[c,e] 2.8686[d,e]
6e	3.1027	3.2533	2.7013	2.8206	2.8293[c]
6f	3.0009	3.1036	2.7160	2.7875	
6g	3.0839	3.3627	2.7734	2.9934	
6h	3.3607	3.3790	3.0355	3.0139	2.8686[c]
6i	3.4912	3.4954	3.1756	3.0823	2.9576[c]
6j	3.5688	3.5791	3.2934	3.2423	3.0995[c,g] 3.0598-3.0225[c,f,g] 3.1058[d,g]
6k	3.6072	3.5912	3.3111	3.2529	
6l	3.9143	3.6323	3.6706	3.2699	
6m	3.9345	4.1893	3.6760	3.5113	
6n	4.1663	3.9798	3.8945	3.6504	
6o	4.1705	3.8797	3.9337	3.5661	

[a]Ref 59. [b]Ref. 72. [c]Ref. 65. [d]Ref. 63. [e]These similar values were averaged and the averaged value is plotted in Figure 4.10. [f]Depending on the solvent. [g]These similar values were averaged and the averaged value is plotted in Figure 4.10.

Table 4.9. ΔE_{H-L} obtained from single point B3LYP calculations and ZINDO calculations for the 7FAR structures optimized with PM3 and COMPASS force field

	ΔE_{H-L} for PM3 structure (eV)		ΔE_{H-L} for FF structure (eV)		
System	B3LYP	ZINDO	B3LYP	ZINDO	Exptl. (eV)
7a	1.5984	2.2802	1.3726	2.0849	
7b	2.0044	2.5696	1.7565	2.3226	
7c	2.3407	2.6541	2.3190	2.6083	
7d	2.3867	2.6314	1.9184	2.2077	
7e	2.5604	2.9811	2.2346	2.6606	
7f	2.5919	2.8428	2.2376	2.4684	
7g	3.0711	3.2648	2.7394	2.9574	2.7911[a]
7h	3.1201	3.2292	2.7454	2.8680	2.8100[a]
7i	2.9655	3.0296	2.6346	2.7375	
7j	3.3073	3.3263	3.0224	3.0527	2.9505[a]
7k	3.1590	3.2695	2.8096	2.9223	
7l	3.6689	3.6199	3.3927	3.3576	2.8819[a]
7m	3.5095	3.4945	3.1990	3.1961	2.9789[a]
7n	4.1853	3.8297	3.9416	3.5841	
7o	3.5685	3.1435	3.9909	2.9343	2.9090[a,d] 3.0131[b,d] 3.0237[c,d] 4.1279[e]

[a]Ref. 65 [b]Ref. 63 [c]Ref. 70 [d]The values that are similar were averaged and the averaged value is plotted in Figure 4.11. [e]Ref. 73.

theory and experiment is observed for the case of the FF//ZINDO calculations. Thus, in the present study, the excited electronic states (including the HOMO–LUMO configuration) of the PAH compounds were calculated using the ZINDO/S[48] method as it is provided in the Gaussian 98 package,[49] and using the COMPASS force field (FF) geometry optimized structures. The difference between the theoretical and experimental data could be due to the fact that the theoretical estimates are derived from a single "frozen" molecule in the gas phase at 0 K without corrections for thermal motion and solvent effects and/or in some cases to an erroneous assignment of the experimental λ_{0-0} band.

3.2.2. HOMO–LUMO Gap in PAHs and Asphaltenes

In this section the calculated HOMO–LUMO gap, and structural arrangement of several PAHs, with 4FAR to 14FAR, are analyzed and compared with the experimental asphaltene fluorescence emission range[7–9] to conclude about the distribution and structure of the fused rings in the aromatic core of the chromophores in asphaltenes. In Figure 4.12 a HOMO–LUMO energy gap (ΔE_{H-L}) vs. the number of fused rings (nFAR) diagram is presented for various PAH systems that have different condensations (catacondensed and pericondensed) and different geometric distributions. The region inside the two horizontal bars corresponds to the experimental asphaltene fluorescence emission range,[7–9] which is

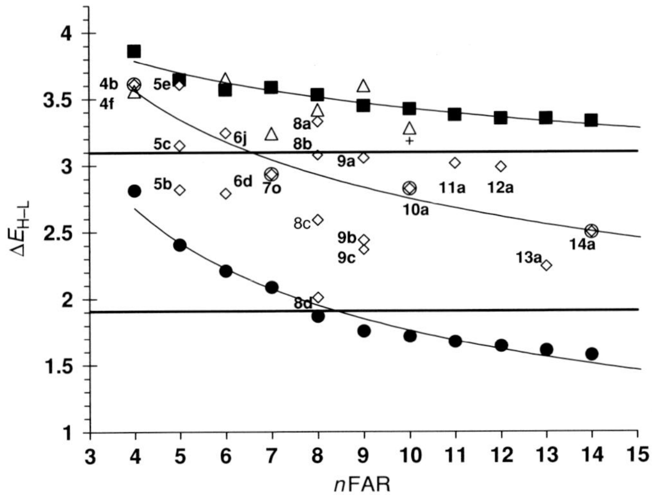

Figure 4.12. Diagram of ΔE_{H-L} vs. the number of fused aromatic rings for several benzenoid PAHs. The solid circles (●) represent the PAH systems with linear structure. The solid squares (■) represent the PAH systems with zigzag structure. The linear and zigzag conformations are not pericondensed or their pericondensation is equal to 0%. These systems are catacondensed. The open triangles (Δ) represent the PAH systems with a full resonant structure (either catacondensed or pericondensed; Figure 4.13). The open circles (◯), with an open diamond inside, represent the PAH systems with a circular structure. These systems have a pericondensation of 100% such as coronene and ovalene. The open diamonds (◇) represent the PAH systems with the highest pericondensation. The labels, which correspond only to the diamond symbol, represent the associated PAH structure. The structures are shown in Figures 4.8–4.11, and in Figure 4.14. The region inside the two horizontal bars correspond to the experimental asphaltene fluorescence emission range.[7] The other curves are the best fit curves for the linear, zigzag, and circular systems. Reproduced with permission from *J. Phys. Chem. A*, 2002, **106**, 11283–11308. Copyright 2002 Am. Chem. Soc.

significant in the range of 400–650 nm (1.9065–3.0981 eV). As it can be seen in Figure 4.12, for all the FAR families the lowest HOMO–LUMO gaps correspond to the linear structures (acenes) that are catacondensed; this means that the acenes are the least stable compounds in the FAR family series. Even though some of the calculated HOMO–LUMO gaps for acenes fall inside the asphaltene fluorescence experimental range (see Figure 4.12), the acenes cannot be considered as structural candidates of asphaltenes due to their low stability, compared with other geometric arrangements. Asphaltenes are chemically stable compounds since they persist for geologic time; thus, they cannot have a linear geometric conformation. The largest HOMO–LUMO gaps correspond to the PAH systems with a zigzag structure, which are also catacondensed, and to the PAH structures, both catacondensed and pericondensed, which are full resonant (see section 3.1, and Figure 4.13). These are the most stable geometric conformations in any PAH family; however, the MO-calculated stability of these PAH compounds is not consistent with the stability of asphaltenes; thus, asphaltenes cannot have an aromatic core that is zigzag or with a geometry of full resonant PAH structure. We found that, in general, asphaltenes are largely pericondensed because the stability of mostly pericondensed PAHs,

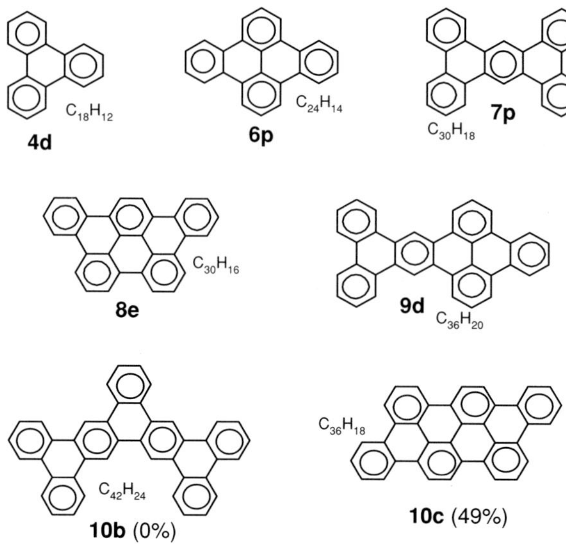

Figure 4.13. Structures of full resonant PAH compounds (FRS), which are mentioned in Figure 4.12.

from 5FAR onward, which are represented with diamond symbols in Figure 4.12, is consistent with the stability of the asphaltenes. This result agrees with the reported experimental XRRS observations. Using carbon K-edge XRRS,[17] it has been concluded that the geometric distribution of the FARs in the aromatic core of asphaltenes is pericondensed.

Groenzin and Mullins concluded, based on the fluorescence result and using size and structure arguments, that asphaltenes have between 4FAR and 10FAR fused rings in the aromatic region.[7,8] The pericondensation of PAHs starts in the 4FAR family with pyrene, which is the only pericondensed 4FAR isomer that exists. We found that the stability of pyrene is not consistent or similar to the stability of asphaltenes. None of the 4FAR arrangements, beside the linear conformation, fall into the experimental range of asphaltenes. We could say that 4 FARs in any arrangement are not good structural candidates for the FAR region in asphaltenes. For the case of the circular systems, with percentage of condensation of 100% (see section 3.1), we conclude that the circular structures 7FAR (coronene, **7o**, Figure 4.11) and 10FAR (ovalene, **10a**, Figure 4.14) are possible structural candidates for the fused ring region in asphaltenes because their HOMO–LUMO gap falls inside the asphaltene experimental range but any other circular structure is not possible (Figure 4.12), because the size (longest dimension) and weight beyond 10FAR do not fulfill the experimental constraints.

In Table 4.10 the calculated size (longest dimension and without considering the hydrogen atoms) of the geometry optimized structures, with the most compact geometry, is presented under the column of molecular size. In this column the size of the most compact (circular) structure is provided and it represents the smallest

Table 4.10. Size and molecular weight for PAH compounds with 4FAR–14FAR in the most compact conformation

*n*FAR	Formula	Highest P_C^a	Number of isomers[c]	Molecular size (Å)[b]	Molec. weight g/mol	Compound
4FAR	$C_{16}H_{10}$	100%	1	7.04	202	**4b**
5FAR	$C_{20}H_{12}$	50%	3	7.49–9.27	252	**5b 5c 5e**
6FAR	$C_{22}H_{12}$	82%	2	7.48–9.24	276	**6d 6j**
7FAR	$C_{24}H_{12}$	100%	1	9.88	300	**7o**
8FAR	$C_{28}H_{14}$	75%	8	9.25–11.56	350	**8a 8b 8c 8d**
9FAR	$C_{30}H_{14}$	89%	3	9.23–9.88	374	**9a 9b 9c**
10FAR	$C_{32}H_{14}$	100%	1	9.861	398	**10a**
11FAR	$C_{36}H_{16}$	83%	20	11.54	442	**11a**
12FAR	$C_{38}H_{16}$	85%	10	11.51	472	**12a**
13FAR	$C_{40}H_{16}$	99.8%	3	12.26	496	**13a**
14FAR	$C_{42}H_{16}$	100%	1	11.51	520	**14a**

[a] P_C Percentage of compactness. It is a measure of the degree of condensation of the PAH structure. The lowest value of P_C is 0% (like in triphenylene, **4f**) and the highest value is 100% (like in coronene, **7o**). [b]Longest dimension without considering the hydrogen atoms. [c]Ref. 28.

size for the pericondensed systems of a given FAR family. In general the *n*FAR families have more than one isomer with the most compact structure. For example, the 8FAR family has eight isomers with the highest percentage of compactness which is $P_C = 79\%$. Some of them are more circular than the others; however, all of them have the same percentage of compactness. In Table 4.10 the number of isomers is reported together with the range in the size, considering all of the isomers, except for 11FAR and onward for which only the size of the most circular isomer is given. The size for 10 fused rings in the most compact circular arrangement is 9.86 Å, and the size of 11 fused rings in the most compact arrangement (**11a**, Figure 4.13) is calculated to be 11.54 Å (Table 4.10). Based only on the experimental size and fluorescence emission criteria, we could conclude that asphaltenes have 5–10 fused rings in their fused ring region, because beyond 10FAR the calculated size is longer than the observed experimental size of ∼11Å, obtained with STM,[14] and of ∼10Å, obtained with HRTM.[15] However, the arrangement of 10FAR would have to be highly compact (with a high percentage of compactness, i.e., 100% pericondensed) to be near the experimental asphaltene size. This agrees with the XRRS experimental results.[17]

If an asphaltene molecule contains a single five fused ring region (5FAR) with 40% of the carbon being part of its ring region (STM results[11]), then the molecular weight of the whole asphaltene would be around 630 g/mol for the most compact circular PAH geometry (see molecular weight column in Table 4.10). On the other hand, if an asphaltene molecule contains a single ten fused ring region (10FAR) with 40% of the carbon being part of its ring region, then the molecular weight of the whole asphaltene would be around 995 g/mol for the most compact circular arrangement. Beyond 10 fused rings the molecular weight would be bigger than the reported experimental molecular weights (see the "Introduction" section).

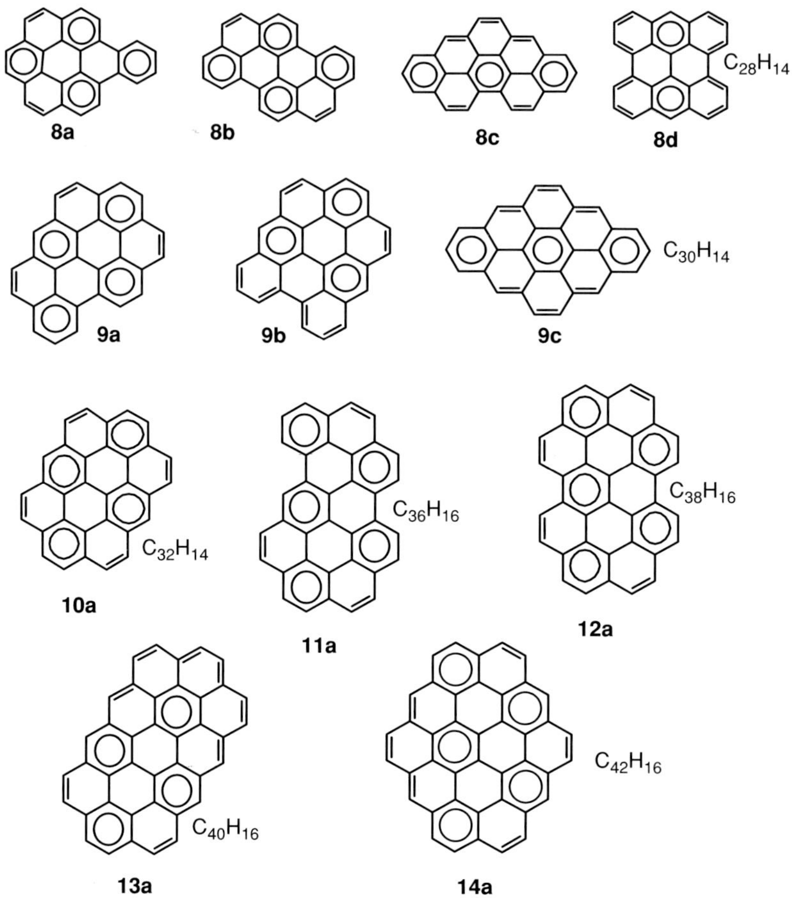

Figure 4.14. PAH structures referred by labels in Figure 4.12.

Thus, by using molecular weight considerations we could say that asphaltenes have between 5 and 10 fused aromatic rings (5FAR–10FAR) which is consistent with the conclusions obtained by size and optical transition considerations. Not all the pericondensed PAHs with 5FAR–10FAR present a stability that is consistent with the asphaltene stability but only some of the pericondensed PAHs have a stability that matches the asphaltene's stability. The stability in PAHs is dictated by the number of Clar resonant sextets present in the structure (see "Introduction"). To figure out the number of resonant sextets that gives a stability that matches the asphaltene stability, we used the Y-rule which predicts, in an easy way, the most likely location and total number of aromatic sextets in pericondensed PAHs. We will continue with the identification of the PAH structural candidates of the aromatic region in asphaltenes in section 3.4 but first we will continue with the presentation and application of the Y-rule, and its validation.

3.3. Aromaticity in PAHs and Asphaltenes: Application of the Y-rule

As we discussed in the Introduction, the establishment of the Clar structure(s) of PAHs that represent the actual chemical stability and reactivity of a given PAH is an open problem that is referred to as the aromaticity in PAHs. The HOMO–LUMO gap magnitude of benzenoid PAHs is related to their aromaticity and stability, that is, to the number of aromatic resonant sextets present in the structure, which in turn is related to the spatial distribution of the hexagonal fused rings. Therefore, the need to know how many resonant rings or sextets are present in pericondensed PAHs (peri-PAHs) arises; i.e., it is necessary to understand the aromaticity in peri-PAHs to be able to understand the aromaticity in asphaltenes and as a consequence to understand their reactivity (see the "Introduction" section for more information). Only benzene has a perfect six π-delocalization, which is intimately connected with its high symmetry, whereas the hexagonal rings of all the annelated systems show varying degrees of π-localizations. In this manner, the practice of drawing a circle inside of each hexagonal ring in PAH systems and in the fused ring region(s) in asphaltenes should be discouraged, and it actually represents a misuse of the circle notation to represent aromatic sextets.[19,23,27] Acree *et al.*[65] were not able to find relationships between point groups (or symmetry elements) and fluorescence probe character for several series of PAHs, and they were unable to predict, from structural considerations, which PAHs will exhibit probe character. We consider that this is due to fact that the presence of localized resonant sextets was not taken into account in the assignment of the molecular symmetries, which in many cases is reduced. There are many benzenoid hydrocarbons that have more than one Clar structure (with the highest number of resonant sextets).[23,27] Which one should one draw? Which one or which ones is or are responsible for the aromaticity? The pioneering work of Clar on the clarification of the aromaticity makes skillful use of the placement of the aromatic sextets through the Clar's rules (see "Introduction"). However, the Clar's rules do not tell us, in an easy way, which is/are the most important Clar structure(s) that represent for any peri-PAH its aromaticity. To solve this problem we have proposed a new method called the Y-rule.

3.3.1. *The Y-Rule in Identifying the Most Important Clar Structures*

As mentioned in section 3.1, the Y-carbons are the internal carbon atoms in the PAH structure with a connectivity of three. We call them Y-carbons because they are in the vertex of what looks like a Y letter. We found, as an observation, that in general the Y-carbons (C_Y) also provide information on the location of the resonant sextets in the pericondensed section in PAHs and consequently in asphaltenes.[19,27] Thus, we came up with a simple rule to draw the most likely localization of the resonant rings (sextets) in pericondensed PAHs and in the fused ring region in asphaltenes. We have called it the *Y-rule* because it involves the Y-carbons. The Y-rule has already been published[19] and validated.[27] However, here we state it again for practical purposes. The Y-rule is phrased as follows: "The resonant sextets are located in the hexagons that contain the C_Y (Y-Carbons).

All the C_Y carbons of the corresponding stoichiometry have to be included, i.e., covered by sextets. When there is more than one possibility to locate the sextets, due to the arrangement of the internal Y-carbons, the possibility that provides the higher symmetry and the higher number of sextets will be the most probable. The structure must contain the highest number of resonant sextets possible. If more than one possibility with the greatest number of resonant sextets and double bonds or most important Clar structure is found, then their superposition provides the π-electronic distribution." The Y-rule has been applied only to planar and concave or distorted pericondensed benzenoid PAHs and to the pericondensed section of the fused aromatic region in asphaltenes. The Y-rule cannot be used for catacondensed PAHs because there are no Y-carbons in these structures. The validity of the Y-rule has already been proven.[27]

Here is a specific example where the Y-rule predicts how the electronic structure is distributed in a PAH where without the Y-rule it was not clear how the π-electronic distribution is in the 8FAR system **8a**, benzocoronene (Figure 4.15). For this system Clar proposed a "lesser" Clar structure in order to account for some observed chemical behavior, instead of the structure having the maximal number of resonant sextets. The "lesser" Clar structure is presented in Figure 4.15, and in this structure there are three localized resonant sextets, five localized double bonds, and one empty ring (the central ring). In the case of **8a** there is a catacondensed region in the structure, which involves the hexagon labeled as 6 (Figure 4.15).

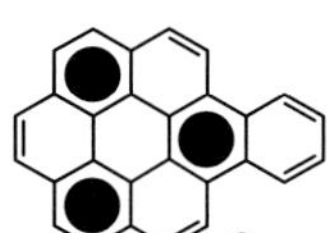

8a "Lesser" Clar structure

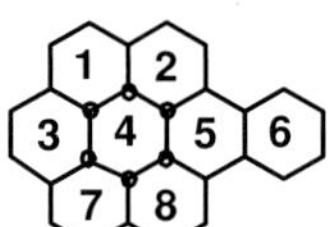

8a σ-backbone

8a Y-rule π-distribution

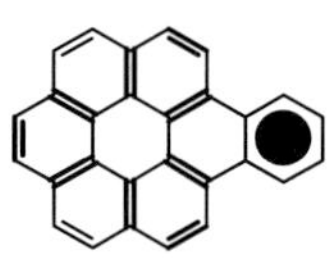

8a-I NICS

Figure 4.15. Comparison of the π-electronic distribution obtained with the Y-rule, the NICS calculation, and the distribution proposed by Clar for the controversial case of benzocoronene. Reproduced with permission from *J. Phys. Chem. A*, 2004, **108**, 10873–10896. Copyright 2004 Am. Chem. Soc.

By applying the Y-rule to the pericondensed region, which resembles the 7FAR coronene (**7o**), it is possible to use all the Y-carbons (which are highlighted in Figure 4.15) by drawing only one resonant sextet occupying the hexagon labeled as 4 (Figure 4.15). But this option does not provide a high number of resonant sextets. The other options are to draw three resonant sextets occupying the hexagons 1, 5, 7 or 2, 3, 8. The later option allows us to draw another resonant sextet in the catacondensed region, i.e., in the hexagon number 6, without overcoming the carbon atoms valence, for a total of four resonant sextets in the structure, two localized double bonds, and two empty rings (hexagons 6 and 5). Thus, this is the distribution of the π-density in **8a**, as determined by the Y-rule, and the structure is presented in Figure 4.15. In this distribution there is one resonant sextet in the hexagon labeled as 6, while in the "lesser" Clar structure there is no resonant sextet in this hexagon.

The performance of the Y-rule has been validated theoretically[19,27] by comparing the π-electronic distribution obtained with it with the π-electronic distribution obtained from NICS[25,34–36] calculations. The NICS is a magnet-based aromaticity index which is related to the magnetic properties of molecules. Molecules with cyclic conjugated π-electron systems, like benzene and PAHs, present a stronger diamagnetic susceptibility as compared to noncyclic conjugated π-systems. The "magnetic anomaly" is due to the ring currents in the π-electron system.[74] The NICS is defined as the negative value of the absolute isotropic magnetic shielding at some selected point in space, e.g., at the center of a ring one is probing, NICS(0), or one angstrom above the geometrical center of the ring that is being probed, NICS(1). Negative and positive NICS values denote aromaticity and antiaromaticity. The NICS values are used to assess the relative aromaticities of the individual rings in small and large PAH compounds. It has already been demonstrated in the literature[23,25,34–36,75,76] that the NICS values represent strong theoretical support for Clar's picture of aromatic π-sextets. However, the advantages of the qualitative Y-rule are that it is of particular importance for the case of large pericondensed PAHs, it takes few minutes to be applied following a very easy methodology, and it provides the same result as the NICS in terms of the location of the resonant sextets, and without having to carry out theoretical quantum chemistry calculations that take time and that can be costly depending on the size of the system. The NICS result for the distribution of the π-density in **8a** is given in the representation **8a-I** in Figure 4.15. This distribution agrees with the π-electronic distribution obtained using the Y-rule. However, neither the π-electronic distribution obtained with the Y-rule nor the NICS calculation of **8a** agrees with the "lesser" Clar structure, and actually the distribution found with the NICS calculation and the Y-rule is the correct one.

The science of PAHs has advanced to the synthesis of large peri-PAHs systems[77] and these extended π-systems create challenging problems for any discussions into π-structure. We have tested the performance of Y-rule for large PAHs by comparing the π-electron distribution obtained with it and the NICS calculations.[27] The system **13b** (Figure 4.16) with 48 π-electrons has six Y-carbons, which are highlighted. After applying the Y-rule, two π-electronic distributions or most important Clar structures are obtained named **I** and **II,** which are equivalent (Figure 4.16). Both distributions present four empty rings in the internal hexagons,

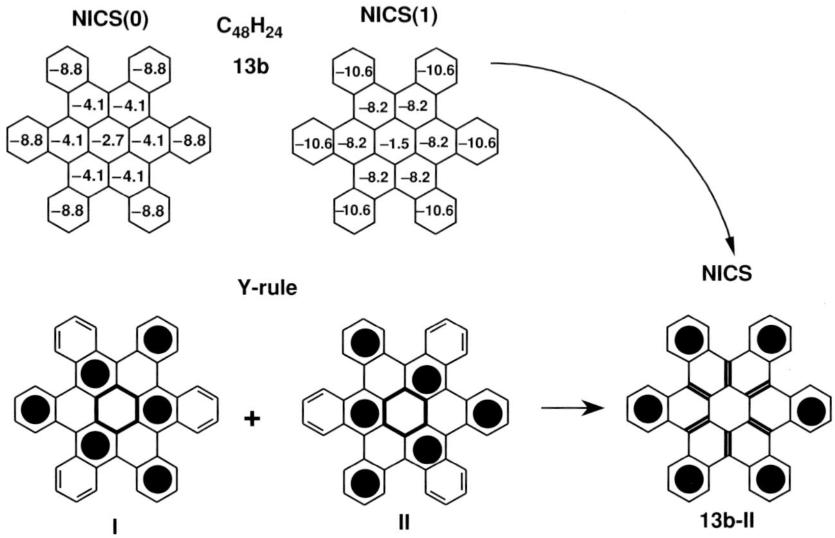

Figure 4.16. Y-rule depiction of aromaticity as Clar structures and calculated NICS(0), and NICS(1) values for the large peri-PAH **13b**. Reproduced with permission from *J. Phys. Chem. A*, 2004, **108**, 10873–10896. Copyright 2004 Am. Chem. Soc.

which resemble triphenylene, and both distributions contain six resonant sextets and six double bonds. As it can been seen all the Y-carbons are covered by sextets in both structures. The calculated NICS(0) and NICS(1) are also given in Figure 4.16. Three types of hexagons are obtained: the external hexagons, which present the highest NICS value; the internal hexagons; and the central hexagon. The NICS analysis for this compound has been done before by Moran *et al.*[25] There are resonant sextets in all the external hexagons, which present the highest NICS value. The calculated NICS is equal for all the internal hexagons, meaning that their electronic environment is the same for all of them. This is possible only if they share a double bond right at the junctions. The central hexagon is an empty ring with a very small NICS value NICS(1) $= -1.5$ ppm (see Figure 4.16). The π-electronic distribution obtained from the NICS calculation is giving as **13b-II** in Figure 4.16. The pictorial sum or superposition of the π-electronic distribution obtained with the Y-rule (**I** and **II**, Figure 4.16) reproduces the π-electronic distribution obtained from the NICS analysis. Randic[24] presents in his paper "On construction of Clar Structures for Large Benzenoids" eighteen Clar structures for hexabenzocoronene (**13b**, Figure 4.16) which have the largest number of resonant sextets (six). Five of the eighteen Clar structures are not related by symmetry (Figure 4.3). Clar discussed the NMR spectra of **13b** in terms of the two structures found by the Y-rule instead of the eighteen structures found by Randic. The superposition of the two structures proposed by Clar and also found with the Y-rule (**I** and **II** in Figure 4.16) gives the structure **13b-II** which represents the actual aromaticity in hexabenzocoronene (**13b**), as it is also supported by the NICS calculation.

We have proven in this section that it is not necessary to find all the Clar valence structures in large benzenoids to understand their aromaticity but it is important to consider only those Clar-type π-electronic distributions obtained with the Y-rule that represent the most important Clar structures, and their superposition, in the case of sextet migration existence. The final π-electronic distribution obtained with the Y-rule is the same as obtained by the NICS calculation, and in the case of very large PAHs, where the NICS calculation is highly costly and time consuming, the best option to elucidate the aromaticity is to use the Y-rule. We suggest that the use of drawing circles in each hexagon of the σ-frame of peri-condensed benzenoid PAHs, to represent the distribution of the π-density, must be discouraged. This practice is being done since long ago[23] due to the fact that it was not possible to know, in an easy way, where the resonant sextets and the double bonds are located. This problem is overcome by using the Y-rule.

3.3.2. Y-Rule and NICS for Asphaltenes

In Figure 4.17 we present the Y-rule π-electronic distribution and the NICS calculation for the aromaticity of a published **asphaltene-1** structure, proposed

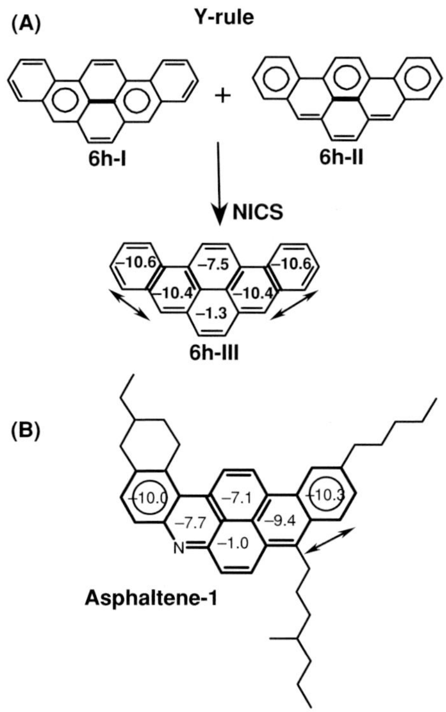

Figure 4.17. (A) Y-rule depiction of aromaticity as Clar structures, and calculated NICS(0) values for the structure **6h** (Figure 4.10). (B) Calculated NICS(0) values for the compound **asphaltene-1** whose associated PAH is **6h**.

by Groenzin and Mullins.[8] In Figure 4.17A we present the electronic distribution obtained with the Y-rule and NICS calculation for the associated free PAH. As it can be seen both distributions agree. The NICS distribution is the result of the superposition of the two distributions found with the Y-rule. The arrows represent the sextet migration between hexagons. In Figure 4.17B we present the calculated NICS of the whole asphaltene structure. We found that in general the aromaticity in the fused ring region in asphaltenes is closely associated with the aromaticity of the corresponding free PAH compound (see Figure 4.17). The presence of heteroatoms (S, N) in the **asphaltene-1** PAH-core has a direct effect mostly on the aromaticity of the adjacent fused ring. In the FAR region of **asphaltene-1** (Figure 4.17B,) the presence of the nitrogen atom stops the migration of the resonant sextet in the left side of the molecule, and the NICS value of the particular hexagon, in which the N atom is located, is decreased.

3.4. The FAR Region in Asphaltenes

In section 3.2.2, it was concluded that asphaltenes have 5–10 fused rings in the fused ring core. The arrangements of the fused ring region in asphaltenes cannot be catacondensed, and they cannot have a full resonant structure. In the case of circular arrangements asphaltenes have a coronene (7FAR $P_C = 100\%$, **7o**, Figure 4.11) and ovalene (10FAR $P_C = 100\%$, **10a**, Figure 4.14) type of structure. In this section we will discuss all the other possible arrangements of the fused ring region in petroleum asphaltenes, i.e., the pericondensed PAH systems with a percentage of compactness different to 100% and that are not of the full resonant type. Some, but not all, of the pericondensed systems with the highest percentage of compactness, different to 100% (points marked with a diamond symbol in Figure 4.12), have a HOMO–LUMO gap that falls into the experimental range of asphaltenes. The isomers with the highest or the lowest number of resonant sextets do not fall into the experimental fluorescence emission range of asphaltenes (Figure 4.12). These isomers are **5e** (Figure 4.9), **6j** (Figure 4.10), and **8a** (Figure 4.14). Our study of PAHs does not include heteroatoms. It is known experimentally that the replacement of carbon in PAH compounds with heteroatoms typically results in a red shift (higher wavelength) of the fluorescence maximum, if there is any spectral effect.[9,59] Thus, those systems with a HOMO–LUMO gap slightly above the upper borderline of the experimental range of asphaltenes, as it is the case of **5c** (Figures 4.12 and 4.9), might fall into the experimental range when a heteroatom is added to the structure.

As it can be seen in Figure 4.12, for a fixed number of FARs or hexagons, the HOMO–LUMO gap increases as the number of resonant sextets increases and vice versa. For example, the HOMO–LUMO gap of the 8FAR systems, **8a** to **8d**, span a range of 1.32 eV (a range of 245 nm) going from two resonant rings (**8d**, Figure 4.14) to four resonant rings (**8a**, Figure 4.14) in the structure. The number of resonant sextets was found by using the Y-rule (see former sections). By comparing the HOMO–LUMO gap between **8b** and **8c,** and between **8c** and **8d,** it is observed that in general the increase of the total number of resonant sextets by one results in

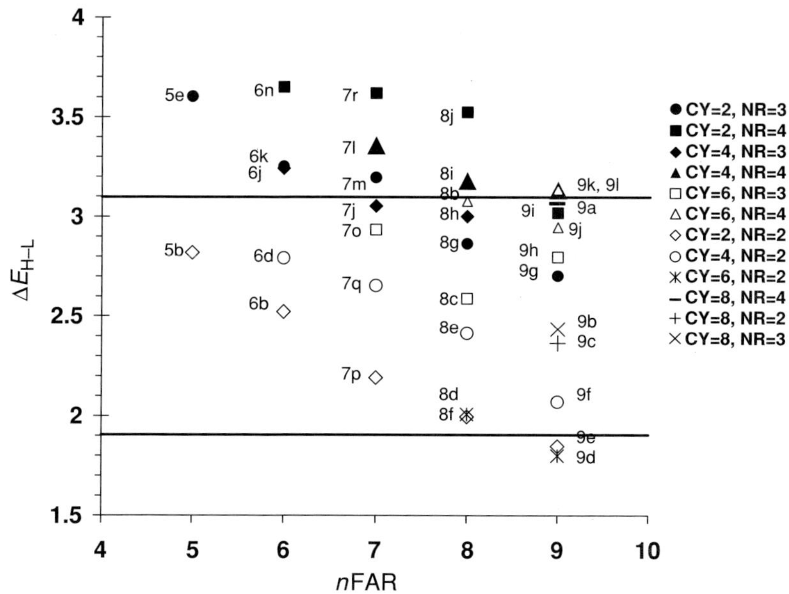

Figure 4.18. Diagram of $\Delta E_{\text{H-L}}$ vs. the number of fused aromatic rings in benzenoid PAHs for several pericondensed PAHs with different combinations of the following structural parameters: the number of internal Y-carbons (C_Y) and the number of resonant sextets (N_R). The labels are associated to the PAH structures which are shown in the following figures: for 5FAR PAH structures see Figure 4.9, for 6FAR PAH structures see Figure 4.10, for 7FAR PAH structures see Figure 4.11 (**7j, 7l, 7m, 7o**) and Figure 4.19 (**7q, 7r,7s**), for 8FAR PAH structures see Figure 4.14 (**8b–8d**) and Figure 4.19 (**8f–8k**), and for 9FAR PAH structures see Figure 4.14 (**9a–9c**) and Figure 4.19 (**9e–9m**). Reproduced with permission from *J. Phys. Chem. A*, 2002, **106**, 11283–11308. Copyright 2002 Am. Chem. Soc.

the opening up of the HOMO–LUMO gap by approximately 0.4–0.5 eV, and this is only the case for a fixed number of fused aromatic rings or hexagons (*n*FARs). In Figure 4.18 the HOMO–LUMO gap vs. the number of fused aromatic rings (*n*FARs) for several pericondensed PAH compounds with a percentage of compactness different to 100% are plotted. The PAH structures associated to the labels are presented in Figures 4.9–4.10, Figure 4.14, and Figure 4.19. There is a relation between the number of resonant sextets (N_R) in the PAH structure and the content of Y-carbons C_Y. Again the two horizontal bars represent the experimental fluorescence emission of asphaltenes. In general for each C_Y and N_R combination there is a decrease in the HOMO–LUMO gap as the number of fused rings (*n*FAR) is increased (Figure 4.18). The systems with a high number of resonant sextets are near the upper experimental borderline of asphaltenes and the systems with a low number of resonant sextets are near the lower experimental borderline of asphaltenes.

In general, the highest number of resonant sextets in the PAH structure produces a HOMO–LUMO gap that does not fall into the experimental result for asphaltenes. The lowest ($N_R = 2$) and the highest ($N_R = 4$) numbers of resonant rings produce a HOMO–LUMO gap that is out of the experimental range for the case of the 9FAR PAHs. On the other hand, the value of $N_R = 2$ is almost out of

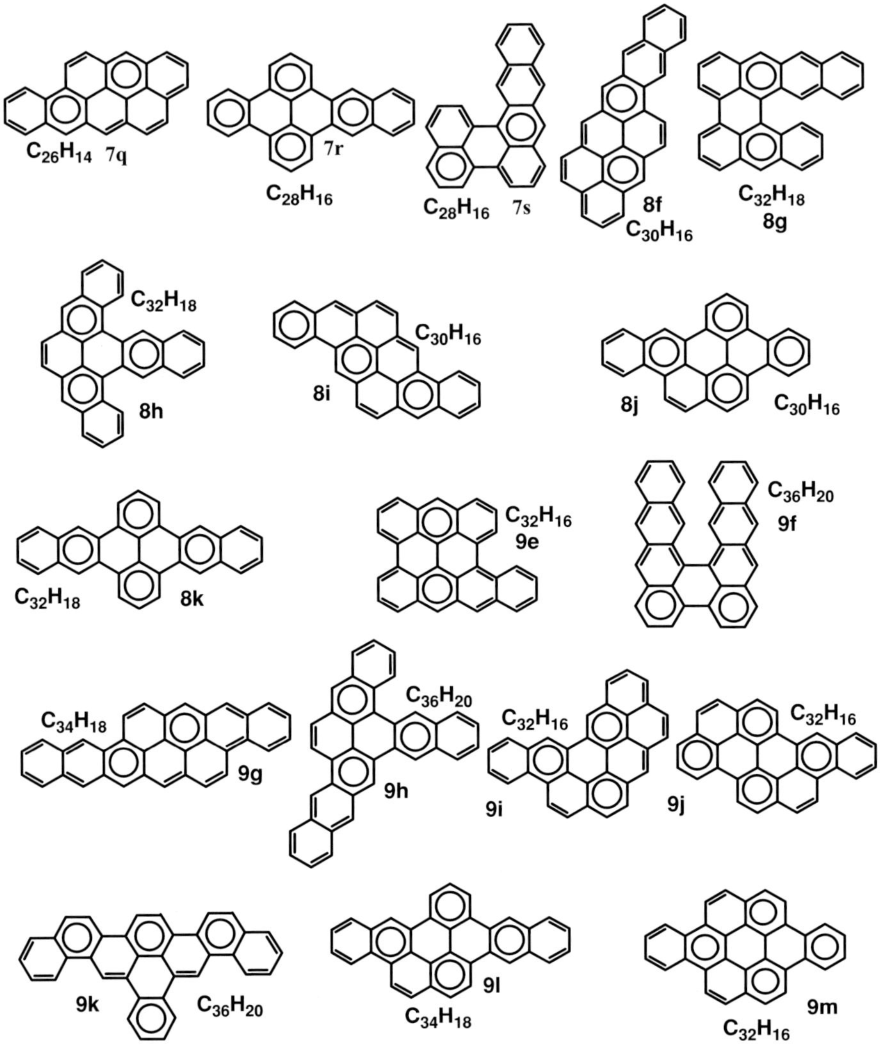

Figure 4.19. Some of the structures referred by labels in Figure 4.18.

the asphaltene experimental range in the case of 8FAR. For 5FAR to 7FAR with the $C_Y = 2, 4$ and $N_R = 3, 4$ combinations, the HOMO–LUMO gap is in general out of the experimental range (Figure 4.18). In Table 4.11, which was obtained from analysis of Figure 4.18, the structural characteristics of the pericondensed PAH systems that are most likely structural candidates of the fused ring region in asphaltenes from 5FAR to 10FAR, including also the number of allowed or most likely number of resonant sextets in asphaltenes, are presented. We have called this table the asphaltene structural parameters table or the ASP-table, for short. The topological characteristics of benzenoid PAHs have previously been extrapolated

Table 4.11. Nonradical benzenoid PAH systems and structural parameters that are allowed as structure of the aromatic region in asphaltenes

*n*FAR[a]	Allowed stoichiometry	Number of isomers	P_C^b	N_R^c	C_Y^d
5FAR	$C_{20}H_{12}$	3	50%	2	2
6FAR	$C_{22}H_{12}$	2	82%	3, 2	4
	$C_{24}H_{14}$	12	33%	3, 2	2
7FAR	$C_{24}H_{12}$	1	100%	3	6
	$C_{26}H_{14}$	9	65%	3	4
	$C_{28}H_{16}$	62	25%	4, 3, 2	2
8FAR	$C_{28}H_{14}$	8	75%	4, 3, 2	6
	$C_{30}H_{16}$	57	45%	4, 3, 2	4
	$C_{32}H_{18}$	295	20%	4, 3, 2	2
9FAR	$C_{30}H_{14}$	3	89%	4, 3	8
	$C_{32}H_{16}$	46	60%	4, 3, 2	6
	$C_{34}H_{18}$	335	33%	4, 3, 2	4
	$C_{36}H_{20}$	1439	16%	4, 3	2
10FAR	$C_{32}H_{14}$	1	100%	4	10
	$C_{34}H_{16}$	34	66%	4, 3	8
	$C_{36}H_{18}$	334	49%	4, 3	6
	$C_{38}H_{20}$	1987	28%	4, 3	6
	$C_{40}H_{22}$	6973	14%	4, 3	6

[a]*n*FAR Number of fused aromatic rings or hexagons. [b]P_C Percentage of compactness. It is a measure of the degree of condensation of the PAH structure. The lowest value of P_C is 0% (like in triphenylene) and the highest value is 100% (like in coronene). [c]N_R Number or resonant sextets in the structure [d]C_Y Number of Y-carbons, internal aromatic carbons with a connectivity of 3.

to the aromatic ring region of asphaltenes[78,79] but the effect of the presence of different number of resonant sextets in the PAH structures was not taken into account in the differentiation of possible structural isomers. In the ASP-table this effect is considered.

3.5. Most Likely PAH Structural Candidates of the FAR Region in Asphaltenes from 5 to 10 Aromatic Rings

In the former section we presented the asphaltene structural parameters table (ASP-table) which contains the topological and electronic characteristics of non-radical benzenoid PAH systems (from 5FAR to 10FAR) that are most likely structural candidates of the FAR region in asphaltenes. In this section we further more analyze this table to reduce even more the number of possibilities of structural candidates and provide PAH structural candidates of the aromatic core in asphaltenes that are pericondensed and that fulfill all the asphaltene size and structure experimental constraints (optical transitions range, STM, HRTM), and that present a π-electron density distribution that agrees with the observed optical transitions (see "Introduction"). In Table 4.11 all the stoichiometries from 5FAR to 10FAR which are pericondensed ($100\% \geq P_C > 0\%$) are presented. These stoichiometries are pericondensed in a greater or lower degree. However, there are stoichiometries that even though they are pericondensed, they have a low percentage of compactness

Table 4.12. Most likely PAH structural candidates of the FAR region in asphaltenes from 5 to 10 aromatic rings. The structures of all the isomer structural candidates are shown in Figures 4.20–4.26 (except Figure 4.23)

*n*FAR[a]	Allowed stoichiometry	Number of isomers	P_C^b	N_R^c	C_Y^d
5FAR	$C_{20}H_{12}$	2	50%	2	2
6FAR	$C_{22}H_{12}$	2	82%	3, 2	4
	$C_{24}H_{14}$	7	33%	3, 2	2
7FAR	$C_{24}H_{12}$	1	100%	3	6
	$C_{26}H_{14}$	5	65%	3	4
	$C_{28}H_{16}$	16	25%	4, 3, 2	2
8FAR	$C_{28}H_{14}$	5	75%	4, 3, 2	6
	$C_{30}H_{16}$	6	45%	4, 3	4
9FAR	$C_{30}H_{14}$	2	89%	4, 3	8
	$C_{32}H_{16}$	6	60%	4, 3	6
10FAR	$C_{32}H_{14}$	1	100%	4	10
	$C_{34}H_{16}$	2	66%	4, 3	8

[a]*n*FAR Number of fused aromatic rings or hexagons. [b] P_C Percentage of compactness. It is a measure of the degree of condensation of the PAH structure. The lowest value of P_C is 0% (like in triphenylene) and the highest value is 100% (like in coronene). [c] N_R Number or resonant sextets in the structure [d] C_Y Number of Y-carbons, internal aromatic carbons with a connectivity of 3.

with a more extended arrangement and less Y-carbons. Due to the larger extension, the size of these systems surpasses the experimental value of asphaltenes ∼10Å (HRTM, fluorescence depolarization)[7,15] and ∼11Å (STM)[14] (see "Introduction"). The most compact isomers, and the isomers with a high percentage of compactness seem to be the most likely structural candidates for the asphaltene aromatic core due to their pericondensed nature and due to size restrictions, and only those with the number of resonant sextets included in the ASP-Table (see Table 4.11) ensure that their λ_{0-0} electronic transition fall into the asphaltene experimental range.

We have optimized and measured the size of most of the isomers considered in Table 4.11 from 5FAR to 10FAR and many of them have been rejected because they do not fulfill all the experimental constraints (see "Introduction"). In Table 4.12 there are presented the stochiometries, structural parameters, and the total number of isomers for the most likely PAH structural candidates of the aromatic core in asphaltenes that are pericondensed, and that fulfill all the asphaltene size and structure experimental constraints (optical transitions range, STM, HRTM, molecular weight), and present a π-electron density distribution that agrees with the optical transitions. In Figures 4.20–4.26 (except for Figure 4.23) all of these isomers are drawn together with their size (longest dimension). The 5FAR pericondensed stoichiometry is $C_{20}H_{12}$ with three isomers (Table 4.11); however, only two of these three isomers are most likely structural candidates of the aromatic core in asphaltenes (see Table 4.12). They are presented in Figure 4.20. Each of these isomers have a total of two Y-carbons ($C_Y = 2$) and two resonant sextets ($N_R = 2$). The third isomer has three resonant sextets in the structure which confers such stability that it is reflected in a HOMO–LUMO gap which falls above the upper limit of the experimental asphaltene range.

Figure 4.20. Structures of the most likely PAH structural candidates for the FAR region in asphaltenes with five fused aromatic rings (5FAR) that fulfill all the experimental constraints. The size (longest dimension) of the optimized geometries as well as the π-electronic distribution obtained with the Y-rule is given. All the Y-carbons are highlighted (bold) or marked in each structure.

The 6FAR pericondensed stoichiometries are $C_{22}H_{12}$ and $C_{24}H_{14}$ with a total of 14 isomers (see Table 4.11). However, only nine isomers are most likely structural candidates of the aromatic core in asphaltenes (Table 4.12). Two of these isomers correspond to the $C_{22}H_{12}$ stoichiometry and they are shown in Figure 4.21. Each of these isomers have a total of four Y-carbons ($C_Y = 4$), which are highlighted, and two and three resonant sextets ($N_R = 2, 3$). On the other hand, not all of the 12 isomers of the $C_{24}H_{14}$ stoichiometry (Table 4.11) are most likely structural candidates of the aromatic core in asphaltenes but only 7 of them

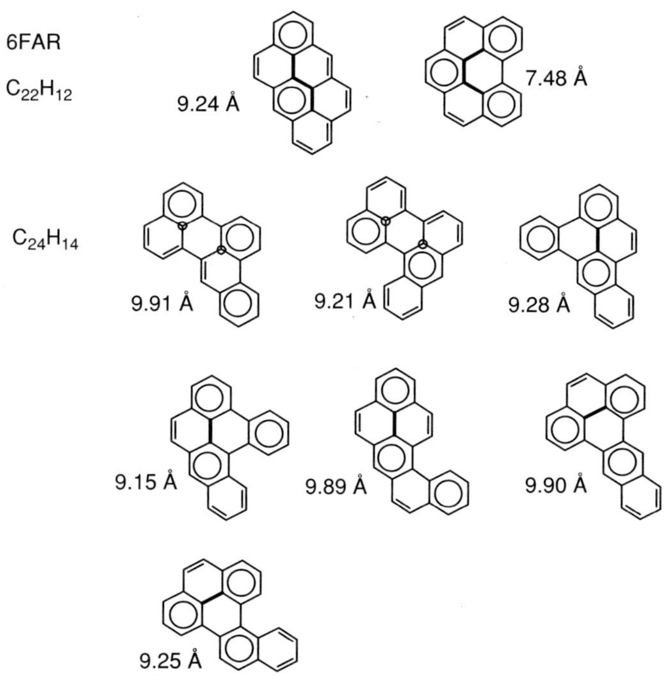

Figure 4.21. Structures of the most likely PAH structural candidates for the FAR region in asphaltenes with six fused aromatic rings (6FAR) that fulfill all the experimental constraints. The size (longest dimension) of the optimized geometries as well as the π-electronic distribution obtained with the Y-rule is given. All the Y-carbons are highlighted (bold) or marked in each structure.

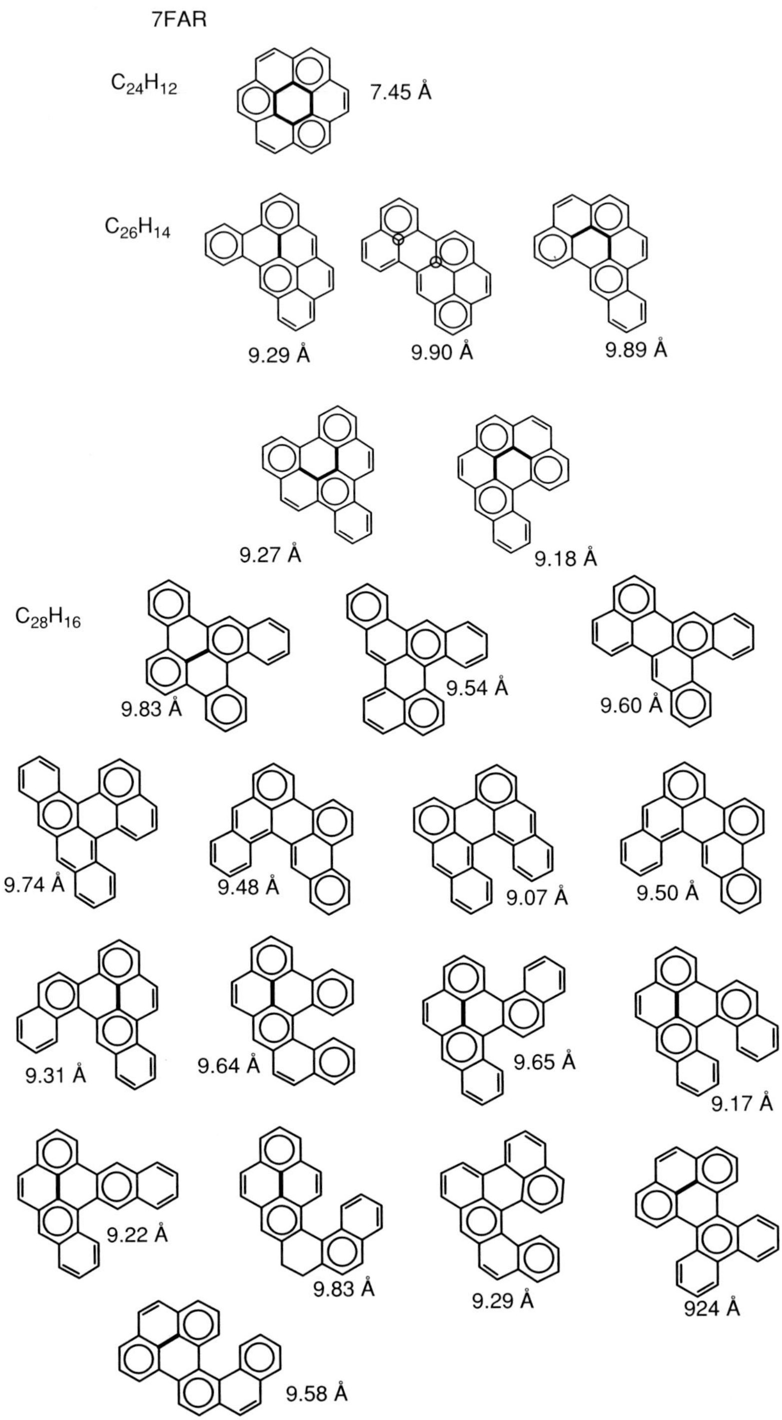

Figure 4.22. Structures of the most likely PAH structural candidates for the FAR region in asphaltenes with seven fused aromatic rings (7FAR) that fulfill all the experimental constraints. The size (longest dimension) of the optimized geometries as well as the π-electronic distribution obtained with the Y-rule is given. All the Y-carbons are highlighted (bold) or marked in each structure.

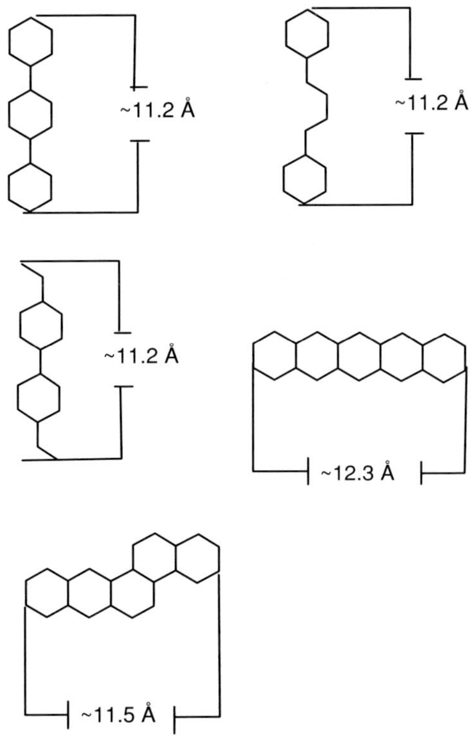

Figure 4.23. Identified sections in PAH isomers that in general have a size or length longer than 11 Å.

(Table 4.12). They are presented in Figure 4.21. Each of these isomers have two Y-carbons ($C_Y = 2$), which are highlighted, and two and three resonant sextets ($N_R = 2, 3$). All the other isomers do not fulfill all the experimental constraints and/or they present a π-electronic distribution that produces a λ_{0-0} transition that do not fall inside the asphaltene experimental range. The 7FAR pericondensed stoichiometries are $C_{24}H_{12}$, $C_{26}H_{14}$, and $C_{28}H_{16}$ (see Table 4.11) for a total of 72 isomers. However, only 22 of these isomers are most likely structural candidates of the aromatic ring region in asphaltenes (see Table 4.12). The structure of these PAH systems are shown in Figure 4.22. The remaining 50 isomers, which are not "good" structural candidates present a size longer than 11 Å (the experimental size for asphaltenes is ~10 Å) and/or present a number of resonant sextets lower or higher to the ones presented in Table 4.11. In Figure 4.23 there are some identified sections of the PAH isomers that in general have a size longer than 11 Å. Most of the 50 isomers, with 7FAR, not considered to be structural candidates of the aromatic core in asphaltenes, present in their structure the identified sections which are longer than 11 Å.

For the case of the 8FAR, 9FAR, and 10FAR families only the pericondensed stoichiometries with a percentage of compactness greater than 45% are most likely structural candidate of the aromatic core of asphaltenes. A lower percentage of

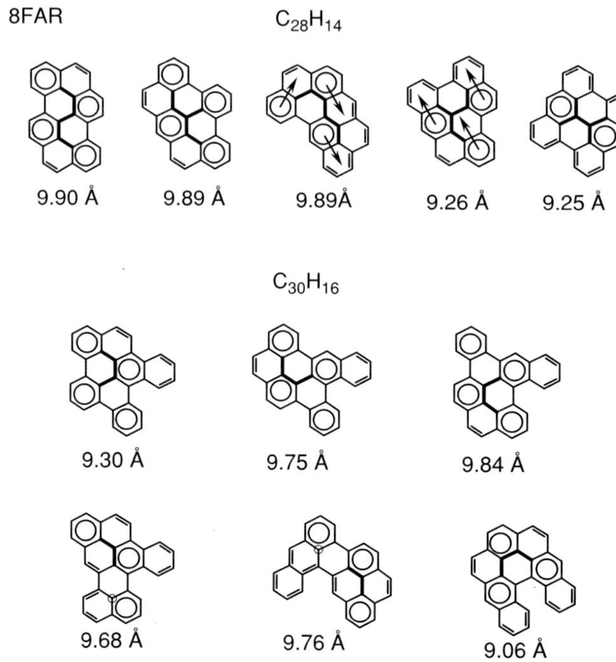

Figure 4.24. Structures of the most likely PAH structural candidates for the FAR region in asphaltenes with eight fused aromatic rings (8FAR) that fulfill all the experimental constraints. The size (longest dimension) of the optimized geometries as well as the π-electronic distribution obtained with the Y-rule is given. All the Y-carbons are highlighted (bold) or marked in each structure. The arrows represent the resonant sextet migration, and only the structures with arrows present sextet migration.

compactness combined with a large number of FARs produce structures that tend to be extended, i.e., with a size longer than 11.2 Å and they present in their structure the identified sections shown in Figure 4.23. Thus, the following stoichiometries are not good candidates for the fused ring region in asphaltenes: $C_{32}H_{18}$ (8FAR); $C_{34}H_{18}$ and $C_{36}H_{20}$(9FAR); $C_{36}H_{18}$, $C_{38}H_{20}$, and $C_{40}H_{22}$ (10FAR) (see Tables 4.11 and 4.12). For the case of 8FAR, only two pericondensed stoichiometries are good as structural candidates of the aromatic region in asphaltenes. These are $C_{28}H_{14}$ and $C_{30}H_{16}$ with only five and six isomers, respectively (Table 4.12). For the case of 9FAR also only two pericondensed stoichiometries are good as structural candidates of the aromatic region in asphaltenes. These are $C_{30}H_{14}$ and $C_{32}H_{16}$ and only a total of eight isomers (see Table 4.12). Finally, for the case of 10FAR only two pericondensed stoichiometries are good as structural candidates of the aromatic region in asphaltenes. These are $C_{32}H_{14}$ and $C_{34}H_{16}$ with only one and two isomers, respectively (Table 4.12). All the other isomers do not fulfill the experimental size and/or the λ_{0-0} gap, due to the π-electronic distribution. The structures of the most likely structural candidates with 8FAR, 9FAR, and 10FAR rings are shown in Figures 4.24–4.26.

In Figures 4.20–4.26 (except Figure 4.23) the Y-carbons are highlighted. As it can be seen the Y-carbons form a “contour diagram” or a “cord” where the total

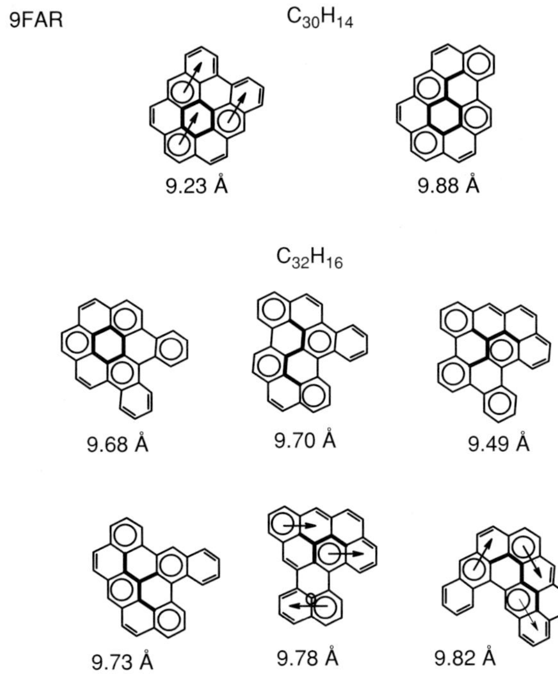

Figure 4.25. Structures of the most likely PAH structural candidates for the FAR region in asphaltenes with nine fused aromatic rings (9FAR) that fulfill all the experimental constraints. The size (longest dimension) of the optimized geometries as well as the π-electronic distribution obtained with the Y-rule is given. All the Y-carbons are highlighted (bold) or marked in each structure. The arrows represent the resonant sextet migration, and only the structures with arrows present sextet migration.

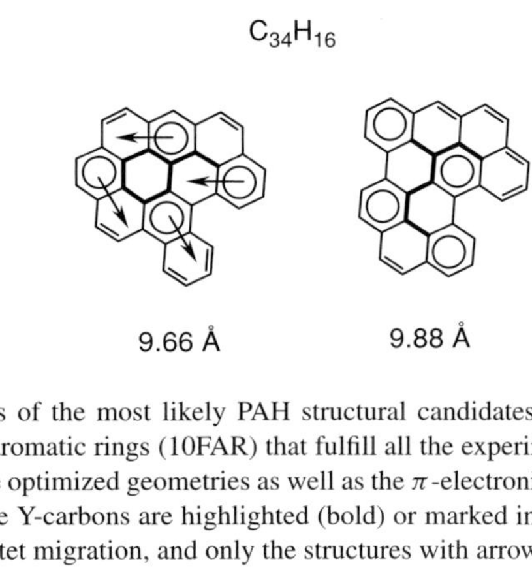

Figure 4.26. Structures of the most likely PAH structural candidates for the FAR region in asphaltenes with ten fused aromatic rings (10FAR) that fulfill all the experimental constraints. The size (longest dimension) of the optimized geometries as well as the π-electronic distribution obtained with the Y-rule is given. All the Y-carbons are highlighted (bold) or marked in each structure. The arrows represent the resonant sextet migration, and only the structures with arrows present sextet migration.

number of points is equal to the number of Y-carbons. Thus, to be able to draw isomers for the same FAR family, it is necessary to know all the possible contour diagrams for the given Y-carbon content. The nonradical benzenoid PAH systems with 5FAR–10FAR fused aromatic rings, which are allowed as structures for the aromatic core region in asphaltenes, have a total number of Y-carbons in the range of $C_Y = 2$ to $C_Y = 10$ (see Tables 4.11 and 4.12). The contour lines or diagrams to draw a total number of Y-carbons from 2 to 10 are presented in Figure 4.27. There are more possibilities for cords; however, those presented in Figure 4.27 fulfill the Y-carbon property of being able to be surrounded by hexagons without creating new Y-carbons, and to be able to apply the Y-rule to the final structure. The most likely structural candidates presented in Figures 4.20–4.26 (except Figure 4.23) have some of the Y-carbon contour diagrams and the other isomers, which are not good as structural candidates, have the same or a different contour diagram. The isomers were constructed by using the contour diagrams in Figure 4.27. For example, to draw an isomer with nine fused aromatic rings (9FAR) and six Y-carbons with a stoichiometry of $C_{32}H_{16}$ (see Tables 4.11 and 4.12), first a contour diagram with six points is chosen out of the eight possibilities given in Figure 4.27. The contour diagram is first drawn and then it is surrounded by hexagons to fulfill the property of the Y-carbons (internal carbons with connectivity of three). The missing hexagons to complete 9FAR are drawn taking care not to exceed the H and C ratio. Finally the resonant sextets are located following the Y-rule (see section 3.3) and the distribution of the π-electrons is finished by placing the remaining

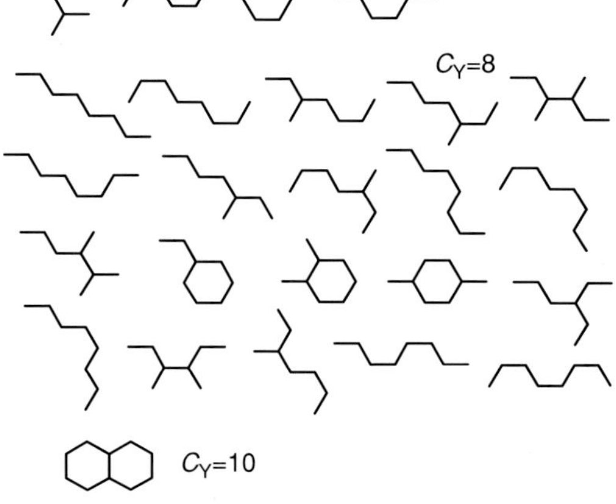

Figure 4.27. Y-carbon contour diagrams.

π-electrons in double bonds taking care of locating the double bonds without exceeding the valence of carbon atoms.

4. Conclusions

Computational chemistry approaches have revolutionized our understanding of the structure and reactivity of molecules, and computation has become the third apex of the triangle representing how we do science with experiment and theory representing the other two apices. Due to the difficult nature and complexity of the asphaltene mixtures, it is impossible to obtain all the answers based only on experimental data and it is almost impossible to get answers only from theoretical chemical approaches. However, as we have shown in this chapter, the combination of computational chemistry and experimental results, such as asphaltene optical absorption and fluorescence emission spectra, is successfully used for the particular problem of the understanding of the size, structure, and geometry of the FAR region in asphaltenes. The structures of the FAR region found with this approach can be used to understand the phenomenon of aggregation of asphaltenes and their interactions with metals and resins, and this information will allow one to design separation systems, to design catalysts for their decomposition as well as to avoid their deposition, and to characterize the heavy fraction of crude oil. Such an understanding is important if we are to get the optimal use from the heavy fraction of any crude oil.

Acknowledgments

Y.R.-M. gratefully thanks Dr. Oliver Mullins and Dr. Marcelo Lozada-Cassou for fruitful and motivating discussions, and gratefully acknowledges the comments and observations of Dr. Jose-Manuel Martinez-Magadan. Thanks are due to Dr. Fernando Alvarez-Ramirez for his assistant with NICS inputs and Y-carbon contour diagrams, and to Dr. Felipe Guevara-Rodriguez for his assistance with Y-carbon contour diagrams and PAH isomers. The assistance of Ms. Mireya Villanueva from Silicon Graphics Inc. with system administration is gratefully acknowledged.

References

[1] Chilingarian, G.V. and T.F. Yen (eds.) (1978). *Bitumens, Asphalts, and Tar Sands*. Elsevier.

[2] Speight, J.G. (1980). *The Chemistry and Technology of Petroleum*. Marcel Dekker, pp. 189–252.

[3] Tissot, B.P. and D.H. Welte (1984). *Petroleum Formation and Occurrence*. Springer-Verlag, Berlin.

[4] Bunger, J.W. and N. C. Li (eds.) (1981). *Chemistry of Asphaltenes*. American Chemical Society, Washington.

[5] Sheu, E.Y. and O.C. Mullins (eds.) (1995). *Asphaltenes, Fundamentals and Applications*. Plenum Press, New York.

[6] Mullins, O.C. and E.Y. Sheu (eds.) (1998). *Structures and Dynamics of Asphaltenes*. Plenum Press, New York.
[7] Groenzin, H. and O.C. Mullins (1999). *J. Phys. Chem. A*. **103**, 11237.
[8] Groenzin, H. and O.C. Mullins (2000). *Energy Fuels* **14**, 677.
[9] Mullins, O.C. (1998). Optical Interrogation of Aromatic Moieties in crude oils and Asphaltenes. In: Mullins, O.C. and E.Y. Sheu (eds.) *Structures and Dynamics of Asphaltenes*. Plenum Press, New York.
[10] Boduszynski, M.M. (1981). Asphaltenes in Petroleum Asphalts: Composition and Formation. In: Bunger J.W. and N.C. Li (eds.) (1981). *Chemistry of Asphaltenes*. Advances in Chemistry Series 195. American Chemical society, Washington.
[11] Boduszynski, M.M. (1988). *Energy Fuels* **2**, 597.
[12] Miller, J.T., R.B. Fisher, P. Thiyagarajan, R.E. Winans, and J.E. Hunt (1998). *Energy Fuels* **12**, 1290.
[13] Rodgers, R.P. and A.G. Marshall (2006). Petroleomics: Advanced Characterization of Petroleum Derived Materials by Fourier Transform Ion Cyclotron Resonance Mass Spectrometry (FR-ICR MS). In: Mullins, O.C., E.Y. Sheu, A. Hammami, and A.G. Marshall (eds.) *Asphaltene, Heavy Oils and Petroleomics*. Springer, New York.
[14] Zajac, G.W., N. K. Sethi, and J. T. Joseph (1994). *Scanning Microsc.* 8, 463.
[15] Sharma, A., H. Groenzin, A. Tomita, and O.C. Mullins (2002). *Energy Fuels* **16**, 490.
[16] Calemma, V., P. Iwanski, M. Nali, R. Scotti, and L. Montanari (1995). *Energy Fuels* **9**, 225.
[17] Bergmann, U., H. Groenzin, O. C. Mullins, P. Glatzel, J. Fetzer, and S.P. Cramer (2003). *Chem. Phys. Let.* **369**, 184.
[18] Seki, H. and F. Kumata (2000). *Energy Fuels* **14**, 980.
[19] Ruiz-Morales, Y. (2002). *J. Phys. Chem. A* **106**, 11283.
[20] Clar, E. (1964). *Polycyclic Hydrocarbons*. Vol. 1, Academic, London.
[21] Clar, E. (1964). *Polycyclic Hydrocarbons*. Vol. 2, Academic, London.
[22] Clar, E. (1972). *The Aromatic Sextet*. Wiley, London.
[23] Randic, M. (2003). *Chem. Rev.* **103**, 3449.
[24] Randic, M., H. Hosoya, and K. Nakada (1995). *Polycyclic Aromat. Compd*. **4**, 249.
[25] Moran, D., F. Stahl, H.F. Bettinger, H.F. Schaefer, and P.v.R. Schleyer (2003). *J. Am. Chem. Soc.* **125**, 6746.
[26] Belloli, R.C. (1983). *J. Chem. Ed.* **60**, 191.
[27] Ruiz-Morales, Y. (2004). *J. Phys. Chem. A*. **108**, 10873.
[28] Dias, J.R. (1994). *Polycyclic Aromat. Compd.* **4**, 87.
[29] Dias, J.R. (1993). *J. Mol. Struct. (THEOCHEM)* **284**, 11.
[30] Dias, J.R. (1991). *J. Mol. Struct. (THEOCHEM)* **230**, 155.
[31] Dias, J.R. (1990). *Theor. Chim. Acta.* **77**, 143.
[32] Dias, J.R. (1999). *Polycyclic Aromat. Compd.* 14–15, 63.
[33] Dias, J.R. (1985). *Acc. Chem. Res.* **18**, 241.
[34] Schleyer, P.v.R., C. Maerker, A. Dransfeld, H. Jiao, and N.J.R.v.E. Hommes (1996). *J. Am. Chem. Soc.* **118**, 6317.
[35] Schleyer, P.v.R., H. Jiao, N.J.R.v.E Hommes, V.G Malkin, and O.L Malkina (1997). *J. Am. Chem. Soc.* **119**, 12669.
[36] Schleyer, P.v.R., M. Manoharan, Z.-X. Wang, B. Kiran, H. Jiao, R. Puchta, and N.J.R.v.E. Hommes (2001). *Org. lett,* **3**, 2465.
[37] Sun, H., P. Ren, and J. R. Fried (1998). *Comput. Theor. Polym. Sci.* **8**, 229.
[38] Sun, H. (1998). *J. Phys. Chem. B*. **102**, 7338.
[39] Molecular Simulation Incorporated (1999). *Cerius2 Modeling Environment*. Release 4.0. Molecular Simulations Incorporated, San Diego, CA.
[40] Eichinger, B.E., D. Rigby, and J. Stein (2002). *Polymer*, **43**, 599.
[41] Kaduk, J.A. and J.A. Hanko (2001). *J. Appl. Crystallogr*, **34**, 720.
[42] Fraaije, J.G.E.M., B.A.C. van Vlimmeren, N.M. Maurits, M. Postma, O.A. Ever, C. Hoffman, P. Altevogt, and G. Goldbeck-Wood (1997). *J. Chem. Phys.* **106**, 4260.
[43] Spyriouni, T. and C. Vergelati (2001). *Macromolecules* **34**, 5306.
[44] Stewart, J.J.P. (1989). *J. Comp. Chem.* **10**, 209.

[45] Stewart, J.J.P. (1989). *J. Comp. Chem.* **10**, 221.
[46] Dewar, M.J.S., E.G. Zoebisch, and E.F. Healy (1985). *J. Amer. Chem. Soc.*, **107**, 3902.
[47] Mora-Diez, N. and R.J. Boyd (2000). *J. Phys. Chem. A*, **104**, 1020.
[48] Zerner, M.C., P. Correa de Mello, and M. Hehenberger (1982). *Int. J. Quantum Chem.* **21**, 251.
[49] Frisch, M.J., G.W. Trucks, H.B. Schlegel, G.E. Scuseria, M.A.; Robb, J.R. Cheeseman, V.G. Zakrzewski, J.A. Jr., Montgomery, R.E. Stratmann, J.C. Burant, S. Dapprich, J.M. Millam, A.D. Daniels, K.N. Kudin, M.C. Strain, O. Farkas, J. Tomasi, V. Barone, M. Cossi, R. Cammi, B. Mennucci, C. Pomelli, C. Adamo, S. Clifford, J. Ochterski, G.A. Petersson, P. Y. Ayala, Q. Cui, K. Morokuma, D.K. Malick, A.D. Rabuck, K. Raghavachari, J.B. Foresman, J. Cioslowski, J.V. Ortiz, B.B. Stefanov, G. Liu, A. Liashenko, P. Piskorz, I. Komaromi, R. Gomperts, R.L. Martin, D.J. Fox, T. Keith, M.A. Al-Laham, C.Y. Peng, A. Nanayakkara, C. Gonzalez, M. Challacombe, P.M.W. Gill, B.G. Johnson, W. Chen, M.W. Wong, J.L. Andres, M. Head-Gordon, E.S. Replogle, and J.A. Pople (1998). *Gaussian 98*, Revision A.7. Gaussian, Inc., Pittsburgh, PA.
[50] Ditchfield, R. (1974). *Mol. Phys.* **27**, 789.
[51] Wolinski, K., J.F. Hinton, and P. Pulay (1990). *J. Am. Chem. Soc.* **112**, 8251.
[52] Becke, A. (1988). *Phys. Rev. A.* **38**, 3098.
[53] Lee, C., W. Yang, and R. G. Parr (1988). *Phys. Rev. B.* **37**, 785.
[54] Chesnut, D.B. (1994). In: G.A. Webb (ed.) *Annual Reports on NMR Spectroscopy*. Vol. 29, Academic Press.
[55] Kutzelnigg, W., U. Fleischer, and M. Schindler (1990). *NMR Basic Principles and Progress*. Vol. 23, Springer-Verlag, Berlin, p. 165.
[56] Foresman, J.B. and A. Frisch (1996). *Exploring Chemistry with Electronic Structure Methods*. Gaussian, Inc., Pittsburgh, PA.
[57] Siegmann, K., H. Hepp, and K. Sattler (1995). *Combust. Sci. Technol.* **109**, 165.
[58] Siegmann, K. and K. Sattler (2000). *J Chem. Phys.* **112**, 698.
[59] Berlman, L.B. (1971). *Handbook of Fluorescence Specta of Aromatic Compounds*. Academic Press.
[60] George, G.A. and G.C. Morris (1968). *J. Mol. Spectrosc.* **26**, 647.
[61] Salama, F. and L. J. Allamandola (1991). *J. Chem. Phys.* **94**, 6964.
[62] Sadtler Research Laboratories Inc. (1975). *The Sadtler Handbook of Ultraviolet Spectra*. Sadtler Research Laboratories, Inc., Philadelphia, PA.
[63] McKay, J.F. and D.R. Latham (1972). *Anal. Chem.* **44**, 2132.
[64] Karcher, W., R.J. Fordham, J. Dubois, P.G.M. Glaude, and J.A.M. Ligthart (eds.) (1990). *Spectral Atlas of Polycyclic Aromatic Compounds*. Dordrech, Reidel.
[65] Acree, W.E., Jr. and S.A. Tucker (1991). *Polycyclic Aromatic Compounds* **2**, 75.
[66] Nakashima, K., S. Yashuda, Y. Ozaki, and I. Noda (2000). *J. Phys. Chem. A.* **104**, 9113.
[67] Roos, B.O., K. Andersson, and M.P. Fülscher (1992). *Chem. Phys. Lett.* **192**, 5.
[68] Garrat, P.J. (1996). *Aromaticity*. John Wiley and Sons, p. 39.
[69] Heinecke, E., D. Hartmann, R. Müller, and A. Hese (1998). *J. Chem. Phys.* **109**, 906.
[70] Halasinski, T.M., D.M. Hudgins, F. Salama, and L.J. Allamandola (2000). *J. Chem. Phys. A.* **104**, 7484.
[71] Waris, R., K.W. Street, Jr., W.E. Acree, Jr., and J.C. Fetzer (1989). *Appl. Spectrosc.* **43**, 845.
[72] Trickey, S.B., G.H.F. Diercksen, and F. Müller-Plathe (1967). *UV Atlas of Organic Compounds*. Plenum.
[73] Canuto, S., M.C. Zerner, and G.H.F. Diercksen (1991). *Astrophys. J.* **337**, 150.
[74] Fleischer, U., W. Kutzelnigg, P.; Lazzareti, and V. Mühlenkamp (1994). *J. Am. Chem. Soc.* **116**, 5298.
[75] Patchkovskii, S. and W. Thiel (2000). *J. Mol. Model.* **6**, 67.
[76] Schleyer, P.v.R. (2001). *Chem. Rev.* **101**, 1115.
[77] Watson, M.D., A. Fechtenkötter, and K. Müllen (2001). *Chem. Rev.* **101**, 1267.
[78] Masuda, K., O. Okuma, T. Nishizawa, M. Kanaji, and T. Matsumura (1996). *Fuel* **75**, 295.
[79] Hamaguchi, H. and T. Nishizawa (1992). *Fuel* **71**, 798.

5

Carbon X-ray Raman Spectroscopy of PAHs and Asphaltenes

Uwe Bergmann and Oliver C. Mullins

1. Introduction

The application of x-ray Raman spectroscopy (XRRS) to gain information about the local structure of carbonaceous systems including complex polycyclic aromatic hydrocarbons (PAHs) and asphaltenes is discussed in this chapter. This novel approach to directly probe carbon type in such systems has become practical only recently with the help of intense new synchrotron x-ray sources and innovation in spectrograph design. XRRS is the energy loss version of x-ray absorption spectroscopy (XAS) a technique well established to characterize local structure and chemistry in an element-specific manner. At the carbon K-edge conventional XAS lies in the so-called soft x-ray region, and its application to numerous systems and experimental conditions encounters severe problems related to the submicron path length of soft x-rays and electrons. In contrast, XRRS is based on hard x-rays (6–10 keV) and provides a means for obtaining bulk carbon XAS with the advantage of a much more penetrating probe ($\sim$mm path length). We will discuss the theoretical and experimental background of XRRS, and will show with the help of several examples how this technique enables understanding of the structure of asphaltenes and other related materials.

The optimal exploitation of hydrocarbon resources mandates proper predictive science. However, predictive science is largely precluded if fundamental structure remains unresolved. Lack of proper prediction in the utilization of crude oil can lead to expensive errors. For example, petroleum is being extracted in large quantities from very remote and expensive locations such as in water depths of 10,000 feet and well depths of 30,000 feet. In such an economic environment, unanticipated fluid complexities can and do result in enormous unforeseen costs; proper prediction is mandated. A major focus of petroleum science is to relate fundamental molecular structures of crude oil to its function, that is, its phenomenological behavior. This agenda is embodied in the new field—Petroleomics. The most enigmatic component of crude oil is the asphaltene fraction; that is our focus herein.[1,2]

Uwe Bergmann • Stanford Synchrotron Radiation Laboratory, P.O. Box 20450, Stanford, California 94309. **Oliver C. Mullins** • Schlumberger-Doll Research, Ridgefield, Connecticut 06877.

In particular, a difficult task is to reveal bulk properties of the aromatic ring systems of asphaltenes. Over the years various techniques have been used to probe the structures of the fused aromatic ring systems in asphaltenes. These include ^{13}C NMR,[3] time-resolved fluorescence depolarization (TRFD),[4,5] scanning tunneling microscopy (STM),[6] and high-resolution transmission electron microscopy (HRTEM).[7] In addition, optical absorption and emission data[8] coupled with molecular orbital calculations have constrained possible aromatic moieties present in crude oils.[9] Three primary issues associated with the aromatic ring systems of asphaltenes are (1) the mean number of aromatic rings per aromatic moiety, (2) the width of this distribution and (3) the geometry of the ring systems. Direct molecular imaging[6,7] as well as TRFD[4,5] have indicated that the mean number of rings in fused ring systems in petroleum asphaltenes is 7. These techniques also generally indicate that the ring distribution spans 4–10 rings for the bulk of asphaltene chromophores. The third topic of importance for aromatic ring systems is the geometry of the ring linkages, generally described in the extremes as pericondensation where three aromatic rings share single bridgehead carbon atoms, or catacondensation where bridgehead carbons are shared by two aromatic rings. Coronene is an example of pericondensation while pentacene is catacondensed. The linear arrangement of rings in pentacene makes it a member of the acene family of catacondensed aromatics. Information on such ring geometries has been more difficult to come by. STM images asphaltene chromophores; however, at most only tens of molecules are investigated making extrapolation to bulk properties difficult. ^{13}C-NMR results are suggestive that pericondensation applies for asphaltenes.[3] However, direct and systematic evidence is needed to establish the geometry of asphaltene ring systems.

Here, we will discuss results to obtain structural information on carbonaceous systems based on a relatively new spectroscopic tool. The technique is referred to as X-ray Raman scattering, nonresonant x-ray Raman scattering, x-ray Raman scattering spectroscopy and more recently, as in this chapter, just as x-ray Raman spectroscopy (XRRS). XRRS is closely related to a better known technique, namely x-ray absorption spectroscopy (XAS). In fact, XRRS is the x-ray energy loss version of XAS, analogous to inner-shell electron energy loss spectroscopy (ISEELS) or EELS. XAS has grown over the past several decades and is now a widely applied tool for element-specific studies of the local molecular and electronic structure. XAS is commonly divided into NEXAFS (near edge x-ray absorption fine structure) or, equivalently, XANES (x-ray absorption near edge structure) and EXAFS (extended x-ray absorption fine structure) regions.[10] In NEXAFS, an incident x-ray photon promotes an inner-shell electron into an unoccupied molecular orbital. The resulting absorption spectrum is sensitive to the bonding of the absorbing atom. In the case of K-edge NEXAFS (as discussed in this chapter), where a 1s electron is ejected, the dipole selection rule requires that predominantly p-type orbitals are probed. In EXAFS, the electron is ejected with more energy and can hence backscatter from the surrounding neighbor atoms. As a result the absorption cross section displays an interference pattern in electron momentum space. By Fourier transform, information of neighbor atom type, number, distance, and disorder can be extracted. Because of the small amplitude of the EXAFS oscillations very good data statistics are needed to obtain an EXAFS spectrum of sufficient quality. Because the cross section for XRRS is very small, it is difficult to obtain EXAFS data in XRRS mode. Therefore, in the current chapter, we will restrict our discussion on XRRS-based NEXAFS studies. It should be noted though, that very recently it has been possible to obtain high quality XRRS mode EXAFS at the oxygen K-edge in a study of water and ice.[11] In the future such EXAFS studies should also be practical for carbonaceous systems.

Conventional NEXAFS has been successfully employed to gain detailed structural and chemical information about hydrocarbons for several years.[12] Characteristic near edge features such as $1s \rightarrow \pi^*$ and $1s \rightarrow \sigma^*$ resonances contain information about the type of bonds[13] and the intramolecular bond lengths.[14] Furthermore, the peaks can be used to unambiguously determine the orientation of aromatic hydrocarbons on, e.g., metal surfaces.[15] In recent years ab initio calculations have been used to interpret detailed NEXAFS features taken at high resolution, as shown in studies on polycyclic aromatic hydrocarbons[16] or dimethyl phthalate isomers.[17] NEXAFS has also been extensively applied to study polymers. Here, the combination of the technique with x-ray microscopy has proven very powerful.[18]

Light-atom NEXAFS spectra are traditionally measured in transmission, using Auger yield, electron yield, sample photocurrent or fluorescence excitation modes.[12] Due to the submicron path lengths of soft x-rays, transmission measurements require very thin samples that are difficult to prepare. For example, the absorption length of carbonaceous material above its K-edge energy is less than 0.1 μm. Furthermore, the samples have to be transversely homogeneous. To simplify sample preparation, the various electron detection methods are often employed. However, these methods have probed depths of less than 50 Å and are thus surface sensitive.[19] In fact, they might provide information about an oxide coating or sorbed atoms rather than about the bulk sample. Fluorescence yield probes deeper into the sample, but in concentrated systems it suffers from strong artifacts due to "saturation effects"[20] which severely limit a quantitative analysis. Finally, for low Z elements variations in fluorescence yield across the absorption edge can lead to spectral distortions.[21]

Summarizing these potential problems shows that there is a large class of systems and experimental conditions where the bulk properties are difficult to probe by these conventional NEXAFS methods. They include heterogeneous concentrated compounds, such as large PAHs, asphaltenes, and coals with their complex structures and sensitivity to surface oxidation, reactive materials, liquids, and systems under extreme pressure or temperature. It is here where XRRS, which uses more penetrating hard x-rays (>6 keV), has crucial advantages. For the study of carbonaceous materials both its "bulk" (~mm) sensitivity and less stringent requirements on the sample environment are favorable. Equipment such as furnaces, in situ chambers, flow tubes, and high-pressure cells can all be used (>6 keV). XRRS is hence a technique that can retain all the experimental advantages of a hard x-ray measurement, while providing the unique structural information that is contained in the soft x-ray NEXAFS.

What is today known as XRRS was first mentioned in 1923 by Smekal, even before Raman introduced similar concepts for inelastic photon scattering accompanying valence electron excitation.[22] It took more than 40 years, however, before in 1967 experiments by Suzuki[23] led to the clear and unambiguous observation of x-ray Raman spectra on elements from Be to C. Earlier in the same year the theoretical work by Mizuno and Ohmura[24] had established the close connection between XRRS and XAS.* The first demonstrations of XRRS for spectroscopic applications appeared in the 1980s when tunable x-rays provided by synchrotron radiation (SR) sources became more widely available.[26,27] Recently, XRRS has been applied in increasing numbers to look at K-edges ranging from Li to F.[25,28–31] This work has shown that XRRS is now clearly beyond demonstration experiments. The potential of XRRS has been recognized, and at the largest third-generation SR facilities such as ESRF, APS, and SPring-8, the technique is performed in increasing numbers. Furthermore, at the newest 3 GeV class third-generation synchrotrons including Stanford's SPEAR3 ring and SOLEIL in France efforts to enable routine XRRS studies are underway.

2. Theory

In the XRRS process an incident photon is inelastically scattered and part of its energy is transferred to excite an inner-shell electron into an unoccupied state. The energy lost in the scattering corresponds to the absorption energy in XAS, and, under the dipole approximation, the resulting XRRS spectral features are identical to those in XAS. For a randomly oriented sample using an unpolarized x-ray beam, the transition probability for XRRS, w, is described by[27]:

$$w = \left\{(4\pi^3 e^4 h)/(m^2 \nu_i \nu_j)\right\}(1+\cos^2\theta)\,|<f|\exp(iqr)|\,i>|^2 \times \delta(E_f - E_0 - h(\nu_i - \nu_j)), \quad (5.1)$$

where $<f\,|$ and $|\,i>$ are the final and initial state wave functions, ν_i and ν_j are incident and scattered x-ray frequencies, θ is the scattering angle, and q is the momentum transfer. The matrix element is essentially identical to that in XAS with q replacing the polarization vector ε in XAS. But, because the process does not involve the annihilation of a photon but rather the scattering, there can be differences when qr is of order one or larger. This fact can be exploited to study unoccupied states with different symmetries. For example, K-edge XAS probes the p-density of unoccupied states (because an s electron is excited) whereas XRRS at large q is sensitive also to other symmetries as e.g., the s-density of states. This additional information can in some cases enhance the understanding of the local bonding and recently several authors have reported such studies.[29] In the work discussed here we will, however, focus on low q studies in the dipole limit. In that case where $qr \ll 1$, the dipole approximation is valid and (also using $|\,k_i\,| \cong |\,k_j|$) the above equation becomes[27]:

$$w = \left\{(64\pi^5 e^4 h)/(m^2 c^2)\right\}(1+\cos^2\theta)\sin^2(\theta/2)\,|<i\,|\,r\,|\,f>|^2, \quad (5.2)$$

*For a brief history regarding the discovery of XRRS see reference 25.

where the matrix element is the same as for dipole XAS.[24] The term $(1 + \cos^2\theta)\sin^2(\theta/2)$ reflects the general angular dependence of XRRS for unpolarized x-rays. For studies using synchrotron radiation, which is most commonly polarized in the horizontal plane w is proportional to $\cos^2\theta\,\sin^2(\theta/2)$ in the case of scattering in the horizontal plane and w is proportional to $\sin^2(\theta/2)$ in the case of scattering in the vertical plane. Hence, ideally XRRS experiments are performed using a vertical scattering plane.

In addition to the angular dependence the XRRS cross section is, like x-ray scattering in general, dependent on the scattering volume, which in turn is dependent on x-ray energy and sample atomic number Z. For a given energy the scattering cross section scales with Z^{-4} indicating that XRRS is most suited for light (low Z) elements. Furthermore, there is a dependence on the x-ray energy E proportional to E^3, suggesting that high energy x-rays are used. This, however, is more than compensated by experimental effects related to analyzer efficiency and resolution. Most XRRS experiments rely on perfect-crystal curved Bragg optics and the efficiency of such devices scales $\sim E^{-3.2}$. In addition, the energy resolution is proportional to E. Therefore, depending on sample Z and required penetration (in case of high pressure or in situ cells) typical x-ray energies to perform XRRS have been in the 6–10 keV range. Finally, the XRRS cross section is inversely proportional to the energy loss ΔE,[32] which favors low Z elements and/or absorption edges with small energies, typically <1 keV.

3. Experiment

As mentioned above, it is the energy difference ΔE between incident x-rays of energy E and scattered x-rays of energy E' that results in the excitation of a core electron into an empty state in XRRS. To obtain this spectral information knowledge of both E and E' is required. On a synchrotron-based x-ray beamline equipped with a monochromator it is straightforward to obtain x-rays with well defined and tunable energy, i.e., knowledge of E. The experimental challenge of XRRS is to provide an efficient means to measure the energy of the scattered photon E'. This is due to the fact that x-ray monochromators are based on perfect crystal Bragg optics that generally only accept a very small solid angle. For the incident well collimated synchrotron beam this is not a problem. The scattered x-rays, on the other hand, leave the sample into essentially all directions, and it requires a special geometry to accept even a small fraction of the total solid angle. In the work described here a multi-crystal spectrometer based on spherically curved Si Bragg crystals aligned in Rowland geometry is used.[33] Figure 5.1 shows the schematic setup of an XRRS experiment and the actual spectrometer used to measure the scattered x-rays.

The details of the instrument are described in reference 33. but three important parameters should be mentioned here: (1) the overall solid angle of the instrument is 0.5% of 4π sr, (2) the total energy resolution (convolution of monochromator and spectrometer resolution) is $\sim$1 eV when the spectrometer is

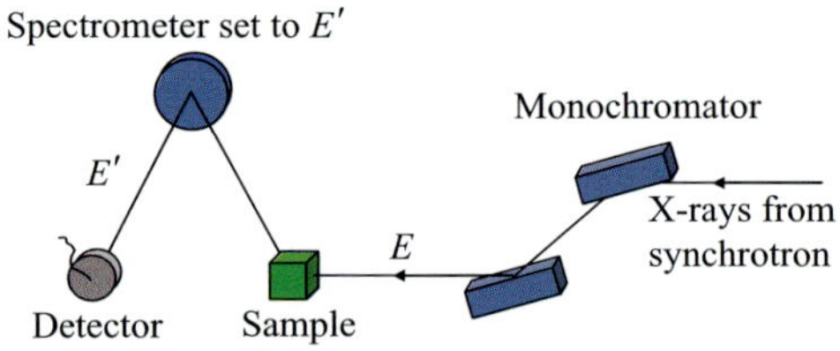

Figure 5.1. Top: Schematic setup of an XRRS experiment. The incident x-ray beam is monochromatized to an energy E, the spectrometer (analyzer) selects the scattering energy E'. A spectrum is obtained by varying E with the monochromator at fixed E'. Bottom: Photo of actual spectrometer used in experiments. The eight analyzer crystals are all set to the same energy E', focusing the beam into the detector. Using multiple analyzers improves the instrument efficiency by capturing a larger solid angle.

operated at $E' = 6.46$ keV and the incident energy E is varied with a Si(111) beamline monochromator, and (3) the q range covered by this multi-crystal spectrometer corresponds to $0.27 < r^*q < 0.5$ for the carbon K-edge. As confirmed in a comparison with ISEELS (performed at very low q) this q range is sufficiently low to fulfill the dipole limit for the studies shown here.[34] Figure 5.2 shows the total scattering spectrum from a graphite sample (left) and the comparison to conventional NEXAFS taken in electron yield (right).[35] All XRRS spectra were taken at BioCAT beamline 18ID at the Advanced Photon Source which is a division of Argonne National Laboratory and the experimental details can be found in reference 31.

The petroleum asphaltenes used here were obtained using n-heptane as the precipitating solvent. Their preparation has been described elsewhere.[2] The coal asphaltene was supplied to us by Professor M. Iino at Tohoku University. This coal asphaltene sample was obtained from a bituminous coal sample, Tanito Harum from Indonesia. The asphaltene sample was prepared from the coal liquifaction residue. The pyridine soluble fraction was isolated, and its toluene soluble fraction was then isolated; the n-hexane asphaltenes of this fraction were then collected. Most of the PAHs were acquired from Aldrich Chemical company and were used without purification. Benzo[c]naphtho[gra]tetracene was obtained from Dr. John Fetzer.

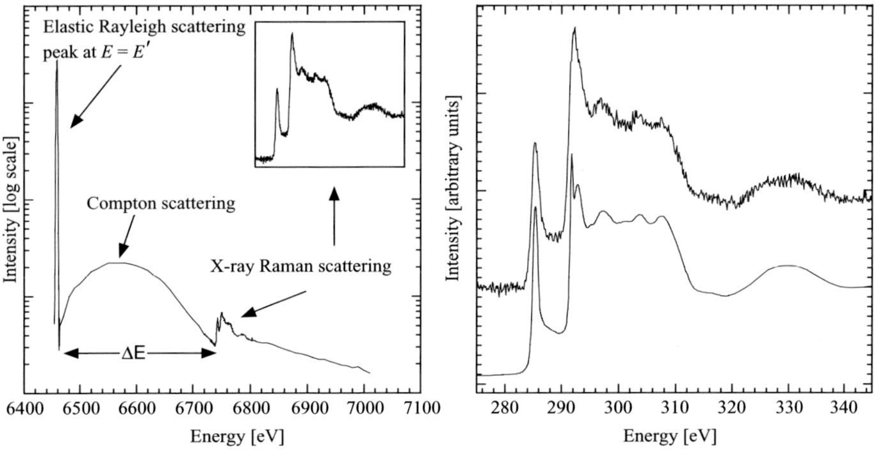

Figure 5.2. Left: Total scattering spectrum from graphite. Plotted is monochromator energy E at fixed analyzer energy E'. The quasi-elastic Rayleigh peak is followed by the Compton spectrum and K-edge x-ray Raman scattering. Right: Comparison of graphite NEXAFS for XRRS (1 eV resolution) versus conventional electron yield NEXAFS (0.15 eV resolution). The spectral features are well reproduced when using the XRRS method. Note that the energy axis electron yield corresponds to ΔE in XRRS.

4. Results and Discussion

As shown in Figure 5.2, XRRS does indeed produce spectra equivalent to conventional NEXAFS. Hence, the interpretation of XRRS can be done identically to that of well established Carbon K-edge NEXAFS.[12] We will now discuss what structural information is contained in the spectra and how this can help to understand asphaltenes. New detailed analyses and some previously published results will be presented.[30,31,34,36]

One of the questions that can be addressed with XRRS is the ratio of aromatic to saturated carbons in asphaltenes. It was shown that for asphaltenes from crude oil this ratio is approximately 1:1.[30] To demonstrate the effect of aromaticity on the NEXAFS spectra, Figure 5.3 (middle) shows a series of spectra of hydrocarbons with varying ratio of aromatic to saturated carbons (top to bottom: coronene, 1-phenyl hexane, solardye, n-octacosane ($C_{28}H_{58}$), and 2,2,4-trimethylpentane). The various features in the spectra include a strong 1s–π^* resonance at 285 eV that is decreasing with decreasing aromatic fraction and absent in the spectrum of fully saturated carbon. There is a sharp absorption edge at ~288 eV which is a signature of saturated carbon. There is a broader feature at ~293 eV assigned to a 1s–σ^* resonance which is present in both aromatic as well as saturated carbons. In addition, the coronene spectrum shows an even broader additional 1s–σ^* resonance between 297 and 311 eV. The assignment of these resonances with respect to intramolecular structure (and extramolecular structure in case of sorbed systems) has been discussed in detail by Stöhr.[12] We will use the spectra in a finger print type analysis for the asphaltenes, and substantiate this analysis by a series of least

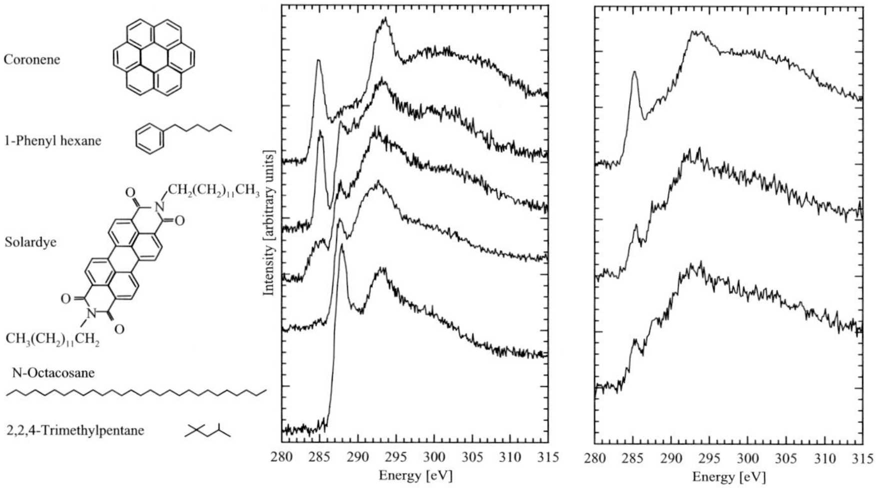

Figure 5.3. Left: Comparison of hydrocarbon spectra with varying degree of aromatic to saturated carbon. Spectra correspond to compounds shown left (top to bottom: coronene, 1-phenyl hexane, solardye, *n*-octacosane, 2,2,4-trimethylpentane). Right: Asphaltene spectra, top: Tanito Harum coal, middle: Venezuela #20 crude, bottom: KUHM crude.

square fits. Figure 5.3 right shows spectra from coal asphaltene (top) and two different crude oil coal asphaltenes (middle: Venezuela #20, bottom: KUHM). Both crude oil asphaltene spectra are very similar whereas the coal spectrum is markedly different. From a comparison with the left figure it is obvious that coal asphaltene does not contain a significant fraction of saturated carbon, its spectrum looks, in fact, rather similar to that of coronene. A more extended comparison with a series of PAHs suggests that coal asphaltene spectra are similar to those of 4–7 predominantly pericondensed ring systems and that the percentage of saturated carbon is no more than $\sim$20% (see detailed analysis below). The crude asphaltene spectra, on the other hand, indicate features typical for saturated as well as aromatic carbons. There is a sharp 1s–π^* resonance at 285 eV indicating the presence of aromatic carbon and the edge feature at 288 eV is a clear signature of saturated carbon.

^{13}C NMR provides unambiguous analyses of the saturate to aromatic fraction content in carbonaceous materials such as asphaltenes. To improve our understanding of the XRRS technique, we analyze the XRRS data on asphaltenes with large differences in this ratio. Petroleum asphaltenes are known to have large fractions of saturated carbon, whereas coal asphaltenes have a much smaller fraction of carbon in saturated groups. This is in accord with the very different alkane content of the source materials for the asphaltenes. To get a quantitative estimate of this ratio based on XRRS, we have performed two component fits using all combinations of the compounds shown in Table 5.1 and in Figure 5.3. *n*-Octacosane turned out to give the best fits as model compound for saturated carbon in the crude asphaltenes. Possibly the high –CH_2– function group is needed to match the alkyl content of asphaltenes which is dominated by this functional group. Table 5.1 shows the fit

Table 5.1. Two-Component Fits for Asphaltene Spectra Using *n*-Octacosane and the PAH Model Compounds[a]

	Venezuela #20		KUHM		Tanito Harum Coal	
Aromatic carbon	Aromatic carbon %	Goodness	Aromatic carbon %	Goodness	Aromatic carbon %	Goodness
Benzene	33.2	2.46	33.9	2.91	61.8	3.85
Naphthalene	41.4	1.85	42.5	2.20	76.1	1.37
Anthracene	41.7	2.26	43.7	2.39	72.8	3.86
Phenanthrene	**42.8**	**1.80**	44.1	2.09	**78.3**	**1.15**
2,3-Benzanthracene	45.9	2.11	48.0	2.26	79.6	2.91
1,2-Benzanthracene	**46.1**	**1.79**	**47.5**	**2.08**	**82.5**	**1.23**
Triphenylene	42.0	1.82	42.9	2.20	76.5	1.40
Benzo[a]pyrene	**43.7**	**1.77**	**45.4**	**1.96**	**78.5**	**1.13**
Benzo[e]pyrene	**45.6**	**1.74**	46.9	2.04	**81.8**	**0.99**
Naphtho[2,3-a]pyrene	45.9	1.83	**48.2**	**1.91**	82.0	1.38
Benzo[ghi]perylene	40.2	2.39	41.6	2.68	77.2	2.16
Coronene	43.4	2.09	44.9	2.34	77.9	2.13
Benzo[e]naphtho[gra]tetracene	43.1	2.22	44.5	2.53	81.4	1.77
Quinoline	35.9	2.24	36.9	2.60	66.4	2.48
Acridine	44.3	2.03	46.6	2.08	77.3	2.72
7,8-Benzoquinoline	40.6	1.89	**42.4**	**2.06**	**75.5**	**1.09**
Carbazole	**48.6**	**1.71**	**50.4**	**1.94**	87.2	1.65
Dibenzothiophene	43.2	1.87	44.6	2.15	71.5	3.86

[a]The percentage of aromatic carbon and the goodness of the fits are shown for each combination. The best five fits for each of the asphaltenes are indicated in bold numbers.

results of *n*-octacosane combined with a series of aromatic compounds. The percentage of aromatic carbon as well as the goodness of the fit is shown in the table. The best five fits of all asphaltenes are indicated in bold letters. Using these five fits, the average percentage of aromatic carbon in Venezuela #20 is 45.4%, in KUHM 46.8%, and coal asphaltene 79.3%. The data also show that neither benzene nor the straight PAH linkages of three or more carbon rings such as anthracene and 2,3-benzanthracene (tetracene) give good fits. Note that the very best fit for coal asphaltene (goodness = 0.85) is obtained by a mixture of 72.8% benzo(a)pyrene with 27.2% solardye. Because solardye has 56% saturated carbon, the resulting percentage of aromatic carbon in coal asphaltene based on this fit is 84.8%. These values of the aromatic fraction obtained by XRRS fitting are close to those obtained by ^{13}C NMR on the same samples; 48% for Venezuela #20, for 53% for KUHM, and 88% for the coal asphaltene.[37]

Not surprisingly the best fits are obtained with PAHs that have a relatively narrow 1s–π^* resonance similar to that in asphaltenes. The exception is benzene where the 1s–π^* resonance is much too intense. In that case only a very small fraction would be sufficient to obtain a 1s–π^* resonance of intensity comparable with that observed in petroleum asphaltenes. This, on the other hand, would lead to a very poor fit of the other spectral features. Benzene is, in fact, the worst model to simulate the aromatic fraction of petroleum asphaltenes (see Table 5.1). It is instructive to consider the 1s–π^* resonance linewidth in detail. The description

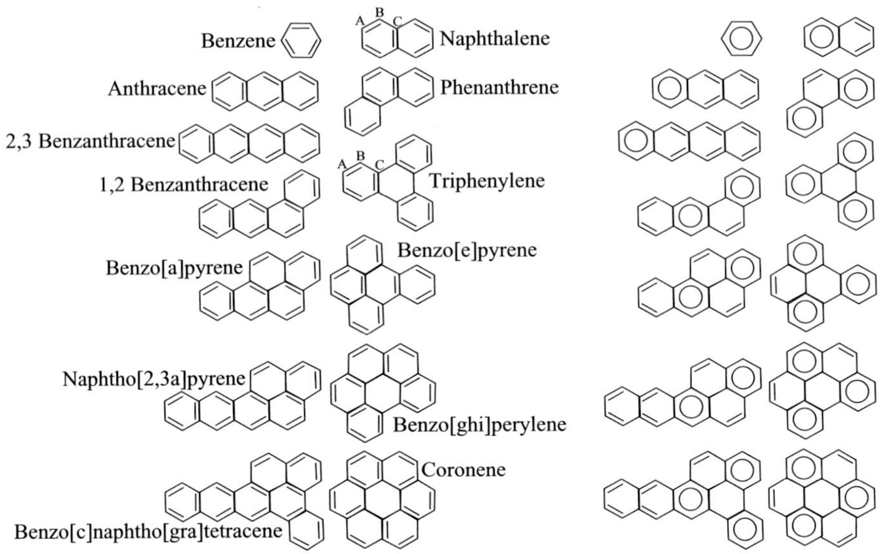

Figure 5.4. Structures of selected 1–7 ring PAHs studied with XRRS. On the right the isolated double bond—aromatic sextet description is shown. The correlation between width of the 1s–π^* resonance and ratio of double bond to sextet carbons strongly supports this description. In a different approach distinct carbon sites such as those shown in naphthalene and triphenylene are used to calculate NEXAFS spectra. See discussion below and reference 34.

of PAHs in terms of isolated double bonds and aromatic sextets was introduced decades ago by Clar.[38] The objective of this description is to provide a simple heuristic to understand the location of π electron density in PAHs and to have some idea of the stability of the PAH. One represents the PAHs as consisting of the maximal number of "benzene rings" within a PAH. All double bonds not consumed by these benzene rings are indicated explicitly as double bonds. Within this description the common representation of PAHs as drawing a Kekule circle in each of the PAH hexagons is incorrect. In this description, no two rings sharing a hexagon side can both be aromatic sextets (see Figure 5.4, right).

This aromatic sextet—isolated double bond description does not have a convenient shorthand notation unless we simply call it Clar's model. But it is useful in providing a very simple framework to understand the location of π electron density in PAHs. For instance, this description immediately predicts a low π electron density in the center hexagon of coronene. Often simple but powerful descriptions become part of undergraduate curricula to empower students' understanding. However, the isolated double bond—aromatic sextet description has remained relatively unknown probably for reasons beyond its awkward name. Indeed this description has been confirmed and its utility expanded by recent work in molecular orbital calculations of PAHs.[9] In addition, general features of PAHs such as the correlation of carbon type (isolated double bond vs. aromatic sextet) with electronic transition energies is known. For example, relatively large pericyclic PAHs are colorless while relatively small catacyclic PAHs (pentacene) are colored. Correspondingly, pericyclic PAHs are relatively more resistant to oxidative degradation. However, finding a direct correlation of a measured physical parameter of PAHs vs. carbon type has proved elusive until recent XRRS investigations of PAHs.[31]

It is known in sulfur NEXAFS that the energy location of the primary spectral peak depends strongly on the sulfur oxidation state.[39,40] The 1s-3p resonance of sulfur varies essentially linearly on oxidation state (from 0 to +6) over an energy range of up to 14 eV. It is also known that the 1s–π^* peaks of aromatic nitrogen depend strongly on nitrogen type.[41,42] Pyridinic nitrogen is basic, the 1s–π^* peak occurs at lower energy, pyrrolic nitrogen shares its lone pairs in the aromatic ring system creating a slightly positive charge on this nitrogen; the 1s–π^* peak in pyrrolic nitrogen is blue shifted relative to pyridinic nitrogen by ~3 eV. The double bond carbon and the sextet carbon are the same oxidation state unlike sulfur for many sulfur compounds. And the resonance effects for the two carbon types do not create large electron density differences the way they do for pyridinic and pyrrolic nitrogen. Thus, the XRRS carbon spectrum cannot be expected to yield two completely distinct 1s–π^* resonances, one for isolated double bond carbon, the other for aromatic sextet carbon. However, one might expect an increasing linewidth as the fraction of isolated double bond carbon increase from zero contribution (e.g., benzene or triphenylene) to some finite contribution (e.g., anthracene or tetracene). We interpret our XRRS data of PAHs in terms of the aromatic sextet—isolated double bond description.[31] When describing PAHs in terms of carbons participating in double bonds and aromatic sextets (Figure 5.4, right), the width of the 1s–π^* resonance is linearly related to the ratio of double bond to sextet carbon.

Compounds with a small ratio have a sharp 1s–π^* resonance. In compounds with a large ratio the resonance splits into several peaks leading to a much increased width. The extreme cases of the systems we measured are benzene and triphenylene

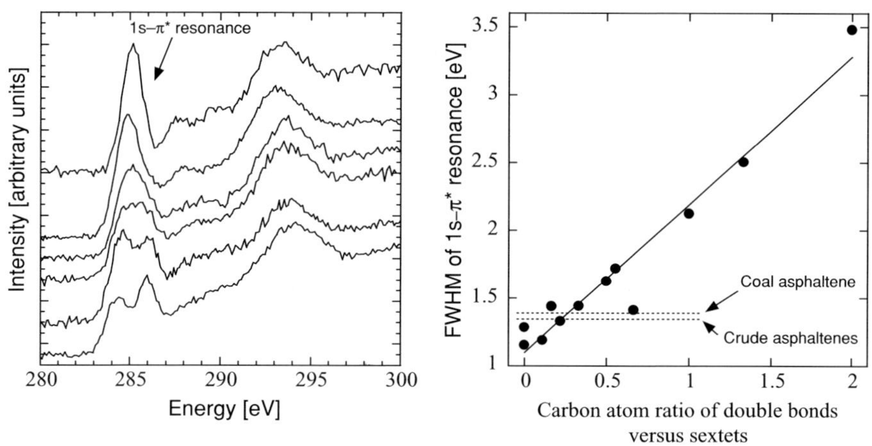

Figure 5.5. Left: Spectra of selected PAH model compounds. From top: triphenylene, benzo[ghi] perylene, benzo[a]pyrene, naphtho[2,3a]pyrene, anthracene, 2,3-benzanthracene. Right: Correlation of 1s–π^* linewidth to the carbon ratio of double bonds versus aromatic sextets and linear least squares fit. The measured 1s–π^* linewidths of coal and crude asphaltenes are shown as dashed lines for comparison.

with a ratio of 0 and 2,3-benzanthracene (tetracene) with a ratio of 2. Figure 5.5 (left) shows some spectra of selected PAH model compounds with an increasing ratio of double bond to sextet carbon (top to bottom). Figure 5.5 (right) shows the correlation and the corresponding linear fit. Also shown in the figure are dashed horizontal lines indicating the measured widths of the 1s–π^* resonance in coal (1.39 eV) and crude asphaltenes (1.34 eV), respectively. This suggests that the aromatic carbons have an average double bond to sextet ratio of order 0.25 in crude oil asphaltenes and slightly more in coals asphaltenes. Such a large degree of sextet carbon generally indicates more pericondensed ring systems.

A different, more detailed, approach to describe how NEXAFS spectra reflect the electronic structure of carbonaceous systems is based on GSCF3 (Gaussian self-consistent field version 3) calculations.[43] GSCF3 is an ab initio code designed specifically for inner-shell excitation and ionization calculations. The program uses the Hartree–Fock–SCF approach to solve for the energies and molecular orbitals of the system under investigation. This method has been used recently to calculate spectra for 1–4 ring PAHs, that were compared to experimental data based on ISEELS and XRRS.[34] In this method a separate calculation is performed for each distinct chemical site in the molecule. For example, both naphthalene and triphenylene have three different carbon sites (indicated A, B, C in Figure 5.4), resulting in three separate spectra. These spectra are then added in the correct proportion to obtain the total NEXAFS spectrum for each molecule.[34] It is interesting to note that in, e.g., triphenylene the calculations at all three sites show a 1s–π^* resonance at the exact same energy position, resulting in a sharp peak for the total spectrum. On the other hand, in naphthalene and even more pronounced in anthracene, these resonances occur at different energies (resulting in split 1s–π^* resonances). Both of these results are consistent with the double bond sextet description, which is based on a very different approach. Future theoretical and experimental studies on a larger set of PAHs including more complex systems will show to what extent and precision the very simple but powerful double bond sextet description of PAHs is applicable.

One of the principle results of XRRS for asphaltenes is that they are dominated by sextet carbon, independent of asphaltene origin. This is not surprising, asphaltenes are born in organic geochemical processes which naturally take geologic time. Stable molecules such as PAHs with a large fraction of sextet carbon are expected to be constituents of asphaltenes; unstable molecules such as linear catacondensed molecules are not expected. For instance, pentacene degrades readily in months in a lab setting—thus is not too likely to be found in a cretaceous crude oil! Knowing the PAH geometry, one is finally ready to utilize MO calculations coupled with known optical properties[8] to predict fused ring numbers in petroleum asphaltenes. Consistency with other direct measurements is found, petroleum asphaltenes consist of 4–10 fused ring systems.[8,9]

It is well known that asphaltenes contain numerous non-carbon elements embedded in the carbon ring systems, most notably nitrogen and sulfur. These compounds are especially important in crude oil utilization. In refining, the aromatic sulfur is much more difficult to remove than the saturated sulfur. Also, the vanadium and nickel content in crude oils is generally chelated by aromatic nitrogen substantially in porphyrin structures. NEXAFS applied to sulfur[39,40] and to

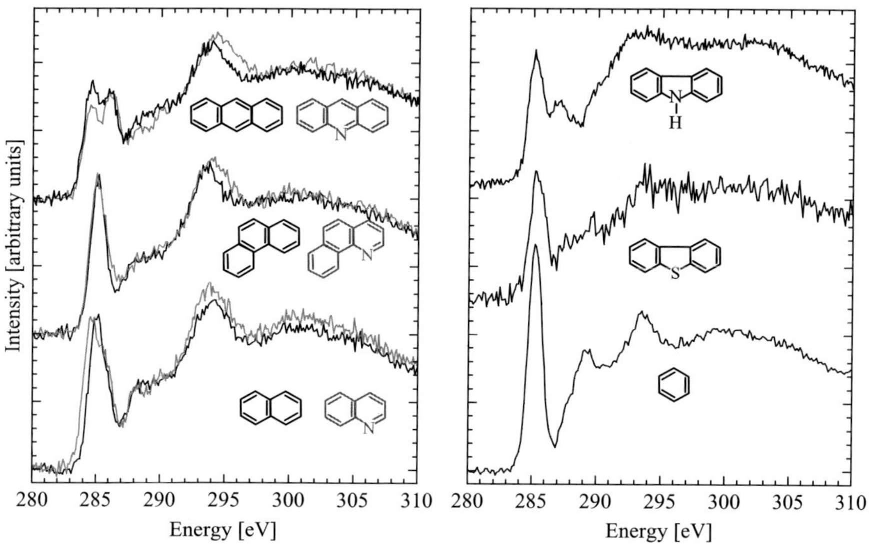

Figure 5.6. Left: Comparison of two and three ring spectra with those where one carbon is replaced by nitrogen (bottom to top black spectra: naphthalene, phenanthrene, anthracene, grey spectra: quinoline, 7,8-benzoquinoline, acridine). Right: Comparison of mixed compounds with two more or less isolated carbon rings with benzene (bottom to top: benzene, dibenzothiophene, carbazole).

nitrogen[41,42] have been very informative in determining their functional groups in asphaltenes. We now address the question of how these two elements affect the carbon NEXAFS spectra of PAHs. Figure 5.6 (left) shows a comparison of two and three ring spectra with those where one carbon is replaced by nitrogen (grey spectra).

As shown in Fig. 5.6, comparing the carbon and homologous N series, the main 1s–σ^* resonance is slightly stronger but comparable in the homologous N series. Furthermore, the broad 1s–π^* resonances are very comparable in width for the carbon and homologous N series. For example, acridine and anthracene have nearly the same 1s–π^* resonance widths. The same applies for phenanthrene and 7,8-benzoquinoline. Consequently, the XRRS data suggest that the Clar model works for 6-membered nitrogen containing aromatic rings.

A different situation applies for 5-membered heteroatom containing rings. As shown in Fig. 5.6, the 1s–π^* resonance of dibenzothiophene (DBT) is very similar to that of benzene. In other words, the two benzosubstitents of DBT can be treated as sextets. This is cosistent with the atomic-like description of XANES spectra of sulfur containing compounds.[39,40] On the other hand, the carbazole spectrum exhibits considerable line broadening of the 1s–π^* resonance. This is consistent with the significant sextet aromaticity of the pyrrole ring as clearly seen in the XANES spectra of corresponding compounds.[41] Thus, carbazole is more similar to anthracene than benzene (cf. Figure 5.6). The XRRS results suggest nitrogen and sulfur containing aromatics fit well within the Clar model. Typically in asphaltenes, sulfur and nitrogen are the most abundant heteroatoms and are a few weight percent thus, on average, an asphaltene molecule possesses one or two heteroatoms. In any event, XRRS is seen to be very sensitive to the specific aromatic nature of complex compounds. And the XRRS data is seen to be complimentary to the K-edge XANES spectroscopy as shown in, e.g., references 39, 40, 41, 44. As discussed above, these conventional soft x-ray methods have unfortunately numerous limitations. It is therefore the hope of the authors that in future, more efficient, spectrometers will make possible the study of these elements with XRRS-based NEXAFS.

5. Conclusion and Outlook

A new technique to study the structure of asphaltenes and other carbonaceous materials as well as PAHs was introduced in this chapter. Previously, conventional NEXAFS has been used extensively to study sulfur and nitrogen in carbonaceous materials, now there is a related technique to study carbon, the element of greatest interest and the defining element of these materials. The technique, XRRS, makes use of the large penetration depth of hard x-rays and the sensitivity of low Z NEXAFS. This combination is beneficial, and in some cases even crucial, to study bulk properties of these economically important systems. Asphaltenes are determined to be dominated by sextet carbon independent of asphaltene source; this is consistent with the known stability of PAHs with a large fraction of sextet carbon. The XRRS results constrain molecular orbital calculations in a crucial way to corroborate previous findings that petroleum asphaltenes consist primarily of ring systems with 4–10 fused rings. These important XRRS results will help guide concept development in understanding nanoaggregate formation in asphaltene solutions due to the importance of these ring systems in intermolecular binding. Indeed, it is likely that in general van der Waals attraction of fused ring systems is a dominating factor in asphaltene nanoaggregate formation. The development of relations between fundamental molecular properties to aggregate growth in asphaltenes is one of the goals of Petroleomics. To understand function, study structure.

The routine application of XRRS is still in its infancy, but existing and future facilities will help to perform such studies on a larger scale around the world. The prospects of using this powerful tool to help optimize the use of the immensely important natural carbon resources are very good indeed. Finally, it should be noted that XRRS is one of the very few spectroscopic techniques that can be applied at future ultra-fast x-ray laser sources such as Stanford's Linac Coherent Light Source (LCLS) (see http://www-ssrl.slac.stanford.edu/lcls/) and the X-ray Free Electron Laser (XFEL) in Hamburg, Germany (see http://xfel.desy.de/content/e169/index_eng.html). Here, extremely intense ~250 femtosecond short x-ray pulses will enable research on the ultra-fast time scale. In a pump probe experiment, for example, a short laser pulse can be used to initiate a chemical reaction and XRRS can be applied to characterize the structural changes, as e.g., the dissociation of a molecule. In this manner a "movie" of the chemical reaction can be recorded. Unlike the energy scanning technique described in this chapter, XRRS instrumentation for such pump-probe experiments is based on wavelength dispersive optics, where the whole NEXAFS spectrum is recorded at once in a position sensitive detector.[45] Being able to probe catalytic reactions in carbonaceous systems on a femtosecond timescale is certainly an exciting prospect.

Acknowledgments

The authors would like to thank Henning Groenzin, Dr. Pieter Glatzel, Prof. Steven P. Cramer, and Dr. John Fetzer for help during sample preparation, experiments and data analysis, and discussion. XRRS experiments were performed at BioCAT at the Advanced Photon Source (APS) and the staff is acknowledged for support. Use of the APS was supported by the U.S. Department of Energy, Basic Energy Sciences, Office of Science, under contract No. W-31-109-ENG-38. BioCAT is a National Institutes of Health-supported Research Center RR-08630. Portions of this research were carried out at the Stanford Synchrotron Radiation Laboratory, a national user facility operated by Stanford University on behalf of the U.S. Department of Energy, Office of Basic Energy Sciences. This research was partially supported by the National Institutes of Health, grants 44891-5, GM 44380, and GM-48145, and by the Department of Energy, Office of Biological and Environmental Research.

References

[1] Chilingarian, G.V. and T.F. Yen (1978). *Bitumens, Asphalts, and Tar Sands*. Elsevier Scientific Pub. Co., New York; Tissot, B.P. and D.H. Welte (1984). *Petroleum Formation and Occurrence*. Springer-Verlag, Berlin; Bunger, J.W. and N.C. Li (eds.) (1984). *Chemistry of Asphaltenes*. American Chemical Society, Washington, DC; Mullins, O.C. and E.Y. Sheu (eds.) (1998). *Structures and Dynamics of Asphaltenes* Plenum, New York.

[2] Sheu, E.Y. and O.C. Mullins (eds.) (1995). *Asphaltenes: Fundamentals and Applications*. Plenum, New York.

[3] Calemma, V., P. Iwanski, M. Nali, R. Scotti, and L. Montanari (1995). *Energy & Fuels* **9**(2), 225; Scotti, R. and L. Montanari (1998). In: Mullins, O.C. and E.Y. Sheu (eds.), *Structures and Dynamics of Asphaltenes*, Plenum, New York.

[4] Groenzin, H. and O.C. Mullins (1999). Asphaltene molecular size and structure. *J. Phys. Chem. A* **103**(50), 11237.

[5] Groenzin, H. and O.C. Mullins (2006). *Asphaltene Molecular Size and Weight by Time-Resolved Fluorescence Depolarization*. Chapter 2, this book.

[6] Zajac, G.W., N.K. Sethi, and J.T. Joseph (1994). *Scanning Microsc.* **8**(3), 463.

[7] Sharma, A., H. Groenzin, A. Tomita, and O.C. Mullins (2002). Probing order in asphaltene and aromatic ring systems by HRTEM. *Energy & Fuels* **16**(2), 490; Sharma, A. and O.C. Mullins (2006). *Insights into Molecular and Aggregate Structures of Asphaltenes Using HRTEM*. Chapter 8, this book.

[8] Mullins, O.C. (1998). Optical interrogation of aromatic moieties in crude oils and asphaltenes, Chapter 2 in O.C. Mullins and E.Y. Sheu (eds.), *Structures and Dynamics of Asphaltenes*. Plenum, New York.

[9] Ruiz-Morales, Y. (2002). *J. Phys. Chem. A* **106**, 11283; Ruiz-Morales, Y. (2006). *Molecular Orbital Calculations and Optical Transitions of PAHs and Asphaltenes*. Chapter 4 in this book.

[10] Koningsberger, D.C. and R. Prins (1988). *X-ray Absorption: Principles, Applications, Techniques of EXAFS, SEXAFS, and XANES*. John Wiley and Sons, New York.

[11] Bergmann, et al. Submitted.

[12] Stöhr, J. (1992). *NEXAFS Spectroscopy*. Springer-Verlag, Berlin, New York.
[13] Francis, J.T., C. Enkvist, S. Lunell, and A.P. Hitchcock (1994). *Can. J. Phys.* **72**, 879–884.
[14] Sette, F., J. Stöhr, and A.P. Hitchcock (1984). *J. Chem. Phys.* **81**, 4906–4914.
[15] Yannoulis, P., R. Dudde, K.H. Frank, and E.E. Koch (1987). *Orientation of aromatic hydrocarbons on metal surfaces as determined by NEXAFS*, 519–28.
[16] Oji, H., R. Mitsumoto, E. Ito, H. Ishii, Y. Ouchi, K. Seki, T. Yokoyama, T. Ohta, and N. Kosugi (1998). *J. Chem. Phys.* **109**, 10409–10418.
[17] Urquhart, S.G., A.P. Hitchcock, A.P. Smith, H.W. Ade, and E.G. Rightor (1997). *J. Phys. Chem. B*, **101**, 2267–2276.
[18] Urquhart, S.G., A.P. Hitchcock, A.P. Smith, H.W. Ade, W. Lidy, E.G. Rightor, and G.E. Mitchell (1999). *J. Electron Spectrosc. Relat. Phenom.* **100**, 119–135.
[19] Abbate, M., J.B. Goedkopp, F.M.F. de Groot, M. Grioni, J.C. Fuggle, and S. Hofmann *et al.* (1992). *Surf. Interface Anal.* **18**, 65.
[20] Goulon, J., C. Goulon-Ginet, R. Cortes, and J.M. Dubois (1982). *J. Phys.* **43**, 539.
[21] deGroot, F.M.F., M.-A. Arrio, P. Sainctavit, C. Cartier, and C.T. Chen (1995). *Physica B* **208–209**, 84.
[22] Smekal, A. (1923). *Naturwissenschaften* **11**, 873.
[23] Suzuki, T., (1967). *J. Phys. Soc. Jpn.* **22**(5), 1139.
[24] Mizuno, Y. and Y. Ohmura (1967). *J. Phys. Soc. Jpn.* **22**(2), 445.
[25] Bergmann, U., P. Glatzel, and S.P. Cramer (2002). *Microchem. J.* **71**(2–3 SI), 221.
[26] Tohji, K. and Y. Udagawa (1987). *Phys. Rev. B (Condensed Matter)* **36**(17), 9410; Schülke, W., U. Bonse, H. Nagasawa, A. Kaprolat, and A. Berthold (1988). *Phys. Rev. B (Condensed Matter)* **38**(3), 2112; Schülke, W., A. Berthold, A. Kaprolat, and H.-J. Güntherodt (1988). *Phys. Rev. Lett.* **60**(21), 2217.
[27] Tohji, K. and Y. Udagawa (1989). *Phys. Rev. B (Condensed Matter)* **39**(11), 7590.
[28] Watanabe, N., H. Hayashi, Y. Udagawa, K. Takeshita, and H. Kawata (1996). *Appl. Phys. Lett.* **69**(10), 1370; Udagawa, Y., N. Watanabe, and H. Hayashi (1997). *J. Phys. Iv* **7**(C2 PT1), 347; Caliebe, W.A., J.A. Soininen, E.L. Shirley, C.C. Kao, and K. Hamalainen (2000). *Phys. Rev. Lett.* **84**(17), 3907; Bowron, D.T., M.H. Krisch, A.C. Barnes, J.L. Finney, A. Kaprolat, and M. Lorenzen (2000). *Phys. Rev. B* **62**(14), R9223; Soininen, J.A., K. Hamalainen, W.A. Caliebe, C.C. Kao, and E.L. Shirley (2001). *J. Phys. (Condensed Matter)* **13**(35), 8039; Galambosi, S., J.A. Soininen, K. Hamalainen, E.L. Shirley, and C.C. Kao (2001). *Phys. Rev. B* **6402**(2), art. no.−024102; Bergmann, U., P. Wernet, P. Glatzel, M. Cavalleri, L.G.M. Petterson, A. Nilsson *et al.* (2002). *Phys. Rev. B* **66**, 092107; Wernet, P., D. Nordlund, U. Bergmann, M. Cavalleri, M. Odelius, H. Ogasawara *et al.* (2004). *Science* **304**(5673), 995.
[29] Krisch, M.H., F. Sette, C. Masciovecchio, and R. Verbeni (1997). *Phys. Rev. Lett.* **78**(14), 2843; Hamalainen, K., S. Galambosi, J.A. Soininen, E.L. Shirley, J.P. Rueff, and A. Shukla (2002). *Phys. Rev. B* **65**(15), 155111; Sternemann, C., M. Volmer, J.A. Soininen, H. Nagasawa, M. Paulus, H. Enkisch *et al.* (2003). *Phys. Rev. B* **68**(3); Feng, Y., G.T. Seidler, J.O. Cross, A.T. Macrander, and J.J. Rehr (2004). *Phys. Rev. B* **69**, 125402.
[30] Bergmann, U., O.C. Mullins, and S.P. Cramer (2000). *Anal. Chem.* **72**(11), 2609.
[31] Bergmann, U., H. Groenzin, O.C. Mullins, P. Glatzel, J. Fetzer, and S.P. Cramer (2003). *Chem. Phys. Lett.* **369**(1–2), 184.
[32] Caliebe, W.A. (1997). PhD Thesis, Physics, University of Kiel, Germany.
[33] Bergmann, U. and S.P. Cramer (1998). *SPIE Int. Soc. Opt. Eng. San Diego Calif.* **3448**, 198.
[34] Gordon, M.L., D. Tulumello, G. Cooper, A.P. Hitchcock, P. Glatzel, O.C. Mullins *et al.* (2003). *J. Phys. Chem. A* **107**, 8512.
[35] Anders, S., J. Diaz, J.W. Ager, R.Y. Lo, and D.B. Bogy (1997). *Appl. Phys. Lett.* **71**(23), 3367.
[36] Bergmann, U., H. Groenzin, O.C. Mullins, P. Glatzel, J. Fetzer, and S.P. Cramer (2004). *Petroleum Sci. Technol.* **22**(7–8), 863.
[37] Buenrostro-Gonzalez, E., H. Groenzin, C. Lira-Galeana, and O.C. Mullins (2001). *Energy & Fuels* **15**(4), 972.
[38] Clar, E. (1964). *Polycyclic Hydrocarbons*. Academic Press, New York; Clar, E. (1972). *The Aromatic Sextet*. John Wiley and Sons, New York.
[39] George, G.N. and M.L. Gorbaty (1989). *J. Am. Chem. Soc.* **111**, 3182.

[40] Waldo, G.S., O.C. Mullins, J.E. Penner-Hahn, and S.P. Cramer (1992). *Fuel* **71**(1), 53.

[41] Mitra-Kirtley, S., O.C. Mullins, J. Van Elp, S.J. George, J. Chen, and S.P. Cramer (1993). *J. Am. Chem. Soc.* **115**(1), 252.

[42] Mullins, O.C (1995). Chapter 2. In: Sheu, E.Y. and O.C. Mullins (eds.), *Asphaltenes: Fundamentals and Applications*, Plenum, New York.

[43] Kosugi, N. (1987). *Theor. Chim. Acta* **72**(2), 149; Kosugi, N. and H. Kuroda (1980) *Chem. Phys. Lett.* **74**(3), 490.

[44] Kirtley, S.M., O.C. Mullins, J. van Elp, S. George, J. Chen, S.P Cramer *et al.* (1992). *Biochim. Biophys. Acta* **0132**(3), 249; Mitra-Kirtley, S., O.C. Mullins, J.F. Branthaver, and S.P. Cramer (1993). *Energy Fuels* **7**(6), 1128; Mullins, O.C., S. Mitra-Kirtley, J. Vanelp, and S.P. Cramer (1993). *Appl. Spectrosc.* **47**(8), 1268.

[45] Bergmann, U. and R. Frahm (2001). TDR XFEL workshop series. In: Hastings, J. and Th. Tschentscher (eds.), *Methods and Instrumentation for the XFEL*, p. 52.

6

Sulfur Chemical Moieties in Carbonaceous Materials

Sudipa Mitra-Kirtley and Oliver C. Mullins

1. Introduction

X-ray Absorption Near Edge Structure (XANES) spectroscopy has been employed to characterize the different chemical structures of sulfur in kerogens, asphaltenes, and coals. Commonalities are found for the sulfur chemistry in these disparate carbonaceous materials. Most of the sulfur is organic, with thiophenic structures typically the most abundant and sulfidic forms also fairly abundant. Oxidized organic sulfur in lesser amounts is found in low rank coals, in the different fractions of one crude oil, and Type I kerogens. In addition, there is pyrite in the coals and pyrite/elemental sulfur in the kerogens. The sulfur chemistry is shown to reflect the carbon chemistry in kerogens and coals. Type II kerogens have a larger ratio than Type I kerogens of aromatic to saturated carbon. Likewise higher rank coals also have a larger ratio of aromatic to saturated carbon than lower rank coals. Here, it is shown that Type II kerogens and higher rank coals similarly have larger fractions of aromatic sulfur. This important result establishes a relationship between the carbon and sulfur chemistry of these materials. The nitrogen chemistry of the carbonaceous materials was also investigated with XANES. In all of the materials, the nitrogen is almost entirely aromatic with pyrrolic nitrogen being the most abundant.

The presence of sulfur in carbonaceous samples is often an impediment in the processing and utilization of fossil fuel resources. Sulfur oxides released into the environment during combustion of these materials have been a matter of concern for decades. A knowledge of the sulfur chemistry in carbonaceous materials helps in the mitigation of sulfur-induced problems in resource utilization. In addition, elucidation of sulfur structures in carbonaceous samples, such as kerogens, helps one understand the complex maturation processes of these materials over geological time. Chemical properties of reservoired hydrocarbon fluids are now being utilized to a much greater degree to address major issues in the production of oil and gas. A new look at the chemistry of these fluids and of their source materials is mandated. Sulfur also affects specific properties such as the solubility

Sudipa Mitra-Kirtley • Rose-Hulman Institute of Technology, Terre Haute, Indiana 47803.
Oliver C. Mullins • Schlumberger-Doll Research, Ridgefield, Connecticut 06877.

characteristics of the various naturally occurring organic compounds such as the different fractions of crude oils. For a multitude of purposes, sulfur chemistry in carbonaceous materials is important to investigate and in some ways is unique in that it is the "other" element of principle concern in organic compounds beside carbon in fossil fuels.

The investigative method of choice for sulfur analysis is X-ray Absorption Near Edge Structure (XANES) spectroscopy. The sulfur XANES method is selective probing the sulfur containing moieties specifically, even if present in small mass fractions, and is insensitive to the presence of other elements. For the most part, the sulfur XANES signatures of different sulfur chemical groups are quite distinct, making the corresponding interpretation quite robust; the primary sulfur XANES spectral signature is based on simple concepts from atomic physics. In large measure, sulfur XANES spectra are equally sensitive to all sulfur moieties due to the invariant photoionization cross-section of the sulfur 1s electrons. XANES methods are nondestructive, which is highly desirable for chemical structure elucidation. XANES spectroscopy utilizes various synchrotrons in public user facilities which are excellently supported. Here, we employ XANES spectroscopy to study the sulfur chemistry in carbonaceous materials. There is a rich history of application of XANES methods to probing sulfur chemistry in carbonaceous materials. We do not attempt a review article here. Rather, it is our intention to outline a unified treatment of sulfur chemistry in disparate carbonaceous materials. We search for common themes in the sulfur chemistry and for possible relations to the carbon chemistry. A broader understanding of the most relevant issues will help direct future investigations for more specific agendas.

It is instructive to treat concurrently a variety of naturally occurring carbonaceous materials in order to elucidate common themes regarding sulfur. A cursory overview of sundry carbonaceous materials is developed particularly as pertains to the sulfur chemistry. In general, carbonaceous sediments are the result of the gradual deposition of material in a hydrological setting. Less than 1% of the total primary product of organic matter is preserved in sediments.[1,2] Bacteria and phytoplankton are responsible for the majority of organic matter in both recent and ancient sediments,[3,4] although a large contribution may also come from higher plant material found in terrestrial environments.[4] Organic carbon may constitute a significant portion of the sediments that we are interested in, such as some shales, organic-rich limestone, and coal.[4] Ultimately, the fossil fuels upon which our society is so dependant evolve from these carbonaceous sediments, although only a small fraction of the original organic material ends up as such fuels. Substantial variations in deposited organic matter result in part from different source materials. For example, the differences between terrestrial (principally swamps and marshes), marine, and lacustrine environments have a pronounced effect on the resulting organic components of the sediments.

After this original organic material is deposited into sediments, those sediments undergo gradual evolution as various biological, chemical, and physical processes occur. The first two stages of these transformations are referred to as diagenesis and catagensis.[4,5] *Diagenesis* is the earlier of the two phases, and it is characterized by relatively low pressure and temperature conditions. Diagenesis involves both biological processes and nonbiological processes. *Catagenesis* is the later stage and is characterized by greater burial depth, higher pressure and temperature. The boundary between these two stages is considered to be when the temperature of the sediment reaches close to 100°[4] (although some sources use a lower temperature threshold.[4,6]) During catagensis, kerogens begin to evolve hydrocarbons and other organic compounds[6] and petroleum formation begins.[4,5,7] In coal formation, the boundary is generally placed at the point in the process of coal maturation between so-called "brown coals" (lignite and subbituminous coal) and "hard coals" (bituminous and anthracite).[5,8]

In this chapter, we briefly outline the deposition and alteration of organic matter, emphasizing particularly the sulfur chemistry. A description of XANES follows, and sulfur XANES results on a variety of carbonaceous samples, including kerogens,[9] asphaltenes,[10] and coals[11] are presented. Relations between sulfur and nitrogen XANES[12,13,14] are highlighted. A discussion of the comparison of the different sulfur structures in these carbonaceous materials concludes the chapter.

2. Carbonaceous Materials

2.1. Production and Deposition of Organic Matter

Productivity refers to the amount of organic matter generated in a given area per unit time. The availability of sunlight and water for photosynthesis greatly affects the amount of biomass produced in a given area. In general, the productivity increases linearly with light intensity until a saturation level is attained.[15] Productivity is also impacted by the availability of nutrients.[16] Certain elements, such as nitrogen, phosphorus, and silicon, are critical to plant growth,[5] and are termed "biolimiting elements".[4] The fact that terrestrial production is deposited in the presence of air allows aerobic bacteria to degrade the organic material before it can be preserved in sediments.[5] Thus, even though terrestrial productivity is of the same order as marine productivity,[17] the vast majority of organic material preserved in sediments is found in sub-aqueous regions.[5] The preservation of organic matter in sediments is dependent upon productivity, depth and oxicity of the water column through which the organic matter must fall,[18–20] water velocity, particle size of the sediment,[21] and oxicity of the sediment itself.[4,5] Surprisingly, of these factors, productivity is not the most important determinant of how much organic material is ultimately preserved in the sediment.[22] Instead, limiting the time that the organic material is exposed to oxygen leads to higher total organic content of the sediment.[23]

In general, the nature of deposited organic matter into sedimentary layers is affected by many factors.[4,5,24] Terrestrial vs. marine organic matter have explicit underlying chemical differences. In marine environments, including open oceans, estuaries, and continental shelves, unicellular organisms are the dominant producers of organic matter.[4,5] The biomass produced in marine environments is almost entirely (ca 95%) attributed to phytoplankton[4,5,25] and marine production accounts for a large portion of the total global production of biomass.[4] However, many economically important deposits originate in terrestrial and lacustrine environments, and their source of production is an important consideration. Terrestrial deposits of concentrated organic matter occur in areas of high productivity of multicellular biological matter,[17] with either still or slow-moving water, such as swamps and rain forests. Lakes contain only a tiny fraction of the water in the hydrosphere, but they can be important precursors of petroleum and coal beds.[4,5,26] The prebiomass that produces organic sediments in lakes is, like marine environments, generally composed of unicellular plants.[17]

The carbonaceous matter in terrestrial sediments is rich in lignins and other resinous material found in the cell walls that make up structural material of higher-order plants.[5] These lignins have a high proportion of aromatic carbon compounds and also significant oxygen content. In addition, the decay of terrestrial plant material usually occurs in the presence of some oxygen, which can partially oxidize reduced carbon, yielding aromatic compounds.[4,5] Thus, the higher aromatic content of the lignins, coupled with the lower hydrogen content due to oxidation, results in terrestrial deposits containing high aromatic carbon content. In addition, the primary heteroatom tends to be oxygen. These terrestrial deposits tend to produce coal. In contrast, marine and lacustrine deposits of organic material typically lack the aromatic-rich lignins derived from higher plant structural material; they are mostly derived from proteins, carbohydrates, and lipids.[5] In addition, these sediments are typically formed in anaerobic or anoxic environments. Thus, they have low aromatic content and low oxygen content. These deposits tend to produce gas and/or petroleum.[5]

2.2. Diagenesis

Diagenesis is the first phase of the processes that act upon the organic material.[5,27] This involves decomposition of the organic material, a process known as *Mineralization*.[4] Typically, physical, chemical, and biological processes all contribute to the process of mineralization. The rate of mineralization due to biological factors in diagenesis is dependent upon the amount of oxygen available, since oxygen is necessary for aerobic bacteria and detritivores to thrive.[4] The presence of oxygen allows aerobic bacteria and fungi to produce extracellular enzymes, which hydrolyze and oxidize insoluble proteins and polysaccharides in the deposited organic material.[4,5] This results in the release of gaseous CO_2 as a by-product of this breakdown.[4,5] As organic matter falls through the water column, biological activity begins to transform it. As that material settles into a sedimentary deposit, and even through early stages of burial, these biological agents continue to act to decompose the organic material.

The process of bioturbation (physical churning of the decomposing material due to burrowing activities of worms and other detritivores) mixes the sediments, and allows oxygen to penetrate deeper into the layer.[4,5] However, once the environment becomes anoxic, detritivores can no longer survive, and bioturbation ceases.[4,5] Further reduction of the oxygen supply will effectively end all aerobic bacterial activity. Biological degradation of organic material will continue in an anaerobic environment, albeit at a much slower rate.[4,5] Anaerobic bacteria continue to oxidize organic matter by using alternative electron-accepting atoms, such as (in decreasing order of net energy release) manganese, nitrate, iron, sulfate, and bicarbonate, in the place of oxygen.[4,28,29] In addition, some natural gas is produced during the earliest phases of diagenesis as a by-product of bacterial respiration; this process is referred to as *biogenic methanogenesis*.[4]

2.3. Sulfur in Carbonaceous Sediments

The chemical environment greatly affects the interaction of the anaerobic microbes that are at work during diagenesis. Typically, marine environments are rich in sulfates,[4,24] and thus in sulfate-reducing bacteria. Since nitrate-reducers and methanogens (bacteria that produce methane in anoxic conditions) compete poorly with the sulfate-reducers,[30,31] the dominant form of carbon oxidation in anaerobic marine environments is due to the sulfate-reducing bacteria.[32] The activity of these bacteria results in the production of sulfide. Sulfide is rapidly combined with iron if present to form iron sulfide.[33] Where iron is not present, hydrogen sulfide and other sulfides are produced. However, sulfur tends to react with reactive iron species far faster than it does with organic material.[34] Thus, the presence of such iron species limits the degree of sulfurization of organic sediments.[4,33] In terrestrial lakes and soils, there is typically far less sulfate present. In these environments, the nitrate-reducers and methanogens will play a more important role.[4] The production of this biomethane in these processes has an economic importance. Petroleum sourced from shales often possesses low quantities of sulfur because the iron present in the shale scavenges the reduced sulfur to form iron sulfides. Petroleum sourced from carbonates is often rich in sulfur due to the lack of iron in many carbonates. There is also a contribution from the sulfur content of the original organic material. Inorganic sulfur can be incorporated into resistant biomacromolecules (e.g., lipids) during early diagenesis.[35] While the amount of sulfur that originates from selective preservation of lipids[4] or carbohydrates[36] varies with the environment of the source organic material, it can be an important source of sulfur compounds in sediments.

Diagenesis involves important nonbiological processes as well. The organic residue left after microbial activities cease often undergoes compaction and condensation reactions as burial depth increases, forming new polymeric material with complex cross-linking chains. These reactions result in complex polycondensed organic materials, called geopolymers, or humic substances.[4–6] By the end of diagenesis, these geopolymers will have evolved into either brown coal or kerogens.[37] When the sediment is overwhelmingly composed of organic matter (as compared to mineral matter), sourced from higher-order plants, the diagenetic pathway typically goes through peat and then brown coal.[4] Marine and lacustrine sediments, sourced from unicellular plants, typically produce kerogen as the end-product of diagenesis.[4,5,22]

It is illustrative to follow the development of brown coal from humic substances. In peat bogs, humic substances continue to undergo diagenesis, and evolve into humic coals.[4,5] The acidity of the sediments generally increases with burial depth.[4,38] This pH gradient limits conditions that are favorable for microbial activity to the near-surface layers.[38] Thus, deeper layers of the peat bog favor large-scale preservation of organic material. Early in diagenesis, invertebrates in the surface layer of the peat aid in the mechanical breakup of plant material in a process called *bioturbation*.[4] Peatification continues as fungi attack the more resistant woody portions, composed of lignins and cellulose.[39] This in turn creates significant amounts of aromatic units, carboxylic acid, and phenolic units.[4] The biochemical transformation then continues, leading to additional depolymerization and the loss of functional groups. Products such as CH_4, NH_3, CO_2, and H_2O, are produced during this phase.[4,7,38,40] As diagenesis progresses, less and less cellulose and lignin remains in the peat, and humic substances increase.[4,38,40] Lipid material, though a minor component of higher plants (mainly found in leaves, spores, pollen, and fruits), is concentrated in peat due to the resistance of such lipids to degradation.[5] By late diagenesis, the lignin content of peat will have been reduced and the loss of oxygen-containing functional groups leaves the residue highly concentrated in carbon and hydrogen. By the end of diagenesis, the peat has been transformed into *brown coal*, with lower atomic H/C and O/C ratio.[5,41] Brown coal is light in color and soft.

2.4. Kerogen Formation

Geopolymers (humic substances) in marine and lacustrine sediments become larger, more complex, and more insoluble as diagenesis proceeds with increasing burial depth.[5] This is associated with the loss of hydrophyilic functional groups.[4,5] Ultimately, after severe condensation reactions, humic substances are transformed into kerogens, which are insoluble in (organic) solvents. Kerogen is the precursor of fluid hydrocarbons, petroleum and natural gas in particular, which are produced from kerogens as a result of thermal degradation during catagenesis. Bitumens are soluble organic compounds primarily containing carbon and hydrogen; some of bitumen is from lipids in the original organic matter.[43] Almost 95% of the ancient sedimentary organic carbon is in the form of kerogen.[5,42] At the end of diagenesis, organic matter in marine and lacustrine sediments is largely kerogen,[4,22] with the remainder bitumens.[4] The primary heteroatoms in this organic matter are sulfur, oxygen, and nitrogen. Included in the complex structure are various hydrocarbon components, such as paraffin chains.

2.5. Coal and Kerogen Macerals

Coal and kerogen contain macerals, fine-grained remains of plant material that are preserved in the sediments.[38] Macerals are divided into types that are physically distinctive and chemically different from one another. The various types of macerals are often characterized by their optical properties (reflectance, transmittance, and/or fluorescence).[4,5,40,44] They are usually subdivided into three broad categories, in order of decreasing reflectance: interinite, vitrinite, and exinite (or liptinite).[4,22,45] In reality, because the composition and evolution of macerals is not easily categorized, these categories often overlap and their boundaries are not precise.[4] The characteristics of the different macerals become progressively more similar with an increase in coal rank.[5,46]

Inertinite is unchanged by heating (i.e., it is thermally inert). Inertinite is thought to be the charred remains of woody plant material (hence its lack of response to heating). It is highly reflective when polished, has high optical absorbance, and does not exhibit fluorescence. Inertinite appears dull brown to black, and is friable. Under microscopic observation, inertinite macerals are opaque and have an angular outline.[45] Vitrinite produces fused carbon residue[4,7] and evolves some gas when heated[4]; the potential for it to produce gas is more than that of the other macerals.[5] Vitrinite is the (uncharred) remains of woody plant material.[40,47] It displays intermediate reflectance and absorbance, and usually does not exhibit much fluorescence. Physically, vitrinite has a lustrous brown to black appearance; under magnification, it displays an angular particulate structure, with cellular outlines sometimes visible.[4,40] Exinite (sometimes also called liptinite) is transformed into tar and evolves gas when heated.[4,7,46] This type of maceral is derived from lipid-rich casings and shells ("exines") of spores and pollen, resin bodies, and algal remains.[4,46] They exhibit low reflectance, low absorbance, and intense fluorescence. They appear translucent, often with red or yellow coloration.[4] Under microscopic examination, these macerals retain the characteristic shape of the materials that formed them (spores are typically flattened spheroids, and resins are translucent ovoids).[4]

Kerogens are commonly classified by several types, often associated with different macerals.[48] Plotting H/C and O/C ratios on a van Krevelen diagram[49] shows similarities between kerogen types and certain coal macerals.[5,37] True Type I kerogen is relatively rare,[5,50] and exhibits a very high (1.25–1.50 or more) atomic H/C ratio and a low (0.03–0.15) O/C ratio.[4,5,51] It contains large lipid portions, significant aliphatic groups, and low proportion of aromatic units.[4,5] This type has the highest oil producing potential compared to other kerogen types.[4,5,51] Some of the characteristics of this type of kerogen match those of liptinite-type coal macerals. The oxygen in Type I kerogen is mostly in ether[4] and ester[4,5] groups. It appears as finely laminated particles, dark and dull to the eye. Type I kerogens originate typically from alginates[4] in low oxygen environments, such as anoxic organic muds in lacustrine environments (i.e., lagoons and shallow lakes).[4,52]

The H/C ratios of Type II kerogen can be as high as 1.3, and O/C ratios as high as 0.2; both values fall as the kerogen matures. These ratios usually correspond with those of exinite macerals.[4,5] Type II is by far the most common type of kerogen. It contains more aromatic units than Type I, but aliphatic structures are still important.[4,5] Esters are important functional groups in this type of kerogen[4,5] Sulfur is often present in significant amounts in Type II kerogens.[4,5,22,51] High-sulfur Type II kerogens are often separated into a sub-type, Type II-S, when the S/C ratio exceeds 0.04.[53] Type II kerogens typically originate in marine sediments,[5] although may be formed in other environments.[4]

Type III kerogen is quite common. It is characterized by a low initial H/C ratio (<1.0) and high O/C ratio (as high as about 0.3).[4,5] These values characterize vitrinite coal macerals. Aromatic carbon content is higher than Type I or II, and polycyclic aromatic compounds, ketones, and carboxylic acids are significant in this type.[4,5] Much of the oxygen in this kerogen type is found in non-carbonyl groups, such as heterocycles, quinones, ethers, and methoxy groups.[5] Type III kerogens originate primarily from vascular plants lacking in lipids or waxy matter, and often contain identifiable plant remains.[4,5] This type of kerogen is less likely to generate oil compared to Types I and II, although it may generate gas at sufficient burial depths.[4,5] High sulfur brown coals ($S/C > 0.04$), where the H/C ratio is low ($\sim$1.00) and O/C ratio high ($>$0.20), have been classified as Type III-S kerogens.[54]

Type IV kerogen has characteristics of inertinite, as well as vitrinite macerals.[4] It has a very low H/C ratio and is composed mainly of polycyclic aromatic compounds.[22] It also has a low oxygen content[4]. It appears black and opaque, and originated in plant material that was oxidized (charred) on land, then transported to a deposition site. It provides little hydrocarbon generating potential, and is often viewed as not a "true" kerogen.[4]

2.6. Catagenesis

Catagenesis is the later phase of the evolution of carbonaceous sediments, and occurs at greater burial depths, higher temperatures, and higher pressures.[4,5] This phase of the conversion of hydrocarbon precursors to fossil fuels is entirely a chemical and physical process—the temperatures are too extreme for biological agents to survive. During catagenesis, kerogen undergoes thermal transformations, releasing hydrocarbons and other organic compounds.[6] Thermogenic natural gas production, and all petroleum production takes place during catagensis.[4–6,51] At these temperatures and pressures coals undergo thermal and chemical maturation, evolving from brown coals through various coal ranks, ultimately to anthracite.[4,5,40,46] Peat and brown coals lose considerable amounts of water starting at about 100°C.[4] As the thermochemical transformations continue, the coals become darker, harder, and shinier. There are several identified stages of coal maturation, known as ranks. In order from low to high rank, there is brown coal (lignite), sub-bituminous coal, high-volitile bituminous coal, medium volatile bituminous coal, low volatile bituminous coal, semi-anthracite, anthracite, then finally meta-anthracite.[4,5,40] As coal progresses through these ranks, the carbon content increases, and the moisture content decreases.[4,5,40] At the final stage, anthracite is very hard and shiny, and has the highest carbon content of all the ranks. Vitrinite reflectance increases throughout this maturation process and is used to identify the coal rank. As coal matures, considerable amounts of methane are generated. This generation of methane has been a major hazard for coal mining operations. Recently, coal bed methane is being increasingly exploited.

Petroleum is generated, primarily from Type II kerogen, when the sediment temperature is in the "oil window" (ca 100–150°C).[4,5,7] Type I kerogen also tends to produce petroleum, but it is relatively scarce when compared to the much more common Type II kerogen.[5] Thermogenic natural gas is the primary hydrocarbon product above these temperatures (up through about 180°C). At even higher temperatures, around 200°C, any remaining organic matter is transformed into graphite and methane (this last phase is sometimes given a separate designation, metagenesis).[6]

2.7. Asphaltene Fractions in Crude Oils

Crude oils are often separated into four fractions: saturates, aromatics, resins, and asphaltenes (SARA). Note that these resins are not plant resins mentioned earlier. Asphaltenes are defined based on their solubility characteristics. The standard solubility definition for asphaltenes is that they are insoluble in *n*-alkanes (such as *n*-heptanes), and soluble in toluene.[55,56] One definition of resins is that they are soluble in n-alkanes (such as n-heptanes), but insoluble in ethyl acetate. These solubility definitions are somewhat chemically selective; asphaltenes derived from different crude oil sources have been found to have similar chemical properties.[55,57,58]

The primary elemental constituents of petroleum asphaltenes are hydrogen and carbon. The H/C ratios of asphaltenes are approximately 1.1:1.[55,57,58] The mean molecular weight of petroleum asphaltenes is about 750 g/mole, and the mean number of fused aromatic rings is about 7.[59,60] ^{13}C nuclear magnetic resonance (NMR) and x-ray studies show that about 40% of the carbon is aromatic, and the rest saturated.[57] Fluorescence depolarization studies have shown that asphaltenes have one or two fused-aromatic ring systems in a molecule.[59,60] Infrared (IR) and NMR spectroscopy studies show that hydrogen is mostly found on saturated carbon.[61] Asphaltene molecules are shaped "like your hand" with the palm representing the core aromatic ring system and the fingers representing the peripheral alkane substituents[Chapter 2, this book].

Although asphaltenes contain a higher percentage of heteroatoms than the original oil, the heteroatomic content in asphaltenes is typically a few atomic percent.[5,62] The heteroatom content of the asphaltenes, especially the polar constituents, partially determine the chemical properties of these fractions. The small fraction of metals found in crude oils is enriched in the asphaltene fraction.[63] The more reactive sites in asphaltenes are often associated with heteroatoms. It is therefore imperative to decipher the heteroatom chemistry in asphaltenes in order to fully understand the structures of asphaltenes.

3. X-Ray Absorption Near Edge Structure (XANES)

XANES is based on the excitation of inner shell or core electrons in an atom to various resonances and to the continuum via x-ray excitation.[64] XANES probes the valence states of the atom to be studied, and thus identifies the different chemical bonding states and chemical structures in which the atom is present in the sample. The resonances in the sulfur K-edge XANES spectra can be explained with an atomic picture in mind. Often, the sulfur XANES spectrum is characterized by one narrow, intense resonance, representing the electronic transition from $1s \rightarrow 3p$ atomic levels, overlaid on a smooth background step representing the photoionization threshold of the 1s electron. It is a bit surprising that the valence shell of chemically bonded sulfur atoms can still be treated within a framework of atomic orbitals, but this appears to apply at least conceptually. Figure 6.1 shows a typical sulfur K-edge XANES spectrum, this of dibenzothiophene. The position

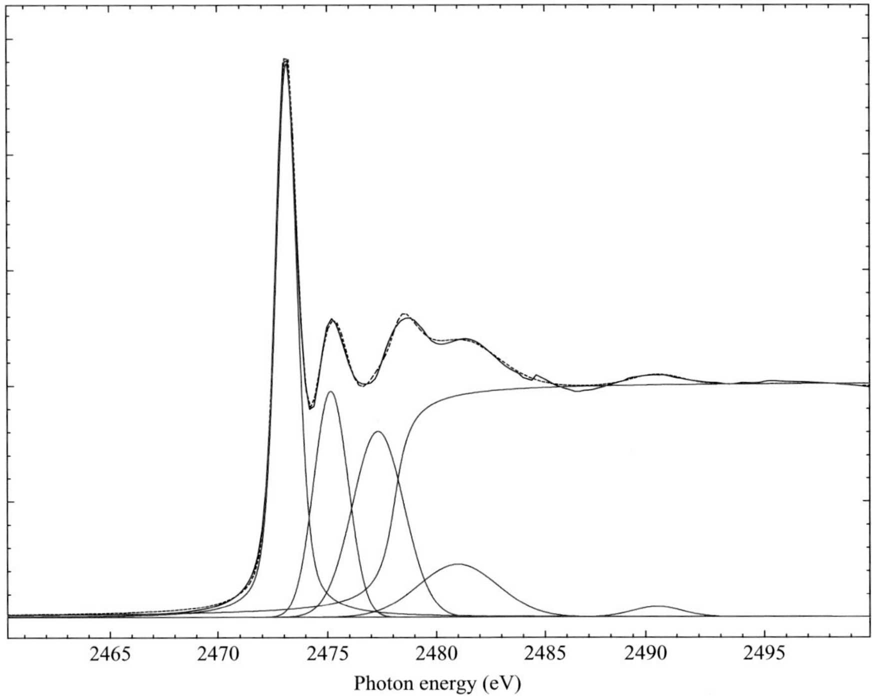

Figure 6.1. Sulfur K-edge XANES spectrum of dibenzothiophene.

and amplitude of the sulfur resonance in dibenzothiophene coincide with the resonance of other thiophene-containing compounds. The primary resonances, along with several smaller resonances, are evident in the figure.

The energy needed to ionize the K-shell electrons of atoms varies as Z^2, where Z is the atomic number of the atom. For photon energies much larger than the ionization energy, the photoionization cross-section decreases as E^3, where E is the photoelectron energy. Large oscillator strength occurs for direct excitation of resonances near the ionization threshold (within 30 eV of the edge).[64]

The probability that x-rays will be absorbed is given by μ, the absorption coefficient. If the beam intensity is reduced by a fraction $\delta I/I$ by a thin layer of the sample of thickness δx, then according to Beer's Law,

$$\frac{\delta I}{I} = -\mu \delta x = -\rho \sigma \delta x. \tag{6.1}$$

Here ρ is the density of the sample, and σ its absorption cross section.

The absorbance, A, is defined as:

$$-\log \frac{I_t}{I_0} = A. \tag{6.2}$$

Here I_t is the transmitted intensity, and I_0 is the incident intensity.

Absorption of the x-ray excites the atom, forming a vacancy in a core electronic level. Within a time frame of about 10^{-15}–10^{-13} seconds,[65,66] the atom de-excites via emission of a fluorescence x-ray or an Auger electron. In the

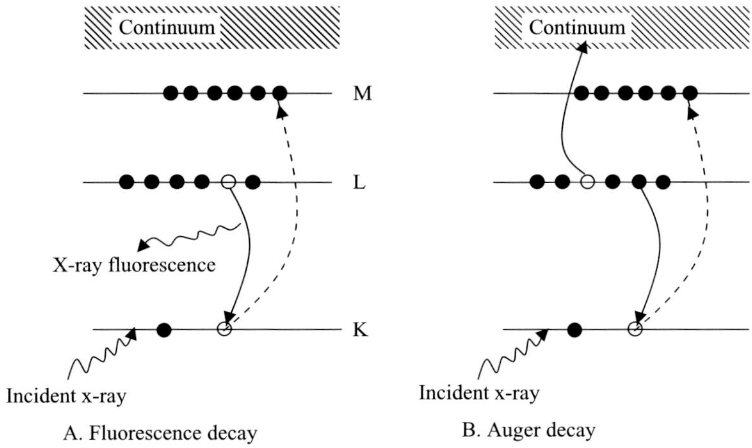

Figure 6.2. Two different decay processes for an atom with inner shell excitation.

fluorescence mode, an electron from a higher energy state de-excites into the vacant hole, with emission of an x-ray photon. In the radiationless Auger process, one electron loses energy by de-exciting to a lower energy state, coupled with another electron being ejected to the continuum. This can be viewed as inelastic scattering between two electrons in the higher energy level (Figure 6.2). With elements of high atomic number x-ray fluorescence can dominate, while for low atomic number elements the Auger process dominates.

For very thin solid samples, uniformity of the sample thickness is often impossible to achieve,[67] the varying thickness of the sample gives rise to a nonlinearly distorted absorbance, A, making interpretation difficult. Thus, determination of A by measurement of transmitted intensity is not preferred. For thick uniform samples, the transmitted x-ray intensity can be very low, and an accurate determination of A is precluded. For solids where the structures to be studied are present in low concentrations (even in the order of parts per million), the fluorescence technique is preferred for determination of A.[68]

The intensity of the fluorescence x-ray signal can be approximated by[69]:

$$I_F = \frac{\phi\, I_0 \sigma_S(E_i)}{\sigma_S(E_i) + \sigma_B(E_i) + \sigma_B(E_F)} \left[1 - \exp - \{\sigma_S(E_i) + \sigma_B(E_i) + \sigma_B(E_F)\}\right]. \tag{6.3}$$

Here, it is assumed that the fluorescence x-rays travel the same path as the incident x-rays, but in the opposite direction. ϕ is the quantum yield, σ_S and σ_B are respectively the absorption cross sections of the selected element and of all other elements of the sample. E_i and E_F are the incident and fluorescence photon energies, respectively.

For very thin samples the following simple linear term is obtained:

$$\frac{I_F}{I_0} \propto A. \tag{6.4}$$

For very thick samples, where the element of concern is present in very low concentrations, thus, the strong background approximation, $\sigma_B \gg \sigma_S$, one considers penetration of the incident beam only into a thin section of the sample. This results in a linear relationship between σ_S and the absorbance A, and Equation (6.4) applies. In the electron yield detection, the background absorption cross section, σ_B, for the emitted electron is larger than for x-ray apsorption, again resulting in a similar linear relationship.

In all the XANES results presented here, the samples have been crushed to obtain small particle sizes and diluted in appropriate solvents so that the distortions to the spectra are at a minimum. The details of the sample preparation are described later in this chapter.

4. Experimental Section

The sulfur and nitrogen XANES experiments described herein were performed at National Synchrotron Light Source, Brookhaven National Laboratory. This section of the chapter describes several aspects of the experiments.

4.1. Synchrotron Beamline

Synchrotron radiation is the intense electromagnetic radiation resulting from curved trajectories of relativistic electrons.[64] This radiation is concentrated only in the forward direction. The orbits of the accelerated electrons are maintained in storage rings, and the paths are curved by means of bending magnets.[70] The properties of synchrotron radiation include wide spectral range, very high intensity, and polarization. The national synchrotron light source (NSLS) is one of the first dedicated storage rings constructed. NSLS operates with two electron storage rings: the vacuum ultra-violet (VUV), and the x-ray rings. A linear accelerator feeds electron bunches into a 750 MeV energy booster, which, in turn, feeds the electron bunches into the two rings. The synchrotron radiation is channeled through tangential beam lines to specific experimental setups. Figure 6.3 shows a typical setup for an XANES experiment with synchrotron radiation.

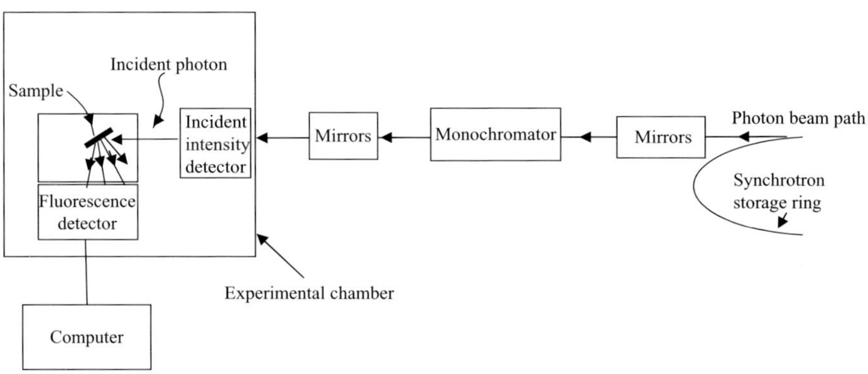

Figure 6.3. Schematic of a typical XANES experimental setup at a synchrotron radiation facility.

All beam paths are maintained at vacuum conditions of 10^{-10}–10^{-11} torr range to avoid collisions with residual gas molecules. Deterioration of the electron beam in the storage ring over time (usually over several hours), requires dumping the residual beam and then injecting a fresh batch of electrons into the ring. The electron beam duration of the hard x-ray ring is typically 20 hr, and of the VUV ring is 6 hr.

NSLS x-ray beamline X19A was used for most of the sulfur work presented here. This beamline spans an energy range of 2.1–17 KeV. Torroidal and spherical mirrors are used to collimate and focus the beam. The monochromator installed at X19A is made of Si[111] crystals, in a double-crystal configuration. As the penetration length of the hard x-rays is large, and the wavelengths are comparable to the lattice spacings of the crystals, dispersion is achieved by Bragg reflection.

For the X19A beamline, a 10 μm thick Beryllium window is the point of exit of the x-ray beam from the beamline. The sample chamber placed next to the window is purged with helium gas to reduce x-ray absorption. A Passivated Implanted Planar Silicon (PIPS) detector, made by Canberra, was used for the kerogen studies, and a Stern-Heald[71] detector was used for the asphaltene and the earlier coal studies. These detectors were used in the fluorescence mode. A Pentium PC with Windows operating system software was used to change beamline parameters when needed, and to collect XANES data.

For the nitrogen XANES work, the VUV sychrotron at the NSLS was employed. The NSLS beamline U4B was built by AT&T Bell Laboratories.[72] A grating monochromator was used for the experiments, with a grating of 600 lines/mm. The sample chamber used for the nitrogen studies was maintained at a pressure of 10^{-9}–10^{-10} torr. A multi-element Ge detector,[73] operating at liquid nitrogen temperature, was used for fluorescence measurements, the total count rates being 1,500 counts/channel. An electron channeltron detector was used for electron-yield data, operating at 1,000–1,500 V. The energy resolution at 400 eV was about 140 meV.

4.2. Samples

Our XANES study on kerogens[9] included the following samples: Indiana (IN) limestone, Bakken shale, Woodford shale, and three Green River Shale samples GR-1, GR-2, and GR-3. The Green River shale kerogens are Type I, and have been fully described by Owen.[74] GR-1 was obtained from the Exxon Colony mine near Parachute, Colorado. GR-2 was obtained from an outcrop near Rock Springs, Wyoming, from the Tipton member of the Green River formation. GR-3 is from a different part of the Mahogany zone than the Exxon Colony mine. The H/C atomic ratio in an immature Green River kerogen is typically 1.5, and the O/C ratio is less than 0.1.[37] The IN sample was a Type II kerogen and was prepared from Indiana limestone, obtained from Indiana Limestone Company. The Bakken and Woodford samples, previously described[75,76] are Type II kerogens.

All the kerogen containing samples were first pulverized, and then treated with aqueous HCl to digest the carbonates. The resulting solids were washed with toluene to remove the bitumen. The bitumen samples were also studied by nitrogen XANES. GR-3, IN limestone, Bakken, and Woodford samples were further treated with HF to digest the silicates, following a standard procedure.[77] The final kerogen samples used in these experiments were powdery, brown to black in color, and nonvolatile. These samples were ground up with mortar and pestle, and smeared over wax paper, which was then positioned in the path of the x-ray beam in the sample chamber.

Our sulfur XANES study on petroleum[10] included asphaltenes extracted from virgin stock tank crude oils, CAL and KUW2 (UG8). To obtain so-called oils, resins and asphaltenes, a solution of 40 cc of pentane per gram of crude oil was prepared. The solution was stirred for 24 hr and filtered, and the filtered residue was washed with pentane until the pentane wash was colorless. The oil fraction was then obtained by evaporating the pentane from the filtrate. The separated solid was dissolved in a small amount of toluene. After heptane was added to this in the volume ratio 40:1, the asphaltenes were filtered. The heptane solution was evaporated to obtain the resins. To obtain the XANES asphaltene spectra, the asphaltenes were ground and smeared on a sulfur-free Mylar film, or dissolved in CCl_4, and evaporated on the film. In addition to the asphaltenes, the resins and oil fractions extracted from the same two oils were also studied.

Waldo *et al.*[78] studied a suite of eight asphaltenes, extracted from crude oils from different regions around the world. These asphaltenes were obtained by *n*-heptane flocculation from various crude oils. The asphaltene content in the crude oils varied from 2 to 15%. Elemental analyses showed that the sulfur content in the asphaltenes varied from 0.68 to 7.61 wt.%.

Eight coals from the Argonne Premium Coal Sample Bank at Argonne National Laboratory,[79] were studied for both sulfur[11] and nitrogen[14] characterization. These samples were from various places in the US, and ranged in rank from lignite to low volatile bituminous. A low volatile bituminous coal was obtained from Pocahontas #3, VA (POC), a medium volatile bituminous coal from Upper Freeport, PA (UF), four high volatile bituminous coals from Pittsburg #8, PA (PITT), Lewiston-Stockton, WV (WV), Blind Canyon, UT (UT), and Illinois #6, IL, (IL), a sub-bituminous from Wyodak-Anderson, WY (WY), and a lignite from Beulah-Zap, ND (ND). The samples were stored in gas-tight, nitrogen filled capsules, and they were mounted on Mylar bags immediately after opening the capsules in air, to ensure negligible air oxidation.

Some of the sulfur model compounds studied[9–11] included pyrite, elemental sulfur, dibenzyl sulfide, diphenyl disulfide, dibenzothiophene, dibenzyl sulfoxide, 1,1′-bi-2-naphthol bis(trifluoromethanesulfonate), dibutyl sulfone, and ferrous sulfate. All of these model compounds were obtained from Aldrich Chemical Company and Alfa Aesar Company. The organic models were diluted in different organic solvents, and the inorganic models were diluted in boron nitride, both to a sulfur concentration of 0.1% by weight. The prepared mixtures were transferred to small sealed Mylar bags, which were mounted on a sample holder. The sample holders, made of poly-methylmethacrylate (PMMA), were designed to fit into the slots of the main sample chamber. All our kerogen and asphaltene spectra were energy calibrated using the elemental sulfur 1s → 3p resonance at 2471.9 eV as the standard. In the sulfur study on coals, the same resonance was shifted to 0 eV, and all other sulfur spectra were calibrated accordingly.[11] After every beam dump, the standard was run with the new beam to monitor of any shift in energy.

For the nitrogen studies on kerogens[12], GR-1, GR-2, GR-3 samples were used. The study also included bitumens extracted from the GR-1 and GR-2 samples, and from New Albany (NAB) shale. The coal sample suite consisted of the same suite of eight coals as in the sulfur study.[14] The asphaltene fractions were obtained from two crude oils, CAL and UG8.[13] The kerogen and coal samples were mounted on nitrogen-free double sticking tape, and the asphaltene samples were dissolved in toluene and evaporated on the tape.

The set of nitrogen model compounds included five pyridine analogues, four pyridone analogues, four pyrrole analogues, two aromatic amines, two porphyrin compounds, two pyridinium analogues, and two saturated amines. These samples were obtained from Aldrich Chemical Company and Alfa Aesar Company. All the model compounds were in powder forms, and were mounted on nitrogen-free double-sticking tape.

4.3. Least Squares Fitting Procedure

A least squares fitting routine is employed to fit the XANES spectra. The fitting procedures were performed using either Apple-based KaleidaGraph software, or Windows-based WinXAS software.[80] First, the background was subtracted from all spectra using a linear fit of pre- and post-XANES contributions. The spectra were then normalized with respect to the individual step heights. The step magnitude, representing excitation of the core electron to the continuum, is independent of the particular chemical identity of the element in the molecule. Each XANES spectrum was fitted with the requisite number of peaks and one arctangent step. The parameters of the arctangent function were held constant for all sulfur functional groups, except in the cases of sulfone, sulfonate and sulfate. For these oxidized samples a high energy edge jump was used. The spectra of known structures were first fitted, and then the spectra from carbonaceous samples were fitted using similar energy positions as the corresponding features in the model spectra. Figure 6.1 shows an example of the fitting procedure, where the spectrum of dibenzothiophene is fitted with several peaks and an arctangent step.

Different relative percentages of the various sulfur and nitrogen structures in the carbonaceous materials were then calculated using the results of the areas from the different resonances in the model spectra. If one known model structure exhibits a secondary resonance in the XANES spectrum, care was taken to subtract off the secondary peak contribution due to this model in the spectrum of carbonaceous materials before calculating the contribution from other molecular structures. All the spectra were analyzed within the energy range where the signature resonances of all the possible model spectra occur. Figure 6.4 shows a spectrum of a kerogen, GR-3, fitted with resonances representing elemental sulfur (or pyrite), sulfide, thiophene, sulfoxide, sulfonate, and sulfate. The widths and

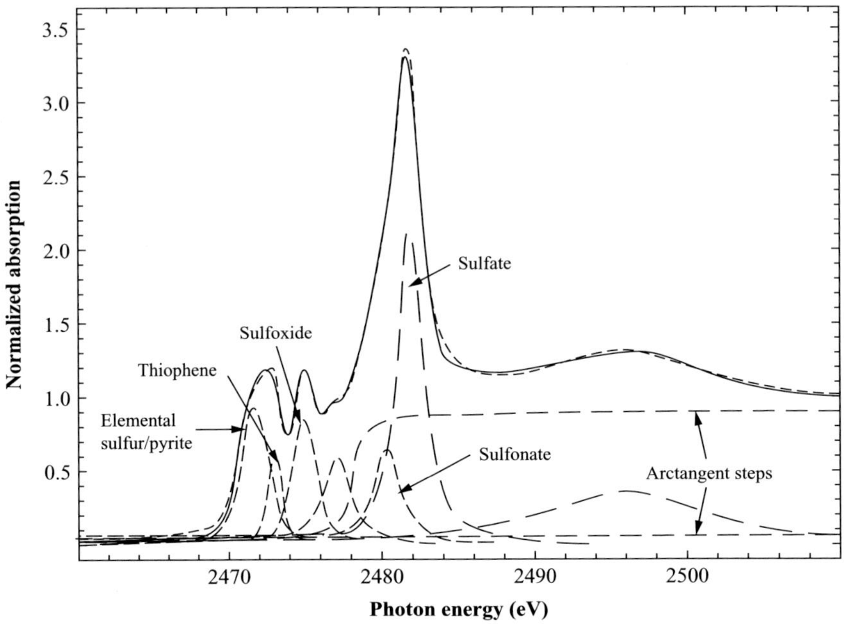

Figure 6.4. Least squares fitted spectrum of a sulfur K-edge XANES spectrum of kerogen. The resonances represent, in order of increasing energy, elemental sulfur/pyrite, sulfide (negligible), thiophene, sulfoxide, sulfonate, and sulfate. The unlabeled resonances represent secondary peaks of reduced sulfur species.

the positions of the resonances and the steps were held similar for all the kerogen spectra. The unlabeled peaks in Figure 6.4 correspond to secondary resonances of reduced sulfur species.

In the sulfur case, there is some amount of uncertainty in the exact amounts of the reduced forms of sulfur. This is due to the small energy separations of the $1s \rightarrow 3p$ resonances between elemental sulfur (and/or pyrite) and sulfides, and between sulfides and thiophenes, making it difficult to resolve the two signature peaks.

5. Results and Discussions

Figure 6.5 shows the sulfur K-edge XANES spectra of several sulfur models, as studied by Waldo *et al.*[78] With an increase in formal oxidation number of sulfur, the position of the strong resonance moves to higher photon energies.[9–11,78,81] In addition, the amplitude of the 1s–3p peak grows with higher sulfur oxidation states due to increasing 3p vacancy. Table 6.1 lists the energy positions of major resonances in the sulfur K-edge XANES spectra of different structures. There is a difference of about 10 eV of energy between the $1s \rightarrow 3p$ resonance energy from pyrite (formal oxidation number of -1) to sulfate (formal oxidation number of $+6$). Spectra of sulfur structures belonging to the same chemical species, comparatively, show little difference in the position of this major resonance. Even though aliphatic sulfide and thiophene have the same formal oxidation state, thiophene

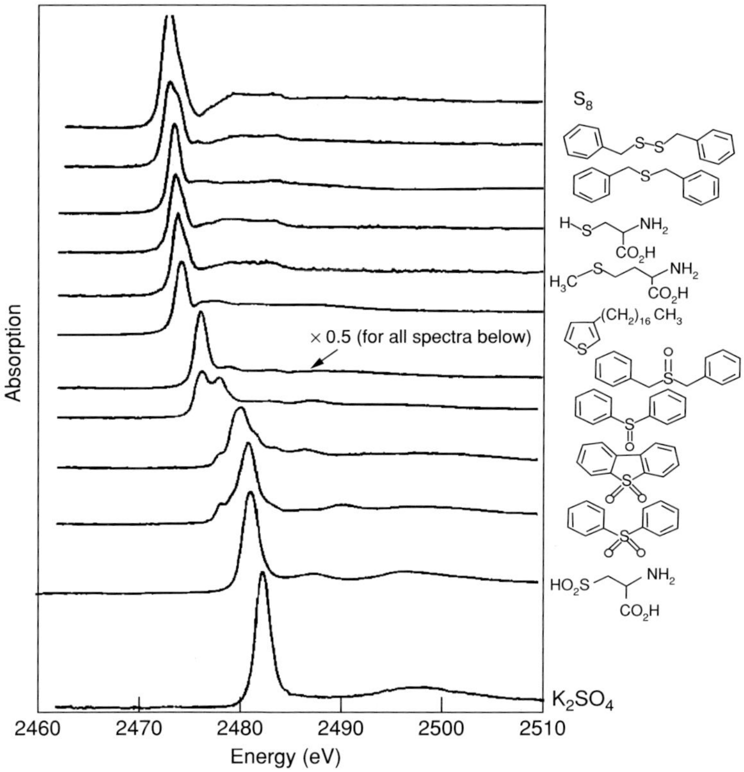

Figure 6.5. Sulfur K-edge XANES for different sulfur model compounds. The 1s → 3p resonance shifts to higher energies as the formal oxidation of sulfur increases to higher values.[78,81]

has higher 1s → 3p energy than sulfide. This can be explained due to aromatic delocalization.[14] The electron density of sulfur in thiophene is reduced probably due to delocalization of the $3p_z$ electrons in the thiophene ring; a similar occurrence has been observed in the case of nitrogen.[12–14]

Table 6.1. Energies of the 1s → 3p Resonances for Different Sulfur Compounds[a]

Compounds	Formal oxidation number of sulfur	1s → 3p position (eV)
Pyrite	−1	2471.5
Elemental sulfur	0	2472.0
Diphenyl disulfide	0	2472.2
Dibenzyl sulfide	0	2472.7
Dibenzothiophene	0	2473.3
Dibenzylsulfoxide	2	2475.4
Dibutyl sulfone	4	2479.5
Sulfonate	5	2481.1
Sulfate	6	2482.1

[a]The Structures are listed in order of the formal oxidation number of sulfur.

5.1. Sulfur XANES on Kerogens

Figure 6.6 shows the sulfur K-edge XANES spectra of six kerogen samples. Numerous variations in the intensities of the spectral features among the different samples are evident, even among the Green River Shale samples, which belong to the same parent formation. The results of a least squares fitting procedure of the relative abundances of the different sulfur chemical structures are tabulated in Table 6.2. There are systematic differences between the functional groups in Type I and Type II kerogens. The outstanding difference is in the thiophenic content between the Type I and Type II kerogens. All the Green River samples, which belong to the Type I category, have a smaller thiophenic fraction than the Woodford, Bakken and IN limestone samples, which belong to Type II. This is in accord with the fact that the carbon aliphatic/aromatic ratio in kerogens decreases from Type I to Type II.[5] The differences in sulfur chemistry, therefore, reflect the differences in carbon chemistry between Type I and II kerogens. Sulfide is abundant in all of the samples except in IN limestone. An earlier XANES study of a Kimmeridge Type IIS (sulfur rich Type II) kerogen showed that thiophene was the dominant sulfur functionality.[82]

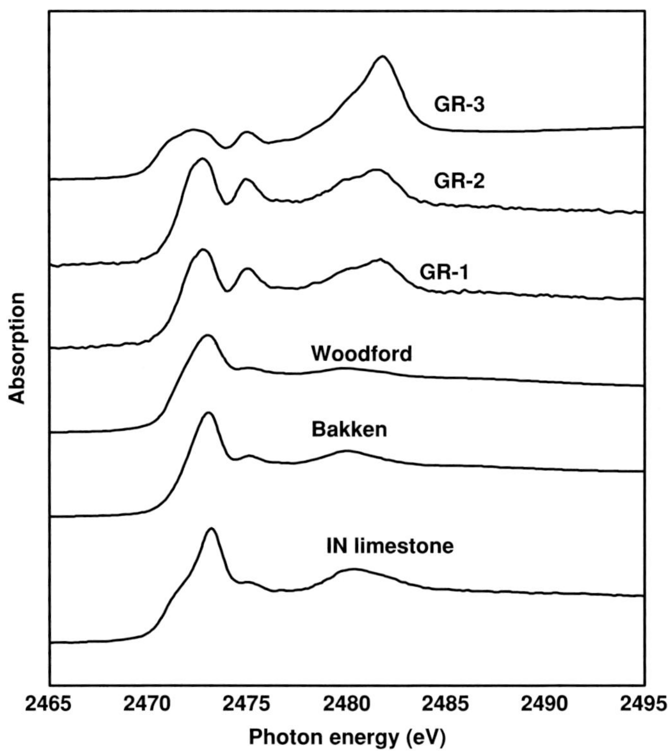

Figure 6.6. Sulfur K-edge XANES spectra of several kerogens samples.

Table 6.2. Relative Abundances of Sulfur Structures Found in Different Kerogens

Kerogen (Type)	Sulfur	Sulfide	Thiophene	Sulfoxide	Sulfonate	Sulfate
IN limestone (II)	29	2	59	2	9	<1
Bakken (II)	11	8	70	2	0	7
Woodford (II)	19	9	71	0	0	0
GR-1 (I)	9	22	39	8	9	11
GR-2 (I)	14	23	39	12	3	9
GR-3 (I)	31	4	22	7	11	24

In IN limestone most of the oxidized form remains in higher sulfur oxides, with negligible amounts (within our error bounds) in the sulfoxide form. All the kerogen samples have reduced inorganic sulfur as well; elemental sulfur and/or pyrite. This is especially evident in IN limestone and GR-3. In GR-3, considerable amounts of sulfate are also found. In an earlier XANES study on humic substances,[83] it was found that sulfides, sulfonates and sulfates were the major contributors, and in the marine samples, both sulfonates, and sulfides were abundant.

This small amount of sulfide estimated to be present in IN limestone may be due to indeterminacy in the analysis; the 1s $\rightarrow$ 3p resonance of sulfide is only 0.2 eV higher than that of elemental sulfur, and there is only a separation of about 0.5 eV between the 1s $\rightarrow$ 3p resonances of pyrite and elemental sulfur. Earlier studies of high sulfur kerogens did not report examining the possibility that there might be elemental sulfur in their samples.[82,84,85] In coals sometimes pyrite gets oxidized into elemental sulfur.[86] In kerogen, the occurrence of elemental sulfur could be a result of oxidation of pyrite, or of either thermochemical reduction or biological reduction of sulfate.[83] This quandary might be resolved with L-edge XANES studies where the reduced forms of sulfur have well separated 1s $\rightarrow$ 3p resonances.

5.2. Sulfur XANES on Oil Fractions

Sulfur K-edge XANES studies on petroleum and its fractions have been very successful over the last two decades.[10,78,81] Figure 6.7 shows sulfur XANES spectra of asphaltene, resin, and oil fractions from two different crude oils, CAL and UG8, as well as several sulfur model compounds. First, it is evident from even

Table 6.3. Relative Abundances of Different Sulfur Species in the Asphaltene, Resin, and Oil fractions from Two Crude Oils

Samples	Sulfide	Thiophene	Sulfoxide	Sulfone	Sulfate
CAL Asph	15	29	50	5	1
CAL Resin	11	27	59	1	1
CAL Oil	24	27	46	1	1
UG8 Asph	40	55	2	1	1
UG8 Resin	40	52	5	1	1
UG8 Oil	45	47	5	1	1

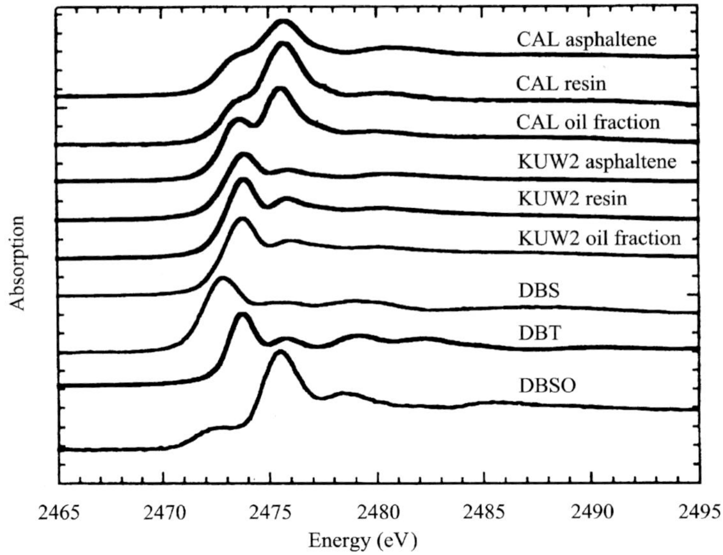

Figure 6.7. Sulfur K-edge spectra of asphaltenes, resins, and oil fractions. These spectra are compared with those of dibenzyl sulfide, dibenzyl thiophene, and dibenzyl sulfoxide.

the raw data that all the sulfur in the asphaltene exists in the form of thiophene, sulfide, and/or sulfoxide. Second, a major difference is seen between the two suites of crude oil components. In the CAL suite, the sulfoxide peak dominates in all the fractions. In contrast, sulfoxide is a minor component in the UG8 samples.

Table 6.3 shows the relative abundances of sulfur forms in the different crude oil components from the two oils. In all three UG8 samples, thiophene is most dominant, followed by sulfide. Likewise the CAL samples are all fairly similar to each other. The oil fractions from both CAL and UG8 show slightly higher sulfide content than the other fractions, perhaps since the carbon is known to be less aromatic in the oil fractions than in the heavier ends. Furthermore, the UG8 samples exhibit smaller fractions of thiophene than the CAL samples. The sulfoxide fraction in CAL is thought to be primarily alkyl sulfoxide.[78] This is consistent with the fact that CAL has low maturity, being reservoired in shallow depths. All three fractions of CAL show large sulfoxide fractions. Even though sulfoxide is very polar, it is not limited to the asphaltene fraction. Mitra-Kirtley *et al.* studied the same asphaltene samples as Waldo *et al.* after almost 7 years and the resulting agreement was very good. No change in the oils or the asphaltenes took place over this span, thereby limiting possible spurious explanations of CAL sulfoxide formation.

5.3. Sulfur K-Edge XANES on Coals

Sulfur K-edge XANES methodology of analyzing coals and coal macerals has been exploited for many years.[11,87,88] Figure 6.8 shows the deconvoluted spectrum of one of the coals,[11] where the zero of the energy is taken at the $1s \rightarrow 3p$ resonance of elemental sulfur. Table 6.4 lists the results of this coal analysis in

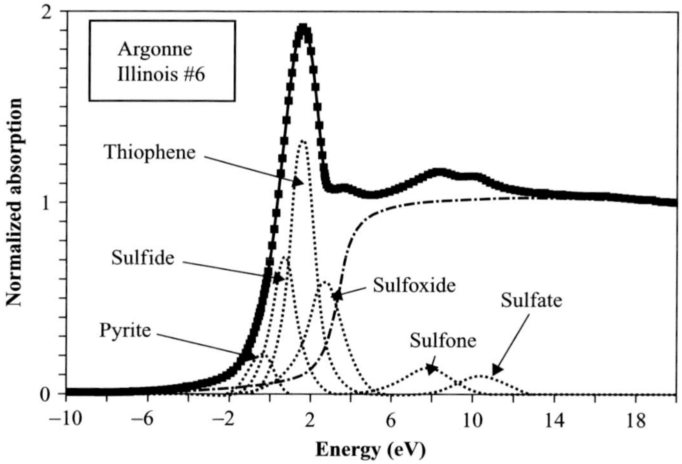

Figure 6.8. Sulfur K-edge spectrum of coal. The fitting procedure used resonances representing, in order of increasing energy, pyrite, sulfide, thiophene, sulfoxide, sulfone, and sulfate.[11]

order of decreasing rank. It is evident that the thiophene fraction of organic sulfur increases and the sulfide fraction decreases with increasing coal rank. This is again similar to the carbon chemistry in that the higher rank coals are more aromatic than lower rank coals. Thiophenic sulfur is the most dominant form of sulfur in all the samples. The coals also exhibit some inorganic sulfur. The pyrite content agreed well with independent Mossbauer spectroscopy results on the same coals.[89] Beulah Zap coal shows a large percentage of sulfate, believed to be present as gypsum.

Several coal macerals were also studied by XANES analysis,[11] and the exinites were found to be richer in sulfides than the vitrinites and the inertinites (Table 6.5).[11] This is significant, as exinites are less aromatic than vitrinites and inerinites, another instance when the sulfur chemistry is in parallel to the carbon chemistry. Many of the maceral separates were found to contain pyrite. It is interesting to note that the inertinites often have a lower thiophene to sulfide ratio

Table 6.4. Relative Percentages of the Different Sulfur Structures Found in Coals Listed In Order of Coal Rank[11]

Coals in decreasing rank	Pyrite	Sulfide	Thiophene	Sulfoxide	Sulfone	Sulfate
Pocahontas No. 3, VA	24	0	75	*	1	*
Upper Freeport, PA	62	6	32	*	1	*
Pittsburgh #8, PA	52	13	35	*	1	*
Lewis Stockton, WV	26	16	56	1	1	*
Blind Canyon, UT	40	15	46	*	*	*
Illinois #6, IL	48	19	33	*	*	*
Wyodak-Anderson, WY	24	29	46	*	*	*
Beulah Zap, ND (fresh)	29	28	30	2	*	12
Beulah Zap, ND (old)	37	24	30	2	*	7

*Negligible quantities, less than 1%.

Table 6.5. Pyrite, Sulfide, Thiophene Percentages in Several Coal Macerals[11]

Samples	Pyrite	Sulfide	Thiophene	Thiophene/Sulfide
PSOC 733 PA, Appale, hvAb				
Exinite		32	62	1.93
Vitrinite		13	76	5.85
Inertinite		17	70	4.12
W. Ky. No. 9 hv Ab				
Exinite	29	20	38	1.90
Vitrinite	29	17	41	2.41
Inertinite	56	12	23	1.92
IL #6				
Floated coal	34	21	38	1.81
Sporinite		43	44	1.02
Vitrinite		23	73	3.17
Inertinite	59	12	23	1.92
PSOC 1110 UT, So-W. SUBC				
Exinite	11	23	37	1.61
Vitrinite	9	23	42	1.83
PSOC 1111 UT So-W. mvb				
Exinite		26	53	2.04
Vitrinite		27	56	2.07
Inertinite	7	23	50	2.17
Texas Lignite				
Floated coal	49	18	30	1.67
Sporinite		46	54	1.17
Vitrinite	12	28	55	1.96
Inertinite	73	10	14	1.40
Wyodak				
Micronized	27	25	45	1.80
Sporinite		41	56	1.37
Vitrinite		27	72	2.67
Inertinite	72	10	16	1.60

hvAb: high volatile A bituminous coal
SUBC: subbituminous C coal
mvb: medium volatile bituminous coal
These coals and maceral separates are not part of the Argonne Premium Coal Sample Bank suite of coals mentioned earlier.

than the vitrinites thus not congruent with the increased carbon aromatization of inertinite. This higher sulfide content in inertinites often accompanies very high pyrite content in the inertinite. Perhaps some inorganic sulfur of the inertinite is reacting with organics producing some sulfide.[35] Some of the macerals showed the presence of sulfur oxygen groups, such as sulfoxide, sulfone, and sulfates.

5.4. Nitrogen XANES

XANES methodology in analyzing the nitrogen content in carbonaceous samples has also been very successful. Mullins[55] has provided an overview of the nitrogen XANES research done on carbonaceous samples. Unlike the sulfur

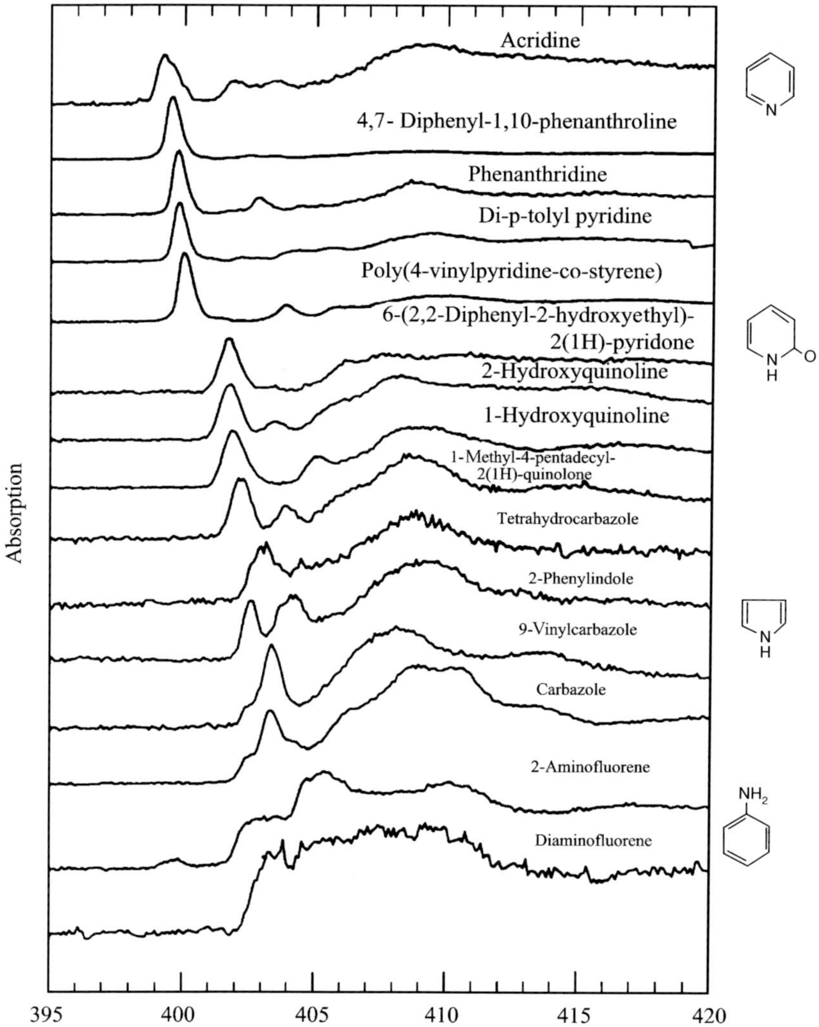

Figure 6.9. Nitrogen XANES spectra of several nitrogen model compounds. The spectra from similar structures are similar, and well separated from those belonging to other structures.

case, all the nitrogen in these samples is found to be in organic forms. The major resonances in the nitrogen XANES spectra are attributed to the different $1s \rightarrow \pi^*$, and $1s \rightarrow \sigma^*$ transitions.[12–14,90] That is, molecular orbitals need to be considered for nitrogen while atomic orbitals largely suffice conceptually for sulfur. In the aromatic molecules, several π^* transitions may be evident, whereas saturated structures exhibit only σ^* transitions. Figure 6.9 shows the nitrogen XANES spectra of several model compounds. The signature peaks of pyridine, pyrrole, and aromatic amine structures are well separated, and this helps in easy analysis of the carbonaceous samples. The nitrogen in pyridine and in pyrrole has the same formal oxidation state. However, the lone pair of electrons on the nitrogen in pyridine is in an sp^2 orbital and not involved in the aromatic bonding. The localization of the

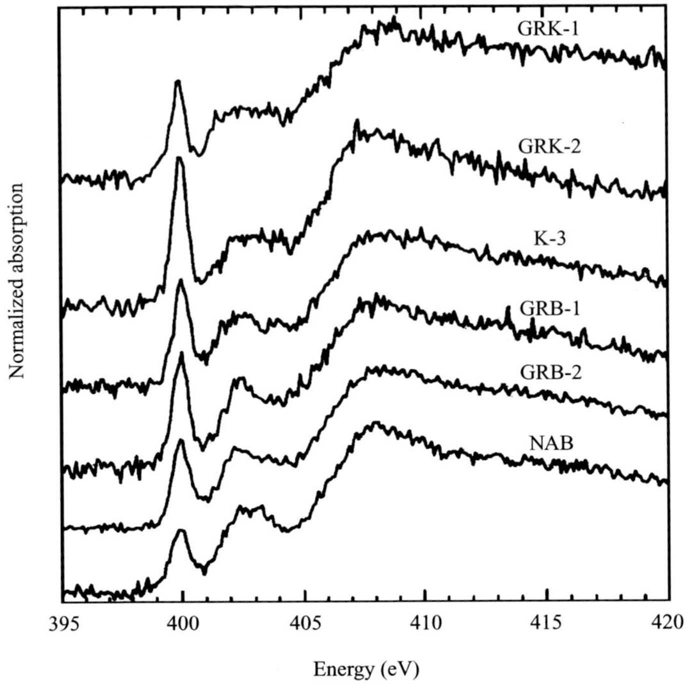

Figure 6.10. Nitrogen XANES spectra of several kerogens and bitumens.

lone pair on the nitrogen, along with the high electronegativity of nitrogen, results in a slightly negative charge on the nitrogen in the pyridine. In contrast, in the pyrrole structure, nitrogen does not possess a formal double bond. The lone pair of electrons on nitrogen in a pyrrole structure is in the p_z orbital, and is present in a 5-membered ring system. The pyrrole nitrogen donates two electrons to the π system thereby reducing its electron density. Thus, the pyridinic nitrogen resonance has a lower energy π^* resonance than pyrrole.[55]

Kerogens and bitumens have also been studied by nitrogen XANES methods.[12] Figure 6.10 shows the nitrogen XANES spectra of three oil shale

Table 6.6. Relative Abundances of Nitrogen Structures in Kerogens and Bitumens

Samples	Pyridine	Pyridone[a]	Pyrrole	Aromatic amine	Saturated amine	% N
Kerogen GRK-1	20	23	42	15	0	1.09
Bitumen GRB-1	27	11	49	11	2	0.83
Kerogen GRK-2	34	11	42	2	11	1.03
Bitumen GRB-2	20	19	49	12	0	1.47
Kerogen K-3	24	20	45	10	0	1.81
Bitumen NAB	14	11	67	8	0	2.60
Bitumen NAB 21% porphyrin[b]	6	11	55	9	0	2.60

[a] Includes pyridinium, private communication O.C. Mullins, S. Mitra-Kirtley, and S.P. Cramer.
[b] Porphyrin fraction estimated from V, Ni content.

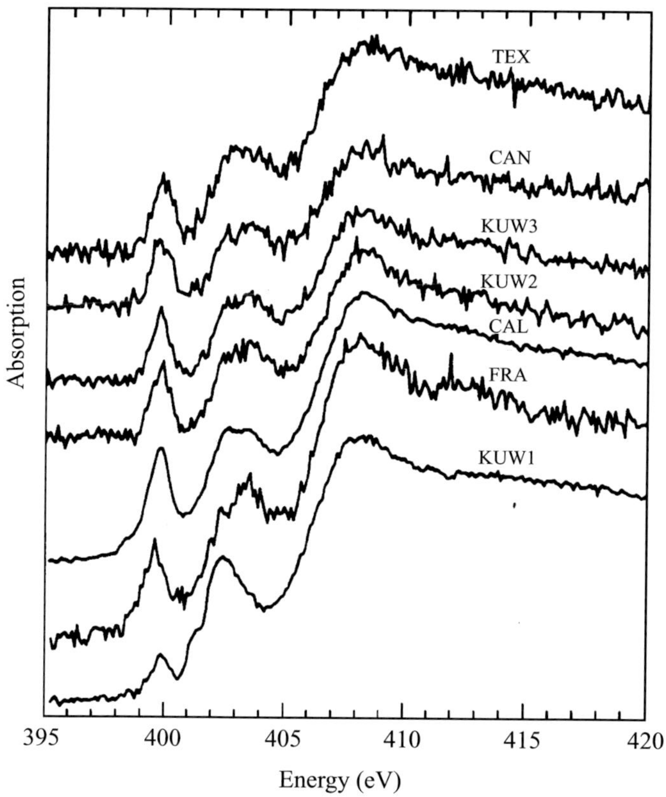

Figure 6.11. Nitrogen K-edge XANES spectra of several asphaltenes.

kerogens and three bitumens. Table 6.6 shows the results of the least squares fitting procedure for kerogens and bitumens. Most of the nitrogen is present in aromatic forms; pyrrole, followed by pyridine, are the most dominant nitrogen structures in the kerogens and bitumens. The pyridone and aromatic amine fractions in these samples are small, and in one sample the amount of saturated amine is noticeable. The pyridone 1s-π^* resonance overlaps the same resonance for pyridinium. The occurrence of quinolones has been found in earlier studies to be present in crude oils and asphalts.[91,92] One of the bitumens, extracted from New Albany shale, has a considerable amount of porphyrin. Porphyrin has half pyrridinic half pyrrolic nitrogen.[13] In this sample, the porphyrin percentage was first estimated using the V and Ni analysis of the sample. It was assumed that all of the V and Ni were complexed with porphyrin structures, and from the nitrogen mass fraction, the amount of nitrogen in porphyrin structures was calculated.

Figure 6.11 shows the nitrogen XANES of seven asphaltenes.[13] Upon comparison of the two figures, Figures 6.9 and 6.11, the pyridine and pyrrole peaks are clearly visible even in the raw data from asphaltenes. The pyrrolic peak widths in the asphaltene spectra are larger than for individual pyrrolic model compounds, probably as a result of the presence of a large number of contributing pyrrole analogues. Table 6.7 tabulates the results of the nitrogen analysis of the asphaltenes.

Table 6.7. Relative Abundances of Nitrogen Molecular Structures in Several Asphaltenes

Asphaltene samples	Pyrrole	Pyridine	Saturated amine
FRA ST1	76	22	3
Kw1 BG5 7% porphyrin[a]	87	2	2
Can Sales	74	26	0
Kw3 W40	77	22	2
Kw2 UG8	79	22	3
Kw2 UG8[b]	71	26	3
Tex Frio	80	18	0
CAL 3% porphyrin[a]	65	30	2

[a] Estimated from the metal content.
[b] Obtained from electron yield detection. All other data by fluorescence yield.

Almost all of the nitrogen found in the asphaltenes studied is aromatic. There is little evidence of saturated amines. Pyrrolic nitrogen in these samples is uniformly predominant, and the smaller pyridine fraction varied considerably. These asphaltenes are the same set for which the sulfur analysis was performed. CAL

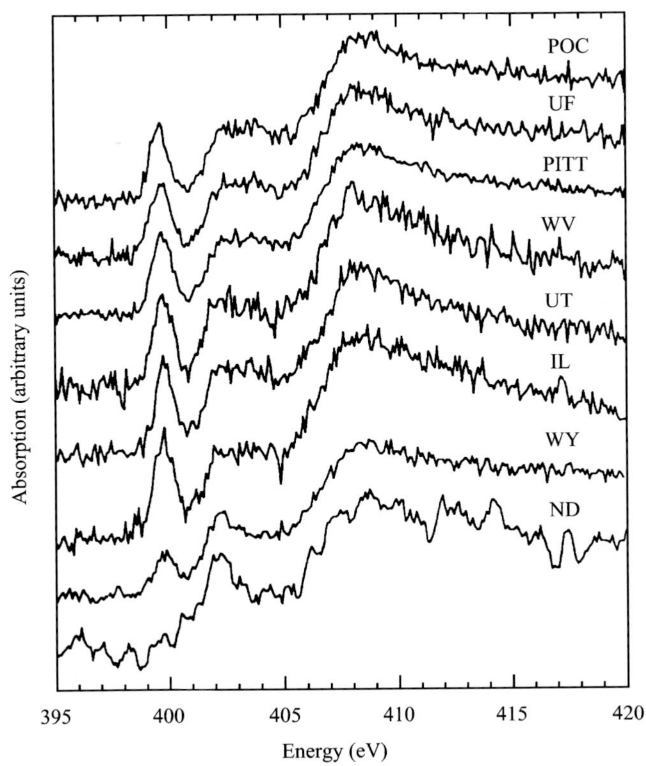

Figure 6.12. Nitrogen XANES spectra of different coals. As the coals increase in rank (bottom to top), the pyridone signature vanishes and the pyridine signature becomes more prominent.

Table 6.8. Relative Amounts of Different Nitrogen Structures Present in the Suite of Eight Argonne Premium Coals

Sample	Pyridine	Pyridone[a]	Pyrrole	Aromatic amine
POC	18	8	66	8
UF	18	8	66	8
PITT	20	6	65	9
WV	20	16	55	9
UT	17	15	60	8
IL	20	19	54	7
WY	10	29	51	10
ND	2	42	50	6

[a] Includes pyridinium.

asphaltene showed the smallest pyrrole and largest pyridine fractions. While some pyridone was found in kerogens and bitumens, none was found in the asphaltenes. This suggests that maturation probably resulted in conversion of pyridone to pyridine in the asphaltenes.

Figure 6.12 shows the XANES spectra of the eight Argonne premium coals[14] with descending rank from the top to the bottom. Table 6.8 shows the relative percentages of the different structures present in these coals. Pyrroles, followed by pyridines were again found to be the most dominant, and some amounts of pyridones and aromatic amines were also found. As the rank of the coals decreased, less pyridine and more pyridone were evident. This is in accord with the fact that with an increase of coal rank, the amount of oxygenated molecules decreases. The inverse correlation between pyridine and pyridone fractions as a function of coal rank is apparent even in the raw XANES data. Small amounts of quarternary nitrogen were also found in coal by an x-ray photoelectron spectroscopy (XPS) study.[94] The XANES π^* resonance of quarternary nitrogen overlaps with that of pyridone, hence this methodology is unable to differentiate between contributions from the two structures with large accuracy.

6. Conclusion

XANES spectroscopy has proved to be a very versatile analysis tool for identifying and quantifying different heteroatom structures present in complex systems. Sulfur and nitrogen molecular structures in coals, macerals, kerogens, bitumens, and asphaltenes have been elucidated by XANES methodology, and provide important information about the formation and maturation of these important carbonaceous samples. In all the three types of carbonaceous samples, kerogens, coals, and oil fractions, the primary organic sulfur structures are in thiophenic and sulfidic forms. A few samples had significant sulfoxide and higher sulfur oxide fractions. In all these samples, the sulfur chemistry mimics the carbon chemistry. The more aromatic samples in each of these categories systematically show larger amounts of thiophenic sulfur. The coals studied have a wider range of thiophenic content than the kerogens or the asphaltenes. Among the coal macerals, sulfidic sulfur is found to be larger in exinites (associated with Type II kerogens) than in the vitrinites (associated with Type III kerogens) and inertinites (associated with Type IV kerogens).

The primary difference in the distribution of sulfur between coals, kerogens, and asphaltenes, is the presence of inorganic sulfur in coals and kerogens and not in asphaltenes. The inorganic sulfur in the coals is mostly in the form of pyrite, and in the kerogens in the form of pyrite, and/or elemental sulfur, and sulfate. Significant variation in these fractions is found for the three Type I GR shale kerogens. One coal has considerable amounts of sulfate, possibly due to the presence of gypsum in the sample.

Nitrogen XANES analysis of carbonaceous samples shows that almost all of the nitrogen in these samples is aromatic and virtually all of the nitrogen is organic. Pyrrolic nitrogen dominates almost all samples. Low maturation coals exhibit a pyridone fraction which is likely converted to pyridine in higher rank coals. In coals and possibly kerogens small amounts of pyridinium are also present.

K-edge XANES analysis methods have been very fruitful for the analysis of sulfur and nitrogen structures in complex carbonaceous samples. This method provides a robust, yet nondestructive tool for both qualitative and quantitative analysis of the heteroatom chemistry in these samples. In the future, third generation synchrotron beamlines, with high resolution and greater photon flux, would make the identification of peaks easier and improve the signal to noise ratio.

References

[1] Hedges, J.I. and F.G. Prahl (1993). Early diagenesis: Consequences for applications of molecular biomarkers. In: M. Engle and S. Macko (eds), *Organic Geochemistry: Principles and Applications.* Plenum, New York.

[2] Eglington, G. (1969). The organic chemist's approach. In: G. Eglington and M.T.J. Murphy (eds.), *Organic Geochemistry*. Springer-Verlag, New York.

[3] Welte, D.H. (1969). Organic matter in sediments. In: G. Eglington and M.T.J. Murphy (eds.), *Organic Geochemistry*. Springer-Verlag, New York.

[4] Killops, S.D. and V.J. Killops (1993). *An Introduction to Organic Geochemistry*, 2nd edn. Longman Scientific and Technical.

[5] Tissot, B.P. and D.H. Welte (1984). *Petroleum Formations and Occurrence*. Springer-Verlag, New York.

[6] Rullkotter, J. (1993). The thermal alteration of kerogen and the formation of oil. In: M.H. Engel, and S.A. Macko (eds.), *Organic Geochemistry: Principles and Applications*. Plenum, New York.

[7] White, W.M. (2005). Organic geochemistry. In: *Geochemistry* (an on-line textbook). Johns Hopkins University Press, in press.

[8] International Committee for Coal petrology (1963). *International Handbook of Coal Petrology*, 2nd edn. 1963, 1st Supplement 1971. Centre National de la Recherche Scientifique, Paris.

[9] Wiltfong, R., S. Mitra-Kirtley, O.C. Mullins, B. Andrews, G. Fujisawa, and J. Larsen (2005). Sulfur speciation in different kerogens by XANES spectroscopy. *Energy & Fuels* **19**, 1971.

[10] Mitra-Kirtley, S., O.C. Mullins, C.Y. Ralston, D. Sellis, and C. Pareis (1998). Determination of sulfur species in asphaltene, resin, and oil fractions of crude oils. *Appl. Spectrosc.* **52**(12), 1522.

[11] Huffman, G.P., S. Mitra, F.E. Huggins, N. Shah, S. Vaidya, and F. Lu (1991). Quantitative analysis of all major forms of sulfur in coal by x-ray absorption fine structure spectroscopy. *Energy & Fuels* **5**, 574.

[12] Mitra-Kirtley, S., O.C. Mullins, J.F. Branthaver, and S.P. Cramer (1993). Nitrogen chemistry of kerogens and bitumens from x-ray absorption near-edge structure spectroscopy. *Energy & Fuels* **7**(6), 1128.

[13] Mitra-Kirtley, S., O.C. Mullins, J. van Elp, S.J. George, J. Chen, and S.P. Cramer (1993). Determination of the nitrogen chemical structures in petroleum asphaltenes using XANES spectroscopy. *J. Am. Chem. Soc.* **115**, 252.

[14] Mullins, O.C., S. Mitra-Kirtley, J. van Elp, and S.P. Cramer (1993). Molecular structure of nitrogen in coal from XANES spectroscopy. *Appl. Spectrosc.* **47**, 1268.

[15] Parsons, T.R., M. Takahashi, and B. Hargrave (1977). *Biological Oceonographic Processes*. Pergamon Press, New York.

[16] Ryther, J.H. (1963). Geographic variations in productivity. In: M.N. Hill (ed), *The Sea.* Wiley, New York.

[17] Geider, R.J. et al. (2001). Primary productivity of planet earth: Biological determinants and physical constraints in terrestrial and aquatic habitats, *Global Change Bio.* **7**, 849.

[18] Hollander, D., F. Behar, M. Vanderbroucke, P. Bertrand, and J.A. McKenzie (1990). Geochemical alteration of organic matter in eutrophic Lake Greifen: Implications for the determination of organic facies and the origin of lacustrine source rocks. *Am. Assoc. Petroleum Geologists Stud. Geol.* **30**, 181.

[19] Canfield, D.E. (1989). *Deep Sea Research* **36**, 121.

[20] Glenn, C.R. and M.A. Arthur (1985). *Chem. Geol.* **48**, 325.

[21] Premuzic, E.T., C.M. Benkovitz., J.S. Gaffney, and J.J. Walsh (1982). *Org. Geochem.* **4**, 63.

[22] Tyson, R.V. (1995). *Sedimentary Organic Matter*. Chapman and Hall, New York.

[23] Hartnett, H.E., R.G. Keil, J.I. Hedges, and A.H. Devol (1998). Influence of oxygen exposure time on organic carbon preservation in continental margin sediments, *Nature* **391**, 572.

[24] Berner, R.A. (1980). *Early Diagenesis*. Princeton University Press, New Jersey.

[25] DeVooys, C.G.N. (1979). Primary production in aquatic environments. In: B. Bolin, E.T. Degens, S. Kempe, and P. Ketner (eds.), *The Global Carbon Cycle*. Wiley, Chichester.

[26] Meyers, P.A. and R. Ishiwatari (1993). The early diagenesis of organic matter in Lacustrine sediments, In: M.H. Engel, and S.A. Macko (eds), *Organic Geochemistry: Principles and Applications*. Plenum, New York.

[27] Bluck, B.J. (1969). Introduction to sedimentology, In: G. Eglington, and M.T.J. Murphy (eds), *Organic Geochemistry: Methods and Results*. Springer-Verlag, New York.

[28] Nedwell, D.B. (1984). The input of mineralization of organic carbon in anaerobic aquatic sediments, *Adv. Microb. Ecol.* **7**, 93.

[29] Nealson, K.H. and D. Saffarini (1994). Iron and manganese in anaerobic respiration: Environmental significance, physiology, and regulation, *Ann. Rev. Microbiol.* **48**, 311–343.

[30] Ward, D.M. and M.R. Winfrey (1985). Interactions between methanogenic and sulfate-reducing bacteria in sediments, *Adv. Aquatic Microbiol.* 3:141.

[31] Widdel, F. (1986). Sulfate-reducing bacteria and their ecological niches in anaerobic bacteria. In: E.M. Barnes, and G.C. Mead (eds), *Habitats Other than Man*, Blackwell, Oxford, p. 157.

[32] Jorgensen, B.B. (1982). Mineralization of organic matter in the sea bed—the role of sulfate reduction, *Nature*, **296** (Apr.), 15.

[33] Hartgers, W.A. J.F. Lòpez, J.S. Sinninghe Damsté, C. Reiss, J.R. Maxwell, J.O. Grimalt (1997). Sulfur-binding in recent environments: II. Speciation of sulfur and iron and implications for the occurrence of organo-sulfur compounds, *Geochim. Cosmochim. Acta.* **61**, 4769.

[34] Gransch J.A. and J. Posthuma (1973). The origin of sulfur in crudes. In: B. Tissot, F. Bienner, (eds), *Advances in Organic Geochemistry*, Editions Technip, Paris.

[35] deLeeuw, J.W. and C. Largeau (1993). A review of macromolecular organic compounds that comprise living organisms and their role in kerogen, coal, and petroleum formation. In: M.H. Engel, and S.A. Macko, (eds.), *Organic Geochemistry: Principles and Applications*. Plenum, New York.

[36] Sinninghe Damaste, J.S. M.D. Kok, J. Koster, and S. Schouten, (1998). Sulfurized carbohydrates: An important sedimentary sink for organic carbon? *Earth Planetary Lett.* **164**, 7.
[37] Durand, B. and J.C. Monin (1980). Elemental analysis of kerogen (C, H, O, N, S, Fe). In: B. Durand, (ed.), *Kerogen: Insoluble Organic Matter from Sedimentary Rocks*. Editions Technip, Paris.
[38] Teichmuller, M. (1982). Origin of petrographic constituents of coal, In: E. Stach, M-TH. Mackowsky, M. Teichmuller, G.H. Taylor, D. Chandra, and R. Teichmuller, (Translation and English revision by D.G. Murchison, G.H. Taylor, and F. Zierke), (eds.), *Stach's Textbook on Coal Petrology*. Gebruder Borntraeger, Berlin, Stuttgart.
[39] Moore, L.R. (1969). Geomicrobiology and geomicrobiological attack on sediment organic matter. In: G. Eglington and M.T.J. Murphy, (eds.), *Organic Geochemistry : Methods and Results*. Springer-Verlag, New York.
[40] Bustin, R.M., A.R. Cameron, D.A. Grieve, and W.D. Kalkreuth (1983). *Coal Petrology Its Principles, Methods, and Applications*. Geological Association of Canada, Short Course Notes.
[41] Durand, B., M. Parratte, and P. Bertrand (1983). *Rev. lust. Fr. Pétr.*, **38**, 709.
[42] Hedges, J.I. (1992). Global biogeochemical cycles: progress and problems, *Mar. Chem.* **39**, 67–93.
[43] Brocks, J.J. and R.E., Summons (2003). Sedimentary hydrocarbons, biomarkers for early life. In: H.D. Holland, and K.K. Turekian (eds.), *Treatise on Geochem.* **8**, p. 63.
[44] Cooper, B.S. and D.G., Murchinson (1969). Organic geochemistry of coal, In: G. Eglington and M.T.J. Murphy, (eds.), *Organic Geochemistry: Methods and Results*. Springer-Verlag, New York.
[45] Senftle, J.T., C.R. Landis, and R.L. McLaughlin (1993). Organic petrographic approach to kerogen characterization. In: M.H. Engel, and S.A. Macko, (eds.), *Organic Geochemistry: Principles and Applications*. Plenum, New York.
[46] Stach, E. (1982). Fundamentals of coal petrology. In: E. Stach, M-TH. Mackowsky, M. Teichmuller, G.H. Taylor, D. Chandra, and R. Teichmuller, (Translation and English revision by D.G. Murchison, G.H. Taylor, and F. Zierke), (eds.), *Stach's Textbook on Coal Petrology*. Gebruder Borntraeger, Berlin, Stuttgart.
[47] Davis, A., W. Spackman, and P. Given (1976). *Energy Sources* **3**(1), 55.
[48] Tissot , B., B. Durand, J. Espitalie, Combaz (1974). Influence of the nature and diagenesis of organic matter in formation of petroleum, *Am. Assoc. Pet. Geol. Bull.* **58**, 499.
[49] van, Krevelen, D.W. (1961). *Coal.* Elsevier, Amsterdam.
[50] Whelan, J.K., and C.L. Thompson-Rizer (1993). Chemical methods for assessing kerogen and protokerogen types and maturity. In: M.H. Engel and S.A. Macko (eds.), *Organic Geochemistry: Principles and Applications.* Plenum, New York.
[51] Vandenbroucke, M., (2003). Kerogen: From types to models to chemical structure, *Oil Gas Sci. Tech.* **58**, 243.
[52] Talbot, M.R. (1988). The origins of lacustrine oil source rocks: Evidence from the lakes of tropical Africa. In: A.J. Fleet *et al.*, (ed.), *Lacustrine Petroleum Source Rocks*. Oxford, Blackwell.
[53] Orr, W.L. (1986). Kerogen/asphaltene/sulfur relationships in sulfur-rich Monterey oils. In: D. Leythaeuser and J. Rullkötter, eds. *Advances in Organic Geochemistry*, **10**, 499.
[54] Sinninghe Damste, J.J., X.F.C. de las Hera, and J.W. de Leeuw (1992). *J. Chromatogr.* **607**, 361.
[55] Sheu, Y. and O.C. Mullins, (eds.) (1995). *Asphaltenes, Fundamentals and Applications*. Plenum, New York, London.
[56] Pfeiffer, J.H. (1950). *The Properties of Asphaltic Bitumen*. Elsevier, Amsterdam.
[57] Mullins, O.C. and E.Y. Sheu (eds.) (1998). *Structures and Dynamics of Asphaltenes*. Plenum, New York.
[58] Chilingarian, G.V. and T.F. Yen (eds.) (1978). *Bitumens, Asphalts, and Tar Sands*. Elsevier Scientific Publishing Co.: New York.
[59] Groenzin, H. and O.C. Mullins (1999). Asphaltene molecular size and structure. *J. Phys. Chem. A* **103**, 11237.
[60] Groenzin, H. and O.C. Mullins (2000). Molecular sizes of asphaltenes from different origin. *Energy & Fuels* **14**, 677.
[61] Scotti, R. and L. Montanari (1998). In: O.C. Mullins, and E.Y. Sheu (eds.), *Structures and Dynamics of Asphaltnees*. Plenum, New York, Ch 3.

[62] Chilingarian, G.V. (1981). In: J.W. Bunger, and N.C. Li (eds.), *Chemistry of Asphaltenes*. American Chemical Society, Washington, D.C.
[63] Yen, T.F. (ed.) (1975). *The Role of Trace Metals in Petroleum*. Ann Arbor Science Inc., Michigan.
[64] Stohr, J. (1992). *NEXAFS Spectroscopy*. Springer-Verlag, Berlin.
[65] Wagner, C., W. Riggs, L. Davis, and J. Moulder (1979). *Handbook of X-ray Photoelectron Spectroscopy*. Perkin Elmer Corporation, Eden Prairie, MN.
[66] Newville, M., *Consortium for Advanced Radiation Sources*. University of Chicago, Chicago, IL.
[67] Pfalzer, P., J.P. Urbach, M. Klemm, S. Horn, M.L. denBoer, A.I. Frenkel (1999). Elimination of self-absorption in fluorescence hard x-ray absorption spectra, *Phys. Rev. B*. **60**, 9335.
[68] Bunker, G. (1988). Biophysics Collaborative Access Team Report: Basic Techniques for EXAFS.
[69] Goulon, J.C. Goulon-Ginet, R. Cortes, J.M. Dubois (1982). *J. Phys.* **43**, 539.
[70] Margaritondo, G. (1988). *Introduction to Synchrotron Radiation*. Oxford University Press, New York.
[71] Lytle, F.W., R.B. Greegor, D.R. Sandstrom, E.C. Marquies, J. Wong, C.L. Spiro *et al.* (1988). *Nucl. Instrum. Methods* **226**, 542.
[72] Chen, C.T. (1987). Concept and design procedure for cylindrical element monochromators for synchrotron radiation, *Nucl. Instrum. Methods Phys. Sect. A*. **256**, 595.
[73] Cramer, S.P., O. Tench., M. Yocum, H. Kraner, L. Rogers, V. Radeka *et al.* (1991). X-ray absorption fine structure. In: S.S. Hassain, (ed.), *Proc. of the 6th International XAFS Conference*. Ellis Horwood, Chichester, p. 640.
[74] Owen, L.B. (1987). DOE Oil Shale Sample Bank, Quarterly Report, July–September 1987, January–March, 1987, Salt Lake City, Utah, Terra-Tek, Inc. Report numbers: TR 87-89 and TR 88-14.
[75] Larsen, J.W., C. Islas-Flores, M.T. Aida, P. Opaprakasit, P. Painter (2005). Kerogen chemistry 2. Low-temperature anhydride formation in kerogens, *Energy Fuels*, **19**, 145.
[76] Zeszotarski, J.C., R.C. Chromik, R.P. Vinci, M.C. Messmer, R. Michels, and J.W. Larsen (2004). Imaging and mechanical property measurements of kerogen via nanoindentation, *Geochim. Cosmochim. Acta*. **68**(20), 4113.
[77] Saxby, J.D. (1970). Isolation of kerogen in sediments by chemical methods, *Chem. Geol.* **6**, 173.
[78] Waldo, G.S., O.C. Mullins, J.E. Penner-Hahn, and S.P. Cramer (1992). Determination of the chemical environment of sulphur in petroleum asphaltenes by X-ray absorption spectroscopy, *Fuel*, **71**, 53.
[79] Vorres, K.S. (1990). The argonne premium coal sample program, *Energy & Fuels* **4**, 420.
[80] WinXAS software, Ressler, T., Fritz-Haber-Institut der MPG, Department of Inorganic Chemistry, Faradayweg, Berlin.
[81] George, G.N. and M.L. Gorbaty (1989). Sulfur K-edge absorption spectroscopy of petroleum asphaltenes and model compounds, *J. Am. Chem. Soc.* **111**, 3182.
[82] Sarret, G., T. Mongenot, J. Connan, D. Sylvie, M. Kasrai, G. Michael Bancroft *et al.* (2002). Sulfur speciation in kerogens of the Orbagnoux deposit (Upper Kimmeridgian, Jura) by XANES spectroscopy and pyrolysis, *Org. Geochem.* **33**(8), 877–895.
[83] Vairavamurthy, M.A., D. Maletic, S. Wang, B. Manowitz, T. Eglington, and T. Lyons (1997). Characterization of sulfur-containing functional groups in sedimentary humic substances by X-ray absorption near-edge structure spectroscopy, *Energy & Fuels*, **11**, 546.
[84] Olivella, M.A., J.M. Palacios, A. Vairavamurthy, J.C. del Rio, and F.X.C. de las Heras (2002). A study of sulfur functionalities in fossil fuels using destructive- (ASTM and Py-GC-MS) and non-destructive- (SEM-EDX, XANES and XPS) techniques, *Fuel*, **81**(4), 405.
[85] Eglington, T., J.E. Irvine, A. Vairavamurthy, W. Zhou, and B. Manowitz (1994). Formation and diagenesis of macromolecular organic sulfur in Peru margin sediments, *Org. Geochem.* **22**, 781.
[86] Buchanan, H., K.J. Coombs, P.M. Murphy, and C. Chaven (1993). *Energy & Fuels*, **7**, 219.
[87] Spiro, C.E., J. Wong, F. Lytle, R.B. Greegor, D. Maylotte, and S. Lampson (1984). *Science* **226**, 48.
[88] Hussain, Z., E. Umbach, D.A. Shirley, J. Stohr, and J. Feldhaus, (1982). *Nucl. Instrum. Methods* **195**, 115.
[89] Shah, N., R.A. Keogh, F.E. Huggins, G.P. Huffman, A. Shah, B. Ganguly *et al.* (1990). *Prepr. Pap. Am. Chem. Soc. Div. Fuel Chem.* **5**(3), 784.

[90] Mitra-Kirtley, S., O.C. Mullins, J. Chen, J. van Elp, S. George, C.T. Chen *et al.* (1992). Nitrogen chemical structure in DNA and related molecules by X-ray absorption spectroscopy, *Biochim. Biophys. Acta*, **1132**, 249.

[91] Copelin, E.C. (1964). Identification of 2-quinolones in a California crude oil, *Anal. Chem.* **36**(12), 2274.

[92] Petersen, J.C., R.B. Barbour, S.M. Dorrence, F.A. Barbour, and R.V. Hel (1971). Molecular interactions of asphalt. Tentative identification of 2-quinolones in asphalt and their interaction with carboxylic acids present. *Anal. Chem.* **43**(11), 1491.

[93] Mitra-Kirtley, S., O.C. Mullins, J. van Elp, and A.P. Cramer (1994). *Prepr. Pap. Am. Chem. Soc. Div. Fuel Chem.* **39**(3), 820.

[94] Kelemen, S.R., M.L. Gorbaty, and P.J. Kwiatek (1994). Quantification of nitrogen forms in argonne premium coals. *Energy & Fuels* **8**(4), 896.

7

Micellization

Stig E. Friberg

1. Introduction

Micellization in aqueous and nonaqueous systems is compared and the difference in the nature of the driving force for the association process is emphasized. In the aqueous system the high interfacial energy between the water molecules and the hydrocarbon chains of the surfactant is the primary factor in the process, while in a nonpolar system the attractive interaction between the polar parts serves as the force governing the association process. Hence, the micellization in aqueous solutions may be perceived as a phase separation that is modified through geometrical restrictions, while the corresponding phenomenon in oils is a chemical equilibrium similar to that of alcohols in such media. The consequence of this difference is that the process in water is highly cooperative and the designation of a critical micellization concentration (cmc) is justified. In the nonpolar environment, on the other hand, the association process is gradual and the term cmc is not warranted. In this case the interactions leading to the micellization are amenable to analysis by spectroscopic methods.

The special case of association of asphaltenes in toluene is not entirely covered by these mechanisms. The recent description as the formation of nanoaggregates by Mullins seems appropriate and the consequences of this approach are reviewed.

"Nonsense, McBain, nonsense," according to rumors, is the only comment from the chairman of the session, at which the young McBain introduced the phenomenon of micelles in aqueous solutions[1] almost a hundred years ago. Today that comment has a touch of comics; micelles have become part of the colloidal area and an important part at that.

However, seen from the perspective of the time of the statement, it becomes, if not polite, certainly reasonable. This was a time in chemistry, when the idea by Arrhenius had been generally accepted that salt ions in aqueous solutions do not associate to molecules. It was realized that the less than expected activity of ions at high concentrations was not due to association of salt ions to molecules, but instead to reduced chemical potential due to interaction between them in the solution. It is easy to understand the doubts of the leading scientists, when suddenly

Stig E. Friberg • Chemistry Department, University of Virginia, Charlottesville, Virginia 22903.

into that scene came young McBain stating that soap ions did not only form small associations like the ones found for weak acids; but actually formed associations of several tens of molecules.

Almost 20 years later the next development; the introduction of the micellization as a critical event[2,3] was also met with less than general enthusiasm at first.

Today the micelles and their formation are perceived as part of the phase science of surfactants[4] and the phenomenon of micellization, although important, is not the only feature of interest in the area.

2. Micelles in Aqueous Solutions

Although the thermodynamic description of the micellization has varied[5,6] all the experimental results indicate a cooperative process. In a narrow range of concentration an increase of the surfactant amount leads to the formation of aggregates of surfactants, which do not separate from the solution.

The most convenient and certainly the most used method to detect the association is to plot the surface tension of an aqueous solution against the logarithm of the surfactant concentration. The sudden change of the direction of the curve is called the critical micellization concentration (cmc). The two branches of the curve form two straight lines and their interception denotes the concentration in question, the critical micellization concentration (cmc) (Figure 7.1).

Prior to describing the results of different analysis of this phenomenon, it is essential to realize the limitations of the approach. It assumes that the activity can be replaced by concentration in the Gibbs adsorption equation

$$\Gamma = -[\mathrm{d}\gamma/\mathrm{dln}\,(a_S)]/RT, \tag{7.1}$$

in which Γ is the surface excess, γ the surface tension, a_S the activity of the surfactant, R the gas constant, and T the temperature.

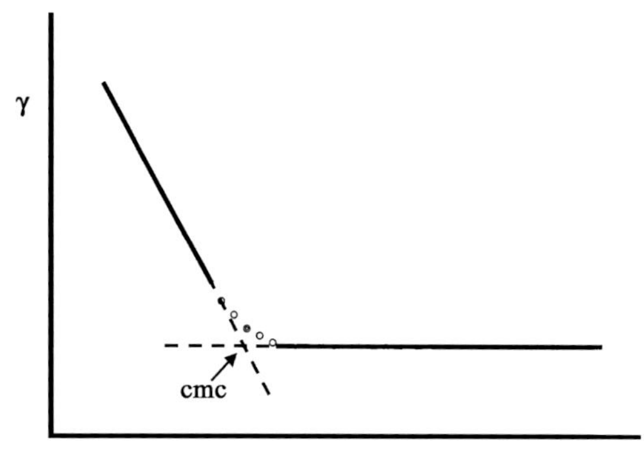

Figure 7.1. Extrapolated lines of the surface tension versus logarithm of surfactant concentration cross at the critical micellization concentration (cmc).

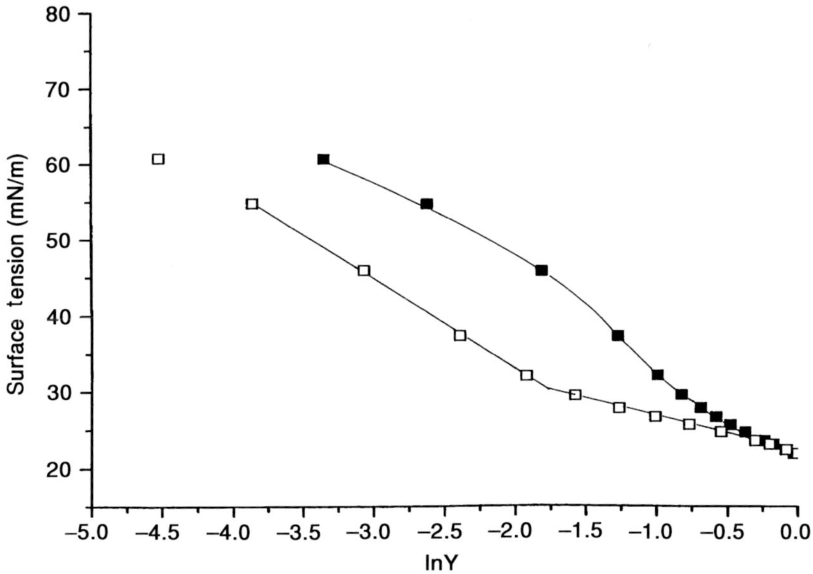

Figure 7.2. Surface tension of aqueous solutions of ethanol plotted against the logarithm of alcohol concentration (open squares) and against logarithm of activity (filled squares).

This is sufficiently accurate for surfactants, whose activity coefficient is approximately constant for the low concentrations below the cmc. However, for short chain amphiphiles, such as hydrotropes[7] this is not true. Neglecting this fact has led to a most embarrassing mistake.[8] The surface tension of a series of alcohols plotted against the logarithm of their concentration and the curves showed knick points as exemplified by the values for ethanol (Figure 7.2). The knick point was interpreted as an indication of the onset of association of alcohols in water.[8] However, a correct plot of the surface tension versus the logarithm of the activity of the alcohol, Figure 7.2, obtained from its vapor pressure[9] shows only a smooth curve with no knick point. Strangely, the mistake was not corrected for a considerable time.[10] It should be mentioned that Srinivas[11] in his attempt to clarify the difference between hydrotrope and surfactant association avoided this mistake by presenting the surface tension curves versus concentration of the amphiphile.[11]

There are a large number of methods to determine the critical micellization concentration other than surface tension. Mukerjee has discussed the optimal application of these for the purpose.[12,13]

The micellization has been treated as an equilibrium process involving consecutive association of surfactant molecules with a series of equilibrium constants[5]

$$S_n + S_1 = S_{n+1};\ K_n. \tag{7.2}$$

The values of the constants are adjusted to experimental results to provide a correct size distribution of the micelles.

However, obtaining all values of K_n would be formidable task and as a first simplification all K:s are put as equal, but such an action removes the tool to limit the size of the micelles. A further simplification directly applies the experimental result that the micelles have certain size and that the size distribution is narrow. Accepting this condition, equilibrium between the unimers of surfactants and micelles with a specific number of surfactants is postulated.

$$nS_1 = S_n \tag{7.3}$$

with an equilibrium constant K. The advantage of this simplification is that it offers an illustration of the fact that the value of n has a most drastic influence on the variation of the fraction of the total number of molecules engaged in the association structure as exemplified in Figure 7.3; in which, for further simplification, the equilibrium constant was chosen as unity. The results in the figure illustrate the most important feature of the micellization.

The association to a structure containing a large number of molecules shows an abrupt change from the amount of surfactant in the association structure being virtually equal to zero at low concentrations to become the dominant feature, when the surfactant concentration exceeds a certain value. Such behavior resembles that experienced during a phase separation and the micellization has been treated as such.[5,14–16] The essential feature of such a treatment is the use of the chemical potential of the molecules in the separated phase, such as in a water/hydrocarbon system, as the standard state, which is correct for a system with a true phase separation. For a hydrocarbon separated from water the chemical potential is conveniently put equal to the chemical potential of the pure hydrocarbon and the chemical potential

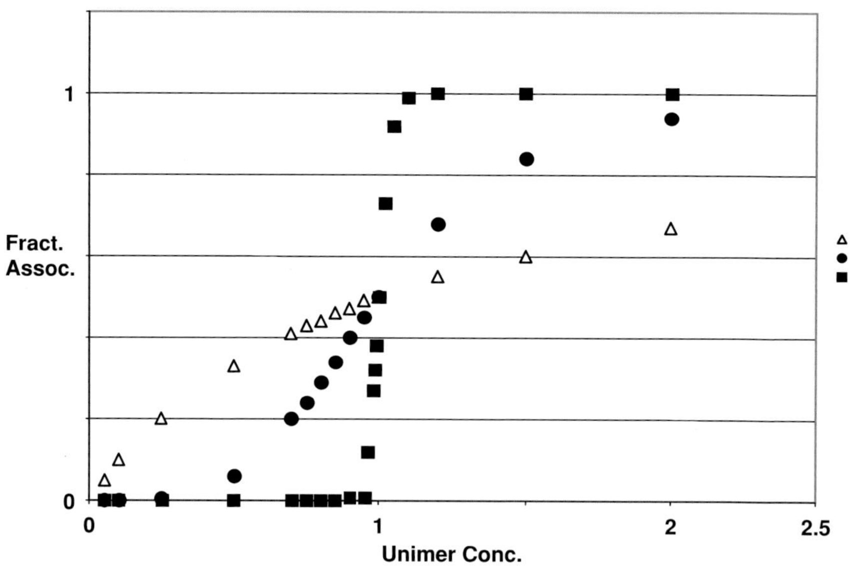

Figure 7.3. The fraction of associated molecules, Fract. Assoc., in the equilibrium; $nS_1 = (S_n)$, for different values of n and with an equilibrium constant equal to one. Open triangles: $n = 2$; Filled circles: $n = 5$; Filled squares: $n = 50$.

of the dissolved molecules becomes

$$\mu_1 = \mu_0 + RT\ln a_1. \tag{7.4}$$

However, such a treatment does not account for the fact that micellization does not include infinite aggregation.

This problem was overcome[17] by considering the repulsion between head groups as a balancing factor. The Australian school[18]introduced geometrical constraints to obtain a finite aggregation number. The geometry of the association structure is monitored by the balance between the volume of the hydrocarbon chain and the area occupied by the polar parts of the surfactant. The ratio

$$R = \nu_H / a_0 l_H, \tag{7.5}$$

in which ν_H is the volume the hydrocarbon chain occupies in the association structure, a_0 the area of the polar part, and l_H the approximate hydrocarbon chain length, determines the allowed geometry of the micelle. For $R < 1/3$, the preferred geometry is a sphere, Figure 7.4, and an estimation of the aggregation number, N, is also obtained from

$$Na_0 < 4\pi l^2 \quad \text{and} \quad N\nu_H < 4\pi l^3/3 \tag{7.6}$$

giving

$$N < 36\pi\nu_H^2/a_0^3. \tag{7.7}$$

Later contributions in the area of aggregation have attempted to present more details of the interaction energy. Nagarajan[19] lists five of these; the hydrophobic effect, the interfacial contribution from the contact between the hydrocarbon core and water, the electrostatic and steric repulsion of the head groups, and the energy from the deviation of the packing of the hydrocarbon chains from their relaxed extended state.

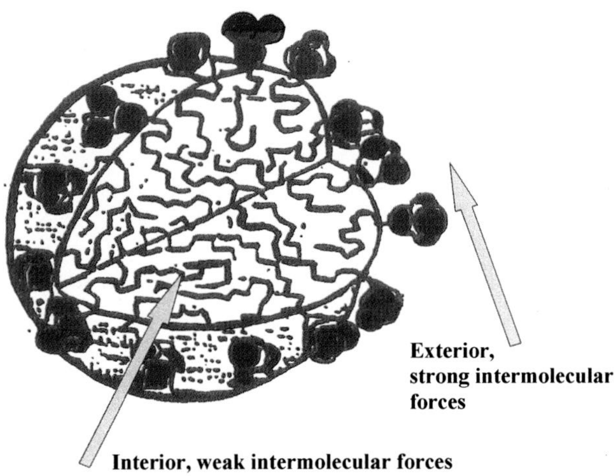

Figure 7.4. A "normal' micelle in aqueous solution with the hydrocarbon chains turned inwards.

As a summary, it may be stated that the micellization in aqueous systems is a consequence of stronger intermolecular forces in the solvent than in the interior of the micelle (Figure 7.4). In short, micellization means a reduction of the high-energy water/hydrocarbon interface by moving the hydrocarbon chains to the interior of the micelle (Figure 7.4). This factor, if acting alone, would cause a phase separation, but the growth of the micellar size is moderated by the repulsion between the polar groups.[17,18]

A most illustrative, but for some reason overlooked, proof of this fact is found in the behavior of surfactants with less repulsion between the head groups such as monoglycerides, lecithin, and polyoxyethylene alkyl ethers with short ethylene oxide chain; less than five oxyethylene groups. For these compounds the association is infinite and, instead of forming aggregates with definite size within the liquid, the association becomes infinite and forms a separate phase.

The association to *inverse micelles* in nonpolar media offers more varied phenomena and a separate section will be devoted to them. At first the traditional inverse micellization of water using common surfactants is not a "critical" event, the association is a stepwise process covering a significant concentration range as a contrast to the conditions in aqueous solutions.

Secondly, the association of asphaltenes in oils is, in turn, a different process. Since the latter phenomenon[20] is of tremendous commercial interest as well as constituting an interesting scientific problem[21]; it will also be treated in a separate section.

3. Inverse Micellization in Nonpolar Media

For these systems there is no relevant reduction in the surface free energy, the hydrophobic effect is nonexistent and early attempts to introduce a critical micellization concentration in such systems were refuted.[6] In fact, it should be observed that the initial addition of surfactant to a hydrocarbon and the consecutive formation of inverse micelles after addition of water[22] results in extremely small changes in surface tension and, furthermore, that this reduction is not due to the molecules per se being surface active. It cannot be overemphasized that surface tension measurements in nonpolar media cannot be expected to leave similar information about micellization as those in aqueous solutions.

Hence, instead of focusing the attention on the properties of the solvent, as is the case for aqueous micelles, the factor to be monitored for the association into inverse micelles is the interaction between the polar groups of the surfactant. As a contrast to the conditions in aqueous micelles, Figure 7.4, the attractive forces between the polar groups now cause the association (Figure 7.5). The solubilization limit of water into the structure is now monitored by the *internal* intermolecular forces.

The importance of this interaction will be illustrated by choosing three examples of surfactants with different solubility in hydrocarbons. The first two examples are concerned with a sodium soap, which itself is virtually insoluble in

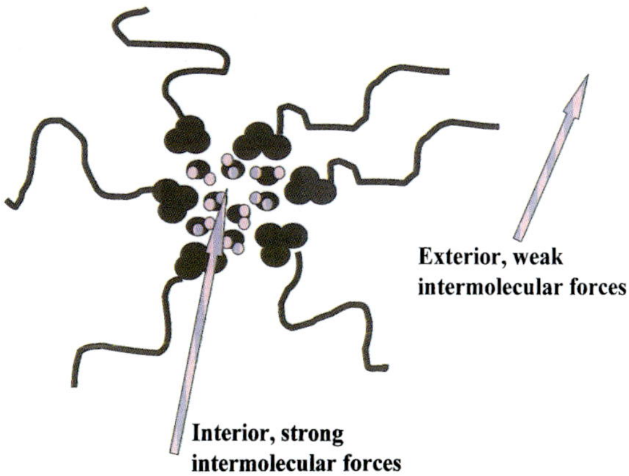

Figure 7.5. An inverse micelle in oil with the hydrocarbon chains turned outwards.

hydrocarbons and the third example is a polyethylene glycol ether, the solubility of which with hydrocarbons is mutually complete.

The soap is chosen, because its combination with a carboxylic acid and with an alcohol is an excellent example of the decisive importance of the intermolecular forces between the polar groups. The soap, when combined with the alcohol shows no solubility in a hydrocarbon, but, when combined with the corresponding carboxylic acid, Figure 7.6, the solubility is extremely high.[23] The reason for this difference is the distinction of the hydrogen bond between the soap carboxylate group and the acid carboxylic group versus the corresponding bond between the soap carboxylate group and the alcohol hydroxy group. The unusual strength of the hydrogen bond in the first case is the cause of the solubility of the soap/acid combination in hydrocarbons, while the interactions in the soap/alcohol combination are not sufficiently strong to confer solubility.

The carboxylate/carboxylic group hydrogen bond was identified by a combination of IR and NMR spectroscopy[24] and the molecular weight determinations[23] gave a composition of four acids and two soaps. The structure suggestion, Figure 7.7, was confirmed by quantum mechanical calculations.[25] The structure in Figure 7.7 offers an intuitive explanation for the solubility in hydrocarbons. The polar parts in the center of the structure are well shielded from the surrounding nonpolar environment by the six hydrocarbon chains extending radially from the center. At a first glance it may be tempting to name the structure an *inverse micelle*, but its size and the bonds involved rather justify the designation pre-micellar aggregate.

Addition of water gives rise to inverse micelles and these show a tremendous capacity to solubilize water (Figure 7.6). The quantum mechanical calculations mentioned,[25] demonstrated the negative free energy change by addition of water to form inverse micelles (Figure 7.8). However, these micelles are formed by gradual addition to the inverse structure in agreement with the general trend in inverse micelles (Figure 7.9).

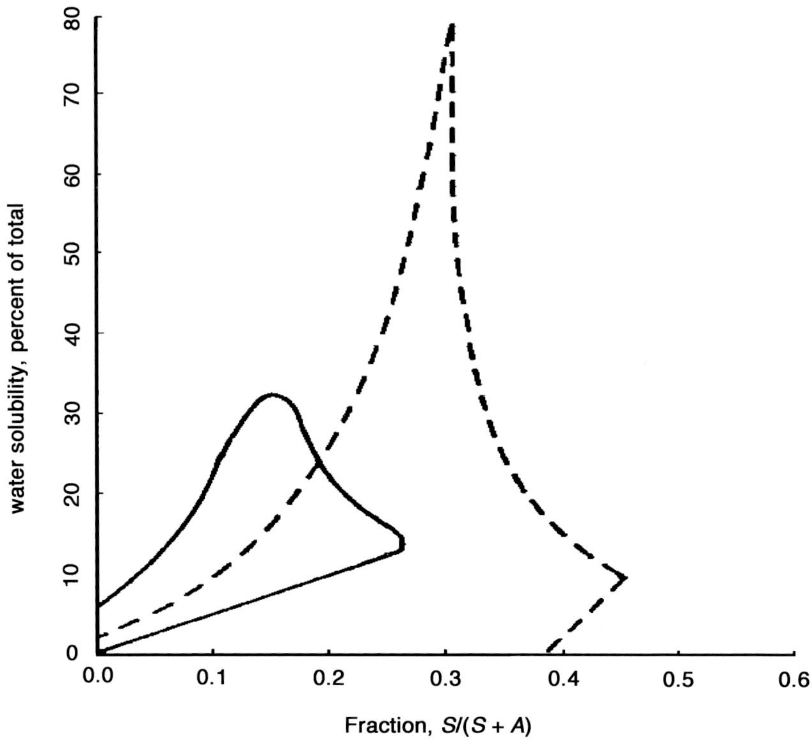

Figure 7.6. The solubility of water and surfactant (*S*) are different, when combined with a carboxylic acid (dashed line) and an alcohol (full line).

In the soap/alcohol combination there is insignificant solubility in a hydrocarbon (Figure 7.6). Solubility is first achieved, when water is added to the combination. The free energy difference between the alcohol/soap/water association complex and the three compounds in pure form has been calculated.[26] The

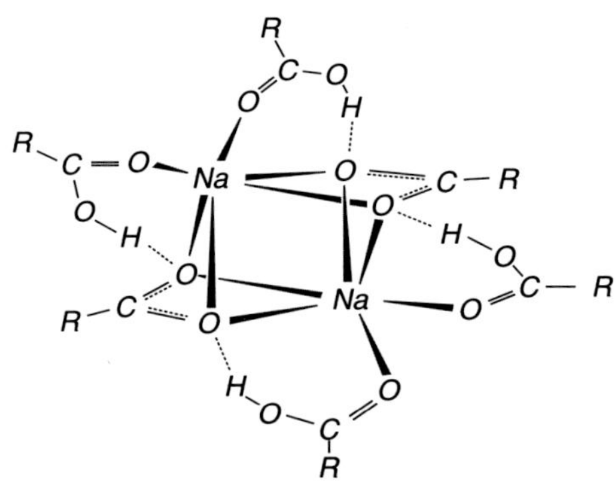

Figure 7.7. A pre-micellar aggregate of two carboxylate ions and four carboxylic acids. The OH-groups of the acids are hydrogen bonded to the ionized carboxylates and the acid carbonyl groups ligand the sodium ion.

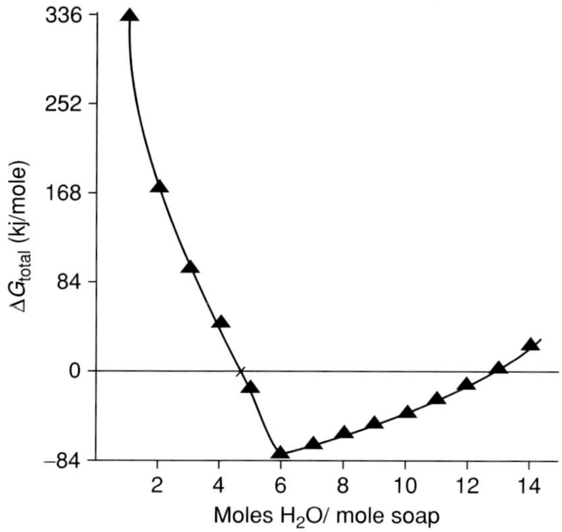

Figure 7.8. The free energy of formation for soap/water association structures in alcohol solution.

results[27] revealed that a minimum of four water molecules are necessary to bring the free energy difference into the negative range, Figure 7.8, in excellent agreement with the experimental results (Figure 7.6). Figure 7.8 indicates the maximum number of water molecules as 13 to retain stability of this primary association. On the other hand, the experimental results, Figure 7.6, reveals the inverse micellar solution with the acid to solubilize huge amounts of water and at a first glance it may appear that the calculations are inadequate.

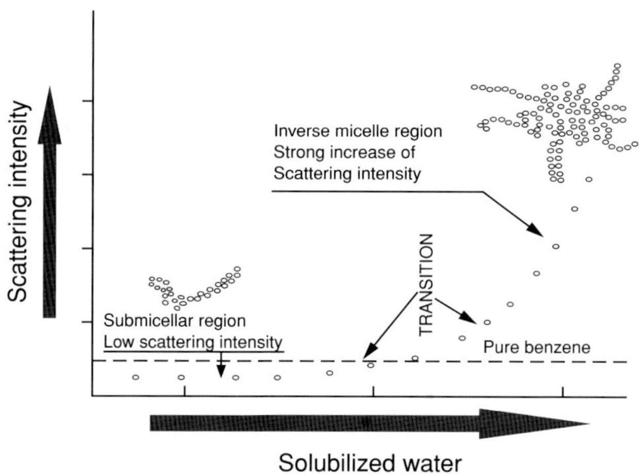

Figure 7.9. Light scattering intensity of inverse micellar solutions with increased water content in the system water, potassium oleate, and pentanol. The pentanol/oleate/ benzene ratio is constant at 1/1/1.

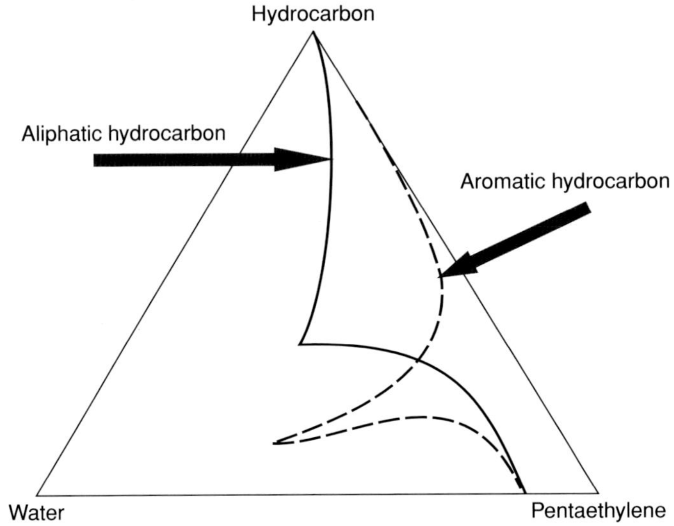

Figure 7.10. The solubilization of water into the inverse micelles stabilized by a nonionic surfactant depends strongly on the structure of the hydrocarbon. (—) cyclohexane, (— — —) benzene.

So is not the case. Unpublished calculations showed most convincingly that a combination of several soap molecules will accept proportionally more water than a single one and the number of 13 water molecules only reflects the stability of an association with *one* soap molecule. As demonstrated in Figure 7.9, with increasing amount of water consecutive association takes place to form colloidal structures, inverse micelles.

Inverse micellization by nonionic surfactants show a similar behavior, but now the weak interaction between the polar groups makes these interactions sensitive to the electronic structure of the hydrocarbon. Figure 7.10 reveals an illustrative example of this feature. The initiation of solubilization of water into a benzene solution of a nonionic surfactant requires an extremely high concentration of the surfactant, more than 50% by weight. As a contrast, when cyclohexane is the nonpolar compound, the solubilization commences at very small concentrations.

The reason for this distinction is the interaction between the aromatic hydrocarbon and the polar group of the surfactant. This surprisingly strong interaction was first discovered by Christenson[28] by NMR investigations and later confirmed by direct calorimetry determinations.[29] As the NMR experiments showed[28] the solubilization into the benzene system cannot begin, before the surfactant molecules have initiated *self-association* without water present. In short, water molecules are accepted into the structure first, when the polar groups are already aggregated. In the cyclohexane system, on the other hand, the addition of water initiates the association prior to the concentration of surfactant to be at the level of self-association. Even in this case the association is gradual and the expression "critical micellization concentration" is not warranted.

4. Asphaltene Association in Crude Oils

A new association structure in nonpolar environment has recently been introduced.[21] It is concerned with asphaltenes, which are a part of crude oils. Their presence significantly adds to the oil properties; and cause most serious problems during recovery and refinery.[30–32] Hence, their structure and properties have been intensely investigated over the years as evidenced by the large numbers of references in this book.

The properties of the crude oil as well as of different model systems strongly indicate association of these compounds in organic media. This association has traditionally been called micellization[33–35] and critical micellization concentrations, cmc:s, have been published[36,37] covering a huge range of values. Such a discrepancy between different results certainly merits a further analysis of the phenomenon.

The reason for this discrepancy has recently been referred to the problems of determining true molecular weights of the asphaltenes.[21] The problem is found in the use of osmotic methods, which by their nature only with difficulty are able to distinguish between monomers and association structures. A large number of methods[21] have been used to resolve this problem and the molecular weight of the asphaltenes may now be considered established at the level of 700 g/mol.

With this result accepted, it seems possible to make some progress in the features of asphaltene self-association in nonpolar solvents. The primary issue is the structure of the associations and the fundamental forces, which give rise to the phenomenon. The second issue is to explain the contradiction that follows from the first issue.

The key factor for the association, as pointed out by Buckley in her chapter in this book; is poor solubility of the asphaltenes. This feature is expected, polyaromatic nuclei display strong solubility changes with the number of benzene rings. So, e.g., is the difference in solubility in aromatic solvents between naphthalene and anthracene pronounced and the polyaromatic nucleus of the asphaltenes is certainly the most essential structural entity for the association.

Accepting this basic thesis, two issues remain to be clarified. At first, why does the lack of solubility of the asphaltenes not lead to phase separation? Secondly, is the poor solubility being caused by the superior intermolecular forces of the solvent, as is the case for the micellization in water in water, Figure 7.4, or is the stronger intermolecular forces of the solubilizate the key feature, as is the case for inverse micelles, Figure 7.5? The question is answered in favor of the latter. The polynuclear aromatics lack solubility in aliphatic hydrocarbons, because they enjoy higher strength of the intermolecular forces and the asphaltene "micelles" in hydrocarbons are of the inverse kind.

However, this conclusion leads to further an unresolved issue. It does not offer an explanation to the cause of the limitation of growth after the association is begun. For "normal" micelles in water the limitation is caused by the repulsion between the polar groups, which prevents the growth into a lamellar structure (Vide retro). The size of traditional inverse micelles, on the other hand, is regulated by the intermolecular forces. As is demonstrated by the results in Figure 7.8, the addition

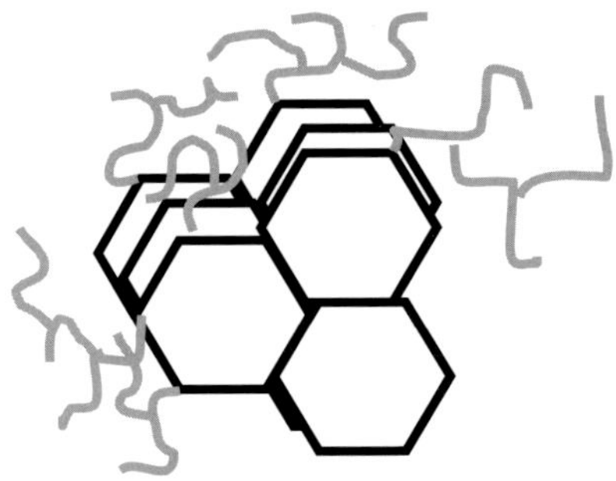

Figure 7.11. Suggested packing of the association structure of asphaltene molecules in hydrocarbons.

of more water molecules leads to a minimum of the free energy difference and the limit to the size of the micelles is now a question of intermolecular forces.

It is obvious that this concept is not applicable to the association of asphaltenes in toluene, since there is no reasonable mechanism that would give a reduction of the intermolecular forces with increased number of molecules added to a stacked association structure.

Hence, a reasonable conclusion is that the asphaltene association in toluene forms an interesting third case. As mentioned earlier the limit to micellar growth cannot be similar to the conditions in aqueous solutions, because the primary cause of the association is attractive intermolecular forces, contrary to the case in aqueous systems.

Hence, the most reasonable conclusion is a limit to association by steric repulsion from the aliphatic chains as recently postulated,[21] (Figure 7.11).

However, accepting this structure and mechanism for the asphaltene association in hydrocarbons leads to an obvious contradiction. In short, if the intermolecular forces between asphaltene molecules are stronger than the toluene–asphaltene interactions, why is then the surface tension of toluene reduced after addition of asphaltenes? Addition of oil soluble surfactant to a hydrocarbon leads to *increased* surface tension as expected. In the asphaltene case *reduced* surface tension has been reported. In addition, the new results of the onset of association show the surface tension to be reduced even after the association process has begun.

As of today there are no final answer to these questions, but the following considerations may be helpful for the future efforts.

At first the knick-point observed in the surface tension curves versus the concentration of asphaltenes may, in fact, be an artifact, caused by the fact that the activity coefficient of the asphaltene molecules is not constant. If the activity coefficients of the asphaltenes could be obtained, the sudden change in the curve may disappear as was the case for alcohol molecules in water.[10]

Furthermore, the reduction in surface tension at concentration in excess of the association concentration may, actually, be a result of the association per se. This feature has been demonstrated in traditional systems and named *conformation induced surface activity*. This reduced surface tension in oil systems is based

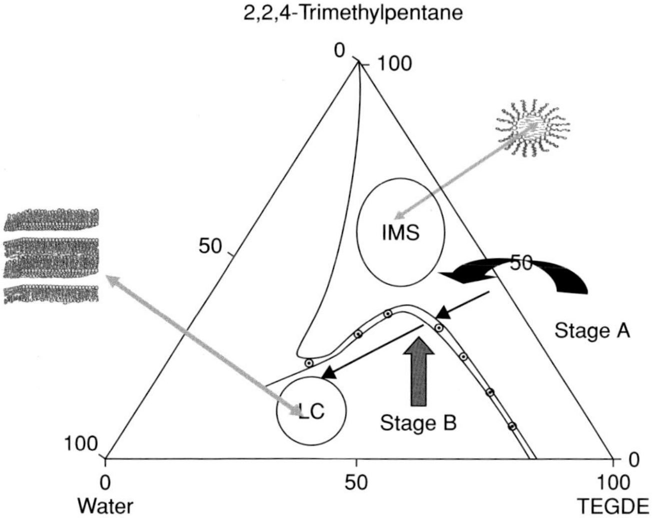

Figure 7.12. Water solubilized into a hydrocarbon/tetraehtylene glycol dodecyl ether (TEGDE) solution initially forms inverse micelles (stage A). With increased amount of water a phase change to a lamellar liquid crystal takes place (stage B).

on a *conformation change* due to the polar interactions. Adding water to liquid hydrocarbon/surfactant solutions, does not lead to a reduction of surface tension. However, after further addition of water and when the maximum water capacity of the inverse micelles is exceeded, a lamellar liquid crystal formed (Figure 7.12). This structure has significantly lower surface free energy in spite of the fact that it contains more water.[22] The reason for this reduction is that the change from inverse micelles to a lamellar liquid crystal, Figure 7.12, means a change in conformation of the surfactant molecules. Instead of having a completely liquid structure with no preferred orientation, the change to the lamellar liquid crystal causes the methyl groups to be preferentially exposed at the surface (Figure 7.12). A layer of methyl groups has a significantly lower surface free energy than a layer of methylene groups[38] and the lamellar liquid crystal has a lower surface free energy than the disordered liquid.

It seems tempting to ascribe the reduced surface tension in the hydrocarbon/asphaltene system to an analogous mechanism. Association of the asphaltene molecules due to the greater intermolecular forces between the polyaromatic nuclei, would expose the aliphatic groups of the asphaltenes, Figure 7.11, in an ordered fashion and create nanosructures of peripheral aliphatic groups and a lowered surface tension in an aromatic environment.

5. Conclusions

The formation of micelles in aqueous solutions is described as a modified phase separation. The significant difference in interaction energy between the water molecules and the hydrocarbon chains of the surfactant promote a complete phase separation. This event is counteracted by the repulsion forces between its polar groups giving rise to association structures of limited size.

As a contrast, the association to inverse micelles in an oil environment is driven by the interattractive forces between the polar groups of the surfactant and with the solubilized water molecules. The phase separation in this case is monitored by the maximum amount of water that can be accommodated in the spherical structures of the inverse micelles, or, for lower content of surfactant by the reduction of intermolecular interaction.

It is shown that the association of asphaltenes in nonpolar environment is different from these two categories and that new mechanisms must be employed to understand their structure and mechanism for association.

Acknowledgments

The author would like to express his gratitude to Drs. Sjoeblom, Mullins, and Dabros for their constructive discussions about these problems.

References

[1] McBain, J.W., M.E. Laing, and A.F. Titley (1919). *J. Chem. Soc.* **115**, 1279.
[2] Ekwall, P. (1927). About the surface activity in solutions of sodium soaps of long chain fatty acids (Diss). *Acta Acad. Abo, Math. Phys.* **4**, 1–83.
[3] Jones, E. (1927). *Phil. Mag.* **4**, 841.
[4] Laughlin, R.G. (1994). *The Aqueous Phase Behavior of Surf*actants. Academic Press, New York.
[5] Lindman, B. and H. Wennerstroem (1980). Micelles aggregation in aqueous solutions. *Topics Curr. Chem.* **87**, 1–84.
[6] Eicke, H.-F. (1980). Surfactants in nonpolar solvents. Aggregation and micellization. *Topics Curr. Chem.* **87**, 85–146.
[7] Friberg, S.E. (2004). Hydrotropes—performance chemicals. *J. Dispersion Sci. Technol.* **25**, 243.
[8] Kahlweit, M., G. Busse, and J. Jen (1991). Adsorption of amphiphiles at water air interfaces. *J. Phys. Chem.* **95**, 5580–5586.
[9] Pemberton, R.C. and C.J. Mash (1978). *J. Chem. Thermodynamics* **10**, 867–888.
[10] Strey, R., Y. Viisanen, M. Aratono, J. Kratohvil, Q.Yin, and S.E. Friberg (1999). On the necessity of using activities in the Gibbs equation. *J. Phys. Chem.* **103**, 9112.
[11] Srinivas, V. and D. Balasubramanian (1998). *Langmuir* **14**, 6658–6661.
[12] Mukerjee, P. (1974). On the experimental determination of cmc. *J. Pharm. Sci.* **63**, 972–976.
[13] Mukerjee, P. and A.Y.S. Yang (1976). Experimental determination of cmc. *J. Phys. Chem.* **80**, 1388–1394.
[14] Shinoda, K. (1980). Surfactant micellization. *Pure Appl. Chem.* **52**, 1195.
[15] Coper, A. (1980). Thermodynamics of surfactant solutions. In: Th.F. Tadros (ed.), Surfactants. Academic Press, New York, pp. 19–52.
[16] Diamant, H. and D. Adelman (2004). Models of gemini surfactants. In: R. Zana and J. Xia (eds.), *Gemini Surfactants.* Marcel Dekker, New York, pp. 37–64.
[17] Tanford, C. (1973). *The Hydrophobic Effect.* Wiley, New York.
[18] Israelachvili, J.N., J.J. Mitchell, and B.N. Ninham (1976). Amphiphilic association structures. Shape factors. *J. C. S. Faraday II* **72**, 1525–1538.

[19] Nagarajan, R. and E. Ruckenstein (1991). Interaction energies in micellization. *Langmuir* **7**, 2944–2969.
[20] Sheu, E.Y. and O.C. Mullins (eds.) (1995). *Asphaltenes: Fundamentals and Applications*. Plenum, New York.
[21] Andreatta, G., N. Bostrom, and O.C. Mullins (2005). *Langmuir* **21**, 2728.
[22] Friberg, S.E., I. Blute, H. Kunieda, and P. Stenius (1986). The stability of hydrophobic foams. *Langmuir* **2**, 659.
[23] Soederlund, G. and S.E. Friberg (1970). Solubility of soap–carboxylic acid complexes in different solvents. I. IR and NMR investigations of sodium caprylate–caprylic acid compounds in carbon tetra chloride. *Z. Phys. Chem.* **70**, 39.
[24] Friberg, S.E., L. Mandell, and P. Ekwall (1969). Solutions of alkali soaps and water in fatty acids. III. IR and NMR investigations. *Kolloid Z. Z. Polymere* **233**, 955.
[25] Bendikson, B., S.E. Friberg, and P.L.M. Plummer (1979). CNDO calculations on the structure of a liquid sodiumcarboxylate–carboxylic acid compound. *J. Colloid Interface Sci.* **72**, 495.
[26] Bendiksen, B. (1981). PhD Thesis, Chemistry Department, University of Missouri at Rolla.
[27] Friberg, S.E. and T. Flaim (1982). Surfactant association structures. *ACS Symposium Series* **177**. American Chemical Society, Washington, DC, pp. 1–17.
[28] Christenson, H. and S.E. Friberg (1980). Spectroscopic investigation of the mutual interactions between nonionic surfactant, hydrocarbon and water. *J. Colloid Interface Sci.* **75**, 276–285.
[29] Nakamura, M., G.L. Bertrand, and S.E. Friberg (1983). Partial molar enthalpies of benzene and water in tetraethylene glycol dodecyl ether–decane solutions. *J. Colloid Interface Sci.* **9**, 516–524.
[30] Havre, T.E. and J. Sjoeblom (2003). *Colloids Surfaces* **228**, 131–142.
[31] Djuve, J., X. Yang, I.J. Fjellanger, J. Sjoeblom, and E. Pelizzetti (2001). *Colloid Polym. Sci.* **279**, 232–239.
[32] Auflem, I.H., T.I. Havre, and J. Sjoeblom (2002). *Colloid Poylm. Sci.* **280**, 695–700.
[33] Mack, C. (1932). *J. Phys. Chem.* **36**, 2901; Nellensteyn, F.J. (1933). *Proc. World Petroleum Congress* **2**, 616.
[34] Pfeiffer, J.P. and R.N.J. Saal (1940). *J. Phys. Chem.* **44**, 139.
[35] Andersen, S.I. and S.D. Christensen (2000). *Energy Fuels* **14**, 38.
[36] Sheu, E.Y., M.M. De Tar, D.A. Storm, and S.J. DeCanio (1992). *Fuel* **71**, 299.
[37] Andersen, S.I. and K.S. Birdi (1991). *J. Colloid. Interface Sci.* **142**, 497.
[38] Zisman, W. (1964). Contact angles of Langmuir–Blodget monolayers. In: R.I. Gould (ed.), *Contact Angle, Wettability and Adhesion*. ACS Advances in Chemistry Series 42. American Chemical Society, Washington, DC, pp. 1–19.

8

Insights into Molecular and Aggregate Structures of Asphaltenes Using HRTEM

Atul Sharma and Oliver C. Mullins

1. Introduction

The recent resolution of the controversy surrounding asphaltene molecular weight coupled with increasing understanding of their molecular structure has enabled the understanding of asphaltene behavior. It has been shown previously that larger ring systems require more alkane substituents to maintain a balance between ring-stacking propensity vs. steric repulsion. Here, stacking and its disruption in asphaltenes and aromatic ring systems are explored using high-resolution transmission electron microscopy (HRTEM). The TEM images are consistent with the presence of aromatic ring systems of ~1 nm diameter for petroleum asphaltenes and 0.7 nm for coal asphaltenes. It is shown that molecularly disparate asphaltenes exhibit stacking invariants. Solubility data herein suggest these stacking invariants naturally follow from the solubility classification of asphaltenes.

"If you want to understand function, study structure," exhorts Francis Crick.[1] However, for systems such as asphaltenes that are defined by an operational solubility classification (e.g., soluble in toluene, insoluble in *n*-heptane[2–5]) as opposed to a standard chemical structural definition, the task of determining the structure, chemical and physical, of asphaltenes is much more difficult. Different techniques such as NMR, IR, XANES, mass spectroscopy, time-resolved fluorescence depolarization spectroscopy (TRFD), x-ray diffraction (XRD), Raman spectroscopy, high-resolution transmission electron microscopy (HRTEM), and a variety of scattering techniques have been applied to study the size and nanoscale arrangement of molecules in asphaltenes. In addition, TEM images can reveal molecular structural information about asphaltenes. However, for 20 years, Crick's sage advice was not applicable to asphaltene science due to the order of magnitude controversy over asphaltene molecular weight. The recent findings regarding this controversy[6,7]

Atul Sharma • Advanced Fuel Group, Energy Technology Research Institute, National Institute of Advanced Industrial Science and Technology, 16-1 Onogawa, Tsukuba 305-8569, Japan. **Oliver C. Mullins** • Schlumberger-Doll Research, Old Quarry Road, Ridgefield, Connecticut 06877.

confirming early mass spectroscopy studies,[8] increasingly supported by a wide variety of techniques,[9–12] has enable Crick's dictate to be followed at long last in asphaltene science. Petroleum asphaltenes have relatively low molecular weights (~750 amu) and coal asphaltenes are even smaller (~500 amu).[6,7] This knowledge coupled with bulk molecular structural information has led to tightly constrained proposed asphaltene structures thereby allowing relations between structure and function to be established.

^{13}C NMR shows that petroleum asphaltenes have approximately 40–50% of their carbon as aromatic.[5,10] Some coal asphaltenes differ significantly here in that ~85% of their carbon can be aromatic.[10] IR studies show that >90% of the hydrogen is substituted on aliphatic groups.[2–5] There is a moderate dependence on specific asphaltene chemical properties with the petroleum source material and on the exact procedure used for asphaltene extraction. There is a significant difference between coal and petroleum asphaltenes that has been very useful in ferreting out structure—function relations. XANES (x-ray absorption near edge structure) and XPS (x-ray photoelectron spectroscopy) studies on sulfur[13,14] show that the sulfur is present in sulfide and thiophene groups with increasing thiophene accompanying greater maturation. Alkyl sulfoxide can also be present[15]; with such a strong dipole moment, the size of the aromatic ring systems are reduced to maintain solubility.[7] That is, the alkyl sulfoxide and the fused ring system present two binding sites in the molecule (bidentate). The large dipole moment of the sulfoxide group requires a reduced fused ring system to keep the total binding energy small enough to maintain solubility. Nitrogen XANES studies show that asphaltene nitrogen is all aromatic with pyrrolic nitrogen dominating.[16] Carbon x-ray Raman spectroscopy has been performed on asphaltenes.[17,18] These new results show that asphaltene ring systems tend to be dominated by sextet carbon[17] and that this arrangement of aromatic ring systems is consistent with formation of the most stable ring systems.[19] The resulting type of petroleum asphaltene structures consistent with this mountain of data is that the bulk of asphaltenes are shaped "like your hand." There is a central core (palm) consisting of a fused aromatic system with possible alicyclic substitution with peripheral alkane constituents (fingers).

With this simple structure, freshman chemistry principles have been shown to be consistent with defining asphaltene solubility, thus asphaltene definition.[10] Fused ring systems have a propensity to stack via van der Waals interaction (and dipolar interactions, etc.) thereby decreasing solubility. Alkane peripheral substituents increase steric interaction thereby increasing solubility. Comparison of coal and petroleum asphaltenes illustrates how these simple principles apply. Petroleum asphaltenes with their substantial alkane substituents possess large ring systems, while (some) coal asphaltenes, with their relative lack of alkanes, possess only small ring systems.[10] That is, large ring systems need alkane substituents to disrupt stacking to maintain solubility in toluene; small ring systems can only have few alkane substituents or else they would be soluble in *n*-heptane. Since the coal source material often contains a smaller fraction of alkanes compared to oil source material, the asphaltene fraction of the corresponding coal will have only small fused ring systems. Of course, the whole coal has large fused ring systems that are not soluble in toluene, thus not asphaltene.

For understanding the nanostructure, HRTEM has been a very successful and preferred technique because of its unique feature to observe the structure directly at an atomic level.[20–31] The advantage of HRTEM over other diffraction techniques is avoidance to large extent of averaging effects. However, as far as the direct imaging of the carbon structure by HRTEM is concerned, some questions arise: for instance, what are the physical limits of a polyaromatic layer, how a stack of layers can be defined (from which misorientation angle and interlayer spacing two layers can be considered as stacked?), how are "single layers" (i.e., nonstacked layers) and "crystallite" or rather "coherent domains" are discriminated? Regarding the layer extent, is the fringe disruption really always considered as the boundary of the polyaromatic layers? The answer is probably yes, when the cycles are terminated by hydrogen, i.e., in the case of disconnected molecules. However, in the case of chars and asphaltenes, the polyaromatic structures are often terminated by alkane substituents. The layers are thus continuous, but more or less distorted (out-of-plane) consisting of individual substituents not planar sheets. A part of the layer can be out of the Bragg angle and disappear in the HRTEM image: the fringe is then interrupted but not the layer. The answers given to such questions appear to be very important for the choice of the procedure of HRTEM image analysis.

Previous TEM direct imaging of asphaltenes and other carbonaceous materials supports the view that stacking disruption is an important issue for asphaltenes.[32–34] Extensive analysis of asphaltenes by HRTEM has shown that long-range order is largely lacking in asphaltenes.[32,33] The aromatic ring systems are readily imaged and are found to associate in stacks comprising typically of only 2–3 fused ring systems. The size of these systems is slightly larger than 1 nm in width and roughly 1 nm for the stack height.[32,33] This size range for the aromatic component of petroleum asphaltenes is consistent with the overall molecular size as determined by fluorescence depolarization.[6,7] The spacing of the ring systems is determined to be slightly larger than the graphitic spacing of 3.35 Å which is expected. If certain carbonaceous materials are subjected to elevated temperatures, graphitization occurs. Because carbon is refractory, temperatures in excess of 1,500°C are required.[32] Graphitization can be monitored using HRTEM to see the growth of long-range order. HRTEM has been shown to be a powerful tool to explore carbonaceous materials particularly with regard to ordering.[20–33]

These simple ideas about stacking and stacking disruption of various samples were tested using HRTEM.[34] This method provides direct imaging of the ordering of molecules in samples providing a stringent test for establishing the importance of stacking in asphaltene identity. HRTEM images of coal asphaltene and petroleum asphaltene have been generated and analyzed. Consistency of these results with findings from previous work utilizing various techniques lends credence to the analyses presented here. The asphaltene images are compared with images of several model compounds including both alkyl substituted and unsubstituted aromatics. Analysis of the effects of alkane substitution on long-range order provides direct support for the importance of fused ring stacking in asphaltene identity.

2. Theory of HRTEM and Image Analysis

2.1. Basics of HRTEM

Millward and Jefferson[20] discussed in detail the theory and history of lattice fringe imaging. Oberlin[32] gave a detailed account of the application of TEM to obtain numerical data from images. The image formation by TEM is governed by three basic principles of electron microscopy (EM).[20]

1. Principle of diffraction or Bragg's law: Only those scattered electrons, which fulfill Bragg's condition can form a diffraction pattern.
2. Kinematic theory of EM or single scattering: Line transfer theory: The sample should act as a weak phase grating. The electrons should be diffracted only once and diffracted beams should be much lesser in amplitude than direct beam. The distribution of intensity in the image plane is governed by phase contrast function.
3. Dynamic theory of EM or multi-slice approach: the thick samples are divided into several thin slices each following the kinematic theory of EM. This theory defines the limit of thickness of sample. For the kinematic theory to be applicable, the thickness must be less than 5 nm for any reliable image.

The correctness of the numerical data from image analysis depends on the reliability of the images. Therefore, it is important to understand the image formation mechanism and various instrumental and theoretical considerations for image formation.

In any image forming system there are two basic processes to consider: the interaction of the incident radiation with the object and the transfer of the scattered radiation to the image plane by the optical system. In general, the interaction of electrons with graphitic carbons involves multiple scattering (dynamic theory) and the subsequent electron optical transfer process is nontrivial. However, if the specimen is sufficiently thin in the direction of the electron beam (5 nm or less) then the relationship of the distribution of the object projection potential and the distribution of image intensity can be treated at least qualitatively in terms of linear transfer theory (kinematic theory). Using this theory the transfer of the information content of the scattered wave to the image plane is described in terms of two successive Fourier transforms (FT) of the object projection potential. The effects of lens aberrations are incorporated into theory by multiplying the first-stage FT (representing the scattered wave amplitude in the back-focal-plane of the objective lens) by a suitable phase contrast transfer function. The most important behavior of this transfer function is dependent upon the spherical aberration coefficient and defect of focus of the objective lens.

Figure 8.1 shows a schematic diagram of image formation by TEM. The scattered beams from the sample (S), which follow Bragg's law, produce a diffraction pattern at the back-focal-plane of the lens (A). This plane is also called the Abbe-image and it corresponds to the diffraction figure of the object (points where each diffracted beam is brought into focus). Because of the very short electron

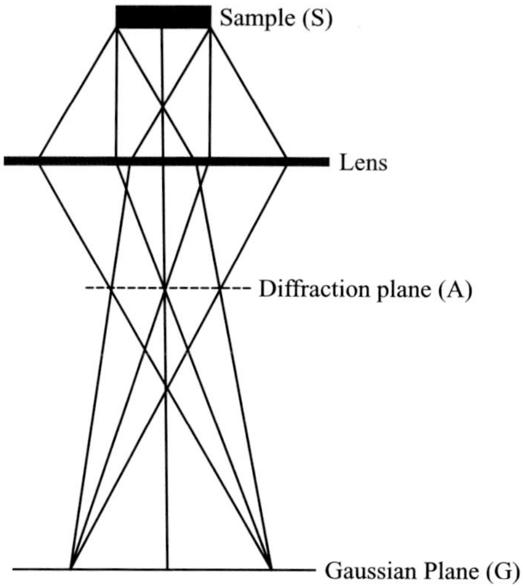

Figure 8.1. Schematic diagram showing image formation by TEM.

wavelength this image may be considered as a plane section of the reciprocal space. It represents the first Fourier Transform of the object which is equivalent to the diffraction pattern obtained by XRD. Each point of the image G shows a perfect correspondence with each point of the object. This means in the Gaussian plane (G) we obtained a second Fourier transform which gives us the real space. Therefore, the objective lens acts as a Fourier transform function to transform the diffraction pattern in reciprocal space to image in the real space. In a TEM, below the Gaussian plane are projector lenses which further magnify the image formed at the Gaussian plane and project it on the fluorescent screen. If the lens is perfect the objective takes into account the phases of the diffracted beams which can interfere with the incident or central beam, so that we may automatically obtain the atomic structure of the object. However, no lens is perfect and it is always more convergent for those rays which are oblique with respect to the optical axis, than for the paraxial rays. This spherical aberration C_s cannot be corrected and causes a phase shift. Any variation in the focal length of the objective lens introduces a phase shift proportional to $(2\theta)^2$. In addition, each diffracted beam has a phase shift of $\pi/2$ with the undeviated beam. So we may write the total phase difference in the Gaussian plane as:

$$\cos\chi = \cos\left(\frac{\pi}{2} - \frac{2\pi}{\lambda}C_s\frac{(2\theta)^4}{4} + \frac{2\pi}{\lambda}\Delta f\frac{(2\theta)^2}{2}\right). \tag{8.1}$$

If all the other lenses of the TEM are focused on the Gaussian plane of the objective lens, we can see on the fluorescent screen an enlarged image of the object and we are in bright field microscopy. The above formula shows that we may compensate C_s for a limited number of diffracted beams and so directly resolve the atomic structure if $\cos\chi = \pm 1$. This can be obtained by choosing a suitable value for Δf, defocus. In carbon samples, the number of diffracted beams able to interfere is too small to use this technique. Therefore, for carbonaceous materials, the lattice imaging technique is widely used. If we focus the projector lens on the Gaussian plane and place an aperture in the back-focal-plane large enough to let through the incident or central beam and an *hkl* diffracted beam, we will obtain in the Gaussian plane a set of dark and bright fringes. Because of the short electron wavelength, the Bragg angles are very small and therefore, the *hkl* planes give diffracted beams only when they are nearly parallel to the incident beam. The lower the crystallinity will be, the larger the interference error could be and intensity of the diffracted beam should be diffuse and faint. For carbon samples, we use the interference between 002 diffracted beam and the incident beam, so that we see the carbon layers edge on. If one uses extremely high magnification, lattice imaging shows only a tiny part of the material and consequently it is not reliable if sample is not homogeneous. Another drawback is that the lattice fringes represent only the projection of lattice planes therefore the carbon layers appear as dark fringes whatever their third dimension may be or whatever their shape may be. We must add that if some graphite-like layers are not grouped or stacked at least in pairs, they cannot be seen by this technique.

2.1.1. Random Small Aromatic Ring Structures

Many carbonaceous materials contain planar aromatic ring structures, more or less piled up in stacks connected to each other with nonaromatic functional groups in a three-dimensional space. In asphaltenes, these liankages can be non-covalent bonding. In addition, the presence of heteroatoms in aromatic sheets or peripheral substitution can introduce nonplanarity in the otherwise planar aromatic sheets. The presence of functional groups or peripheral substituents makes the aromatic sheets present randomly to each other by distorting the orientation of the aromatic sheets. The size of aromatic sheets, presence of aliphatic groups and heteroatoms are mainly dependent on the C/H ratio, heat treatment history, and nature of the precursor. To begin with, it is necessary to understand how diffraction patterns are formed by randomly orientated small aromatic ring structures. Figure 8.2A shows the TEM diffraction pattern from two randomly orientated aromatic layers. These layers will only produce diffraction spots from the 10 and 11 planes. When these layers are orientated to each other at a distance that is equal to characteristic d-spacing of turbostratic carbons, the 002 diffraction spot will also appear in addition to the 10 and 11 as shown in Figure 8.2B. The above results show that irrespective of whether layers are orientated at d_{002} or not, 10 and 11 reflections are always present. The 10 and 11 reflections give the layer size or length of the aromatic layer. On the image plane the 10 and 11 diffraction spots will form a lattice fringe image as a line or spots with a d-spacing of 0.241 and 0.121 nm, respectively, depending on the resolution of the TEM. Figure 8.3 shows the effect of orientation on the results interpreted by TEM images. Aromatic sheet B and aromatic sheet C belong to the same molecule but are connected by an aliphatic group which distorted the orientation of the two sheets. When observed by TEM, sheets A and B which are part of different molecules will appear as stacked layers as they fulfill the (002) diffraction conditions while sheet C will appear as a

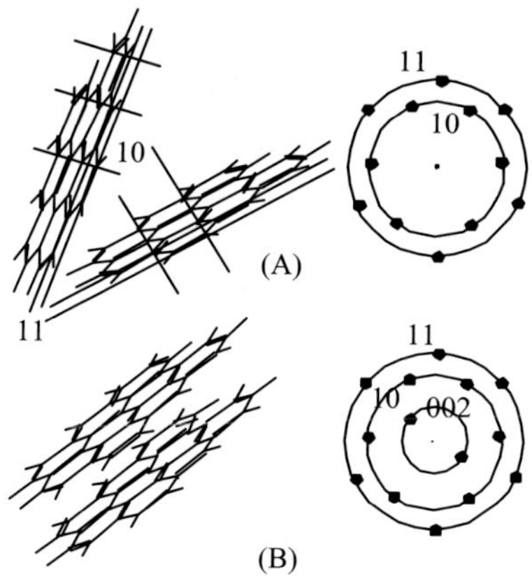

Figure 8.2. Model showing formation of electron diffraction pattern from (A) two randomly orientated layers and (B) two ordered layers at d_{002} spacing.

separate single layer because it will only produce 10 and 11 diffraction. Although it is not impossible, it must emphasized here that sheet C appearing as single layer is rather difficult to observe because of sample thickness, multiple diffractions, and presence of other randomly orientated layers. However, as first approximation, if one interprets the size of the fringe as the size of the molecule, the result will be erroneous. The molecule consists of two aromatic sheets, B and C, disorientated to each other as they are connected by an aliphatic group. Since the aliphatic group does not appear in the fringe image, the molecule appeared as two different fringes which only show the size of the aromatic sheet and not the molecule size. TEM provides information about the extent of orientation and the size of the aromatic

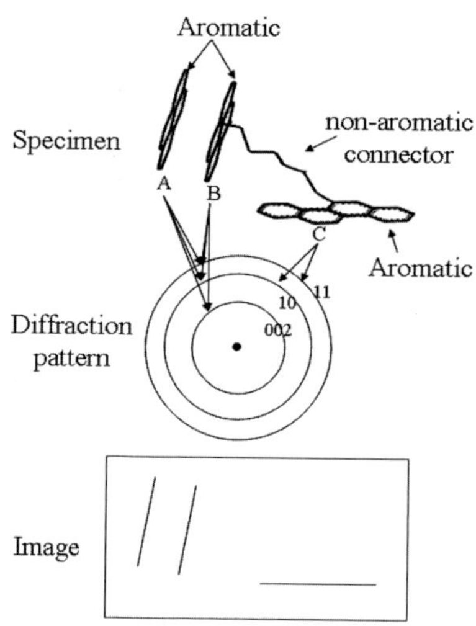

Figure 8.3. Model depicting fringe image of a small aromatic–aliphatic structure system.

sheet from the projected fringes. It is common to consider as a first approximation, the length of the fringe as the size of the aromatic sheet after assuming a simple geometry or shape of aromatic structure such as the pyrene or coronene series. However, it must be kept in mind that the fringe length is a two-dimensional information of a three-dimensional structure projected orthogonally on the screen. The size of the aromatic sheet calculated from the fringe length in many cases may not be the actual size of the sheet. The aromatic sheet size could be bigger than the size interpreted from the fringe layer length. Therefore, it is more common to report the fringe length as layer length rather than layer size. In addition, the TEM does not give any idea about the size and nature of the aliphatic group. Therefore, it is not possible to estimate the size of molecules by only TEM results. To estimate the size of a molecule, it is necessary to correlate information from TEM (aromatic sheet size and degree of orientation), NMR (ratio and nature of aromatic and aliphatic group), mass spectroscopy (distribution of molecular mass), and chemical analysis (C, H, N, O, and S fraction). None of the above techniques used alone can provide an insight into the molecular and physical structure of a carbonaceous material. For example, Aso *et al.*[30,31] employed TPO (chemical analysis) technique, TEM technique, and XRD results to discuss the molecular size, structure and structural transformation on heating of pure carbons, and anthracites. They estimated the molecular size including aliphatic groups by evaluating C, H, N, O fractions from TPO analysis coupled with the layer size obtained by TEM and XRD technique. They reported that the size determined by TPO method gave a much bigger molecular size than that obtained by TEM and XRD. The results were expected as the TEM and XRD give a calculated size only of an aromatic sheet. They recommended that different techniques should be used in conjunction with TEM and XRD to understand the carbon structure.

2.2. Quantitative Information from TEM Images

The image analysis technique consists of (1) filtration of TEM micrographs for noise reduction without losing appreciable information, (2) identification and reconnection of the fringe layers which were disconnected during the filtration step to obtain the extracted image for statistical analysis, and (3) the statistical analysis to evaluate the quantifiable structural parameters. We developed a new filtration technique and an image analysis computer algorithm to obtain the structural parameters as described in the following sections.

2.2.1. Image Processing

The first step in image analysis is digitization of the TEM micrographs. The digitized image is then subjected to a filtration procedure. All filtration procedures use essentially the same methodology of Fourier transform (FT) of the raw image, noise reduction followed by inverse FT, of which the most important step is the noise reduction. A comparison of conventional and our new method has been shown in Figure 8.4. The conventional method uses a square filter to take out the frequencies in the frequency domain image which fall in the square

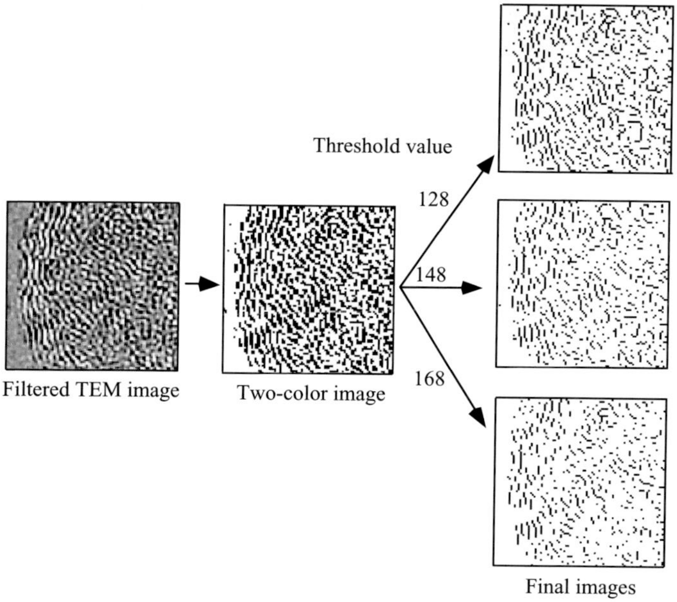

Figure 8.4. Conventional method to separate fringes based on variable threshold value criterion.

filter range, for example, 0.30/nm to 0.50/nm range in FT domain. The new filter that we have used is a step-filter which has only the lower limit which in our case is 0.3/nm in FT domain. The advantage of step filter over square filter is as follows.

Figure 8.2A shows TEM diffraction pattern from two randomly orientated aromatic layers. These layers will only produce diffraction spots from 10 and 11 planes. When these layers are orientated to each other at a distance that is equal to characteristic d-spacing of turbostratic carbons, the 002 diffraction spot will also appear in addition to 10 and 11 as shown in Figure 8.2B. If one applies a square filter with a frequency width of 0.30/nm to 0.50/nm in Fourier domain, the reflections from 10 and 11 appearing at 0.242/nm and 0.121/nm will be filtered out. As a result, information concerning single layers is lost along with the noise when square filter is used. However, we used a step-filter that filters out frequencies of 0.3/nm and lower, thereby retaining the reflections from 10 and 11.

2.2.2. Fringe Separation Technique

The filtered image is then inverse Fourier-transformed and shows a network of fringes connected by Y and/or T shape links. A gray scale TEM image contains pixels with value from 1 to 256 (or 0 to 255), where a pixel with value 1 is white and a pixel with value 256 is black. This gray image must be converted to two-color black and white image if one wishes to do image analysis. This is done by selecting a pixel value between 1 and 256 as cutoff value or threshold value (THV). The THV means all pixels in the image with values less than THV becomes 1 or white and pixels with value equal to or greater than THV becomes 255 or black. Thus by selecting a THV, a gray scale image can be converted to two-color black and white image which can be used for further processing, however, the selection of the THV remains a subjective criterion. In general, in the absence of any defined criterion, the THV is generally taken as 128, a mid-value between 1 and 256 as default value. The conventional method uses the variable threshold value criterion to separate the lattice fringes. This can better be understood from the fact that different extracted images can be obtained from the same filtered TEM image by changing the threshold value. Figure 8.4 shows three different images obtained from the same TEM image by using different threshold values viz; 128, 148, and 168. In this process not only small fringe layers are lost but also the layer size of large fringe layers is reduced. Thus the statistical analysis of all these extracted images will give different structural parameters though they all come from the same TEM image.

We developed a new unique method that not only retains the small layers but also keeps the size of large fringe layers intact while separating the fringes. This is done by first separating the fringe layers followed by reconnection of disconnected layers. The separation procedure is shown in Figure 8.5. The filtered gray scale TEM image (Figure 8.5A) is first converted to two-color image (Figure 8.5D) by selecting the threshold value of 128 and no efforts were made to separate the fringes. The two-color image is then made into a binary image for skeletonization as shown in Figure 8.5F and several layers can be seen connected to each other.

In the next step a duplicate of this skeletonized binary image (Figure 8.5F) is subjected to node extraction program. Figure 8.5G shows the node image. The images in Figure 8.5F and 8.5G are both binary and can be mathematically subtracted. The result of subtraction is shown in Figure 8.5H. Figure 8.5H shows that all the fringes are separated. Once the fringe layers are separated, we reconnected the disconnected layers using the geometrical parameters of these layers and positions of the nodes that were removed. The reconnected image becomes the final image for layer identification and characterization process. However, it must be clearly mentioned and emphasized at this point that reconnecting the neighboring disconnected fringes leads to the increase in the continuous layer thereby unfortunately introducing defects.

2.2.3. Image Analysis: Statistical Image Analysis Algorithm

The extracted image shows fringes as black lines. Some of these fringes are curvilinearity in shape. The NIH Image software is able to characterize these layers by first identifying a layer and then fitting an ellipse to the layer. The mid-point coordinates of the major axis of the fitted ellipse are assigned as the X, Y-coordinates and angle of inclination of the major axis to the X-axis is assigned as the angle of inclination of the layer. By counting the number of pixels, the actual length of the layer can be obtained. The total number of layers (N_L), angle of inclination (θ), X, Y-coordinates (x, y), major and minor axes (r_x, r_y) of the fitted ellipse and layer length (L) become the input data for the image analysis algorithm for statistical analysis. The algorithm makes use of four parameters; aspect ratio,

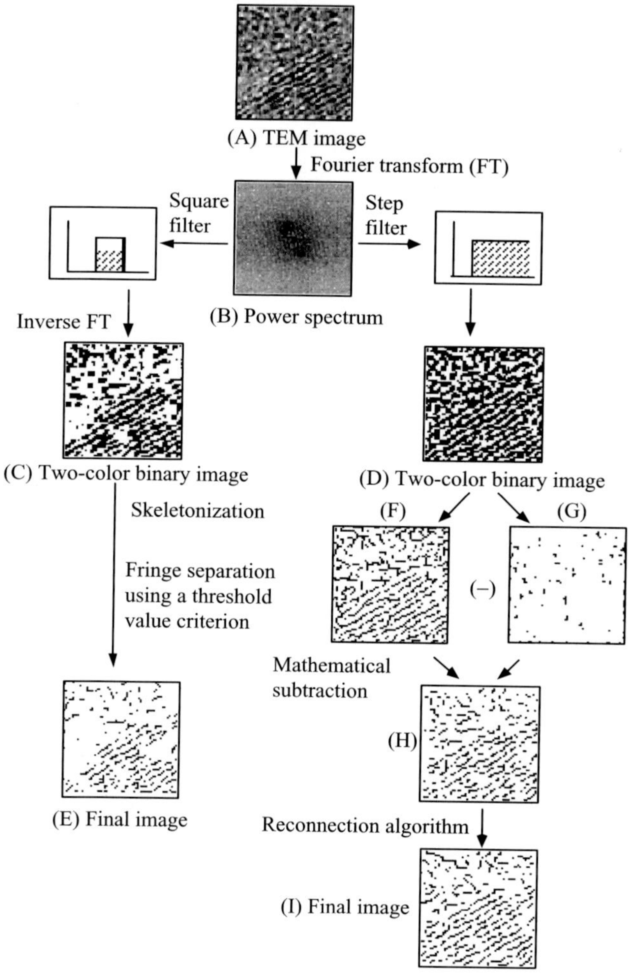

Figure 8.5. Comparison of conventional and new HRTEM image processing methods.

parallelism, overlap view parameter, and the interlayer spacing to evaluate the structural parameters as shown in Figure 8.6.

1. Aspect ratio: The aspect ratio has been defined as the ratio of major axis to minor axis of the fitted ellipse. This parameter defines a limit for curvilinearity in the fringes for calculation purposes. We took fringes with aspect ratio of 2 or more into consideration as constituent of carbon structure.
2. Parallelism: This parameter identifies layers that are parallel to the reference layer. A layer with an angle of inclination within ± 10 degree is considered as parallel to the reference layer. Perpendicular distance between two layers can only be obtained if they satisfy this criterion.

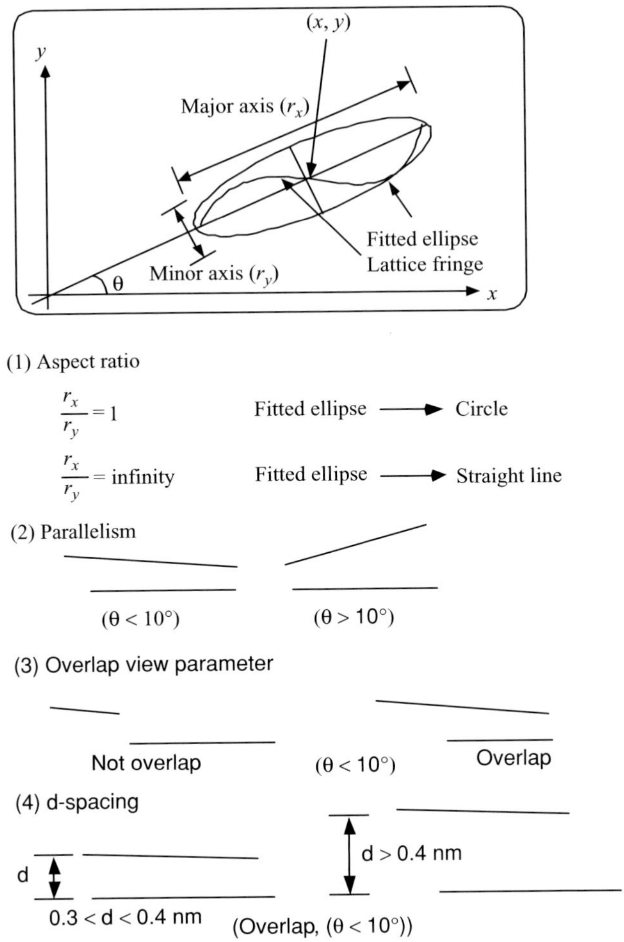

Figure 8.6. Parameters used to identify and classify fringes as stacked or single.

3. Overlap view parameter: This parameter identifies the layers that can form a stack with the reference layer. This is done by drawing perpendicular lines from the two edges of the reference layer and selecting the layers, which are encountered by either of these perpendicular lines.
4. Interlayer spacing: This parameter identifies layers whose perpendicular distances are close to that of typical turbostratic carbon, i.e., 0.344 nm. We selected those layers whose interlayer spacing falls between 0.3 and 0.4 nm for our calculations.

The flow diagram of the statistical image analysis algorithm (SIAA) is shown in Figure 8.7. The source code is written in FORTRAN 77. The validity of algorithm has been established by using a rather simple image and comparing the parameters obtained manually and that from the algorithm; a good

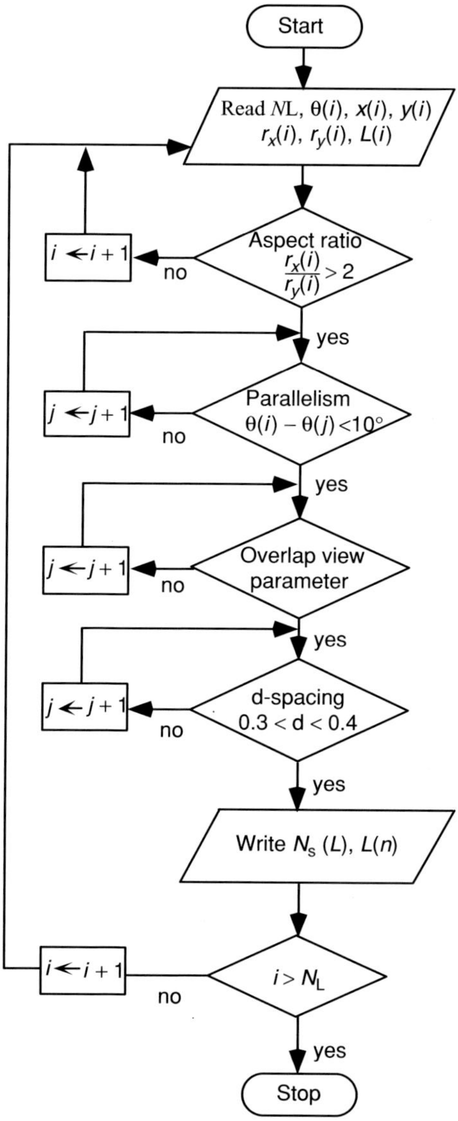

Figure 8.7. Flow diagram of statistical image analysis algorithm.

agreement between the computed parameters and those counted manually was obtained. Using this algorithm and the computerized imaging system it was possible to analyze many pictures to obtain general information from TEM images. The method has been applied to study the structure of various carbonaceous materials including asphaltenes and to quantify the transformations obtained by different processes such as pyrolysis, gasification, thermal treatment, coking, and graphitization.

3. Experimental Section

3.1. Samples

The petroleum asphaltene samples: Ven20 from a Venezuelan crude oil (API gravity = 10), UG8 from a Kuwaiti crude oil (API gravity = 26), and BG5 from a Kuwaiti crude oil (API gravity = 29) were prepared as described previously.[6] These are n-heptane asphaltenes prepared by addition of 40 cc n-heptane to 1 g of crude oil. After 24 hr, the solution was filtered and the precipitate was washed with hot n-heptane. These samples were all individually redissolved in toluene and reprecipitated, no effect of this extra purification step on any of our data was observed.

The Arabian Medium Heavy asphaltene vacuum resid was obtained from Dr. Eric Sheu and was separated as an n-heptane asphaltene. The procedure was described previously.[4] The bituminous coal sample was Tanito Harum (TH) from Indonesia. The TH coal asphaltene was obtained from Professor M. Iino as an n-hexane asphaltene from the coal liquefaction product from Tanito Harum coal. The liquefaction residue was first extracted with pyridine. The pyridine soluble fraction was isolated and dissolved in toluene. And the n-hexane asphaltenes were obtained.[5]

3.2. HRTEM Method

HRTEM fringe imaging requires thin samples that partly transmit the electron beam. For poorly ordered structures, the thickness of the sample is the most likely cause of errors as it is very difficult to eliminate the superimposition of lattice fringes.[24–32] In the present study, the samples were hand-ground to fine powder in ethanol and sprayed over a copper microgrid for TEM observation. TEM observation was performed with a 200 kV transmission electron microscope (JEOL, JEM-2010) and several pictures were taken for each sample from different spots to get a general view. The transmission electron microscope was equipped with an anticontamination trap, a computerized imaging system and EDS (energy dispersive spectroscopy) for elemental analysis. For disordered structures or small crystallite sizes, spherical aberration influences on TEM lattice images must be considered. The phase transfer function was calculated for $\lambda = 0.00251$ nm (electron beam wavelength) and $C_S = 0.5$ mm (spherical aberration coefficient) which are the conditions for the present observations. The transfer function is used to obtain the defocus position at which smoothness in contrast was guaranteed for 0.3 nm and higher basal spacing. In all cases, sub-micron size particles were first examined at moderate magnification to locate the wedge-shaped particles that are optically thin at the edge. The diffraction pattern was taken and elemental analysis was performed. A number of such regions were then imaged at high magnification ($\times$500 k). The TEM images were then subjected to image analysis for semiquantitative information such as interlayer spacing d, stacking distribution and layer size distribution.

4. Results and Discussion

Figures 8.8 A–H and 8.9 show the TEM and skeletonized images of four petroleum asphaltenes and a coal asphaltene, respectively. In these and subsequent images, the light regions correspond to electron transmission and the dark to electron scattering. As it has been shown previously,[32] the aromatic ring systems are readily imaged and are seen as dark lines while the alkanes are not readily observed. These micrographs exhibit some very local order and long-range disorder and are in agreement with extensive previously published work.[32,33] These images illustrate that the ring systems often occur with two or three stacking together. This observation has been repeatedly observed in carbonaceous materials, the small stacks have been referred to as basic structural units.[32,33] The structural parameters, stacking number, average layer size, and d-spacing obtained from these images using image analysis are presented in Table 8.1. The average number of layers per stack for the three petroleum asphaltenes and the TH coal asphaltene is nearly the same, ~2.3 and the layer spacing for petroleum asphaltenes is also nearly the same, 3.7 Å. This spacing is larger than the graphite spacing of 3.35 Å indicating that steric disruption does impact spacing distances. Smaller ring systems than graphite may cause the spacing to increase as well. The layer size for the petroleum asphaltenes is much larger than for the TH coal asphaltene, 1 nm for the former, 0.7 nm for the latter. These results support previous conclusions[10] that asphaltene structures are determined by the interplay of stacking vs. steric hindrance. The smaller ring systems for the coal asphaltene are consistent with the smaller alkane fraction. Here, we see for disparate asphaltenes of significantly different molecular size and alkane content, the stacking parameters (number of molecules in stack, layer spacing) are nearly constant. That is, the invariant defining characteristic of the asphaltenes is not molecular size or aromatic to saturate fraction. The invariant of asphaltenes is the balanced interplay between stacking propensity and steric hindrance to stacking.

The fact that asphaltenes routinely exhibit a lack of long-range order indicates that the lack of order is not some peculiarity of sample preparation or peculiar to certain asphaltenes. This is a general finding. Here, we have examined many samples of several asphaltenes and consistently observed that the length scale of the aromatic ring systems is ~1 nm for the petroleum asphaltenes and is 0.7 nm for the TH coal asphaltene. Smaller ring systems for the coal asphaltene are clearly shown by fluorescence spectroscopy and by fluorescence depolarization (FD) spectroscopy.[7] TRFD obtains the molecular size from the rotational correlation times in solution, a very different technique than HRTEM. For UG8 asphaltene a molecular diameter of 2.1 nm was obtained and for the TH coal asphaltene, 1.1 nm, respectively. TRFD is sensitive to the entire molecule, not just the aromatic ring portion, and so gives somewhat larger sizes. In addition, the FD result corresponds to a hydrodynamic diameter. Reasonable agreement between two very different methods especially as to the relative sizes of the petroleum and coal asphaltene is very encouraging.

To examine the effects of alkanes on aromatic ring stacking, we collected HRTEM images of several model compounds consisting of aromatic ring systems.

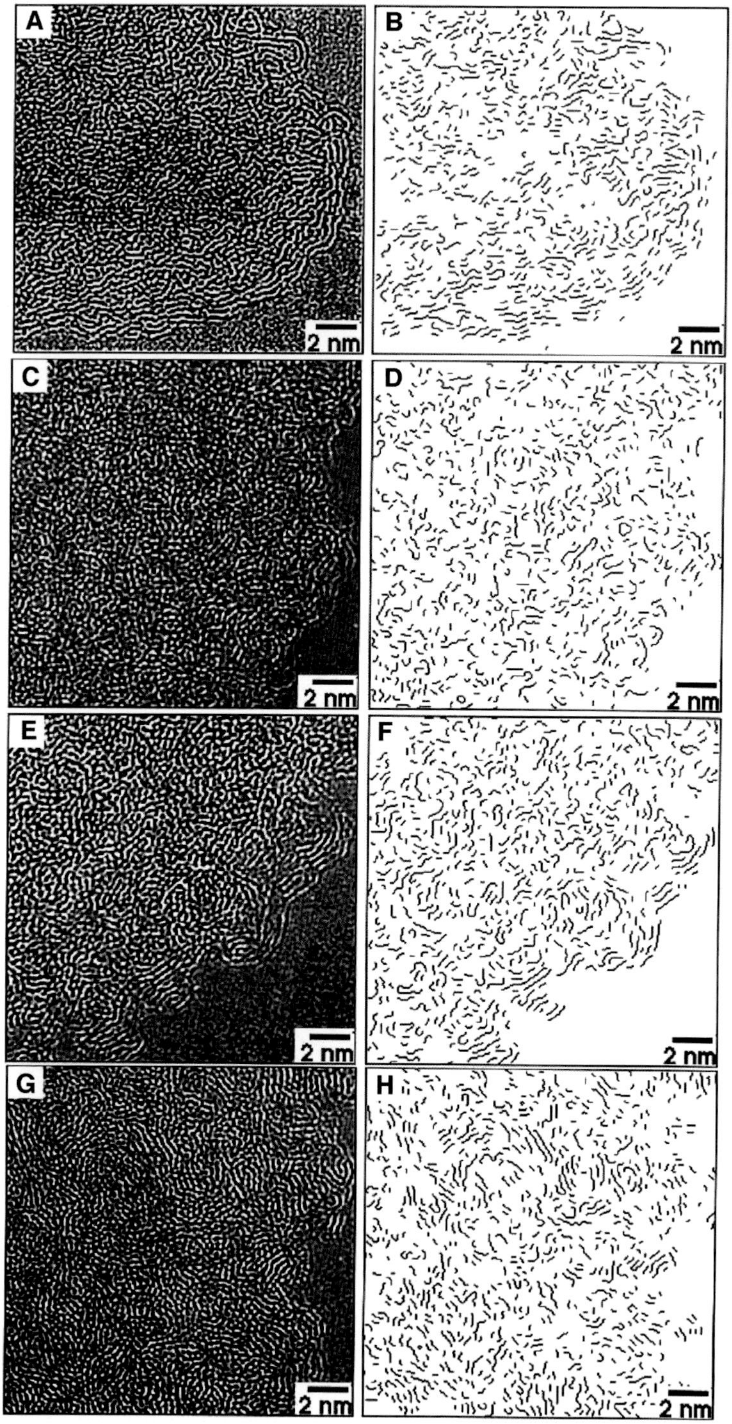

Figure 8.8. HRTEM and skeletonized images of asphaltenes (A, B) BG5, (C, D) Ven20, (E, F) UG8, and (G, H) Arab resid.

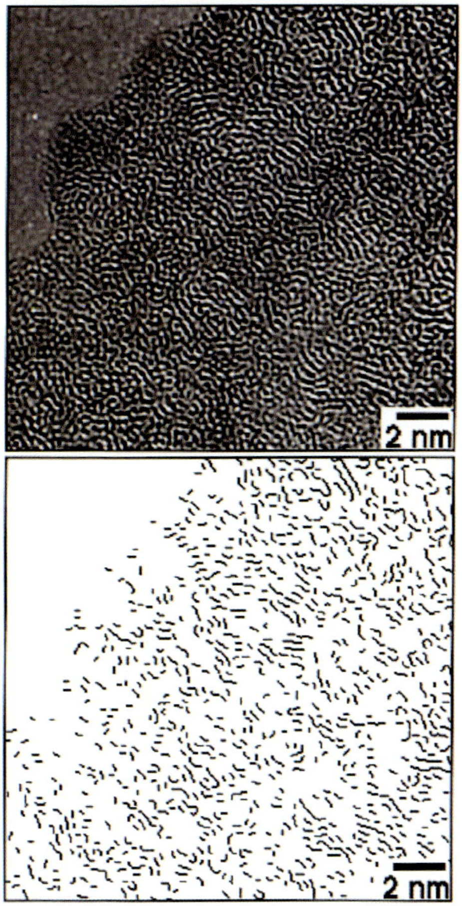

Figure 8.9. HRTEM and skeletonized image of TH coal asphaltene.

Their structures are shown in Figure 8.10. Figure 8.11 shows the HRTEM images obtained for an unsubstituted aromatic compound, naphtho[2,3-*a*]pyrene. Long-range order and aromatic ring stacking are quite evident. There is an overall curvature which is quite evident for this sample. HRTEM requires samples that

Table 8.1. Structural Parameters of Asphaltenes

	Structural parameters[a]		
Sample	n	L (nm)	d_{002} (nm)
BG5	2.1	1.1	0.37
Ven20	2.2	1.0	0.36
UG8	2.3	1.0	0.38
Arab resid	2.4	1.0	0.37
TH coal asphaltene	2.3	0.7	0.37

[a] n is the average number of molecules in a stack; L is the length of the molecule (fringe); d_{002} is the fringe spacing in a stack.

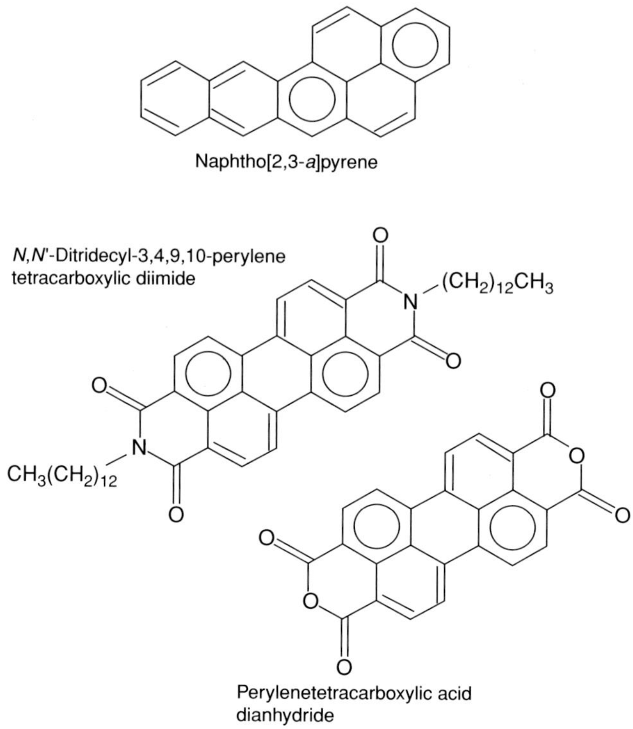

Figure 8.10. Chemical structures of model compounds.

are very thin in order to obtain little distortion of the electron beam, typically a few nm. Consequently, sample edge effects may be important. In Figure 8.11, the curvature might be associated with edge effects and not present in bulk crystals. Nevertheless, the important feature for our purposes is the occurrence of long-range order in this sample.

Previous work implies that the presence of alkanes can disrupt stacking in aromatic ring systems.[10] To test this, we use two similar compounds except that one has long alkane chains, the other does not. Figures 8.12 and 8.13 show the HRTEM images of perylenetetracarboxylic dianhydride and N,N'-ditridecyl-3,4,9,10-perylenetetracarboxylic diimide, respectively. From the perspective of ordering, the primary difference between these two compounds is the presence or absence of the two n-C_{13} carbon chains; Figure 8.10 shows the structure of these two compounds. The image of the aromatic system without the alkane chains (Figure 8.12) exhibits a great deal of long-range order. For the alkylated molecule, the ordering is disrupted. This result corroborates previous work utilizing TRFD, ^{13}C NMR, and IR results indicating that alkane substitutents disrupt stacking.[10] Melting point data of alkylbenzenes, alkyl naphthalenes, and alkyl anthracenes indicate the same thing. The HRTEM results are a direct observation of this disruption.

In addition to melting point data, we found that the alkylated compound is >100 times more soluble in toluene than the unsubstituted compound. Figure 8.14

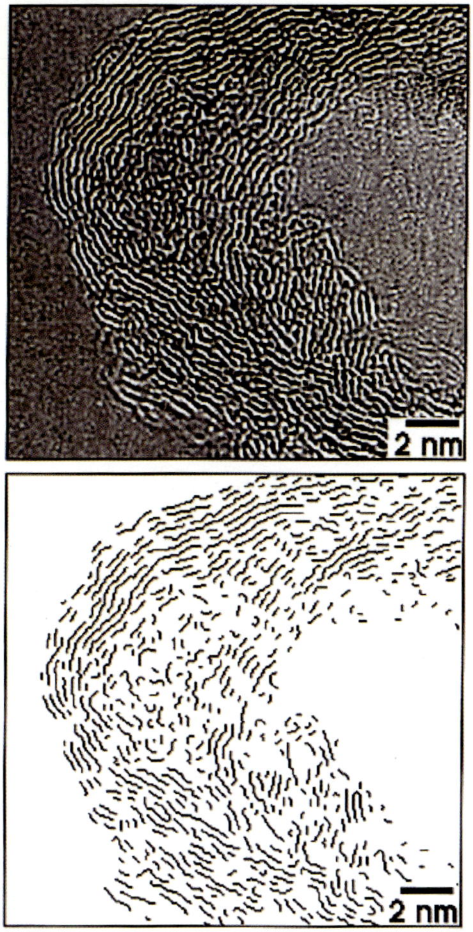

Figure 8.11. HRTEM and skeletonized image of naphtho[2,3-*a*]pyrene.

shows the absorption spectrum for two toluene solutions of these two dyes. The unsubstituted compound is barely detectable in the toluene solution, whereas the alkylated compound produces strong absorption. The absorption coefficients of the two compounds are comparable as shown by diffuse reflection spectroscopy. This is exactly the point we are attempting to make about asphaltenes. If the intermolecular interaction is too strong, then the molecule would not dissolve (in toluene). By definition, asphaltenes are soluble in toluene. We view that the solubility classification dictates the stacking behavior, 2–3 molecules per stack. These constraints dictate a class of molecular structures that scale. Large aromatic ring systems necessitate substantial alkyl substitution; small ring systems necessitate little alkyl substitution (or they would dissolve in *n*-heptane).

Because HRTEM measures very thin edges, sample preparation or other spurious effects can impact the images. For example, the extent of grinding or

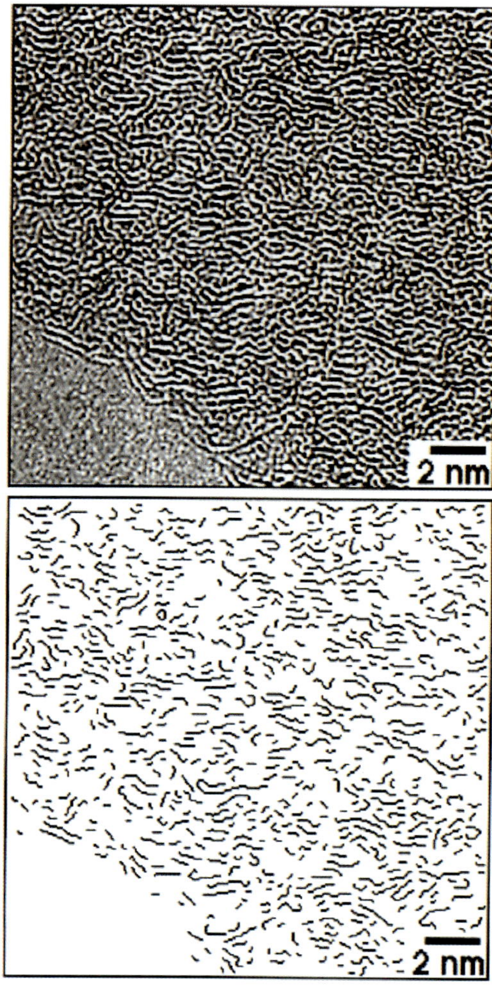

Figure 8.12. HRTEM and skeletonized image of perylenetetracarboxylic dianhydride.

mechanical disruption can have a large effect on the data. As one would expect, it is easy to obtain spurious disorder. Nevertheless, all samples we examined that had alkyl substitution exhibited disordered images. Octaethyl porphyrin complexes are an example, these were chosen because they are about the same size as the TH coal asphaltene.[7] The TH coal asphaltene possesses very little alkane and always shows disorder, so it is not surprising that the alkyl-substituted porphyrins exhibit disorder as well. We found that unsubstituted phthalocyanine did not exhibit order, which was surprising. Perhaps edge effects dominate. Two alkyl-substituted phthalocyanines did not show order as expected.

The salient conclusion is that the asphaltene solubility class precludes significant long-range order. For petroleum asphaltenes with their significant alkanes, long-range order is precluded by the alkanes even for relatively large ring systems,

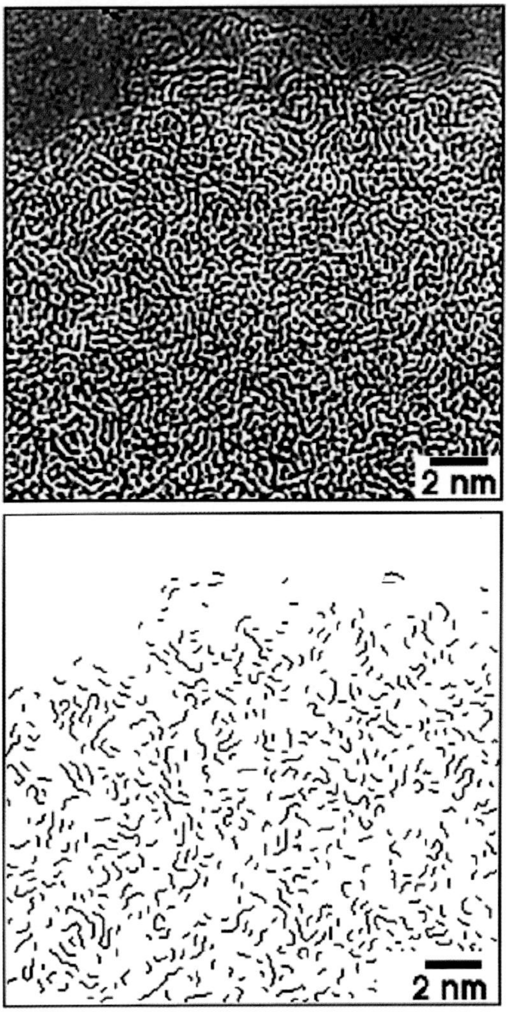

Figure 8.13. HRTEM and skeletonized image of N,N'-ditridecyl-3,4,9,10-perylenetetracarboxylic diimide.

about seven rings on average.[7,19] The HRTEM images clearly show aromatic ring systems for petroleum asphaltenes that are ~1 nm in diameter. This corresponds rough to seven fused ring systems in a pericyclic molecule. For instance, coronene is just shy of 1 nm in (in-plane) diameter. Thus, the HRTEM images of asphaltene ring systems are consistent with many other measurements including TRFD and direct imaging via scanning tunneling microscopy (STM)[35]. In addition, aromatic ring systems in coal asphaltenes are found by HRTEM to be smaller than those in petroleum asphaltenes, ~0.7 nm. This is also known from a comparison of fluorescence emission spectra of these two types of asphaltenes. For coal asphaltenes which have very little alkane substitution, long-range order is precluded

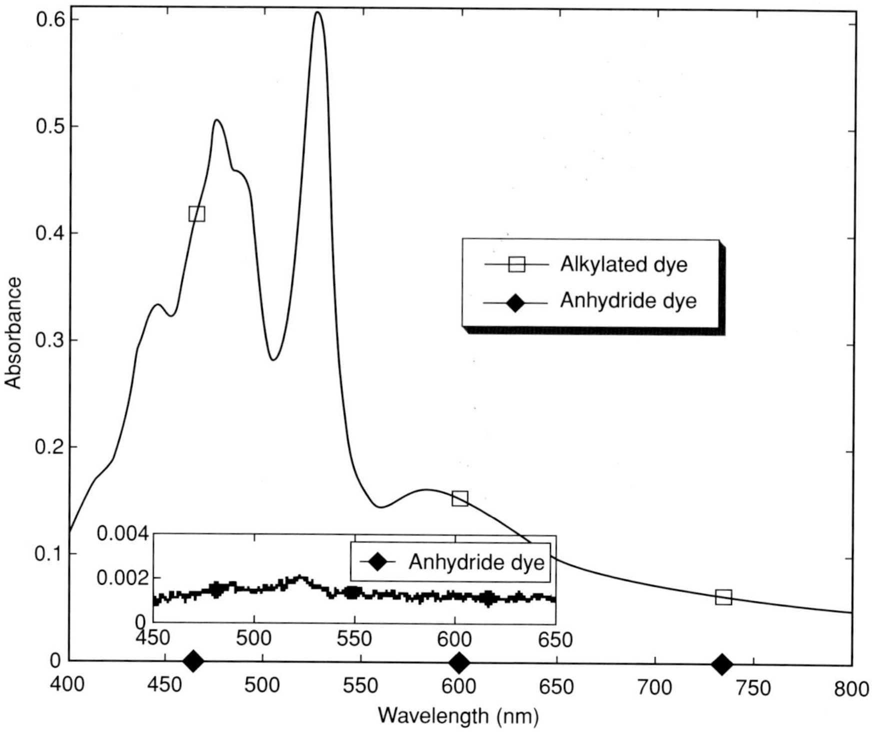

Figure 8.14. The optical absorption spectrum of saturated solutions of the two "perylene" dyes. The alkylated dye is quite soluble, the unsubstituted dye is barely detectable in toluene.

by consisting of ring systems sufficiently small that intermolecular binding is weak. Because van der Waals interaction of aromatic ring system scales with the number of rings, coal asphaltenes can only possess small ring systems, about four rings on average. Of course, both coal and petroleum asphaltenes possess a significant width of the distribution of ring sizes. Nevertheless, the governing principles of the relations between structure and solubility still apply. It is possible that the petroleum asphaltenes with their long alkane chains stack somewhat better in nanoaggregates in solution or in crude oil. In the solid the alkane chains will tend to form spheres distorting stacking further, whereas with nanoaggregate petroleum asphaltenes in solution or in crude oil, the alkane chains will have some tendency to distend into the continuous hydrocarbon (or toluene) phase thereby removing some of the steric interaction responsible for stacking disruption. For those coal asphaltenes with very small alkane fractions, this effect might be quite small.

The TH coal asphaltene has been shown by ^{13}C NMR to have a very small fraction of saturated carbon (15%) whereas the BG5 is ~60% saturated carbon fraction. The XRD profile shown in Figure 8.15 confirms this result. For BG5, a large peak is seen at 20° that is known to originate from saturated carbon. The TH coal asphaltene does not show this peak. Agreement on ring system size

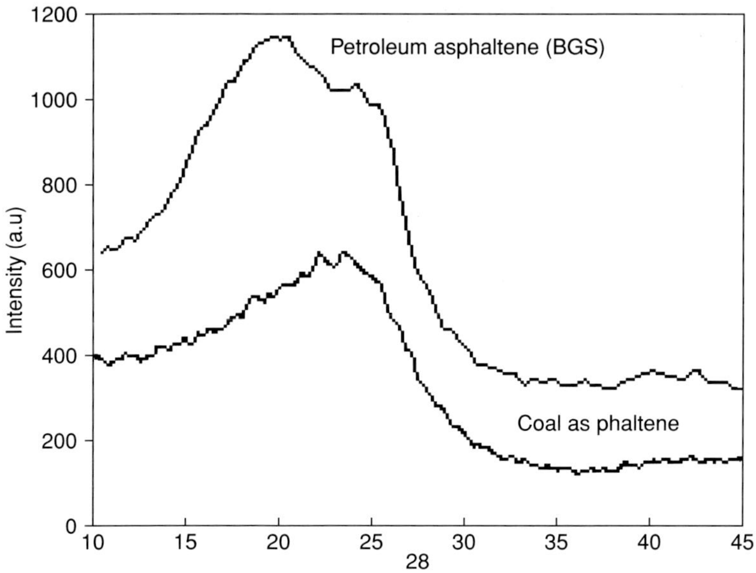

Figure 8.15. X-ray diffraction profile of petroleum asphaltene (BG5) and TH coal asphaltene.

between these HRTEM measurements and TRFD measurements[10] strengthen all conclusions.

5. Conclusions

The present results from HRTEM corroborate the ideas that simple chemical principles govern the identity of asphaltenes; steric repulsion competes with π-bond stacking to establish asphaltene molecular identity. The solubility classification of asphaltenes mandates certain invariants in the stacking behavior of asphaltene molecules, the average intermolecular spacing and the average number of molecules in the stack. In turn, these invariants require a balance between intermolecular stacking of aromatic ring systems vs. steric disruption induced by alkanes. To achieve these invariants, larger ring systems mandate larger alkane chains; likewise smaller ring systems mandate smaller alkane chains. The ability to relate simple chemical principles to asphaltene identity is crucially dependent on the solution of the 20 year, order of magnitude controversy over asphaltene molecular weight. In addition, the HRTEM results are consistent with the presence of fused ring systems of $\sim$1 nm in diameter in petroleum asphaltenes. This direct molecular imaging is in accord with conclusions from many other measurements. HRTEM is seen to provide vital information about asphaltene molecular structure and stacking behavior. Furthermore, HRTEM shows that stacking invariants follow from known simple principles and HRTEM confirms the differing sizes of the fused aromatic ring systems in different asphaltenes. These results provide a sturdy foundation for understanding asphaltenes

Acknowledgments

This study was a part of a collaborative research of O.C. Mullins and H. Groenzin from Shlumberger Doll Research, USA, and Prof. Tomita and A. Sharma from Tohoku University, Japan. Authors on behalf of all group members wish to thank Professor Iino of Tohoku University for supplying the TH coal asphaltene sample and Dr. Eric Sheu of Vanton Research Laboratory for supplying the vacuum resid asphaltene.

References

[1] Crick, F. (1988). *What Mad Pursuit, a personal View of Scientific Discovery*, Basic Books, New York.
[2] Chilingarian, G.V. and T.F. Yen (eds.) (1978). *Bitumens, Asphalts, and Tar Sands*. Elsevier Scientific Publishing, New York.
[3] Bunger, J.W. and N.C. Li (eds.), (1984). *Chemistry of Asphaltenes*. American Chemical Society, Washington, DC.
[4] Sheu, E.Y. and O.C. Mullins (eds.) (1995). *Asphaltenes: Fundamentals and Applications*. Plenum, New York.
[5] Mullins, O.C. and E.Y. Sheu (eds.) (1998). *Structures and Dynamics of Asphaltenes*. Plenum, New York.
[6] Groenzin, H. and O.C. Mullins (1999). Asphaltene molecular size and structure, *J. Phys. Chem. A*. **103**, 11237.
[7] Groenzin, H. and O.C. Mullins (2000). Molecular sizes of asphaltenes from different origin, *Energy Fuels* **14**, 677.
[8] Boduszynski, M.M. (1988). Composition of heavy petroleums. 2. Molecular characterization, *Energy Fuels* **2**, 597.
[9] Miller, J.T., R.B. Fisher, P. Thiyagarajan, R.E. Winans, and J.E. Hunt (1998). Subfractionation and characterization of mayan asphaltene, *Energy Fuels* **12**, 1290.
[10] Buenrostro-Gonzalez, E., H. Groenzin, C. Lira-Galeana, and O.C. Mullins (2001). The overriding chemical principles that define asphaltenes, *Energy Fuels* **15**, 972.
[11] Hortal, A.R., B. Martinez-Haya, M.D. Lobato, J.M. Pedrosa, and S. Lago (2006). On the determination of molecular weight distributions of asphaltenes and their aggregates in laser desorption ionization experiments, *J. Mass Spec.* **41**, 960–968.
[12] Sheu, E.Y., M.M. De Tar, and D.A. Storm (1991). Rheological properties of vacuum residue fractions in organic solvents, *Fuel* **70**, 1151.
[13] George, G.N. and M.L. Gorbaty (1989). Sulfur K-edge x-ray absorption spectroscopy of petroleum asphaltenes and model compounds, *J. Am. Chem. Soc.* **111**, 3182.
[14] Kelemen, S.R., G.N. George, and M.L. Gorbaty (1990). Direct determination and quantification of sulphur forms in heavy petroleum and coals : 1. The X-ray photoelectron spectroscopy (XPS) approach, *Fuel* **69**, 939.
[15] Waldo, G.S., O.C. Mullins, J.E. Penner-Hahn, and S.P. Cramer (1992). Determination of the chemical environment of sulfur in petroleum asphaltenes by X-ray absorption spectroscopy, *Fuel* **71**, 53.
[16] Mitra-Kirtley, S., O.C. Mullins, J. van Elp, S.J. George, J. Chen, and S.P. Cramer (1993). Determination of the nitrogen chemical structures in petroleum asphaltenes using XANES spectroscopy, *J. Am. Chem. Soc.* **115**, 252.

[17] Bergmann, U., H. Groenzin, O.C. Mullins, P. Glatzel, J. Fetzer, and S.P. Cramer (2003). Carbon K-edge X-ray Raman spectroscopy supports simple yet powerful description of aromatic hydrocarbons and asphaltenes, *Chem. Phys. Lett.* **369**, 184.

[18] Gordon, M.L., D. Tulumello, G. Cooper, A.P. Hitchcock, P. Glatzel, O.C. Mullins, S.P. Cramer, and U. Bergmann (2003). Inner shell excitation spectroscopy of fused aromatic molecules by electron energy loss and X-ray Raman techniques, *J. Phys. Chem. A.* **107**(41), 8512.

[19] Ruiz-Morales, Y. (2002). HOMO-LUMO gap as an index of molecular size and structure for polycyclic aromatic hydrocarbons (PAHs) and asphaltenes: a theoretical study, *J. Phys. Chem. A.* **106**(46), 11283.

[20] Millward, G.R. and D.A. Jefferson (1978). In: P.A.Thrower (ed.), *Chemistry and Physics of Carbon*. Marcel Dekker, New York, Vol. 14 pp. 1–78.

[21] Furuta, T., Y. Yamashita, and M. Shiraishi (1989). *Tanso* **140**, 241–247.

[22] Davis, K.A., R.H. Hurt N.Y.C. Yang, and T.H. Headley (1995). *Combust. Flame* **100**, 31–40.

[23] Palotás, Á.B., L.C. Rainey, A.F. Sarofim, J.B.V. Sande, and P. Ciambelli (1996). Effect of oxidation on the microstructure of carbon blacks, *Energy Fuels* **10**, 254–259.

[24] Sharma, A., T. Kyotani, and A. Tomita (1999). A new quantitative approach for microstructural analysis of coal char using HRTEM images, *Fuel*, **78**, 1203–1212.

[25] Sharma, A., T. Kyotani, and A. Tomita (2000). Comparison of structural parameters of PF carbon from XRD and HRTEM techniques, *Carbon* **38**, 1977–1984.

[26] Sharma, A., T. Kyotani, and A. Tomita (2000). Direct observation of layered structure of coals by a transmission electron microscope, *Energy Fuels* **14**, 515–516.

[27] Sharma, A., T. Kyotani, and A. Tomita (2000). Direct observation of raw coals in lattice fringe mode using high-resolution transmission electron microscopy, *Energy Fuels* **14**, 1219–1225.

[28] Sharma, A., T. Kyotani, and A. Tomita (2001). Quantitative evaluation of structural transformations in raw coals on heat-treatment using HRTEM technique, *Fuel* **80**, 1467–1473.

[29] Sharma, A., H. Kadooka, T. Kyotani, and A. Tomita (2002). Effect of microstructural changes on gasification reactivity of coal chars during Low temperature gasification, *Energy Fuels* **16**, 54–61.

[30] Aso, H., K. Matsuoka, A. Sharma, and A. Tomita (2004). Evaluation of size of graphene sheet in anthracite by a temperature-programmed oxidation method, *Energy Fuels* **18**, 1309–1314.

[31] Aso, H., K. Matsuoka, A. Sharma, and A. Tomita (2004). Structural analysis of PVC and PFA carbons prepared at 500–1000 °C based on elemental composition, XRD, and HRTEM, *Carbon* **42**, 2963–2973.

[32] Oberlin, A. (1989). In: P.A. Thrower (ed.), *Chemistry and Physics of Carbon*. Marcel Dekker, New York, Vol. 22, p. 1.

[33] Oberlin, A., S. Bonnamy, and P.G. Rouxhet (1999). In: P.A. Thrower and L.R Radovic (eds.), *Chemistry and Physics of Carbon*. Marcel Dekker, New York, Vol. 26, p. 1.

[34] Sharma, A., H. Groenzin, O.C. Mullins, and A. Tomita (2002). Probing order in asphaltenes and aromatic ring systems by HRTEM, *Energy Fuels* **16**(2), 490.

[35] Zajac, G.W., N.K. Sethi, and J.T. Joseph (1994). Molecular imaging of petroleum asphaltenes by scanning tunneling microscopy, *Scan. Micros.* **8**, 463.

9

Ultrasonic Spectroscopy of Asphaltene Aggregation

Gaelle Andreatta, Neil Bostrom, and Oliver C. Mullins

1. Introduction

High-Q high-resolution ultrasonic spectroscopy is used to detect the onset of aggregation in asphaltene solutions and micelle formation with standard surfactants. This technique allows determination of the speed of sound in solution to a few parts in a million. Aggregation is accompanied by a change in compressibility enabling this ultrasonic technique to determine concentrations of aggregate formation. The ability to detect the critical micelle concentration (CMC) for different standard surfactants with CMCs varying over two orders of magnitude establishes high-Q ultrasonics as a sensitive probe. Asphaltene in toluene is shown to have a critical nanoaggregate concentration (CNAC) of $\sim$100 mg/l which is a much lower concentration than previous reports using other techniques. The strong tendency of asphaltenes to aggregate explains why asphaltene "molecular" weights determined by vapor pressure osmometry are always well in excess of accurate asphaltene molecular weights. Findings herein are consistent with the Yen model with the restriction that the asphaltene molecules are relatively small having mean molecular weights of 750 g/mole. This restriction on molecular structure enables identification of key dynamics of asphaltene behavior, thereby considerably extending the Yen model.

The Yen model consists of a hierarchy of aggregation for asphaltene solutions.[1] Different hierarchies of aggregation correspond to different energies of interaction. There has been considerable uncertainty as to what concentrations correspond to what aggregation. Reports utilizing surface tension measurements[2] and microcalorimetry[3] indicate that primary aggregation or critical micelle concentration occurs at the grams per liter concentration of asphaltenes in toluene (the presence of dispersed water in toluene affects this result[4]). These concentrations seem rather high and there is a question as to whether these techniques have requisite sensitivity and applicability for detecting the primary aggregation of asphaltenes. The uncertainties regarding aggregation vs. concentration are so great as to cast doubt on the Yen model itself. Exacerbating this situation is that

Gaelle Andreatta, Neil Bostrom, and Oliver C. Mullins • Schlumberger-Doll Research, Ridgefield, CT 06877.

controversies surrounding asphaltene molecular weight and molecular architecture preclude the concept of tracing the Yen aggregation hierarchy back to first principles of intermolecular interaction.

Fortunately, the situation has changed dramatically. We would modify the Yen model in that an additional constraint needs to be incorporated. The basic building block of the Yen model—the polydisperse asphaltene molecules—are now largely understood. First, the asphaltene molecular weight has been established beyond doubt. There is broad agreement that virgin crude oil asphaltene molecular weights have a 750 g/mole centroid with an "FWHM" of 500–1000 g/mole. In pioneering work, M.M. Boduszynski originally obtained this result using field-ionization mass spectroscopy (FIMS).[5] A series of time-resolved fluorescence depolarization studies (TRFD) obtains this result by determination of molecular rotational diffusion constants and by the dispersion of these diffusion constants with wavelength.[6–9] These diffusional constant studies are in total agreement with translational diffusion constant studies made by Taylor dispersion using optical absorption detection.[10] More recently, mass spectral studies using ESI-FT-ICR-MS (electrospray ionization fourier transform ion cyclotron resonance),[11] atmospheric pressure photoionization (APPI-MS),[12] and atmospheric pressure chemical ionization (APCI-MS)[13] all agree. The only mass spectral technique that gives inconsistent results involves laser desorption which is now understood to be sensitive to many artifacts.

The TRFD studies also show that the primary intermolecular attraction increases with increasing size of the polycyclic aromatic hydrocarbon (PAH) systems and decreases with increasing alkyl substitution.[6,9] These results suggest that the primary intermolecular attraction is van der Waals interaction of π electrons in more or less a molecular stack. There is clearly an affect on intermolecular interactions from polar groups such as sulfoxide as well,[8] but sulfoxides are often present in small concentrations.[14] The PAH systems in asphaltenes have been directly imaged by scanning tunneling microscopy (STM)[15] and by high-resolution transmission electron microscopy (HRTEM).[16] STM directly images fused rings in individual asphaltene chromophores and finds an average size of asphaltene of 1 nm for the PAH ring systems.[15] HRTEM determines that virgin crude oil asphaltenes have chromophores of $\sim$1 nm in size which is expected for $\sim$7 fused rings.[16] These results are consistent with TRFD studies of rotational diffusion in comparison to known chromophores.[6–9] ESI-FT-ICR-MS studies show that the number of aromatic ring systems in asphaltenes varies from 2 to 12 rings—with an average of 7 rings per molecule.[11] Given the low molecular weights of asphaltenes, and the relatively large fused ring system, the asphaltene molecules have one or sometimes two ring systems per molecule.

The molecular size and molecular architecture of asphaltenes can be used to understand primary aggregate formation of asphaltenes. HRTEM studies image small (e.g., 2 or 3 molecules in a stack) stacks of PAHs at the graphitic sheet separation distance.[16] In particular, the intermolecular attractive and repulsive forces of asphaltenes are short range with likely increasing steric hindrance with greater aggregation thereby implying a size limit on primary aggregation. A picture emerges that primary asphaltene aggregation results when the high energy PAH ring systems are accessible to stacking. The alkane substituents are then subject to a restricted volume to avoid interfering with stacking. Additional high energy PAH surface is complexed until steric repulsion of the alkyl substituents precludes close approach of further PAH ring systems. Since steric repulsion is short range, the expectation is that this size limit is reached with a small number (5–10) of molecules. Vapor pressure osmometry (VPO) studies, carried out at high concentrations, are often in error for molecular weight determination by this range.

These dynamic expectations built upon understanding asphaltene molecular structure can now be tested. First, primary asphaltene aggregation should be at fairly low concentration due to the relatively high binding energy of stacked, large PAH systems. Second, growth of nanoaggregates should largely cease due to increasingly restricted access to PAH systems with increasing aggregate size. A sharp limit to the size of these nanoaggregates might be found. Cluster formation of nanoaggregates should be much lower energy as the high energy interactions are consumed in the primary aggregation process. Thus cluster formation should not occur until much higher concentrations. The question remains, what technique has provable requisite sensitivity to detect primary asphaltene aggregation?

High-Q, high-resolution ultrasonic spectroscopy is one of the most direct and sensitive methods to detect the formation of micelles. The speed of sound is a direct probe of the bulk and so is not sensitive to surface issues, and one can easily exclude transients in the measurements. High-Q ultrasonic measurements have been used successfully to monitor many types of phase transitions in solution. Here, high-Q ultrasonic measurements are performed on aqueous and toluene solutions containing standard surfactants and are compared against known literature values when available to validate the methodology. The governing equations for the micelle phase equilibrium model are given, and all data presented here are interpreted within this framework. The surfactants used included SDS in water, C_{16}TAB in water, Tween 80 in water and separately in toluene, and Brij 35 in toluene. CMC determinations via ultrasonic spectroscopy are shown to agree well for known surfactants over a broad range of CMCs. In particular, surfactants with very small values of CMCs are treated without diffculty using high-Q ultrasonic measurements. We employ these ultrasonic techniques to study several asphaltene–toluene systems up to concentrations of several grams per liter. In addition, with density measurements, the ultrasonic results provide a direct measure of monomer and micelle compressibilities. For all of these solutions, density measurements were made enabling the determination of micellar or nanoaggregate compressibilities in solution and, in some case, monomer compressibilities in solution. Comparisons are emphasized between standard surfactants and asphaltenes.

2. Ultrasonic Spectroscopy

The word spectroscopy is often associated with electromagnetic waves. Indeed techniques such as UV, IR, and visible spectroscopy, fluorescence, light scattering and so on are well established and widely used. Acoustic waves can also be applied to fluid or material analysis. Ultrasonic spectroscopy employs high frequency acoustical waves in the frequency range of 20 kHz to several GHz, to determine different properties of the material.

Ultrasonic spectroscopy allows fast and nondestructive analysis of small samples, in our case, from 1 to 2 mL. The amplitudes of deformations in the ultrasonic waves employed are extremely small insuring no damage to the sample. Moreover, the ultrasonic waves can propagate through most materials, including optically opaque materials.

Ultrasonic propagation is characterized by velocity and attenuation. High-Q ultrasonic spectrometry gives precise frequencies and bandwidths for a series of resonances. From these data, sound speed and attenuation are obtained. If a phase transition happens during a titration, the ultrasonic velocity and attenuation will reflect this change. Changes such as sedimentation, aggregation, and micellization can be detected using this technique.

2.1. Ultrasonic Resonances

Different ultrasonic methods have been developed during the past decades. For instance, time of flight of an ultrasonic pulse through a sample gives the velocity and amplitude reduction corresponds to the attenuation. The Helmoltz resonator technique and the cylindrical resonator technique have also been employed.[17] In our work, we use a plane-wave resonator technique. The resonance cavity is a plane parallel resonator (Figure 9.1).

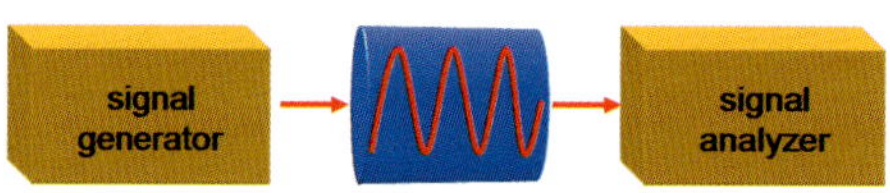

Figure 9.1. Ultrasonic resonance cell. The ultrasonic waves are compressional, thus longitudinal.

The cavity has two ultrasonic transducers, an emitter and a receiver. Ultrasonic waves launched into the cell experience interference from reflections at impedance interfaces (e.g. cell-fluid interface). At certain frequencies corresponding to a whole number of half wavelength in the ultrasonic cell, a transmission resonance occurs. At resonance, there is a large increase of the amplitude measured by the second transducer. The resonant technique builds standing waves at eigenfrequencies inside a cell filled with the solution of interest.

The resonant technique has less signal-to-noise problems than the pulse-echo technique since the signal is of longer duration. The resonant technique can also be more accurate since it is easier to measure frequency very accurately although it does take longer, on the order of seconds, to determine the speed of sound. Using the slower measurement could cause problems with materials that change on a shorter time scale. Since the samples here do not change during the experimental times (from 5 to 30 min in most cases), this problem is not an important inconvenience.

2.2. Plane Wave Propagation

In liquids, linear ultrasonic waves generally produce longitudinal deformations and these waves are characterized, in a linear case, by their velocity and their attenuation. The fundamental quantities obtained from a continuous ultrasonic wave experiment are ultrasonic attenuation and ultrasonic velocity. Attenuation is determined by the energy losses in the compressions and rarefactions in ultrasonic waves, and includes absorption and scattering contributions. The density ρ and the elasticity E of the medium determine ultrasonic velocity u:

$$u = \sqrt{\frac{E}{\rho}}. \tag{9.1}$$

The elasticity E is equal to the bulk modulus in liquids and to the Young modulus for solids. E is extremely sensitive to the molecular organization and intermolecular interactions in the medium and is very sensitive to temperature.

An *ideal one-dimensional isolated resonator in which plane waves are propagated* is relevant for the resonator cell used here. Diffraction and coupling resonances will be ignored in this case. It will be shown later that these assumptions are only true in certain frequency ranges and that the coupling to other acoustic modes cannot be overlooked in other frequency ranges. In the ideal case, the amplitude of the deformation due to the ultrasonic wave propagating in the z-direction in liquid is given by

$$A = A_0 \exp(-\alpha z) \cos\left(\frac{2\pi z f}{u} - 2\pi f t\right), \tag{9.2}$$

where α is the attenuation of the wave, f is the frequency and u the velocity of the wave, and A_0 the magnitude of deformation where l_s is the path in the solution.[18]

For resonance, f_n the frequency of the nth longitudinal resonance is

$$f_n = \frac{nu}{2l_s}. \tag{9.3}$$

Since l_s is constant, we can deduce the relationship between the relative changes in frequency and in velocity:

$$\frac{\delta u}{u} = \frac{\delta f_n}{f_n}. \tag{9.4}$$

Each of the standing wave resonances is characterized by a resonant frequency f_n and a quality factor:

$$Q = \frac{f_n}{\Delta f_n}, \tag{9.5}$$

where Δf_n is the full width at maximum divided by $\sqrt{2}$.

The peak number can be determined using the frequencies of two adjacent modes:

$$\frac{f_{n+1} - f_n}{f_n} = \frac{1}{n}. \tag{9.6}$$

2.3. Experimental Section

The ultrasonic measurements were performed on the HRUS 102 high-resolution ultrasonic spectrometer from Ultrasonic Scientific Ltd. The speed of sound is determined using the resonance technique in the range of frequencies between 2 and 20 MHz. We used ~5 MHz for our experiments; the spectrometer can measure the speed of sound to 5 parts in 10^6. The measurements are made using two identical cells filled with a volume of 1–2 mL, one filled with the experimental solution and the other with the solvent (water or toluene). Both cells are mounted together in the same block and are thermostated at 25 ± 0.1°C, enabling small differences in ultrasonic velocity to be determined.[19] Each cell is a resonance cavity, the cavity walls are a glass chamber built with two lithium niobate transducers on two opposite sides of the cell; one transducer is used as the signal source, the other is the receiver[19] (see Figure 9.1).

Two main factors determine the resolution of the measurements: the quality of the resonance (including a large quality factor Q and the absence of satellites of the resonance peaks) and the stability of the resonances. The first factor requires a high precision in the parallel alignment of the cells and the quality of the lithium niobate piezotransducers. To have strong interference, a high impedance contrast at the boundaries is needed.[19]

Frequencies can be measured to very high accuracies.[20] In a titration, the conversion of frequency to sound speed is performed using Eq. (9.4). A typical ultrasonic spectrum of water is shown in Figure 9.2, several of the sharp ultrasonic resonances are shown in Figure 9.3 at the frequency range used for this study. Figures 9.4 and 9.5 show the comparable spectra for toluene. In these figures, the amplitude of the output signal is shown as a function of the frequency of the acoustic signal. In addition to the narrow acoustic cell resonances, Figures 9.2 and 9.4 show broad resonances (at ~4 MHz, 7 MHz, 10 MHz, for example) that are associated with ultrasonic resonances in the glass walls of the cells and with the transducers. These spectral regions were avoided in all of our experiments. For distilled water at 25°C, the speed of sound is $u_0 = 1496.7$ m/s with a resolution of 0.0075/m. In toluene at 25°C, the speed of sound is $u_0 = 1307.1$ m/s with a resolution of 0.0065 m/s.

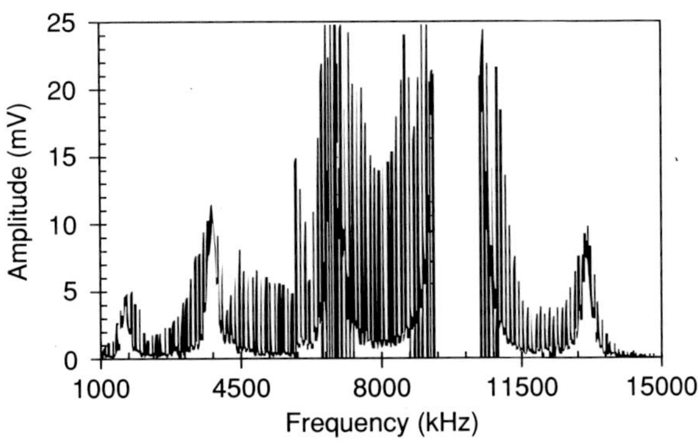

Figure 9.2. Ultrasonic spectrum of water at 25°C showing acoustic cell resonances

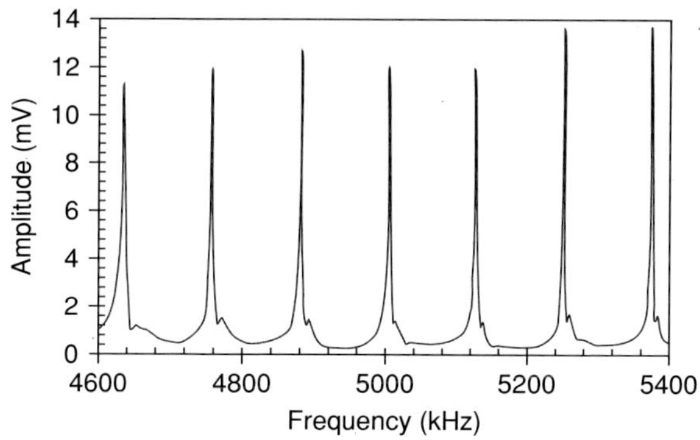

Figure 9.3. Several acoustic cell resonances in the ultrasonic spectrum of water at 25°C

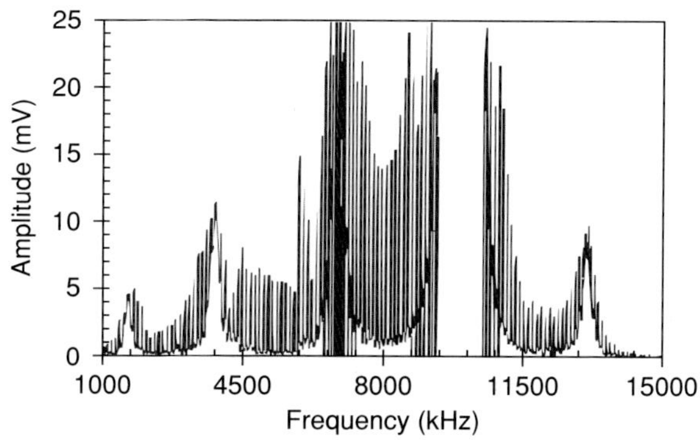

Figure 9.4. Ultrasonic spectrum of toluene at 25°C showing acoustic cell resonances

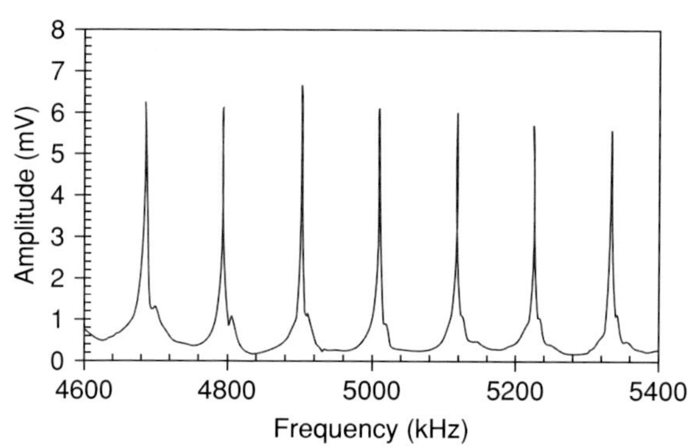

Figure 9.5. Several acoustic cell resonances in the ultrasonic spectrum of toluene at 25°C

2.4. Compressibility of Liquids and Ultrasonic Velocity

The isentropic compressibility κ_s is defined by:

$$\kappa_s = -\frac{1}{V}\left(\frac{\partial V}{\partial P}\right)_S, \tag{9.7}$$

where V is the volume, P the pressure.

If the mass of the considered system is constant,

$$\kappa_s = \frac{1}{\rho}\left(\frac{\partial \rho}{\partial P}\right)_S, \tag{9.8}$$

where ρ is the density of the liquid.

In a liquid, the speed of sound can be linked to the density and the adiabatic compressibility (also see Eq. (9.1)):

$$u^2 = \frac{1}{\rho \kappa_s}. \tag{9.9}$$

For ultrasonic spectrometry, the adiabatic compressibility is used because the compressions and decompressions in ultrasonic waves are too fast for heat dissipations.

The measurements of both the solution density and the ultrasonic velocity allow determination of the solution compressibility; thus, these two different experiments (measuring the density and measuring the ultrasonic velocity) are often performed together.[20–26] Densities are essentially integral quantities and are not very sensitive to variations in aggregation. Compressibilities are differential quantities and are thus much more sensitive to variations in aggregation. Measurement of ultrasonic velocity is a sensitive probe primarily due to the dependance of ultrasonic velocity on compressibility.

3. Micellar Aggregation Model

Ultrasonic spectrometry can be seen as a very useful technique for the study of colloidal solutions and of processes such as aggregation, gelation, flocculation, etc. Micellization[21–27] is a critical component of the science of surfactants and has been studied with great interest by a large number of methods.[28] Micellar aggregation can be linked with the speed of sound in the colloidal solution and the critical micellar concentration (CMC) can be determined very precisely using ultrasonic spectrometry.[21–26,29] Furthermore, both direct and inverse micelles can be studied by this technique.[29,30]

3.1. Theory

Different thermodynamical models have been developed for the study of micellar systems, using the fact that the micellization process can be seen either as a chemical equilibrium described by the law of mass action or as a phase equilibrium.[31]

For dilute surfactant solutions, the ultrasonic velocity and the density of the solution can be expressed as functions of the concentration of surfactants.[21] In general, surfactant molecules in solution are present in monomeric and micellar forms. Here we follow the treatment given in Zielinski *et al.*[21] Consider a volume V, if there are w_0 grams of solvent, w grams of surfactant, w_1 grams of surfactants in the monomeric form and w_m grams of surfactant in the micellar form,

$$w = w_1 + w_m. \tag{9.10}$$

For a solution of volume V, if v_0 is the specific volume of the solvent, $\tilde{v}_1 = \partial V/\partial w_1$ is the apparent specific volume of the monomeric form and $\tilde{v}_m = \partial V/\partial w_m$ is the apparent specific volume of the micellar form, then.

$$V = w_0 v_0 + w_1 \tilde{v}_1 + w_m \tilde{v}_m. \tag{9.11}$$

The mass of the surfactant solution is equal to

$$\rho V = w_0 + w_1 + w_m. \tag{9.12}$$

The density of the solution ρ is then

$$\rho = \rho_0 + (1 - \tilde{v}_1 \rho_0) c_1 + (1 - \tilde{v}_m \rho_0) c_m, \tag{9.13}$$

where ρ_0 is the density of the solvent ($\rho_0 = 1/v_0$), c_1 is the mass concentration of surfactant in the monomeric form and c_m is the mass concentration of surfactant in the micellar form. c is the total mass concentration of surfactant $c = c_1 + c_m$.

If assumed that the phase equilibrium model is valid here,[23] then

- For $c < cmc$, $c_1 = c$ and $c_m = 0$
- For $c > cmc$, $c_1 = cmc$ and $c_m = c - cmc$,

where c is the mass concentration of the solution and *cmc* is the numerical value at the critical micellar concentration (CMC). That is, at CMC, $c_m = 0$ but any new increase in surfactant concentration corresponds to increasing the concentration of micelles c_m.

For $c < cmc$

$$\rho = \rho_0 + (1 - \tilde{v}_1 \rho_0) c. \tag{9.14}$$

And for $c > cmc$

$$\rho = \rho_0 + (\tilde{v}_m - \tilde{v}_1) \rho_0 cmc + (1 - \tilde{v}_m \rho_0) c. \tag{9.15}$$

To obtain the adiabatic compressibility of the solution κ_s as a function of the concentrations of surfactants in the solution, the density is differentiated with respect to pressure P at constant entropy:

$$\kappa_s = \frac{1}{\rho} \left(\frac{\partial \rho}{\partial P} \right)_S. \tag{9.16}$$

Differentiating Eq. (9.13):

$$\left(\frac{\partial \rho}{\partial P} \right)_S = \left(\frac{\partial \rho_0}{\partial P} \right)_S + \left(\frac{\partial (1 - \tilde{v}_1 \rho_0)}{\partial P} \right)_S c_1 + (1 - \tilde{v}_1 \rho_0) \left(\frac{\partial c_1}{\partial P} \right)_S + \left(\frac{\partial (1 - \tilde{v}_m \rho_0)}{\partial P} \right)_S c_m + (1 - \tilde{v}_m \rho_0) \left(\frac{\partial c_m}{\partial P} \right)_S. \tag{9.17}$$

Assuming that the concentration of monomers c_1 and the concentration of micelles c_m change with pressure only through the changes in the volume of the solution, it follows that:

$$\left(\frac{\partial c_1}{\partial P}\right)_S = c_1\kappa \tag{9.18}$$

$$\left(\frac{\partial c_m}{\partial P}\right)_S = c_m\kappa. \tag{9.19}$$

The adiabatic compressibility of the solvent is defined by

$$\kappa_0 = \frac{1}{\rho_0}\left(\frac{\partial \rho_0}{\partial P}\right)_S \tag{9.20}$$

The apparent adiabatic compressibility of the surfactant in the monomeric form is defined by

$$\tilde{\kappa}_1 = -\frac{1}{\tilde{v}_1}\left(\frac{\partial \tilde{v}_1}{\partial P}\right)_S. \tag{9.21}$$

And the apparent adiabatic compressibility of the surfactant in the micellar form is defined by

$$\tilde{\kappa}_m = -\frac{1}{\tilde{v}_m}\left(\frac{\partial \tilde{v}_m}{\partial P}\right)_S. \tag{9.22}$$

Considering Eq. (9.17) and the previous definitions:

$$\begin{aligned}\rho\kappa &= \rho_0\kappa_0 + \kappa[(1-\tilde{v}_1\rho_0)c_1 + (1-\tilde{v}_m\rho_0)c_m] \\ &\quad + \rho_0 c_1\tilde{v}_1(\tilde{\kappa}_1-\kappa_0) + \rho_0 c_m\tilde{v}_m(\tilde{\kappa}_m-\kappa_0),\end{aligned} \tag{9.23}$$

where κ is the isentropic compressibility of the solution.

From Eqs. (9.13) and (9.23):

$$\kappa = \kappa_0 + (\tilde{\kappa}_1-\kappa_0)\tilde{v}_1c_1 + (\tilde{\kappa}_m-\kappa_0)\tilde{v}_mc_m. \tag{9.24}$$

Equation (9.9) relates the ultrasonic velocity in liquid to density and adiabatic compressibility. For dilute solutions ($c_1 \ll 1$ and $c_m \ll 1$):

$$u = u_0 + \frac{u_0}{2}\left[\tilde{v}_1\left(2-\frac{\tilde{\kappa}_1}{\kappa_0}\right) - v_0\right]c_1 + \frac{u_0}{2}\left[\tilde{v}_m\left(2-\frac{\tilde{\kappa}_m}{\kappa_0}\right) - v_0\right]c_m. \tag{9.25}$$

For $c < cmc$,

$$u = u_0 + \frac{u_0}{2}\left[\tilde{v}_1\left(2-\frac{\tilde{\kappa}_1}{\kappa_0}\right) - v_0\right]c. \tag{9.26}$$

And for $c > cmc$,

$$\begin{aligned}u = u_0 &+ \frac{u_0}{2}\left[\tilde{v}_1\left(2-\frac{\tilde{\kappa}_1}{\kappa_0}\right) - \tilde{v}_m\left(2-\frac{\tilde{\kappa}_m}{\kappa_0}\right)\right]cmc \\ &+ \frac{u_0}{2}\left[\tilde{v}_m\left(2-\frac{\tilde{\kappa}_m}{\kappa_0}\right) - v_0\right]c.\end{aligned} \tag{9.27}$$

Equations (9.26) and (9.27) show that the ultrasonic velocity can be modeled by two straight lines in the plot of ultrasonic velocity vs. concentration; one line segment below CMC, the other, above.[21–23,28] The apparent compressibilities of the monomeric form and the micellar form can be deduced from ultrasonic velocity and mass density measurements.[21–23,28,31,32]

3.2. Experimental Results on Surfactants

The different surfactants used here are sodium dodecylsulfate (SDS) from Sigma-Aldrich Chemicals, purity >99%; hexadecyltrimethylammonium bromide ($C_{16}TAB$) from Sigma-Aldrich, purity >99%; polyoxyethylene 23 laurylether (Brij 35) ($C_{12}E_{23}$) from Acros Chemicals, purity 99%; polyoxyethylene sorbitan mono-oleate (Tween 80) from Acros Chemicals, purity >99%. The aqueous solutions were made in Milli-Q water for the density measurements and in distilled water for the ultrasound measurements. The organic solutions were prepared in reagent grade toluene 99.8% from Acros and Sigma-Aldrich. The chemical structure of Tween 80 is shown in Figure 9.6.

All ultrasonic spectra were acquired by diluting solutions from the highest concentrations, with stepwise concentration reductions. Each curve consisted of approximately 15–25 points. For each point, a quantitative dilution was performed; the solution was stirred and allowed to equilibrate for 10–15 minutes prior to recording the ultrasonic frequency for that concentration. The duration of a single run was typically 4–6 hr. No difference in the spectrum was observed if the equilibration time was increased or decreased by a factor of two. The reproducibility of the measurements was checked and found to be very good. All the ultrasonic titrations exhibit a break between two straight-line segments in the curves, as expected from the Eqs. (9.26) and (9.27). The critical micelle concentrations of the different surfactants were given by the intersection of two straight-line segments.

The densities were measured at $25 \pm 0.1°C$ with an Anton Paar DMA 4500 densitometer with a resolution of $5 \cdot 10^{-5}$ g/cm^3. Each measurement was done twice and averaged. The densities of Milli-Q water and toluene were found to be close to the expected value: 0.99704 g/cm^3 and 0.86222 g/cm^3, respectively. Apparent specific volumes can be calculated from the slope of the density vs. concentration graph utilizing Eq. (9.15). The apparent specific volumes of the monomer and of the micelle of SDS, $C_{16}TAB$, and Tween 80 in water and Tween 80 and Brij 35 in toluene were calculated using a straight-line segment of the points at concentrations above the CMC, which are more reliable than the points

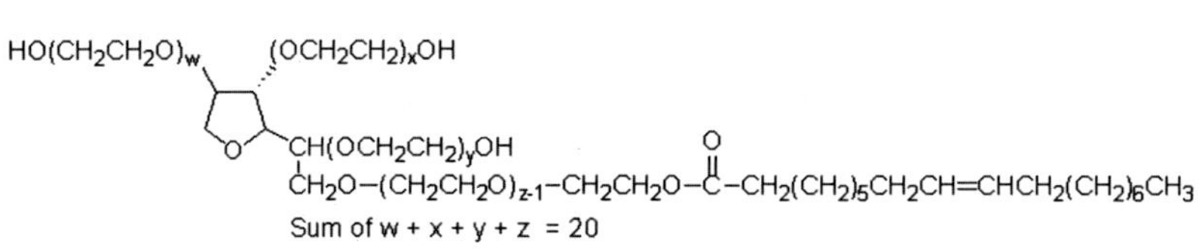

Figure 9.6. Chemical structure of Tween 80

Table 9.1. Speeds of Sound, Densities, and Adiabatic Compressibilities at the Temperature of 25°C of the Solvents Used in This Study

Solvent	u (m/s)	ρ(g/cm^3)	v_0 (cm^3/g)	κ_0 (10^{-5} bar^{-1})
Water	1496.7	0.99704	1.00297	4.48
Toluene	1307.1	0.86222	1.1598	6.79

at concentrations lower than CMC due to the resolution of the densitometer. For Tween 80 in water, the apparent specific volume of the micelle was determined but the apparent specific volume of monomer was not calculated here due to insuffcient measurement resolution at the very low concentrations.

Table 9.1 presents the ultrasonic speeds of sound and densities measured for our two solvents, water and toluene. The derived compressibilities from Eq. (9.9) are also given.

3.2.1. Ionic Surfactants in Aqueous Solution

Two well-known ionic surfactants were studied here in aqueous solutions: SDS (anionic) and C_{16}TAB (cationic). The CMCs were deduced from the ultrasonic velocity vs. concentration plots and taken as the intersections of the two straight-line segments (above and below CMC) and were found to be close to the CMC given in the literature.

Figure 9.7 gives the solution ultrasonic velocity vs. SDS concentration (top) and the solution density vs. SDS concentration (bottom). In all plots, the solid points were used for data fitting (and not the open points).

Figure 9.8 gives the solution ultrasonic velocity vs. C_{16}TAB concentration (top) and the solution density vs. C_{16}TAB concentration (bottom). Table 9.2 presents the CMCs obtained here at 25°C for the ionic surfactants SDS and C_{16}TAB and also gives literature values for these CMCs. Note the agreement.

The density measurements provide apparent specific volumes; these combined with the ultrasonic velocity measurements, allow us to derive the apparent adiabatic compressibilities of monomer and micelle. Our apparent specific volumes and derived compressibilities for SDS and C_{16}TAB are given in Table 9.3 and compared with literature values.

The results are in excellent agreement with the references 22–25, 33. $\tilde{\kappa}_1$ is large and negative for the ionic surfactants SDS and C_{16}TAB in water [21,23–25] while $\tilde{\kappa}_m$ is large and positive.[21,23–25,33] Table 9.3 shows that specific volumes are not greatly affected by micelle formation while apparent compressibilities are greatly affected. This high molar compressibility for the surfactants in the micellar state is probably due to the compressibility of the internal core of micelles and is close to the adiabatic compressibility of pure hydrocarbon liquids with the same length of hydrocarbon chain.[24] For example, Table 9.1 shows that the compressibility of toluene is $6.79 \cdot 10^{-5}$ bar^{-1}.

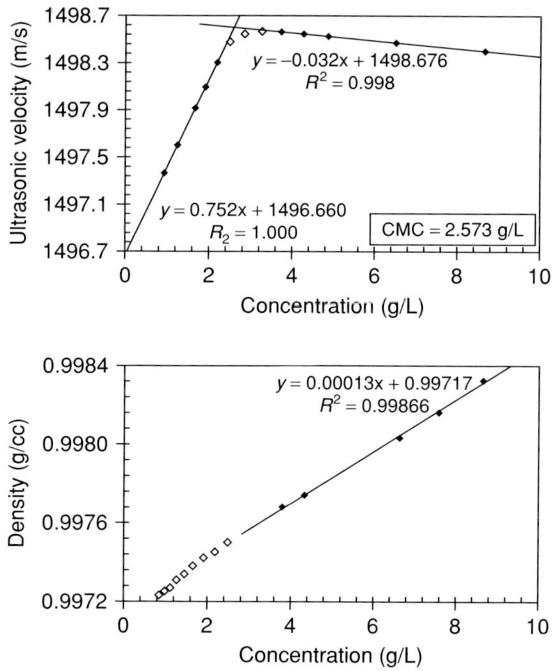

Figure 9.7. Ultrasonic sound velocity vs. concentration of SDS in aqueous solution at 25°C (top) and density measurements of SDS in aqueous solution at 25°C (bottom)

3.2.2. Nonionic Surfactants in Aqueous and Organic Solvents

Tween 80 (see Figure 9.6) is a nonionic amphiphile composed of 20 oxyethylene groups on an oxocyclopentane core which are the hydrophilic part of the molecule, while the hydrocarbon chain is the hydrophobic part of the molecule. Tween 80 was studied here both in water and in toluene. Brij 35 is a polyoxyethylene dodecyl ether, which is a nonionic surfactant with a C_{12} hydrophobic alkyl chain and a hydrophilic chain of 23 polyoxyethylene subunits. Brij 35 was studied in toluene. Figure 9.9 shows the solution ultrasonic velocity of Tween 80 in water (top) and the corresponding density curve (bottom). At the very low concentration of 8 mg/L, Tween 80 exhibits a clear change in the ultrasonic velocity curve. Thus, the ultrasonic method for CMC determination is established over 2.5 orders of magnitude in concentration.

Two nonionic surfactants were run in toluene; Figure 9.10 shows the solution ultrasonic velocity vs. Tween 80 concentration (top) and the corresponding density curve (bottom). Figure 9.11 shows the solution ultrasonic velocity vs. Brij 35 concentration (top) and the corresponding density curve (bottom). For these nonionic surfactants in toluene, it is not surprising that there is not a clear break in the velocity curves (even though we fit sections of the curve with straight lines.) We interpret the changes in slopes in the ultrasonic velocity curves as effective CMCs for Figures 9.10 and 9.11. Table 9.4 lists the CMCs determined here for standard

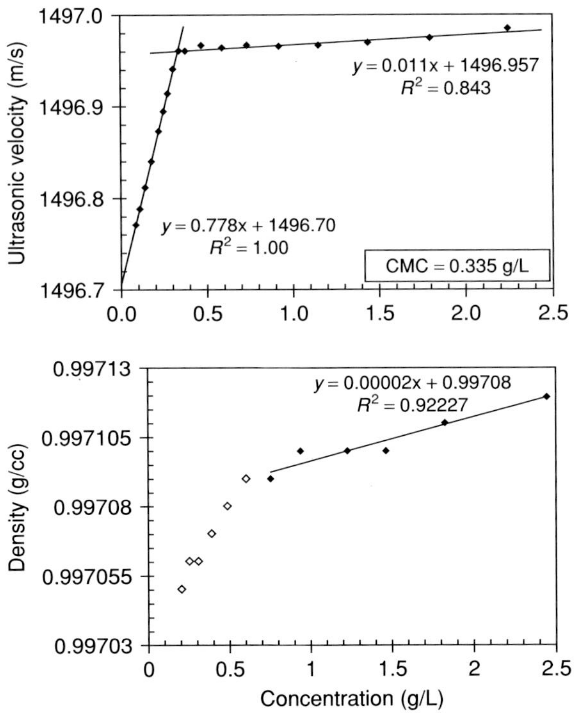

Figure 9.8. Ultrasonic sound velocity vs. concentration of C_{16}TAB in aqueous solution at 25 °C (top) and density measurements of C_{16}TAB in aqueous solution at 25 °C (bottom).

Table 9.2. Values of the Critical Micelle Concentrations at 25°C of the Ionic Surfactants Used in This Study

Surfactants	CMC in this study (g/L)	CMC reference (g/L)
SDS	2.573	2.393 (ref 28) 2.408 (ref 29)
C_{16}TAB	0.335	0.328 (ref 28) 0.334 (ref 24)

Table 9.3. Specific Volumes and Compressibilities of the Monomeric and Micellar Forms of the Ionic Surfactants Used in This Study at Temperature of 25°C

Surfactants	$\tilde{v}_1$ (cm^3/g)	$\tilde{v}_m$ (cm^3/g)	$\tilde{\kappa}_1$ (10^{-5} bar^{-1})	$\tilde{\kappa}_m$ (10^{-5} bar^{-1})
SDS (this work)	0.822	0.873	−1.97	4.02
SDS (ref)	0.813 (ref 33)	0.854 (ref 33)	−1.93 (ref 23)	4 (ref 23)
				4.3 (ref 22)
C_{16}TAB (this work)	0.864	0.983	−1.64	4.33
C_{16}TAB (ref)	0.962 (ref 24)	0.989 (ref 24)	−0.039 (ref 24)	4.28 (ref 24)
	0.964 (ref 25)	0.988 (ref 25)	−1.25 (ref 25)	4.24 (ref 25)
		1.002 (ref 33)		

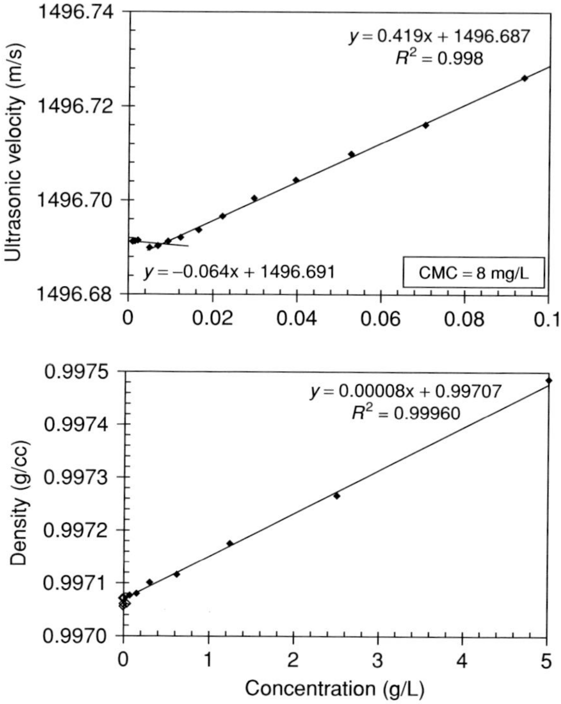

Figure 9.9. Ultrasonic sound velocity vs. concentration of Tween 80 in water at 25°C (top) and density measurements of Tween 80 in water at 25°C (bottom)

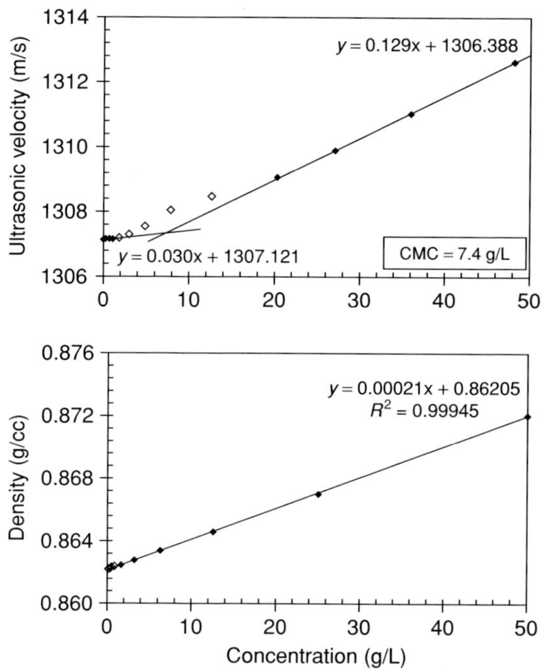

Figure 9.10. Ultrasonic sound velocity vs. concentration of Tween 80 in toluene at 25°C and density measurements of Tween 80 in toluene at 25°C

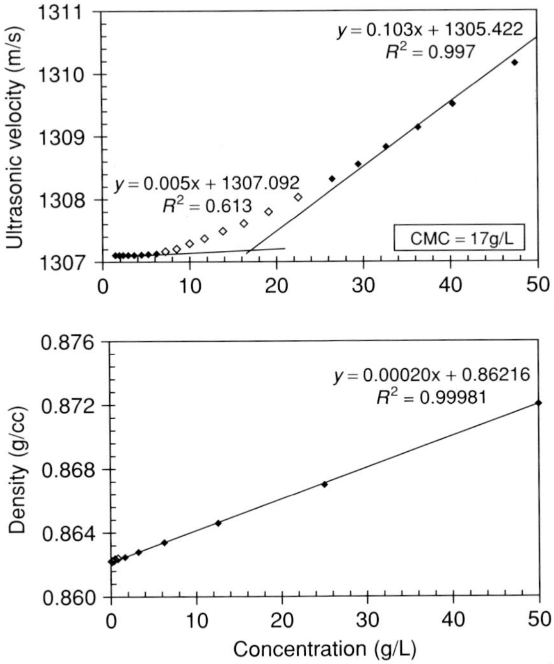

Figure 9.11. Ultrasonic sound velocity vs. concentration of Brij 35 in toluene at 25°C (top) and density measurements of Brij 35 in toluene at 25°C (bottom)

nonionic surfactants. Using measured densities, apparent specific volumes and apparent compressibilities are obtained and are listed in Table 9.5.

Comparison of the ultrasonic curves for ionic surfactants in water vs. nonionic surfactants in toluene shows very different behavior. The differential quantity, the compressibility, is much more sensitive and thus accounts for this change much more than the integral quantity, the density. The ionic surfactants in water have much different apparent compressibilities than the nonionic surfactants in toluene. In particular, the ionic surfactants exhibit negative apparent compressibilities in water for the monomeric form with a very large increase in apparent compressibilities in the micelle. These micelles have a nonpolar core which is anticipated to be rather compressible. On the other hand, the nonionic surfactants in toluene exhibit very large positive apparent compressibilities in toluene for the monomeric form and

Table 9.4. Values of the Critical Micelle Concentrations at 25°C of the Nonionic Surfactants Used in This Study

Surfactants	Solvent	CMC (this study) (g/L)	CMC (ref) (g/L)
Tween 80	Water	0.008	0.013 (ref 34)
Tween 80	Toluene	7.4*	
Brij 35	Toluene	17*	

*Rough approximation.

Table 9.5. Specific Volumes and Compressibilities of the Monomeric and Micellar Forms of the Nonionic Surfactants Used in This Study at Temperature of 25°C

Surfactants	$\tilde{v}_1$ (cm^3.g)	$\tilde{v}_m$ (cm^3.g)	$\tilde{\kappa}_1$ (10^{-5} bar^{-1})	$\tilde{\kappa}_m$ (10^{-5} bar^{-1})
Tween 80 in water		0.9227		1.37
Tween 80 in toluene	0.8895	0.9162	4.38	3.52
Brij 35 in toluene	0.9269	0.9309	5	3.94

show a reduction in apparent compressibilities upon micelle formation. The core for these micelles is polar and is anticipated to be more rigid. Simple heuristics account for these systematics and are useful for comparison with asphaltene results.

The nonionic surfactants do not exhibit a single slope for ultrasonic velocity vs. concentration upon micelle formation. The likely explanation is that there is not a single micelle structure for nonionic surfactants in toluene, there are no charges and toluene surface tension is low. Consequently, there is not a well-defined limit to the micelle's size. The curved slope of ultrasonic velocity vs. concentration indicates that micelles of various sizes form, and that the exact micellar description is a function of concentration.

4. Experimental Results on Asphaltenes

4.1. Background

The molecular structure that has emerged is that the bulk of asphaltene molecules are shaped "like your hand" with an aromatic core (palm) with associated alicyclic rings, and with alkyl groups hanging off the periphery (fingers). One concludes that competing intermolecular interactions are dominant for asphaltenes; van der Waals binding via stacking of aromatic ring systems vs. steric repulsion associated with alkane chains. These standard chemical interactions are utilized by the dye industry; to make an aromatic dye more soluble, alkane substituents are often added. The increased steric repulsion can dramatically increase solubility. A comparison of coal vs. crude oil asphaltenes illustrates this expectation. Coals are much more aromatic and have a much smaller alkane fraction than petroleum. Thus, asphaltenes derived from coals also possess much smaller alkane fractions than petroleum asphaltenes.[9] Correspondingly, coal asphaltenes are subject to much less steric repulsion than petroleum asphaltenes. To maintain the same solubility (the definition of asphaltene), coal asphaltenes must have smaller van der Waals attraction to maintain the balance of intermolecular attractive and repulsive forces. Therefore, coal asphaltenes must possess smaller fused ring systems. This has been shown by fluorescence emission and TRFD measurements,[9,35] as well as direct molecular imaging of coal and petroleum asphaltenes.[36] This important result relates molecular structure with function—in this case solubility.

There are potentially different stages of aggregation in asphaltenes, for instance as proposed in the Yen model.[1] The question arises as to the role of possible hierarchical aggregation structures of asphaltenes in solution. There is recent data suggesting that asphaltene molecules associate in toluene. Laser thermal lensing in asphaltene solutions shows an extremum at roughly 60 mg/L.[37] Recent fluorescence measurements of intensity and red shift indicate that asphaltenes start to associate at 60 mg/L in toluene.[38] High-Q ultrasonic spectroscopy clearly shows asphaltene aggregation at ~100 mg/L.[32,39] It is plausible that asphaltene dimer formation initiates at ~60 mg/L as seen by fluorescence measurements and that nanoaggregates formation is complete at ~150 mg/L as shown by ultrasonic spectrometry.

Using some micellar formalisms for the primary asphaltene aggregation is plausible from a molecular structural point of view. In particular, the concept that nanoaggregate growth shuts off after reaching a certain small size due to steric hindrance is consistent with substantial observations; it needs to be tested. Surface tension measurements have been employed to measure "asphaltene CMC" in pyridine.[40] A clear break in the surface tension data occurred at ~400 mg/L. Other studies have reported "CMCs of asphaltene in toluene" to be much higher in concentration, in the grams per liter range, by Calorymetry[3] and by surface tension.[2,41] There are kinetic issues associated with surface tension measurements that may help explain the large range of values reported for asphaltene CMC. It is standard to determine the CMC of a surfactant in water by measuring surface tension vs. concentration. The surface tension decreases with increasing concentration until the surface is fully saturated with surfactant. Surfactant added at concentrations higher than the CMC form micelles, and the surface tension no longer changes. Similar experiments have been performed with asphaltenes in toluene. However, these experiments are fundamentally flawed.[27] The surface tension of water is high, 71 dynes/cm, and surfactant molecules at the surface lower the surface tension. However, the surface tension of toluene is low. A strongly interacting surfactant molecule such as asphaltene would increase, not decrease the surface tension. In any event, classic measurements of surface tension should not yield the "CMC" of asphaltenes in toluene.

4.2. Ultrasonic Determination of Various Asphaltenes Aggregation Properties

n-Heptane asphaltenes UG8 and BG5 from Kuwaiti crude oils were used, the extraction procedure is described elsewhere.[8] The organic solutions were prepared in reagent grade toluene 99.8% from Acros and Sigma-Aldrich.

UG8 asphaltenes have been used by essentially every technique we have of analyzing asphaltenes. Their properties are common, thus they represent typical virgin crude oil asphaltenes. The top figure of Figure 9.12 shows the solution ultrasonic velocity vs. concentration for UG8 asphaltenes in toluene.[32] A clear break in this curve is observed; this gives a "critical nanoaggregate concentration" (CNAC) for asphaltenes of 0.164 g/L. The velocity vs. concentration curve for asphaltene is similar to that of other nonionic surfactants in toluene strengthening the micellar interpretation for asphaltenes.[32] However, unlike nonionic surfactants in toluene, the ultrasonic velocity vs. concentration is straight not curved

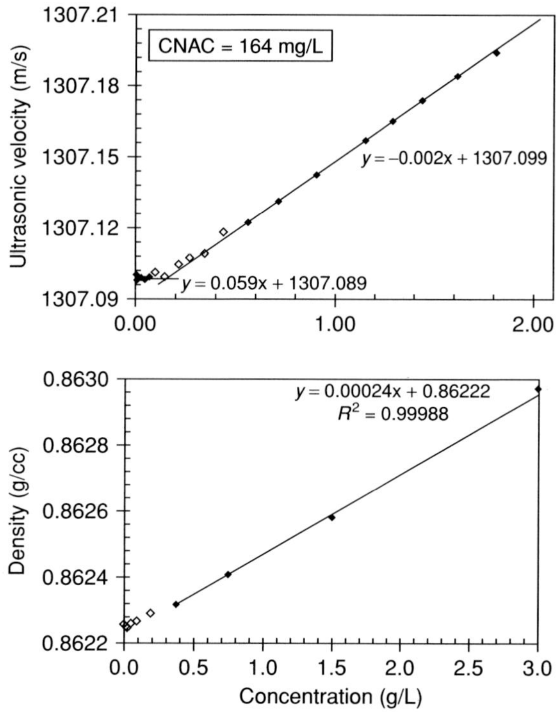

Figure 9.12. Ultrasonic sound velocity vs. concentration of asphaltenes UG8 in toluene at 25°C (top) and density measurements of asphaltenes UG8 in toluene at 25°C (bottom)

at concentrations higher than the CNAC. This result indicates that asphaltenes have only one size of nanoaggregates. Increasing the asphaltene concentration increases the number, not the size, of asphaltene nanoaggregates. Furthermore, at concentrations higher than the CNAC, there is not another break in the curve even up to concentrations of 2 g/L asphaltene in toluene. Either there is no other change in aggregates up to this concentration, or any further change in aggregation has no effect on ultrasonic velocity (because the binding energy is too low to change the compressibility). From the density measurements (bottom figure of Figure 9.12), we can calculate the apparent specific volume of asphaltene nanoaggregates from the slope of the density vs. concentration graph but we cannot get the apparent specific volume of monomer since the concentrations are too low for the accuracy of the densitometer. The apparent adiabatic compressibility in the nanoaggregate form is then calculated from the ultrasonic data with Eq. (9.27). The results are presented in Table 9.6. The apparent compressibility of the asphaltene nanoaggregate is close in magnitude to that of nonionic surfactants again lending credence to the CNAC interpretation for the ultrasonic data of the asphaltenes.[32]

BG5 asphaltenes were obtained from Kuwait Burgan5 crude oil. They too have been subjected to many different kinds of interrogation and they too have typical characteristics. Asphaltenes from UG8 and BG5 might be called "plain

Table 9.6. Values for Asphaltene Nanoaggregates for the Critical Nanoaggregate Concentration, the Apparent Specific Volumes and the Apparent Adiabatic Compressibilities in Toluene at 25 °C

Asphaltenes	CNAC (g/L)	$\tilde{v}_m$ (cm^3/g)	$\tilde{\kappa}_m$ (10^{-5} bar^{-1})
UG8	0.164	0.8814	3.95
BG5	0.048	0.8582	3.45

vanilla" asphaltenes. The top figure of Figure 9.13 shows the solution ultrasonic velocity vs. concentration for BG5 asphaltenes. A break in this curve is evident; this gives a CNAC of 0.048 g/L.[32] The CNAC for BG5 is lower than that of UG8 asphaltenes; nevertheless, a CNAC is evident in both cases. Again, no other change is observed in ultrasonic velocity up to 3 g/cc.[32] As in the case of UG8 asphaltenes, we can calculate the apparent specific volume of the nanoaggregate from the slope of the solution density vs. concentration (Figure 9.13, bottom figure) but we cannot get the apparent specific volume of the monomer because the concentration is too low. From the ultrasonic data, we can calculate the apparent adiabatic compressibility in the micellar form. The results are presented in Table 9.6. Again, we get

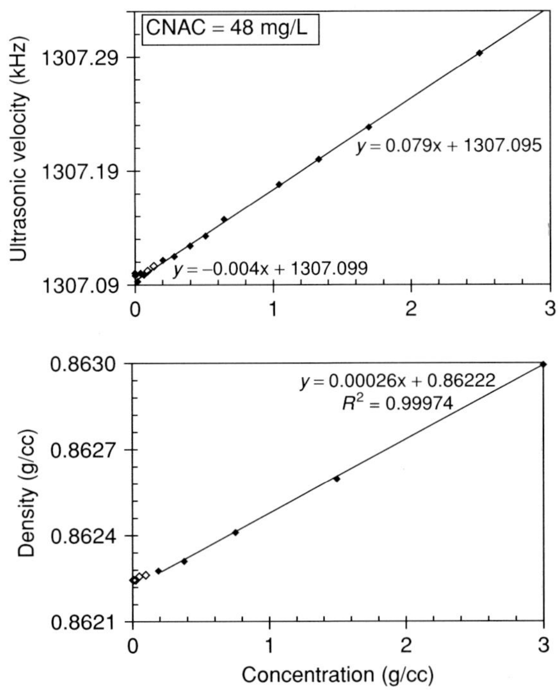

Figure 9.13. Ultrasonic sound velocity vs. concentration of asphaltenes BG5 in toluene at 25°C (top) and density measurements of asphaltenes BG5 in toluene at 25°C (bottom)

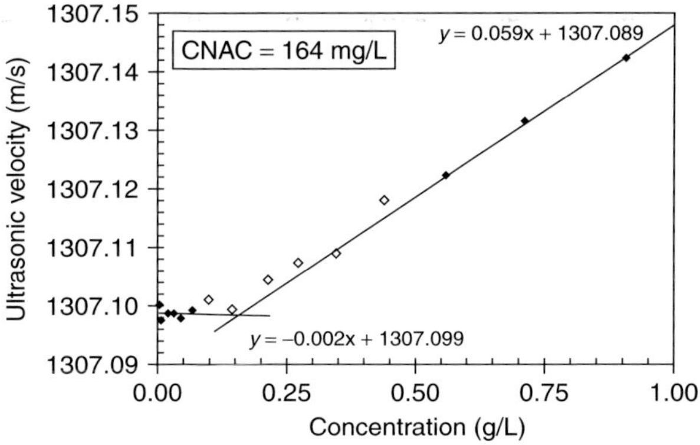

Figure 9.14. Ultrasonic sound velocity vs. concentration of asphaltenes UG8 in toluene at 25°C

agreement between asphaltene micelle apparent compressibilities with those of other nonionic surfactants.

Figures 9.14 and 9.15 show an expanded scale of the ultrasonic velocity vs. concentration for asphaltenes to make clear the CNAC. Figure 9.14 expands the low concentration range of Figure 9.12 while Figure 9.15 expands the low concentration range of Figure 9.13. The break in the ultrasonic velocity curve is quite clear; we interpret this break to be the CNAC. At higher concentrations than the CNAC, there is no change in the ultrasonic slope. This indicates that the nanoaggregates are not changing at these concentrations, just that there are more of them at higher concentration. That is, nanoaggregate growth shuts off. We also note that while a clear break is evident at the CNAC, the ultrasonic data cannot rule out formation of dimers or trimers at concentrations below that CNAC.

Different asphaltenes exhibit CNACs at similar concentrations but with some variability in the exact value; CNACs ~50 to 150 mg/L.[32,39] The apparent compressibilities of the asphaltene nanoaggregates are similar to each other and similar to other apparent micelle compressibilities for other nonionic surfactants in

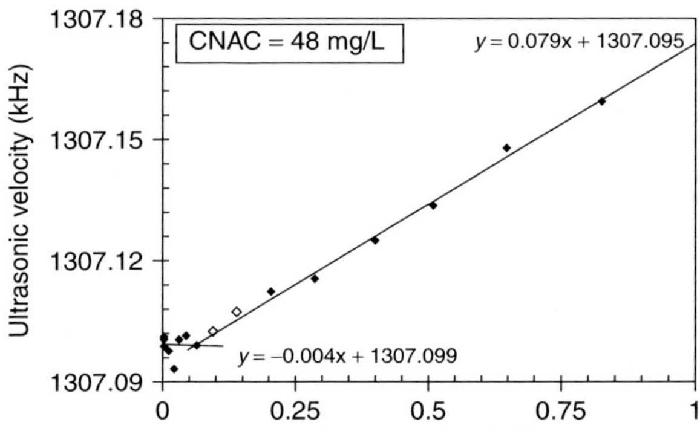

Figure 9.15. Ultrasonic sound velocity vs. concentration of asphaltenes BG5 in toluene at 25°C

toluene. The CNACs determined here are 1–2 orders of magnitude lower than literature reports for asphaltene–toluene systems obtained by other techniques. In our view, the other techniques are not recording proper CNACs. They may be recording some higher-level aggregation phenomenon. Many techniques do not measure an explicit parameter such as apparent nanoaggregate compressibility that can then be checked against known surfactants as we do. Rather, some other techniques only interpret a change in some property as the CNAC. If these techniques are not suffciently sensitive to detect CNACs at 100 mg/L, then corresponding data will be subject to misinterpretation.

The governing chemical principles of asphaltenes that determine solubility and thus define asphaltenes have been shown to be van der Waals (and polar) attraction of aromatic ring systems vs. steric repulsion of their alkane chains[9]; essentially asphaltene molecules are shaped with a core made of a polycyclic aromatic ring system and the alkyl chains in the periphery of the ring system. We believe that the same forces are operative in determining the asphaltene nanoaggregate. The idea is that the first few asphaltene molecules to associate have fairly clear access to intermolecular interaction of the fused ring system. However, subsequent aggregation of more molecules becomes constrained by steric repulsion of alkane substituents thereby impeding further stacking. At some point, the interaction between the nanoaggregate and an additional molecule becomes rather weak, probably due to steric repulsion. At this point new nanoaggregates form upon increasing concentration. These ideas are central to the Yen model[1]; our data are in general in agreement with this well-known model. A possibility for an asphaltene nanoaggregate can be seen in Figure 9.16. Our results extend the Yen model by 1) including the restriction of small molecular size of asphaltene molecules and 2) showing the effect of asphaltene intermolecular interactions on the dynamics of aggregate formation.

The low values of the CNACs help explain why colligative techniques such as vapor pressure osmometry (VPO) always record "molecular" weights that are too high. VPO and other colligative methods are performed at concentrations that are significantly in excess of the asphaltene CNAC; consequently, VPO provides an aggregate weight. VPO is often in error by a factor of ~5 for molecular weight determination; consequently, VPO along with our asphaltene CNAC results here imply that the aggregation number in an asphaltene micelle is ~5.

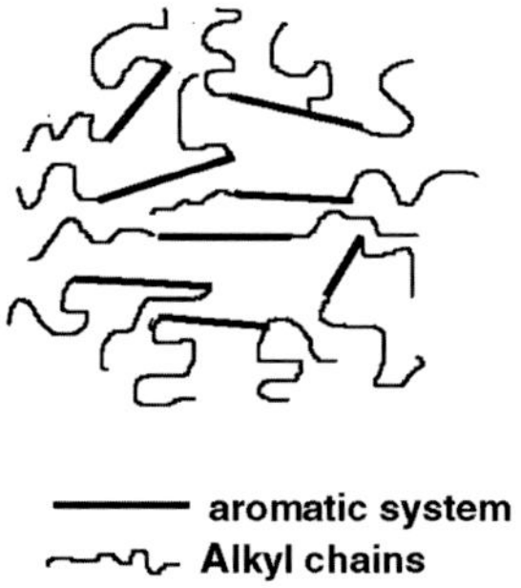

Figure 9.16. Hypothetic structure for the asphaltenes nanoaggregate

4.3. Comparison of Experimental Results on UG8 Asphaltenes and Maltenes

The principal classes of constituents of asphalt and related carbonaceous material are defined by their solubility properties. Asphaltenes can be isolated from various carbonaceous sources (petroleum, bitumen, coal). Asphaltenes are, as stated earlier, the fraction from the crude oil or the coal which are heptane insoluble and toluene soluble. Maltenes are defined as the heptane soluble fraction from the carbonaceous source.

This clear break in the UG8 asphaltene curve (as shown in Figure 9.14) is in contrast to the behavior of UG8 maltenes in toluene shown in Figure 9.17 and expanded in Figure 9.18. It would be diffcult to confuse the maltene behavior with the asphaltene behavior; the maltenes reduce the speed of sound in toluene solutions thereby exhibiting opposite trends compared to asphaltenes.[39] In addition, maltenes do not exhibit even a hint of a break in the speed of sound curve (UG8

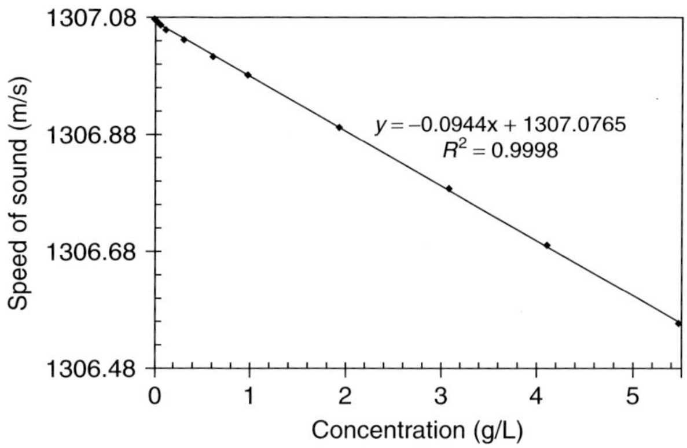

Figure 9.17. Ultrasonic velocity vs. concentration of UG8 maltene in toluene

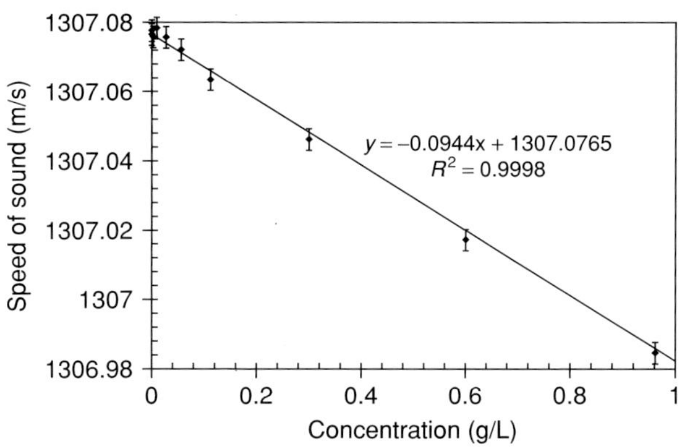

Figure 9.18. Expended region of the ultrasonic velocity vs. concentration of UG8 maltene in toluene as a function of concentration

is a 25 API oil). Figure 9.14 and Figure 9.18 cover similar concentration ranges indicating that the high-Q ultrasonic technique can detect the presence and absence of aggregation.[39] To obtain a detailed understanding of why the ultrasonic slopes are different for maltenes and asphaltenes, we would need accurate density data.

4.4. Differences Between Coal and Petroleum Asphaltenes

Figure 9.19 shows the same curve for the Iino coal asphaltene (we thank Professor Iino for this Tanito Harum coal asphaltene sample) while Figure 9.20 expands the curve. Although the break is more difficult to detect than for the petroleum asphaltene, it is still evident, especially in Figure 9.20, at approximately 180 mg/L.[39] This asphaltene has been shown to be much different in terms of chemical structure than petroleum asphaltenes.[9] However, the intermolecular

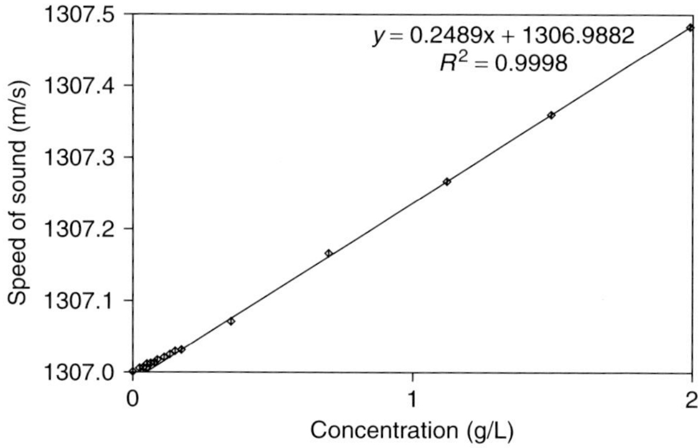

Figure 9.19. Ultrasonic velocity of the Iino coal asphaltene sample as a function of concentration. A break occurs at 180 mg/L indicating the CNAC

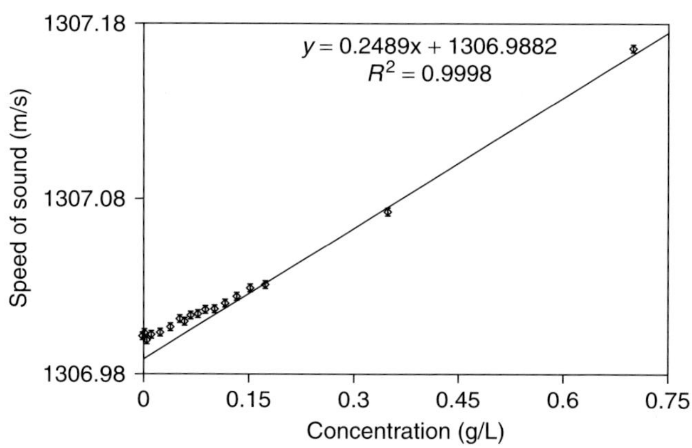

Figure 9.20. For the Iino coal asphaltene in toluene, an expanded region of the previous figure near the CNAC

structural features in the solid where found to be very similar using high-resolution transmission electron microscopy.[16,36] Thus, it is not surprising to find similar aggregation tendencies of coal and petroleum asphaltenes.

5. Conclusion

High-Q ultrasonic spectroscopy has proven to be a very valuable tool in the characterization of micelle formation for known surfactants and of nanoaggregate formation of asphaltenes. CMCs of ionic and nonionic surfactants are easily measured in high and low concentration ranges. For standard surfactants, we obtain excellent agreement between our measurements and literature values of CMCs, apparent specific volumes, and apparent compressibilities. Asphaltenes in toluene exhibit CNACs at ~100 mg/L. The phase equilibrium model for micelles applies readily to all data presented here, surfactant and asphaltene data alike. Furthermore, derived parameters such as the compressibility of nanoaggregates of asphaltenes and of micelles of nonionic surfactants are very similar, thereby strengthening the conclusion that asphaltenes are nonionic surfactants that form nanoaggregates and exhibit CNACs. The literature reports with much larger concentrations for asphaltene CNACs (or CMCs) are not measuring CNACs but perhaps some higher-order aggregation. Our asphaltene CNACs explain why VPO measurements of asphaltene molecular weights are consistently too high; VPO measures aggregate weight. The growth and its termination of nanoaggregates can be understood directly from molecular structural considerations. The formation of nanoaggregates essentially consumes high energy binding sites. The resulting nanoaggregates resembles a "hairy tennis ball" with alkanes surrounding the outside. These nanoaggregates are stable and do not grow; the absence of available high energy binding sites prevent flocculation at moderate asphaltene concentrations. If the least soluble fraction of asphaltenes is isolated, then aggregate growth proceeds unhindered yielding low observed solubilities. The polydispersity in a standard asphaltene sample ensures nanoaggregate initiation from the least soluble molecular fraction and nanoaggregate termination from the most soluble molecular fraction, thereby creating a stable (nano)colloidal suspension. Since the crude oils are even more polydisperse, this implies that the same nanoaggregate formation and stabilization takes place in crude oils. In large measure, these results are in concert with the basic premise of the Yen model introduced so many years ago. These results extend the Yen model illustrating structure-function relations in aggregate formation.

References

[1] Yen, T.F. (1990). *Am. Chem. Soc. Dic. Petroleum Chem. Prepr.* **35**, 312.

[2] da Silva Ramos, A.C., L. Haraguchi, F.R. Nostripe, W. Loh, and R.S. Mohamed (2001). Interfacial and colloidal behavior of asphaltenes obtained from Brazilian crude oils. *J. Pet. Sci. Eng.* **32**, 201–216.

[3] Andersen, S.I. and S.D. Christensen (2000). The critical micelle concentration of asphaltenes as measured by calorimetry. *Energy & Fuels* **14**, 38–42.

[4] Andersen, S.I., J.M. del Rio, D. Khvostitchenko, S. Shakir, and C. Lira-Galeana (2001). Interaction and solubilization of water by petroleum asphaltenes in organic solution. *Langmuir* **17**, 307–313.

[5] Boduszynski, M.M. (2001). Composition of heavy petroleums. 2. Molecular characterization. *Energy & Fuels* **2**, 597–613.

[6] Groenzin, H. and O.C. Mullins. Asphaltene molecular size and weight by time resolved fluorescence depolarization, Chapter 2. This Book.

[7] Groenzin, H. and O.C. Mullins (1999). Asphaltene molecular size and structure. *J. Phys. Chem. A* **103**, 11237–11245.

[8] Groenzin, H. and O.C. Mullins (2000). Molecular size and structure of asphaltene from various sources. *Energy & Fuels* **14**, 677–684.

[9] Buenrostro-Gonzalez, E., H. Groenzin, C. Lira-Galeana, and O.C. Mullins (2001). The overriding chemical principles that define asphaltenes. *Energy & Fuel* **15**, 972–978.

[10] Wargadalam, V.J., K. Norinaga, and M. Iino (2002). Size and shape of a coal asphaltene studied by viscosity and diffusion coefficient measurements. *Fuel* **81**, 1403–1407.

[11] Rodgers, R.P. and A.G. Marshall (2006). Petroleomics: Advanced characterization of petroleum-derived materials by Fourier transform ion cyclotron resonance mass spectrometry (FT-ICR MS). Chapter 3, This book.

[12] Merdrignac, I., B. Desmazieres, P. Terrier, A. Delobel, and O. Larevote (2004). *Proc. Heavy. Org. Dep.*, Los Cabos, Baja Cal. Sur. Mexico, Nov. 14–19.

[13] Cunico, R.I., E.Y. Sheu, and O.C. Mullins (2004). Molecular weight measurement of UG8 asphaltene using APCI mass spectroscopy. *Petroleum Sci. Technol.* **22**, 787–798.

[14] Waldo, G.S., O.C. Mullins, J.E. Penner-Hahn, and S.P. Cramer (1992). Determination of the chemical environment of sulphur in petroleum asphaltenes by X-ray absorption spectroscopy. *Fuel* **71**, 53–57.

[15] Zajac, G.W., N.K. Sethi, and J.T. Joseph (1994). Molecular imaging of petroleum asphaltenes by SCM, *Scan. Micros.* **8**, 463–470.

[16] Sharma, A. and O.C. Mullins. Insights into molecular and aggregate structures of asphaltenes using HRTEM, Chapter 8, This book.

[17] Eggers, F. and U. Kaatze (1996). Broad-band ultrasonic measurement techniques for liquids. *Meas. Sci. Technol.* **7**, 1–19.

[18] Bolef, D.I. and J.G. Miller. High-frequency continuous wave ultrasonics. In: W.P. Mason and R.N. Thurston (eds.), *Physical Acoustic*, Vol. VIII. Academic Press, New York.

[19] Buckin, V. and C. Smyth (1999). High resolution ultrasonic resonator measurements for analysis of liquid. *Sem. Food Anal.* **4**, 113–130.

[20] Freyer E. B., J.C. Hubbard, and D.H. Andrews (1929). Sonic studies of the physical properties of liquids. *J. Am. Chem. Soc.* **51**, 759–770.

[21] Zielinski, R., S. Ikeda, H. Nomura, and S. Kato (1987). Adiabatic compressibility of alkyltrimethylammonium bromides in aqueous solutions. *J. Colloid Interface Sci.* **119**, 398–408.

[22] Bloor, D.M., J. Gormally, and E.J. Wyn-Jones (1984). Adiabatic compressibility of surfactant micelles in aqueous solutions. *Chem. Soc. Faraday Trans.* **1**(80), 1915–1923.

[23] Suárez, M.J., J.L. López-Fontán, F. Sarmiento, and V. Mosquera (1999). Thermodynamic study of the aggregation behavior of sodium *n*-hexyl sulfate in aqueous solution. *Langmuir* **15**, 5265–5270.

[24] Kudryashov, E., T. Kapustina, S. Morrissey, V. Buckin, and K. Dawson (1990). The compressibility of alkyltrimethylammonium bromide micelles. *J. Colloid Interface Sci.* **203**, 59–68.

[25] De Lisi, R., S. Milioto, and R.E.J. Verrall (1990). Partial molar volumes and compressibilities of alkyltrimethylammonium bromides. *Solution Chem.* **19**(7), 665–692.

[26] Mosquera, V., J.M. del Rio, D. Attwood, M. Garcia, M.N. Jones, G. Prieto, M.J. Suarez, and F. Sarmiento (1998). A Study of the aggregation behavior of hexyltrimethylammonium bromide in aqueous solution. *J. Colloid Interface Sci.* **206**, 66–76.

[27] Friberg, S., O.C. Mullins, and E.Y. Sheu (2005). Surface activity of an amphiphilic association structure. *J. Dispers. Sci. Tech.* **26**, 513.

[28] Mukerjee, P. and K.J. Mysels (1971). Critical micelle concentrations of aqueous surfactant systems. *NSRDS-NBS 36*. US Department of Commerce, Washington, DC.

[29] Priev, A., S. Zalipsky, R. Cohen, and Y. Barenholz (2002). Determination of critical micelle concentration of lipopolymers and other amphiphiles: Comparison of sound velocity and fluorescent measurements. *Langmuir* **18**, 612–617.

[30] Amararene, A., M. Gindre, J.-Y. Le Huerou, C. Nicot, W. Urbach, and M. Waks (1997). Water confined in reverse micelles: Acoustic and Densimetric studies. *J. Phys. Chem. B* **101**, 10751–10756.

[31] Blandamer, M.J., P.M. Cullis, L.G. Soldi, J.B.N.F. Engberts, A. Kacperska, N.M. Van Os, and M.C.S. Subha (1995). Thermodynamics of micellar systems: Comparison of mass action and phase equilibrium models. *Adv. Colloid Interface Sci.* **58**, 171–209.

[32] Andreatta, G., N. Bostrom, and O.C. Mullins (2005). High Q-ultrasonic determination of the critical nanoaggregate concentration of asphaltenes and the CMC of standard surfactants. *Langmuir* **21**(7), 2728–2736.

[33] Corkill, J.M., J.F. Goodman, and T. Walker (1967). *Trans. Faraday. Soc.* **63**, 768.

[34] http://www.sigmaaldrich.com/img/assets/17541/Detergent_table2edited.pdf.

[35] Badre, S., C.C. Goncalves, K. Norinaga, G. Gustavson, and O.C. Mullins (2006). Molecular size and weight of asphaltene and asphaltene solubility fractions from coals, crude oils and bitumen. *Fuel*, **85**, 1–11.

[36] Sharma, A., H. Groenzin, A. Tomita, and O.C. Mullins (2002). Probing order in asphaltenes and aromatic ring systems by HRTEM. *Energy & Fuel* **16**, 490–496.

[37] Acevedo, S., M.A. Ranaudo, J.C. Pereira, J. Castillo, A. Fernandez, P. Perez, and M. Caetano (1999). Thermo-optical studies of asphaltene solutions: Evidence for solvent—solute aggregate formation. *Fuel*, **78**, 997–1003.

[38] Goncalves, S., J. Castillo, A. Fernandez, and J. Hung (2004). Absorbance and fluorescence spectroscopy on the aggregation behavior of asphaltene—toluene solutions. *Fuel*, **83**, 1823–1828.

[39] Andreatta, G., C.C. Goncalves, G. Buffin, N. Bostrom, C.M. Quintella, F. Arteaga-Larios, E. Perez, O.C. Mullins (2005). Nanoaggregates and structure-function relations in asphaltenes. *Energy & Fuels* **19**, 1282.

[40] Sheu, E.Y. (1996). Physics of asphaltene micelles and microemulsions— theory and experiment. *J. Phys. Condens. Matter* **8**, A125–A141.

[41] Bouhadda, Y., D. Bendedouch, E.Y. Sheu, and A. Krallafa (2000). Some preliminary results on a physico-chemical characterization of a hassi messaoud petroleum asphaltene. *Energy & Fuels* **14**, 845–853.

10

Asphaltene Self-Association and Precipitation in Solvents—AC Conductivity Measurements

Eric Sheu, Yicheng Long, and Hassan Hamza

1. Introduction

Techniques used for investigating asphaltene self-association are reviewed. The principles, fundamental differences, and limits of each technique are briefly discussed. A new approach using AC conductivity measurement for detecting asphaltene self-association is proposed and demonstrated using Alberta bitumen-derived asphaltene as a model system. Preliminary results show that the AC conductivity measurement is sensitive to subtle capacitance change arising from asphaltene self-association but only within a certain frequency range. A percolation model with parallel capacitor–resistor circuit is adopted to establish the theoretical basis for this approach. This model predicts the functional behavior of the AC conductivity and exhibits phase transition-like behavior upon asphaltene self-association. The conductivity measurements show functional forms similar to the predicted ones and exhibit discontinuity near 120 mg/L in toluene where self-association is believed to occur. This value agrees (on the same order of magnitude) with earlier surface tension,[1] laser thermal lensing,[2] and ultrasonic[3] measurements. In addition to detecting asphaltene self-association, AC conductivity is also applicable to characterization of asphaltene precipitation in toluene upon addition of non-solvent such as heptane. The sensitivity is high and the method is simple. These two experiments suggest that AC conductivity method can be a good option for measuring flocculation, precipitation, and phase separation of petroleum complex fluids, provided the right frequency range is chosen. Further validation of this method is needed for other complex fluids.

Asphaltene is a heavy end component of petroleum material commonly defined as the solvent class that is soluble in toluene but insoluble in aliphatic solvent (e.g., heptane, pentane, etc.). It is an undesired component in many petroleum processes (production, transportation, refining) and engine operation using heavy oils.

Eric Sheu • Vanton Research Laboratory, Inc., 7 Olde Creek Place, Lafayette, California 94549. **Yicheng Long and Hassan Hamza** • CANMET Energy Technology Center–Devon, 1 Oil Patch Drive, Devon, Alberta, Canada T9G 1A8.

It is largely due to its propensity of flocculation and precipitation. In conventional operations, precipitation is the obvious phenomenon to be prevented. The precursor of precipitation is often the flocculation. Thus, much attention has been paid to studying flocculation, hoping to provide earlier warning before precipitation occurs. This is crucial in production where pressure is continuously reduced as depth decreases, which may drive the liquids into the precipitation-envelope and clot the down-pipe. Flocculation can usually be detected using simple laser transmission signal that is associated with a pressure cell.[4]

Ideally, it is even more advantageous to detect the precursor of flocculation, which is the formation of microscopic particles. These particles are generally believed to originate from asphaltene molecular self-association. They serve as the elemental particles for either Oswald ripening-like process or as nucleation centers that eventually prompt precipitation. In either case, the asphaltene self-association is an important phenomenon to investigate and understand. In 1940, Pfeiffer and Saal[5] propose a hypothetical model to describe a possible scenario of an in situ asphaltene containing petroleum liquids. In their hypothesis, asphaltene molecules are peptized by resins, which has smaller polynuclear aromatic cores and/or longer aliphatic chains. Because of the peptized resin molecules around asphaltenes, the asphaltene molecules maintain dispersed in oil. Later, Yen[6] proposed a progressive model that explicitly describes the evolution of the particle size from nanosize aggregates to macroscopic particles observed in precipitates. Yen's model is based on length scale and the elemental particles, as Yen named it, are nanoscale aggregate arising from molecular self-association (or self-assembly). This aggregation step forms the nano size precursors that flocculate later and is the point of discussion of this chapter.

It is generally accepted that asphaltene molecules aggregate in solvent when concentration exceeds a threshold value.[1] This is very much similar to a surfactant system undergoing micellization. Techniques used for determination of the critical micelle concentration (CMC) in aqueous solutions include surface tension, osmotic pressure, high frequency conductivity, equivalent conductivity, interfacial tension, density variation, and detergency.[7] These techniques can be categorized into surface techniques and bulk techniques. Surfactants are highly surface active, thus, surface tension energy is very sensitive to the formation of micelles. This is why surface tension is widely used for determination of CMC.[7,8] It measures the surface tension force, which relates to the surface coverage of the molecules through Gibb's isotherm equation. From this equation, the molecular coverage area and molecular weight can be determined.

In a surfactant system, the molecules are usually well defined. The aggregate size and shape of the micelles can thus be accurately described once the aggregation energies, (hydrophobic, packing, and entropic energies, etc.) are accurately modeled. Because surfactant micellization is primarily dominated by the hydrophobic energy, accurate models have been established using only the hydrophobic energy and entropic energy. The CMC, aggregate size, and phase transition are routinely determined for many industrial systems, either through analysis of experimental data or from the molecular structures. Comprehensive description of the hydrophobic energy, its relation with surfactant polar head(s) and hydrophobic tail(s) can be found in Tenford's "hydrophobic effect"[9] or the review article by Israelachvili.[10]

In the case of asphaltene aggregation, techniques applicable to surfactant systems may not apply, simply because asphaltenes are much less surface active. In addition, the energies lead to asphaltene molecular aggregation is not dominated by the hydrophobic energy. Instead, Van der Waal energy may be a dominating factor though there are still debates on this point. Techniques that have been applied to asphaltene aggregation detection include surface tension,[1] calorimetry,[11,12] laser thermal lensing,[2] and ultrasounic compressibility.[3]

In order to apply the surface techniques, one needs to understand that the effect of the aggregation energy on the surface parameters, such as surface tension, is much smaller than in a hydrophobic energy-driven surfactant aqueous solution. Nevertheless, this energy may still be adequate to modify the surface tension energy to a detectable level if the surface tension contrast between the surface tension of the solvent and asphaltene is sufficiently high. Microscopically, the surface parameter should undergo a phase transition at the point of aggregation onset for the technique to be applicable. In the case of surface tension measurement, it requires the surface sublayer to first being covered by asphaltene molecules and saturate at the point of aggregation, much similar to the surfactant system. In addition, the surface tension of the solvent surface and the asphaltene-covered surface should be significantly different for a surface tensiometer to pick up the signal.

Based on this scenario, one approach is to select a solvent with surface tension higher than the asphaltene surface tension. As asphaltene molecules are added, they are adsorption to the surface sublayer with the "hydrophilic" portion in the solvent to reduce the surface tension energy, and thus the system free energy. As asphaltene molecules saturate the surface, they can move to the bulk as a free molecule to maximize entropy, but other free energies would increase. As concentration continues to increase, at certain point the entropic energy is no longer advantageous over the other positive free energies while self-association is the best option to reduce the free energy. This is the point when asphaltene molecules self-associate and is also the point when surface tension becomes independent of concentration (or much less dependent on concentration).

In order to choose the right solvent, here are the points of consideration. If one takes asphaltene as consisting of a polynuclear aromatic core with short aliphatic chain attached to it, the surface tension is likely between a pure aliphatic molecule and an aromatic molecule. Using benzene and hexane as the examples, then, asphaltene surface tension is likely between 29 dyne/cm (benzene) and 18 dyne/cm (hexane).[13] Because benzene is a good solvent for asphaltene, one expects the surface tension of asphaltene to be close to benzene's surface tension. Therefore, if one chooses hexane, it will give enough surface tension contrast. Unfortunately, asphaltene will not reside at the interface between hexane and air because both air and hexane are much more hydrophobic than asphaltene. As a result, asphaltene can only precipitate after it reaches the solubility limit in hexane. This leaves the surface tension nearly unchanged and undetectable by the surface tension technique. This is why one should select a solvent with the surface tension higher than the surface tension of asphaltene to drive the "aliphatic" part of asphaltene out of the solvent–air interface and into the air while the polar portion resides below the interface. This is to say that the solvent should have a surface tension higher than 29 dyne/cm. Using this principle we chose pyridine for Ratawi resid[1] where the surface tension contrast is about 8 dyne/cm between asphaltene (assuming 30 dyne/cm) and pyridine (38 dyne/cm).[13] Using this solvent, we found the onset concentration to be between 350 and 500 mg/L.

It is rather difficult to find a right condition to accurately determine asphaltene self-association using surface tension technique. When an inappropriate condition is selected, the change of the surface tension upon aggregation may be undetectable until another event occur that changes the surface tension to a detectable level. This event could be flocculation or even bigger object formation. There are aggregation onset concentrations reported in toluene, which does not have enough surface tension contrast.[10] The aggregation onset concentrations obtained in toluene were approximately 10 times higher then what was measured in pyridine.[11,12,14–16] This onset concentration is likely the onset of further agglomeration between asphaltene aggregates, a phenomenon proposed by Yen.[6,17,18]

Onset concentration detected by microcalorimetry method is also about 10 times higher than the pyridine results we obtained. Microcalorimetry measures the heat change, which, in the case of asphaltene aggregation, is very small and is hard to detect. It makes this technique difficult to use for detecting molecular aggregation. One may want to use a modulated heating ramping to see if it is possible to increase the sensitivity. Unfortunately, the modulated heating only provides the dynamic response of the material and the difference between an aggregate and a free molecule is too small to be detected unless the aggregate has a very different morphology. This is not necessary applicable to asphaltene case because asphaltene aggregation is not a first order phase transition and the dynamics is rather slow as we pointed it out in our time-dependent surface tension measurement.[1]

Recent ultrasonic measurements[3] showed that it is possible to pick up compressibility difference between the aggregated state and the nonaggregated state. The onset concentration was found to be between 50 and 250 mg/L. It measures the bulk properties directly, which is advantageous over surface tension technique. However, compressibility is an integrated parameter, which may be inaccurate due to the polydispersity effect on the structure factor at the zero momentum transfer. This, together with the second virial coefficient effect (likely negligible at the asphaltene onset concentration) limits its capability in measuring higher concentration systems.

Other bulk technique used in surfactant CMC determination is the equivalent conductivity measurement (see reference 8, pp. 284–285). DC conductivity meter is usually used for this measurement. In the classical concept, conductivity is related to the number of charges available in the solution and their movement in the media. The conductance is then measured according to the Ohm's law and corrected for the distance between electrodes and the electrode area to get conductivity. Obviously, conductivity signal depends on the movement of the charge particles. One expects the diffusivities of N-monomers and an N-mer are very different. This is true because the total charges available in the solution is the same (if charge condensation is not taken into account) while their movement are vastly different. As a result, the equivalent conductivity (normalized by concentration) would follow very different slope upon increasing concentration. Typical surfactant solutions often show a large slope before CMC and much smaller slope above CMC. This is how one determines CMC using equivalent conductivity method.

The accuracy of equivalent conductivity measurement relies on the difference in movement of the charge carriers, i.e., the difference in the hydrodynamic radius between N monomers and an N-mer assuming they carrier the same total charges. This argument is legitimate if the monomer has a much smaller hydrodynamic radius than the aggregates. However, if the aggregates are small and the monomer is structurally very asymmetric, their hydrodynamic radius can be on the same order and the diffusivity may be too close to be distinguished by the conductivity measurement. Under this circumstance, the equivalent conductivity may not be the right choice for CMC measurement. Unfortunately, this is precisely the case of asphaltene solution. In addition, aggregation of asphaltene can be slow because of the structural arrangement at the later stage of aggregation as seen in simulation[19] and in the evidence of reaction limited aggregation process.[20] This further jeopardizes the possibility of using conventional DC conductivity measurement method.

In order to overcome this hurdle, a new approach is proposed using alternating current of various frequencies to detect the equivalent conductivity as a function of the asphaltene concentration. One immediate advantage of this technique is that the alternating current can effectively eliminate the charge build-up near the electrodes as long as the half-cycle is shorter than the relaxation time of the equivalent RC circuit. In fact, this factor is not severe in the case of asphaltene/solvent system because it is not a highly conducting system compared with an ionic surfactant system. However, one should still be cautious about its effect. In this work, three actions were taken to completely eliminate this effect. First, platinum black electrode was used to enlarge the total surface area. This has been taken by many reported studies.[21–26] Second, low voltage was applied to reduce the driving force, which is directly proportional to the charge movement. Finally, high enough frequency AC current was used to avoid charge build-up. With all three factors taken care of, there was no observable charge build-up near the electrode. In fact, experiment using non-platinum black electrode appeared to be sufficient for conductivity nearly 10 times of the values obtained here.[27]

There is more important advantage associated with using AC conductivity measurements. By using AC potential, the equivalent conductivity represents a derivative quantity of the equivalent RC circuit of the system. In the case of asphaltene solution this derivative quantity is largely dictated by the capacitance change rather than the resistance change because the media are organic solvents, which have high resistance. If we hypothesize that changing asphaltene concentration is essentially changing the capacitance of the equivalent RC circuit, then a relationship between the conductivity and the concentration can be derived and experimentally evaluated. If asphaltene aggregation initiates a discontinuity of the capacitance change, one may be able to detect it by simply measuring the equivalent conductivity (or equivalent conductance) at an appropriate frequency that is sensitive to the size of the aggregate. This is essentially the hypothesis of this work.

We used asphaltene derived from Alberta bitumen to demonstrate that the equivalent conductivity at proper AC frequency range can detect the asphaltene aggregation in toluene and precipitation upon heptane addition. We believe that this idea is sound. Experimental data obtained based on this idea show that the Alberta bitumen-derived asphaltene has an aggregation onset concentration at about 120 mg/L, well within the range reported by our earlier surface tension work, and the recent laser thermal lensing and ultrasonic work. We believe this method is legitimate, the experimental procedure is reliable and the results are creditable.

In Section 2, detailed experimental procedure is described including instrumentation calibration using impedance and toluene dielectric constant. This is followed by a brief discussion of the theory we adopted in Section 3. Section 4 gives the results for both asphaltene aggregation in toluene and precipitation in toluene/heptane mixture. Section 5 discusses the results and justification of the AC conductivity technique.

2. Experimental

2.1. Sample

Asphaltene used in this work was derived from Alberta bitumen using conventional separation technique. A 40:1 (volume:weight) ratio of pentane to bitumen was mixed at ambient temperature under constant agitation for 4 hr. This is followed by filtration using 0.25 μm pore-size filtration paper. The filtered solid phase was dried under nitrogen until a constant weight was obtained.

Prior to conductivity measurement, the powder-like asphaltene was redissolved in the selected solvents (toluene, heptane or their mixtures). All solvents used are reagent grade from Sigma-Aldrich.

2.2. Instrument

Low frequency conductivity measurements were conducted using Hewlett-Packard LF4192 impedance analyzer. A custom-designed cell made of Teflon and Pyrex as the outer shield was used to conduct these measurements. The lengths of the electrode wires were reduced to their minimum to minimize the capacitance effect. The electrode is a four-plate platinum black electrode with 1 mm gap and the total surface area is 9 cm^2 and the cell constant is 0.001, suitable for oil-like systems. The capacitance contributed by the electrode wires was compensated by the standard open–close measurement as part of the calibration. The other factor comes into play is the field inhomogeneity due to the clamp of the sample holder. To avoid its effect, a cylindrical aluminum shield was placed in a symmetric manner to define a field boundary. There were no localized metal parts used in the vicinity of the cell within the field boundary. Toluene sample was used as the calibration curve from 5 Hz to 13 MHz to ensure a constant capacitance across this frequency range under the measuring cell configuration.

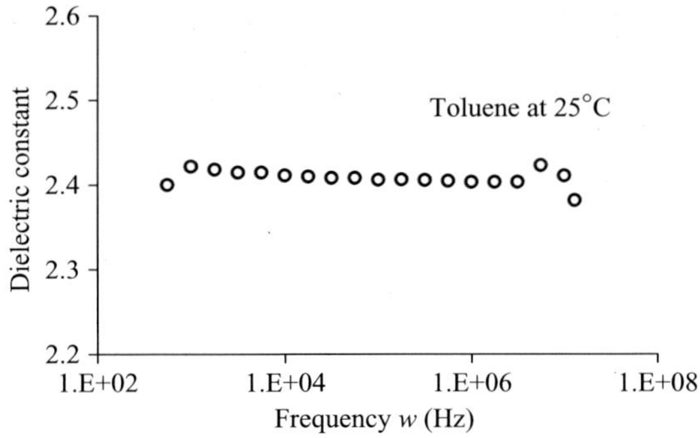

Figure 10.1. Dielectric constant of toluene as a function of frequency ranging from 562 Hz to 13 MHz. The published values for toluene dielectric constant is 2.438 (reference 10, pp. E51–E53).

2.3. Measurement

The instrument was set at parallel circuit mode because the resistance is over 25 kΩ above which the sensitivity on serial mode starts to decline while a parallel model provides much better sensitivity. In addition, a parallel mode is suitable for modeling a percolation system, such as an asphaltene solution.

The potential applied across the electrodes was set at 0.5 V and the frequency was calibrated from 562 Hz to 13 MHz using toluene as the standard solution. Before measurement, the resistance and capacitance contribution from the electrodes was compensated using the standard open–close circuit method. First, the electrodes were open with air load for capacitance measurement and multiplied by the cell constant. This measurement is stored to the LF4192 unit for automatic capacitance compensation. Secondly, the electrode is closed for the instrument to measure the resistance contribution, again, stored to the LF4192 unit for resistance compensation. Following the electrodes compensation measurement, toluene was measured from 562 Hz to 13 MHz. The dielectric constant obtained was within 3% of the published value. Figure 10.1 shows the result. The other calibration curve needed was the impedance curve. It is shown in Figure 10.2. All measured sample curves were normalized by the toluene contribution based on the parallel circuit mode.

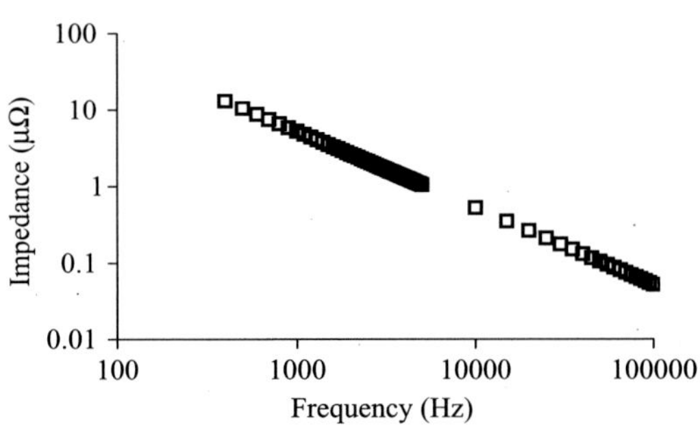

Figure 10.2. Impedance of toluene at 25°C.

3. Theory

For a composite material, modeling the frequency dependent electrical responses often starts with two basic circuits, namely the series and the parallel model as depicted in Figure 10.3.

Take the conductivity of a capacitor C as $j\omega c$, the complex conductivity $G(\omega)$ can be expressed as

$$G(\omega) = \frac{\left(R_1 + R_2 + \omega^2 R_1 R_2^2 C_p^2\right) + j\omega R_2^2 C_p}{(R_1 + R_2)^2 + (\omega R_1 R_2 C_p)^2} \tag{10.1}$$

for the parallel circuit (see Figure 10.3) and

$$G(\omega) = \frac{\omega[\omega(R_1 + R_2)C_s^2 + jC_s]}{1 + \omega^2 \left(R_{11}^2 + R_2^2\right) C_s^2} \tag{10.2}$$

for the series circuit (see Figure 10.3). In Eqs. (10.1) and (10.2) R_1 represents the resistance of the solvent and R_2 the resistance of the dispersed component and C_p and C_s are the insulation effect arising from the nonconducting portion of the system for parallel and series circuits, respectively.

In a 2D random network model, if one assumes the doped material represent a resistor (conducting material) while the insulator (solvent) is a capacitor and the probability of a site occupied by a capacitor (the disperse component) is P, then a frequency dependent conductivity can be simulated.[28]

The main issue in applying AC conductivity for asphaltene aggregation is how to represent an asphaltene system by a proper equivalent circuit. The above Eqs. (10.1) and (10.2) provide the basic of a composite material. However, it is important to understand what role an asphaltene molecule plays in the solution versus an asphaltene aggregate. In the following a structural modeling is described as the basis for using Eq. (10.1) as oppose to Eq. (10.2).

In an asphaltene solution, the main component is the solvent molecules, i.e., toluene, and asphaltene are "foreign" molecules dispersed in the "sea" of toluene molecules. Because toluene molecules has very low conductivity but connect from one electrode to the other (it is the major component), it should be modeled as a resistor. On the other hand, the asphaltene molecules are dispersed and assumed nonconnected with each other. This is equivalent to having an "obstacle" blocking movement of electrons from one side of the asphaltene molecule to the other side of the asphaltene molecule unless it moves. If we apply a high enough frequency AC

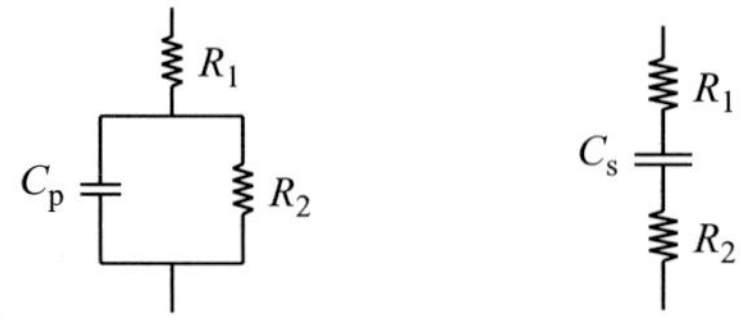

Figure 10.3. Elementary circuits used for modeling composite electrical behaviors.

to the system, the asphaltene molecules can then be assumed stationary, thereby becoming insulators blocking electron movement. This is similar to adding non-conducting microparticles to a conducting system. If one envisions this model, a series circuit appears to fit the description. However, there are continuous "channels" (percolation) where toluene molecules connect themselves from one electrode to the other. Therefore, it should be modeled as a "percolation" parallel circuit as shown in Figure 10.3.

In order to detect asphaltene aggregation using the equivalent circuit concept, it still requires a functional change of the conductivity along an experimentally controllable parameter. In a DC conductivity experiment, one can only control the applied potential and the asphaltene concentration. Since the organic solvents has very low conductance and dielectric constant, DC signal can only come from the asphaltene movement, which is too slow to be significant and the charges asphaltene molecules carry may be too weak to be detected. Therefore, DC conductivity has very limited sensitivity.

As one uses AC conductivity within appropriate frequency range the asphaltene molecules are essentially stagnant and the dependence of equivalent conductivity upon frequency is describable by Eq. (10.1). Moreover, it dependence on asphaltene concentration is a monotonically decreasing function, similar to an electrolyte solution. The main question for detecting asphaltene aggregation is whether the equivalent conductivity is more dependent on frequency or on asphaltene concentration. This is to say that which parameter can provide more sensitive measurement. We chose concentration axis based on the following reason. As asphaltene concentration increases, but below the self-association onset, the equivalent conductivity continue to decrease because there are more and more "individual insulators" introduced. This will slow down the electron movement because of more and more capacitors in parallel with the resistor as an individual unit circuit. However, when they "stack" (or aggregate) the number of insulator decreases leaving a smaller resistance, thus, slow down the decrement of the equivalent conductivity. We thus predict a slope change along the concentration axis when asphaltene concentration increases from below a critical concentration to above it where asphaltene monomers aggregate.

If one choose the frequency axis, both Eqs. (10.1) and (10.2) have their real part behave like $\sim\omega^2$ and the imaginary as $\sim\omega$ at substantially low frequency. As a result, conductivity measurement using either DC or AC as a function of frequency is not applicable for detecting aggregation behavior. Apparently, the concentration axis is more sensitive, provided an appropriate frequency is chosen so that the movement of asphaltene can be neglected. If the frequency is too high, hopping of the electrons from the Fermi level starts to happen within asphaltene molecules leading to a power-law dependence of the conductivity as a function of the frequency ω. This certainly complicates the equivalent conductivity because it becomes a conductor (asphaltene) in parallel with resistor and the conductivity of the asphaltene at this frequency, whether in molecular form or in aggregate form, has the same power law dependence. It thus cannot distinguish the monomer and aggregate states. This more or less suggests that the the conductivity at too high of frequency range will not carry parameters that are aggregation relevant.

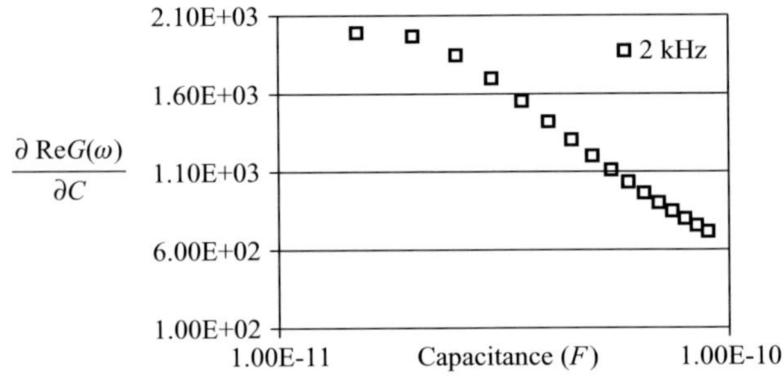

Figure 10.4. Calculated change of conductivity as a function of capacitance at low frequency.

In order to model the equivalent conductivity, we take the partial derivative of the percolating circuit (Figure 10.3 parallel) with respect to C (capacitance, the asphaltene concentration in our case). A simple algebraic functional form is obtained representing the rate of change of the real part of the AC conductance (see Eq. (10.3)). For the simplicity, we assume $R_1 = 0$ and $R_2 = R$,

$$\frac{\partial \mathrm{Re}G(\omega)}{\partial C} = \frac{2R\omega^2 C}{1 + \omega^2 R^2 C^2}. \tag{10.3}$$

Equation (10.3) shows that at high frequency, this function behaves as $1/C$ while at lower frequency range it is ω dependent. Equation (10.3) can be approximated experimentally by the equivalent conductivity by simply dividing the measured conductivity by the asphaltene concentration. The only difference is that it is an AC equivalent conductivity as oppose to the conventional one. We used Eq. (10.3) to detect asphaltene aggregation and phase separation upon addition of nonsolvent.

Figures 10.4 and 10.5 show the theoretical plots of Eq. (10.3) at two frequencies assuming there is no aggregation or phase transition occur within the capacitance (or concentration) range. The curves in Figures 10.4 and 10.5 were calculated based on $R = 25\ \mathrm{M}\Omega$ (resistance measured at 352 mg/L bitumen-derived asphaltene in toluene) as a function of capacitance from 10 to 100 pF. This covers a range of typical asphaltene/toluene systems (~25–50 pF). It is important to pick

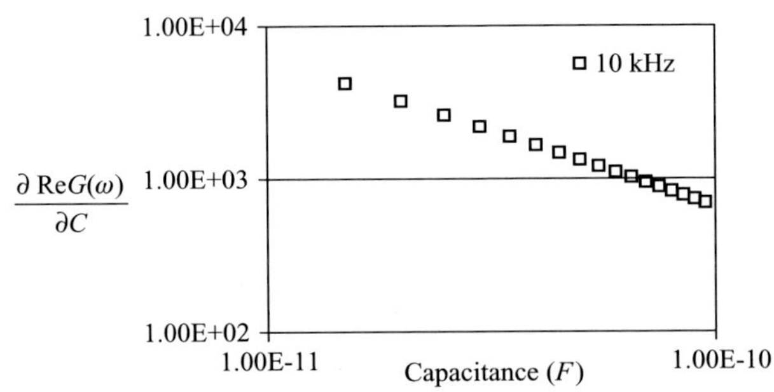

Figure 10.5. Calculated change of conductivity as a function of capacitance at high frequency.

the right range; otherwise, the double-layer formation may occur. Moreover, the breaking point due to aggregation may be overlooked.

One can see the different functional form at 2 kHz and at 10 kHz. These curves do not take into account of the aggregation, which may change the functional form completely. If one assumes asphaltene molecules do not significantly change the bulk resistance, then Eq. (10.3) will only represent the capacitance effect. In this case, when asphaltene molecules aggregate, the capacitance of the equivalent circuit should decrease, resulting in slowing down of the decreasing rate of the conductivity in Figures 10.4 and 10.5 with increasing capacitance.

4. Results

Figure 10.6 shows the real part of the conductivity of the bitumen asphaltene as a function of the AC frequency at various asphaltene concentrations. There are two basic trends observed. One is that the conductivity increases as a function of both frequency and concentration. The other is that the conductivity at low frequency range increases with concentration but becomes more or less independent of the concentration at high frequency. The frequency dependence is consistent with what Eqs. (10.1) and (10.2) predict regardless of percolation or nonpercolation models. As for the conductivity behavior at the low frequency range, it is not obvious that one can extract information using frequency as the primary parameter as expressed in Eqs. (10.1) and (10.2). If one uses Eq. (10.3) and approximate it by the equivalent conductivity, an obvious discontinuity is observed at about 120 mg/L as illustrated in Figure 10.7.

The curve in Figure 10.7 follows Eq. (10.3) rather well, at least qualitatively. Above 120 mg/L, the slope is nearly constant. This is a strong evidence that

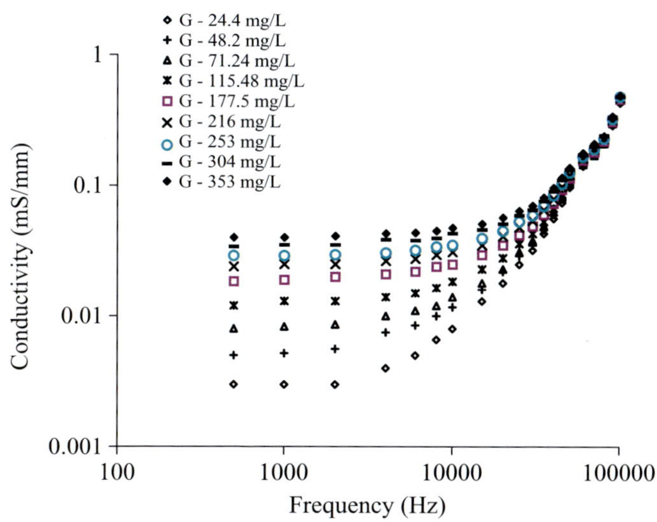

Figure 10.6. Conductivity as a function of frequency at various concentrations.

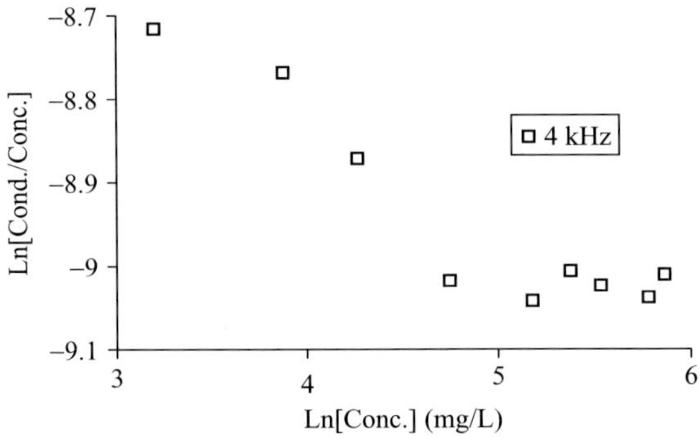

Figure 10.7. Normalized conductivity as a function of concentration at 4 kHz.

asphaltene molecule aggregates at this concentration. Note that this is for 4 kHz. According to Eq. (10.3), one expects the break at 120 mg/L to gradually disappear as frequency increases.

Figures 10.8–10.11 show this trend. At 100 kHz (Figure 10.11), there is no break observed and the functional behavior is exactly what Eq. (10.3) predicts. This suggests that one should be careful in designing experiments in order to pick up the aggregation signal. It is easy to overlook the aggregation if one uses any frequency higher than 100 kHz, which is frequently used in many AC conductivity measurements.

The second series of measurements was for detecting the phase separation (or flocculation induced precipitation). This was done by gradually adding heptane into a 1% asphaltene in toluene as indicated in the phase diagram (Figure 10.12).

Figures 10.13–10.17 show the equivalent conductivity as a function of the asphaltene concentration in the mixed solvents. Note that the lower the concentration the higher the heptane content. It is the same as diluting a fixed asphaltene/toluene (1%) system along the heptane line in the ternary phase diagram shown in Figure 10.12. From these figures, it is clear that there is a phase separation occurred at ~4 g asphaltene per liter of mixed solvent with toluene to heptane volumetric ratio between 4 and 5.

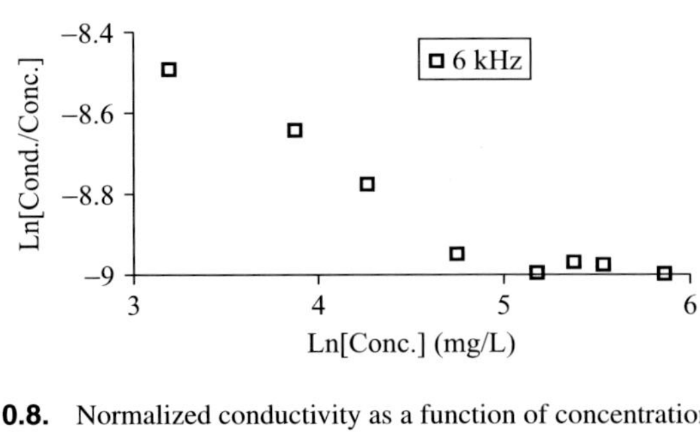

Figure 10.8. Normalized conductivity as a function of concentration at 6 kHz.

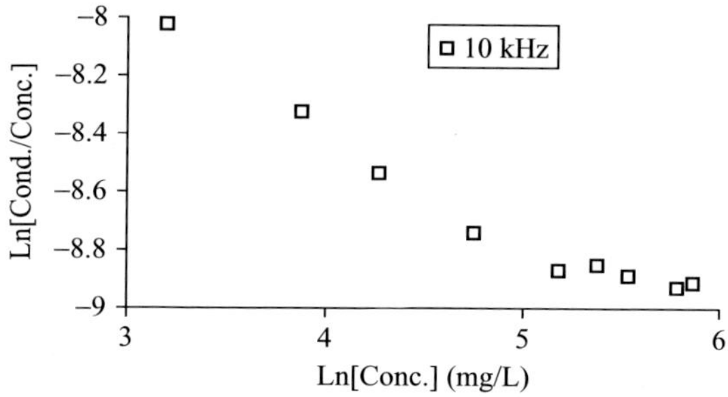

Figure 10.9. Normalized conductivity as a function of concentration at 10 kHz.

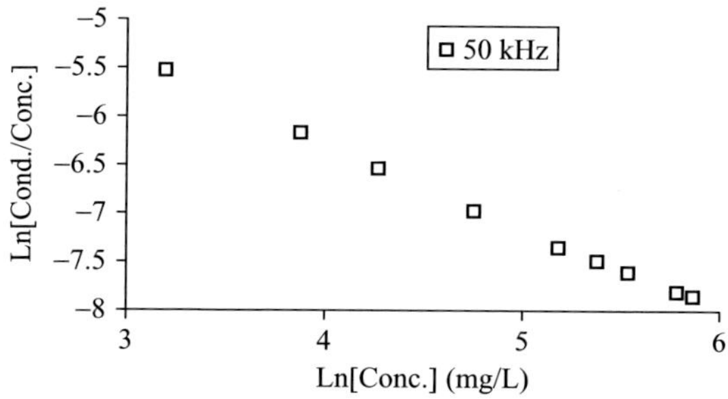

Figure 10.10. Normalized conductivity as a function of concentration at 50 kHz.

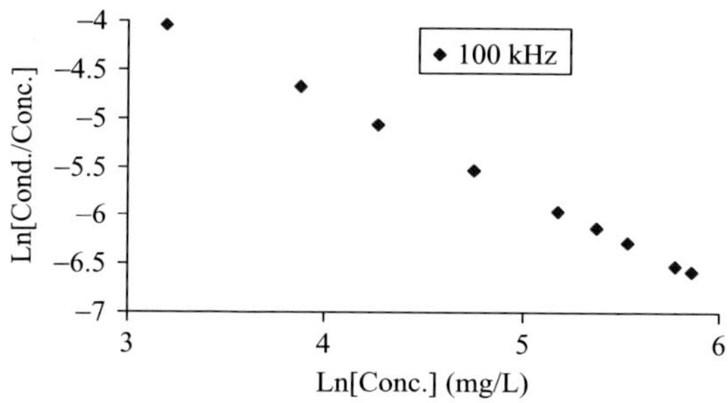

Figure 10.11. Normalized conductivity as a function of concentration at 100 kHz.

The striking points of these equivalent conductivity curves are that the window for detecting the phase transition is much wider than in the case of asphaltene aggregation (see Figures 10.7–10.11). In the case of aggregation detection, one can use only up to 10 kHz. On the other hand, one can practically use 4–100 kHz to detect the phase separation and the signal is rather strong. The other point worth noting is the evolution of the slopes (see Figure 10.13) of the concentration-normalized conductivity in the two-phase region (from B toward the heptane corner in Figure 10.12). Figure 10.18 depicts the slopes of both the isotropic phase (one phase region) and the separated phase region (two-phase region). These slopes

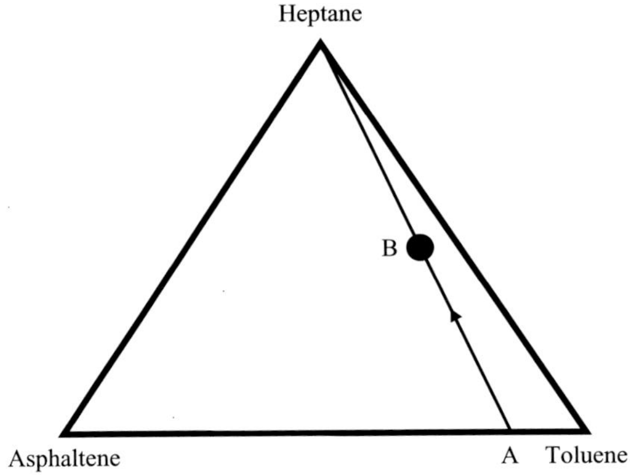

Figure 10.12. A three-component asphaltene ternary system. Stock solution is at point A (1% asphaltene in toluene), diluted along the AB line. The system undergoes phase separation at B (~4 g/L).

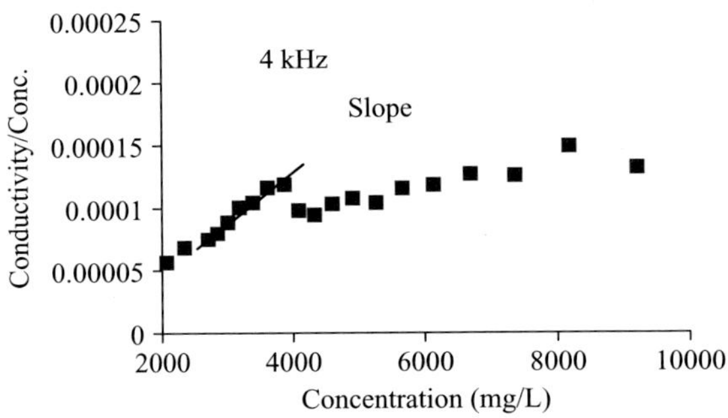

Figure 10.13. Normalized conductivity as a function of asphaltene concentration via hepatne addition (see Figure 10.12 for dilution line).

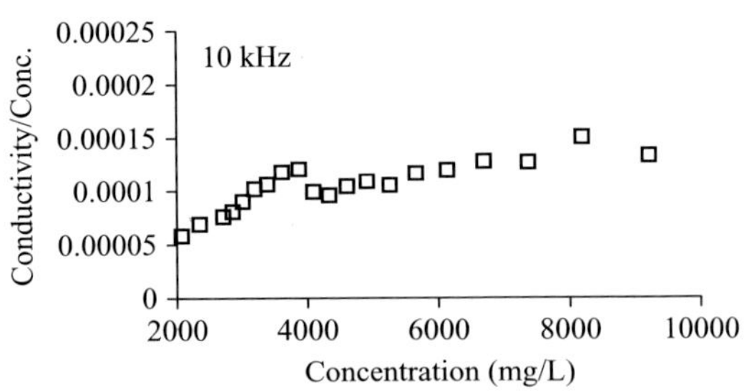

Figure 10.14. Same as Figure 10.13 at 10 kHz.

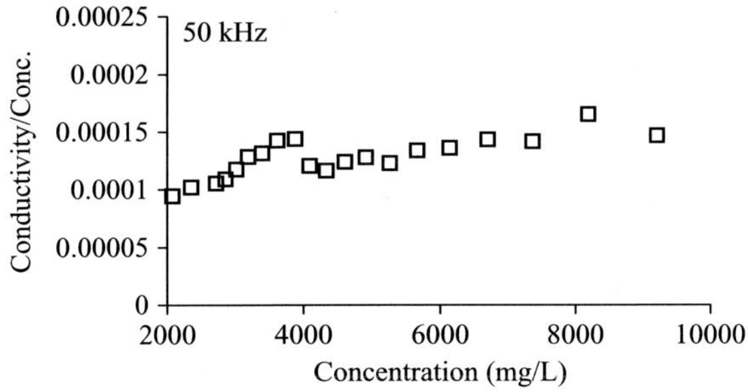

Figure 10.15. Same as Figure 10.13 at 50 kHz.

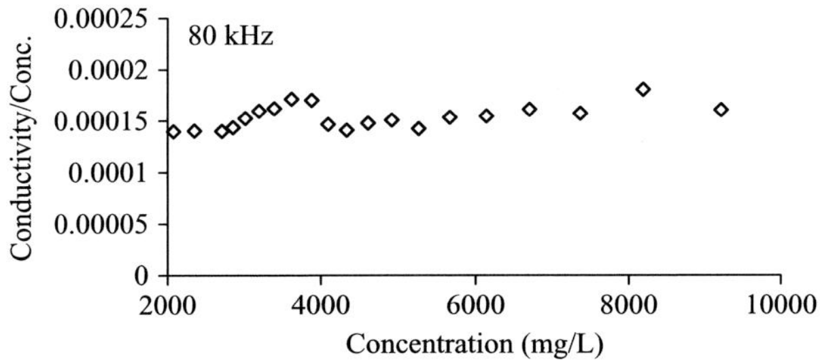

Figure 10.16. Same as Figure 10.13 at 80 kHz.

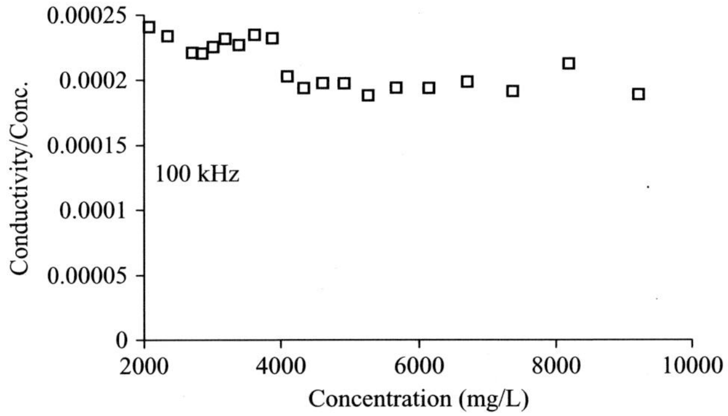

Figure 10.17. Same as Figure 10.13 at 100 kHz.

are not physically meaningful unless one argues that they are similar to a critical phenomenon observed in binary fluids.[29] Nevertheless, it can serve as an indicator for choosing a right frequency range for AC conductivity measurements.

From Figure 10.18, it is clear that the slopes evolve when frequency varies. Similar to the aggregation study, we would like to find a frequency range appropriate for phase separation study. Based on Figures 10.13–10.17, this range can be from 4 to 100 kHz. However, if one uses Figure 10.18 to select the frequency

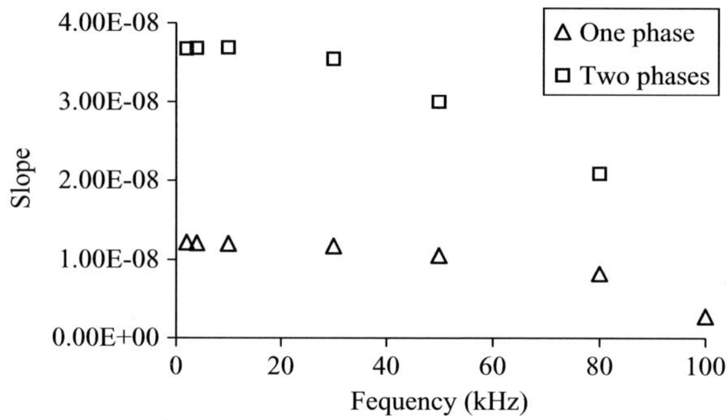

Figure 10.18. Slopes of the two regions. One phase region is at high asphaltene concentration (>4 g/L or between A and B in Figure 10.12) with high toluene content and the two-phase region (<4 g/L or between B and heptane corner) is with high heptane content.

range, 4–30 kHz may be preferred because there is enough difference between the slope of the one-phase region and that of the two-phase region.

5. Discussion and Conclusion

There are two essential issues to be discussed. The first one is the justification of the conductance measurement. Argument is likely on the potential double-layer formation when a potential is applied to the electrode. This is a classical concern and was used as the guideline for setting the electrode–electrode distance and the cell constant. The key parameter that dictates this phenomenon is the RC constant. It indicates how fast a parallel plate electrode will accumulate charges. In our case, the $R \geq 25$ MΩ (25 MΩ is the resistance at 350 mg/L asphaltene) and C is about 30 pF. With this RC constant, the minimum frequency needed to prevent accumulation corresponds to $1/RC \sim 1{,}300$ Hz. This is based on the assumption that there is no surface roughness introduced. Since the cell used in this work is a platinum black which creates much more surface area, the minimum usable frequency should be much lower than 1,300 Hz. Yet, we chose 4 kHz to avoid any possible accumulation. Thus, the charge accumulation at the chosen frequency range (>4 kHz) should not be an issue.

The second issue is related to the assumption that increasing asphaltene concentration is equivalent to increasing capacitance. This assumption was based on the fact that asphaltene is a foreign object as far as the toluene is concerned. Therefore, we argue that both R and C should change upon asphaltene addition. Since the concentration range we study is relatively dilute, we anticipate R to be dominated by toluene, which has a much higher R than asphaltene but the toluene molecules connect from one electrode to the other. As a result, R should be more or less linearly proportional to the asphaltene concentration unless there is electron hopping. However, it is unlikely for electron to hop between asphaltene molecules given the fact that the concentration is very dilute. Moreover, the hopping phenomenon should not happen until the frequency is higher than the critical frequency.[30]

On the other hand, the physical existence of asphaltene molecules in between toluene molecules makes them behave more like a capacitor than a resistor (see Section 3). As a result, the nonlinear change of the rate of change of the conductivity (Eq. (10.3) should be largely from the change of capacitance upon asphaltene addition. Based on this argument, we did not really enforce asphaltene concentration to be equivalent to the capacitance. Instead, we measured equivalent conductivity to observe the functional behavior when asphaltene is added. What we observed was a monotonically decreasing function as predicted by the percolation model (Eq. (10.3)) when frequency is between 2 and 10 kHz. When we further increase the frequency, the electron hopping gradually sets in and overcomes the capacitance effect as illustrated in Figures 10.6 and 10.11 where we demonstrate the concentration independence of the conductivity for frequency above ~75 kHz.

In order to reveal the dependence of the conductivity on asphaltene concentration, we investigated the frequency range only from 4 to 10 kHz. By plotting the equivalent conductivity, we observed aggregation-like behavior. We believe this is a true phenomenon and can only be observed when frequency used is in the right range.

Another point worth noting is the popularity of equivalent conductivity for CMC determination if one follows the CMC work.[7,8] In our case, it was used nearly the same way except we use AC conductivity rather than a standard DC conductivity measurement. DC conductivity measurement often suffers charge deposition near the electrode particularly when conductivity is high. This is why many conductivity cell manufactures coat the platinum electrode to make rough surface.[21] The electrode used here has a cell constant of 0.001, which is calibrated for electrolyte solution with application range well cover the asphaltene solution investigated here. The reason conductivity measurement for CMC determination is not as popular as the surface tension technique is because surfactants are very surface active, which make surface tension a sensitive technique. On the contrary, asphaltene is not as surface active as one can see from our early pyridine work.[1] Thus, one may not be able to correlate the surface tension measurement to what happens in the bulk. This is to say that one may not measure substantial surface tension transition when asphaltene molecules aggregate in the bulk. Therefore, one should look for bulk techniques to detect changes in the bulk or select a right solvent that has enough surface tension contrast for asphaltene to be more surface-active in that particular system.

Final point of discussion is about the VPO measurement of molecule weight of asphaltene. Most VPO work reported at several thousand to several hundred thousands. Because VPO can only operate at a concentration much higher than 120 mg/L, we speculate that the VPO-measured molecular weight is the average molecular weight of an aggregate rather than an asphaltene molecule.

In conclusion, we evaluated the possibility of using equivalent AC conductivity for measuring asphaltene aggregation and phase separation. It is a bulk technique, suitable for detecting changes in the bulk. The results obtained make us believe that it is an appropriate technique for detecting asphaltene aggregation and for solvent initiated phase transition. However, the frequency range should be carefully selected.

6. Future Perspective

The AC conductivity technique should be further evaluated for other asphaltene and petroleum systems. However, we believe this is a good start and can potentially benefit petroleum research community where simple characterization techniques are always demanded. One important note for using this technique is to find the right frequency range and perform accurate system calibration.

References

[1] Sheu, E.Y. (1995). Colloidal properties of asphaltenes in organic solvents In: E. Sheu and O.C. Mullins (eds.), *Asphaltene—Fundamentals and Applications.* Plenum, New York.

[2] Acevedo, S., M.A. Ranaudo, J.C. Pereira, J. Castillo, A. Fernandez, P. Perez *et al.* (1999). Thermo-optical studies of asphaltene solutions: Evidence for solvent solute aggregate formation *Fuel* **78**, 997.

[3] Andreatta, G., N. Bostrom, and O.C. Mullins (2006). *Ultrasonic spectroscopy on asphaltene aggregation.* In: O.C. Mullins, E.Y. Sheu, A. Hammami, and A.G. Marshall (eds.), *Asphaltene, Heavy oils and Petroleomics.* Springer Academic Press, New York.

[4] Ferworn, K. and W. Svrcek (1998). Characterization and phase behavior of asphaltenic crude oils. In: O.C. Mullins and E.Y. Sheu (eds.), *Structures and Dynamics of Asphaltenes.* Plenum, New York.

[5] Pfeiffer, J.P. and R.N. Saal (1940). Asphaltic Bitumens as a colloidal system *J. Phys. Chem.* **44**, 139.

[6] Yen, T.F. (1988). In: M. Grayson and J.I. Krochwitz (eds.), *Encyclopedia of Polymer Science and Engineering*, 2nd edn., Vol. 1. Wiley, New York.

[7] Rosen, M. (1989). *Surfactant and Interfacial Phenomena*, 2nd edn. John Wiley and Sons, New York.

[8] Hiemenz, P.C. (1977). *Principle of Colloid and Surface Chemistry.* Marcel Dekker, New York, pp. 284–285.

[9] Tanford, C. (1980). *The Hydrophobic Effect*, 2nd edn. Wiley, New York.

[10] Israelachvili, J.N., D.J. Mitchell, and B.W. Ninham (1976). Theory of self-assembly of hydrocarbon amphiphiles into micelles and bilayers. *J. Chem. Soc., Faraday Trans. II* **72**, 1525–1568.

[11] Andersen, S.I. and S.D. Christensen (2000). The critical micelle concentration of asphaltenes as measured by calorimetry. *Energy Fuels* **14**, 38.

[12] Andersen, S.I., J.M. del Rio, D. Khvostitchenko, S. Shakir, and C. Lira-Galeana (2001). Interaction and solubilization of water by petroleum asphaltenes in organic solution *Langmuir* **17**, 307.

[13] CRC table. (1989–1990). *Handbook of Chemistry and Physics*, 70th edn. Robert C. Weast, David R. Kide, Melvin Astle, and William Beyer, CRC press, Boca Raton, FL, pp. F33–F35.

[14] Loh, W., R.S. Mohamed, and A.C. Ramos (1999). Aggregation of asphaltenes obtained from a Brazilian crude oil in aromatic solvents *Pet. Sci. Technol.* **17**, 147–163.

[15] Ramos, A.C.D., L. Haraguchi, F.R. Notrispe, W. Loh, and R.S. Mohamed (2001). Interfacial and colloidal behavior of asphaltenes obtained from Brazilian crude oils *J. Petroleum Sci. Eng.* **32**, 201–216.

[16] Bouhadda, Y., D. Bendedouch, E. Sheu, and A. Krallafa (2000). Some preliminary results on a physico-chemical characterization of a Hassi Messaaoud petroleum asphaltene. *Energy Fuels* **14**(4), 845–853.

[17] Yen, T.F. (1972). Present status of the structure of petroleum heavy ends and its significance to various technical applications. *Am. Chem. Soc., Div. Petrol. Chem. Prepr.* **17**(1), F102–114.

[18] Yen, T.F. (1981). Structural differences between asphaltenes isolated from petroleum and from coal liquid. In: Chemistry of Asphaltene. Advance in Chemistry seris 195. American Chemical Society, New York.

[19] Brandt, H.C.A., E.M. Hendriks, M.A.J. Michels, and F. Visser (1995). Thermodynamic modeling of asphaltene stacking. *J. Phys. Chem.* **99**, 10430.

[20] Yudin, I.K., G.L. Nikolaenko, E.E. Gorodetskii, V.R. Melikyan, E.L. Markhashov, V.A. Agayan, M.A. Anisimov, and J.V. Sengers (1998). Crossover kinetics of asphaltene aggregation in hydrocarbon solutions. *Physica A*, **251**, 235–244.

[21] Radiometer analytical technical note. (2004). *Conductivity—Theory and Practice* D61M002, Radiometer Analytical SAS, France 2004-05B.

[22] Jalali, F., M. Shamsipur, and N. Alizadeh (2000). Conductance study of the thermodynamics of micellization of 1-hexadecylpyridinium bromide in (Water + Cosolvent). *J. Chem. Thermodynamics* **32**, 755–765.

[23] Sui, G.P., S.R. Coppen, E. Dupont, S. Rothery, J. Gillespie, D. Newgreen *et al.* (August 2003). Impedance measurements and connexin expression in human detrusor muscle from stable and unstable baldders. *Br. J. Urol., Int.* **92**(3), 297.

[24] Armstrong, C.M. (March 1999). Distinguishing surface effects of calcium ion from pore-occupancy effects in Na^+ channels. *Physiology. Proc. Natl. Acad. Sci.* **96**, 4158–4163.

[25] Pourghobadi, Z., F. Seyyed-Majidi, M. Daghighi-Asli, F. Parsa, A. Moghimi, M.R. Ganjali *et al.* (2000). Synthesis of a new triazine derived macrocycle and a thermodynamic study of its complexes with some transition and heavy metal ions in acetonitrile solution. *Polish J. Chem.* **74**, 837–846.

[26] Ogata, A., Y. Tsujino, and T. Osakai (2000). Selective hydration of alkylammonium ions in nitrobenzene. *Phys. Chem. Chem. Phys.* **2**, 247–251.

[27] Kang, K., H. Kim, K. Lim, and N. Jeong (2001). Mixed micellization of anionic ammonium dodecyl sulfate and cationic octadecyl trimethyl ammonium chloride *.Bull. Korean Chem. Soc.* **22**(9), 1009.

[28] Sen, A. and A. Gupta (1998). Frequency-dependent (AC) conduction in disordered composites: A percolative study. *J. Cond. Matt. Phys.* **2**(6), 282.

[29] Fisher, M.E. (1967). The theory of equilibrium critical phenomena. *Progr. Phys.* **30**(Part II), 615.

[30] Sheu, E.Y. and O.C. Mullins (2004). Frequency-dependent conductivity of utah crude oil asphaltene and deposit. *Energy Fuels* **18**(5), 1531–1534.

11

Molecular Composition and Dynamics of Oils from Diffusion Measurements

Denise E. Freed, Natalia V. Lisitza, Pabitra N. Sen, and Yi-Qiao Song

1. Introduction

We discuss examples and methods for using NMR diffusion measurements to obtain information about molecular sizes, their distributions, and dynamics. Scaling relationships between chain lengths and diffusion constants are derived and tested on diffusion measurements of many samples, including crude oils that are high in saturates. The diffusion constants of asphaltenes are also measured as a function of asphaltene concentration, indicating the formation of asphaltene aggregates at a concentration of approximately 0.2 g/L, and the sizes of the individual asphaltene molecules and aggregates are obtained. The examples and methods discussed in this paper can become the basis for in situ characterization of crude oils.

Crude oils are complex mixtures of molecules encompassing a broad range of shapes and sizes.[1–3] They include molecules ranging from alkanes, which are chain-like and relatively simple, to asphaltenes, which are complex and may interact strongly with one another.[4] The composition determines the properties of crude oils, such as their viscosity and phase behavior. These properties are very important in the production of the oils. For example, the heavy oil components may precipitate and clog the formations and wells, depending on how oils are being lifted to the surface.

There are several reasons why it is also important to characterize the composition of the oil in situ. First, many properties of the fluid depend critically on temperature and pressure, so it can be advantageous to make the measurements downhole. In some cases, oil samples even undergo irreversible changes as they are extracted from the well and transferred to the laboratory for analysis. Second, the fluid composition in a reservoir can exhibit large heterogeneity, and strong compositional gradients have been reported.[5] Because downhole measurements

Denise E. Freed, Natalia V. Lisitza, Pabitra N. Sen, and Yi-Qiao Song • Schlumberger-Doll Research, 36 Old Quarry Road, Ridgefield, Connecticut 06877.

can be made at many different locations, they can be used to better characterize the reservoir. Many analytical techniques, such as high resolution NMR spectroscopy, gas chromatography, and mass spectrometry, require delicate instrumentation and are currently not suited for field application. Instead, NMR diffusion measurements are an attractive method for fluid typing[6] because they are noninvasive and are already used in well-logging.[7–9] In this paper, we show several applications for using diffusion to characterize composition.

It is well-known that the molecular diffusion constant (D) is dependent on molecular size.[10] In a mixture, small molecules generally diffuse faster than large ones. The diffusion constant of all molecules also depends on the common fluid environment contributed from all molecules. Thus, the size distribution of the molecules in the mixture should be reflected in the distribution of diffusion constants, although the relation may not always be simple. In addition, when the molecular size of the species changes, for example, during the self-aggregation of molecules, this change should also be reflected in the diffusion constants. Therefore, the distribution of diffusion constants and its variations can also be used to describe the dynamical processes between species in a mixture.

In this paper, we focus on two aspects of molecular diffusion in fluid analysis. First, we show how distributions of diffusion constants in oils high in alkanes can be used to determine the composition. Namely, we extract the chain length distribution from diffusion data. Second, we show how the distributions of diffusion constants and their changes can reflect the dynamics of self-aggregation of asphaltene molecules in solution.

2. General Theory of Molecular Diffusion

The diffusion constant of a molecule depends both on its size and on the properties of the surrounding fluid. For example, for a hard sphere in a solvent with viscosity η_s, the relation is given by the Einstein–Stokes relation,[10]

$$D = \frac{k_B T}{6\pi \eta_s r}, \tag{11.1}$$

where r is the radius of the sphere. This relation implies that by measuring the diffusion constant, the radius of the diffusing molecule can be found, provided the viscosity of the solvent is known.

Equation (11.1) is applicable to the case where the spheres are dilute and much larger than the solvent molecules. Even as the size of the molecule is decreased, so that, for example, it is the same as the solvent molecules, this relation often still holds once the denominator is multiplied by a microviscosity factor. Similarly, as the spheres are deformed, again the denominator in Eq. (11.1) will be modified by a factor that depends on the shape of the molecule, but otherwise it is still valid.

Once the molecules have internal degrees of freedom, the diffusion constant still depends on the size of the molecule, but the relation in Eq. (11.1) may no longer hold. For example, polymers, which are long, flexible, chain molecules, are known to exhibit a scaling relation between D and the chain length.[11–15] This relation is given by

$$D \propto N^{-\kappa} \tag{11.2}$$

with κ ranging from $1/2$ to 2, depending on whether hydrodynamic effects are significant and on whether the chains are entangled. Pure alkanes also show a similar scaling relation, with $\kappa \approx 2$ (references 16–19).

In mixtures, the dependence of the diffusion constant on the size of the molecule becomes more complex. In addition to the diffusion coefficient scaling with the size of the molecule as in Eqs. (11.1) and (11.2), the diffusion constant also depends on the viscosity of the mixture, which, in turn, depends on the composition. Even within mixtures, though, the self-diffusion coefficient of a molecule can be divided into two parts. The first part depends on the properties of the molecule, including its shape, size, and stiffness. For polymers, this first part of the diffusion constant gives rise to a power law in the chain length, N. For example, in the free-draining limit, it is inversely proportional to N (reference 15), and in the presence of hydrodynamic effects, instead it is inversely proportional to the radius of gyration of the molecule,[20] which also follows a power law in N. Even for shorter chains, such as the alkanes, the diffusion coefficient will scale inversely with the chain length, as long as the only interaction between the segments in the molecule is through a translationally invariant potential.

The second part of the diffusion constant depends on the bulk fluid properties, such as the density and the friction coefficient for a monomer, ξ. It should be the same for all the molecules within the mixture. This suggests the following ansatz for the diffusion constant of the ith component D_i in the mixture:

$$D_i = N_i^{-v} g(\{N_i\}), \tag{11.3}$$

where $g(\{N_i\})$ is a function of all the components in the mixture and is related to the viscosity. The scaling behavior within a mixture reflects directly the single chain dynamics, while any change in D_i for a given component in different mixtures reflects the bulk properties.

We note that Eq. (11.3) is applicable both to hard spheres, as in the Einstein–Stokes equation, and to polymers. In the case where the diffusing particles are dilute and the viscosity of the solvent is known, the diffusion constant directly gives the size of the particles. In Section 5, we shall use this relation to determine the size distribution of asphaltene molecules and aggregates dissolved in toluene. For mixtures, such as melts or oils, the scaling behavior for different molecules within a mixture still reflects directly the single chain dynamics. In addition, the change of D_i for a given component in different mixtures reflects the bulk properties, which depend on all the constituents in the mixture. In Section 4, we shall use these two relations to determine the chain length distribution in mixtures of alkanes and in crude oils.

3. Experimental Method

The measurement of the diffusion constants has generally been done with pulsed field gradient (PFG) NMR.[21] It is noninvasive and capable of studying optically opaque samples. The PFG NMR measures the displacement of molecules as a function of diffusion time, t. The mean-squared displacement $\langle |\mathbf{r}|^2\rangle$ due to Brownian motion is linear in time t and proportional to the self-diffusion constant D of the molecule:

$$\langle |\mathbf{r}|^2\rangle = 6Dt. \tag{11.4}$$

The NMR experiment for measuring diffusion is sketched in Figure 11.1. It utilizes the stimulated-echo pulse sequence[22] with two magnetic field gradient pulses applied: one between the first two 90° pulses, and one after the third pulse. These two gradient pulses are identical in amplitude, G, and duration, δ, and they are separated by a time Δ. The function of the first gradient pulse is to dephase magnetization according to the position of the molecules in the sample. During the subsequent Δ period, the molecules are allowed to diffuse; the second gradient pulse is applied to refocus the phase and produce an echo. The spins that have diffused to a new location do not get refocused completely at the end of the Δ period, and therefore, the echo signal is attenuated. The relationship between the signal amplitude I in the presence of a gradient of amplitude G and the diffusion constant D along the gradient direction is given by[22]

$$\frac{I}{I_0} = \exp\left[-D(\gamma G\delta)^2\left(\Delta - \frac{\delta}{3}\right)\right], \tag{11.5}$$

where I_0 is the signal amplitude at zero gradient, γ is the gyromagnetic ratio ($2.675 \times 10^8\ T^{-1}s^{-1}$ for protons). In a common implementation of this sequence, Δ and δ are kept fixed, while G is varied. The attenuation of the echo signal is then measured as a function of G; and fitting the echo amplitude to Eq. (11.5) gives the diffusion coefficient.

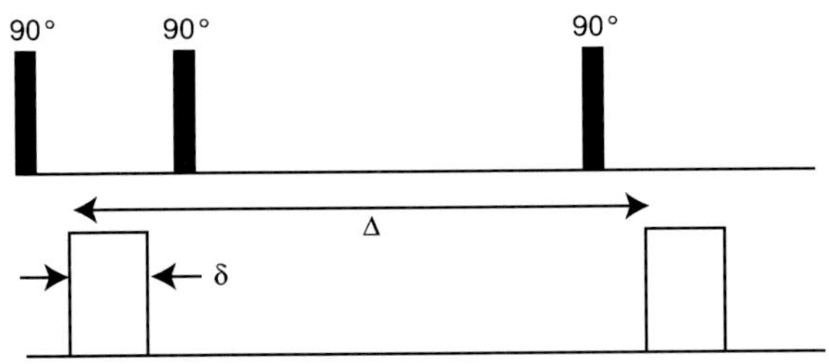

Figure 11.1. Pulse sequence used for NMR diffusion measurements. The duration of the gradient pulse δ and the period Δ were 1.5 and 33 ms, respectively. The diffusive echo attenuation was measured as function of the gradient strength G.

In a mixture of independently diffusing species, the total signal is a sum of all components, so that Eq. (11.5) becomes

$$\frac{I}{I_0} = \int dDp(D)\exp\left[-D(\gamma G\delta)^2\left(\Delta - \frac{\delta}{3}\right)\right], \tag{11.6}$$

where $p(D)$ is the distribution of the diffusion constants. The diffusion weighting is often defined to be $\mathcal{B} = (\gamma G\delta)^2(\Delta - \delta/3)$, so that Eq. 11.6 can be rewritten as

$$I(\mathcal{B})/I(0) = \int dDp(D)\exp(-D\mathcal{B}). \tag{11.7}$$

In this case, a multi-exponential decomposition is required to analyze the data. Generally, it is done by Laplace inversion. In these measurements, we used a specially designed diffusion probe (Bruker Biospin), which allows the application of magnetic field gradients as high as 1200 G/cm (Bruker Biospin).

The Laplace inversion is an ill-conditioned problem since its solution is not unique, and thus, it is quite sensitive to the noise in the input data. A common method for solving the ill-conditioned problem is to use a numerical technique called regularization[23] and such algorithms have been used in NMR.[24–28] The regularization method can provide a stable inversion for a given signal-to-noise ratio. For a given dataset and noise, a limit exists on the smallest resolvable structure (or separation of structures) in the Laplace inversion spectrum.[28,29] It is important to be aware of the spectral resolution in order to interpret properly the results of Laplace inversion.

4. Mixtures of Alkanes

The hard sphere model and the Einstein–Stokes equation (Eq. 11.1) are not adequate for describing diffusion in oils. One example of the failure of the hard sphere model is evidenced by the measurements of diffusion and viscosity in alkanes and oils: In plots of log D versus log kT/η, the data all lie on a single line, regardless of the molecules' radii.[9,30,31] This is in disagreement with Eq. (11.1), which implies that the intercept depends on the radius of the molecule.

In this section, we will instead model oils as mixtures of alkanes, and consider the self-diffusion constant D_i of a molecule in such mixtures. This alkane mixture model has been briefly discussed in reference 32. An alkane can be described by a chain similar to a polymer, only shorter. The number of segments in the chain N can be taken equal to the number of carbon atoms in the alkane. Each segment of the chain interacts with its neighbors and is subject to the Brownian forces of the surrounding fluid. It may also be subject to hydrodynamic interactions.

As described in the Introduction, the self-diffusion coefficient of a molecule can be divided into two parts, the one that depends on the properties of the molecule, and the one that depends on the bulk fluid. The diffusion constant then has the form given by $D_i = N_i^{-\nu} g(\{N_i\})$. We will first address the scaling behavior between the components within a mixture, given by $N_i^{-\nu}$, and then discuss the dependence on the bulk properties of the fluid g later.

Crude oils also contain gases such as methane and ethane. These small molecules are more appropriately described by hard spheres, with the caveat that the diffusing particles are the same size or smaller than the solvent molecules. In that case, we will still assume that the diffusion constant scales inversely with the radius of the molecule, as in the Einstein–Stokes equation, and that the gas molecules are subject to the same internal viscosity function $g(\{N_i\})$ as the other components in the mixture. The diffusion constant for these gas components can then be written as

$$D_{\text{gas}} = r_{\text{gas}}^{-1} \cdot g, \tag{11.8}$$

where g is short for $g(\{N_i\})$ and r_{gas} is a dimensionless parameter proportional to the radius of the molecule.

4.1. Chain-Length Dependence

According to Eqs. (11.3) and (11.8), we expect that within a mixture $D_i N_i^{\nu}$ and $D_{\text{gas}}\, r_{\text{gas}}$ are equal for all components. We call these factors the scaled diffusion constants (SDCs). By requiring the SDCs to be equal for all components within a mixture, we can fit for the value of v, r_{methane}, and r_{ethane}. We analyzed the diffusion data for many binary and ternary mixtures of alkanes. These mixtures included molecules with chain lengths from $N = 1$ to 30, and also benzene and squalene, for a total of 207 data points. The data consist of 12 different pairs or triplets of components, with about four concentrations for each set of components and about six temperature and pressure conditions for each concentration for most of the samples. With a single set of parameters, $\nu = 0.7$, $r_{\text{methane}} = 1.64$ and $r_{\text{ethane}} = 2.32$, we found that the SDCs fall close to the diagonal line with the SDCs ranging by a factor of 25, as demonstrated in Figure 11.2.

It is interesting to contrast the exponent ν found for alkanes with that for polymer melts.For example, it has been established that $D \propto N^{-2}$ for long polymers with entanglements,[13,14] and $D \propto N^{-1}$ for melts of shorter polymers.[15] For these shorter polymers, the scaling behavior has been explained by the Rouse model[15] which we will review briefly. In the Rouse model,[15] a polymer is considered as a chain of N segments. There is a Gaussian distribution of bond lengths, which leads to a spring–like interaction between adjacent segments. In addition, each segment undergoes Brownian motion due to its interaction with the surrounding fluid. In this model, $D \propto \xi^{-1} N^{-1}$. Although alkanes are too short to be Gaussian chains, their diffusion constant will follow the Rouse scaling as long as the only interaction between segments is a translationally invariant potential. The deviation from $\nu = 1$ found for alkanes is then due to correlated motion of the segments, such as the hydrodynamic interaction, that does not come from a translationally invariant potential. In that case, the value of ν reflects the equilibrium configurations of the chain due to the molecular properties such as the stiffness of the chain and

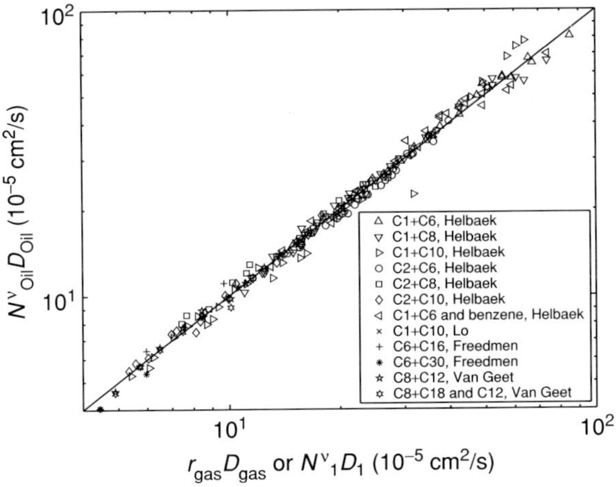

Figure 11.2. Demonstration that $N_i^{\nu} D_i$ and $r_{gas} D_{gas}$ are constant in binary and ternary mixtures. The SDC of the gas component $r_{gas} D_{gas}$ or the lighter component $N_1^{\nu} D_1$ is plotted against that of the heavier component. The prediction of Eq. 11.3 is shown by the black line. The pressure ranges from 0.1 to 60 MPa and the temperature ranges from 25 to 60°C. The data for mixtures with methane and ethane (open symbols) are from reference 34. The data for C_1–C_{10} mixtures (crosses) are from reference 35, those for C_6–C_{16} or C_6–C_{30} mixtures are from reference 9, and those for C_8–C_{12} mixtures are from reference 36.

excluded volume effects.[20] A value of $\nu = 0.7$ is then consistent with the presence of hydrodynamic interactions and the chain being stiffer than a Gaussian chain.

As the chains get longer, one might expect that the behavior will approach that of the Rouse model and $\nu \to 1$. However, for mixtures of C_{12} and C_{60},[33] we determined that $\nu \approx 0.75$ for C_{60} with $\nu = 0.70$ for C_{12}. Although the exponent is increasing for the longer chain, it is still well within the regime of partial screening of the hydrodynamic interactions. In the other limit, as $N \to 1$, the molecules become stiffer and lose their segmental motion. As a result, the radii of methane and ethane are expected to be larger than the extrapolation from the longer molecules and thus $r_{methane} > 1$. This effect is the strongest for methane, weaker for ethane, and within the experimental error for pentane and hexane.

4.2. Dependence on Mean Chain Length and Free Volume Model

Next, we consider the function $g(\{N_i\}) = \text{SDC}$. In Figure 11.3 we plot the SDCs as a function of $\bar{N}$ for many mixtures to show that g is, in fact, only a function of $\bar{N}$ in mixtures. This was first observed in reference 36 for mixtures of C_8 with C_{12} and C_{18}. We find this remains the case for a wide range of mixtures including mixtures with methane or ethane at elevated pressures, as shown in Figure 11.3B. In both Figures 11.3A and 11.3B, the SDCs collapse to a single curve as a function of $\bar{N}$.

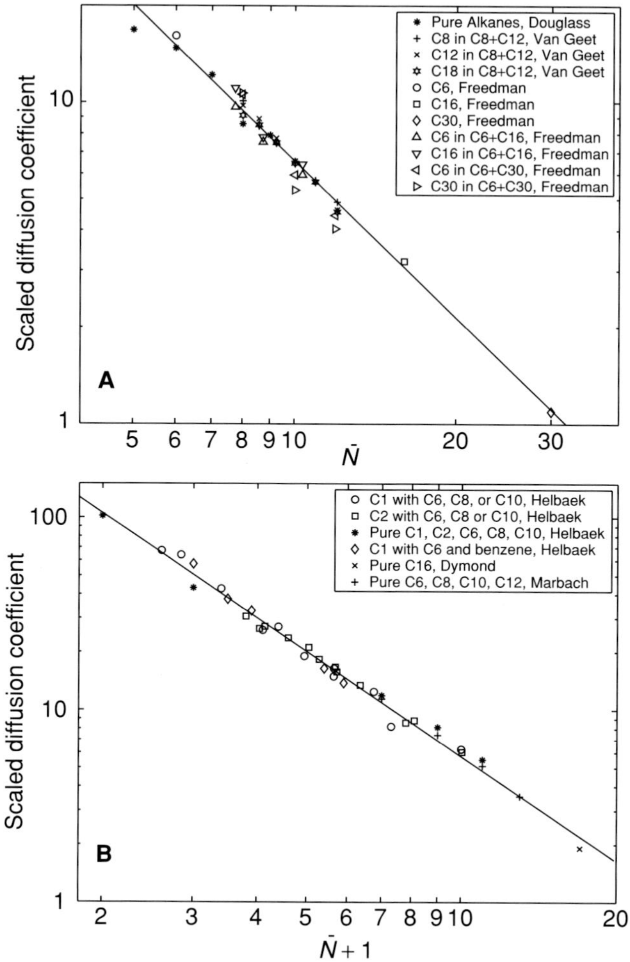

Figure 11.3. (A) Scaled diffusion constants for pure alkanes and mixtures as a function of mean chain length $\bar{N}$. All data are at 25–30°C and at atmospheric pressure. The solid black line shows the fit for the pure alkanes from C_6 to C_{10} and the binary mixtures of C_8 and C_{12} to a power law dependence on the mean chain length. The data for the mixtures are the same as in Figure 11.1. The data for pure alkanes (stars) are from reference 16 and those for pure C_6, C_{16}, and C_{30} are from reference 9. (B) Scaled diffusion coefficients of the gas component as a function of $\bar{N}+1$. The data are at 25–30°C and 30 MPa. The solid black line shows the fit for the binary mixtures to a power law dependence on $\bar{N}+1$. The data for the mixtures are the same as in Figure 11.1. The data for pure C_1, C_2, C_6, C_8, and C_{10} are from reference 34, those for pure C_{16} are from reference 37, and those for pure C_6, C_8, C_{10}, and C_{12} are from reference 38.

This dependence on the mean chain length can be explained by taking into account the end effects of the chains in the free volume model.[11] In this model applied to alkanes,[19] the diffusion constant is given by

$$D_i = A' N_i^{-\nu} \exp\{-E_a/kT\} \exp\{-B/f(\{N_i\})\}, \tag{11.9}$$

where the activation energy E_a for segmental motion and the overlap function $B \approx 1$ are considered to be independent of chain length, and A' is a constant.[19] von Meerwall *et al.*[19] used the Rouse value of $\nu = 1$, but we will use the experimental value of 0.7 for these alkane mixtures. In Eq. (11.9), the only dependence of the SDC on the composition of the mixture is through the free volume fraction $f(\{N_i\})$, where $f(\{N_i\}) =$ free volume/total volume. The volume per molecule υ_T for alkanes depends on the volume per segment υ_s and the extra free volume per end υ_e. To a very good approximation, both of these volumes are independent of chain length.[11,19,39,40] Then the total volume per molecule is given by

$$\upsilon_T = 2\upsilon_e + N\upsilon_s. \tag{11.10}$$

The free volume per segment υ_{sf} is also considered to be independent of chain length for polymers.[11] Thus, in a mixture, the average volume per molecule, $\bar{\upsilon}_T$ and the average free volume per molecule $\bar{\upsilon}_f$ are given by

$$\bar{\upsilon}_T = 2\upsilon_e + \bar{N}\upsilon_s, \qquad \bar{\upsilon}_f = 2\upsilon_e + \bar{N}\upsilon_{sf}, \tag{11.11}$$

where $\bar{N}$ is the molar average of the chain lengths. Hence, $f(\{N_i\})$ and thus $g(\{N_i\})$ depend only on $\bar{N}$.

For polyethylene and pure alkanes, it is well established that the diffusion constant D scales as $N^{-\kappa}$ with $\kappa \approx 2$.[17–19,41] This means that for pure alkanes, the scaled diffusion constant $DN^{-\nu} = g(N)$ must follow a power law, too. In other words, because κ does not equal ν, the internal viscosity function must also follow a power law in N. Since the SDCs in a mixture depend only on $\bar{N}$, this implies that within a mixture $D_i N_i^{\nu}$ should follow a power law in $\bar{N}$. Figure 11.3A shows that to a good approximation,

$$D_i N_i^{\nu} = A\bar{N}^{-\beta}. \tag{11.12}$$

For the data in Figure 11.3A at 25–30°C and atmospheric pressure, we find that $\beta = 1.62$ and $A = 2.73 \times 10^{-3}$ cm^2/s. For mixtures with a large amount of methane and ethane, Eq. (11.12) does not fit well when $\bar{N}$ is less than 3. Instead, as shown in Figure 11.3B, a power law in $\bar{N} + 1$ works quite well:

$$D_i N_i^{\nu} = D_i r_i = A(\bar{N} + 1)^{-\beta}. \tag{11.13}$$

The second equation, with appropriate values of A and β will also fit the data with larger $\bar{N}$ because when $\bar{N} \gg 1$ this again approaches a power law in $\bar{N}$ as in Eq. (11.12).

4.3. Comparison with Experiments

The parameters A and β are independent of composition, but can depend on temperature and pressure. We can use known mixtures to obtain A and β at the desired temperature and pressure. More importantly, once A and β have been calibrated, we can obtain the mean chain length and the chain length distribution of any mixture of alkanes directly from the measured diffusion distribution. In particular, according to Eq. (11.12), the mean chain length is given by

$$\bar{N} = (A^{1/\nu}\overline{D^{-1/\nu}})^{\nu/\nu+\beta}, \tag{11.14}$$

where $\overline{D^{-1/\nu}}$ is the molar average of the diffusion constant raised to the $-1/\nu$ power. It can be directly calculated from the diffusion distribution. The diffusion distribution from NMR measurements is usually weighted by the proton number $p(D_i)$, which is very close to the weight fraction as long as N_i is not too close to 1. In that case, the molar average of $D^{-1/\nu}$ can be expressed in terms of the proton number as follows:

$$\overline{D^{-1/\nu}} = \frac{\sum p_i}{\sum p_i D_i^{1/\nu}}. \tag{11.15}$$

These equations for the mean chain length, combined with the relation between D_i and N_i given by Eq. (11.12), can be used to determine the composition of *any* mixture from the distribution of the diffusion constants.

We have applied the scaling model to analyze crude oil samples and the results for two samples are shown in Figure 11.4. The distributions of diffusion constants $p(D)$ were measured by nuclear magnetic resonance experiments using the conventional pulsed-field gradient spin echo technique,[21] as described in Section 3. The echo signal was measured for a series of 32 gradient values, and Laplace inversion[27] was applied to obtain $p(D)$.

In Figure 11.4, the measured diffusion constant distributions $p(D)$ for two different crude oils are shown on the left-hand side. Both oils contain a relatively

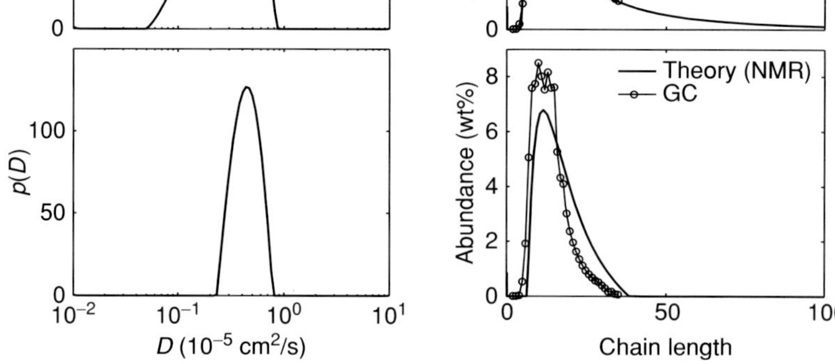

Figure 11.4. Diffusion distributions (left column) and chain length distributions (right column) for two crude oil samples that are high in saturates. The chain length distributions calculated from the diffusion constant using the scaling theory are compared with those measured by gas chromatography (GC). Note that the GC data extends only to C_{36}.

large amount of saturates, over 85%, so the alkane model should be applicable. On the right-hand side, the chain length distributions obtained from $p(D)$ are shown and they reflect quite well the main features of those measured by gas chromatography. The difference between the narrow and broad distributions is clearly reflected in the chain length distributions found from the NMR diffusion data. We note that the gas chromatography data gives more detail than the NMR chain length distributions, but only extends to C_{36}. Instead, the NMR chain length distribution covers the whole range of chain lengths and thus gives information about both the light and heavy ends of the oil.

4.4. Viscosity

Lastly, we estimate the viscosity of alkanes and their mixtures. We shall use the expression for viscosity in the polymer models with hydrodynamic effects, the Zimm model, because the self-diffusion constant of the alkanes is consistent with that of a polymer with some hydrodynamic effects. For comparison, we shall also calculate the viscosity in the Rouse model, or free-draining limit, even though, strictly speaking the models are for chains that are considerably longer than the alkanes.

In the Rouse and Zimm models, the viscosity is related to the rotational diffusion constant D_R by[11,12]

$$\eta = b' \frac{c}{N} \frac{kT}{D_R}. \tag{11.16}$$

In this equation, c is the number of segments per unit volume and is related to the density ρ by $c = \rho N/M$, where M is the mass of the chain. The constant b' depends on whether the Rouse or Zimm model is used. For both models, in the absence of excluded volume effects, the rotational and translational diffusion constants are related by

$$D_R \propto \frac{D}{Nl^2}, \tag{11.17}$$

where l is the effective segment length. Again, the constant of proportionality depends on which model is used. Combining Eqs. (11.16) and (11.17) gives the relation between the viscosity and the translational diffusion constant:

$$\eta = cl^2 bkT/D, \tag{11.18}$$

where for the Rouse model $b = 1/36$ and for the Zimm model $b = 0.0833$.

Note that the product $\eta D/T$ is independent or nearly independent of chain length. This is what is observed for both alkanes, refined oils and crudes.[9,30,31] This would not be the case for hard spheres, where one would expect the product to scale with the chain length. Instead, in the polymer models the chain length scaling drops out due to the "anomalous" dependence on chain length of both the translational and rotational diffusion constants.

We can check these equations more quantitatively by comparing the predictions for the values of $\eta D/T$ from the polymer models with those found experimentally. For alkanes and refined oils, $\eta D/T$ was found to be 3.90×10^{-8} cp cm^2/sK in reference 31 and for alkanes and crude oils, it was found to be 5.05×10^{-8} cp cm^2/sK in references 30 and 9. By fitting to the data for pure alkanes in references 16 and 36, we find that $\eta D/T = 3.8 \times 10^{-8}$ cp cm^2/sK, in agreement with the data for alkanes and refined oils.

For comparison, for very long chains, one would expect the Rouse model to be valid. In that case, we can take the density to be $\rho \approx 0.8$ g/cm^3 and the effective segment length to be $l = \sqrt{6.67} \times 1.54$ Å.[42] The Rouse model then gives $D\eta/T = 2.1 \times 10^{-8}$ cp cm^2/sK. This is almost a factor of two smaller than the experimental value. Instead, for the alkanes, one would expect the Zimm model to be more appropriate. For chain lengths around 10, the effective distance between segments is better given by[42] $l \approx \sqrt{4} \times 1.54$ Å, and the density is closer to $\rho \approx 0.75$ g/cm^3, in which case the Zimm model gives $D\eta/T = 3.6 \times 10^{-8}$ cp cm^2/sK. This agrees very well with the experimental values of $D\eta/T$ for the alkanes. This is somewhat surprising given the simplicity of the model and that alkanes are too short to be fully described by the Zimm model.

In a mixture, according to the polymer models,[11] the viscosity is just a sum of the viscosity of each component in the mixture, weighted by the number of molecules of that component per unit volume. Thus the total viscosity is

$$\eta = \sum_i \frac{\#\text{ of } i\text{th molecule}}{\text{unit volume}} \frac{kT}{(D_\mathrm{R})_i}. \tag{11.19}$$

The relation between the translational and rotational diffusion coefficients then gives

$$\eta = bcl^2kT \sum_i y_i/D_i, \tag{11.20}$$

where y_i is the weight fraction of the ith component. Finally, the viscosity can be expressed in terms of the chain lengths of the constituents in the mixture via Eq. (11.12) for the diffusion constant:

$$\eta = \frac{l^2bckT}{A} \bar{N}^\beta \sum_i y_i N_i^\nu. \tag{11.21}$$

A similar equation in terms of r_{gas} can be used if the mixture contains methane or ethane.

In Figure 11.5, the viscosity calculated from Eq. 11.21 is compared to the experimental values. We have used the Zimm value for b and a density of 0.75 g/cm^3. The parameters A and β were obtained for the specific temperatures of the data used in the figure. For simplicity, we have not included the chain length dependence of ρ and l. The agreement with the experimental data is quite good, especially considering the simplicity of the model. The agreement might be improved by including the N-dependence of ρ and l. Notice that at low viscosity, pentane and the mixture of C_1 and C_{10} are nearing their boiling or bubble points, while at the high end, hexadecane is nearing its freezing point. Thus, some deviations from the solid line are expected at the high and low ends.

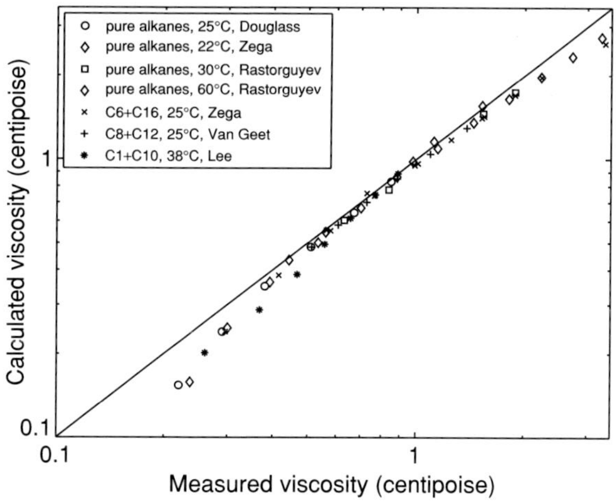

Figure 11.5. Comparison of theoretical and experimental values of viscosity for pure alkanes and mixtures. For pure alkanes, the data at 25°C are from reference 16, the data at 22°C are from references 43 and 44, and the data at 30 and 60°C are from reference 45. For mixtures, the data for C_6 and C_{16} are from reference 43, those for C_8 and C_{12} are from reference 36, and those for C_1 and C_{10} are from reference 46.

4.5. Discussion

In conclusion, we have shown that the diffusion and viscosity of mixtures of alkanes follow simple scaling laws based on the chain size of the components. We have demonstrated that these scaling laws can be used to determine the viscosity and chain sizes in a mixture from the distribution of the diffusion constants. These scaling laws also work for finding the chain lengths in live oils.

There are several limitations of this technique. One is the inherent diffculty in obtaining the distribution of diffusion coefficients from the raw data. For example, the regularized inverse Laplace transform will give a distribution of chain lengths, even in the case when there is only one or two diffusion coefficients. This is due to the limited resolution of Laplace inversion at finite signal-to-noise ratio.[29] Judicious choices for gradient or echo spacings can improve the calculated distribution, and inverting the raw data directly for the chain lengths may be useful. Relaxation time distributions may also be an attractive alternative to diffusion measurements.

The second limitation of the method in this paper, as applied to mixtures of crude oils, is that it has only been justified rigorously for linear chains and gases. Crude oils can contain branched molecules, aromatics and asphaltenes, among other things. For these different types of molecules, and also possibly for very long alkanes, the relation between the radius of gyration R_g and chain length can be altered, which can change the value of ν. Also, the amount of free volume within a mixture can be altered and the flexibility and ease of motion of these molecules can differ significantly from the linear chains, as seen, for example, in reference 47. In the next section, we will turn our attention to the diffusion of asphaltenes in solution. These molecules behave differently from the alkanes, not only for the reasons given above, but also because they can associate and form aggregates.

5. Dynamics Of Asphaltenes In Solution

Asphaltenes are complex organic compounds found in crude oils and coals, defined as "insoluble in *n*-alkanes, such as *n*-pentane and *n*-heptane, and soluble in toluene under certain conditions."[4] Asphaltene molecules consist of an aromatic core with aliphatic side-chains attached to it; and individual molecules can self-associate, forming aggregates.[4] Significant progress has been made in determining both structural and aggregation properties of asphaltenes,[4,48–58] but the understanding of self-association phenomena, especially on a molecular level, is still incomplete. The previous studies of asphaltene aggregation were done on the basis of small-angle x-ray,[53,59–61] neutron[62] and light[52,63] scattering, viscosity[54,64] and conductivity[62] measurements, fluorescence depolarization techniques,[49–51] and ultrasonic measurements.[48,65] NMR measures dynamical characteristics (diffusion constants and relaxation times) and thus, potentially, can be more representative in describing aggregation phenomenon. The other advantages of NMR are that it is a noninvasive technique and it can be used to study optically opaque samples, such as high concentration asphaltene solutions.

Molecular diffusion can be a direct probe of the molecular sizes via the Einstein–Stokes equation. Aggregation of molecules affects their mobility; this makes the molecular diffusion an excellent tool for exploring asphaltene aggregation. A number of studies have measured the diffusion constants of asphaltenes[66–68] and used them to describe the behavior of asphaltenes in mixtures with other compounds,[68,69] as well as the flocculation phenomenon.[70,71] In this work we focus on using molecular diffusion as a probe to study the aggregation of asphaltenes.

Here, we consider asphaltenes in dilute solutions. We show that the distribution of diffusion constants can be used to determine the molecular sizes of asphaltenes and their aggregates, while changes in the diffusion constants reflect the dynamics of aggregation.

5.1. The Proton Spectrum of Asphaltene Solutions

Asphaltene solutions were prepared by dissolving solid asphaltene samples in deuterated toluene-d_8. All solutions were equilibrated for several hours prior to experiments. The solid asphaltene samples were obtained by precipitating crude oil in *n*-heptane and were also used in reference.[48] Figure 11.6A shows the ^{1}H chemical shift spectrum of an asphaltene solution in toluene-d_8. The sharp peaks at 7.5 and 2.2 ppm are due to the protonated toluene and the peak at 0.7 ppm is due to impurities in the solvent, as can be seen from the comparison with the spectrum of the toluene-d_8 solvent. The broad spectral feature ranging from 0.9 to 2.2 ppm is missing in the solvent spectrum and, therefore, is related to the asphaltene sample. According to its chemical shift, this feature represents the aliphatic protons. According to the proposed structure of asphaltenes, this signal

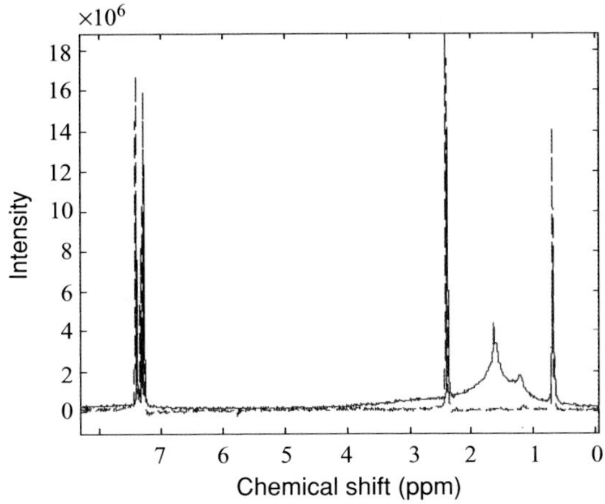

Figure 11.6. ^{1}H NMR chemical shift spectra of the asphaltene (2 g/L) and deuterated toluene solution (solid line) and the deuterated toluene solvent (dashed line). The sharp peak at about 0.7 ppm in the solvent spectrum is due to impurities.

is due to the aliphatic protons of the side chains attached to the aromatic ring.[4] The precise assignment of resonances is not straightforward because of the spectral broadening. We used the integrated intensity of this feature in our diffusion analysis and refer to it as the "asphaltene signal" in the following discussion.

5.2. The Diffusion Constant and Diffusion Spectrum

The diffusive decay of the asphaltene signal reveals multi-exponential behavior (Figure 11.7A); Figure 11.7B represents the distribution of diffusion constants

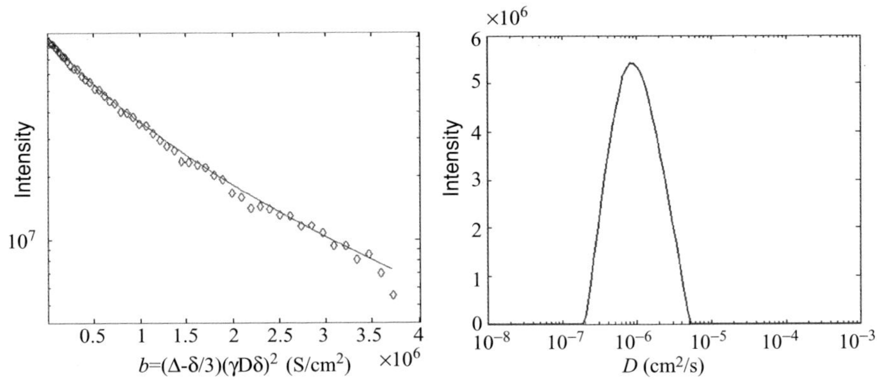

Figure 11.7. (A) The diffusive decay of the integral of the asphaltene signal as a function of diffusion weighting factor, $\mathcal{B}$. The line is a fit for the data obtained using Laplace inversion. (B) The distribution of diffusion constants extracted from this decay.

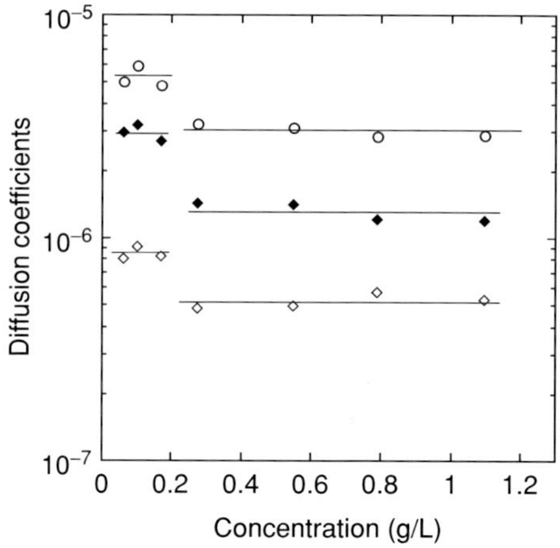

Figure 11.8. The diffusion constant of the fast (circles) and slow (open diamonds) components, and the average (solid diamonds) as a function of the concentration of asphaltenes in the toluene solution. Lines are a guide to the eye.

extracted from this decay. This distribution was obtained by the numerical inverse Laplace transform, described elsewhere.[27] We will refer to this distribution as the "diffusion spectrum" in our discussion.

The diffusion decay can also be fit by a double-exponential decomposition. Since these two methods fit the data equally within statistical error, we cannot determine the exact shape of the diffusion spectrum. We will take the double-exponential fit as the range of diffusion constants in the following discussion. The D values obtained from the double-exponential fit for the 2.1 g/L asphaltene sample are different by almost an order of magnitude, namely, 2.0×10^{-6} cm^2/s and 5.3×10^{-7}cm^{2}s. We will define them as the "fast" and "slow" components, respectively. In Figure 11.8 the diffusion constants of the fast and slow components and the average diffusion constant are plotted as a function of the concentration of asphaltenes. The average diffusion constant was obtained by a single exponential fit to the initial diffusion decay (Figure 11.7A).

The striking features of this plot are the sudden change in the diffusion constant D of the average and the two components at a concentration of ~0.2 g/L and the uniformity of D above and below that concentration. The concentration at which the sudden change in D occurs is consistent with that of the kink in the compressibility measured by ultrasound velocity and attributed to the aggregation of asphaltene molecules.[48] The diffusion constants of both components are shown to be constant above and below this concentration, suggesting that above and below this concentration there is no significant change in the molecular size of the species.

5.3. Discussion

On the basis of previous studies, asphaltene solutions at very low concentrations contain single asphaltene molecules. At high concentrations, molecular aggregates may form. The molecular size and thus the diffusion constant are different for aggregates and single asphaltene molecules. Therefore, measuring the diffusion constant at different asphaltene concentrations can provide information about the aggregation of the molecules.

5.3.1. Very Low Concentrations

At low concentrations (<0.2 g/L) the solution is dominated by single asphaltene molecules. The diffusion spectrum reflects the size distributions of these species; the size distribution of the single molecules is mostly determined by sample heterogeneity. If we assume that the asphaltene molecules are spherical, then the Einstein–Stokes relation, Eq. (11.1), gives the radius of the molecule in terms of its diffusion constant. We find that the molecular diameter obtained from the center of the spectrum ($D = 3 \times 10^{-6}$ cm^2/s) is 26 Å. The fast and slow components set a range for the diffusion constants, which gives molecular diameters ranging from 12 to 76 Å.

Our estimate of the minimum asphaltene size is consistent with previous estimates[49–51] of the minimum asphaltene diameter of about 12 Å. However, our estimate of the average size and the large end is somewhat larger. For example, Groenzin and Mullins[51] used fluorescence depolarization to obtain the rotational correlation times of asphaltenes. They derived the range of asphaltene sizes to be 12–24 Å with a peak at 19 Å from the shape of the fluorescence spectrum with a wavelength range of 410–635 nm. This discrepancy is possibly due to the higher sensitivity of NMR for larger proton-rich molecules compared to fluorescence spectroscopy. The sensitivity of fluorescence spectroscopy is proportional to the fluorescence quantum yield which was found to decrease for larger asphaltene molecules.[72] On the other hand, the NMR signal is proportional to the number of hydrogen atoms in a molecule so that the larger molecules are heavily weighted. It is also possible that there are some asphaltene aggregates, such as dimers and trimers, forming even at the lowest concentration of our samples.[51]

We note that asphaltene molecules are often considered to be oblate ellipsoids, instead of spheres.[51,73] In that case, the radius of the major axis is slightly larger than the radius found for spherical molecules. For example, if the ratio of the major to minor axis is 2, the major axis is only 21% larger than the radius found for a sphere, and, in the limit of an infinitely thin disk, it is only 57% larger.

5.3.2. Intermediate and High Concentrations

Above 0.2 g/L, we observe a rapid reduction of the diffusion constant by approximately a factor of two, for both the average diffusion constant and the fast and the slow components. At this concentration, we find that the relative populations of the two species changes very little. This data unambiguously indicates a slowing down of the molecular dynamics above the concentration 0.2 g/L, which coincides with the change in ultrasonic measurements.[48] For this range of concentration, the change in viscosity is negligible.[54] According to the Einstein–Stokes equation, this reduction of D then corresponds to a doubling of the molecular diameter and thus an eightfold increase in volume. This data suggests that the asphaltene aggregates formed at concentrations just above 0.2 g/L are those with eight asphaltene molecules, consistent with the aggregate size obtained in the fluorescence depolarization measurements.[49–51] Again, we are assuming that both the asphaltene molecules and the aggregates are spherical. For other shapes, the radii and volumes will be modified accordingly. In addition, we are assuming the asphaltene molecules are densely packed. If there is space between the molecules, then, in the case of spherical molecules, the number of molecules in an aggregate will be somewhat less than eight.

For higher concentrations, 0.2–1.2 g/L, no significant change in the diffusion constants of the components occurs. This indicates that the eight-molecule aggregates are stable in this concentration range and that the increasing concentration of asphaltenes results in an increase in the number of aggregates in the solution without significantly changing the molecular dynamics.

6. Conclusions

We have shown that NMR measurements of translational diffusion provides information about molecular sizes, the distributions of sizes, and molecular dynamics. Many traditional analytical techniques, such as gas chromotography, high field NMR spectroscopy, and mass spectrometry may be diffcult to implement in field applications. NMR diffusion measurements on the other hand do not require high magnetic fields and can be made cheaper, compact and portable[7,74] and are already in use in NMR well-logging. The examples and methods discussed in this paper can become the basis for in situ characterization of crude oils during oil exploration.

Acknowledgment

We thank Gaelle Andreatta for providing the original solid sample and Oliver Mullins for discussions.

References

[1] Brooks, B.T., C.E. Boord, S.S. Kurtz Jr., and L. Schmerling (eds.) (1954). *The Chemistry of Petroleum Hydrocarbons*, Vol. 1. Reinhold Publishing, New York.

[2] Speight, J.G. (1980). *The Chemistry and Technology of Petroleum.* Marcel Dekker, New York.

[3] Tissot, B.P. and D.H. Welte (1984). *Petroleum Formation and Occurrence*. Springer-verlag, Berlin.

[4] Mullins, O.C. and E.Y. Sheu (1988). *Structure and Dynamics of Asphaltenes.* Plenum Press, New York.

[5] Betancourt, S., G. Fujisawa, O.C. Mullins, A. Carnegie, C. Dong, A. Kurkjian *et al.* (2003). "Analyzing hydrocarbons in the borehole", *Oilfield Rev.* **15**, 54.

[6] Hürlimann, M.D., L. Venkataramanan, and C. Flaum (2002). "The diffusion-spin relaxation time distribution as an experimental probe to characterize fluid mixtures in porous media", *J. Chem. Phys.* **117**, 10223.
[7] Kleinberg, R. (1995). Well logging. In: D.M. Grant and R. K. Harris (eds.), *Encyclopedia of Nuclear Magnetic Resonance.* John Wiley and Sons, New York.
[8] Hürlimann, M.D., L. Venkataramanan, C. Flaum, P. Speier, C. Karmonik, R. Freedman *et al.* (2002). Diffusion-Editing: New NMR Measurement of Saturation and Pore Geometry. In: *43rd Annual SPWLA Meeting.* Oiso, Japan, p. paper FFF.
[9] Freedman, R., A. Sezginer, M. Flaum, A. Matteson, S. Lo, and G. J. Hirasaki (2000). "A New NMR Method of Fluid Characterization in Reservoir Rocks: Experimental Confirmation and Simulation Results", Paper SPE 63214 presented at the *2000 SPE Annual Technical Conference and Exhibition.*
[10] Einstein, A. (1905). " Über die von der molekularkinetischen Theorie der Wärme geforderte Bewegung von in ruhenden Flüssigkeiten suspendierten Teilchen", *Ann. Phys.* **17**, 549.
[11] Ferry, J.D. (1980). *Viscoelastic Properties of Polymers.* John Wiley and Sons, New York.
[12] Doi, M. and S.F. Edwards (1998). *The Theory of Polymer Dynamics.* Oxford University Press, New York.
[13] de Gennes, P.G. (1971). "Reptation of a polymer chain in the presence of fixed obstacles", *J. Chem. Phys.* **55**, 572.
[14] Klein, J. (1978). "Evidence for reptation in an entangled polymer melt", *Nature* **271**, 143.
[15] Rouse, P.E. (1953). "A theory of the linear viscoelastic properties of dilute solutions of coiling polymers", *J. Chem. Phys.* **21**, 1272.
[16] Douglass, D.C. and D.W. McCall (1958). "Diffusion in paraffin hydrocarbons", *J. Phys. Chem.* **62**, 1102.
[17] Bachus, R. and R. Kimmich (1983). "Molecular weight and temperature dependence of self-diffusion coefficients in polyethylene and polystyrene melts investigated using a modified N.M.R. field gradient technique", *Polymer* **24**, 964.
[18] Vardag, T., N. Karger, and H.-D. Lüdemann (1991). "Temperature and pressure dependence of self-diffusion in long liquid n-alkanes", *Ber. Bunsenges. Phys. Chem.* **95**(8), 859.
[19] von Meerwall, E., S. Beckman, J. Jang, and W.L. Mattice (1998). "Diffusion of liquid n-alkanes: Free-volume and density effects", *J. Chem. Phys.* **108**(10), 4299.
[20] Zimm, B.H. (1956). "Dynamics of polymer molecules in dilute solution: Viscoelasticity, flow birefringence and dielectric loss", *J. Chem. Phys.* **24**, 269.
[21] Callaghan, P.T. (1993). *Principles of NuclearMagnetic Resonance Microscopy.* Oxford University Press, New York.
[22] Hahn, E. (1950). "Spin echoes", *Phys. Rev.* **80**, 580.
[23] Tikhonov, A.N. and V.Y. Arsenin (1977). *Solutions of Ill-Posed Problems.* JohnWiley and Sons, New York.
[24] Venkataramanan, L., Y.-Q. Song, and M. D. Hürlimann (2002). "Solving fredholm integrals of the first kind with tensor product structure in 2 and 2.5 dimensions", *IEEE Trans. Signal Proc.* **50**, 1017.
[25] Song, Y.-Q., L. Venkataramanan, M. Hürlimann, M. Flaum, P. Frulla, and C. Straley (2002). "T1-T2 correlation spectra obtained using a fast two-dimensional laplace inversion", *J. Magn. Reson.* **154**, 261.
[26] Kroeker, R.M. and R. M. Henkelman (1986). "Analysis of biological NMR relaxation data with continuous distributions of relaxation times", *J. Magn. Reson.* **69**, 218.
[27] Fordham, E.J., A. Sezginer, and L.D. Hall (1995). "Imaging multiexponential relaxation in the (y, logT1) plane, with application to clay filtration in rock cores", *J. Magn. Reson. Ser. A* **113**, 139.
[28] Borgia, G.C., R.J.S. Brown, and P. Fantazzini (1998). "Uniform-penalty inversion of multi-exponential decay data", *J. Magn. Reson.* **132**, 65.
[29] Song, Y.-Q., L. Venkataramanan, and L. Burcaw (2005). "Determining the resolution of laplace inversion spectrum", *J. Chem. Phys.* **122,** 104104.
[30] Lo, S.-W., G.J. Hirasaki, W.V. House, and R. Kobayashi (2000). Correlations of NMR Relaxation Time with Viscosity, Diffusivity, and Gas/Oil Ratio of Methane/Hydrocarbon Mixtures.

In: *Proceedings of the 2000 Annual Technical Conference and Exhibition.* Society of Petroleum Engineers.

[31] Straley, C. (2002). *The Low Field NMR of Refined Oils at Elevated Temperature and Pressure* (unpublished).

[32] Freed, D.E., L. Burcaw, and Y.-Q. Song (2005). "Scaling laws for diffusion coefficients in mixtures of alkanes", *Phys. Rev. Lett.* **94**, 067602 .

[33] von Meerwall, E. and E.J. Feick (1999). "Diffusion in binary liquid n-alkane and alkane-polyethylene blends", *J. Chem. Phys.* **111**(2), 750.

[34] Helbæk, M., B. Hafskjold, D.K. Dysthe, and G.H. Sørland (1996). "Self-diffusion coefficients of methane or ethane mixtures with hydrocarbons at high pressure by NMR", *J. Chem. Eng. Data* **41**, 598.

[35] Lo, S.-W., G.J. Hirasaki, R. Kobayashi, and W.V. House (1998). "Relaxation time and diffusion measurements of methane and n-decane mixtures", *The Log Analyst* **39**, 43.

[36] Van Geet, A.L. and A.W. Adamson (1964). "Diffusion in liquid hydrocarbon mixtures", *J. Phys. Chem.* **68**(2), 238.

[37] Dymond, J.H. and K.R. Harris (1992). "The temperature and density dependence of the self-diffusion coefficient of n-hexadecane", *Mol. Phys.* **75**(2), 461.

[38] Marbach, W. and H.G. Hertz (1996). "Self- and mutual diffusion coefficients of some n-alkanes at elevated temperatures and pressures", *Zeitschrift f`ur Physikalische Chemie, Bd.* **193**, 19.

[39] Van Geet, A.L. and A.W. Adamson (1965). "Prediction of diffusion coefficients for liquid n-alkane mixtures", *Industrial Eng. Chem.* **57**(7), 62.

[40] Kurtz, S.S. Jr. (1954). Physical properties and hydrocarbon structure. In: B.T., Brooks, C.E. Boord, S.S. Kurtz Jr., and L. Schmerling (eds.), *Chemistry of Petroleum Hydrocarbons.* Reinhold Publishing Corporation, New York, pp. 275–331.

[41] McCall, D.W., D.C. Douglass, and E.W. Anderson (1959). "Diffusion in ethylene polymers. IV", *J. Chem. Phys.* **30**(3), 771.

[42] Flory, P.J. (1969). *Statistical Mechanics of Chain Molecules.* John Wiley and Sons, New York.

[43] Zega, J.A. (1988). Spin lattice relaxation in pure and mixed alkanes and their correlation with thermodynamic and macroscopic transport properties. Master's Thesis, Rice University.

[44] Zega, J.A., W.V. House, and R. Kobayashi (1989). "A corresponding-states correlation of spin relaxation in normal alkanes", *Physica A* **156**, 277.

[45] Rastorguyev, Y.L. and A.S. Keramidi (1974). "Experimental study of the coefficient of dynamic viscosity of n-alkanes at high pressures and various temperatures", *Fluid Mechanics Sov. Res.* **3**, 156.

[46] Lee, A.L., M.H. Gonzalez, and B.E. Eakin (1966). "Viscosity of methane–n–decane mixtures", *J. Chem. Eng. Data* **11**(3), 281.

[47] Bachl, F. and H.-D. Lüdemann (1986). "p,T-dependence of self-diffusion in simple alkanes", *Physica* **139, 140B**, 100.

[48] Andreatta, G., N. Bostrom, and O.C. Mullins (2005). "High-Q ultrasound determination of the critical nanoaggregate concentration of asphaltenes and the critical micelle concentration of standard surfactants", *Langmuir* **21**, 2728.

[49] Buenrostro-Gonzalez, E., H. Groenzin, C. Lira-Galeana, and O.C. Mullins (2001). "The overriding chemical principles that define asphaltenes", *Energy Fuels* **15**, 972.

[50] Groenzin, H. and O.C. Mullins (2000). "Molecular size and structure of asphaltenes from various sources", *Energy Fuels* **14**, 677.

[51] Groenzin, H. and O.C. Mullins (1999). "Asphaltene molecular size and structure", *J. Phys. Chem.* **103**, 11237.

[52] Ralston, C.Y., S. Mitra-Kirtley, and O.C. Mullins (1996). "Small population of one to three fused-aromatic ring moieties in asphaltenes", *Energy Fuels* **10**, 623.

[53] Tanaka, R., E. Sato, J. Hunt, R. Winans, S. Sato, and T. Takanohashi (2004). "Characterization of asphaltene aggregates using X-ray diffraction and small-angle X-ray scattering", *Energy Fuels* **18**, 1118.

[54] Evdokimov, I., N. Eliseev, and B. Akhimetov (2003). "Initial stages of asphaltene aggregation in dilute crude oil solutions: studies of viscosity and NMR relaxation", *Fuel* **82**, 817.

[55] Waldo, G.S., O.C. Mullins, J.E. Penner-Hahn, and C.P. Cramer (1992). "Sulfur speciation in heavy petroleum: determination of the chemical environment of sulfur in asphaltenes", *Fuel* **71**, 53.

[56] Sheu, E. (1992). "Polydispersity analysis of scattering data from self-assembled systems", *Phys. Rev. A.* **45**, 2428.

[57] Ravey, J., G. Decouret, and D. Espinat (1988). "Asphaltene macrostructure by small angle neutron scattering", *Fuel* **67**, 1560.

[58] Sheu, E., M.M. De Tar, D.A. Storm, and S. DeCanio (1992). "Aggregation and kinetics of asphaltenes in organic solvents", *Fuel* **71**, 299.

[59] Espinat, D., D. Fenistein, L. Barré, D. Frot, and Y. Briolant (2004). "Effects of temperature and pressure on asphaltenes agglomeration in toluene. A light, X-ray, and neutron scattering investigation", *Energy Fuels* **18**, 1243.

[60] Fenistein, D. and L. Barré (2001). "Experimental measurement of the mass distribution of petroleum asphaltene aggregates using ultracentrifugation and small-angle X-ray scattering", *Fuel* **80**, 283.

[61] Siddiqui, M., M. Ali, and J. Shirokoff (2002). "Use of X-ray diffraction in assessing the aging pattern of asphalt fractions", *Fuel* **81**, 51.

[62] Sheu, E. (1996). "Physics of asphaltene micelles and microemulsions-theory and experiment", *J. Phys.: Condens. Matter* **8**, A125.

[63] Yudin, I.K., G.L. Nikolaenko, E.E. Gorodetskii, V.I. Kosov, V.R. Melikyan, E. Markhashov *et al.* (1998). "Mechanisms of asphaltene aggregation in toluene–heptane mixtures", *J. Petroleum Sci. Eng.* **20**, 297.

[64] Priyanto, S., G.A. Mansoori, and A. Suwono (2001). "Measurement of property relationships of nanostructure micelles and coacervates of asphaltene in a pure solvent", *Chem. Eng. Sci.* **56**, 6933.

[65] Andreatta, G., C.C. Goncalves, G. Buffin, N. Bostrom, C.M. Quintella, F. Arteaga-Larios *et al.* (2005). "Nanoaggregates and structure-function relations in asphaltenes", *energy euels* **19**(4), 1282–1289.

[66] Norinaga, K., E.Wargardalam, V. Takasugi, S. Iino, and S. Matsukawa (2001). "Measurement of self-diffusion coefficient of asphaltene in pyridine by pulsed field gradient spin-echo 1H NMR", *Energy Fuels* **15**, 1317.

[67] Östlund, J.-A., S.-I. Andersson, and M. Nydén (2001). "Studies of asphaltenes by the use of pulsed-field gradient spin echo NMR", *Fuel* **80**, 1529.

[68] Östlund, J.-A., M. Nydén, and P. Stilbs (2004). "Component-resolved diffusion in multicomponent mixtures. A case study of high-field PGSE-NMR self-diffusion measurements in asphaltene/naphthenic acid/solvent systems", *Energy Fuels* **85**, 531.

[69] Östlund, J.-A., M. Nydén, I. Auflem, and J. Sjblom (2003). "Interactions between asphaltenes and naphthenic acids", *Energy Fuels* **17**, 113.

[70] Östlund, J.-A., P. Wattana, and M. Nydén (2004). "Characterization of fractionated asphaltenes by UV–vis and NMR self-diffusion spectroscopy", *J. Colloid Int. Sci.* **271**, 372.

[71] Östlund, J.-A., J.-E. Löfroth, K. Holmberg, M. Nydén, and H. Fogler (2002). "Flocculation behavior of asphaltenes in solvent/nonsolvent system", *J. Colloid Int. Sci.* **253**, 150.

[72] Mullins, O.C. (1988). Optical interrogation of aromatic moieties in crude oils and asphaltenes. Chap. II. In: O.C. Mullins and E.Y. Sheu (eds.), *Structures and Dynamics of Asphaltenes*, Plenum, New York, pp. 21–77.

[73] Groenzin, H. and O.C. Mullins (2003). "Molecular size of asphaltene solubility fractions", *Energy Fuels* **17**, 498.

[74] Eidmann, G., R. Savelsberg, P. Blümler, and B. Blümich (1996). "The NMR mouse, a mobile universal surface explorer", *J. Magn. Reson. A* **122**, 104.

12

Application of the PC-SAFT Equation of State to Asphaltene Phase Behavior

P. David Ting, Doris L. Gonzalez, George J. Hirasaki, and Walter G. Chapman

1. Introduction

A method to characterize crude oil including asphaltenes using the perturbed chain form of the statistical associating fluid theory (PC-SAFT) is presented. The theory accurately predicts the bubble point, density, and asphaltene precipitation onset for the oil. Examples show that the theory predicts asphaltene instability due to changes in pressure, temperature, and fluid composition. Further work demonstrates the effect of asphaltene polydispersity and resins on the phase behavior of asphaltenes. The approach demonstrates that laboratory and field observations of asphaltene phase behavior can be explained based only on molecular size and van der Waals interactions.

This chapter provides an application of the statistical associating fluid theory (SAFT)[1–4] equation of state (EOS) to model the effects of pressure, temperature, and composition on the phase behavior and stability of asphaltenes in crude oil. SAFT is a versatile molecular model capable of predicting the effects of molecular shape, van der Waals forces, polar interactions, and association on the thermodynamic properties and phase behavior of fluids. The approach we have taken is to use the minimum number of (physically relevant) parameters possible to describe phase behavior of asphaltenes by including only essential interactions.

Given the high degree of complexity of crude oil and its large variability in composition due, in part, to differing source rock properties and migration history, one may question the relevance of a molecular-based model in light of our current understanding of asphaltenes. It is our belief that the bulk phase behavior of asphaltenes can be accurately described if we can correctly account for its major physical attributes and its interactions with other species in oil. There are many advantages in using a predictive molecular-based EOS model. For instance,

---

P. David Ting, Doris L. Gonzalez, George J. Hirasaki, and Walter G. Chapman • Department of Chemical Engineering, Rice University, Houston, Texas 77005.

the sensitivity of asphaltene phase behavior to the effects of temperature, pressure, and asphaltene polydispersity can be quickly and confidently modeled. An EOS framework can also be more readily implemented into existing reservoir and thermal–hydraulic simulators used in industry. At the other end of the scale, the EOS approach, when compared to more rigorous molecular simulations, has the advantage of speed and ease of use. This comes, of course, at the expense of detailed descriptions at the molecular level.

With the molecular basis of the SAFT approach, it is necessary to define the system to be modeled. Our viewpoint is described and justified in Section 1 of this chapter. In Section 2, we present in detail the characterization of an oil with SAFT including applications to asphaltene phase behavior on reservoir depressurization and gas injection. Conclusions are provided in the final section.

1.1. Asphaltene Properties and Field Observations

The currently accepted definition of asphaltenes is an operational one based on solubility (i.e., asphaltenes are insoluble in heptane or pentane and soluble in toluene). As such, it reveals little about the structure of asphaltenes. Most researchers agree that asphaltenes are a polydisperse mixture of molecules containing polynuclear aromatic, aliphatic, and alicyclic moieties with small amounts of dispersed heteroelements such as oxygen, sulfur, nitrogen, vanadium, and nickel. When compared to other crude oil components, asphaltenes are the heaviest fraction of a distribution (in molecular weight as well as aromaticity) of compounds that include aromatics and resins in the lower molecular weight sub-fractions. The accepted definition for asphaltenes is, in essence, an arbitrarily divided subfraction of this distribution. Asphaltenes are more aromatic (low H/C ratio) than most other oil fractions, are larger in molecular weight, and have higher solubility parameter.[5]

The interesting phase behavior of asphaltenes in oil can be deduced by studying a few representative examples of field experiences with asphaltene problems.[6–11] These examples not only elucidate asphaltene behavior typically seen in the field and in PVT laboratories but also help us gain a better understanding of the type of molecular interactions between asphaltenes and other oil species. From these experiences, it can be concluded that asphaltenes are usually stable in heavier oils well above the bubble point pressure under reservoir conditions. Light oils with little asphaltenes are the most susceptible to asphaltene problems. Field operators have reported that asphaltenes are unstable over a range of pressures during reservoir depressurization. More specifically, the effect of pressure on asphaltene phase separation is most pronounced for light oil near the bubble point. Asphaltenes are stable at high pressure and at pressures well below the bubble point, but, in many cases, they are unstable at pressures somewhat above and below the bubble point. Far below the bubble point, asphaltenes tend to be stable since most of the precipitants (methane, ethane, nitrogen, etc.) have escaped from the liquid. Compositional changes such as oil blending or miscible flooding sometimes result in asphaltene precipitation. Interestingly, temperature changes may result in either asphaltene precipitation or solubilization. For instance, in the propane deasphalting process, asphaltenes become increasingly unstable with temperature increase.[12] However, for n-alkane (n-C_{5+}) titrations, asphaltene stability improves with increasing temperature.[12] These examples highlight the need, by such disciplines as flow assurance and production engineers, for a predictive model that is capable of explaining these observations.

A distinction needs to be made between thermodynamic asphaltene stability (which is the focus of this work) and asphaltene-related field problems. Just because asphaltenes may become unstable during production does not necessarily mean that deposition will be encountered. Factors such as (1) the properties of the asphaltene that precipitated including its "stickiness," (2) the amount of the asphaltene that precipitated, and (3) the flow pattern in the production system (flowline, tree, riser, inside topside facility, etc.) all play a role in determining whether the precipitated asphaltenes will result in field problems. However, the ability to accurately model asphaltene as a polydisperse system will help in elucidating its deposition tendencies. Finally, it is important to note that asphaltene from one region (source) may have very different properties than asphaltene from another region.

1.2. The Two Views of Asphaltene Interactions

In the last 50+ years, two views to describe the phase behavior of asphaltenes have emerged. In one view, which we will call the "molecular solution" approach, asphaltenes are treated as molecules that are solubilized by the oil. Asphaltene precipitation is treated as liquid–liquid or solid–liquid equilibria. The hypothesis in this framework has been that molecular size and van der Waals interactions, which are related to molecular polarizability, dominate asphaltene phase behavior in reservoir fluids; the more polarizable components (the resins and aromatics) solubilize the asphaltenes and the less polarizable components (saturates) destabilize the asphaltenes. Proponents of this view have used such approaches as Flory–Huggins theory[13] or equations of state to model asphaltene phase behavior.[14–18] In this chapter we report on the application of the SAFT equation of state to model asphaltene phase behavior.[19,20] Although the SAFT approach and its extensions can account for polar and association effects, we find that the effects of molecular size and van der Waals attractions are sufficient to explain reported observations of asphaltene phase behavior.

The application of Flory–Huggins regular solution theory to describe asphaltene phase behavior was first proposed by Hirschberg *et al.*[15] While the Flory–Huggins regular solution-based approaches have been used with varying success to model asphaltene solubility with n-alkane titrations under ambient conditions, it is difficult to extend the approach to model asphaltene solubility under reservoir conditions. In a sense, the theory is not a "complete" equation of state; it requires the molar volumes and solubility parameters under reservoir conditions. These values must be obtained from an equation of state or estimated from empirical correlations. For example, in the work of Chung *et al.*,[21] the Flory–Huggins regular solution model is combined with the Peng–Robinson cubic equation to model asphaltene solubility in oil. And in the work of Burke *et al.*,[22] the Flory–Huggins regular solution model parameters were obtained from the Zudkevitch–Joffe–Redlich–Kwong equations. Some success in predicting asphaltene precipitation at reservoir conditions from Flory–Huggins parameters fit to ambient titration data has been reported using an extrapolation due to Wang and Buckley.[14, 23]

Another limitation of the Flory–Huggins model is its inability to predict certain classes of phase behavior. As an example, the model requires temperature dependent binary interaction parameters to show lower critical solution behavior, phase behavior known to occur in systems with large size differences between molecules.

Another popular classical thermodynamics approach to model asphaltene behavior is to use cubic equations of state. In the method proposed by Nghiem *et al.*,[24] the C_{31+} heavy end of crude oil is first divided into nonprecipitating and precipitating sub-fractions. Different interaction parameters (between these sub-fractions and light ends) are then assigned to reproduce experimental results. In another example, Akbarzadeh *et al.*[16] modified the Soave–Redlich–Kwong cubic equation by adding an additional aggregation size parameter to asphaltenes. The cubic equations have relatively simple functional forms and are easy to implement into existing reservoir simulators because cubic equations have been used extensively to describe the thermodynamic behavior of reservoir fluids. However, the major shortcoming of cubic equations of state is that they cannot describe the phase behavior of systems with large size disparities[25] and that they cannot accurately describe liquid densities. Accurate modeling of liquid density is essential for an equation of state to predict liquid–liquid equilibria over a range of conditions. While the calculated densities can be modified using volume translation techniques, volume translations do not affect phase equilibria calculations.

In our approach, asphaltene instability is modeled as liquid–liquid equilibria. We have chosen to base our model on the SAFT equation of state because of its ability to accurately model fluid densities as well as phase behavior for mixtures with substantial size asymmetry.[20, 25] SAFT and its extensions also explicitly account for association and polar interactions. The details of the model are discussed in Section 2.

As noted above, the "molecular solution" approach is not the only approach used in literature to describe asphaltene phase behavior in oil. In this other approach generically called the "micellar" approach, the structural characteristics of asphaltenes and resins are emphasized—asphaltenes are viewed to be stabilized by resins via polar–polar interactions. The basis of this viewpoint is that asphaltenes and resins are the most polar fractions of crude oil because they contain heteroatoms of various proportions. When resins are added, less asphaltenes will precipitate. And when *n*-alkanes are added, asphaltenes will precipitate because of the dilution of resins in the mixture. The argument is that resins stabilize asphaltenes in a similar way to surfactants stabilizing a micro-emulsion in an oil/water mixture. Thermodynamic models that take the micellar view include the solid–asphaltene colloidal model proposed by Leontaritis and Mansoori,[26] the reversible micellization model proposed by Victorov and Firoozabadi,[27] and the McMillan–Mayer–SAFT-based theory proposed by Wu.[12, 28, 29] Interestingly, the SAFT framework is versatile enough that it has been used to develop models for both points of view.

1.3. Our View and Approach

The underlying hypothesis of our approach in modeling asphaltene phase behavior is that molecular size and nonpolar van der Waals interactions dominate asphaltene phase behavior in reservoir fluids. We find that the phase behavior described in field and laboratory experiences are similar to those seen in oligomer and polymer systems. For instance, the pressure, temperature, and compositional behavior of asphaltenes in reservoir fluid described in Section 1.1 is qualitatively similar to the behavior of polystyrene in a mixture of cyclohexane and CO_2.[30] In such systems, the phase behavior can be predictively modeled by only considering molecular size and van der Waals interactions. Note that the existence of other types of intermolecular interactions are not being discounted or trivialized in our framework; we are simply taking the approach that the behavior of asphaltenes in crude oil systems can be described to a large extent by accounting for molecular size and van der Waals interactions only.

Our hypothesis is supported by other evidence. For instance, an investigation of asphaltene solubility in over 40 polar and nonpolar solvents by Wiehe[31] shows that asphaltenes are soluble in solvents with high field force solubility parameters (which is a measure of nonpolar, van der Waals interaction strength) and insoluble in solvents with moderate and high complexing solubility parameters (which is a measure of hydrogen bonding and polar interaction strengths). While this observation is consistent with the idea that van der Waals interactions determine asphaltene phase behavior, it does not readily fit in the micellar framework.

As another example, consider that some relatively nonpolar molecules of similar size and structure can be either precipitants or solvents for asphaltenes. Toluene ($C_6H_5CH_3$) is a good solvent for asphaltenes while *n*-heptane (C_7H_{16}) is a precipitant. Similarly, both carbon disulfide (CS_2) and carbon dioxide (CO_2) are weakly polar and of similar molecular structure, but CS_2 is a good solvent for asphaltenes while CO_2 is a precipitant. In these examples, the more polarizable molecules (such as toluene and carbon disulfide, molecules with the stronger van der Waals interactions as shown by their solubility parameters) are the better solvents for asphaltenes.

In some situations, the role of polar or hydrogen bonding interactions is important. For example, asphaltenes may aggregate on the water–oil interface and stabilize water emulsions.[32] Also, addition of a sufficiently large amount of alkyl–benzene derived amphiphiles (such as dodecyl benzene sulfonic acid) can inhibit asphaltene aggregation.[33] While polar interactions may play a role, our hypothesis is that asphaltene phase behavior in the reservoir, such as on reservoir depressurization, is shaped to a larger extent by the nonpolar interactions in the oil.

Recent structural investigations on asphaltene behavior[34] suggest that multiple levels of interactions may be occurring in fluids containing asphaltenes. In all except very good solvents and under near-infinite dilution concentrations, asphaltenes exist as molecular aggregates (of several units, each ~500–800 amu in size) and each aggregate behaves as if it is a single molecule. Our approach at this point is to assume that asphaltenes have pre-aggregated in crude oil systems and that the asphaltenes in our model would, in fact, exhibit characteristics of the molecular aggregate so that the molecular weight in our model represents the MW of the aggregate. This approach is justified since oil systems (and the model systems investigated in this work) are neither infinitely dilute in asphaltenes nor very good solvents for asphaltenes.

2. Introduction to SAFT

We adopt the SAFT equation of state for our study of asphaltenes because of its demonstrated ability to accurately describe and predict the effects of large molecular size differences and association on phase behavior of complex mixtures.[25, 30, 35, 36] This has been seen in numerous applications of SAFT to polymer solutions and hydrogen bonding fluids. For example, SAFT has become an important tool in predicting polymer phase behavior to prevent fouling in polymer processing.[37]

SAFT was developed by Chapman *et al.*[1–3] based on extensions and simplifications of Wertheim's theory for associating fluids.[38–40,41] In SAFT, molecules are modeled as chains of bonded spherical segments. As shown in Figure 12.1, SAFT determines the free energy of a fluid as the sum of the free energy for a collection of spherical "segments" (from which molecules are constructed) plus the change in free energy on "bonding" these spherical "segments" in a prescribed manner to form the molecules of interest. This change in free energy can be calculated from Wertheim's theory. Finally, if the molecules have association sites, the change in free energy due to these directional interactions are explicitly accounted for using Wertheim's theory.

A theory like SAFT that is based in statistical mechanics offers several advantages. The first advantage is that each of the approximations made in the development of SAFT such as the chain and association terms has been verified against molecular simulation results.[1] In this way, the range of applicability and the shortcomings of each term in the equation of state have been assessed. A second advantage is that the SAFT parameters have physical meaning. For example, a chain molecule is characterized by the diameter or volume of a segment, the number of segments in the chain, and the segment–segment van der Waals attraction. These parameters are, in general, fit to saturated liquid densities and vapor pressures for the pure components. Since the equation of state parameters are physical, they

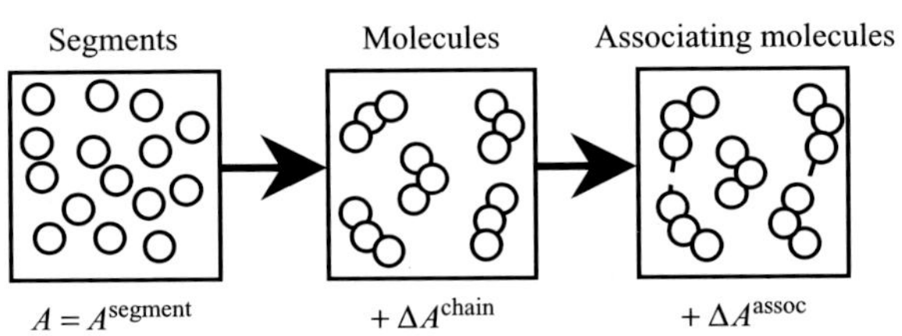

Figure 12.1. Contributions to the SAFT equation of state for an associating polyatomic fluid.

behave systematically within a homologous series. Furthermore, parameters for new systems can be estimated from those of previously modeled systems. In this way, parameters for saturates, aromatics, and resins can quickly be determined to model a crude oil.

Numerous forms of the SAFT equation of state have been proposed.[1–4, 42–44] These forms differ only in the segment term used to account for the van der Waals attraction between molecules; all use the same chain and association terms as introduced in the original SAFT papers by Chapman *et al.*[1–3, 42] Because each of these SAFT versions shares the same basic form of the equation of state, they each give similar results. In this work, we report results using the Perturbed Chain version of SAFT or PC-SAFT due to Gross and Sadowski.[4] We expect qualitatively similar results if another version of SAFT is used instead.

2.1. PC-SAFT Pure Component Parameters

For each non associating species in SAFT, the equation of state requires the specification of three physical parameters: σ, the diameter of each molecular segment, m, the number of segments in the molecule, and ε/k, the interaction energy (van der Waals attraction) between each molecular segment. A list of PC-SAFT parameters for compounds of interest to this work is given in Table 12.1.

An important feature of SAFT is that the fitted pure component parameters (σ, m, and ε/k) behave in a systematic manner with molecular weight for different classes of compounds[4, 46] (see Figure 12.2). Furthermore, species with both aromatic and aliphatic characteristics have EOS parameters that lie in between the aromatic and n-alkane correlations in a systematic manner depending on their degree of aromaticity or aliphaticity. For instance, the parameters for benzene derivatives and cycloalkanes approach the pure component parameters of the alkanes as the degree of aliphaticity increases. These "well-behaved" correlations between the pure component parameters and molecular weight have three implications: (1) SAFT parameters for crude oil components or lumps of components (pseudo-components) can be estimated from their average molecular weights; (2) SAFT parameters can be estimated for substances whose vapor pressures and/or liquid densities are difficult to measure. In the case of asphaltenes, the correlations of aliphatic and aromatic SAFT parameters with molecular weight provide the upper and lower bounds of the asphaltene SAFT parameters; and (3) The "well-behaved" pure component SAFT correlations imply that the EOS can easily be extended to model the effects of polydispersity.

Note that the systematic behavior shown in Figure 12.2 has been observed for a number of SAFT implementations (see previous section). Discussions from this point on will focus on the use of PC-SAFT.

2.2. PC-SAFT Characterization of a Recombined Oil

To use the SAFT equation of state to model live oil systems, we need to be able to account for the fluid's (PVT) phase behavior using a minimum number of real components and "realistic" pseudo-components. In this section, we

Table 12.1. PC-SAFT Pure Component Parameters[a]

Substance	MW (g/mol)	m	σ (Å)	ε/k (K)	T range (K)	AAPD[b] P^{sat}	AAPD[b] ρ^{liq}
n-Alkanes							
Methane	16.04	1.0000	3.7039	150.03			
Ethane	30.07	1.6069	3.5206	191.42			
Propane	44.09	2.0020	3.6184	208.11			
Butane	58.12	2.3316	3.7086	222.88			
Pentane	72.15	2.6896	3.7729	231.20			
Hexane	86.18	3.0576	3.7983	236.77			
Heptane	100.20	3.4831	3.8049	238.40			
Octane	114.23	3.8176	3.8373	242.78			
Nonane	128.25	4.2079	3.8448	244.51			
Decane	142.29	4.6627	3.8384	243.87			
Undecane	156.31	4.9082	3.8893	248.82			
Dodecane	170.34	5.3060	3.8959	249.21			
Tridecane	184.37	5.6877	3.9143	249.78			
Tetradecane	198.39	5.9002	3.9396	254.21			
Pentadecane	212.42	6.2855	3.9531	254.14			
Hexadecane	226.45	6.6485	3.9552	254.70			
Eicosane	282.55	7.9849	3.9869	257.75			
Cycloalkanes							
Cyclopentane	70.13	2.3655	3.7114	265.83			
Methyl-cyclopentane	84.16	2.6130	3.8253	265.12			
Ethyl-cyclopentane	98.18	2.9062	3.8873	270.50			
Cyclohexane	84.15	2.5303	3.8499	278.11			
Methyl-cyclohexane	98.18	2.6637	3.9993	282.33			
Ethyl-cyclohexane	112.22	2.8256	4.1039	294.04			
Cycloheptane	98.19	2.6870	3.9352	296.15	300–570	0.44	0.21
Cyclooctane	112.22	2.9222	4.0028	304.67	300–620	0.71	1.06
Polynuclear aromatics							
Benzene	78.11	2.4653	3.6478	287.35			
Naphthalene	128.17	3.0915	3.8333	348.40	373–633	1.45	0.77
Anthracene	178.23	3.5291	4.0922	402.13	500–830	1.81	1.57
Phenanthrene	178.23	3.4890	4.1053	403.06	330–780	1.51	1.39
Naphthacene	228.29	4.6432	3.8942	407.60	660–940	0.88	4.23
Chrysene	228.29	5.1201	3.8400	385.73	580–940	0.81	3.61
Pyrene	202.26	3.6847	4.1151	427.35	450–850	1.37	3.76
Aromatic and polynuclear aromatics derivatives							
Toluene	92.14	2.8149	3.7169	285.69			
Ethylbenzene	106.17	3.0799	3.7974	287.35			
Propylbenzene	120.19	3.3438	3.8438	288.13			
Butylbenzene	134.22	3.7662	3.8727	283.07			
Tetralin	132.21	3.3131	3.8750	325.07			
Biphenyl	154.21	3.8877	3.8151	327.42			
1-Methylnaphthalene	142.20	3.4064	3.8961	345.71	393–673	0.29	0.50
1-Phenylnaphthalene	204.27	4.7634	3.8582	336.53	330–780	1.51	1.39
m-Terphenyl	228.29	5.6273	3.7967	329.18	420–840	1.52	0.26

Table 12.1. (*Continued*)

Substance	MW (g/mol)	m	σ (Å)	ε/k (K)	T range (K)	AAPD[b] P^{sat}	AAPD[b] ρ^{liq}
			Gases				
Nitrogen	28.01	1.2053	3.3130	90.96			
Carbon dioxide	44.01	2.0729	2.7852	169.21			
Carbon disulfide	76.14	1.6919	3.6172	334.82			

[a] m is the number of segments that make up a molecule, σ is the segment diameter in angstroms, ε/k is the interaction energy between a pair of segments in K. For substances fitted by Ting,[20] the AAPDs and temperature range of the experimental data are given.[45] All other PC-SAFT parameters are from Gross and Sadowski.[4]

[b] Average absolute percent deviation $= \frac{1}{ndp} \sum_{ndp} \left| \frac{\text{Calculated} - \text{Experimental}}{\text{Experimental}} \right| 100\%$.

will outline (1) a method to model live oil systems using a small number of real components and pseudo-components and (2) a method to characterize asphaltenes in the SAFT model. Both (1) and (2) take advantage of the systematic SAFT parameter–molecular weight behavior discussed in Section 2.1.

Consider the application of PC-SAFT to a recombined "live" oil (separator oil plus its associated gas) for which the bubble point, GOR, stock tank oil density (API gravity), density above saturation, and asphaltene precipitation onsets have been measured.[47] To model the phase behavior of the recombined oil, the oil is

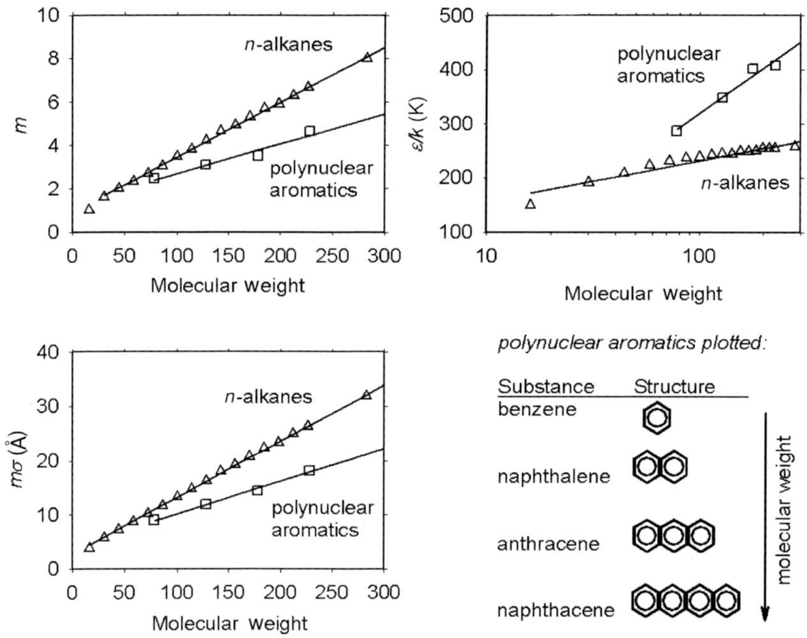

Figure 12.2. Plots of PC-SAFT pure component parameters as a function of molecular weight for saturates and polynuclear aromatics.

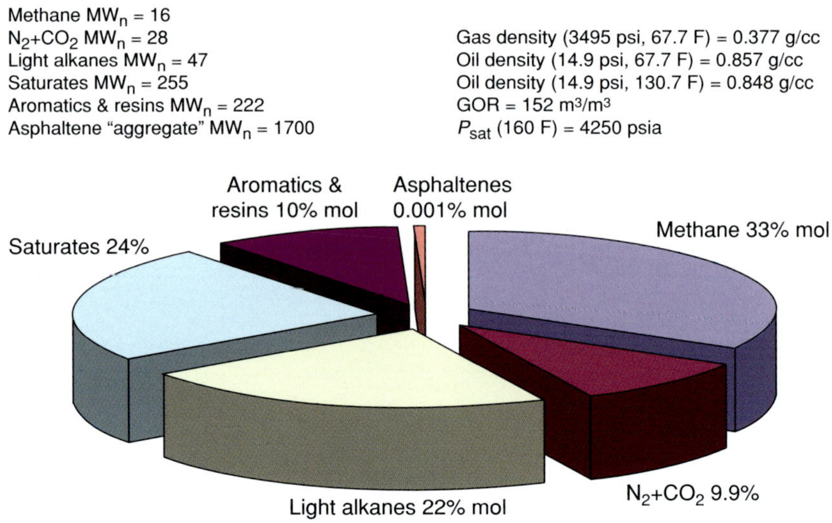

Figure 12.3. Representation of a recombined oil with six pseudo-components.

treated as a mixture of six pseudo-components (Figure 12.3). Three sub-fractions (methane, $N_2 + CO_2$, and light n-alkanes) are used to describe the separator gas and three sub-fractions (saturates, aromatics + polynuclear aromatics, and asphaltenes) are used to describe the stock tank oil.

The relative amount of each sub-fraction given in Figure 12.3 can be calculated based on composition data from gas chromatography (GC), SARA fractionation data, and GOR data. The lumping of N_2 with CO_2 was appropriate in this case because very little CO_2 was present. More generally, we treat N_2 and CO_2 as separate components.

Note that the proposed lumping scheme introduces a small error in the description of the fluid properties. This is because there are nonnegligible amounts of light n-alkanes such as butane and pentane partitioned in both the "light n-alkane" pseudo-components and the "saturates" pseudo-component. However, the effect of this error on the fluid phase behavior was found to be small.

2.2.1. Characterization of Separator Gas

The three PC-SAFT parameters (σ, m, and ε/k) for each pseudo-component are related to the average molecular weight of that pseudo-component. Parameters for the pseudo-components can be interpolated/extrapolated from correlations shown in Figure 12.2. For instance, the PC-SAFT parameters for the "light n-alkane" pseudo-component lie on the n-alkane correlations in Figure 12.2 with an average molecular weight of 47. For a real component, PC-SAFT parameters are taken from fits to the component's saturated liquid densities and vapor pressures. Parameters for the separator gas real- and pseudo-components are given in Table 12.2.

Table 12.2. Fractionation of the Recombined Oil with GOR = 152 m^3/m^3 into a 6-Component Mixture

Component	Mole fraction	Average MW	m	σ (Å)	ε/k (K)
Methane	0.3300	16	1.0000	3.7039	150.03
Nitrogen and carbon dioxide	0.0985	28	1.2053	3.3130	90.96
Light alkanes	0.2234	47	1.9854	3.6917	206.12
Saturates	0.2445	255	7.3765	3.9635	257.12
Aromatics and resins	0.1023	212	5.3351	3.7637	323.70
Asphaltenes	0.0013	1700[a]	29.5000	4.3000	395.00

[a] Molecular weight of pre-aggregated asphaltene.

To accurately describe the van der Waals attraction between unlike pairs of molecules, the energies of interaction are often modified in an equation of state via the use of binary interaction parameters. The binary interaction parameters used in this work were fit to vapor–liquid equilibria data, with the values listed in Table 12.3.

The PC-SAFT calculated density of 0.3708 g/cm^3 obtained from the three pseudo-component treatment of the separator gas compares well with the measured density of 0.3773 g/cm^3 at 241.3 bar and 293 K.

2.2.2. Characterization of Stock Tank Oil

The stock tank oil is treated as a mixture of three pseudo-components (saturates, aromatics + resins, and asphaltenes). This particular division was set up

Table 12.3. Binary Interaction Parameters Used in This Study

Component	Methane	$N_2 + CO_2$	Light alkanes	Saturates	Aromatics and resins	Asphaltenes
Methane	0	0[a]	0.01[b]	0.03[c]	0.03[d]	0.03[e]
$N_2 + CO_2$	0[a]	0	0.07[f]	0.01[g]	0.13[h]	0.13[e]
Light alkanes	0.01[b]	0.07[f]	0	0.006[i]	0.02[k]	0.02[e]
Saturates	0.03[c]	0.01[g]	0.006[i]	0	−0.01[j]	0
Aromatics and resins	0.03[d]	0.13[h]	0.02[k]	−0.01[j]	0	0
Asphaltenes	0.03[e]	0.13[e]	0.02[e]	0	0	0

[a] based on methane–N_2 binary data of Kidnay *et al.*[48]
[b] Based on methane–propane binary data of Reamer *et al.*[49]
[f] Based on N_2–propane binary data of Grauso *et al.*[50]
[c] Based on methane–decane data of Reamer *et al.*[51]
[d] Based on methane–toluene binary data reported in this work and from Elbishlawi and Spencer.[52]
[g] Based on N_2–decane binary data of Azarnoosh and McKetta.[53]
[i] Based on propane–hexadecane binary data of Joyce.[54]
[h] Based on N_2–benzene binary data of Miller and Dodge.[55]
[k] Based on propane–benzene binary data of Glanville *et al.*[56]
[j] Based on toluene–dodecane and toluene–hexadecane binary data of Messow and Engel.[57]
[e] The *kij*s between asphaltenes and separator gas sub-fractions are assumed to be the same as the *kij*s between aromatics + resins and separator gas sub-fractions.

partly to take advantage of the SARA fractionation data. Unlike SARA, which breaks down an oil into saturates, aromatics, resins, and asphaltenes, we grouped aromatics and resins into one single pseudo-component because they form a part of the same distribution in molecular weight and aromaticity (that actually extends out to the asphaltenes). The following assumptions are applied to the oil composition data to determine the composition of each sub-fraction given in Figure 12.3.

- We assume that saturates can be distinguished from aromatics below the C_{10} carbon cut.
- Since 72.3% mass of the stock tank oil are made up of saturate components (from SARA fractionation analysis of the example oil in this work), 72.3% mass of the materials from the higher carbon cuts (the C_{10} cut and higher) are assumed to be saturate components. The remaining 27.7% (mass) of the materials are assumed to be either aromatics + resins (up to, and including the C_{29} cut) or aromatics + resins and asphaltenes (in the C_{30+} cut).
- The asphaltene fraction accounts for 2.5% mass (from SARA analysis) of the stock tank oil, and all asphaltenes are found in the C_{30+} sub-fraction.
- The molecular weight distribution (and thus the average molecular weight) of the C_{30+} saturates follows an exponential distribution in mole amount
- The components in the saturates sub-fraction are all normal alkanes.

Based on these assumptions, the average molecular weights were first determined and the PC-SAFT parameters subsequently obtained (from correlations in terms of average molecular weight in Figure 12.2) for the saturates pseudo-components. The parameters are presented in Table 12.2. Binary interaction parameters are presented in Table 12.3 based on fits of vapor–liquid equilibria data for saturates. SAFT is one of the few equations of state capable of accurately predicting liquid–liquid equilibria based on binary interaction parameters fit to vapor–liquid equilibria data.

In the case of the aromatics + resins pseudo-component, the PC-SAFT parameters depend not only on the average molecular weight but also on the average aromaticity of the sub-fraction. This is because the aromatics + resins sub-fraction is a mixture of aromatics, polynuclear aromatics (PNA), PNA derivatives, and resins. Since the PC-SAFT parameters for compounds in each of these classes follow different parameter correlation curves, it is important to quantify this "degree of aromaticity." An aromaticity parameter was defined to interpolate between parameters for PNA's and those for aromatic derivatives. For a stock tank oil, the degree of aromaticity was adjusted (1) to match the stock tank oil density and (2) so that a "model" n-C_7 insoluble asphaltene is completely soluble in a fluid composed of only the aromatics + resins fraction. Hence, the aromatics + resins pseduo-component possesses the minimum degree of aromaticity necessary to dissolve asphaltenes. The binary interaction parameter between the saturates and the aromatics + resins pseudo-components are set to −0.01 based on the optimal *kij* for toluene–dodecane and toluene–hexadecane VLE data[57] at 353 and 333 K.

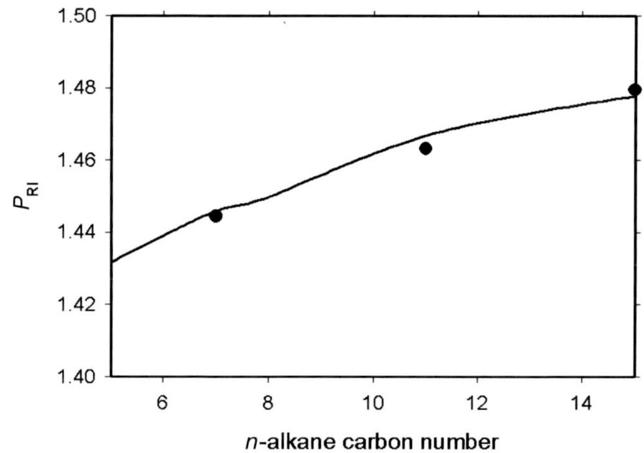

Figure 12.4. Refractive index at the point of asphaltene precipitation (P_{RI}) from an equi-volume mixture of stock tank oil and alpha-methyl naphthalene on titration with n-alkane. The curve shows results from the PC-SAFT equation of state and the data are from Wang and Buckley (J.S. Buckley and J. Wang, Private Communication, 2003).

At this point, all PC-SAFT parameters have been determined except for the parameters for the asphaltenes. We fit the PC-SAFT parameters for asphaltenes to data for asphaltene precipitation on n-alkane titration of the stock tank oil (see Figure 12.4). For the oil illustrated in this work, the refractive index of the mixture at the initial point of asphaltene precipitation (P_{RI}) was measured by Wang and Buckley (J.S. Buckley and J. Wang, Private Communication, 2003). The P_{RI} is related to the composition and density of the oil; it can also be correlated to the solubility parameter of the oil. In measuring the P_{RI}, the stock tank oil was initially mixed with an equal volume of alpha-methyl naphthalene to re-dissolve any precipitated material. For a given oil/precipitant pair, the addition of asphaltene solvents to the oil shifts the amount of precipitant needed to induce asphaltene precipitation but does not significantly alter the refractive index at precipitation onset.[58] By simulating these experiments using PC-SAFT, we are able to fit the three PC-SAFT parameters for asphaltenes to reproduce the P_{RI} behavior. In these calculations, to minimize the number of adjustable parameters, the binary interaction parameters between asphaltenes and the other stock tank oil components (and alpha-methyl naphthalene) were set to zero. Using these asphaltene parameters PC-SAFT predicts a solubility parameter for the asphaltene of 21.85 $MPa^{0.5}$ that is consistent with values reported in the literature.[5, 15, 22]

2.3. Comparison of Results and Analysis of Asphaltene Behavior

The fit of asphaltene parameters to titration data is shown in Figure 12.4 in terms of the refractive index of the mixture at the initial point of asphaltene precipitation (P_{RI}). Although the asphaltene parameters were fit to ambient condition

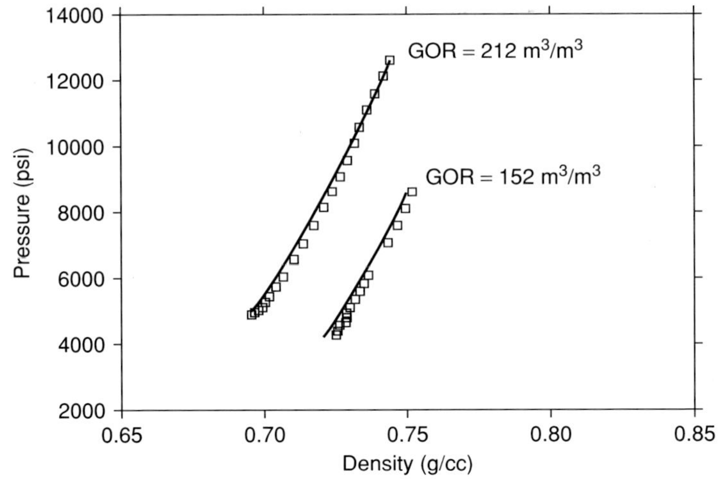

Figure 12.5. Experimental[47] and PC-SAFT-predicted (lines) one-phase recombined oil densities with two different gas–oil ratios at 71.1°C.

alkane titration experiments (for an oil blended with an equal volume of alpha-methyl naphthalene), our experience is that these same asphaltene parameters are applicable for asphaltenes in uncontaminated oil at reservoir conditions. PC-SAFT parameters for the asphaltene component are given in Table 12.2. This asphaltene has a PC-SAFT calculated solubility parameter of 21.85 $MPa^{0.5}$ and a molar volume of 1,437 cm^3/mol. Given that the density of asphaltenes is estimated to be between 1.13 and 1.20 g/cm^3, the SAFT-calculated stock tank oil asphaltene has an implied *MW* between 1,624 and 1,724. Because the asphaltenes are assumed to be pre-aggregated in the oil, this reflects the *MW* for the pre-aggregated asphaltenes.

A comparison of PC-SAFT-predicted and experimental densities for the one-phase recombined oil with a gas–oil ratio of 152 m^3/m^3 (for the reservoir fluid) and a GOR of 212 m^3/m^3 (for the reservoir fluid with additional separator gas added) is shown in Figure 12.5. Although PC-SAFT slightly under-predicts the recombined oil densities, the predicted densities are in good agreement with experimental values.

The PC-SAFT predicted asphaltene instability onset and bubble point pressures for the example recombined oil at reservoir temperature (71.1°C) are plotted versus experimental results in Figure 12.6 as a function of separator gas composition (corresponding to GORs' of 152 m^3/m^3 and 212 m^3/m^3). PC-SAFT accurately predicts both the bubble points and the asphaltene instability onset points for this oil in the composition range investigated. According to the equation of state calculations, we would expect asphaltene precipitation problems to occur at higher pressure for higher separator gas concentrations.

Consider what is predicted to happen on reservoir depressurization in Figure 12.6. If we imagine a reservoir fluid at 0.2 mass fraction of gas and 10,000 psia, the asphaltenes are soluble in the oil. On reservoir depressurization to just

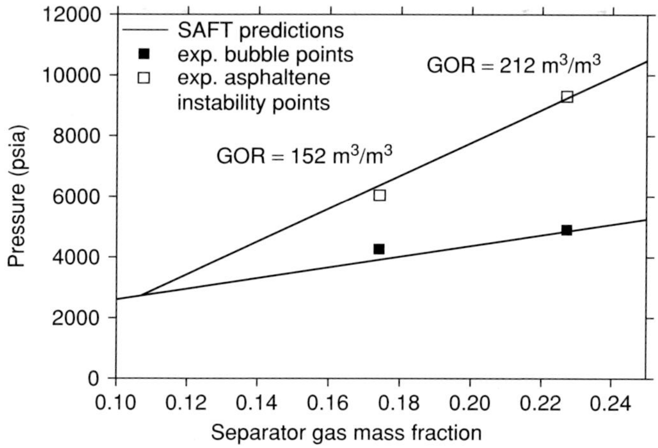

Figure 12.6. PC-SAFT-predicted and measured[19, 47] asphaltene instability onset and mixture bubble points for the recombined oil at 71.1°C.

below 8,000 psia, asphaltenes begin to come out of solution. On further depressurization, we reach the bubble point of the oil where the light ends begin to come out of solution. As the pressure is decreased further, more light ends leave the solution until at just below 3,000 psia, the oil becomes stable for asphaltenes again. This explains the field observation that asphaltenes precipitate only in a certain pressure range. Of course, the effect of asphaltene re-dissolution kinetics, which may be significant, is not considered in the thermodynamic model.

We find similar phase behavior on N_2 injection for an oil studied by Jamaluddin *et al.*[59] In this system, the parameters for the components in the oil are determined by the same approach as previously explained in this chapter. One difference is that N_2 and CO_2 are treated as separate components because each is present in reasonably high concentration. Also, the parameters for the asphaltene component are fit to the asphaltene instability pressure prior to N_2 injection. Details of the calculations are presented in reference 60. A comparison of PC-SAFT results with experimental results is given in Figure 12.7. The injection of 5, 10, and 20 mole% of N_2 strongly increases the asphaltene instability onset. The oil is originally unstable probably due to its initial high content of CO_2. The difference between the asphaltene onset pressure and the bubble point pressure ($P_{onset}-P_{bbp}$) increases with the amount of injected N_2. The SAFT predictions closely follow the experimental findings.

For this same system before N_2 injection, the pressure/temperature isopleth projection of the phase diagram has been measured. In Figure 12.8, we compare the asphaltene onset as a function of temperature with experimental data.[59] PC-SAFT predicts, in qualitative agreement with experiment, asphaltene instability at low temperature and at high temperature. This type of phase boundary is commonly seen in mixtures of components with large size asymmetry. At low temperature, differences in van der Waals interactions and molecular size between the asphaltene

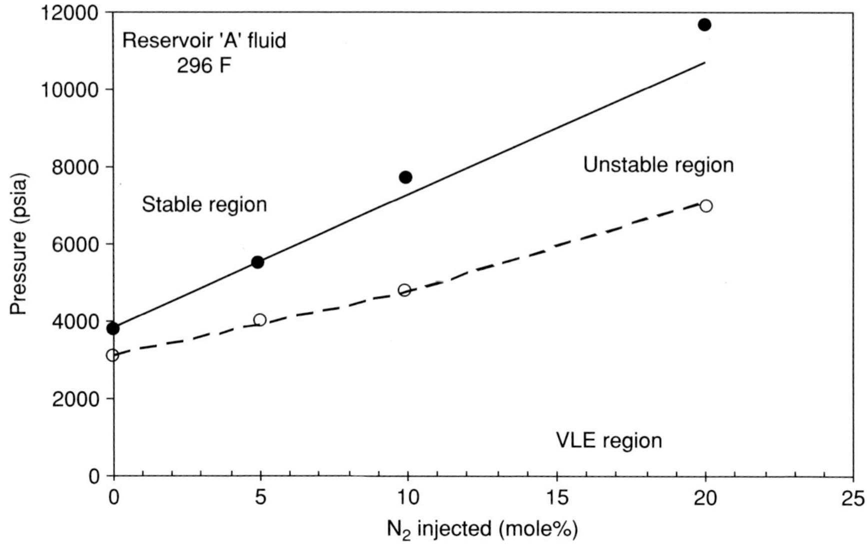

Figure 12.7. PC-SAFT-predicted and measured[59] asphaltene instability onset and mixture bubble points as a function of nitrogen concentration in a recombined oil at 296° F.

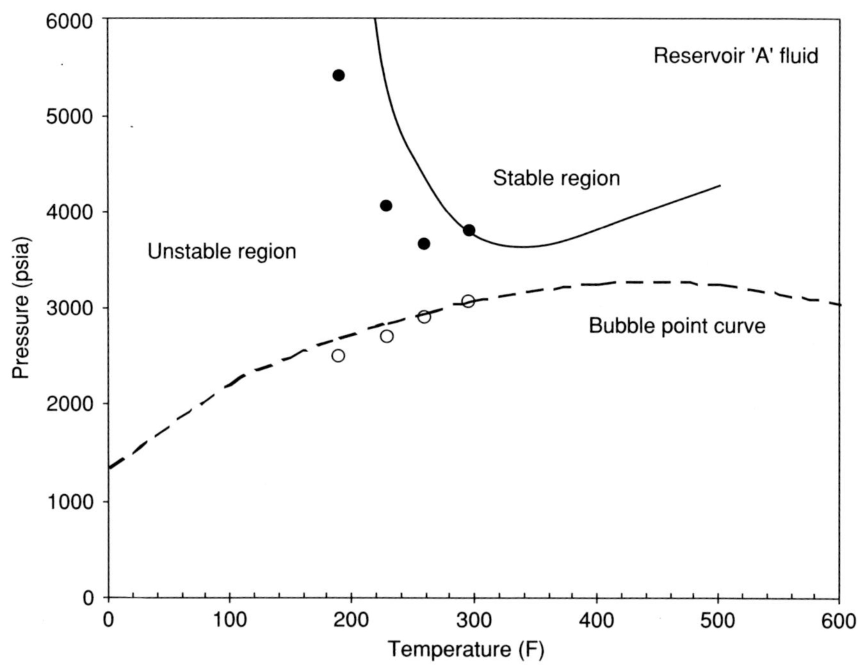

Figure 12.8. The temperature dependence of the asphaltene instability curve and bubble curve predicted by PC-SAFT and experimental measurements[59] for the reservoir fluid prior to nitrogen injection.

molecules and solvent (crude oil) cause phase separation. At high temperature, the oil becomes a poor solvent and the solution demixes because the light components cause the oil to expand with increase in temperature.

2.4. Effect of Asphaltene Polydispersity on Phase Behavior

Results thus far have shown[19, 20, 60] that PC-SAFT can adequately describe asphaltene phase behavior in a recombined oil under reservoir conditions. In these calculations the asphaltenes were treated as a single, monodisperse component in oil. Since asphaltenes are, in actuality, a polydisperse class of compounds with resins as their lowest molecular weight sub-fraction, the effects of asphaltene polydispersity need to be considered.

Here we examine the effect of asphaltene polydispersity on asphaltene's thermodynamic phase behavior in oil. We will show that the lowest molecular weight asphaltenes (including the resins) can stabilize, via nonpolar interactions, the higher molecular weight asphaltenes. We also show, via modeling analyses, qualitative differences in the behavior of the various asphaltene (and resin) sub-fractions. The polydispersity of asphaltenes has been observed in the laboratory, both in terms of the amount of asphaltenes precipitated with various alkane precipitants and in the characterization of the precipitated asphaltenes as "hard" or "soft" asphaltenes. Since deposition tendencies onto pipeline surfaces have often been associated with variations in the morphology and the composition of the precipitated asphaltene phase, an understanding of the molecular weight distribution of polydisperse asphaltenes in equilibrium phases should help us gain a better understanding of the various asphaltene sub-fraction's deposition tendencies.

In this section of the work, polydisperse asphaltenes are represented as four pseudo-components in SAFT. These pseudo-components are denoted n-C_{i-j}, representing the asphaltene fraction insoluble in the n-alkane of carbon number n-C_i, but soluble in the n-alkane n-C_j. Thus, the pseudo-components are denoted n-C_{3-5}, n-C_{5-7}, n-C_{7-15}, and n-C_{15+}. Hence, n-C_{15+} asphaltenes are the asphaltenes fraction insoluble in pentadecane. This assumes, for example, that no n-C_{7-15} asphaltene will precipitate on addition of pentadecane. We will see that there is some co-precipitation, leading to an iterative division of pseudo-components. Since we are interested in the qualitative effects of polydispersity at this point, we consider these pseudo-components suitable for our purposes. The reader should keep in mind that, traditionally, the n-C_{3-5} fraction is called resins and that conventional asphaltene extraction techniques generally identify asphaltenes as the n-C_{5+} or n-C_{7+} insoluble fractions of heavy organics. All asphaltene sub-fractions in this work are soluble in aromatic solvents (assessed via the PC-SAFT equation of state).

2.4.1. Selection of PC-SAFT Parameters for Polydisperse Asphaltenes

In experiments performed by Wang,[14] the asphaltenes from an oil were separated into various solubility fractions using excess n-pentane, n-heptane, and n-pentadecane precipitants; these sub-fractions are called n-C_5 insoluble asphaltenes, n-C_7 insoluble asphaltenes, and n-C_{15} insoluble asphaltenes, respectively. The asphaltene instability onsets with n-alkane titrations were measured for mixtures of toluene with each isolated asphaltene sub-fraction (at ambient conditions and with an asphaltene/toluene ratio of 1 g/100 mL toluene). The experimental asphaltene fractionation and titration data for Lagrave oil from Wang[14] were used to fit parameters for each asphaltene pseudo-component.

Using this model, we could then study how polydispersity and resins affect the stability of asphaltenes. Although each sub-fraction, including the resins, is polydisperse, we model asphaltenes as either three or four pseudo-components (depending on whether the resin fraction, which is the n-C_{3-5} asphaltene sub-fraction, was included): an n-C_{15+}sub-fraction, an n-C_{7-15} sub-fraction, an n-C_{5-7} sub-fraction, and an n-C_{3-5} sub-fraction. The method used to fit the SAFT parameters for each asphaltene pseudo-component was similar in principle to the monodisperse SAFT asphaltene characterization procedure discussed above and given in Ting *et al.*[19, 20]. Succinctly:

1. SAFT parameters were fit for the n-C_{15+} asphaltene sub-fraction to reproduce the experimental data on the minimum volume fraction of alkane precipitant needed to induce asphaltene instability (ϕ_v^{ppt}) for mixtures of n-C_{15} insoluble asphaltene, toluene, and various n-alkanes (see Figure 12.9).
2. The asphaltene made from the combination of n-C_{15+} and n-C_{7-15} sub-fractions were assumed to represent the n-C_7 insoluble asphaltenes; a second set of SAFT parameters were fit for the n-C_{7-15} sub-fraction to reproduce (together with the previously fitted n-C_{15+} sub-fraction) the experimental ϕ_v^{ppt} data for a mixture of n-C_7 insoluble asphaltene, toluene, and n-alkane as shown in Figure 12.9.
3. A third set of SAFT parameters were fit for the n-C_{5-7} sub-fraction so that the combination of the n-C_{15+} (fitted in Step (1)), n-C_{7-15} (fitted in Step

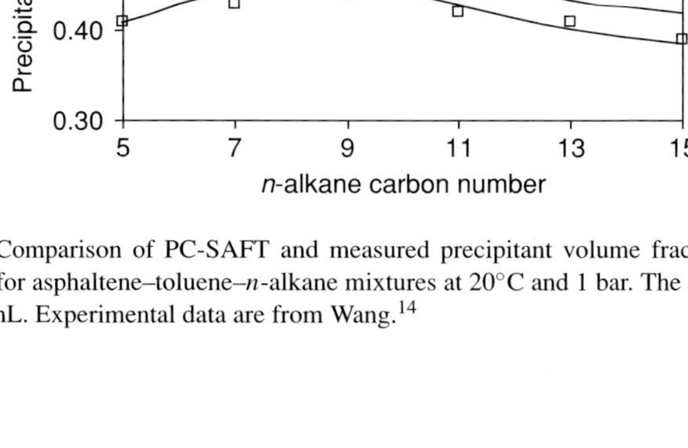

Figure 12.9. Comparison of PC-SAFT and measured precipitant volume fraction at asphaltene instability onset for asphaltene–toluene–n-alkane mixtures at 20°C and 1 bar. The asphaltene/toluene ratio is 1 g/100 mL. Experimental data are from Wang.[14]

(2)), and n-C_{5-7} sub-fractions represented the n-C_5 insoluble asphaltenes and reproduced the experimental ϕ_v^{ppt} data for a mixture of n-C_5 insoluble asphaltene, toluene, and n-alkanes as shown in Figure 12.9.

4. Due to lack of precipitation data, the PC-SAFT parameters for the resin (n-C_{3-5}) sub-fraction were obtained by decreasing the PC-SAFT parameters (m and ε/k) of the n-C_{5-7} asphaltene sub-fraction until a set of PC-SAFT parameter was obtained that would make the resins insoluble in propane (tested at 10 bars) and soluble in n-pentane.

Because there is insufficient data to uniquely fit all of the model parameters for polydisperse asphaltenes, certain approximations and relationships have to be made:

- The molecular weights of all PC-SAFT asphaltene sub-fractions were set to be linearly dependent on chain length, m. This was done because the experimental molecular weight of each asphaltene sub-fraction is not known and because the SAFT chain length is roughly linearly dependent on molecular weight for polynuclear aromatics. The constant of proportionality ($MW = m/0.0216$) used in this work was set to give the n-C_{15+} asphaltene sub-fraction a molecular weight of 2,500.
- The segment diameters of all asphaltene sub-fractions were set to 4 Å. This is an average value of the segment diameters for most polynuclear aromatics and polynuclear aromatics derivatives.

A comparison of the equation-of-state fitted and the experimental ϕ_v^{ppt} data is shown in Figure 12.9, with the fitted PC-SAFT asphaltene parameters listed in Table 12.4. As seen in the figure, the agreement between PC-SAFT calculated and measured ϕ_v^{ppt} is qualitative. PC-SAFT is able to describe the change in magnitude (and to a lesser extent, the curvature) of the ϕ_v^{ppt} vs. n-alkane carbon number curve between n-C_{15} insoluble and n-C_5 insoluble asphaltenes. For the precipitation onsets with n-C_7 and n-C_5 extracted asphaltenes, PC-SAFT underestimates the

Table 12.4. PC-SAFT Parameters for the Various Asphaltene Sub-Fractions (Including Resins)

			SAFT parameters			
Asphaltene sub-fraction	MW	m	σ (Å)	ε/k (K)	δ(MPa$^{0.5}$)	ρ(g/cm^3)
n-C_{15+}	2500[a]	54	4.00	350.5	22.17	1.150
n-C_{7-15}	1852[a]	40	4.00	340.0	21.52	1.137
n-C_{5-7}	1806[a]	39	4.00	335.0	21.25	1.133
Resin	556	12	4.00	330.0	20.41	1.103
Monodisperse – n-C_5	2080[a]	46	4.00	350.5	22.13	1.120

[a] Molecular weight of pre-aggregated asphaltene.

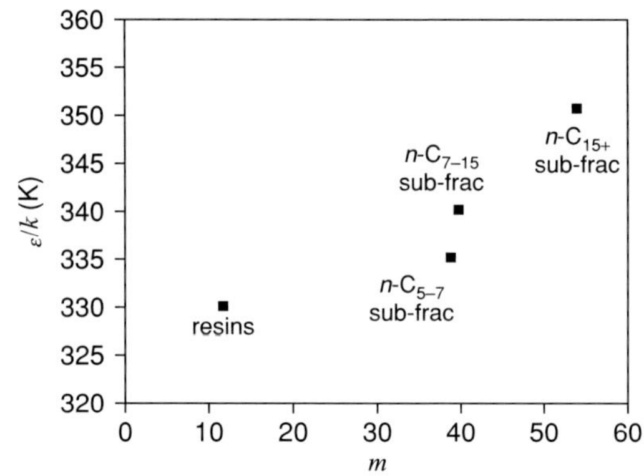

Figure 12.10. Plot of ε/k vs. m for the various PC-SAFT asphaltene sub-fractions and resin.

experimental ϕ_v^{ppt} data in cases where larger n-alkanes (undecane and higher) are used to induce asphaltene precipitation. As seen in the figure, PC-SAFT predicts a maximum in the volume fraction of precipitant at a carbon number of about 9, in agreement with previous experimental observations.

A plot of the SAFT parameters ε/k vs. m for the various SAFT asphaltene sub-fractions and resin (the n-C_{3-5} sub-fraction) shows a well-defined trend between ε/k and m (Figure 12.10). The asphaltenes precipitated by the lower molecular weight n-alkanes tend to be smaller in size and have lower segment energy. It is also interesting to note that despite having the same segment size, the segment energy of the "heavier" asphaltene sub-fractions seem to be larger. The asphaltene sub-fractions have a segment energy close to that for naphthalene and alpha-methyl naphthalene (Table 12.1 and Figure 12.2); however, the chain lengths, m, of the asphaltenes are much larger. Finally, the chain length (m) for the resin sub-fraction seems to be much smaller than the other asphaltene sub-fractions. Equation of state parameters for the asphaltene and resin fractions are given in Table 12.4.

2.4.2. Effects of Asphaltene Polydispersity and Resin Addition

To investigate the roles resins and asphaltene polydispersity play on asphaltene phase behavior in oil, we will compare the phase behavior of four model oil mixtures containing monodisperse or polydisperse asphaltenes. In all model systems, toluene is the model oil and the asphaltenes have a fixed concentration of 7.5 g asphaltenes to 100 mL of toluene. The properties and asphaltene/resin contents of these systems are listed in Table 12.5 and are discussed in the following paragraph. For these model system investigations, the focus is on the qualitative trends/behaviors and all binary interaction parameters between all species are set to zero (following Ting *et al.*[19, 20]).

The biggest difference between the various systems in Table 12.5 is that asphaltenes in Systems 1 and 2 are monodisperse, while the asphaltenes in Systems 3 and 4 are polydisperse. More specifically, the asphaltene used in Model System

Table 12.5. Description of Four Representative Model Oils Tested to Study the Effects of Asphaltene Polydispersity and Resin Addition[a]

Model system	Included asphaltene fractions
System 1	Monodisperse–n-C_{15+} sub-fraction only
System 2	Monodisperse–n-C_5
System 3	n-C_{15+} : n-C_{7-15} : n-C_{5-7} with mass ratios of 4.5 : 2.0 : 3.5
System 4	n-C_{15+} : n-C_{7-15} : n-C_{5-7} with mass ratios of 4.5 : 2.0 : 3.5 and resins 10 g/100 mL toluene (20°C, 1 bar)

[a] All model oils contained 7.5 g of asphaltene to 100 mL of toluene (20°C, 1 bar).

1 is monodisperse and was fit to experimental ϕ_v^{ppt} data for the n-C_{15} insoluble Lagrave asphaltenes[14] (also called the n-C_{15+} asphaltene fraction in this work). The asphaltene used in Model System 2 is monodisperse and was fit to experimental ϕ_v^{ppt} data for the n-C_5 insoluble Lagrave asphaltenes. The asphaltene used in Model System 3 is polydisperse and the SAFT parameters for each asphaltene sub-fraction were fit to the experimental ϕ_v^{ppt} data of the fractionated asphaltenes. Model System 4 is similar to Model System 3 with the exception that 10 g of resin/100 mL toluene (approximately 1–2 moles resin/100 moles toluene) is added to the system. The amount of resin added (10 g resin per 100 mL toluene) is arbitrary.

The effects of n-alkane addition on the amount of asphaltenes precipitated (at 20°C and 1 bar) for the four model oil mixtures are shown in Figures 12.11 and 12.12. For systems containing monodisperse asphaltenes (Figure 12.11), the change in asphaltene solubility is dramatic: asphaltenes go from completely soluble to almost completely insoluble in the model oil when the volume fraction of the n-alkane precipitant is increased slightly past the asphaltene instability onset point.

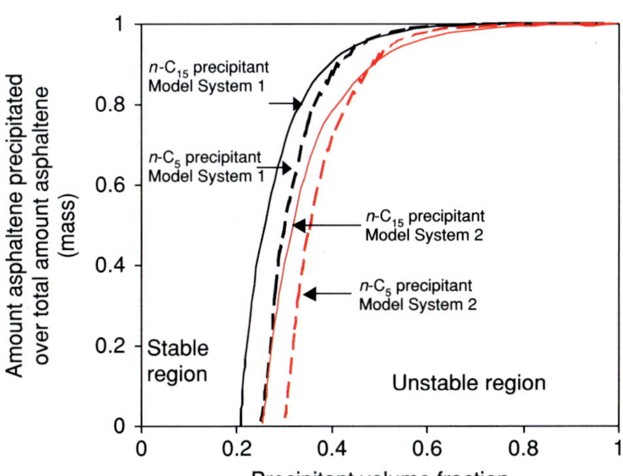

Figure 12.11. Solubility of monodisperse asphaltenes in model oil (7.5 g asphaltene/100 mL toluene) mixed with n-alkanes at 20°C and 1 bar.

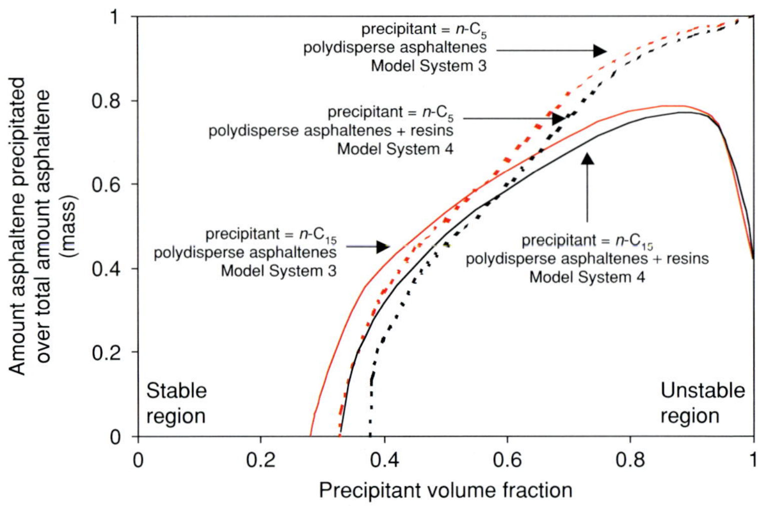

Figure 12.12. Solubility of polydisperse asphaltenes (with or without resins) in model oil (7.5 g total asphaltene/100 mL toluene) mixed with n-alkanes at 20°C and 1 bar.

As expected, the lower molecular weight asphaltenes (the monodisperse asphaltene fit to the n-C_5 insoluble asphaltenes ϕ_v^{ppt} data) are more soluble than the higher molecular weight asphaltene (the monodisperse, n-C_{15} insoluble asphaltene) in terms of the amount of precipitant needed to induce asphaltene instability. When sufficiently large amount of n-alkanes are added to the model oil, all asphaltenes will precipitate.

A large change in the amount of precipitated asphaltenes vs. precipitant volume fraction can be seen when the effect of asphaltene polydispersity (and resin addition) is taken into consideration (Figure 12.12). By treating asphaltene as a polydisperse specie, the amount of asphaltenes precipitated increases much more gradually with precipitant addition. A significant amount of asphaltenes will stay in solution even at high precipitant volume fractions, and more asphaltenes can be precipitated using lower molecular weight n-alkanes. It is interesting to note that when n-C_{15} is used as the precipitant, SAFT predicts the existence of a solubility minimum around $\phi_v^{ppt} = 0.9$.

A comparison of the SAFT-predicted behavior of polydisperse asphaltenes with and without resins show that the presence of resins will increase the amount of precipitant needed to induce the onset of asphaltene instability (Figure 12.12). Furthermore, at lower precipitant volume fractions in the oil, the amount of asphaltenes that will precipitate is less when resins are present. Even though only dispersion interactions are considered in these PC-SAFT models, the lower molecular weight asphaltenes and especially resins will stabilize the heavier asphaltenes in the oil. It can be seen in Figure 12.12 that the effects of resins on asphaltene phase behavior in the oil become less pronounced as the oil is diluted more with precipitants.

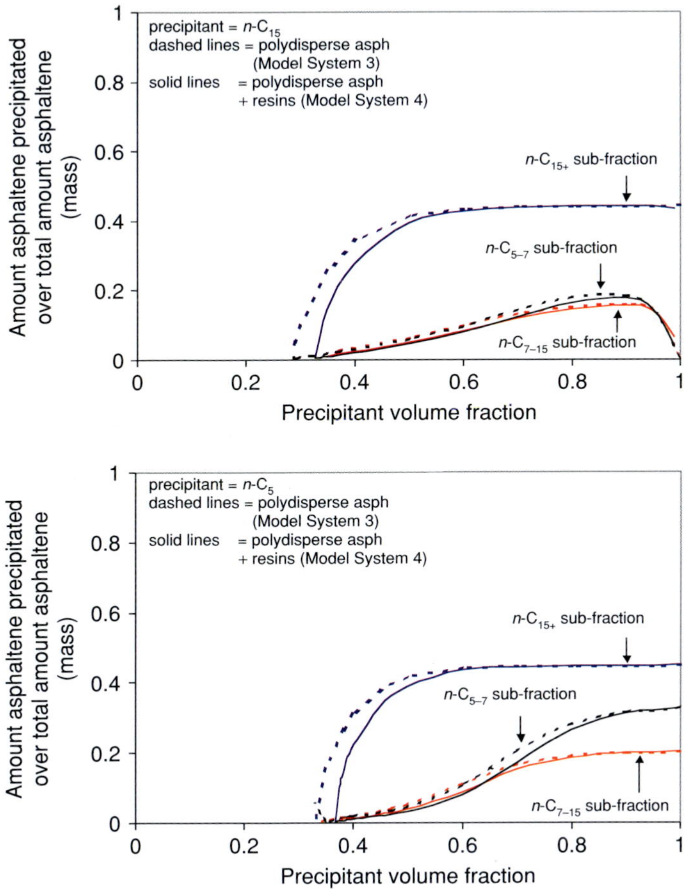

Figure 12.13. Normalized distribution of the asphaltene sub-fractions in the precipitated phase as a function of volume fraction precipitant in the model oil mixtures.

A plot of the mass distribution of the asphaltene sub-fractions as a function of precipitant volume fraction is shown in Figure 12.13. As seen in the figure, near the initial asphaltene instability onset, the precipitated phase is composed mostly of the heaviest asphaltene fractions (in this case, the n-C_{15+} sub-fraction). As the amount of precipitant is increased further, more and more lower molecular weight asphaltenes will precipitate.

3. Summary and Conclusions

The effect of pressure, temperature, and composition on the phase behavior of asphaltenes in crude oil systems can be explained in terms of van der Waals interactions between molecules using the PC-SAFT equation of state. A method was introduced to characterize the recombined oil including asphaltenes as a six-component mixture using the PC-SAFT equation of state. The real components and pseudo-components were chosen based on saturates–aromatics–resins–asphaltenes (SARA) fractionation, gas chromatography, and gas–oil ratio information. Equation of state parameters for each component (except asphaltenes) were determined from generalized correlations in terms of molecular weight. Binary interaction parameters were fit to vapor liquid equilibria data. Asphaltene parameters were determined by modeling asphaltene precipitation upon titration of oil with n-alkanes at ambient conditions. PC-SAFT was found to accurately predict the density, the bubble point curve and the asphaltene instability region for the recombined oil over a range of temperatures, pressures, and compositions.

The PC-SAFT equation of state has proved useful in explaining laboratory and field observations of asphaltene phase behavior. For example, when asphaltene precipitates near the bubble point, the model predicts asphaltene stability at high pressure and at pressures well below the bubble point. On titration of the oil with various n-alkanes, we show that the volume fraction of precipitant required for asphaltene precipitation as a function of carbon number of the alkane precipitant reaches a maximum at about nonane. The model also predicts that asphaltenes can become unstable as temperature is decreased or increased depending on the composition of the oil and the temperature. The model has also proved to be predictive in describing the effect of gas injection (nitrogen and methane) and of alpha-methyl naphthalene on asphaltene stability.

We have investigated the effect of polydispersity and resins on asphaltene phase behavior for a model system. We find that resins delay the onset of asphaltene precipitation, but the total amount of asphaltene precipitated is unaffected by the resins. SAFT calculations show that the lower molecular weight asphaltenes and resins play a large role in stabilizing higher molecular weight asphaltenes in oil. Resin's stabilizing effects on polydisperse asphaltene is greatest in the region of incipient asphaltene instability; when sufficiently large amounts of n-alkane precipitants are added, similar amounts of asphaltenes would precipitate regardless of the presence of resins in the oil. An analysis of the mass distribution of the asphaltene sub-fractions in the precipitated phase shows that the largest asphaltenes will precipitate first, followed by the precipitation of smaller asphaltenes upon further oil dilution. In our study, the heaviest asphaltene sub-fraction precipitated first.

The amount and type of data presently available for onset of asphaltene precipitation can be adequately modeled without including the effect of polar groups or association. Although extensions of the PC-SAFT equation of state can account for these effects, presently available data is insufficient to justify the additional parameters needed to include these effects.

Acknowledgments

We gratefully acknowledge the Department of Energy, DeepStar, Chevron-Texaco, DB Robinson, and the Consortium of Processes in Porous Media at Rice

University for their financial support. We also thank Jeff Creek and Jill Buckley for many helpful discussions.

References

[1] Chapman, W.G., G. Jackson, and K.E. Gubbins (1988). Phase equilibria of associating fluids—Chain molecules with multiple bonding sites. *Mol. Phys.* **65**, 1057–1079.

[2] Chapman, W.G., K.E. Gubbins, G. Jackson, and M. Radosz (1989). SAFT: Equation-of-state model for associating fluids. *Fluid Phase Equil.* **52**, 31–38.

[3] Chapman, W.G., K.E. Gubbins, G. Jackson, and M. Radosz (1990). New reference equation of state for associating liquids. *Ind. Eng. Chem. Res.* **29**, 1709–1721.

[4] Gross, J. and G. Sadowski (2001). Perturbed-chain SAFT: An equation of state based on a perturbation theory for chain molecules. *Ind. Eng. Chem. Res.* **40**, 1244–1260.

[5] Wiehe, I.A. and K.S. Liang (1996). Asphaltene, resins, and other petroleum macromolecules. *Fluid Phase Equil.* **117**, 201–221.

[6] Haskett, C.E. and M. Tartera (1965). Practical solution to problem of asphaltene deposits—Hassi Messaoud field, Algeria. *J. Pet. Technol.* **17**, 387–391.

[7] Tuttle, R.N. (1983). High-pour and asphaltic crude oils and condensates. *J. Pet. Technol.* **35**, 1192–1196.

[8] Adialalis, S. (1982). *Investigation of Physical and Chemical Criteria as Related to the Prevention of Asphalt Deposition in Oil Well Tubings.* PhD Thesis, Imperial College.

[9] Chavez, A.J.L. and M.A. Lory (1991). Estudio sobre la depositacion de material organico en instalaciones de produccion del area marina de Campeche. *Revista del Instituto Mexicano del Petroleo* **22**, 55–67.

[10] Escobedo, J. and G.A. Mansoori (1995). Asphaltene and other Heavy-Organic Particle Deposition during Transfer and Production Operations. In: *SPE Annual Technical Conference and Exhibition Production Operations and Engineering*, Dallas, TX, pp. 343–358.

[11] von Albrecht, C., W.M. Salathiel, and D.E. Nierode (1977). Stimulation of asphaltic deep wells and shallow wells in Lake Maracaibo, Venezuela. Advances in Methods of Increasing Well Productivity and Injectivity. *Can. Pet. Technol., J., Oil Sands* **PD7(1)**, 55–62.

[12] Wu, J. (1998). *Molecular Thermodynamics of Some Highly Asymmetric Liquid Mixtures.* PhD Thesis, University of California at Berkeley.

[13] Prausnitz, J.M., R.N. Lichtenthaler, and E.G. de Azevedo (1986). *Molecular Thermodynamics of Fluid-Phase Equilibria*, 2nd edn. P T R Prentice-Hall, New Jersey, U.S.A.

[14] Wang, J.X. (2000). *Predicting Asphaltene Flocculation in Crude Oils.* PhD Thesis, New Mexico Institute of Mining & Technology.

[15] Hirschberg, A., L.N.J. deJong, B.A. Schipper, and J.G. Meijer (1984). Influence of temperature and pressure on asphaltene flocculation. SPEJ, *Soc. Pet. Eng. J.* **24**, 283–293.

[16] Akbarzadeh, K., S. Ayatollahi, M. Moshfeghian, H. Alboudwarej, W.Y. Svrcek, and H.W. Yarranton (2002). Regular solution model for asphaltene precipitation from bitumens and solvents. In: *American Institute of Chemical Engineers Spring National Meeting.* New Orleans, LA, pp. 218–224.

[17] Cimino, R., S. Correra, and P.A. Sacomani (1995). Thermodynamic Modeling for Prediction of Asphaltene Deposition in Live Oils. In: Proceedings *Soc. Pet. Eng. Intl. Symp. on Oilfield Chem.* San Antonio, TX, pp. 499–512.

[18] Nghiem, L.X., M.S. Hassam, R. Nutakki, and A.E.D. George (1993). Efficient Modeling of Asphaltene Precipitation In: Proceedings *Soc. Pet. Eng. Annual Tech. Conf. and Exhib.* Houston, TX. pp. 375–384.

[19] Ting, P.D., G.J Hirasaki, and W.G. Chapman (2003). Modeling of asphaltene phase behavior with the SAFT equation of state. *Pet. Sci. Tech.* **21**, 647–661.

[20] Ting, P.D. (2003). *Thermodynamic Stability and Phase Behavior of Asphaltenes in Oil and of other highly Asymmetric Mixtures.* PhD Thesis, Rice University.

[21] Chung, F., P. Sarathi, and R. Jones (1991). Modeling of Asphaltene and Wax Precipitation, NIPER-498, DOE Topical Report, Bartlesville, OK.

[22] Burke, N.E., R.E. Hobbs, and S.F. Kashou (1990). Measurement and modeling of asphaltene precipitation. *J. Pet. Technol.* **42**, 1440–1446.

[23] Wang, J.X., J.S. Buckley, N.A. Burke, and J.L. Creek (2003). A Two-Component Solubility Model of the Onset of Asphaltene Flocculation in Crude Oils. Paper 15254 presented at the *2003 SPE Offshore Technology Conference*, Houston, May 5–7; Wang, J.X. and J.S. Buckley (2001). *Energy Fuels* **15**, 1004–1012.

[24] Nghiem, L.X. and D.A. Coombe (1997). Modeling asphaltene precipitation during primary depletion. *Soc. Pet. Eng. J. SPE* 36106, **2**, 170–176.

[25] Ting, P.D., P.C. Joyce, P.K. Jog, W.G. Chapman, and M.C. Thies (2002). Phase equilibrium modeling of mixtures of long-chain and short-chain alkanes using peng-robinson and SAFT. *Fluid Phase Equil.* **206**, 267–286.

[26] Leontaritis, K.J. and G.A. Mansoori (1987). Asphaltene flocculation during oil production and processing: A thermodynamic colloidal model. In: *Society of Petroleum Engineers of AIME*. San Antonio, TX, pp. 149–158.

[27] Victorov, A.I. and A. Firoozabadi (1996). Thermodynamic micellization model of asphaltene precipitation from petroleum fluids. *Am. Inst. Chem. Eng. J.* **42**, 1753–1764.

[28] Wu, J., J.M. Prausnitz, and A. Firoozabadi (1998). Molecular-thermodynamic framework for asphaltene-oil equilibria. *Am. Inst. Chem. Eng. J.* **44**, 1188–1199.

[29] Wu, J., J.M. Prausnitz, and A. Firoozabadi (2000). Molecular thermodynamics of asphaltene precipitation in reservoir fluids. *Am. Inst. Chem. Eng. J.* **46**, 197–209.

[30] Bungert, B. (1998). Modeling Polymer Systems Using the Perturbed-Chain Statistical Associating Fluid Theory Equation of State. Komplexe Phasengleichgewichte von Polymerlösungen. *Dissertation*, Technische Universität Berlin, Germany. Gross, J. and G. Sadowski (2002). *Ind. Eng. Chem. Res.* **41**, 1084–1093.

[31] Wiehe, I.A. (1996). Two-dimensional solubility parameter mapping of heavy oils. *Fuel Sci. Tech. Intl.* **14**, 289–312.

[32] Moore, E.W., C.W. Crowe, and A.R. Hendrickson (1965). Formation effect and prevention of asphaltene sludges during stimulation treatments. *J. Pet. Technol.* **17**, 1023–1028.

[33] Chang, C.L. and H.S. Fogler (1993). Asphaltene Stabilization in Alkyl Solvents using Oil-Soluble Amphiphiles. In: *Soc. Pet. Eng. Intl. Symp. on Oilfield Chem.* New Orleans, LA, pp. 339–349.

[34] Groenzin, H. and O.C. Mullins (1999). Asphaltene molecular size and structure. *J. Phys. Chem. A* **103**, 11237–11245.

[35] Economou, I.G. (2002). Statistical associating fluid theory: A successful model for the calculation of thermodynamic and phase equilibrium properties of complex fluid mixtures; Economou, I.G. (2002). Molecular Modelling of Materials Laboratory, Institute of Physical Chemistry, National Research Center for Physical Sciences "Demokritos", GR-15310 Aghia Paraskevi Attikis, Greece. *Ind. Eng. Chem. Res.* **41**, 953–962.

[36] Muller, E.A. and K.E. Gubbins (2001). Molecular-based equations of state for associating fluids: A review of SAFT and related approaches. *Ind. Eng. Chem. Res.* **40**, 2193–2212.

[37] Jog, P.K., W.G. Chapman, S.K. Gupta, and R.D. Swindoll (2002). Modeling of liquid-liquid-phase separation in linear low-density polyethylene-solvent systems using the statistical associating fluid theory equation of state. *Ind. Eng. Chem. Res.* **41**, 887–891.

[38] Wertheim, M.S. (1984). Fluids with highly directional attractive forces. I. Statistical thermodynamics. *J. Stat. Phys.* **35**, 19.

[39] Wertheim, M.S. (1984). Fluids with highly directional attractive forces. II. Thermodynamic perturbation theory and integral equations. *J. Stat. Phys.* **35**, 35–47.

[40] Wertheim, M.S. (1986). Fluids with highly directional attractive forces. III. Multiple attraction sites. *J. Stat. Phys.* **42**, 459–476.

[41] Wertheim, M.S. (1986). Fluids with highly directional attractive forces. IV. Equilibrium polymerization. *J. Stat. Phys.* **42**, 477–492.

[42] Chapman, W.G. (1990). Prediction of the thermodynamic properties of associating lennard-jones fluids: Theory and simulation. *J. Chem. Phys.* **93**, 4299–4304.

[43] Kraska, T. and K.E. Gubbins (1996). Phase equilibria calculations with a modified SAFT equation of state. 1. Pure alkanes, alkanols, and water. *Ind. Eng. Chem. Res.* **35**, 4727–4737.

[44] GilVillegas, A., A. Galindo, P.J. Whitehead, S.J. Mills, G. Jackson, and A.N. Burgess (1997). Statistical association fluid theory for chain molecules with attractive potentials of variable range. *J. Chem. Phys.* **106**, 4168–4186.

[45] DIPPR Database (2003). Design Institute for Physical Properties, Brigham Young University, UT.

[46] Huang, S.H. and M. Radosz (1990). Equation of state for small, large, polydisperse, and associating molecules. *Ind. Eng. Chem. Res.* **29**, 2284–2294.

[47] Ting, P.D. (2003). *Thermodynamic Stability and Phase Behavior of Asphaltenes in Oil and of other highly Asymmetric Mixtures.* PhD Thesis. Rice University.

[48] Kidnay, A.F., R.C. Miller, W.R. Parrish, and M.J. Hiza (1975). Liquid–vapour phase equilibria in the N2–CH4 system from 130 to 180 K. *Cryogenics* **15**, 531–540.

[49] Reamer, H.H., B.H. Sage, and W.N. Lacey (1950). Phase equilibria in hydrocarbon systems: Volumetric and phase behavior of the methane-propane system. *Ind. Eng. Chem.* **42**, 534.

[50] Grauso, L., A. Fredenslund, and J. Mollerup (1977). Vapour-liquid equilibrium data for the systems C2H6+N2, C2H4+N2, C3H8+N2, and C3H6+N2. *Fluid Phase Equil.* **1**, 13–26.

[51] Reamer, H.H., R.H. Olds, B.H. Sage, and W.N. Lacey (1942). Phase equilibria in hydrocarbon systems. *Ind. Eng. Chem.* **34**, 1526–1531.

[52] Elbishlawi, M. and J.R. Spencer (1951). Equilibrium relations of two methane-aromatic binary systems at 150F. *Ind. Eng. Chem.* **43**, 1811–1815.

[53] Azarnoosh, A. and J.J. McKetta (1963). Nitrogen-n-decane system in the two-phase region. *J. Chem. Eng. Data* **8**, 494–496.

[54] Joyce, P.C. (1999). Vapor-Liquid Equilibria for the Hexane+Tetracosane and Hexane+ Hexatriacontane Systems at Elevated Temperatures and Pressures. *Vapor–Liquid Equilibria for Long-Chain Hydrocarbons in Supercritical Alkane Solvents.* PhD Thesis, Clemson University; Joyce, P.C., J. Gordon, and M.C. Thies (2000). *J. Chem. Eng. Data* **45**, 424–427. Joyce, P.C. and M.C. Thies (1998). Vapor-liquid equilibria for the hexane + hexadecane and hexane + 1-hexadecanol systems at elevated temperatures and pressures. *J. Chem. Eng. Data* **43**, 819–822.

[55] Miller, P. and B.F. Dodge (1940). The system benzene-nitrogen: Liquid-vapor phase equilibria at elevated pressures. *Ind. Eng. Chem.* **32**, 434–438.

[56] Glanville, J.W., B.H. Sage, and W.N. Lacey (1950). Volumetric and phase behavior of propane-benzene system. *Ind. Eng. Chem.* **42**, 508.

[57] Messow, U. and I. Engel (1977). Thermodynamische Untersuchungen an Losungsmittel/n-Paraffin-Systemen. *Z. Phys. Chem.* **258**, 798.

[58] Buckley, J.S., G.J. Hirasaki, Y. Liu, S. von Drasek, J. Wang, and B.S. Gill (1998). Asphaltene precipitation and solvent properties of crude oils. *Pet. Sci. Technol.* **16**, 251–285.

[59] Jamaluddin, A.K.M., N. Joshi, F. Iwere, and O. Gurnipar (2001). An investigation of asphaltene instability under nitrogen injection. Presented at the Society of Petroleum Engineers International Petroleum Conference and Exhibition, Villahermosa, Mexico, *SPE 74393*.

[60] Gonzalez, D.L., P.D. Ting, G.J. Hirasaki, and W.G. Chapman (2005). Prediction of asphaltene i nstability under gas injection with the PC-SAFT equation of state. *Energy Fuels* **19**, 1230–1234.

13

Application of Isothermal Titration Calorimetry in the Investigation of Asphaltene Association

Daniel Merino-Garcia and Simon Ivar Andersen

1. Introduction

In biochemistry the use of isothermal titration calorimetry (ITC) is abundant in the investigation of interactions between different components. In the present chapter, we describe the application of this technique to the studies of petroleum compounds especially asphaltenes and resins and their interaction.

This work reports our experiences with ITC in the investigation of association and interactions of asphaltenes in solution. First, we discuss the use of ITC on well-described compounds and the detection of different association mechanisms relevant to the discussion of asphaltenes, followed by examples of different experiments possible with asphaltenes in toluene solution.

The work deals especially with the analysis of the concept of critical micellar concentration (CMC) and its application to asphaltene solutions. First, the proposed technique (isothermal titration calorimetry, ITC) is tested with well-known chemicals that have been reported to undergo micellization. Then, the technique is further applied to asphaltene solutions in toluene. As previously reported, the data obtained by ITC show that the existence of an asphaltene CMC could not be verified by this direct method and no signs of micellization were detected even when the concentration was lowered as much as 34 ppm. Instead, a stepwise mechanism of association is believed to explain better the aggregation behavior of asphaltenes in organic solutions, as shown in the second part of this chapter. An overview of the different studies that can be carried out by isothermal titration calorimetry is as well presented. In particular, this can be used in combination with, e.g., spectroscopic techniques and chemical alterations to study specific types of interactions and the influence and weight of these in overall association and reaction behavior. The aim is in this case specifically to understand the role of the hydrogen bond in

Daniel Merino-Garcia • Consultant, Pedro Barruecos 2 4C, 47002 Valladolid, Spain. **Simon Ivar Andersen** • Center for Phase Equilibria and Separation Processes, Department of Chemical Engineering, Bldg. 229, Technical University of Denmark, DK-2800 Kgs. Lyngby, Denmark.

asphaltene stability, revealing the result that naturally occurring hydrogen bonding sites may be very important in enhancing natural stability and the interaction with inhibitor components. Hence, we state that molecular flexibility and the ability to enter into specific interactions with smaller molecules lead to more stable asphaltenes in petroleum.

2. The Concept of Micellization

It is a well-known phenomenon that surfactants form micellar aggregates in liquid solution when the solute concentration exceeds a specific threshold value. This threshold, depends on the nature and structure of the surfactant, as well as the solvent and the presence of any other compounds, even in very small quantities. Basically, the surfactant structure consists of a lipophilic and a hydrophilic part, and, in a simplistic view, the surfactant is often seen as a lipophilic hydrocarbon chain and a hydrophilic head group. In reality, several more complex molecules exist which possess similar features, in the sense that they orientate at the water/oil interface and will form micellar aggregates when dissolved either in water or in an oil phase. Depending on the solvent, the number of associating molecules is highly different. Typically, the aggregation number (or number of compounds per micelle) is high in water but low in nonaqueous solution. As well as the molecular orientation in the two different solvents of course is different, so is the structure of the aggregates formed and the degree of polydispersity. From a thermodynamic perspective, the formation of micelles is not a true phase transition, since no macroscopic phase is formed.

However, the phase transition approach has been used in order to describe the phenomenon. One of the most important parameters is the critical micelle concentration (CMC). The CMC is the concentration region beyond which the surfactant will start to form micelles by association with other surfactants in order to diminish the energy of the system. Hence, the CMC is actually the monomer concentration in a solution with a concentration $C >$ CMC, as all surfactant molecules added beyond this magnitude will enter into micelle formation. CMC may also be defined as the total concentration of the surfactant at which a very small fraction is in the associated state, or where the concentration of micelles/aggregates becomes zero upon dilution. The micellization process is a dynamic process and, therefore, micelles will form and dissociate rapidly in the mixture. Besides, the CMC is more a concentration range than a fixed concentration. For some compounds a second CMC at a higher concentration has been reported as well related to a change in micellar shape (i.e., spherical to rod-like cylindrical micelles).[1]

The CMC is experimentally observed by a change in a given measured macroscopic property as a function of concentration. A number of methods exists for the determination of the critical micelle concentration of a molecule with self-assembly properties in a solution. The most abundant methods applied comprise measurements of conductivity, dielectric constant, surface or interfacial tension, osmotic pressure, calorimetric heat of dilution, and size exclusion/gel permeation chromatography.[2–4] All these methods may have their deficiency in detecting a clear CMC, depending on the surfactant species involved. Besides, in aqueous solutions where the aggregation number is large for most species, the detection is easier by most methods. However, in nonaqueous solution, the aggregation number tends to be small and hence not all methods are able to determine clearly the CMC.[5] In addition to this, two types of procedures exist for conducting the experiment: (1) measurement of physical property of series of solutions of varying concentration, which is a static test; and (2) a dynamic test where a solution is diluted while the measurement is performed.

Isothermal titration calorimetry (ITC) is a dynamic technique, a dilution or addition process where the heat of any ongoing process (exothermic or endothermic) is recorded. Therefore, ITC is a direct technique, but when several different processes take place at the same time, their contributions to the total heat developed cannot be separated unless blind tests are carried out.

3. Experimental

3.1. Asphaltene Separation

In the following, a standard procedure has been used, which is based on the IP143 method with modifications.[6] The ratio of solvent-to-oil is 30 mL heptane/g and the mixing is performed using ultrasound. Following this, precipitation is performed at constant temperature, which in general is room temperature. The final filtration is performed using vacuum filtration on membrane filters (0.5 μm). The filter is washed with *n*-heptane and the asphaltenes are extracted by dissolution in hot toluene. Excess solvent is removed by rotavaporization and the complete drying is done under a nitrogen stream. The raw asphaltenes this way obtained are then washed by small amounts of heptane using again ultrasound for mixing, centrifugation and decantation. This is done till the washings appear colorless. The final drying is performed in a vacuum oven. In large batch separations, the first separation step is performed in a centrifuge followed by similar treatment, but scaled to the amount of material.

The calorimeter was a VP-ITC 2000 from Microcal. It was kept in a controlled environment glove box to enhance baseline stability, humidity, etc. Synchronous fluorescence spectroscopy was performed using a PC interfaced MPF-3 Perkin-Elmer fluorescence spectrometer with a wavelength difference of 20 nm between excitation and emission wavelength. Using the synchronous mode allows for more detailed spectra or finger printing type of spectra than ordinary excitation spectra. The fairly large difference of 20 nm is recommended for complex mixtures such as crude oil fractions. Infrared spectra was recorded in liquid cells on a Perkin-Elmer Paragon 1000 FTir spectrometer. Dried toluene was used as solvent and subtracted in background. Details are given along the text or in referenced works by the authors.

In order to avoid effects related to the presence of trace water the toluene solvent used in ITC and IR was dried using molecular sieves and contact with metal sodium. In the literature reversed micellization of surfactants is indeed stated to be driven by the presence of trace water. Simple calculations indicate that in the normal concentration range applied in asphaltene investigations the water-to-asphaltene molecular ratio is above 1 and could reach 20 for very low asphaltene concentrations as the water concentration in bottled toluene may reach 200 ppm as received. Hence, in the present work the association of asphaltenes investigated reflects the system with a minimum of water, whereas systems report in the literature often reflects the system asphaltene–water–solvent in which water is an important component driving the association.

4. Application of ITC to Surfactants

Aqueous surfactant solutions have been studied in numerous occasions and the existence of critical micelle concentration (CMC) has been reported using a variety of techniques, as mentioned above. Calorimetry has also been applied to the determination of CMCs of surfactants, mainly in aqueous solutions. While surface tension measures an interfacial property to infer the behavior of the bulk, calorimetry is based on the direct measurement of bulk properties such as the heats of dilution and dissociation. The experiments consist of the sequential injection of a solution of surfactant into pure solvent. The concentration of the injected solution is high enough to assure that the surfactant exist in micellar state. In the first injections, the solution is diluted to a concentration below the CMC. Thus, the heat measured is a combination of both the heat of dilution of the monomer and the heat of dissociation of the micelles. After a certain number of injections, the resulting concentration in the cell is above the CMC and only the heat of dilution of micelles is measured. The data collected by the calorimeter are displayed as peaks. Each of them represents one injection and positive and negative peaks represent endothermic and exothermic processes, respectively. The integration of the area between the peaks and the baseline gives the heat developed per injection. Plots of heat of mixing versus final concentration may therefore be used to measure CMC. The typical plot would consist of three regions: a first region of high heat developed, resulting from both demicellization and dilution of monomers, a transition region (CMC region) and a third region with less heat developed, which corresponds to the dilution of the micelles. Birdi[7] determined CMC using a mixing calorimeter in which the heat of dilution was measured in batch single dilution experiments and showed the changes in CMC for mixed surfactants. The technique is tedious, as it requires large sample volumes and each single end-point concentration requires an entire experiment. The latter is optimized by the application of isothermal titration calorimetry.

Figure 13.1 shows the data obtained in the titration of an aqueous solution of sodium dodecyl sulphate (SDS) and sodium cholate. The first chart displays the raw data. The positive peaks indicate that the de-micellization is endothermic at this temperature. The integration of the area below each peak gives the heat developed per injection, as shown in the second chart. As explained above, the heat developed decreases sharply when the concentration in the cell surpasses the CMC. As expected, the CMC is not a single concentration but a region or

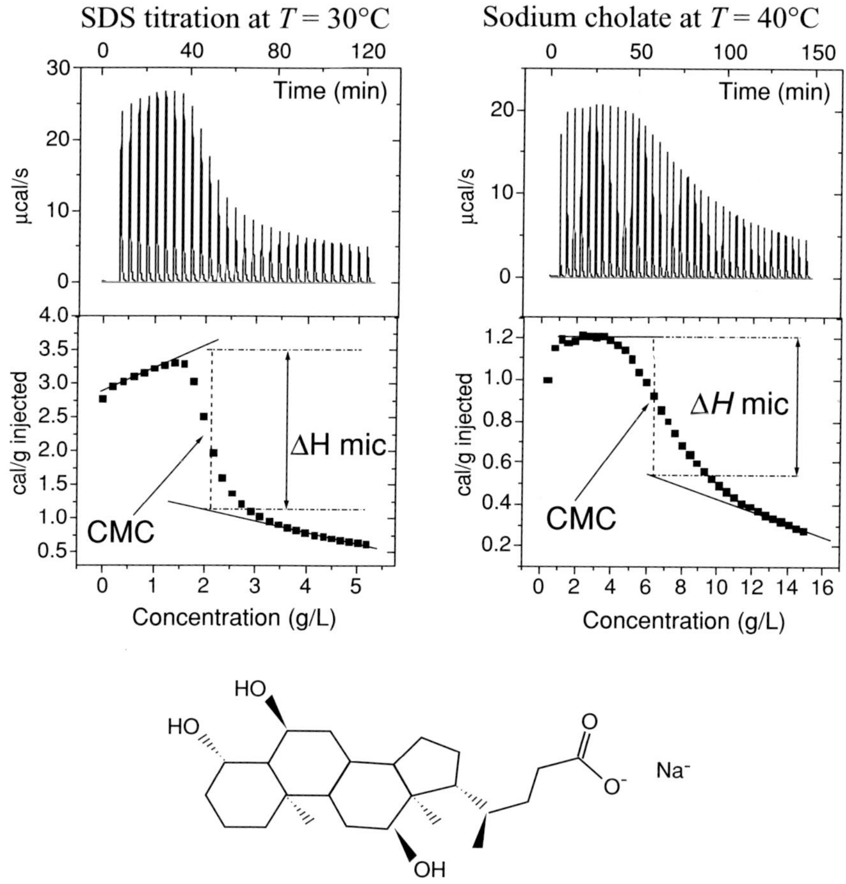

Figure 13.1. (A) Titration of 30 g/L of SDS into water at 30°C. (B) Titration of 80 g/L of sodium cholate into buffered water (pH = 7) at 40°C. Structure of sodium cholate.

transition zone. As can be seen, some degree of extrapolation is required especially for the structural more complex cholate. CMC can also be determined from the first derivative of the curve in which the CMC is displayed as a minimum. The plot of cumulated heat versus concentration may also be applied but these graphs are difficult to interpret if the heat beyond CMC does not reach an almost constant value. The value of CMC of SDS using ITC from Figure 13.1A is 2.4 g/L, while Paula *et al.*[8] and Andersen and Christensen[9] both reported 2.5 g/L at 30°C, using similar techniques. The CMC calculated with ITC is also in good agreement with other techniques such as fluorescence probing and surface tension. The calculated heat of micellization is −2.7 kJ/mol, while Paula *et al.*[8] reported –2.5 kJ/mol and Sharma *et al.*[10] reported –2.28 kJ/mol, using isoperibolic calorimetry.

While SDS is a typical surfactant with a hydrophilic and a hydrophobic part with an aggregation number in the range of 50–60, sodium cholate is a much more

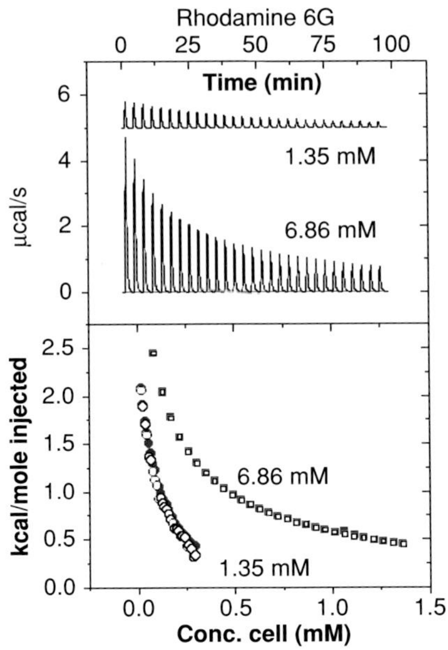

Figure 13.2. Titration of 8.86 mM Rodhamine G6 into water at 30°C.

complex molecule showing a much lower aggregation number and, therefore, also lower heat of micellization. As can be seen in Figure 13.1A, final plateau value of low heat is not reached, and hence CMC is difficult to determine. Contrary to surfactants, dyes such as Rhodamine 6G do not require a critical concentration to control the association in aqueous solutions. They associate following a stepwise mechanism, which implies that the aggregation number is increased gradually as the concentration increases. An example of the ITC charts (enthalpograms) of the dye Rhodamine 6G injected in two different concentrations is shown in Figure 13.2. No initial plateau exists in the first injections that would indicate the complete dissociation of the aggregates into monomers, as observed in the surfactant experiments. The heat developed depends on the syringe concentration, which is due to the gradual change in the aggregation number of the dye with increasing concentration. The enthalpogram shows that, as the concentration in the cell increases, the heat developed has an exponential decay, due to the fact that the solution in the cell becomes more similar to the one in the syringe and less aggregates dissociate.

4.1. Nonaqueous Systems

The formation of micelles by surfactants in nonaqueous or nonpolar solvents has been a matter of dispute. The formation of these reversed micelles, in which the hydrophilic part is situated in the center of the micelle while hydrocarbon chains extend into the solvent phase, has been proven to be very much dependent on the presence of water in the solvent, even at a trace level.[11] Hence, the envisioned picture that emerges shows a micelle that has a core of water surrounded by the surfactant molecules. This core of water would consist of molecules, not droplets like in microemulsions. Ferrari[12] showed a dependence of the reaction energetics going from exothermal to endothermal with the concentration of water for AOT (sodium bis(2-ethylhexyl) sulfosuccinate) in various solvents. In the authors' laboratory, drying with both metallic sodium and molecular sieves, values down to 10 ppm of water in toluene have been obtained after drying from approximately 200 to 300 ppm of water in commercial toluene. Therefore, total nonaqueous media could not be produced and water is present in all cases. This hinders greatly the discussion about the role of water in micelle formation.

5. ITC Experiments with Asphaltene Solutions: Is There a CMC?

The existence of a critical micelle concentration of asphaltenes in organic solvents and, hence, also inferred to exist in neat crude oil, has been the subject of numerous works in the literature. Rogacheva *et al.*[13] and later Sheu *et al.*[14] followed by others used surface tension to point at possible values of CMC, as is commonly done with surfactants. However, this technique only reports indirectly what is going on in the bulk phase, as the measurement is directed towards the interface between air and the solution. It is indeed correct that, for many asphaltene samples, a drop in surface tension is observed with increasing concentration, followed by a constant magnitude. This could be seen as a clear indication of CMC, if there was an analogy between the behavior of asphaltenes and surfactants. On the other hand, given the very low aggregation number of asphaltenes as given both in the original Yen model in the range of 4–5 units (and confirmed recently using other techniques by Yarranton *et al.*[15]) the observed behavior could just be a result of surface saturation or a higher affinity towards the asphaltenic solution than towards the interface. Other techniques have also been used[16] and many reach by relating observations to the existence of CMC magnitudes in the range of 1–5 g/L, if toluene is used as solvent.[17, 18]

Given the universal type of measurement that ITC is able to give, the existence of CMC for asphaltenes in toluene was examined. It was observed that the heat developed was dependent on the concentration of asphaltenes in the syringe (Figure 13.3). This indicates a different aggregation state of the asphaltenes, even if in all cases the concentration is above the estimated CMC. Besides, it was not possible to find an initial plateau or a break point in any experiment, as observed in experiments with surfactants. In Figure 13.4, two analysis procedures to determine CMC are compared. In Figure 13.4B, the use of cumulated heat is presented, as applied by Andersen and Birdi.[19] As can be seen, this approach can lead to bias as the researcher may be led to fitting linear curves at low and high concentration for the asphaltenes and obtain a CMC in the intersection of the lines although the curve displays a smooth gradual change. Hence, it is obvious that the heat/injection versus concentration plot is much less prone to errors of this kind and it is considered as the preferable approach.

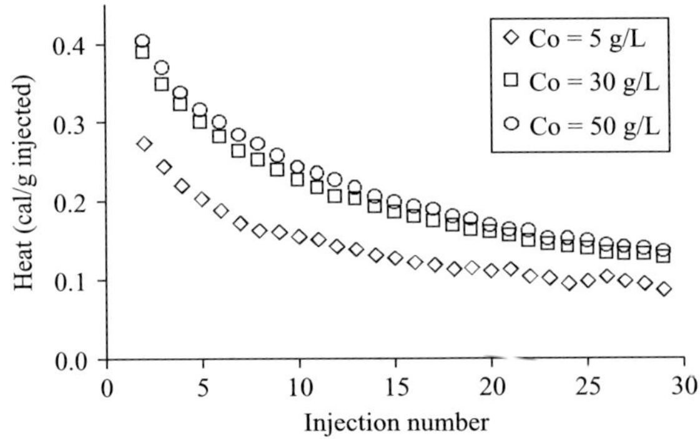

Figure 13.3. Influence of initial concentration of asphaltenes in injected solution. Sample OMV asphaltenes.

Figure 13.5A shows a comparison of the heat developed between tests with and without asphaltenes. The heat obtained in the reference test is subtracted prior to any analysis of the data. Figure 13.5B shows the titration of 8 different asphaltenes injected at a concentration of 5 g/L. The lowest concentration reached in all experiments is 0.07 g/L. None of them presents a plateau at low concentrations that would indicate the presence of a CMC. Therefore at no point do the dilution process go from associated to monomeric state and the heat measured will reflect the process of going from one association state to another equilibrium aggregation state.

The chart looks more like the one of Rodhamine 6G, leading to the conclusion that the association of asphaltene also occurs stepwise as was also pointed out by Acevedo *et al.*,[20] by measurements of thermal diffusivities. One could always argue that a multi-component mixture like asphaltenes could exhibit a multitude of CMCs leading to a smoothening of the curve such that no CMC could be observed using our technique.

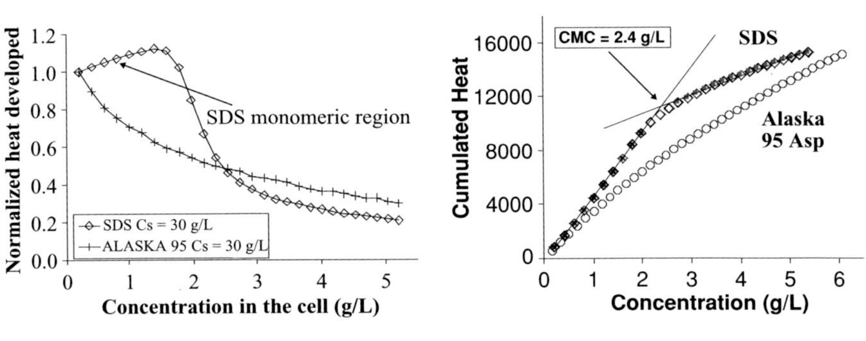

Figure 13.4. Analysis of calorimetric data for presence of CMC. (A) Enthalpograms of SDS and asphaltene Alaska 95, normalized heat. (B) Cumulated heat analysis of same data. Heat of asphaltenes multiplied by 6 for comparison.

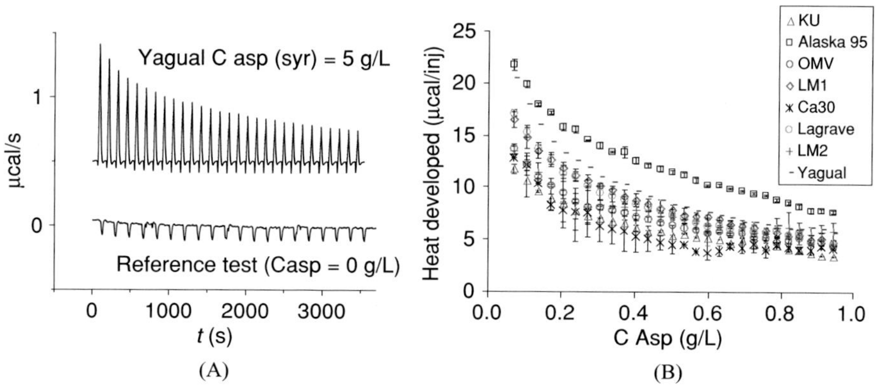

Figure 13.5. (A) Reference data and one example of raw data of an asphaltenes test. (B) ITC experiments of 8 asphaltenes. Injection concentration 5 g/L into dried toluene at 30°C.

That hypothesis cannot be validated and it is actually of little use when developing models to describe the behavior of asphaltenes. The micellization approach can be written into equations when the CMC is unique, not when it tends to infinite, and in any case the modeling would have to approach a stepwise mechanism.

The conclusion, based on analysis of ITC experiments and analogies to aqueous solutions, is that the concept of CMC is not applicable to petroleum asphaltene in solution because:

1. Asphaltenes present a similar behavior to dyes, that is to say, they seem to associate stepwise.
2. The aggregation number is too small to consider asphaltene aggregates as typical micelles.

However, the authors acknowledge the evidence of aggregation and association of asphaltenes in solution, which is best illustrated by the quenching and red shift of the fluorescence of asphaltene solutions seen in Figure 13.6. In the fluorescence spectroscopic experiment, one may look at the fluorescence intensity as a function of concentration and wavelength (often in the 5–100 ppm region). A decrease in intensity (quenching) as a function of concentration indicate strong molecular interaction.[21] The example in Figure 13.6 was observed for four investigated asphaltenes all having significant quenching starting between 10 and 50 ppm solutions in agreement with recent results by Goncalves *et al.*[22] Similar results were reported by Andersen[23] who also reported that in 5 ppm solutions the fluorescence signal was affected by composition of the solvent changing this between 100% toluene and 10% toluene-in-heptane. The latter indicates that at this very dilute conditions the solvent still affects the solute–solute interactions.

We do accept that a large quantity of data do indicate that physical properties of solutions indeed change in the region of 0.5–5 g/L in toluene, but that

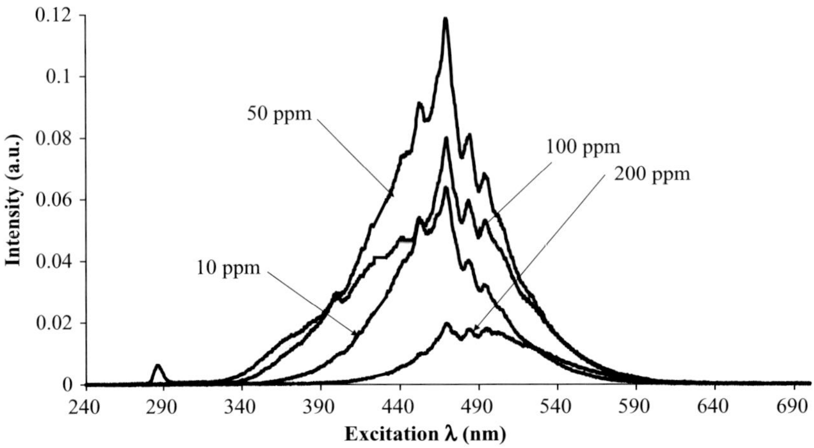

Figure 13.6. Synchronous fluorescence ($\Delta\lambda = 20$ nm) of Yagual asphaltenes in toluene conc. from 10 to 200 ppm.

it is different from the CMC mechanism and more likely associated to a limiting effect on the extent of aggregation. The stepwise mechanism indicates that at a given concentration aggregates will exist but the ratio of monomers to aggregated species will change. Hence, one cannot apply a concept of a critical concentration above which the monomer concentration is constant. However, we may in dilution experiments reach extreme dilute conditions where the number and size of the aggregates is so small that experimental data will reflect the "monomeric" state. The fluorescence quenching given above may reflect an increased aggregate–aggregate interaction or the consolidation of the aggregated state at elevated concentration.

6. Modeling ITC Experiments

In order to derive meaningful information from the heat-traces, apart from that of direct qualitative nature, it is necessary to apply a model. Since no CMC can be deduced, the stepwise mechanism is the preferred approach. The model developed herein is based on the chemical theory. It assumes that the formation of dimers, trimers and so on is represented by an equilibrium and it considers that all the heat developed is due to the modification of the equilibria upon the addition of more asphaltenes after each injection. This type of model can be applied to both the asphaltene self-association as well as to the interaction with resins or other components, such as inhibitors. In this last case, ITC could give insight into the efficiency of these substances by evaluation of binding energies. In the present work, nonylphenol has been used as a model compound both for resins and for typical inhibitor chemistry, as shown in the next section. The goal of the modeling is not only to correlate the data but also to derive thermodynamic variables such as the enthalpy of association, which can be used in comparison of material from different sources, as well as in the prediction of phase behavior of asphaltenes in order to solve industrial problems. This approach hence adapts directly what in biochemistry is known as biocalorimetry where, i.e., protein–ligand interaction is investigated.[24]

The basis of the model is the equilibrium between interacting species and the aggregates formed, following the same approach as usually applied to polymer growth. It is however important to keep in mind that this fact does not imply that the growth of asphaltenes is assumed to be linear:

$$A_n + A_1 \leftrightarrow A_{n+1} \Rightarrow [A_{n+1}] = K_{n+1}[A_n][A_1].$$

In this case, all external influences such as friction losses, etc. should of course be removed before the analysis. In order to do so, a reference experiment is carried out, in which the syringe is filled with pure solvent and the heat developed is subtracted from all experiments that involve asphaltene solutions.

For the sake of simplicity, it is assumed that the enthalpy of association of all molecules is the same (an average for the multitude of molecular species in the involved). The equilibrium constants allow the calculation of the concentration of monomers, dimers and so on, and the number of bonds dissociated is calculated from the differences in concentration between the moments before and after each injection. The experimental heat developed is then fitted with the following equation:

$$\text{Heat developed} = \text{Number of association sites broken (mol)} \times (-\Delta H).$$

The fitting parameters are the equilibrium constant K_i and the average enthalpy of association ΔH.

Different model approaches have been investigated, depending on the degree of complexity required.[25] In asphaltene self-association studies, four assumptions were applied:

1. DIMER model: It is assumed that only dimers are formed. Therefore, only one equilibrium constant and one enthalpy ΔH are used as fitting parameters.
2. EQUAL K model: More species can react to form larger aggregates. In order to keep the same number of parameters, it is assumed that all equilibrium constants are equal.
3. ATTENUATED K model: To simulate the steric effect, the equilibrium constants of the formation of species with grater aggregation numbers are reduced, following the simple relationship: $K = K_{2/2} = \cdots = K_{i/i}$. The number of fitting parameters is again only two.
4. TERMINATOR model: The fraction is divided into two types of molecules, those that allow the continuation of the growth of the aggregate, as they contain more than one association site (Propagators) and those that act as Terminators, limiting the size of the final aggregate. A third parameter is added, which is the ratio of terminators to propagators (T/P) is the asphaltene fraction.

In the above models, the enthalpy of association ΔH is considered to be the same in all reactions, since the objective is to keep the number of fitting parameters as low as possible.

The modeling of nine individual asphaltenes showed the same tendency. Dimer models had a very good fit, while ATTENUATED K had the best fit of the more complex models. The calculated values of ΔH are small (in the range of -2 to -7 kJ/mol), a bit smaller than the usual hydrogen bond magnitudes (-8 to -40 kJ/mol)[26] and also smaller than the stacking of some pure aromatic compounds, such as pyrene (-15 kJ/mol).[27] It is not possible to determine which is the main driving force for association based on these experiments, as both mechanisms suggested (stacking and hydrogen bonding) present values of the same order of magnitude and would compete with each other.[28] These values, however, should be considered with caution and only as a qualitative indication of the range of average enthalpies, since the ΔH depends strongly on the assumed MW of asphaltenes. In the present case, $MW = 1000$ g/mol was arbitrarily chosen for all asphaltene molecules. An increase in this magnitude resulted in a linear increase in ΔH_a. Moreover, taking the heat of interaction between phenol molecules as a reference value ($\Delta H_{\text{formation}} = -16.6$ kJ/mol), the heat developed in these experiments is rather low, suggesting that a fraction of asphaltenes do not participate in these tests. This is to be expected considering the nonspecific forces involved in the precipitation of asphaltenes by n-alkane solvents. SEC analysis of heptane–toluene fractions of asphaltenes also indicated that a substantial fraction could be inactive in self-association. They would precipitate after the addition of n-heptane because of the size difference, not because of aggregation issues.[18] The underestimation of the enthalpy of self-association may as well be related to all the processes involved in the association that are not taken into account explicitly, such as de-solvation of aggregates, solvation of monomers, tangling of branches, and conformational changes.

However, at the current stage no model has been developed that can take polydispersity in MW and structure into account. Aggregation numbers derived in this work indicated that the concentration of aggregates of $n > 5$ was negligible.

7. Application of ITC to Various Aspects of Asphaltene Association and Interaction with Other Substances

In the previous section, ITC was used to investigate asphaltene self-association in organic solvents. In this section, the versatility of the technique is demonstrated on four different issues: (1) the investigation of subfractions of asphaltenes; (2) the effect of blocking the hydrogen bonding functionalities in asphaltenes by methylation; (3) the interaction with a model resin, namely nonylphenol; and finally (4) the interaction with native resins. In order to make a somewhat uniform investigation, only results from toluene solutions are reported, although some investigations in other solvents have been performed as part of our research.[29] The issue of the effect of water on the association will be discussed as part of the above investigations.

Similar to ordinary surfactants in apolar/nonaqueous solutions experimental evidence exist that indicate that asphaltenes association is as well dependent on the presence of water.[30] This effect has been observed in different degrees by different techniques. Even if it is still a matter of research, the adsorption of asphaltene molecules to the oil–water interface and the following oil in water emulsion stability is obviously a direct evidence of the presence of hydrophilic interactions. Solidification of asphaltenes occurs in a more ordered structure when water is present.[31] Therefore, it was decided to work with controlled humidity and drying of solvents to the possible extent, as explained above. This is done in order to avoid any doubts related to the presence of significant amounts of water such as the usual 100–200 ppm of water in aromatic solvents. In the application to petroleum reservoirs, water is abundant and must be included in future considerations in connection with model development.

7.1. Investigation of Asphaltene Subfractions

Many aspects of asphaltene chemistry in the literature have focused on the standard asphaltenes, precipitated by either *n*-heptane or *n*-pentane. However, in reality, problems occurring during handling of petroleum are often caused by a minor fraction of these asphaltenes. Therefore, it is of great importance to understand the properties of different asphaltene subfractions. This is a very important topic, and the number of reported investigations looking at fractions and properties of fractions is indeed fortunately increasing. The most popular fractionation procedure has been the use of different ratios of the two defining solvents, heptane and toluene.[18, 32, 33] This procedure, however, leads to very small apparent differences in chemical composition of the material although indications from size exclusion chromatography did indicate a difference in association affinity. The insoluble fraction showed consistently a larger degree of association.[18] Stronger and polar solvents, such as acetone, have also been applied to obtain greater differences among the fractions.[33–35] In particular, this procedure is expected to lead to a more polar soluble fraction where hydrogen bonding could dominate relative to the insoluble fraction. For instance, Takanohashi *et al.*[36] showed that coal asphaltenes fractionated using acetone and pyridine resulted in a soluble fraction in acetone that was the least aromatic.

In the present work, *n*-heptane asphaltenes (using our standard method described above) were subfractionated in mixtures of acetone and toluene at room temperature. The asphaltenes were placed in the extracting fluid and ultrasonicated for 60 min, left overnight and centrifuged to obtain an insoluble fraction (INS) and a soluble (SOL) fraction. Two asphaltenes were analyzed, called LM1 and KU. In order to give good yields, the solvent mixture was selected to give a 1:1 fractionation based on weight, meaning that the solvent ratios in terms of toluene to acetone were 30:70 and 60:40 (volume) for LM1 and KU, respectively. It was observed that around 10% of the INS fraction of KU was not totally soluble in toluene. A fact that has been observed in previous experiments on asphaltene extraction.[35, 37] It is expected that, after fractionation, some asphaltenes are not soluble in toluene, as we remove some of the molecules that may act as co-solvents. If it is assumed that no chemical alteration takes place during the fractionation, this is a proof of the delicate balance between the petroleum constituents where the final stability depends on the presence of other molecules and specific interactions. Andersen and Speight[37] found that a fraction of asphaltenes turned into insoluble in toluene after extensive washing in 40% toluene/heptane, and this change in solubility was followed by a remarkable change in the hydrogen bonding region of the infrared spectrum, indicating that hydrogen bonding leads to stability through interaction with lighter compounds.

For the two different asphaltenes investigated the heat developed from solubles is much higher than for insolubles. In order to investigate the possibility of chemical alteration upon interaction with acetone, the fractions were mixed in the same proportion as in the original asphaltenes and the heat of dissociation of the mixture (SOL + INS) was found to match the original asphaltenes.[29] This indicates that no chemical alteration took place during the fractionation. For KU, SOL developed again more heat, and, in this case, it was also demonstrated that water has a pronounced effect on the heat traces.

In Figure 13.7, the titration is performed in dried toluene and in a mixture having approximately 300 ppm water. The presence of water apparently leads to an exothermic reaction lowering the otherwise dominating endothermic reaction of dissociation. Injection of toluene into solutions of 10 g/L of KU INS and SOL showed that only SOL gave endothermic peaks which could be related to dissociation. INS only showed peaks that could be related to the friction heat while the toluene was injected into the cell.

There is a number of possibilities to explain the low heat of dissociation of INS fractions: (1) INS represents molecules of low degree of association or (2) the interactions are so strong that dilution does not lead to dissociation. Investigation of the fractions using infrared and fluorescence spectroscopy did indicate that the hydrogen bonding capacity (derived from IR) of INS and SOL fractions of LM1 was similar, whereas the fluorescence showed relatively less intensity for INS. Although the techniques applied showed a difference in the association of subfractions of asphaltenes, the subject needs indeed further investigation. These evidences become even more important if it is hypothesized that the insoluble

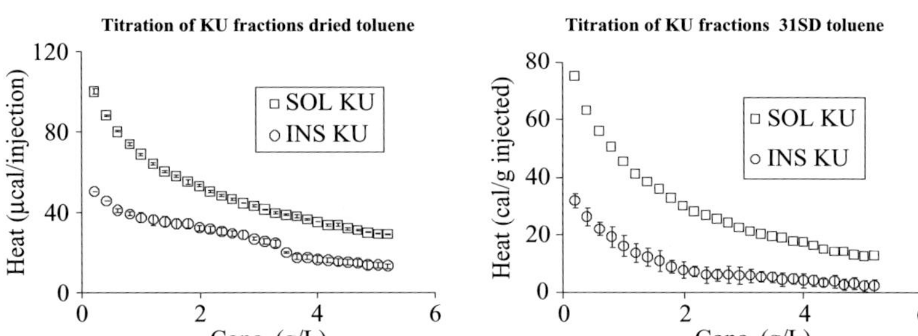

Figure 13.7. Investigations of subfractions of heptane asphaltene KU; 30 g/L injected at 30°C. (A) Dried toluene. (B) Toluene with water content of ca. 300 ppm.

fractions are more in the same family of molecules precipitating or depositing during the processing of petroleum.

7.2. Effect of Methylation of Asphaltenes

Understanding the molecular structure and concentrations of different oxygen, nitrogen, and sulfur species is especially important in asphaltenes, as these have been implicated in oxidation, degradation, and molecular associations.[38, 39] They affect properties of asphaltenes by participating in intermolecular associations through hydrogen bonding and formation of secondary structures.[40–44]

Since the ITC analysis of nonmodified asphaltenes (hereafter called "raw asphaltenes") does not allow discerning between types of interaction, it was decided to alter the asphaltenes. By comparing the ITC results of modified and raw asphaltenes, it would be possible to determine the relative importance of hydrogen bonding and stacking of aromatic regions of the molecular structures. There are some examples of asphaltene alteration in the open literature: Gould *et al.*[45] showed that removal of hydrogen bonding sites lead to a decrease in viscosity of solutions of asphaltenes. The use of methylation by phase transfer catalysis combined with spectroscopy has been used for quantitative determination of acidic functions in petroleum materials.[46, 47] Hence, the combination with ITC may enhance the understanding of the balance of interactions by simply removing quantitatively these effects.

Herein, the effect of methylation in asphaltene self-association is described. In this reaction, acidic hydrogen, present in functionalities such as –OH, –COOH, –SH, and –NH, is substituted with a methyl group without affecting the hydrocarbon structure. Hence, one could assume that aromatic stacking would not be affected. The procedure by Desando and Ripmeester[48] was followed: the material is deprotonated with an organic base (tetra-*n*-butylammoniumhydroxide) and the resulting anion reacted with and alkyl halide (methyl iodide). For LM2 asphaltenes, the reaction led to an almost total removal of the infrared broad band between 3,600 and 3,100 cm^{-1}, indicating a quantitative removal of acidic hydrogen (Figure 13.8). Peaks at 3,050 cm^{-1} are related to the aromatic structure.

Elemental analysis of this asphaltene indicated an increase in H/C from 1.15 to 1.23, as expected from the introduction of the CH_3 groups. By assuming a molecular weight of 1,000 units, easy algebra leads to the number of sites that have been affected by the reaction. This calculation gives 7.9 sites affected by methylation in LM2 asphaltenes. ITC experiments show that the methylation of asphaltenes leads to a significant decrease in the heat developed (Figure 13.9). The blockage of potential hydrogen bonding sites decreases very significantly the capacity of self-association of asphaltenes.

Table 13.1 collects the results obtained with five heptane asphaltenes, in terms of variation in the heat developed in ITC experiments, as well as the variation of the hydrogen bonding index in IR spectroscopy. This index gives an idea of the hydrogen bonding capacity and is defined as:

$$I\,(\mathrm{HB}) = \frac{\mathrm{Absorbance}\ \left(3500 - 3100\ \mathrm{cm}^{-1}\right)}{\mathrm{Absorbance}\ \left(3500 - 2740\ \mathrm{cm}^{-1}\right)}$$

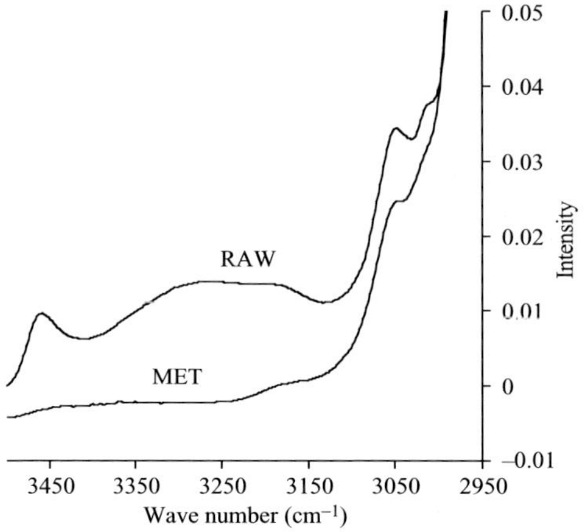

Figure 13.8. FTir of hydrogen bonding region results for LM2 raw heptane asphaltene, methylated asphaltenes.

The relative small variation in Yagual is due to low degree of hydrogen bonding in the original asphaltene. The lack of direct proportionality between the two methods might be due to the fact that the infrared spectral index also includes the area where the introduced methyl group is observed in the strong stretching

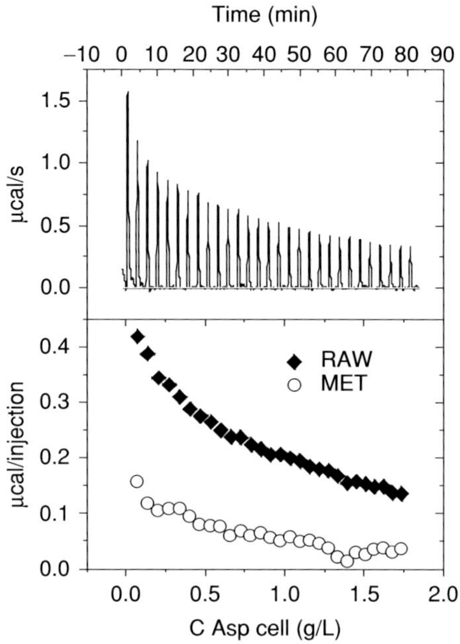

Figure 13.9. ITC titration results for LM2 asphaltenes raw and MET-hylated at 30°C and 30 g/L in dried toluene.

Table 13.1. Effect of Methylation on Heat Developed During ITC and Infrared Spectroscopic Analysis in Toluene Solutions

	Percent variation caused by Methylation	
	In ITC experiments	In IR index HB
Alaska 95	−52	−74
Yagual	−27	−9
Ca30	−27	−43
LM2	−62	−75
Lagrave	−26	−62

vibrations. This could indicate that not only the number of sites but also the position plays a role.

In Figure 13.10, the effect of methylation on the fluorescence spectra is given. Methylation of Alaska 95 asphaltenes led to a significant shift of the band to shorter wavelengths (blue shift), while only a small change is observed for Yagual. Emission at shorter λ would in principle imply the presence of smaller aromatic rings,[49] but the reaction does not alter the core of asphaltene molecules. If some molecules emit at short wavelengths in the methylated samples it is because they did as well in the raw asphaltenes. These small molecules that emit at short λ would be bonded through hydrogen bonding in the original raw asphaltenes, as association is believed to move the bands to longer wavelengths (red shift).[50]

In the spectrum of raw asphaltenes, the emission of these species is reabsorbed by the neighbor molecules due to strong interaction within the material (even at concentrations of 2 ppm in toluene). Yagual has a lower H-bonding capacity and the blue shift is not seen. This also indicates that the asphaltenes from this particular crude may have a different structure in terms of aromatic condensation. Interestingly, Yagual asphaltenes come from a very instable crude oil. This leads to a supporting evidence that the hydrogen bonding capacity of asphaltenes plays a significant role in the *stabilizing* mechanisms of asphaltenes—such that a large hydrogen bonding capacity gives rise to more stable asphaltenes as the interaction with other petroleum constituents (resins and polars) counteracts the asphaltene–asphaltene interactions.

Therefore, one may assume that although asphaltene may self-associate through H-bonding, this is also the only favorable way that these molecules can interact with other smaller species. In this, it is assumed that the aromatic–aromatic core interaction often mentioned as a driving force for asphaltene–asphaltene interactions may not be very pronounced between resin-like (small polar aromatics) and asphaltenes.

7.3. Interaction of Asphaltene with Other Compounds

The study of asphaltenes in toluene solutions is only a step in the understanding of these species and may look very remote from the real situation in the

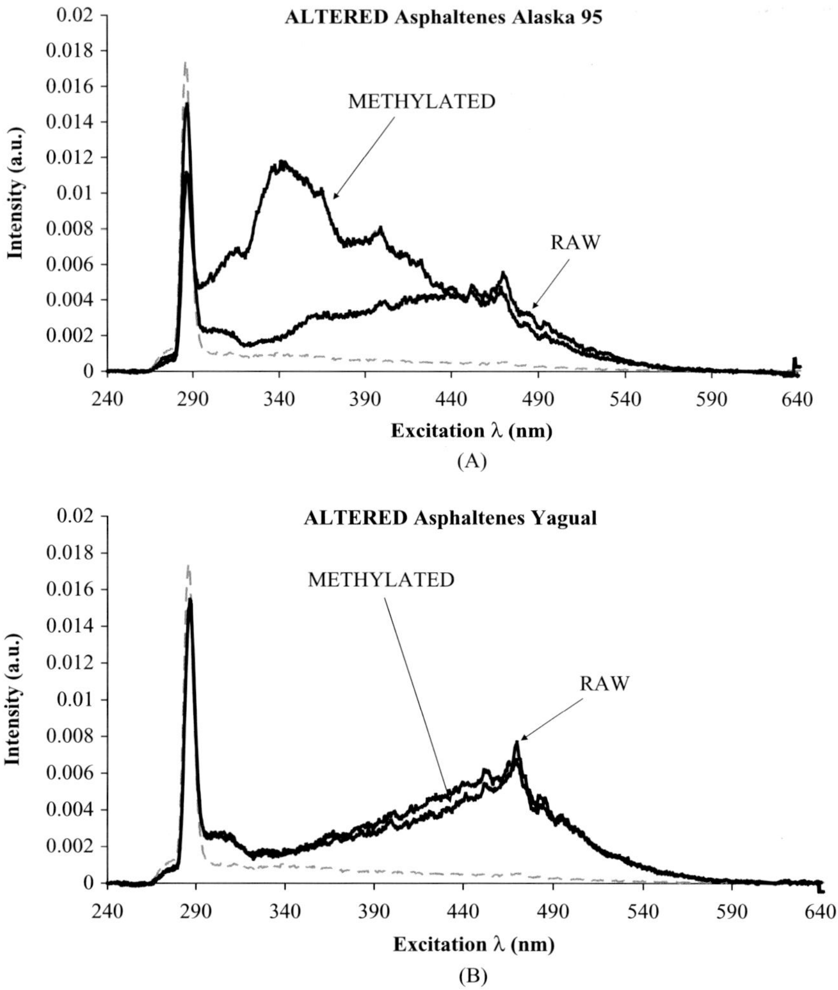

Figure 13.10. Synchronous fluorescence ($\Delta\lambda = 20$ nm) of raw and methylated asphaltenes from two sources (2 ppm toluene solutions).

crude oil. Even though this may give rise to an important insight in the mechanisms of molecular association, it has long been known that stability issues include the entire oil and the interactions among all components in it. In order to take ITC investigations a step forward towards the real problem, the interaction between asphaltene and other components was investigated.

As a first approach, the interaction with a model compound was studied.[51] Nonylphenol was chosen for several reasons: it is a well-known amphiphile, it has been successfully applied as an inhibitor of asphaltene aggregation and it has mechanisms of association similar to the ones resent in oil resins, namely hydrogen bonding and aromatic π–π interactions.

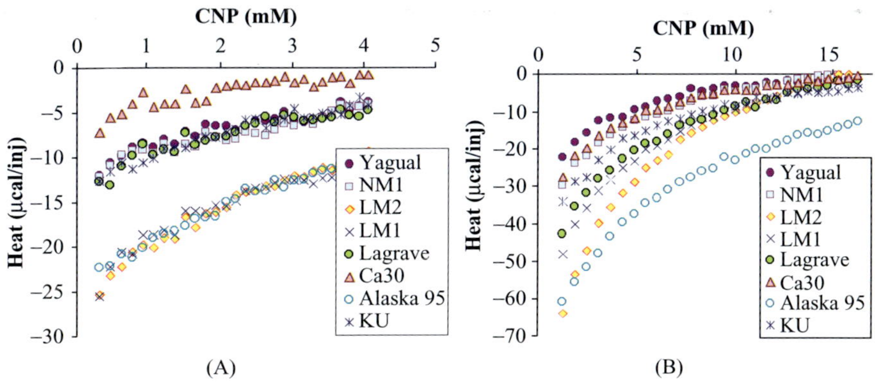

Figure 13.11. Heat developed in the injection of nonylphenol solutions into 1 g/L ASP: (A) 5 g/L of NP; (B) 20 g/L of NP.

Experiments consisted of the injections of nonylphenol into an asphaltene solution in toluene. Nonylphenol was injected at several concentrations in toluene, ranging from 5 to 100 g/L. The heat developed in this kind of experiment would contain several contributions, including the dissociation of nonylphenol aggregates, the interaction of nonylphenol with asphaltenes and also friction losses. In order to individuate the heat developed in the interaction between nonylphenol and asphaltenes, a reference test is performed. In it, a solution of the same nonylphenol concentration is injected into pure solvent, and the heat developed in this test is subtracted from the data obtained when asphaltenes are present. It is assumed that the remaining heat is only due to the interaction between the compounds of interest.

Figure 13.11 shows that *stable* asphaltenes such as Alaska 95, LM1, and LM2 have more heat developed in the interaction with nonylphenol than the *instables* (Lagrave, Ca30, NM1, and Yagual). KU is a stable crude but does not interact with NP as much as the other stable asphaltenes. Alaska 95 is the asphaltene with greater capacity of interaction with NP. This suggests that there may be a relationship between the capacity of interaction with nonylphenol and the stability of the asphaltenes in the crude. Since the main mechanism of interaction is hydrogen bonding, this would imply that asphaltenes with a high hydrogen bonding capacity would become more stable in the crude, by interaction with the surrounding maltenes. This is in agreement with the evidences presented in the section about chemical alteration of asphaltenes.

Experiments with higher nonylphenol concentrations, in the range of 100 g/L, allow the study of the saturation of asphaltene sites. As shown in Figure 13.12, the heat developed in the interaction reaches zero at a certain nonylphenol concentration, indicating that the newly added nonylphenol does not interact any longer with asphaltenes. The concentration of NP (C^*) at which the sites become saturated has been calculated by drawing the trend line in the linear region of the curve. If it is assumed that all nonylphenol molecules in the cell would

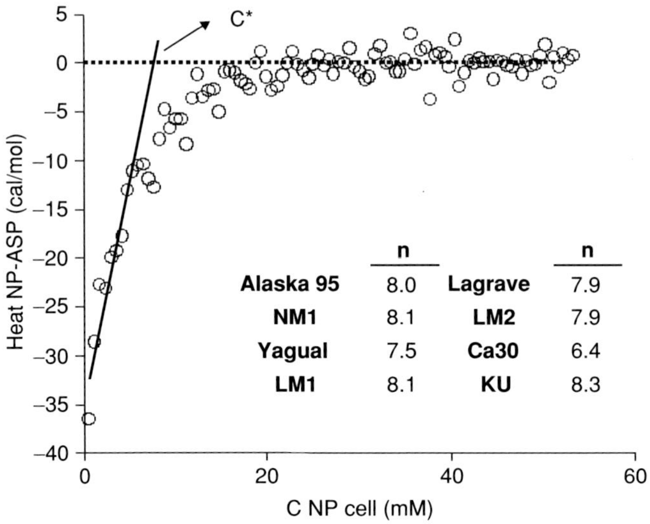

Figure 13.12. Calculation procedure for maximum number of sites per molecule in LM2 asphaltenes. Table gathers experiments with asphaltenes from other sources.

interact with asphaltenes, C* gives the number of sites n available for interaction. n varies from 6 to 8 depending on the asphaltene. In reality, not all nonylphenol molecules are attached to asphaltenes, so these values can be considered an upper limit. It is, however, interesting to compare this result with the number of sites affected by methylation, in the case of LM2 asphaltenes (see above). The agreement is very good, indicating that Nonylphenol attaches to asphaltenes by means of hydrogen bonds: the same bonds that are blocked upon methylation. The magnitude found for all asphaltenes investigated was surprisingly high considering the normal depicted average molecular structure from which probably not more than 4–5 sites could be expected.

In experiments with native resins, the same methodology was followed. Solutions of high resin concentration were injected into asphaltene solutions and the heat developed was measured.[52] Reference experiments were carried out and the heat developed was subtracted to determine the heat evolved in the interaction between resins and asphaltenes.

Asphaltenes are believed to associate stepwise, and in the previous sections the self-association of asphaltenes has been successfully modeled with polymerization-type reactions:

$$A_n + A_1 \leftrightarrow A_{n+1} \Rightarrow [A_{n+1}] = K_{n+1}[A_1][A_n] \tag{13.1}$$

$$A_n + R \leftrightarrow A_n R \Rightarrow [A_n R] = K_{Rn}[R][R_n] \tag{13.2}$$

$$K = K_2 = K_3 = \cdots = K_{n+1} = K_{R1} = K_{R2} = \cdots = K_{Rn} \tag{13.3}$$

$$\Delta H_a = \Delta H_{a2} = \Delta H_{a3} = \cdots = \Delta H_{an+1} \tag{13.4}$$

$$\Delta H_i = \Delta H_{i1} = \Delta H_{i2} = \cdots = \Delta H_{in} \tag{13.5}$$

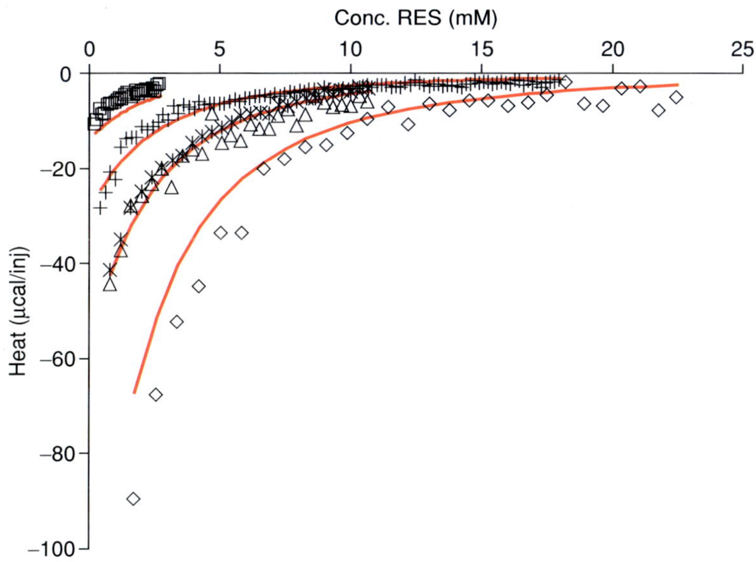

Figure 13.13. Predicted fits by TERM model of Alaska 95 experiments with 1 g/L ASP, using the average ΔH and K. (−) Fit of model; Conc. RES = 9 g/L (□); 35.8 g/L (∗); 35.9 g/L (Δ); 60 g/L (+); 75.3 g/L (◇).

To simplify the approach, it is considered that the equilibrium constants are the same as those of the propagation reactions, but the resin–asphaltene interaction is modeled with a different value of ΔH. Asphaltene constants have been taken from the fit of asphaltene self-association experiments.

Figure 13.13 shows that average values of ΔH and K are able to fit successfully in all experiments of Alaska 95 asphaltenes of the same asphaltene concentration. In spite of all the assumptions made, it is possible to model all resin concentrations with one set of parameters ($\Delta H_{A-R} = -3.2$ kJ/mol and $K = 377$ l/mol). The enthalpies are in the same range as the ones reported for asphaltene interaction with nonylphenol. They are one order of magnitude lower than the typical hydrogen bonding (−10, −40 kJ/mol) and in the lower limit of permanent dipole interactions (−4, −20 kJ/mol). The values reported here are in agreement with modeling data obtained by Buenrostro-Gonzalez *et al.*[53] They applied SAFT-VR (variable range) equation to model the precipitation of Maya asphaltenes, obtaining an enthalpy of interaction of −3.3 kJ/mol, as a fitted parameter. Even if the energies measured by ITC may seem small, it must be taken into account that they account not only for association but also the energies developed in the conformational changes of the molecules to accommodate for binding. Solvation effects have as well been disregarded. It is not clear, either, that all molecules in both fractions will be equally active in the interaction. It is practically impossible to develop a model that accounts for all these effects in a system of such a complexity as asphaltene and resin fractions. For the sake of simplicity, the heat developed is assigned to association, but it is necessary to keep in mind that the binding energies may be underestimated. Nevertheless, these experiments can provide data to state-of-art models, which do not consider the polydispersed nature of asphaltenes.

8. Conclusions

In the present chapter, we have shown that isothermal titration calorimetry can successfully be applied to the investigation of association of asphaltenes to get further insight into the various bonding mechanism involved in both stability and instability of asphaltenes. Using different approaches it appears that the hydrogen bond is important in the stability of asphaltenes: apparently asphaltenes derived from stable oils have a higher degree of hydrogen bonding capability. Besides, blocking of sites using methylation did in some cases lower the solubility of the asphaltenes in toluene. Hence, it is envisioned that hydrogen bonding sites are used to avoid or diminish asphaltene–asphaltene association by interaction with smaller molecules.

One of the very important conclusions of this work is that no single specific critical micelle concentration has been detected and hence this concept will not be applicable in asphaltene science. Instead, a stepwise aggregation mechanism is proposed to account for the well-known association of asphaltenes into nanostructures. Data presented in the literature for dilute asphaltene solutions can be approached along the same lines. However, the analogy often made with aqueous solutions in the interpretation of data to find CMCs cannot be recommended. The region above say 5–10 g/L is probably dominated by a limiting growth effect in which interparticle repulsion is dominating more than further growth of particles. The findings indicate that no single CMC was detectable down to approximately 50 ppm of asphaltenes in toluene. What happens at even lower concentrations are still open for discussion.

Furthermore, the ITC technique has proven to be a powerful tool to investigate interactions between additives and asphaltenes, and may as such be developed for screening of asphaltene inhibitors. For resins, it was shown how trends could be modeled based on average interaction parameters and chemical-theory-based models. In terms of this, the technique is also capable of defining magnitude ranges for parameters in association-based models and hence helps in bringing these closer to a predictive state.

Given the ease of performing this type of experiment after proper training and development, the technique can be applied in deriving standard input values for modeling or for screening of interacting components.

Acknowledgments

The authors thank the Danish Technical Science Council (STVF) for financial support under the talent project. The skilful chemical alteration of asphaltenes by Dr. Priyanka Juyal is highly appreciated. The KU asphaltene sample was kindly supplied by Dr. J.M. del Rio, IMP, Mexico.

References

[1] Pérez-Rodríquez, M., G. Prieto, C. Rega, L.M. Varela, F. Sarmineto, and V. Mosquera (1998). *Langmuir* **14**, 4422.
[2] Birdi, K.S. (1988). *J. Chromatrogr. Library* **40**, 399.
[3] Majhi, P.R. and S.P. Moulik (1998). *Langmuir* **14**, 3986.
[4] Goodwin, J. (2004). *Colloid and Interfaces with Surfactants and Polymers*—An Introduction, Wiley, New York.
[5] Eicke, H.F. (1980). *Topics in Current Chemistry* Vol. 87. Springer-Verlag, New York, p. 85.
[6] IP 143/90. (1985). *Standards for Petroleum and its Products*. Institute of Petroleum, London.
[7] Birdi, K.S. (1985). In: D.O. Shah (ed.), *Macro and Microemulsions—Theory and Applications*, ACS Symp. Series 272. American Chemical Society Washington, DC, pp. 67–74.
[8] Paula, S., W. Sus, and A.J. Blume (1995). *J. Phys. Chem.* **99**, 11742.
[9] Andersen, S.I. and S.D. Christensen (2000). *Energy Fuels* **14**, 38.
[10] Sharma, V.K., R. Bhat, and J.C. Ahluwalia (1987). *J. Coll. Interface Sci.* **115**(2), 396.
[11] Bey Temsamani, M., M. Maeck, and I. ElHassani, H.D. Hurwitz (1998). *J. Phys. Chem. B* **102**, 3335.
[12] Ferrari, M. (2002). MSc Thesis, Uni. Degli Studi di Trieste, Facolta di Ing., Italy.
[13] Rogacheva, O.V., R.N. Rimaev, V.Z. Gubaidullin, and D.K. Khakimov (1980). *Colloid J. USSR* (Translated version) **42**(3), 586.
[14] Sheu, E.Y., M.M. DeTar, D.A. Storm, and S.J. DeCanio (1992). *Fuel* **71**, 29.
[15] Yarranton, H.W., H. Alboudwarej, and R. Jakher (2000). *Ind. Eng. Chem. Res.* **39**, 2916.
[16] Deo, M.D. (2002). In: C. Lira-Galeana (ed.), 2002. *International Conference on Heavy Organic Deposition*, Mexico.
[17] Loh, W., R.S. Mohamand, and A.C.S. Ramos (1999). *Pet. Sci. Technol.* **17**, 147.
[18] Andersen, S.I. (1994). *J. Liquid Chrom.* **17**, 4065.
[19] Andersen, S.I. and K.S. Birdi (1991). *J. Coll. Interface Sci.* **142**, 497.
[20] Acevedo, S., M.A. Ranaudo, J.C. Pereira, J. Castillo, A. Fernández, P. Pérez *et al.* (1999). *Fuel* **78**, 997.
[21] Groenzin, H. and O.C. Mullins (1999). *J. Phys. Chem.* **103**, 11237.
[22] Goncalves, S., J. Castillo, and A. Fernández, and J. Hung (2004). *Fuel* **83**, 1823.
[23] Andersen, S.I. (1900). PhD Thesis, Technical University of Denmark, Dept. Phys. Chem. (1900).
[24] Bladamer, M.J. (1998). In: J.E. Ladbury, and B.Z. Chowdhry (eds.), *Biocalorimetry Application of Calorimetry in the Biological Sciences.* Wiley, New York.
[25] Merino-Garcia, D., J. Murgich, and S.I. Andersen (2004). *Petroleum Sci. Technol.* **22**(7/8), 735.
[26] Prausnitz, J.M., R.N. Lichtenthaler, and E.G. Azevedo (1999). *Molecular Thermodynamics of Fluid Phase Equilibria.* Prentice Hall, New Jersey.
[27] Martin, R.B. (1996). *Chem. Rev.* **96**(8), 3043.
[28] Murgich, J. (2002). Petroleum *Sci. and Tech.* **20**(9/10), 983.
[29] Merino-Garcia, D. (2004). PhD Thesis, Technical University of Denmark, Dept. Chem. Eng.
[30] Andersen, S.I., J.M. del Rio, D. Kvostitchenko, S. Shakir, and C. Lira-Galeana (2001). *Langmuir* **17**, 307.
[31] Czarnecki, J. (2003). In: J. Sjöblom (ed.), *Proceedings of the 4th International Conference on Petroleum Phase behavior and Fouling.* Trondheim, Norway.
[32] Andersen, S.I., A. Keul, and E.H. Stenby (1997) *Pet. Sci. Technol.* **15**(7/8), 611.
[33] Buenrostro-Gonzalez, E., M. Espinosa-Pena, S.I. Andersen, and C. Lira-Galeana (2001). *Energy Fuels* **16**(3), 732.
[34] Hillmann, E.S. and B. Barnett (1934). *ASTM Proc.* **37**, 559.

[35] Acevedo, S., G. Escobar, M.A. Ranaudo, J. Piñate, A. Amorin, and M. Diaz *et al.* (1997). *Energy Fuels* **11**, 774.
[36] Takanohashi, T., M. Iino, and K. Nakamura (1998). *Energy Fuels* **12**, 1168.
[37] Andersen, S.I. and J.G. Speight (1992). *Prepr. Am. Chem. Soc. Div. Fuel* **38**(3), 1335.
[38] Speight, J.G. and S.E. Moschopedis (1981). Chemistry of asphaltenes. In: J.W. Bunger, and N.C. Li (eds.) *Advances in Chemistry Series.* American Chemical Society, Washington, DC, p. 1.
[39] Snape, C.E. (1989). In: L.D. Field, and S. Sternbell (eds.), *Analytical NMR.* Wiley, New York, p. 65.
[40] Leon, O., E. Rogel, A. Urbina, A. Andujar, and A. Lucas (1999). *Langmuir* **15**, 7653.
[41] Murgich, J., J.A. Abanero, and O.P. Strausz (1999). *Energy Fuels.* **10**, 278.
[42] Fotland, P. and H. Anfindsen (1996). *Fuel Sci. Tech. Inter'l.* **14**, 101.
[43] Acevedo, S., B. Mendez, A. Rojas, I. Layrisse, and H. Rivas (1985). *Fuel* **64**, 1741.
[44] Xu, Y., Y. Koga, and O.P. Strausz (1995). *Fuel* **74**, 960.
[45] Gould, K.A., M.L. Gorbaty, and J.D. Miller (1978). *Fuel* **57**, 510.
[46] Liotta, R. (1979). *Fuel* **58**, 724.
[47] Rose, K.D. and M.A. Francisco (1987). *Energy Fuels* **1**(1), 233.
[48] Desando, M.A.A. and J.A. Ripmeester (2002). *Fuel* **81**, 1305.
[49] Ralston, C.Y., S. Mitra-Kirtley, and O.C. Mullins (1996). *Energy Fuels* **10**, 623.
[50] Pesce, A.J., C.G. Rosen, and T.L. Pasby (1971). *Fluorescence Spectroscopy.* Marcel Dekker, New York, p. 54.
[51] Merino-Garcia, D. and S.I. Andersen (2004). *Langmuir* **20**(4), 1473.
[52] Merino-Garcia, D. and S.I. Andersen (2004). *Langmuir* **20**(11), 4559.
[53] Buenrostro-Gonzalez, E., A. Gil-Villegas, J. Wu, and C. Lira-Galeana (2002). In: C. Lira-Galeana, (ed.), *Proceedings 2002 International Conference on Heavy Organic Deposition.* Mexico.

14

Petroleomics and Characterization of Asphaltene Aggregates Using Small Angle Scattering

Eric Y. Sheu

1. Introduction

Petroleum is a mixture of organic material consisting of a serious of molecules with increasing molecule weight but with decreasing carbon to hydrogen ratios. This monotonic trend leads to distinctive properties of each class, cut by solvents. Asphaltene is a class soluble in toluene but not in heptane. Importance of asphaltene lies in its relevance with petroleum operations. Many properties of petroleum liquids are due to the interplay between asphaltene with other co-exist components. These complex interactions impact on petroleum phases, and thus the operations. The so-called petroleomics is a scheme to link the molecular structures of the most relevant components in the petroleum liquid to it overall properties, similar to the proteomics widely accepted in biological sciences. However, the asphaltene molecular structure and compositions, though relevant to the macroscopic properties of petroleum liquids, their aggregates on the colloidal length scale could be more relevant to the properties of the petroleum mixtures. In this regard, there is a need to thoroughly characterize these aggregates using advanced techniques, such as Small angle X-ray scattering (SAXS) and small angle neutron scattering (SANS) to bridge the molecular structures of asphaltenes and the operational parameters that are commonly applied in the fields.

The intent of petroleomics has been to address large length scale physical and chemical properties of petroleum liquids and solids using molecular structure of the components comprising them. This route of approach is similar to using statistical mechanical theory for describing the macroscopic properties of a system, such as using the inter-particle potential and structure factor to characterize viscosity.[1–3] It can only be taken when adequate molecular information is available. In surfactant chemistry, it is possible to achieve such a goal because in most cases the molecular structures are known. Modeling of their aggregate structures at various thermodynamic conditions is much easier. To adopt this philosophy

---

Eric Y. Sheu • Vanton Research Laboratory, Inc., 7 Olde Creek Place, Lafayette, California 94549.

and technology for complex systems like asphaltene and petroleum liquids, detail molecular information should be available as the starting point. However, for the most refractory and complex components of petroleum material, asphaltene, it has been difficult to unambiguously identify their molecular structures until recently.[4,5]

Since molecular information is available for asphaltene to date, it appears that the time is right for looking into the petroleomics, in spite of many stumbling blocks ahead. This development is in many ways similar to the human genomic project where genes were identified and characterized. In order to apply the knowledge about genes to molecular pharmaceutical and biotechnology, there is a need to map out the correlation between genes and diseases. In many major diseases (diabetes, cardiovascular diseases, etc.), more than one gene is involved, making it difficult to quantitatively link the characteristics of each involved gene to the disease via a simple integration process. The combined effect that leads to many diseases were found to be highly nonlinear because the "contribution" from each gene may be vastly different. This expected yet unexpected outcome had slowed down the gene treatment and drug development pace.

What makes it even more complex is that each disease has to be deconvoluted, in order to identify how a relevant gene is involved. Mathematically, this is equivalent to extracting a set of parameters from their integrated value. It is theoretically not possible to find a single set of parameters, as we know that the number of eigen functions are often infinitive. Thus, using gene and the molecular structures to link to any integrated effect, such as a disease symptom is seemingly unreachable.

Knowing this chilling fact in human genome project and proteomics, it becomes obvious that if petroleomics is to be practical one day, it is necessary to develop linkages from the smallest length scales, i.e., molecular length scale, to the macroscopic scale that engineers deal with in the field.

If there are designated linkages of the petroleum molecules to the overall petroleum liquid characteristics, it is possible to completely describe the properties of petroleum liquids using molecular structures. Similar type of linkage was the hope in the human genome project. As mentioned earlier, the advantage of the human genome project is that most biological reactions are specific and structural controlled. In the case of asphaltene and petroleum liquids, the interactions between components are not as well defined as the key-and-lock reactions. This makes petroleomics more difficult than the human genome-linked disease cases or the proteomics case. Under this circumstance, a set of integrated parameters that have designated linkage with the molecular structures may be more useful. These integrated parameters would serve as the fundamental parameters. As long as these integrated parameters are physically meaningful and are directly related to the field parameters, the linkage from molecular structures to the filed is essentially established. Thus, it is important to identify such a set of parameters.

In the petroleomics, asphaltene is a central component because its phases have directly impact on petroleum operations. Through the twentieth century, many prestigious works have been reported on asphaltene properties. The results unanimously showed that there is aggregation propensity in many classes of petroleum materials. It is particularly true for asphaltene. It however only forms a portion of petroleum liquids and the overall characteristics of petroleum liquids may not be driven by asphaltene structures or structures of other components alone. In this situation, integrated parameters approach becomes important though they are not the primary parameters. One such group of nonprimary parameters includes size, shape, structural factor, and polydispersity of asphaltene aggregates. One may be able to use them as the fundamental parameters to derive the rest of the field and operational parameters. To do so, two conditions are necessary. One is these parameters should be fully describable by asphaltene molecular structures and the other is these parameters can be related to the overall properties of the petroleum liquids in the field.

To identify and characterize these colloidal parameters, techniques that measure the right colloidal length scale are needed. Colloids usually have length scales from about 10 to 1000 Å. Small angles X-ray scattering (SAXS) and small angle neutron scattering (SANS) are two well-known and accurate techniques applicable for length scale of this range. They have been applied to many colloidal systems with good success.[6–11] Both SAXS and SANS are microscopic techniques suitable for extracting information such as the colloidal form factor and structure factor. In principle, structures-and dynamics-related information is available in the data generated by these techniques. However, the data treatment and analysis are nontrivial. Recent advancement on statistical mechanical theory has made it easier to obtain crucial information.

In this chapter, a review of the small angle scattering work on asphaltene systems reported in the past will be presented and discussed. The limits, spectroscopic configurations, and data analysis schemes will be described. Instruments available for this type of work and the characterization of the instruments are provided. In Section 2, the fundamentals of asphaltene aggregation, colloidal particle formation, and aggregation mechanism are reviewed. This is followed by description of the small angle scattering theories and data analysis schemes in Section 3. Section 4 lists the SAXS and SANS instruments available in the world. Both small-scale rotating anode X-ray source or synchrotron source are described, along with the configurations of the available instruments. The specific spectrometers for the work presented here will be detailed. Section 5 gives the results and analyses of the SANS and SAXS measurements of the selected asphaltene systems in this work. In Section 6, discussion on the scattering curves and how to unambiguously determine the colloidal parameters using scattering data is given. Conclusions are given in Section 7 followed by future perspective in Section 8.

2. Asphaltene Aggregation

Asphaltene aggregation has been a subject of interest for many decades since it was identified in the early twentieth century, partly because of its significance in petroleum processing and partly because of the parallel development in colloidal science. To date, the energies that drive asphaltene aggregation are still a research topic. The importance of asphaltene aggregation research, from the petroleomics point of view, is to relate the asphaltene molecule structures and the aggregation energies to the properties of their aggregates. This is the most crucial step because it is the linkage between molecular structures and their first integrants.

The techniques used for detecting asphaltene aggregation can be categorized into surface techniques and the bulk techniques.[12] The bulk techniques are more convenient if the aggregation is to be further analyzed. This includes understanding thermodynamic properties, such as osmotic pressure, excess internal energy[13] and aggregate dynamics. These thermodynamic properties are related to the asphaltene aggregates via the structure, size and polydispersity of the aggregates. Unfortunately, there are no known techniques that can simultaneously detect asphaltene aggregation onset and in the mean time characterize the properties of the aggregates. More than one technique is usually needed. Moreover, it requires multiple theories to construct the relationship between aggregation data and the thermodynamics of the system. As a result, a specific set of theories is needed to obtain some thermodynamic properties from the aggregation onset data. While another set of theories is needed for analyzing their aggregates to obtain other (or similar) thermodynamic properties of the aggregates and compare to those extracted from the aggregation onset measurements. Small angles X-ray scattering (SAXS) and small angle neutron scattering (SANS) are the two techniques most suitable and powerful for colloidal structure measurements, certainly fit for this study. For analysis, several statistical mechanical theories have been developed for extracting microscopic information from SAXS and SANS data. Combining the right spectrometer and data analysis schemes, it is possible to accurately characterize asphaltene aggregates and obtain the parameters needed for linking to operational parameters.

3. SAXS and SANS

SAXS makes use of the electron density difference to identify and measure the particle size, shape, and polydispersity. For example, asphaltene aggregates and its surrounding, either well characterized solvents or petroleum liquids, may differ substantially in their electron density. In 1940, Pfeiffer and Saal proposed the resin-peptized model to explain suspension of asphaltene in petroleum liquids.[14] If this argument is correct, one should be able to detect the suspended asphaltene particles as long as the asphaltene particle has different electron density from resins and other light components. This electron density contrast requirement is a restriction of SAXS technique. Fortunately, most asphaltene molecules contain some degree of heteroatoms that make their electron densities slightly higher than the commonly used solvents or lighter components in petroleum liquids. However, the contrast is small, thus long data acquisition time may be needed to collect statistically significant data. This requires the instrument to be stable and configuration well selected.

SANS applies similar principle to SAXS except the contrast for neutrons to detect the object in the system is the scattering length densities, which depends on the difference of the nuclei that comprise the molecules. In asphaltene systems, it is the difference of the neutron scattering cross section between the asphaltene molecules and their surrounding molecules. In the case of asphaltene system, there is no enough contrast for neutron to distinguish asphaltene aggregates form their surrounding. Fortunately, hydrogen and deuterium are very different in their scattering length densities. One thus can dissolve asphaltene in deuterated solvents to enhance contrast for SANS measurements. Other than the contrast difference (SAXS is based on electron density and SANS on nuclei) SAXS and SANS applied the exact same scattering theory arising from the Born approximation which assumes no energy loss (elastic scattering) in the scattering process.

For an asphaltene system with aggregates suspended in solvent molecules of different scattering contrasts, an incident photon (SAXS) or neutron (SANS) of zero impinging angle form a de Broglie plane wave is scattered by the suspended scatterers (asphaltene and solvent) in a different manner such that the scattered spherical wave will leave the system at different angles. Figure 14.1 shows this process where I_0 is the incident intensity of photons (or neutrons). Taking the incident photon or neutron as wave traveling to the system at zero angle, this wave will leave the system at different angle depending on what it hits. As mentioned earlier, the scattering theory used here is based on the Born approximation. It requires the incoming wave to be scattered only once (no multiple scattering) and that the scattering process does not involve energy loss or gain (elastic scattering). If the incident wave hits a solvent molecule, it has a probability P_θ to leave the system at an angle θ. This probability distribution (scattering distribution function) is expected to be different from the probability distribution if it hits an asphaltene molecule whether the asphaltene molecule is a free molecule in the solvent or in an asphaltene aggregate. Therefore, if one subtracts the probability distribution of the solvent scattering from the solution scattering, the resulting scattering distribution is contributed from asphaltene only. This scattering intensity distribution can then be analyzed to extract information about these scatterers, such as their sizes, shapes, polydispersity, and the interactions among them. Because these basic properties are governed by the thermodynamics of the system we can also learn about the

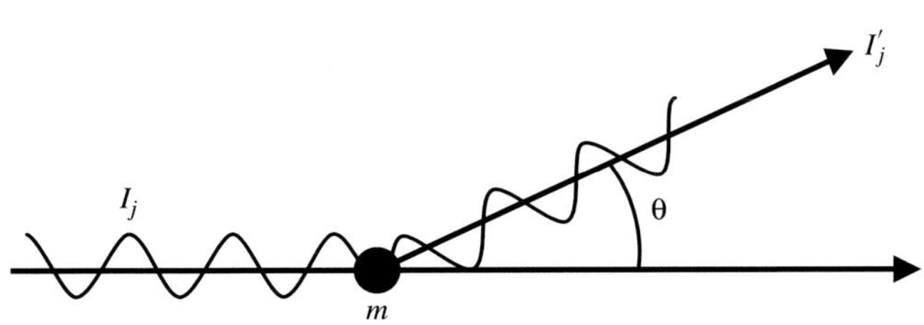

Figure 14.1. Schematic of a scattering process at point m assuming m has no size.

state of the system (e.g., internal energy, enthalpy, and energies involved in the aggregation, etc.)

The above description of scattering process is based on an assumption that the scatterer, S, has no size. Mathematically, this scattering distribution function can be expressed as

$$f_m(\vec{Q}) = \rho_m(\vec{r}_m)e^{i\vec{Q}\cdot\vec{r}_m}, \tag{14.1}$$

where Q is the scattering vector (or momentum transfer from the initial state to the final state) with a value of

$$\left|\vec{Q}\right| = \frac{4\pi}{\lambda}\sin\frac{\theta}{2} \tag{14.2}$$

for an incident wave of wavelength λ and $\vec{r}_m$ is the position vector where the scatter resides at the time when it interacts with the incident wave.

In reality the scatterers do have sizes. In the asphaltene case, they are the sizes of the aggregates. For simplicity we assume the composition is uniformly distributed throughout the aggregate and that there is an average electron density (SAXS) or scattering length density (SANS). This allows us to assign an average electron density for the aggregate. With this setting, there will be interference between the waves scattered from one side of the aggregate and the ones scattered from the other side. This interference is from the waves that are scattered from the same particle. Because the interference depends on the structure of the particle, it is called the intra-particle structure factor and it contains information of the particle size and shape. Figure 14.2 demonstrates this intra-particle scattering process. From the scattering pattern, it is in principle possible to extract the particle size and shape.

As one can see from Figure 14.2, the scattered waves, I'_j and I'_k are not in-phase as they are before being scattered. This correlation carries the size and shape

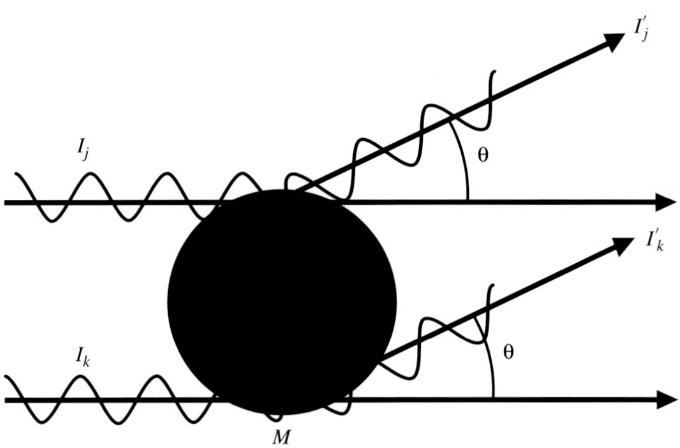

Figure 14.2. Phase shift between two incident waves after scattered by an object M carries the particle size and shape information in the form factor.

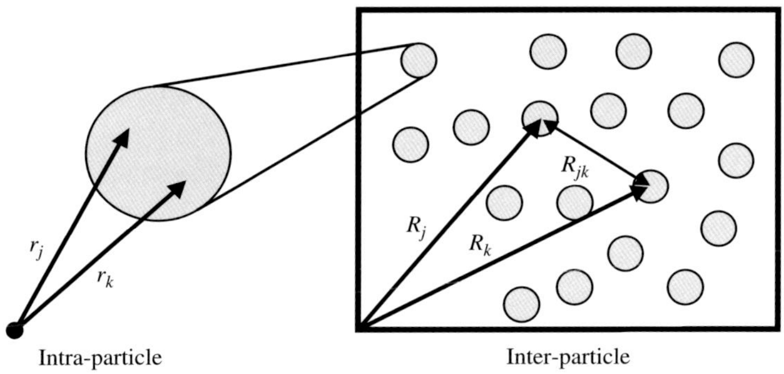

Figure 14.3. Intra- and inter-particle interactions.

information and is in the scattering distribution function. Equation (14.1) can thus be extended to express the scattering function of this nonzero size scatterer with intra-particle interactions as

$$f_M(\vec{Q}) = \int_M [\rho_j(\vec{r}_j) - \rho_{\text{solvent}}]e^{i\vec{Q}\cdot\vec{r}_j}d\vec{r}_j. \tag{14.3}$$

Essentially, Eq. (14.3) is an integration of Eq. (14.1) over the particle M that has size effect on scattering. Once the scattering function is known, the expected value of the scattering function can be obtained:

$$P(|\vec{Q}|) = \left\langle \left| f_M(\vec{Q}) \right|^2 \right\rangle. \tag{14.4}$$

Equation (14.4) is the scattering intensity distribution function from one nonzero size particle. This is known as the form factor. It carries the size, shape, and polydispersity information of the particles in the system but in a coupled form. In order to get the size, shape, and polydispersity information individually, a proper decoupling scheme is needed (see Figure 14.3).

In addition to the size effect, the form factor, the scattering intensity distribution function is also affected by the interactions among particles. That is to say that the scattering of the wave also depends on the interactions strength among the particles in the system. When there are more particles in unit volume the interactions become stronger, thereby, affecting more on the scattered waves. Figure 14.3 shows the schematic of the system and the relationship between form factor (intra-particle interactions) and the inter-particle interactions. The overall scattering function taking into account both interactions can be expressed as

$$I(\vec{Q}) = \frac{1}{V}\sum_{k=1}^{N_p}\left\langle \left| f_k(\vec{Q}) \right|^2 \right\rangle + \frac{1}{V}\left\langle \sum_{k=1}^{N_p}\sum_{\substack{k=1 \\ j\neq k}}^{N_p} f_k(\vec{Q})f_k^*(\vec{Q})e^{i\vec{Q}\cdot(\vec{R}_k-\vec{R}_j)} \right\rangle, \tag{14.5}$$

where N_p is the number of particles and V is the total volume of the system. The first term of Eq. (14.5) is the contribution from the form factor as given in Eq. (14.4) and the second term is from the interactions among particles, known as the structure factor because it carries information about how the particles are arranged in the system.

If the particles are monodisperse in size, Eq. (14.5) can be simplified as

$$\begin{aligned} I(\vec{Q}) &= n_{\mathrm{p}} \left\langle \left| f_k(\vec{Q}) \right|^2 \right\rangle \left\{ 1 + \left\langle \sum_{k=1}^{N_{\mathrm{p}}} \sum_{\substack{k=1 \\ j \neq k}}^{N_{\mathrm{p}}} f_k(\vec{Q}) f_k^*(\vec{Q}) e^{i\vec{Q}\cdot(\vec{R}_k - \vec{R}_j)} \right\rangle \right\} \\ &= n_{\mathrm{p}} P(\vec{Q}) S(\vec{Q}) \end{aligned} \tag{14.6}$$

with $S(\vec{Q})$ being the structure factor. If the system is isotropic, $S(\vec{Q})$ can be expressed by averaging over the angle as

$$\left\langle S(\vec{Q}) \right\rangle = S(Q) = 1 + 4\pi n_{\mathrm{p}} \int_0^\infty [g(r) - 1] \frac{\sin Qr}{Qr} r^2 dr, \tag{14.7}$$

where $g(r)$ is the so called pair distribution function representing the probability of finding a particle at distance r when there is a particle at origin. It is essentially the local number density of the particles. $g(r)$ is the most essential function that links the thermodynamics of the system to the structure factor which can be determined from a scattering measurement. Figure 14.4 shows the physical meaning of $g(r)$ qualitatively.

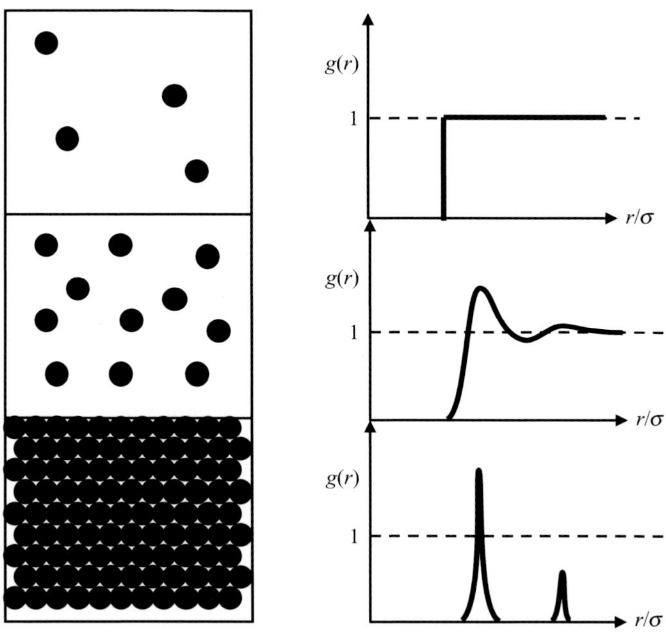

Figure 14.4. Pair correlation function $g(r)$ represents the local number density surrounding a particle. It is correlated to how the particles are arranged and transition from crystalline to dilute solution where there are no particle–particle interactions.

The above description clearly indicates that there are two major factors to be determined in any scattering experiment. They are the form factor $P(Q)$ and the structure factor $S(Q)$. $S(Q)$ is equal to unity if one deals with a very dilute solution with short-range interactions. This happens for volume fraction less than 0.1 as illustrated in the Stoke–Einstein equation and the Einstein viscosity equation as examples.[15] For asphaltene solutions, the media are organic with very low dielectric constants. Thus, it is plausible to assume that the inter-particle interactions are short-ranged and are negligible. In this case, only $P(Q)$ needs to be dealt with.

Further simplification of the scattering intensity distribution function is to apply approximation for particle size and shape characterization. For example, the $P(Q)$ can be approximated by the radius of gyration using the Guinier plot,

$$I(Q) = I_o e^{-Q^2 \frac{R_g^2}{3}}. \tag{14.8}$$

Using Eq. (14.8) one can plot the logarithm of $I(Q)$ versus Q^2 to evaluate the radius of gyration R_g. Once the radius of gyration is determined, the dimension of the particle can be calculated if the shape of the particle is known. Unfortunately, the Guinier plot only provides the radius of gyration, not the shape of the particles. To determine the shape of the particle, other analyses are needed. There are three approaches to determine particle shape. One is to use the Porod plots[9] to look for the linear regions in a cylindrical plot or a flat particle plot. For examples, if one obtains linear behavior when plotting $QI(Q)$ versus Q^2 in the medium Q range, the object is likely a long cylinder. The short dimension can be calculated by the slope obtained from the linear section. This particular plot is for long cylinder. On the other hand, if one plots $Q^2I(Q)$ versus Q^2, then it is for analyzing flat particles through analyzing the linear region at the high Q for the thickness and the low Q for the horizontal dimension. Comprehensive review of this approach along with the concept of invariant (volume to surface area ratio) can be found in reference 8.

Another approach to determine the particle shape is to presume the shape of the particles. Using this presumed shape the $P(Q)$ can be rigorously calculated with the dimensional parameters built-in. This $P(Q)$ is then used to fit the scattering data, from which the dimension of the particle can be determined. However, one needs to check if the shapes of the particle presumed are justifiable. To justify the presume particle shapes, one needs another axis with a requirement that there is a well-defined functional behavior along that particular axis. In this way, one can use the extracted particle dimension and the presumed shape to evaluate this parameter along that particular axis and see if the parameter indeed shows the functional behavior it supposes to. In a previous report[16] we show this method using concentration as the axis and the contrast as the parameter.

Finally, one can also determine the shape by presuming the particle shape for data fitting and then calculate the invariants to see if the presumed shape is correct. This method was discussed in details in a previous report.[17] Another approach is to directly invert the scattering function to obtain the distance distribution function, $p(r)$, which represents the distribution of the scattering density. It provides direct information about the scattering object and shape can be mapped out this way. Brunner-Popela and Glatter give a comprehensive review of this approach.[18]

4. SAXS and SANS Instruments

An SAXS spectrometer is composed of three parts—X-ray source, collimator and geometric configuration, and detector. The X-ray sources can be a simple X-ray tube, a rotating-anode using various metals as targets or a synchrotron source. An X-ray tube conventionally uses tungsten wire as the filament to generate electrons, which were subsequently accelerated and bombard to a metal coating (can be tungsten, copper or copper–graphite mixture). This simple process creates continuous energy X-ray from approximately 120 eV to 120 keV. The corresponding electromagnetic radiations have a wavelength range, λ, from $\sim$0.01 to 10 nm. The X-rays generated are largely white radiations (or Bremsstrahlung) with occasional characteristic beams depending on the target used. In order to use a specific energy of X-ray, a collimator is used to select a particular X-ray wavelength. As a result, most energy in Bremsstrahlung range is abandoned in the collimation process. This limits the X-ray's intensity for application to colloidal systems where $\lambda \sim$ 0.5–3 nm are most frequently used. One stronger source is a rotating anode source with metal targets. Frequently, copper is used as the target. To generate X-ray, a cathode-like filament generates electrons when heated. These electrons are accelerated to several thousand volts before bombarding to a metal target. The energy of these electrons is high enough to excite the K_α electrons. Following the decay of the excited states, photons are generated at X-ray energy range. The X-ray so generated is exclusively characteristic beam with a very narrow photon energy distribution, so the X-ray intensity of the same wavelength is much higher. In the case of a copper target, the wavelength of the photon is 1.54 Å (8 keV energy). A collimator is still needed for the rotating anode source to sharpen the resolution.

Synchrotrons produce the highest intensities of X-rays. A synchrotron source consists of a large dimension storage ring. These very large closed ring accelerates electrons bunches and constrained them by high power magnetic fields. These electrons can be accelerated to a few GeV energy range. At these energies the particles are relativistic with velocities very close to the speed of light and as they are bent by magnetic fields they emit synchrotron radiation tangentially. This synchrotron radiation is used as a source of electromagnetic radiation. Their wavelengths extend from the infrared through the visible and ultraviolet to high-energy X-rays ($\sim 10^{-4}$ to $\sim 10^2$ keV). Due to its high intensity synchrotron is a valuable source for characterizing dilute systems and the surface properties of monolayer.

However, there is a concern in using synchrotron radiations for investigation of asphaltene. Most asphaltene solvents are relatively low in boiling points (80°C for benzene, 110°C for toluene, and 69°C for hexane). When they are exposed to synchrotrons, the samples may be heated rapidly or even evaporated in the sample cell. It is particularly true when solvent-like pentane is used with Kapton window from which pentane molecules can penetrate. In this regards, a rotating anode X-ray spectrometer is probably the best choice for studying asphaltene aggregates. Synchrotron source will be useful if the electron density contrast between asphaltene and the solvent molecule is small and a rotating anode cannot provide statistically significant data. However, one needs to deal with the sample evaporation issue before using a synchrotron source.

Once the energy source is selected, the second part of the spectrometer is the collimation and configuration. For selection of wavelength in an X-ray tube or a rotating anode, either flat or curved graphite is used. After the energy band is selected the beam is further focused by a set of double bounced (or triple bounced) mirrors followed by a series of pinholes to define the beam size before impinging onto the sample. Usually, the beam path is in a vacuum metal tube to avoid leakage and air scattering, which may diverge the beam. In many cases, a pinhole is used just before the sample cell, known as the beam size defining pinhole. It determined the beam size and thus the scattering volume. The third component is the detector. To collect the scattering radiations either a line-detector or a 2D position sensitive detector detects the scattered photons. The Q range of a spectrometer is the most important configuration that needs to be determined prior to experiment to ensure the length scale range of interested is covered by the configuration. The radius of the detector and the sample-to-detector distance (SDD) determines the Q range (see Eq. (14.2) where $\sin\theta$ is a function of SDD and the detector radius). Because the sample-to-detector distance can be changed in many spectrometers, the Q range can be selected at most spectrometers.

Laboratory rotating anode are relatively common, however, majority of these instruments are for institutional use only while synchrotron sources are readily available for the public.[19] Each X-ray source differs from its X-ray energy band to the resolution (i.e., $\Delta\lambda/\lambda$).

The SAXS work presented here was performed at the 10-m small angle X-ray facility at the Oak Ridge National Laboratory. The X-ray source is a rotating anode at 4 kV and 100 mA. It is a copper target with a take-off angle of 6°. The monochromator for wavelength selection was a flat graphite to select the $\lambda = 1.54$ Å photons. The sample holder is a liquid cell with double Kapton window. A 2D detector was used with continuous wire divided into 64×64 pixels. A mechanical pump was used to maintain 10^{-4} torr of vacuum in the spectrometer during measurements. Fe-55 (24.4% 5.9 keV and 2.9% 6.5 keV) was used to calibrate the detector pixel-to-pixel sensitivity and the sample transmission was determined using a with-carbon and without-carbon process. The scattering spectrum obtained was in absolute scale of cm^{-1} representing the differential scattering cross section of the sample. It is important to obtain the absolute intensity because it helps determine the particle shape and is usually not possible to fit the data if the model is incorrect. This further assures the unambiguity of the data analysis.

SANS spectrometer are usually available at national laboratories because it requires either a spallation source or a nuclear reactor. Table 14.1 is a short list of small angle neutron scattering facility available in the world. A small angle neutron source is usually used for characterizing particles of colloidal sizes. In order to generate enough neutron intensity for characterizing this size range, cold sources, either by liquid helium or liquid hydrogen, are used to generate neutron wavelength from ~1 to 30 Å. Typical wavelength used for colloidal systems is around 5 Å,

Table 14.1. A Short List of SANS Spectrometers Available to the Public in the World

Place	Source	Power	Moderator	Date available
Bombay	India, DHRUVA	100 MW	liquid CH_4	1986
Brookhaven	USA HFBR	60 MW	liquid H_2	1977
Grenoble	France RHF/ILL	57 MW	liquid D	1972, 1985, 1987
Julich	Germany FRJ 2	23 MW	liquid H_2	1972, 1985, 1987
Gaithersburg	USA NIST	20 MW	sol D_2O, liquid H_2	1987, 1995
Tokai Mura	Japan JRR-3	20 MW	liquid H_2	1988
Budapest	Hungary KFKI	15 MW	liquid H_2	1989
Chengdo	China HWRR	15 MW	liquid H_2	1988
Saclay	France ORPHEE/LLB	14 MW	liquid H_2	1980
Leningrad	Russia VVR-M	10 MW	liquid H_2 + liquid D_2	1985
Berlin	Germany BER 2	10 MW	gas H_2	1988
Riso	Denmark Pluto	10 MW	gas H_2	1975
Rutherford	GB ISIS	Pulsed	gas H_2, liquid CH_4	1985
Argonne	USA IPNS	Pulsed	sol, liquid CH_4	1986
Los Alamos	USA LANSCE	Pulsed	liquid H_2	1986
Tsukuba	Japan KENS-1	Pulsed	sol CH_4	1987

which, in most reactor source spectrometers, ends up with $\sim 10^6$ neutron/cm^2 flux before entering sample. Other than the source, SANS spectrometer is similar to an SAXS spectrometer. The detectors are different from SAXS detectors but are also position sensitive consisting of pixels.

Two spectrometers were used to generate the data to be discussed here. One was the spallation source neutron facility called intense pulsed neutron source (IPNS) at Argonne National Laboratory (ANL). It has a Q range from 0.008 to about 0.3 Å^{-1}, equivalent to a spatial resolution of few angstroms to several thousand Å particle sizes. The other spectrometer used was the 30-m NG7 spectrometer at the National Institute of Standards and Technology (NIST). It is a reactor source SANS spectrometer. The wavelength of the cold neutron is from 5 to $\sim$30 Å. Because it is a 30-m long spectrometer, the sample-to-detector distance can be adjusted from 1 m to nearly 20 m, which make the Q range much wider (from 0.002 to 0.7 Å^{-1}). Quartz cells were used to hold samples since neutron has high transmission in quartz.

5. SAXS and SANS Experiments and Results

There are several sets of measurements reported here; some uses SAXS and others use SANS. Asphaltenes used for measurements were from Ratawi vacuum residue or Arabian Medium Heavy vacuum residue using standard heptane extraction process. Briefly, 1 g of vacuum residue asphaltene is mixed with 40 mL of heptane and mixed for 24 hr before being filtered by Whatman No. 5 paper to separate the insoluble fraction (asphaltene) from the rest. The insoluble fraction was then dried under nitrogen until constant weight is obtained. The heptane soluble fraction was further cut by the same procedure but using pentane as the

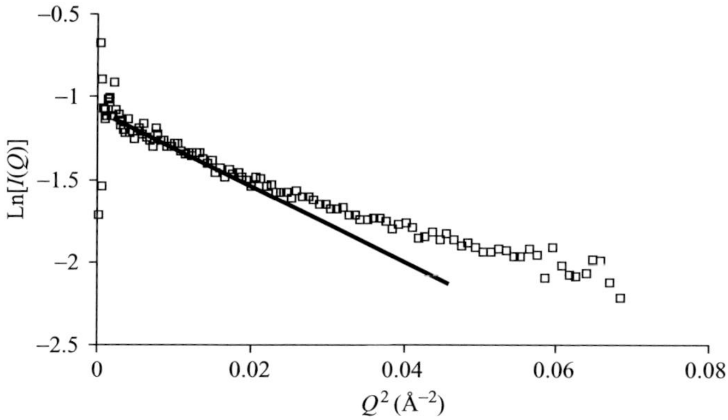

Figure 14.5. Guinier plot of the SAXS intensity distribution function of Ratawi resin (heptane soluble and pentane insoluble fraction) in deasphalted oil (C5S). The radius of gyration calculated is 7.8 Å.

solvent. After this process, three fractions were obtained—asphaltene (C7I), resin (heptane soluble but pentane insoluble), and pentane soluble (C5S).

5.1. SAXS Measurement on Ratawi Resin and Asphaltene

This experiment was performed at Oak Ridge National Laboratory using 1.54 Å wavelength from a copper target rotating anode X-ray spectrometer with Q range from 0.007 to 0.4 $Å^{-1}$. The temperature was kept at 25°C. Figure 14.5 shows the scattering intensity distribution function. Simple Guinier plot yields a radius of gyration of 7.8 Å. Assuming it is a spherical object,[17] then the radius R is about 10 Å. Taking 1.25 Å as the carbon–carbon bond length, this is equivalent to less than 7 carbon bond lengths. Compared with the recent asphaltene molecular structure proposed,[4,5] it is reasonable to argue that this is not an aggregated. It is more of the average size of a resin molecule.

We argue that the SAXS-derived dimension of a resin molecule is reasonable. First, the $QR_g < 1$ for the range we used to derive the R_g. It is a requirement of the Guinier theory. Secondly, the intra-molecular structure may rotate in a way the scattering process captures a spherical-like object. Thus, an assumption of a spherical object to derive R from R_g is a plausible process. The true dimension should be accurate within the first order of approximation. We do not emphasize that the molecules are spherical but do believe an exercise using a spherical object to get R is acceptable. Fluorescence emission[4] showed similar dimension for UG8 resin. This agreement cannot be accidental when two mechanisms are vastly different, one by photon–electron interaction resulted scattering pattern analysis while the other by relaxation mechanism. The agreement should be a reflection of the true dimension of this class of material.

Figure 14.6 shows a 20.7 Å radius of gyration. Again, one assumes a spherical model to get 26.7 Å. This radius is considerably larger than an asphaltene molecular model and should be considered an aggregate. Fluorescence emission shows 19.7

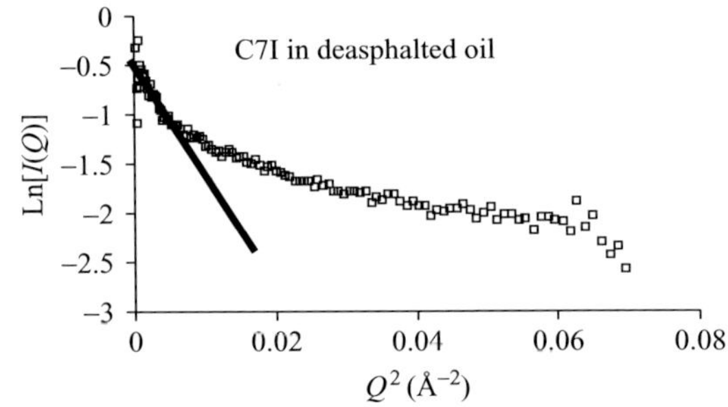

Figure 14.6. Guinier plot of the SAXS intensity distribution function of Ratawi asphaltene (heptane soluble and pentane insoluble fraction) in deasphalted oil (C5S). The radius of gyration calculated is 20.7 Å.

in diameter.[4] If we take this number as the asphaltene diameter, the asphaltene molecule radius can be in the range of 10–12 Å if the blue wing of the fluorescence wavelength is taken into account. Using these numbers, the volume ratio between an aggregate and a molecule is about 11–19. This is a rough estimate of the aggregation number assuming compact packing, which is not the case. Therefore, this is just a hand-waving argument but should be accurate to the first order of approximation. If we take the void of the packing into account, the aggregation number can be up to 30% lower.[17] Other report showed a range of ~25–45 Å size asphaltene particles.[11] While the individual asphaltene molecule may not vary as much the aggregate size can have more variation due to the different composition. Therefore, a 26.7 Å of radius obtained here is well within the range.

Instead of using the Guinier plot, one can establish a form factor with structural parameters built in. Details about the form factor with polydispersity had been discussed in a previous report.[17] Using form factor and an appropriate polydispersity model one can fit the scattering intensity distribution function and extract the radius and the degree of polydispersity. Figure 14.7 shows the radii extracted from such a fitting scheme using Schultz distribution function as the polydispersity function.[16] As one can see, the asphaltene aggregates from different sources consistently have radii between 25 and 45 Å. The 100-wt% concentration shown here is defined as the asphaltene concentration in the vacuum reside. The actual asphaltene concentration is about 20% in the vacuum residue. The lower concentrations were made by diluting the 100-wt% system by deasphalted oil. The other message delivered by this plot is that the aggregate size does not increase rapidly like a micellar system. It is statistically unchanged if the polydispersity is taken into account.

The relative independence of aggregate size as a function of the concentration leads to the structural evolution Yen[20,21] proposed many years ago. Yen proposed formation of elementary particles upon aggregation as the first step. These elementary particles do not heavily depend on asphaltene concentration. However, these particles may further aggregate to form much bigger particles in which an elemental particle maintains its own integrity and intra-particle structures. This second

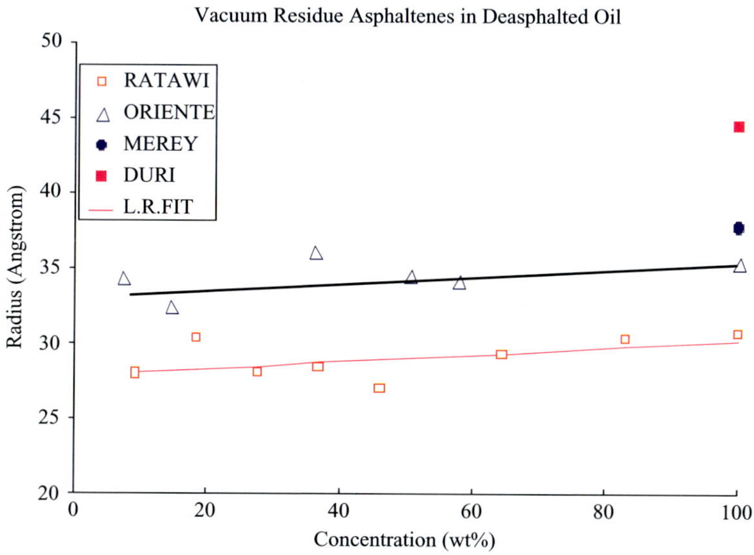

Figure 14.7. SAXS-derived radii of Ratawi asphaltene aggregates in vacuum residue (100 wt%) and in deasphalted oil (diluted from vacuum residue) and the radii of aggregates of various asphaltenes in their vacuum residue state.

step of aggregation is more of flocculation than aggregation. If these secondary aggregates are much bigger than the elementary aggregates, they may not be detected by the SAXS we performed due to the limited Q range. A much smaller Q is needed to detect these particles if they exist. Some SANS facilities can reach lower Q than the SAXS spectrometer used. In the following two SANS experiments are to be discussed where some indication of these large particles can be observed.

5.2. SANS Measurement on Asphaltene Aggregation, Emulsion, and Dispersant Effect

Figure 14.8 shows the 1% (wt) Ratawi asphaltene in deuterated toluene/pyridine mixtures. The curves were vertically shifted for clarity. As one can see the particle sizes are similar in all mixtures and are in the nano range. They are likely the elementary particles described in Yen's model, or the smallest aggregates.

It should be noted that the radii of gyration obtained from Figure 14.8 are from $Q = 0.015$ to 0.023 Å^{-1}. With 30 Å as the radius of gyration one gets $R_g Q = 0.45$–0.69 which is smaller than 1. So the Guinier approximation applies. For lower Q (i.e., less than 0.015 Å^{-1}) the intensity increases rapidly, indicative of much larger objects. Guinier analysis shows they are about 120 Å. This is likely the further agglomeration of the elemental particle of 40 Å. The message from these curves and Guinier analysis is that the particle size remains nearly the same from toluene to pyridine. The other important result is that the particle size does

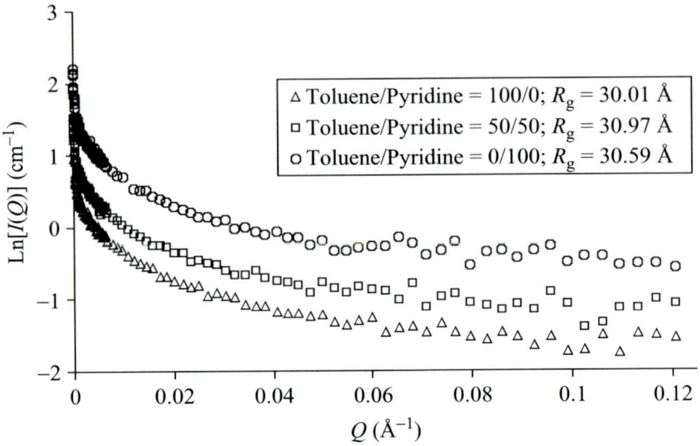

Figure 14.8. One percent Ratawi asphaltene in toluene/pyridine mixtures.

not increase upon increasing asphaltene concentration as indicated in a previous report.[11]

Because SANS uses deuterated solvents (D-toluene and D-pyridine) to enhance scattering contrast. There is always a need to check the effect of the deuterated component, the isotope effect. This was achieved by mixing protonated and deuterated solvents to see if the results will change—a technique known as contrast variation. Figure 14.9 shows a series of measurements using the mixed solvents. The scattering intensities spectra appear similar except their intensities because of the contrasts.

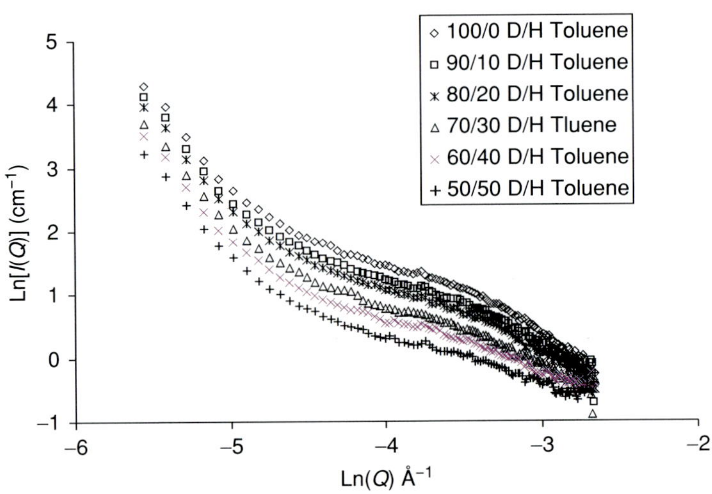

Figure 14.9. Contrast variation measurements for series of D-toluene/H-toluene mixtures. The asphaltene concentration is 1%.

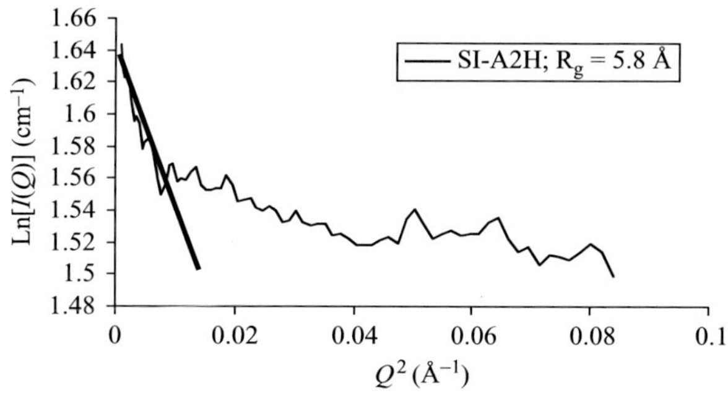

Figure 14.10. Contrast variation experiment. See text for the composition and details.

Knowing that the deuterated solvent effect is negligible, the next question will be the morphology of the aggregates. It is obvious that the aggregate will not be similar to surfactant systems where molecules have tight packing in an aggregate. In asphaltene aggregates, it is expected to have voids because of the wide spread of the molecule structures that form asphaltene aggregates. The reasonable questions to ask for answering the packing questions are the roughness of the aggregate surface, the "core" size and their morphology, if it can be answered by SANS. The approach is again the contrast variation technique. Here we demonstrate how to use this technique to answer some details of the aggregate structure.

Figure 14.10 shows 5% asphaltene solutions in different environments. SI-A2H is a mixture of 4:1 of 5% asphaltene in protonated toluene and $pH = 2$ deuterated water. Because the major structure is asphaltene aggregates but is "masked" by the protonated toluene in the bulk, thereby only the deuterated water region shows neutron scattering contrast with respect to the environment. Thus, the scattering is mainly from the water region. The "water core" was found to be 5.8 Å in the radius of gyration, approximately one asphaltene molecule dimension. Note that the scattering intensity is very low.

The next two systems are SII-A2H: same concentration and oil:water ratio but the toluene is deuterated and water protonated; SIII-A2H: deuterated toluene and deuterated water. As one can see the particle size extracted for SII-A2H and SIII-A2H2 are similar as expected. Physically, it means that the surfaces are not too rough and it is hydrophobic in nature. This is to say that the cores are more polar where water molecules prefer to stay to minimize the water–toluene contact. In this series, SANS demonstrates its unique capability of studying the core and surface morphology. Later, similar strategy is applied to vacuum resid instead of a solvent system and similar conclusion can be drawn (see Figure 14.13).

After studying the morphology, structures of aggregates, it is natural to ask the next question and that is how to prevent aggregation. One way is to introduce dispersant. If the dispersant successfully prevent aggregate formation we should see particle size being subdued. Here we investigated a simple surfactant, the sodium dodecyl sulfate (SDS) and applied SANS to evaluate the structural change before and after adding SDS. Figure 14.12 illustrates the scattering intensity distributions.

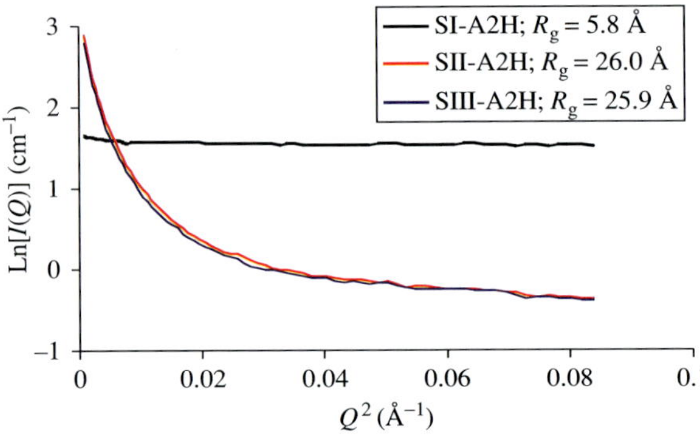

Figure 14.11. Contrast variation study with SI-A2H = asphaltene/H-toluene/D-water, SII-A2H = asphaltene/D-toluene/H-water, and SIII-A2H = asphaltene/D-toluene/D-water.

Apparently, SDS does have effect on the structure of the aggregates. The striking point is that the effect is much more on the low Q region than the high Q region. It clearly shows that the SDS does reduce clustering of the elemental particles but does not affect the structure of the elemental particles. This is a very important result. It suggests that the energy involves in the clustering of the elemental particles is much smaller and can be dispersed by adding a relative weak dispersant like SDS. However, the elemental particles that formed by the asphaltene molecules have much stronger aggregation-energy thus will not be dispersed by SDS.

Having discussed the results from solvent systems, it is interesting to know if SANS can be applied to measuring asphaltene structure in vacuum residue with small amount of deuterated solvent added. Figure 14.13 is such a study. Similar to the solvent case, water appears to reside in the core rather than with or between the large aggregates (the clustering of elementary particles). The scattering intensity distributions in the low Q range appear to be nearly unchanged. The sizes remain similar. However, contrast in the higher Q range increase drastically making the

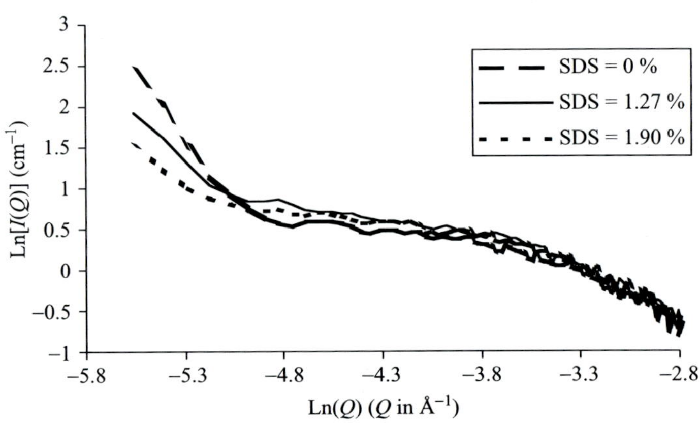

Figure 14.12. Five percent asphaltene in toluene with added SDS.

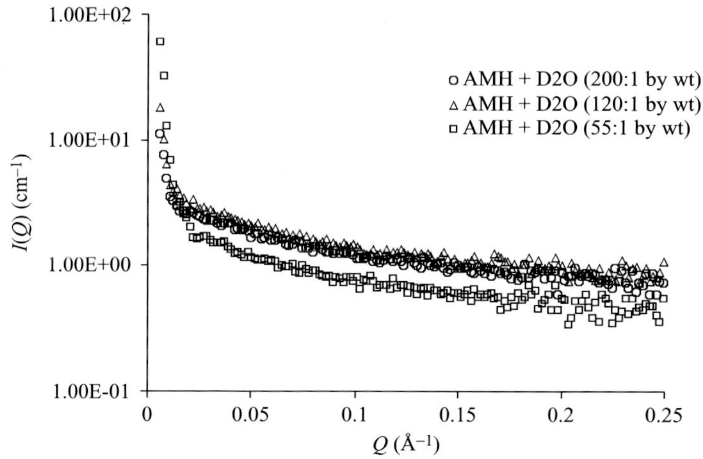

Figure 14.13. SANS of Arabian medium heavy (AMH) asphaltene in vacuum residue with added deuterated water.

incoherent scattering much smaller. This is a direct indication that water molecules are associated with the asphaltene core of the elemental particles. Moreover, the scattering intensity distribution functions are practically unchanged meaning that the water molecules are filling the void of the elementary particles only. They do not change the status of the structure of the aggregates.

6. Discussion

Small angle scattering is a sophisticated technique with obvious advantage that it is a true microscopic technique and information it carries include statistical mechanical parameters such as intra-particle and inter-particle interaction. This allows one to unambiguously determine the pair distribution function, $g(r)$. It represents the local number density of particle and is a unique quantity that can be linked to thermodynamic properties such as excess internal energy. Small angle scattering basically carries all information we need to learn about a system, from microscopic to macroscopic properties. However, the scattering intensity distribution $I(Q)$ is an integrated quantity coupled by intra-particle (form factor) and inter-particle (structure factor) scattering spectra. In order to extract information one needs to decouple $I(Q)$ into the two factors.

There are forward and backward methods to achieve this goal. The forward method is to setup two functions, one for the intra-particle and one for inter-particle and then combine them to compare with the experimental measurement. This is a modeling approach (modeling form factor and structure factor) and requires a fitting process with preset adjustable parameters. The drawback of this approach is obviously the modeling and the fitting process. In the modeling, one needs to presume a particle size, shape, and possible polydispersity distribution. This will involve at least three to four parameters. In addition, inter-particle interaction often involves at least two parameters. A total of five to six adjustable parameters are too many and can produce ambiguous results. Thus, it is necessary to minimize the number of adjustable parameters and to perform fitting with restriction. Application of fitting restriction is nontrivial and can be misleading. In previous reports, we propose several methods to justify the fitting when the modeling approach was taken.[16,17] Most of the efforts in scattering study go to developing appropriate data analysis schemes, particularly when modeling and fitting are used.

The other approach is to apply model independent analyses. The simplest one is the Guinier plot, which is applicable in the Q range where QR_g is less than unity with R_g being the radius of gyration of the particle. The advantage of this approach is that it is model independent and R_g can be accurately determined. However, there are many systems with $QR_g > 1$, mainly because of the limitation in instrument. In this case the Guinier analysis cannot be applied. Moreover, when a system is polydisperse, an average R_g will be obtained but some of the particles in the polydispersed system may not meet the $QR_g < 1$ requirement which make the Guinier analysis for a polydisperse system questionable.

Other approach is the invariant method, which is related to the surface-to-volume (S/V) ratio. This approach has an obvious merit of being able to identify the particle shape more effectively. Its restriction is that the data should be integrated to get the S/V and in many cases, the data collected do not extend to a level where intensity is close to zero. In this case, there may be error involved in the calculation of S/V and jeopardize the justification of the shape determined. In a previous work, we combined the modeling and invariant method to identify particle shape and size.[17]

Inverse Fourier transformation method is also applied to decouple the form factor and structure factor.[9,18] Xu *et al.*[22] applied this method to identify asphaltene aggregates to be spherical-like. This is a good method but may suffer similar drawback to the S/V approach because Fourier transformation requires data to practically decay to zero for integration. This is to say that the contrast and spectrometer should be well tuned and configured to meet the requirements. If there is inhomogeneity in the system or within the scattering particles, this method will have to be abandoned.

Although model independent methods have several issues to deal with, it is still a much better method to use whenever possible. This is particularly important for asphaltene research because it is a mixture system and one expected some degree of inhomogeneity from one aggregate to the other. Therefore, it is more important to obtain a statistical average than to model an individual particle. In addition, most field applications require only the statistical parameters, which can be obtained using simple analysis. To this end, the requirements become more instrument related than data analysis related.

7. Conclusion

We describe the importance of petroleomics and its relevance to the proteomics, from the application point of view. We then introduce the small angle X-ray and neutron scattering techniques for characterization of asphaltene systems.

This includes basic theory, instrumentation, sample preparation, and data analysis. Examples from Ratawi asphaltene, Arabian medium heavy asphaltene, and vacuum residues were used for demonstration. X-ray scattering data collected at the Oak Ridge National Laboratory and neutron scattering data from Argonne National Laboratory and National Institute of Standards and Technology were presented to illustrate various behaviors of these asphaltenes. Discussion on merits and drawbacks of these techniques was given in details for reader to judge what techniques may be useful for a particular system.

8. Future Perspectives

SAXS and SANS provide structural information that are related to thermodynamics and the equation of state. These techniques can potentially be used for determination of the phases of petroleum liquids and solids. It is important to identify the relevant parameters by which crucial operational parameters can be quantitatively determined and later on controlled. In the next decade or so, the central role of petroleum production will shift from sweet crude to heavy oils where flocculation, precipitation, sedimentation and other kinetically unstable situations may dominate the operations. It is thus essential to establish a simple yet accurate method for tracking the phases of petroleum liquids and solids. On the other hand, the control of the phases requires understanding from the molecular level and the colloidal level. While petroleomics starts from the molecular properties, the colloidal length scale is expected to play much more important role, at least for now, because it can link to phase separation, miscibility and the transport properties through statistical mechanical theory. Therefore, techniques such as SAXS and SANS are expected to take major responsibility for helping development and maturation of petroleomics.

Acknowledgments

I am indebted to many co-workers, students, and laboratory assistants during the process of the scattering work, which spanned more than 5 years. Technical supports from Argonne National Laboratory, Oak Ridge National Laboratory, and National Institute of Standards and Technology are especially thankful. Many thanks go to Ms. De Tar who prepared many samples and performed numerous measurements.

References

[1] Liu, Y.C. and E.Y. Sheu (1996). Low shear viscosity of a dense ionic micellar solution. *Phys. Rev. Lett.* **76**, 700.

[2] Pätzold, G. and K. Dawson (1996). Connection of microstructure to rheology in a microemulsion model. *Phys. Rev. E* **54**, 1669.

[3] Pätzold, G. and K. Dawson (1996). Rheology of self-assembled fluids. *J. Chem. Phys.* **104**, 5932.

[4] Groenzin, H. and O.C. Mullins (2000). Molecular size and structure of asphaltenes from various source. *Energy Fuels* **14**, 677.
[5] Ruiz-Morales, Y. (2002). HOMO-LUMO gap as an index of molecular size and structure for polycyclic aromatic hydrocarbons (PAHs) and asphaltenes: A theoretical study. *J. Phys. Chem.* 11283.
[6] Chen, S.H. and R. Rajagopalan (eds.) (1990). *Micellar solutions and microemulsions—Structure, Dynamics, and Statistical Thermodynamics*. Springer-Verlag, New York.
[7] Sköld, K. and D.L. Price (eds.) (1986). *Neutron scattering*. In: *Method of Experimental Physics*, Vol. 23. Academic Press, Orlando.
[8] Feigin, L.A. and D.I. Svergun (1987). *Structure Analysis by Small Angle X-ray and Neutron Scattering*. Plenum, New York.
[9] Glatter, O. and O. Kratky (eds.) (1982). *Small Angle X-ray Scattering*. Academic Press, New York.
[10] Pilz, I. (1982). Proteins. In: O. Glatter and O. Kratky (eds.), *Small Angle X-ray Scattering*. Academic Press, New York.
[11] Sheu, E.Y. (1995). Colloidal properties of asphaltenes in organic solvents. In: E.Y. Sheu and O.C. Mullins (eds.), *Asphaltene—Fundamentals and Applications*. Plenum, New York.
[12] Andreatta, G., N. Bostrom, and O.C Mullins (2006). Ultrasonic spectroscopy on asphaltene aggregation. In: O.C. Mullins, E.Y. Sheu, A. Hammami, and A.G. Marshall (eds.), *Asphaltene, Heavy Oils and Petroleomics*. Springer Academic Press, New York.
[13] Velázquez, E.S. and L. Blum (1999). Variational mean spherical scaling approximation for non-spherical molecules: The case of dimers. *J. Chem. Phys.* **110**(22), 10931.
[14] Pfeiffer, J.P. and R.N. Saal (1940). Asphaltic bitumens as a colloidal system. *J. Phys. Chem.* **44**, 139.
[15] Hiemenz, P.C. (1977). *Principle of Colloid and Surface Chemistry*. Marcel Dekker, New York, pp. 284–285.
[16] Sheu, E.Y., K.S. Liang, S.K. Sinha, and R.E. Overfield (1992). Polydispersity analysis of asphaltene solutions in toluene. *J. Coll. Int. Sci.* **153**, 399.
[17] Sheu, E.Y. (1998). Self-association of asphaltenes: structure and molecular packing. In: O.C. Mullins and E.Y. Sheu (eds.), *Structures and Dynamics of Asphaltenes*. Plenum, New York.
[18] Brunner-Popela, J. and O. Glatter (1997). Small-angle scattering of inter-acting particles. I. Basic principles of a global evaluation Technique. *J. Appl. Cryst.* **30**, 431–442.
[19] http://www-als.lbl.gov/als/synchrotron_sources.html
[20] Yen, T.F. (1972). Present status of the structure of petroleum heavy ends and its significance to various technical applications. *Am. Chem. Soc., Div. Petrol. Chem. Preprint* **17**(1), 102–104.
[21] Yen, T.F. (1981). Structural differences between asphaltenes isolated from petroleum and from coal liquids. In: J. Bunger and N.C. Li (eds.), *Chemistry of Asphaltene*. Advance in Chemistry series 195. American Chemical Society, New York.
[22] Xu, Y.N., Y. Koga, and O.P. Strausz (1995). Characterization of athabasca asphaltenes by small-angle X-ray scattering. *Fuel* **74**(7), 960.

15

Self-Assembly of Asphaltene Aggregates: Synchrotron, Simulation and Chemical Modeling Techniques Applied to Problems in the Structure and Reactivity of Asphaltenes

Russell R. Chianelli, Mohammed Siadati, Apurva Mehta, John Pople, Lante Carbognani Ortega, and Long Y. Chiang

1. Introduction

Increased understanding of the structure and chemistry of asphaltenes is essential to developing ways of mitigating the effects of asphaltenes, destroying them or finding new uses for them. The chemical structure and physical structure of the asphaltenes are unique and much has been learned about their physics and chemistry.[1] However, there are still fundamental questions regarding the origin and structure of asphaltenes that remain to be answered. In this report, new synchrotron WAXS (wide angle x-ray scattering data) and SAXS (small angle x-ray scattering data) for Venezuelan and Mexican asphaltenes are reported showing the ubiquitous presence of the "asphaltene particles" with sizes in the 3–5 nm ranges. The particles exist both as correlated packets in the precipitated asphaltene and in the parent crude oil as individual particles. Furthermore, in the second section of this report the self-assembly of the "asphaltene" particles from model compounds is reported. That the "asphaltene particles" can self-assemble indicates the basic stability of the particles and generates interesting questions regarding the origins of petroleum.

Increasingly, heavy crudes are becoming a major source of petroleum hydrocarbons as lighter crudes become scarce. One major difference between a light crude and a heavy crude is the asphaltene content of the crude. The asphaltene fractions contain most of the metals in the crude and generally more sulfur and

Russell R. Chianelli and Mohammed Siadati • Materials Research and Technology Institute, University of Texas at El Paso, El Paso, Texas. **Apurva Mehta and John Pople** • Stanford Synchrotron Radiation Laboratory, Stanford, California. **Lante Carbognani Ortega** • Consultant, Caracas, Venezuela, Present address: University of Calgary, Alberta, Canada. **Long Y. Chiang** • University of Massachusetts, Lowell, Massachusetts.

nitrogen than the rest of the crude. Asphaltenes are complex mixtures of polyaromatic molecules containing large amounts of sulfur, nitrogen, and metals. They are difficult to convert to lighter fractions and contain metals that foul catalysts and present a disposal problem. They also affect viscosity, causing plugging problems in production and transportation operations.[2] Currently, the asphaltenes can either be converted with great difficulty and expense by addition of hydrogen at high temperature and pressure or they can be destroyed by coking or other means of disposal. The future will require either better means of conversion of asphaltenes or new uses to make the asphaltene more valuable than fuel. Both these approaches require deeper knowledge of the asphaltene chemistry and physics, despite much has been learned about them.[3]

Asphaltenes are thought to be the remains of biological molecules from which the petroleum was formed. Therefore they contain metals like vanadium and nickel in phorphyrinic ring-like structures reminiscent of biological molecules.[4] However, their biological origins and how they assemble into the structures that we see in petroleum is still open to discussion. Also open to debate is the existence of petroleum micelles from a classical colloidal point of view. "Aggregates" are the structures believed to describe better the observed "micellar" behavior of crude oils.[5]

Asphaltenes occur as colloidal suspension in hydrocarbon liquids. The composition of the crude and physical parameters such as pressure determines whether the asphaltenes remain in solution. As a result of production, transportation, and refining of crude oils the composition of the crude changes. These changes often cause the asphaltenes to precipitate resulting in plugging or incompatibility problems. Understanding heavy crudes requires that the phase behavior of the crudes is understood under all conditions relevant to production, transportation, and refining processes. Many techniques have been applied to develop understanding of the complex asphaltene systems. Techniques applied include NMR,[6] STM (scanning tunneling microscopy),[7] and others. Optically, anisotropic structures are also seen in asphaltene containing petroleum residues indicating a degree of molecular order within the asphaltene. Generally, these techniques describe the average degree of aromatic condensation to be approximately seven rings. For example, in the cited references, Maya asphaltenes were imaged by STM in dilute solutions of THF on highly oriented pyrolytic graphite. The sizes and structures of the asphaltenes were observed in the STM. Asymmetric structures were observed with dimensions averaging 1.04 ± 0.19 nm. Studies such as these establish the dimensions of the aromatic cores of asphaltenes.

It is, however, the existence of larger structures occurring in asphaltenes that are the subject of this report and the ability of these structures to "self-assemble." X-ray and neutron scattering studies have shown the existence of discrete particles of approximately 3–5 nm in crudes. These and larger structures (aggregates $\sim$20 nm and super aggregates) have been postulated to explain scattering data in various flocculated asphaltenes and their literature has been thoroughly reviewed.[8] New synchrotron WAXS and SAXS for Venezuelan and Mexican asphaltenes are reported showing the ubiquitous presence of these "asphaltene particles."

Furthermore, in the second section of this report the self-assembly of the asphaltene particles from model compounds is reported. That the asphaltene particles can self-assemble indicates the basic stability of the particles and underlines that the original model of Yen is essentially correct.[9]

2. WAXS Synchrotron Studies and Sample Preparation

A WAXS study on a Mayan crude oil was performed to demonstrate the information obtained from this technique. X-ray scattering techniques performed at synchrotron facilities can determine heavy crude phase behavior directly in either the liquid or solid state. The sample studied was the Mayan petroleum crude oil, *Miguel Hildago* that is a typical Mexican heavy crude. Both the crude oil and the heptane insoluble fraction of the crude, which is by definition the asphaltene fraction of the crude, were studied. Scattering techniques are complimented by other synchrotron techniques such as XAFS (x-ray adsorption fine structure). XAFS determines chemical state and molecular structure of the Ni and V in asphaltenes under various conditions.

The synchrotron studies were performed at SSRL (Stanford Synchrotron Radiation Laboratory) from 2000 to 2003 under a grant from the DoE (Department of Energy) BES (Basic Energy Sciences). WAXS data were taken on beamline 2–1 with 10 keV radiation. Some studies were performed at an earlier time as indicated in the cited references. Standard sample preparation techniques were used and can also be found in the cited references.

Information that can be extracted from WAXS asphaltene data is indicated schematically in Figure 15.1. There are three essential features:

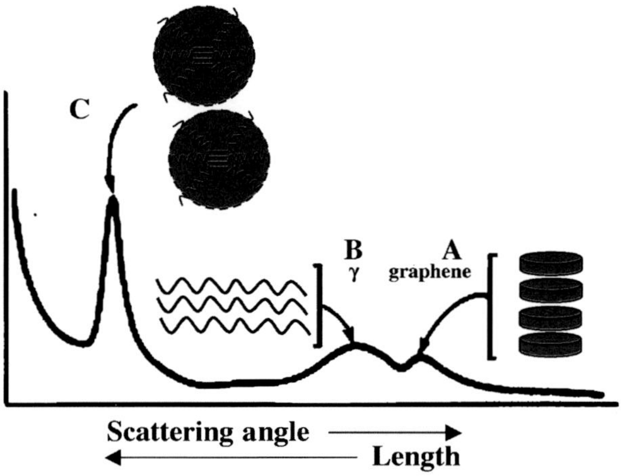

Figure 15.1. Schematic of information obtained from WAXS (wide angle x-ray scattering) of asphaltenes. Region A: aromatic stacking (the graphite 002 often called the graphene peak), Region B: coherence of paraffin interaction (the γ of Yen), and Region C: micellar aggregation diameter.

- Low angle peak (*C*) indicating asphaltene aggregation (30–40 Å).
- Peak (*B*) typically occurring between 4.0 and 5.0 Å. This peak is the so-called γ peak that reflects coherence between paraffin chains in the asphaltene.
- Peak (*A*) that occurs between 3.4 and 3.5 Å (L_g = graphitic stacking). This peak is the so-called graphene peak that reflects the graphitic stacking within the asphaltene core.

Peaks at higher angles are the graphitic peaks 10 (100) and 11 (110) that reflect the in plane order parameter for the graphitic sheets and generally give a "diameter," measured by line broadening analysis, for the aromatic core of between 8 and 17 Å (L_a = diameter of aromatic stacks).

The WAXS data were collected at SSRL on beam line 2–1. The vertical collimation and high brightness of the synchrotron beam allowed use of an Si (111)-based detector suitable to resolve lattice changes of the order of 0.1%. The size of the focused beam was 2 × 1 mm and approximately 10^{11} photons/s are incident on the sample. The WAXS patterns were collected as close as possible to the direct beam. Generally, the direct beam interfered at $2\theta = 2°$. The data collection scan continued to $2\theta = 120°$ at the Zn K-edge (9.659 keV).

Quantitative information could be obtained using the x-ray scattering intensity for a collection of atoms:

$$I_{eu} = \sum_m \sum_n f_m f_n e^{iS \cdot R_{mn}},$$

where f_m is the x-ray atomic scattering factor of m-type atoms, S is the x-ray scattering wave vector with $S = |\mathrm{S}| = 4\pi \sin\theta/\lambda$, and the vector R_{mn} connects atom m and atom n. Assuming a random (powder) arrangement of the structure with respect to the incoming x-ray beam, a spherical average gives the Debye scattering equation:

$$I_{eu} = \sum_m \sum_n f_m f_n \frac{\sin SR_{mn}}{SR_{mn}}.$$

The full widths at half-maximum (FWHM) of the (002) peaks were measured directly from the x-ray patterns in order to approximate crystallite dimensions of the *graphene* sheets in the *c*-axis direction using the Debye–Scherrer relation:

$$D_{002} = \frac{k_{002}\,\lambda}{\beta_{002}\cos\theta},$$

where D_{002} is the dimension of the particle along the stacking direction, λ is the wavelength of the x-rays ($\lambda = 1.2836$ Å), θ is the diffraction angle, and β_{002} (or FWHM) is the angular line width. The shape factor k_{002} depends on the shape of the particle and is equal to 0.76 for random layer lattice, which can be used for asphaltene structure.[10] The resulting graphitic stacks or graphene sheets are part of the asphaltene core as described by Yen.

The crystalline-order along the basal direction can be evaluated using the Debye–Scherrer equation applied to the widening of the (110) diffraction peak.

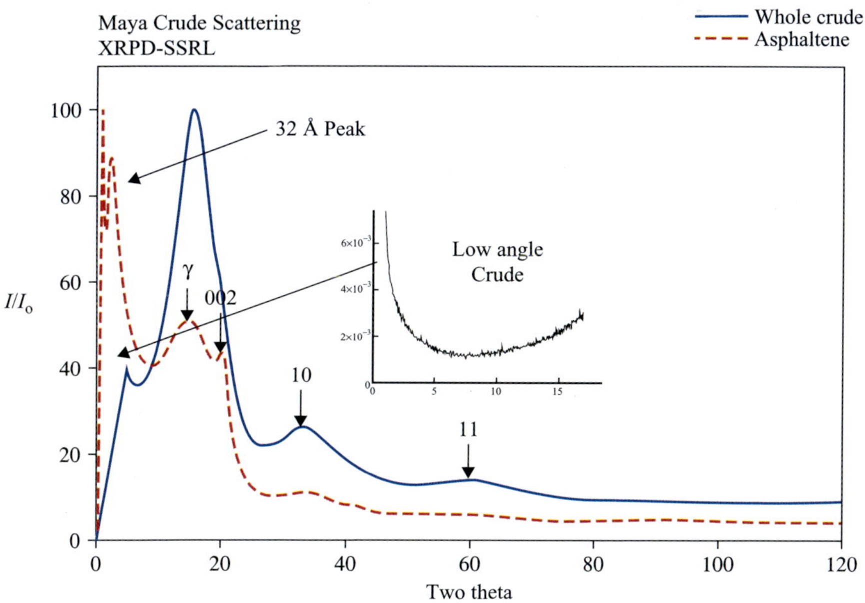

Figure 15.2. WAXS from Maya whole crude and asphaltene, insert is background near the direct beam for the whole crude showing absence of the Bragg reflection.

As with the (002) peak, the (110) peak is not influenced by imperfect stacking or bending/folding of the layers. In that case, the shape factor k_{110} varies with the β_{110} angular line width, but it can be determined following the values reported by Liang *et al.* using computer calculations of the scattered x-ray intensity for model-layered lattice structures. According to the experimental angular line widths measured in the present study, k_{110} values vary between 1.42 and 1.56. Details of the random layer lattice scattering analysis can be found in Perez De la Rosa *et al.*[11]

The data for Maya crude and asphaltene are shown in Figure 15.2. The measured data are shown in Table 15.1. All parameters are consistent with Yen's original survey of asphaltenes in the previously cited references. One aspect that is important is that the low angle (32 Å) peak is not seen in the whole crude. This indicates that the asphaltene micelles are not correlated in the whole crude if they

Table 15.1. WAXS Data Summary for Maya Asphaltenes

WAXS peak	d (graphitic)	$\omega \cos\theta$	L	#Repeat
Low angle	$d = 32$	0.0174	111 Å	~3
Saturate	$d_\gamma = 4.78$	0.336	4.2 Å	~1
Graphene	$d_{(\text{grahene})} = 3.53$	0.052	14.0 Å	~4
100	$d_{(100)} = 2.03$	0.2485	9.17 Å	~4
110	$d_{(110)} = 1.23$	0.2563	9.00 Å	~7

are present. The next section discussing small angle x-ray scattering (SAXS) gives further evidence that they do exist in the whole crude.

3. SAXS

Small angle x-ray scattering is used to study structures of size about 10 Å or larger. This technique is particularly applicable to providing information regarding the micellar structure of asphaltenes as described previously.

A brief description of the technique and its application to the study of asphaltenes follows. A more in depth description of the techniques can be found in reference 12. The SAXS intensity of the investigated material (q) is recorded as a function of the angle of scattering (2θ), where q is the reciprocal space scattering vector and is related to the real space geometry as:

$$q = (4\pi/\lambda)\sin\theta.$$

Considering the Bragg law:

$$\lambda = 2d\sin\theta,$$

where d is the real space distance. The inverse relation between q and d is:

$$qd = 2\pi.$$

Monochromatic x-rays are scattered from the sample and collected on a CCD camera. The differential scattering cross-section is expressed as a function of the scattering vector q. The value of q is proportional to the inverse of the length scale (Å^{-1} or nm^{-1}). Whenever the sample contains a scattering length density inhomogeneity of dimension larger than ~10 Å, scattering becomes observable in the small-angle region, and its study requires the technique of SAXS. Information on such relatively large-scale structures is contained in the intensity patterns of the scattered x-rays at small angles, typically at 2θ less than 2°. The reciprocity between size of the scattering object and q means that information on relatively large sizes is contained in $I(q)$ at small q.

Guinier law: When the sample contains particles of unknown shape, or when the shape is irregular and not describable in simple terms, the scattering function in the limit of small q is given by

$$I(q) = (\rho_o v)^2 \exp(-q^2 R_g^2/3),$$

where $I(q)$ is the intensity of independent scattering by a particle. This relation is known as the Guinier law and allows determination of the radius of gyration R_g of a particle of unknown shape and size from small-angle scattering measurement. Based on the Guinier law, when the logarithm of $I(q)$ is plotted against q^2 the initial slope gives $R_g^2/3$. Radius of gyration is the root mean-square distance of all points in the particle from its center of mass.

Porod Law ($\ln I[q]$ vs. $\ln q$): As q increases, the curve falls off rapidly for spheres and less so for disks and rods and, the asymptotic form of the intensity curves at large q can be represented by

$$I(q) \propto q^{-a}.$$

At large q, the most important theoretical result in accord with the Porod law is the prediction that $I(q)$ should decrease as $\sim q^{-4}$. An exponent of 4 in ln $I(q)$ vs. ln$q$ plot indicates a 3D spherical particle with smooth surface. Values of 2 and 1 indicate 2D thin disks and 1D thin rods, respectively. Therefore, the power-law exponent at large q reflects the dimensionality of the scattering object.

3.1. Fractal Objects

At large q the intensity $I(q)$ of the scattering from a sphere decay as q^{-4}, from a thin disk as q^{-2}, and from a thin rod as q^{-1}. The power-law exponent at large q is therefore seen to be related to the dimensionality of the scattering object. There are, however, many cases in which the intensity varies as unexpected or even fractional power of q. The inverse power-law exponents that differ from 1, 2, or 4 can be explained in terms of the concept of a *fractal*. Mandelbrot promulgated the description of complex patterns in nature in terms of *fractal geometry*.[13] The concept has been applied to the study of increasing numbers of irregular objects in all branches of science.[14–16] A well-known example of a fractal is the length of a coastline, that increases in length as the yardstick with which it is measured is made smaller. Other examples are the irregular aggregates of tiny silica or soot particles, the pattern of dendritic growth of crystals, the trace left by an electric discharge starting from a point in a dielectric, and the shape of a polymer coil, etc. A fractal possesses *dilation symmetry*, that is, it retains a *self-similarity* under length scale transformations. In other words, if we magnify part of the structure, the enlarged portion looks just like the original. In a mathematically defined fractal object, this self-similarity extends from an infinitesimally small to an infinitely large scale, but in an object occurring in nature, there is an upper bound imposed by the largest dimension of the object and a lower bound due to the size of the basic building blocks of the structure. A fundamental characteristic of a fractal is its *fractal dimension*. For example, if a sphere of radius r is drawn around a point in the object, then the fractal object is a:

line, if the mass $M(r)$ within the sphere is proportional to r
sheet, if the mass $M(r)$ within the sphere is proportional r^2
solid, 3D object, if the mass $M(r)$ within the sphere is proportional to r^3.

Therefore, in a fractal the following general relation is obeyed:

$$M(r) \propto r^d,$$

where the fractal dimension (d) is a number between 1 and 3. An illustration for hypothetical asphaltenes is indicated in Figure 15.3. Typically, as discussed below asphaltenes occur in disc-like, elliptical or spherical forms depending on their origin or their subsequent treatment.

The fractal dimension (d) can also be fractional. The smaller the value of d, the more open the structure is, and as d is reduced to 1, the object becomes a line

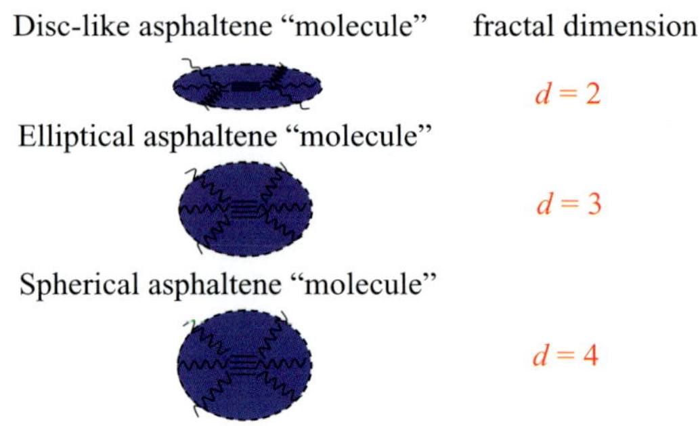

Figure 15.3. Fractal dimension for disc-like, elliptical, and spherical asphaltene.

if it remains singly connected. Since the volume of the sphere is proportional to r^3, the density (r) of actual material embedded in it is

$$\rho(r) \propto r^{d-3},$$

this shows that the density is no longer a constant of the object but rather decreases as the size of the volume being considered is increased. The fractal object discussed in the preceding paragraph is called a *mass fractal*. Some objects possess a surface that is rough and exhibit fractal properties. Such an object is called a *surface fractal*. The moon pockmarked with craters of all sizes and a clump of cauliflower are both examples of a surface fractal. An island with a fractal coastline is an example of a surface fractal in 2D space.

The 2D (surface fractal) is easier to visualize than a 3D one. Imagine we cover an island completely with square tiles of edge length l, and we mark those tiles that at least partially overlap the coastline. Suppose $N(l)$ is the number of tiles are so marked. If the coastline is smooth and nearly straight, $N(l)$ will be proportional to l^{-1} as we use tiles of different size l. If the coastline is irregular and fractal, the number $N(l)$ of marked tiles depends more strongly on l, and is proportional to l^{-d_s} where d_s is a number larger than 1. The length $L(l)$ of the coastline is then:

$$L(l) \propto l^{1-d_s}.$$

The number d_s is the *fractal dimension* of this 2D surface fractal. The fractal dimension of a 3D surface fractal can be defined in a similar manner. In particular, if $S(r)$ is the surface area measured with a measuring tool of characteristic area r^2, then:

$$S(r) \propto r^{2-d_s}.$$

The value of d_s ranges from 2 to 3 for a surface fractal in 3D space. It is equal to 2 when the surface is perfectly smooth and approaches 3 when the surface is so folded that it almost completely fills the space (such as a tightly crumpled napkin).

3.2. Scattering from Mass Fractal Objects

If we consider a mass fractal object as a distribution of mass points, the normalized correlation function $y(r)$ is the probability of finding a mass point a distance r apart from an arbitrary mass point selected within the object. We construct a spherical shell of radius r and thickness dr around the selected point. Based on $M(r) \propto rd$, the number of mass points enclosed in the shell is proportional to $4\pi r^{d-1}dr$. Since the volume of the shell is equal to $4\pi r^2 dr$, the correlation function is

$$\gamma(r) \propto r^{d-3}$$

whose range of validity is $R \gg r \gg a$, where R is the overall dimension of the object ($\sim R_g$), and a is the size of the basic building block of the structure, which could be as small as an atom or a molecule. Valid for $1/R \ll q \ll 1/a$, the scattering intensity $I(q)$ is

$$I(q) \propto q^{-d}.$$

This relation indicates that the intensity of scattering from a mass fractal decays with q with an exponent between -1 and -3.

3.3. Scattering from a Surface Fractal Object

We regard the system obeying an ideal two-phase model, and that its interface boundary, instead of being smooth, is now fractal. Often the second phase is simply a vacuum, we draw a sphere of radius r from every point on the phase boundary. The larger the radius r, the smoother the surface will be. Its area $S(r)$ is given by

$$S(r) = S_0\, r^{2-d_s},$$

where S_0 is a constant, which is the surface area itself when $d_s = 2$ (smooth surface). The scattering intensity $I(q)$ is

$$I(q) \propto q^{-(6-d_s)}.$$

A log–log plot of $I(q)$ against q will therefore give a straight line, with the slope equal to $(6 - d_s)$. Since d_s for a 3D surface fractal is between 2 and 3, the exponent of q is to be between -3 and -4 (the latter limit corresponds to the *Porod law* for a smooth interface boundary).

4. SAXS Studies of Venezuelan and Mexican Asphaltenes

SAXS data on selected asphaltenes described below was performed on beamline 1–4 of the Stanford Synchrotron Radiation Laboratory (SSRL) at the Stanford Linear Accelerator Center (SLAC), in Stanford, CA. Beamline 1–4 focused x-ray source with a flux of 10^{10} photons on a spot size of 0.5 mm (vertical) $\times$ 1 mm (horizontal). The radiation is monochromatic, reflected from a [111] Si crystal (which is also bent to provide horizontal focusing) to a wavelength of $\lambda = 1.488$ Å. The

Table 15.2. Asphaltenes Used in the SAXS Study

Crude	Sample	Comments
Boscan	Whole crude oil	Heavy oil
GU	FEEDGU	Vacuum resid
GU	HCKGU	Hydrocracked vacuum resid
S1-O	S1-0	Unstable oil
Maya	Miguel Hildago	Heavy oil

samples were mounted in a holder manufactured at SSRL with two 25 μm thick KAPTON windows with an active path length of sample material ~1 mm. The 16 bit SAXS data were collected at room temperature on a cooled CCD-based area detector. These data were corrected for background scattering and scattering from the sample cell windows. One-dimensional profiles of the data were acquired by radial integration routines. The q range sampled was $q_{\text{initial}} < q < q_{\text{final}}$ Å^{-1}, (where q is the scattering vector: $q = 4\pi \sin\theta/\lambda$ for x-ray photons of wavelength λ scattered through an angle of 2θ).

Four Venezuelan and one Mexican asphaltenes were studied using the SAXS technique. The Venezuelan asphaltenes have been extensively characterized by one of the authors.[17, 18] The asphaltenes studied are summarized in Table 15.2. It can be noted that the studied hydrocarbon fractions are representative of the heaviest and most difficult to convert feedstocks. The Maya crude oil contains 3.5% sulfur and consists of 33.9% 1050 + (fraction boiling above 1050°F). The Boscan crude oil contains 5.5% sulfur and consists of 55.5% 1050+ according to the PetroPlan assay list.[19]

The small q and the large q scattering regions are seen in Figures 15.4 and 15.5, respectively. The *Guinier* plots [ln (I) vs. q^2] are shown in Figures 15.6 and 15.7. The *Porod* plots [ln(I) vs. ln(q)] are shown in Figures 15.8 and 15.9. The *Guinier* plots yield information regarding the radius of gyration (R_g) in the large and small real space distances. The Venezuelan asphaltenes have cores in the 3–5 nm size range and larger aggregates in the 50 nm region. The *Porod* data indicate that based on the fractal dimension previously described the 3–5 nm cores are disc-like or elliptical in nature; while the 50 nm aggregates are spherical in nature. We further notice in Figure 15.10 that three of the Venezuelan asphaltenes also show Bragg reflections that occur in the region of 3.2–4.1 nm in good agreement with the *Guinier* analysis. The presence of the Bragg peaks in the asphaltenes indicates that the cores are correlated in all the samples except the hydrocracked asphaltene. The correlation of asphaltene cores is discussed further below. These data are summarized in Table 15.3 and shown schematically in Figures 15.11 and 15.12. In Table 15.3 and Figure 15.11, we see that the data analysis indicates that there is an elliptical core with the fractal dimension (d) varying from 2.01 to 2.81 nm. According to the analysis described above the asphaltene cores can be described as mass fractals ($1 < d < 3$). As the fractal dimension decreases the structures are described as more porous. Thus, the Boscan heavy crude asphaltene is more porous than the more aromatic asphaltene from a vacuum resid suggesting

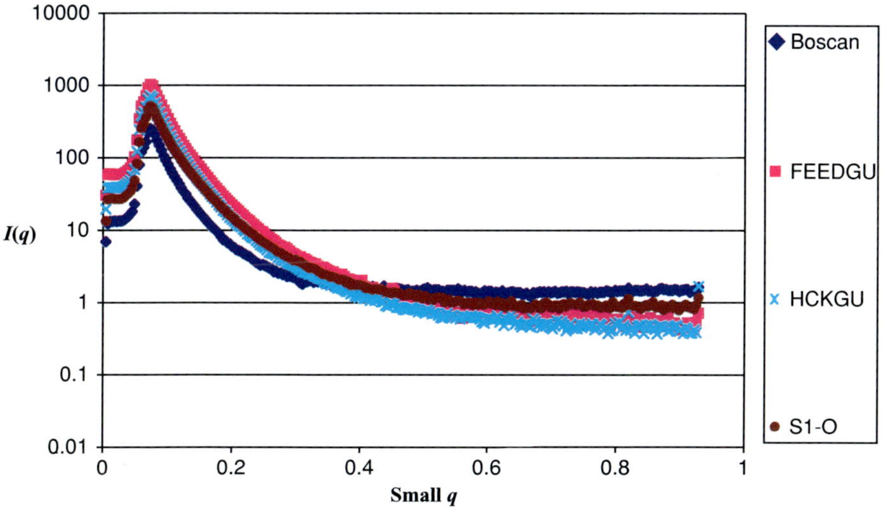

Figure 15.4. $I(q)$ vs. q for Venezuela asphaltenes in the small q region.

further changes during treatment. More work is required to see if the fractal analysis described above gives real information regarding structural changes in asphaltenes as they are processed. However, the previously mentioned agreement between the Porod analysis and the Bragg reflection data is a solid result indicating the core existence and correlation.

In Table 15.3, we also see the longer range analysis that indicates that there is an association in the 25.6–25.8 nm range. In this case the fractal dimension is in the range from 3.6 to 4.0. In this range as indicated in the previous section the aggregates are described as surface fractals ($3 < d < 4$). This indicates that the

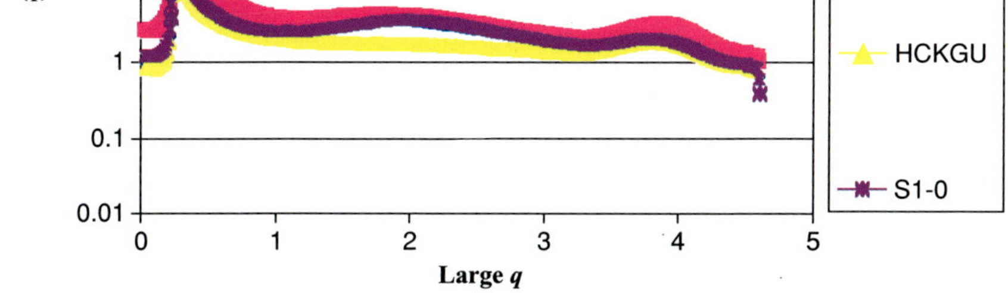

Figure 15.5. $I(q)$ vs. q for Venezuela asphaltenes in the large q region.

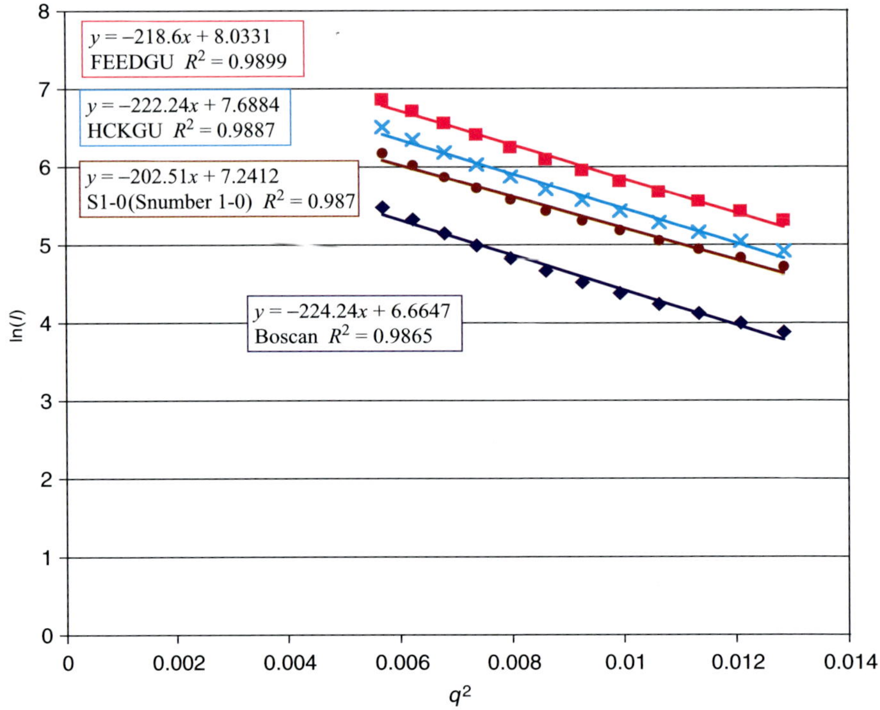

Figure 15.6. *Guinier* plot $\ln(I)$ vs. q^2 for Venezuelan asphaltenes in the small q region.

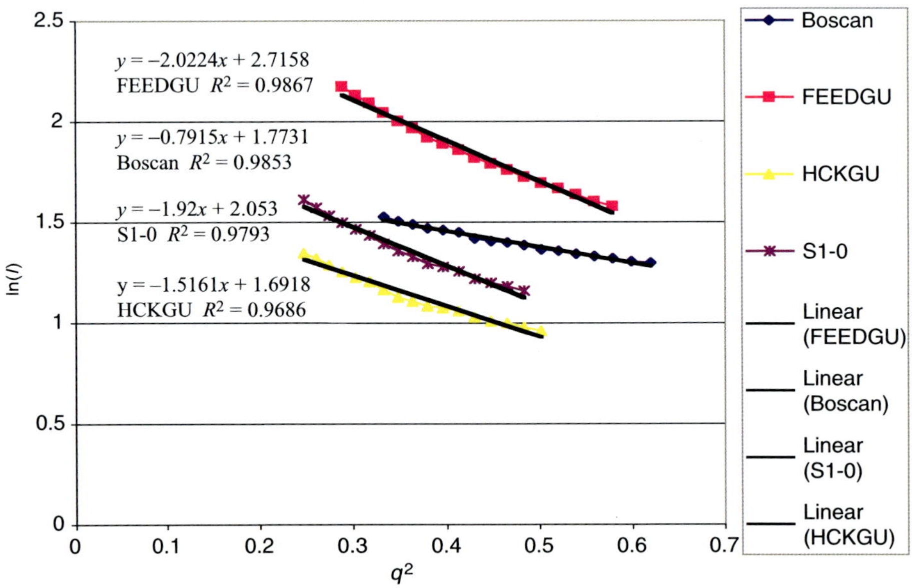

Figure 15.7. *Guinier* plot $\ln(I)$ vs. q^2 for Venezuelan asphaltenes in the large q region.

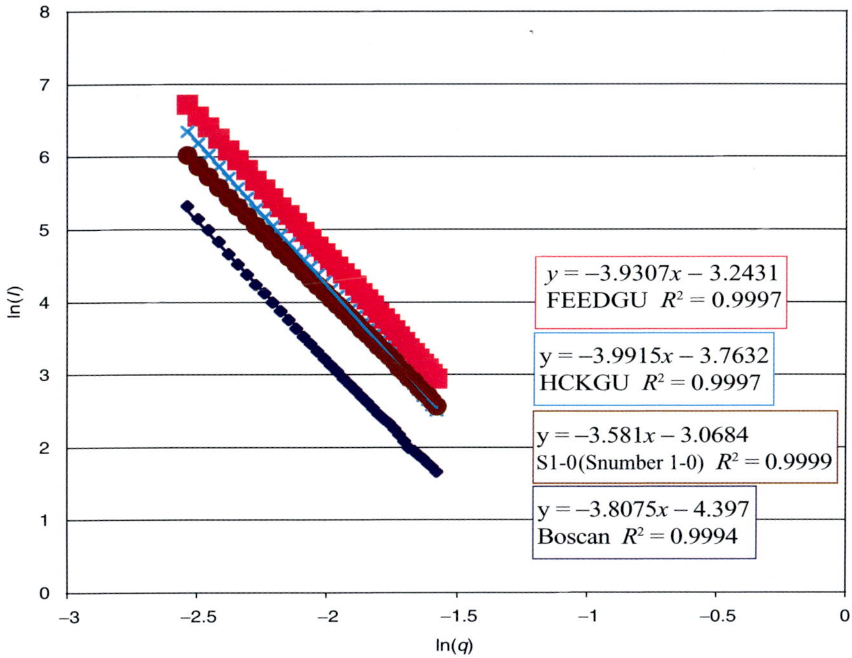

Figure 15.8. *Porod* plot ln(I) vs. ln(q) for Venezuelan asphaltenes in the small q region.

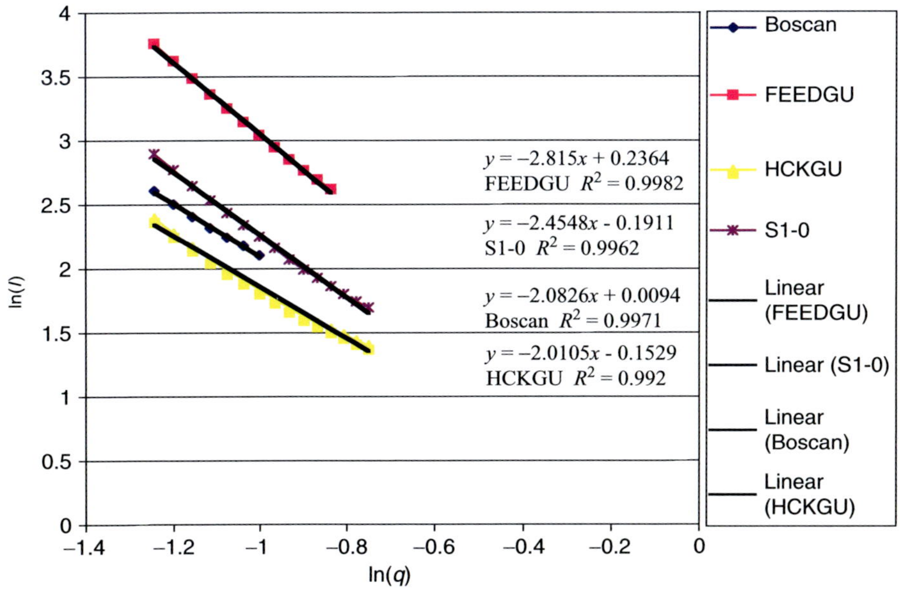

Figure 15.9. *Porod* plot ln(I) vs. ln(q) for Venezuelan asphaltenes in the large q region.

Table 15.3. SAXS Data Summary for Venezuelan Asphaltenes

Asphaltene	q^2 Guinier R_g (nm) large	ln(q) Porod fractal dim.	q^2 Guinier R_g (nm) small	ln(q) Porod fractal dim.	Bragg peak (nm)
Boscan heavy crude	25.9	3.8	1.54	2.08	4.06
Feed GU vacuum resid	25.6	3.9	2.46	2.81	3.44
Unstable S1 0 crude	24.6	3.6	2.4	2.46	3.17
Hydrocracked GU vacuum resid	25.8	4.0	2.13	2.01	—

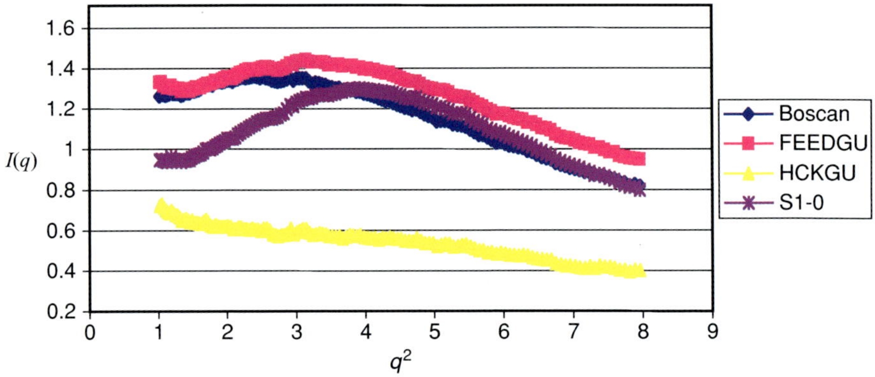

Figure 15.10. *Bragg* region ln(I) vs. q^2 for Venezuelan asphaltenes indicating micelle formation.

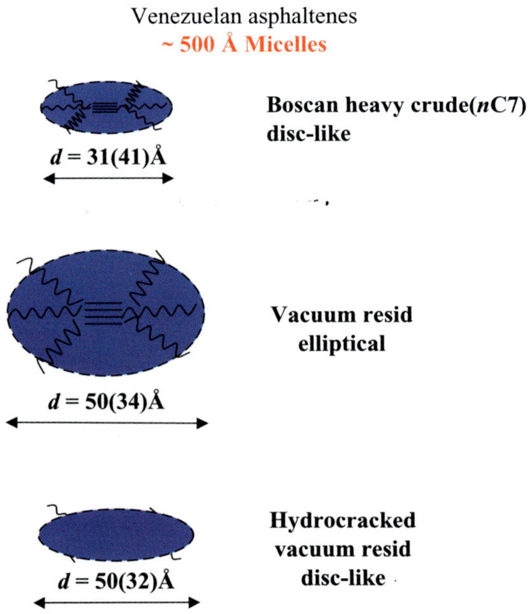

Figure 15.11. Properties of Venezuelan asphaltenes.

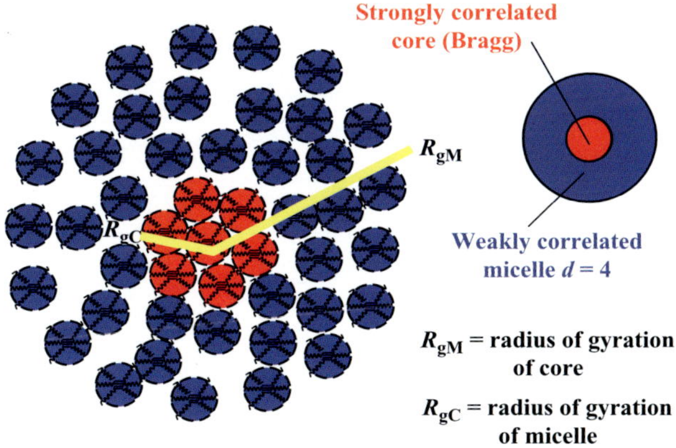

Figure 15.12. Schematic of Venezuelan asphaltene aggregation.

aggregates are porous or rough, with the porosity increasing as the fractal dimension approaches 3. The analysis could also be done in terms of d_s as previously indicated. However, this information is not contained in Table 15.3 for simplicity. The results of this analysis are shown schematically in Figure 15.12.

A similar situation can be seen in analysis of the data from the Maya asphaltenes. In this case the original crude oil was also analyzed. Figures 15.13 and 15.14 show the data for small and large q. Figures 15.15 and 15.16 show the *Guinier* plots and the linear regression fits. Figures 15.17 and 15.18 show the *Porod* plots. The analysis is summarized in Table 15.4. The *Porod* analysis indicates that Maya asphaltene has a core with a diameter of 3.76 nm in reasonable agreement

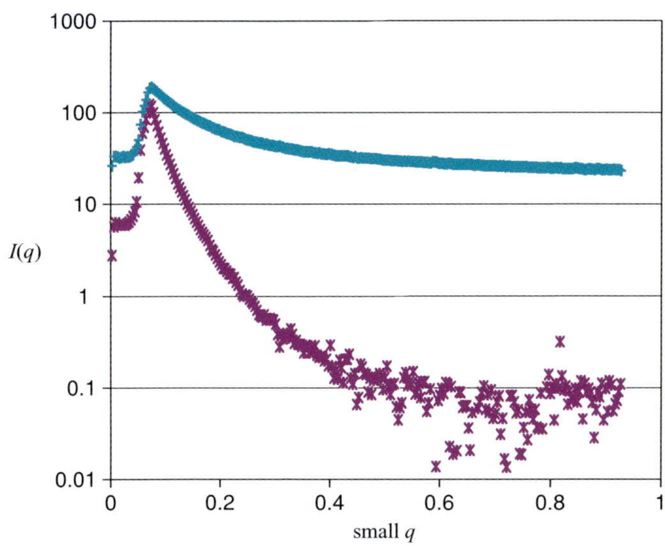

Figure 15.13. $I(q)$ vs. q for Maya asphaltenes/crude in the small q region.

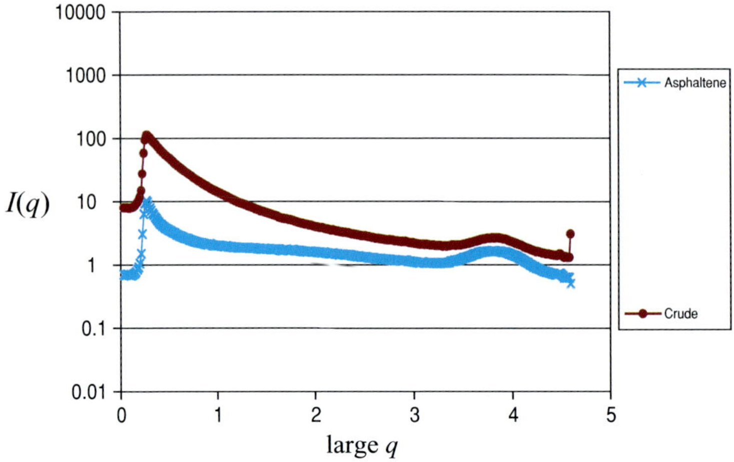

Figure 15.14. $I(q)$ vs. q for Maya asphaltenes/crude in the large q region.

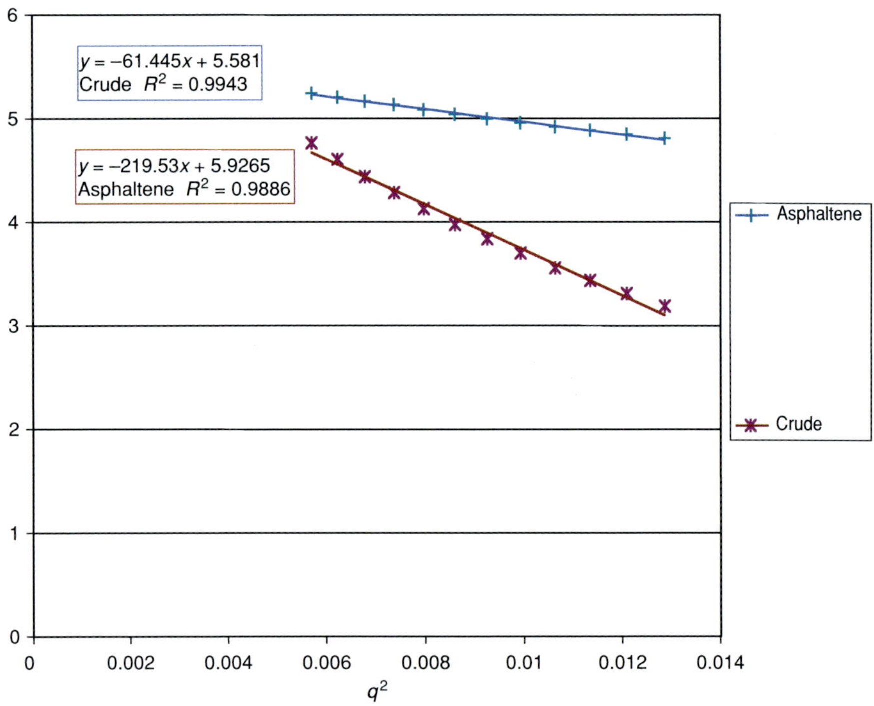

Figure 15.15. *Guinier* plot $\ln(I)$ vs. q^2 for Maya asphaltenes/crude in the small q region.

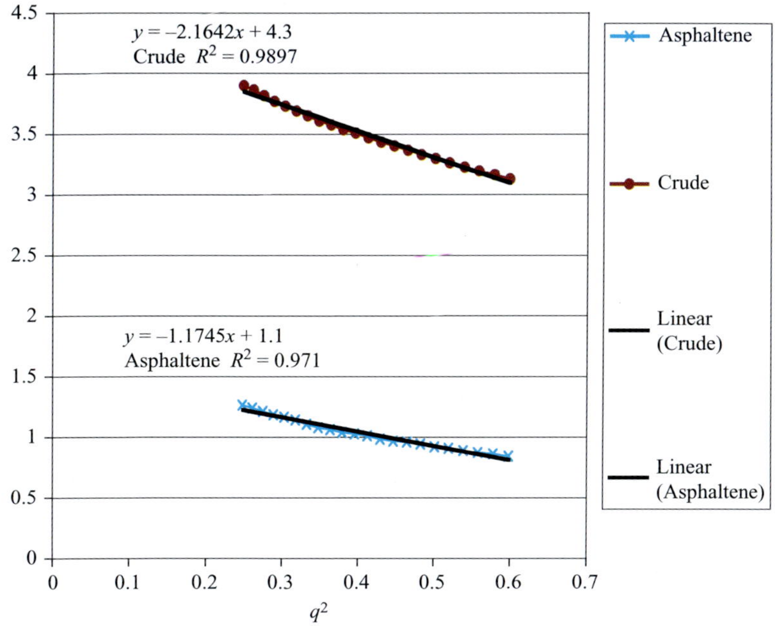

Figure 15.16. *Guinier* plot ln(I) vs. q^2 for Maya asphaltenes/crude in the large q region.

with the position of the Bragg reflection at 3.22 nm. The core is also correlated as in the case of the Venezuelan asphaltenes. Finally, the *Porod* analysis for the original crude oil shows unambiguously the presence of the asphaltene cores at 5.10 nm. In their suspended state in the crude oil they are uncorrelated as would be expected. The meaning of the secondary aggregation and the very low fractal dimension is not known at this writing and further work is required. Nevertheless, the power of the SAXS analysis is clearly indicated and the presence of the asphaltene core is clearly demonstrated in all the cases studied and in particular in the whole crude.

We see a further possible interpretation of the aggregated asphaltene in Table 15.5. In this table the approximate number of repeated aggregate units is presented. The coherence length indicated in the table is calculated by applying the standard Debye–Scherrer equation previously described. The number of repeats is calculated by dividing the coherence length by the core diameter. For

Table 15.4. SAXS Data Summary for Maya Asphaltenes/Crude

Asphaltene	q^2 Guinier R_g (nm) large	ln(q) Porod fractal dim.	q^2 Guinier R_g (nm) small	ln(q) Porod fractal dim.	Bragg peak (nm)
Maya crude	13.6	1.1	2.55	1.76	—
Miguel hildago					
Maya asphaltene	6.2	4	1.88	1.73	3.2 (2)

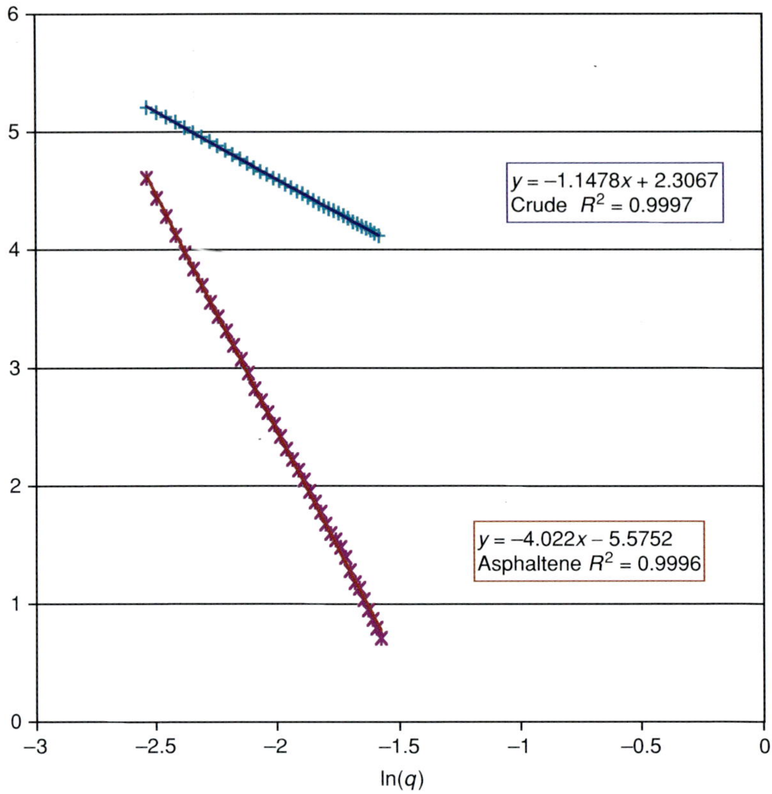

Figure 15.17. *Porod* plot ln(I) vs. ln(q) for Maya asphaltenes/crude in the small q region.

the Maya asphaltene the average number of repeats is 2 and the model indicated in Figure 15.12 is a fair representation of asphaltene aggregation as understood by application of x-ray scattering techniques. In the Venezuelan asphaltenes the repeat numbers are significantly higher indicating a higher degree of aggregation. In the case of the vacuum residuum asphaltene (FEEDGU) the number of repeats actually indicates that the asphaltene cores are correlated to a much greater extent than indicated in Figure 15.12. Further work is required to extend the interpretation of the WAXS and SAXS data in understanding asphaltene structure. However,

Table 15.5. Approximate Repeats for Asphaltene Cores

Crude	Sample	Coherence length, D (nm)	Average repeat
Boscan	Heavy Crude	11.3	~3–4
GU (Vac. Resid)	FEEDGU	22.6	~5–6
Instable Medium Crude oil	S1-0	11.3	~3–4
Maya	Miguel Hildago	6.0	~2

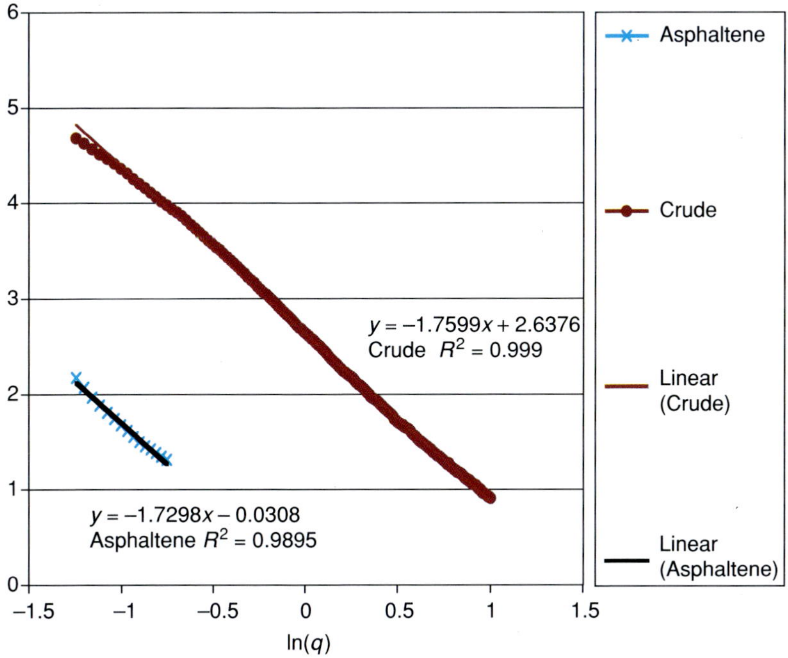

Figure 15.18. *Porod* plot $\ln(I)$ vs. $\ln(q)$ for Maya asphaltenes/crude in the large q region.

it is interesting to note that the heavier Boscan crude oil has a higher degree of asphaltene aggregation than that of the "lighter" Maya heavy crude oil.

5. Self-Assembly of Synthetic Asphaltene Particles

The ubiquitous nature of the asphaltene particles was discussed in the previous sections. Asphaltene particles in the 30–40 nm range occur with great frequency. These structures appear to be highly stable structures that appear during the process of transformation of kerogen to petroleum hydrocarbon. Actual chemical compositions of naturally occurring asphaltenes are rather complex and varied in their derivative sources. To facilitate our understanding of the process of asphaltene formation and the physical characteristics of synthetic asphaltenes, simplified polyaromatic discotic molecules were used in a thermal simulation model of "synthetic asphaltene." The thermal model produces polycondensed structures in the presence of reactive discotic polyaromatics and aliphatic oil intermediates that possess chemistry in a close resemblance to conditions of the asphaltene formation in deep underground environments.

The following section describes the synthetic formation of asphaltene particles by thermolyzing discotic liquid crystal-like multialkylated aromatics. Thermolysis products of the discotic liquid crystals show amazingly similar physical properties to real asphaltenes.[20] The general formula for these highly oriented

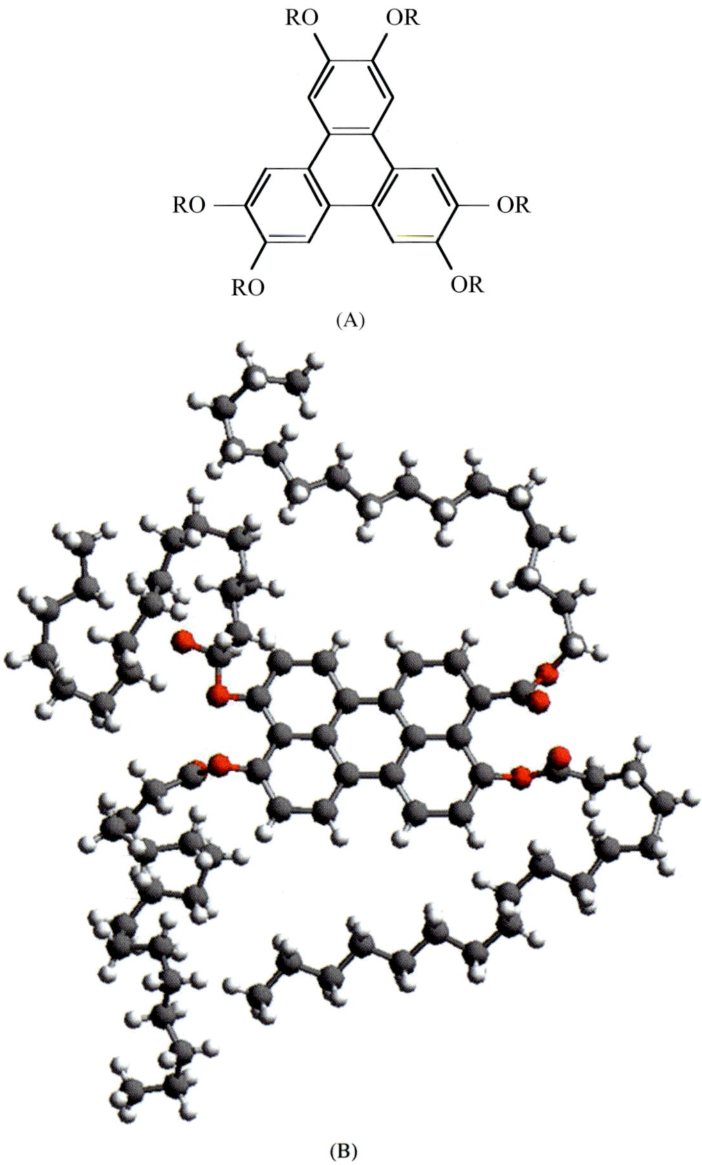

Figure 15.19. (A) Structure of discotic liquid crystal where $R =$ undecyl, heptyl, hexyl, undecanoyl. (B) Force Field (Cerius2 Accelrys Corp.) relaxed simulation of the TOCP precursor molecule. The molecule in this configuration is disc-like with a 3 and 5 nm diameter.

discotic liquid crystalline compounds is shown in Figure 15.19A and a relaxed molecular simulation of the TOCP (tetra octyl carboxylate perylene) is shown in Figure 15.19B. These compounds because of the polycondensed aromatic core form oriented liquid crystal mesophases at their melting temperatures. In the case of TOCP the melting temperature is 106°C. The liquid crystalline material can then be further pyrolyzed by heating in the absence of air to form products that

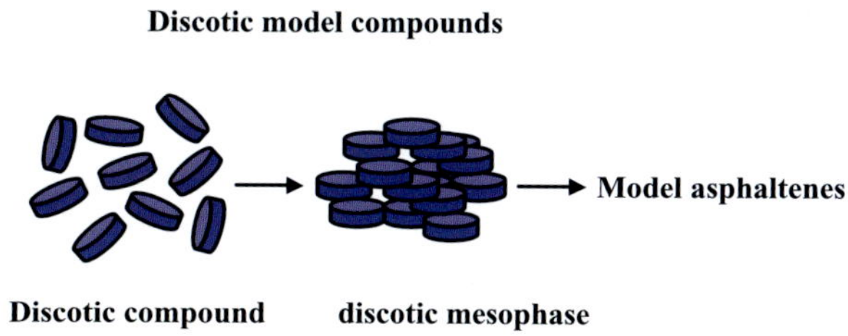

Figure 15.20. Schematic of transformation of discotic molecule to discotic mesophase to self-assembled asphaltene.

spontaneously assemble synthetic asphaltene particles as shown schematically in Figure 15.20.

Schematic molecular thermal simulation of 1,6,7,12-tetra(octadecyl) perylene-tetracarboxylate ester TOCP (tetra octadecanoxy carboxy perylene) is shown in Figure 15.21. Synthesis of TOCP was carried out by tetraesterification

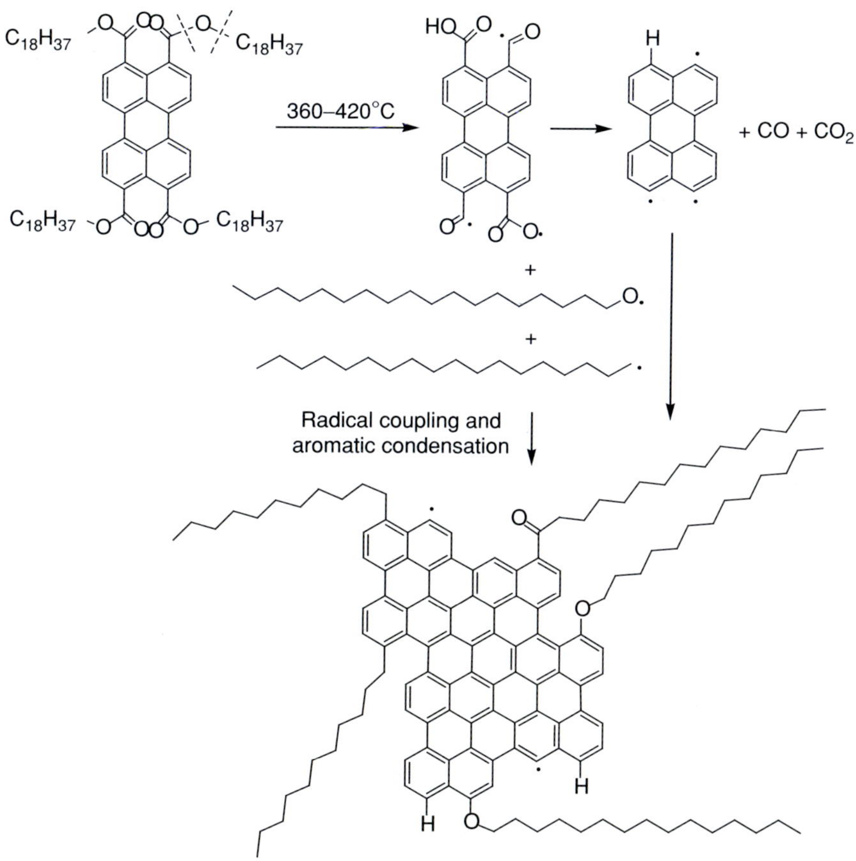

Figure 15.21. Schematic presentation of aliphatic and polyaromatic thermal products upon the heat treatment of TOCP at 360–420°C involving radical coupling and aromatic condensation.

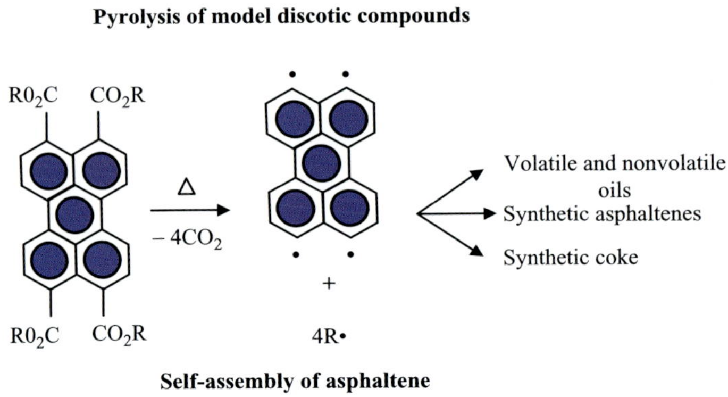

Figure 15.22. Schematic of pyrolysis of TOCP followed by self-assembly of asphaltenes from a sea of free radicals.

of potassium perylene tetracarboxylate salt with 1-octadecylbromide. The planar moiety of condensed polyaromatic perylene core gave high tendency of the molecules to assemble into oriented liquid crystalline mesophases at temperatures below their melting transition, via aromatic–aromatic interactions. In the case of TOCP the melting transition occurs at 106°C. The perylene core is highly thermal stable. During the thermal treatment process with the increase of temperature to above 350°C, the aliphatic ester center of TOCP consisting of two carbon–oxygen bonds CO–O and O–C becomes the weakest region for thermal cleavage to occur. Thermal degradation of these two bonds gave the corresponding perylenyl carbonyl radical and perylenyl carboxyl radical, as shown in Figure 15.21. This is also shown schematically in Figure 15.22.

Subsequent decarbonylation and decarboxylation, respectively, of these intermediate radicals afforded perylenyl radicals with the production of CO and CO_2. The release of aliphatic octadecyl chains from TOCP in the thermal treatment resulted in reactive octadecyl radicals and octadecanoxyl radicals. At temperatures above 420°C in the absence of air, complete thermal conversion of TOCP ester moieties into a "sea of free radicals" containing aromatic cores and long-chain alkyl fragments was achieved. At this stage the aliphatics–polyaromatics mixtures with incorporation of reactive free radicals generated in situ provided appropriate reaction environment mimicking reactive intermediates involved in the condensation transformation of natural petroleum components. Further radical coupling among perylenyl radicals, octadecyl radicals, and octadecanoxyl radicals led to various combinations of aliphatics–polyaromatics condensates in close resemblance to the composition of asphaltene. The resulting thermal products are denoted "synthetic asphaltene" accordingly. Spontaneous assembly of polyaromatic moieties of synthetic asphaltene into particle aggregates was expected to occur during the thermal transformation.

Interestingly, a close resemblance in optical absorption characteristics of synthetic asphaltene to that of natural asphaltene was observed in infrared spectroscopic (IR) measurements, as shown in Figure 15.23. A clear chemical conversion

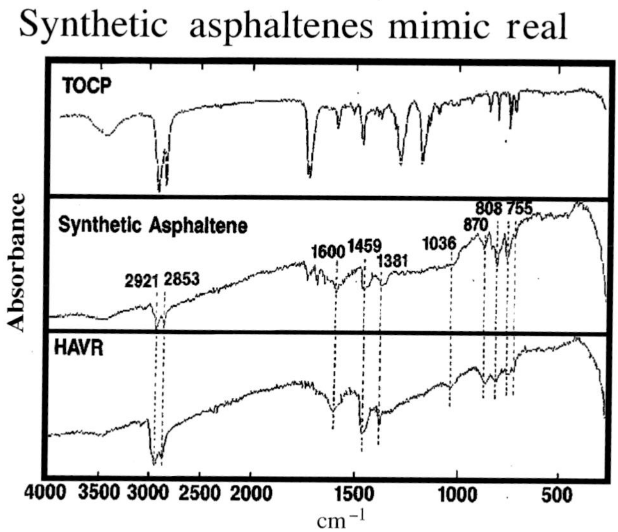

Figure 15.23. Infrared spectra of the TOCP precursor, synthetic asphaltene and HAVR (heavy Arab vacuum resid) asphaltene indicating the remarkable similarity between the synthetic and the HAVR asphaltene.

of the starting material TOCP (Figure 15.22) to synthetic asphaltene was substantiated by a large intensity loss of the band centered at 1720 cm^{-1} corresponding to the optical absorption of carbonyl groups, indicating a near quantitative loss of ester groups and their involvement in thermal cleavage reactions. Broadening of many IR bands in the spectrum of synthetic asphaltene revealed a multicomponent mixture of this synthetic thermal residual. Surprisingly, absorption bands of the overall spectrum match well with the corresponding band position and intensity of the IR spectrum derived from HAVR (heavy Arab vacuum residuum). This confirmed the accurate chemical modeling of particular natural asphaltene formation by using a different or perhaps a mixture of discotic compounds in an appropriate carbon to hydrogen ratio coupled with the polycondensation reaction mechanism for the growth of aliphatics–polyaromatics condensates. The synthetic asphaltene is remarkably similar to the HAVR asphaltene. This further confirmed by the comparison shown in Table 15.6. It appears that any particular natural asphaltene could be chemically modeled by using a different or perhaps a mixture of these compounds with the appropriate carbon to hydrogen ratio.

Figure 15.24 shows the results of WAXS studies during the systematic thermal conversion of TOCP molecules from the melting transition, maltene formations, and asphaltene formations, to coke formations. The first WAXS pattern of the discotic liquid crystalline mesophase was collected at 106 °C showing a large ordering peak of aliphatic chains with a relatively smaller aromatic–aromatic stacking ordering in a scattering angle range similar to those of the peaks derived from asphaltene particles. As the pyrolytic temperature increased to 340°C for the mal-

Table 15.6. Comparison of Some Properties of the Synthetic and HAVR Asphaltenes

	Chemical properties of synthetic asphaltenes and HAVR			
	H/C	%*C* Aromatic	%*C* Aliphatic	M_W
Synthetic asphaltenes	1.06	58	42	2,750
HAVR asphaltenes	1.10	33–45	55–65	1,000–10,000

tene formation, progressive development of typical asphaltene WAXS peaks was detected with the increase in intensity of asphaltene particle peak in the 3.0–4.0 nm region along with the paraffin peak and the graphene peak. The polyaromatics–polaromatics (graphenic) ordering peak and the asphaltene particle peak became highly enhanced when the pyrolytic temperature increased to 420°C for the asphaltene formation and the subsequent coke formation. The results substantiated that a polycondensed aromatic core for the graphenic polyaromatics formation during the pyrolysis process is essential for the self-assembly of the asphaltene into a particle form. As the comparison using a similar small discotic molecule BH8 without containing a polycondensed aromatic core, its pyrolytic transformation does not produce the corresponding synthetic asphaltene. Figure 15.25 focuses on the graphene/paraffin WAXS region. The figure indicated the coking process that occurs with the synthetic and the HAVR asphaltene. Again the similarity is

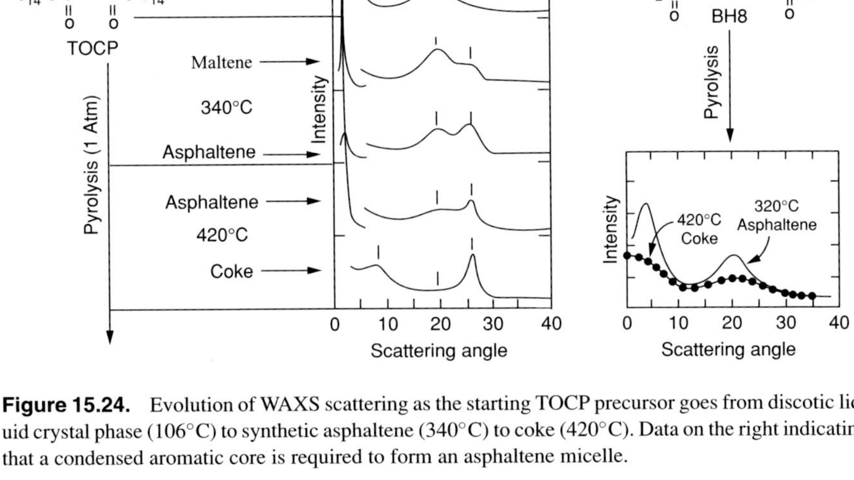

Figure 15.24. Evolution of WAXS scattering as the starting TOCP precursor goes from discotic liquid crystal phase (106°C) to synthetic asphaltene (340°C) to coke (420°C). Data on the right indicating that a condensed aromatic core is required to form an asphaltene micelle.

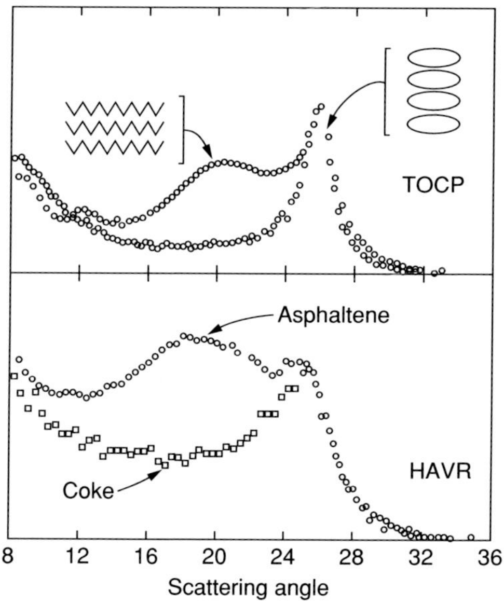

Figure 15.25. WAXS scattering data for synthetic and HAVR asphaltenes in the paraffin correlation and aromatic stacking regime showing the similarity as coking occurs.

remarkable with the disappearance of the paraffin peak observed as the coking process proceeded to strip the aliphatic chains from both the synthetic and the natural asphaltene.

6. Conclusions

The existence of asphaltene particles with cores in the 3.0–4.0 nm region and the aggregation of these cores into larger structures in the 25 nm region is confirmed by WAXS and SAXS studies. This work confirms the earlier work of T.F. Yen. These asphaltene structures are ubiquitous and stable. The self-assembly of asphaltenes from model compounds described above seems a significant step for understanding the origins of the structure of asphaltenes. Future work will allow modeling and simulating asphaltenes from any petroleum source. This work may lead to a better understanding of asphaltenes from their biological origins to their production and use in the petroleum industry.

Acknowledgments

We would like to acknowledge Exxon Research and Engineering Co., Atofina Corp. USA, the Robert A. Welch Foundation, and the DoE Stanford/SSRL "Gateway Program" for supporting this work. We would like to thank Miguel José Yácaman for providing the Maya asphaltene samples. Gathering of Venezuelan asphaltenes was possible thanks to funding provided by Petroleos de Venezuela during the 80–90's.

References

[1] Yen, T.F. and G.V. Chilingarian (2000). Asphaltenes, Asphalts Vol. 2. Elsevier, N.Y.

[2] Speight, J.G. (1980). *The Chemistry and Technology of Petroleum*. Marcel Dekker, New York.

[3] Sheu, E.Y. and O.C. Mullins (eds.) (1995). *Asphaltenes Fundamentals and Applications*. Plenum, New York.

[4] Yen, T.F. (1975). *The Role of Trace Metals in Petroleum*. Ann Arbor Science, MI-USA.

[5] Merino-Garcia, D. and S.I. Andersen (2005). Calorimetric evidence about the application of the concept of CMC to asphaltene self-association. *J. Disp. Sci. Technol.*, **26**, 217–225.

[6] Calemma, V., P. Iwanski, M. Nali, R. Scotti, and L. Montanari (1995). Structural characterization of Asphaltenes of different origins. *Energy Fuel* **9**, 225–230.

[7] Zajac, G.W., N.K. Sethi, and J.T. Joseph, (1994). Molecular imaging of petroleum asphaltenes by scanning tunneling microscopy: Verification of structure from 13C and proton nuclear magnetic resonance data. *Scanning Microscopy* **8**, 463–470.

[8] Mullins, O.C. and E.Y. Sheu (eds.) (1998). *Structure and Dynamics of Asphaltenes*. Plenum, New York.

[9] Yen, T.F., J.G. Erdman, and S.S. Pollack (1961). Investigation of the structure of petroleum asphaltenes by x-ray diffraction. *Anal. Chem.* **33**(11), 1587–1594.

[10] Liang, K.S., R.R. Chianelli, F.Z. Chien, and S.C. Moss (1986). Computer calculation of scattering intensity for disordered molybdenum disulfide. *J. Non-Cryst. Solids* **79**, 251.

[11] Perez De la Rosa, M., S. Texier, G. Berhault, A. Camacho, M.J. Yácaman, A. Mehta, and R.R. Chianelli (2004). Structural studies of catalytically stabilized model and industrial-supported hydrodesulfurization catalysts. *J. Catal.* **225**, 288–299.

[12] Ryong-Joon Roe. (2000). In: *Methods of X-Ray and Neutron Scattering in Polymer Science*. Oxford University Press, Oxford.

[13] Mandelbrot, B.B. (1983). *The Fractal Geometry of Nature*. Freeman, San Francisco.

[14] Martin, J.E. and A.J. Hurd (1987). Surface and mass fractals in vapor-phase aggregates. *J. Appl. Crystallogr.* **20**(2), 61–78.

[15] Liu, S.H. (1986). *Solid State Phys.* **39**, 207.

[16] Schmidt, P.W. (1989). In: D. Avnir (ed.), *The Fractal Approach to Heterogeneous Chemistry*. Wiley, New York, p. 67.

[17] Carbognani, L. and E. Rogel (2002). Solvent swelling of petroleum asphaltenes. *Energy Fuels* **16**(6), 1348–1358.

[18] Carbognani, L., E. Contreras, R. Guimerans, O. Leon, E. Flores, and S. Moya (2001). Physico-chemical characterization of crudes and solid deposits as guideline for optimizing oil production. In: *Proceedings of the SPE International Symposium on Oilfield chemistry*. Houston, Texas (paper SPE 64993), Feb 13–16.

[19] http://home.flash/~celjure/engineering/petroplan/assay/index.htm

[20] Chiang, L.Y., N.A. Clark, K.S. Liang, A.N. Bloch (1985). Highly oriented fibers of discotic liquid crystal. *J. Chem. Soc. Chem. Commun.* **11**, 695–696.

16

Solubility of the Least-Soluble Asphaltenes

Jill S. Buckley, Jianxin Wang, and Jefferson L. Creek

1. Introduction

The key to understanding many asphaltene-related phenomena is a quantitative description of the solubility conditions at which the least-soluble asphaltenes begin to flocculate from a crude oil, often referred to as the onset of flocculation. Models that treat asphaltene flocculation as a liquid–liquid phase separation of large solute molecules dispersed in a solvent composed of much smaller molecules can successfully describe experimental observations in which solubility conditions vary due to changes in pressure and composition. Formation of small, well-dispersed asphaltene aggregates of colloidal dimensions (on the order of nanometers) does not invalidate the thermodynamic approach to modeling asphaltene phase behavior.

The parameters needed to describe asphaltene phase behavior are solubility parameters and molar volumes of asphaltic and nonasphaltic portions of the oil. There are experimental barriers to accurate measurement of these important parameters, especially for the asphaltenes. We review several approaches to estimation of the solubility parameters of stock tank oil (STO) and mixtures with flocculating agents at the onset conditions, including the use of refractive index to estimate solubility parameters. We discuss the minimum data requirements for quantifying and predicting asphaltene instability from experiments with liquid alkane nonsolvents that define an asphaltene instability trend (ASIST) and we demonstrate application of STO ASIST data to prediction of asphaltene instability during depressurization of live oil. Finally, we apply the thermodynamic model to predict asphaltene instability in mixtures of petroleum fluids.

Asphaltenes are defined, based on standardized tests, as the materials in petroleum products that are insoluble in *n*-heptane or *n*-pentane, but soluble in benzene or toluene (e.g., ASTM D2007). Asphaltene characterization techniques can be divided into two main groups: those based on determination of the *amount* of asphaltene using the standardized tests and those based on observations of

Jill S. Buckley and Jianxin Wang • Petroleum Recovery Research Center, New Mexico Tech, Socorro, New Mexico. **Jefferson L. Creek** • Chevron Energy Technology Co., Flow Assurance Team, 1500 Louisiana St., Houston, Texas.

the *onset* of asphaltene insolubility, as suggested by Oliensis.[1] Tests in the first group specify extreme conditions of very poor solubility (e.g., mixing 40 parts *n*-heptane with 1 part oil) to produce the maximum amount of asphaltene. Asphaltene problems, however, can occur at solubility conditions that are much less extreme. Information is needed about the solubility conditions at which the least soluble asphaltenes first begin to form a separate phase and about how those solubility conditions change with temperature, pressure, and oil composition. In this chapter, we focus on quantifying the solubility of the least soluble asphaltenes at the onset of asphaltene flocculation.

1.1. Importance of the Least-Soluble Asphaltenes

Stable crude oils are those in which asphaltenes are well dispersed. At very low concentrations asphaltenes may exist as molecules, but in most oils the asphaltenes probably form small aggregates with dimensions on the order of a few nanometers.[2] The first appearance of aggregates that are large enough to scatter light and to be seen with the aid of an optical microscope is often referred to as the onset of asphaltene flocculation or precipitation. Similar onset conditions can be defined by filtration and other techniques. The onset is a useful reference point that correlates with changes in the impact of asphaltenes in a variety of situations of practical interest. Some of the phenomena associated with the onset of asphaltene flocculation include:

- formation of asphaltene deposits[3–6];
- stabilization of water-in-oil emulsions[7,8];
- poisoning of catalysts (reference 9 and references cited therein);
- fouling of hot metal surfaces[10];
- extent of wettability alteration.[11]

In some reservoirs, destabilization of asphaltenes can occur during production as a result of changes in pressure, temperature, and/or composition. Stable oils can be destabilized by mixing with injection or lift gas. In some cases, mixing of two stable oils can result in destabilization of asphaltenes. Chemical reactions including cracking and oxidation can change the stability of existing asphaltenes or create new asphaltenes from species in the oil that originally were soluble in heptane.

Asphaltene stability in a particular crude oil system can be viewed as having two aspects. The first is a property of the asphaltene fraction itself, which has some inherent stability that depends on the chemical composition and distribution of molecular properties of the material in the asphaltene fraction. The second aspect is the influence of the nonasphaltene fraction on asphaltene stability.

Asphaltene onset titration results are often reported in terms of the volume (or mass) of nonsolvent that must be added to initiate asphaltene aggregation. This volume is a function of both the inherent asphaltene stability and the solvent quality of the nonasphaltene portion of the oil. Values of solubility parameters of oil and onset mixtures cannot be obtained from such volumetric data without some further characterization of the starting material, as discussed later in this chapter. Often the crucial information that would permit assessment of onset titrations in terms of solubility parameters has not been reported.

1.2. Detection of the Onset of Asphaltene Instability

Onset conditions can be detected optically with or without the aid of a microscope,[1] by light scattering,[12] by conductivity,[13] by filtration,[14] or by viscosity measurements.[15] Onset detection methods were recently reviewed by Correra *et al.*[16] All of these techniques should provide similar information, but the methods are not identical. Some require dilution of the oil with an asphaltene solvent. They may not detect asphaltenes at the same stage of aggregate growth, either because of the measurement principle employed or because of differences in the amount of time permitted for asphaltene flocs to grow, a process that can be very slow, especially near the onset conditions.[17] Interference from the presence of wax crystals, gas bubbles, emulsified water, and inorganic particulates can also contribute to the uncertainty of some onset measurements.

In any process that involves mixing of poor asphaltene solvents with oil, spurious results can be obtained if local concentrations of the poor solvent exceed the onset conditions. Major differences can be expected between the onset conditions as determined by adding a poor solvent to a stable dispersion and the point at which the last asphaltenes disappear upon addition of a good solvent to flocculated asphaltenes. Wiehe and Kennedy[18] point out that the importance of the order of mixing extends to cases where two crude oils are mixed if asphaltenes are not stable in mixtures of all proportions.

Differences in results between laboratories can be ascribed in part to real differences in sensitivity, responses to interference, and flocculation kinetics for specific experimental protocols. Nevertheless, it should be possible to obtain comparable results with any of the common methods for detection of asphaltene flocculation, provided care is taken to avoid the problems discussed above. Details of the onset method used in this work are presented in Appendix I.

The major problem that has impeded comparisons between laboratories is not onset detection, but the absence of a quantitative description of solubility conditions at the onset and of the solvent quality of the starting material. The importance of independent measures of oil and onset solubility parameters will be illustrated later in this chapter.

1.3. Asphaltenes as Colloidal Dispersions

For years, asphaltenes defied many of the best characterization efforts of petroleum chemists. In addition to the analytical difficulties associated with complex mixtures generally, asphaltenes have a strong tendency to self-associate and likely exist as unassociated molecules only in very dilute solutions in good solvents. Estimates of molecular weight based on colligative properties give results up to hundreds of thousands of Daltons, depending on the details of the analytical procedure.[19] Size exclusion measurements have been applied, but are inaccurate because interactions between asphaltenes and the column material cannot be eliminated.[20,21]

With the aid of small angle neutron and x-ray scattering techniques, continuous increases or decreases in aggregate size have been demonstrated that depend on changes in solvent quality.[22,23] Although quantification of aggregate size and shape from these experiments is model dependent,[24,25] it seems clear that aggregates of colloidal dimensions can exist. This fact complicates phase behavior calculations since forces between colloidal particles depend on factors beyond simple equilibrium thermodynamics.

Nellensteyn[26] inferred a colloidal structure for bitumen from observations such as the Tyndall effect, Brownian motion of particles viewed at high magnification, dialysis, and ultrafiltration. He further observed that an asphalt-like dispersion could be prepared from asphalt-base oil using a dispersion of finely divided elemental carbon. From these observations, he developed a model of asphaltenes as a lyophobic sol (to which he applied the term micelle) in which a graphitic lyophobe is stabilized by lyophilic "protective bodies" similar to resins.[27] Pfeiffer and Saal[28] elaborated upon this conceptual model to explain rheological observations consistent with the colloidal nature of bitumen. In particular, they identified insufficient resin coating as the cause of flocculation or formation of a gel network. This model has been adopted by succeeding generations of asphaltene researchers with little further examination and its influence on subsequent research would be difficult to overstate.

To be colloidal, a system must have one dimension in the size range from nanometers to micrometers. Surface area-to-volume ratio is high for such a system so surface forces are significant and gravity is less important than it would be for larger objects of the same materials. The forces between colloidal particles depend on their size, shape, and separation distance as well as their material properties. Because of the existence of colloids, the system cannot be treated as a homogeneous, true solution. Beyond that, the implications of the existence of colloidal-sized particles can be very different, depending on the colloidal material, the continuous phase, their interactions, and the means of colloidal stabilization. Thermodynamic considerations apply to colloidal systems, but application can be complicated since interaction energies depend on size and shape of colloidal particles. Some colloids act as a separate phase (lyophobic colloids) while others can be treated as part of the continuous phase (lyophilic colloids). Although micelles (structures formed by amphiphilic molecules to minimize their free energy) are colloidal, not all colloids form micelles.

Analogies to micelle-forming surfactants[29] are probably not appropriate to asphaltenes in hydrocarbon dispersion. The hypothetical structure of asphaltene micelles was critically examined in a review by Cimino *et al.*[5] They asked why resins should be considered an essential part of the model, given that asphaltenes are soluble in aromatic solvents in the complete absence of resins. Why a resin coating should disassociate from asphaltene cores in the presence of paraffinic diluents has never been adequately explained, nor does the asphaltene/resin micelle model help to explain asphaltene flocculation as a function of pressure. Because of their tendency to self-associate, asphaltenes exist as colloidal-sized particles in crude oils, bitumens, and heavy petroleum products, but that fact has been widely misinterpreted as implying many things about asphaltenes that have little or no basis in experimental observations.

1.4. Asphaltenes as Lyophilic Colloids

Polymer science presents many well-studied example of lyophilic colloids and there has been reasonable success in adapting thermodynamic calculations by accounting explicitly for differences in size between solvent and colloidal-sized solutes. The Flory–Huggins approach, developed to describe solubility of polymers, was first applied to asphaltenes by Hirschberg *et al.*[30] There are, however, obstacles to implementation of this approach. Covalently bonded polymers have reasonably well-defined distributions of molecular weights. Asphaltene association produces aggregates whose size is affected by most molecular weight measurements and is therefore only poorly defined. In view of the complexity of the fluids involved, simplifying assumptions are essential. The effect of those assumptions on the accuracy of model predictions is a source of some potential problems. Nevertheless, progress has been made toward understanding and describing asphaltene solubility by treating asphaltenes as lyophilic colloids.

1.5. Solubility of Large Molecules

The Flory–Huggins approach has been the basis for several methods of predicting asphaltene instability.[5,30,31] Because polymers are much larger than the surrounding solvent molecules, the regular solution theory equation for the free energy of mixing was adapted to account for the effects of differences in size. Application of that approach to asphaltene aggregates yields Eq. (16.1)

$$\Delta G_{\text{mixing}} = RT(\eta_{\text{m}} \ln \phi_{\text{m}} + \eta_{\text{a}} \ln \phi_{\text{a}} + \eta_{\text{m}}\phi_{\text{a}}\chi) \tag{16.1a}$$

$$\text{where } \chi = \frac{v_{\text{m}}}{RT}(\delta_{\text{a}} - \delta_{\text{m}})^2 \tag{16.1b}$$

and η is number of moles, ϕ is volume fraction, v is molar volume, δ is solubility parameter, R is the universal gas constant, and T is absolute temperature. The subscripts a and m refer to *asphaltene* and to the *mixture* of all components except the asphaltenes, respectively. Solubility depends on concentrations, molar volumes, and the solubility parameters of components a and m. Increasing the Flory–Huggins interaction parameter (χ) by the absolute value of the solubility parameter difference increases the free energy of mixing (ΔG_{mixing}), corresponding to a decrease in solubility. The free energy of mixing is also increased by increasing the molar volume of the oil mixture.

In reality, there are species in oil with a broad range of molecular sizes and the distinction between the "asphaltene" and "mixture" fractions is an artificial one. Nevertheless, this approach captures the essential features of asphaltene onset behavior.

1.6. Solubility Parameters

1.6.1. Solubility Parameters of Pure Components

The solubility parameter (δ) of a compound with a measurable energy of vaporization and molar volume can be calculated from experiment, since δ is defined as

$$\delta = \sqrt{\frac{\Delta U}{v}}, \tag{16.2}$$

where ΔU is the energy of vaporization to the ideal gas state.

The solubility parameter of a mixture is the volume weighted average of the solubility parameters of the individual components[32]:

$$\delta_{\text{mixture}} = \sum_i \phi_i \, \delta_i, \tag{16.3}$$

where ϕ_i and δ_i are volume fraction and solubility parameter of species i, respectively. The units of solubility parameter are $\text{MPa}^{1/2}$ or $(\text{cal/cm}^3)^{1/2}$ where $1\ (\text{cal/cm}^3)^{1/2} = 2.0455\ \text{MPa}^{1/2}$.

1.6.2. Solubility Parameter Estimates from Solubilization/ Precipitation Experiments

Solubility parameters can be estimated by mixing a material whose solubility properties are not known with materials of known solubility parameters. Mitchell and Speight[33] measured the amount of precipitate produced by addition of liquids of varying solubility parameter to Athabasca bitumen. Burke *et al.*[34] estimate the solubility parameter of an asphaltene sludge to be about 20.5 $\text{MPa}^{1/2}$. Wiehe[35] published an extensive set of two-dimensional solubility measurements with heavy oil fractions. One dimension corresponds to what Wiehe called "field forces" (non-polar forces such as van der Waals that have no preferred directionality), the other to complexing forces (polar or orienting forces such as hydrogen bonding). Solubility of heptane asphaltenes from Cold Lake bitumen was highest in solvents with the field-force components of solubility parameter in the range from 17 to 20 $\text{MPa}^{1/2}$ and complexing components from 0 to about 5 $\text{MPa}^{1/2}$. Since the square of the total solubility parameter equals the sum of the squares of the components,[36] these asphaltenes were soluble in mixtures with solubility parameters from 18.8 to 21.3 $\text{MPa}^{1/2}$ or an average of 20.1 $\text{MPa}^{1/2}$. In similar experiments with saturate, aromatic, and resin fractions, solubility regions generally spanned the entire range of field-force components of available test solvents, making estimation of oil solubility parameter difficult or impossible by this technique. Such experiments are time consuming, limited in accuracy, and are not routinely performed with petroleum fluids. It is unusual to find estimates of solubility parameters of either the oil or its asphaltenes in standard oil characterization.

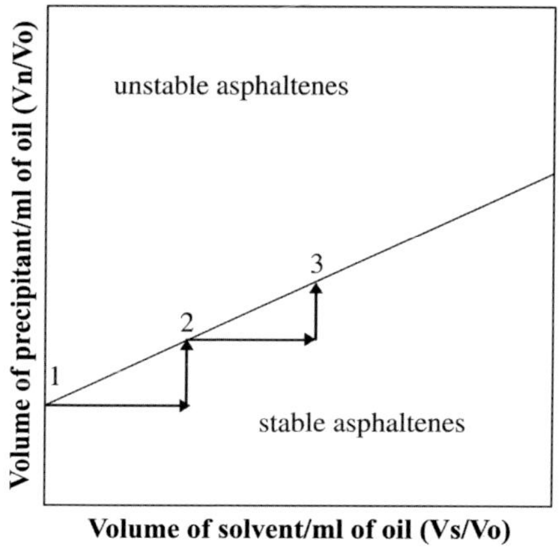

Figure 16.1. Schematic illustration of the typical relationship between amounts of solvent and nonsolvent at the onset conditions (adapted from Mertens).[37] At point 1, $V_{s1} = 0$ and V_{n1} is the minimum amount of precipitant required to initiate asphaltene flocculation. At points 2 and 3, solvent has been added to the oil, increasing the amount of precipitant required to initiate flocculation. The onset points 1, 2, and 3 define a straight line above which asphaltenes are unstable.

1.6.3. Solubility Parameter Estimates from Dilution Experiments

Numerous studies have found a simple linear relationship between the amounts of solvent and nonsolvent at the onset condition. These amounts can be expressed as volumes (or weights) per unit volume (or weight) of oil sample, as illustrated in Figure 16.1. Equation (16.4) describes the empirical relationship between onset values of V_n/V_o and V_s/V_o (where V is volume, the subscripts o, s, and n denote oil, solvent, and nonsolvent, respectively, S is the slope and I the intercept of a straight line through the experimental data):

$$\frac{V_n}{V_o} = S\frac{V_s}{V_o} + I. \tag{16.4}$$

Similar linear relationships have been reported for a wide variety of solvent/nonsolvent pairs.[5,12,37–39]

Mertens[37] first suggested interpretation of the linear relationship between volumes of solvent and nonsolvent in terms of a critical solubility parameter, δ_{Cr}, an approach that has recently been revived.[18,40] These analyses are based on the assumption that there exists a critical solubility parameter at the onset of asphaltene flocculation that is unaffected by dilution. We will return to this "critical solubility parameter" assumption after introduction of another, more direct method of estimating solubility parameters from measurements of mixture refractive indices.

A summary of many historical reports of titration data, showing similarities between interpretations from many different authors, was published by Donaggio *et al.*[41]; an expanded summary is given in Appendix II.

1.6.4. Solubility Parameter Estimates from RI

Refractive index (RI) measurements were suggested by Buckley *et al.*[42] as an alternative method of characterizing solvent conditions in oil and onset mixtures. In systems where interactions are dominated by London dispersion forces, it can be shown that the strength of those interactions is related to the difference in refractive indices between two materials (assuming the materials have similar absorption frequencies).[43] According to Lorenz–Lorentz equation, the RI of a medium measured at visible light frequencies is related to the electronic polarizability of the medium, α_0, by:

$$\frac{n^2-1}{n^2+2} = \frac{\alpha_0}{3\bar{v}\varepsilon_0}, \tag{16.5}$$

where n is RI measured with the sodium-D line at 20 °C, ε_0 is the permittivity of vacuum, and $\bar{v}$ is the volume occupied per molecule. For a medium with density ρ and molecular weight M,

$$\bar{v} = \frac{M}{\rho N_0} \tag{16.6}$$

where N_0 is Avogadro's constant. Thus, Eq. (16.5) can be rewritten as:

$$\frac{n^2-1}{n^2+2} = \frac{\alpha_0 \rho N_0}{3M\varepsilon_0} \tag{16.7}$$

In the visible frequency range, n^2 is approximately equal to the dielectric constant ε. The electronic polarizability α_0 is an intrinsic property of a molecule that represents the extent of induced dipole moment resulting from the displacement of electron clouds in a molecule by an external electric field E. A more common expression for Eq. (16.7) is

$$\frac{n^2-1}{n^2+2} = \frac{\rho}{M}R, \tag{16.8}$$

where $R = \alpha_0 N_0/(3\varepsilon_0)$ is the molar refraction of the material and is independent of temperature and pressure. Thus, for a pure substance, refractive index is a function of density. It has been shown by laboratory measurements that Eq. (16.7) is accurate at room temperature for either nonpolar or polar molecules.[43]

For nonpolar species, the cohesive energy is roughly proportional to $[(n^2-1)/(n^2+2)^{3/4}]^2$.[43] An even simpler empirical relationship can be demonstrated between δ and F_{RI}, where $F_{RI} = (n^2-1)/(n^2+2)$, as shown in Figure 16.2.[44] For nonpolar materials, RI can be converted to solubility parameter at ambient temperature using Eq. (16.9):

$$\delta(\text{MPa}^{1/2}) = 52.042F_{RI} + 2.904. \tag{16.9}$$

This technique is useful for estimating solubility parameter of liquid mixtures such as STOs for which RI can readily be measured. Taylor *et al.*[47] have shown that RI can be estimated for heavy oils by extrapolation of measurements with diluted solutions. RI of mixtures with light components can be calculated from the

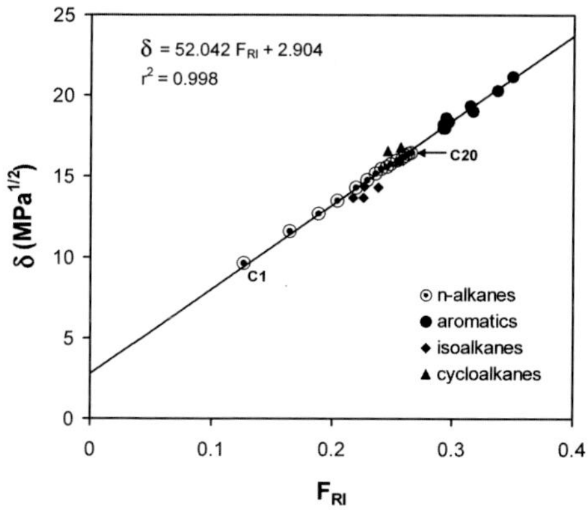

Figure 16.2. Empirical correlation between δ and F_{RI} (after Buckley and Wang[44]). RI for methane, ethane, and propane were calculated based on their partial molar volume,[45] other data are from Barton.[46]

composition using published values of molar refraction.[48] If the gas composition is known, RI of live oil can be calculated from measured RI of STO using conventional PVT measurements of gas–oil ratio and formation volume factor as a function of pressure.[42]

1.6.5. The "Critical Solubility Parameter" Assumption

Experimental results that follow the linear trend shown in Figure 16.1 have been observed by many laboratories, including ours. While the experimental observations are indisputable, the assertions that this linearity implies that there is a critical solubility condition at the onset of asphaltene flocculation and that the onset condition is not affected by dilution are open to question.

Equation (16.4), which expresses the linear relationship between the volumes V_n and V_s, can be rewritten as

$$V_n - SV_s - IV_o = 0. \tag{16.10}$$

Dividing through by V_t (where $V_t = V_o + V_s + V_n$) gives

$$\phi_n - S\phi_s - I\phi_o = 0, \tag{16.11}$$

where ϕ_i is the volume fraction of component i and

$$\phi_n + \phi_s + \phi_o = 1. \tag{16.12}$$

According to Eq. (16.3), the onset solubility parameter is the volume-weighted average of the component solubility parameters at the onset:

$$\delta_{onset} = \delta_n\phi_n + \delta_s\phi_s + \delta_o\phi_o. \tag{16.13}$$

Equations (16.11) through (16.13) can be combined to give an equation of the form

$$\delta_{onset} = C_1\phi_s + C_2, \tag{16.14a}$$

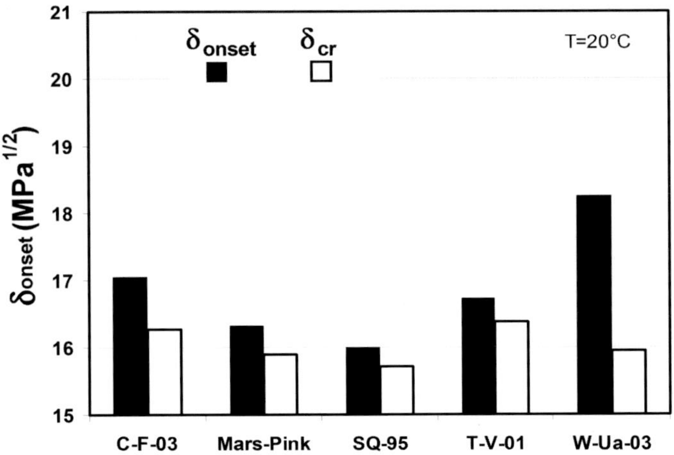

Figure 16.3. Assuming constant solubility parameter at onset with varying dilutions results in an overly optimistic estimate of the onset condition ($\delta_{Cr} < \delta_{onset}$) for oil/toluene/*n*-heptane mixtures.

where

$$C_1 = \delta_s - \frac{1+S}{1+I}\delta_o + \frac{S-I}{1+I}\delta_n \tag{16.14b}$$

$$C_2 = \frac{\delta_o + I\,\delta_n}{1+I}. \tag{16.14c}$$

As Eq. (16.14) shows, the observed linear relationship (Eq. 16.4) does not mean that δ_{onset} is constant, except in the special case where $C_1 = 0$. δ_{onset} is a function of the extent of dilution; it can either increase or decrease with dilution, depending on the sign of C_1. Note that δ_{onset} depends on the solubility parameter of the oil, δ_o, which must be determined independently. When there is no dilution with toluene ($\phi_s = 0$), δ_{onset} can be calculated from C_2, if δ_o is known. Only if ϕ_o is set to zero in Eqs. (16.11) through (16.13) (corresponding to infinite dilution) does the equation for δ_{onset} reduce to that obtained by assuming the existence of δ_{Cr}.[18,40] The differences between onset solubility parameters obtained by these two approaches are compared in Figure 16.3 for five crude oil samples, with toluene as the solvent and *n*-heptane as the nonsolvent. In all cases, δ_{Cr} is lower than δ_{onset}. The calculated oil solubility parameters ($\delta_{o,Cr}$) corresponding to δ_{Cr} are much lower than those estimated from measured RI values using Eq. (16.9) ($\delta_{o,RI}$), as shown in Figure 16.4.

Additional experimental evidence shows that the solubility parameter (based on RI measurements) at the onset of flocculation varies significantly with dilution.[49,50] An example that clearly illustrates a decrease in onset solubility parameter (δ_{onset}) with increasing toluene dilution is shown in Figure 16.5.

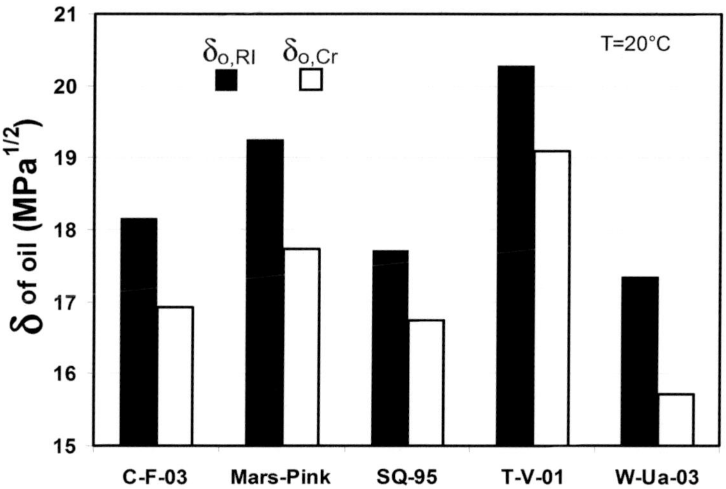

Figure 16.4. Assuming constant solubility parameter at onset with varying dilution leads to significant underestimation of the oil solubility parameter ($\delta_{o,Cr} < \delta_{o,RI}$) for oil/toluene/$n$-heptane mixtures, where $\delta_{o,Cr}$ is the solubility parameter calculated for the oil assuming a constant value of δ_{Cr} and $\delta_{o,RI}$ is calculated from RI measurements.

1.6.6. Solubility Parameter Estimates from EOS Calculations

Although not a standard option, δ can be calculated using an equation of state. Solubility parameters were calculated for experimentally determined

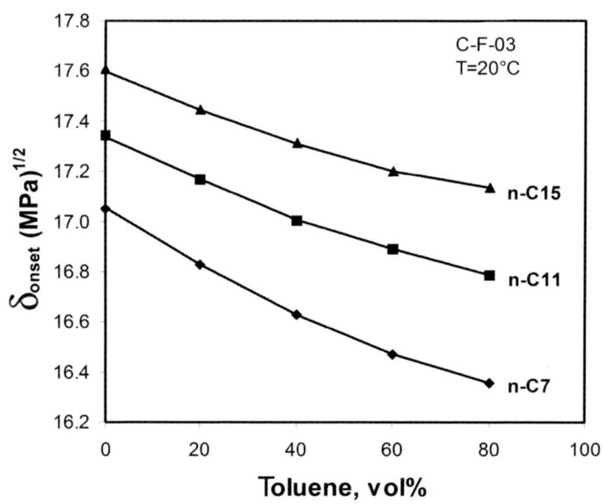

Figure 16.5. Onset solubility parameter (δ_{onset}) decreases with addition of toluene for mixtures of C-F-03 STO, toluene, and three nonsolvents, each of which was tested at 20°C. (Note: solubility parameters were calculated from measured values of RI, as described in the previous section.)

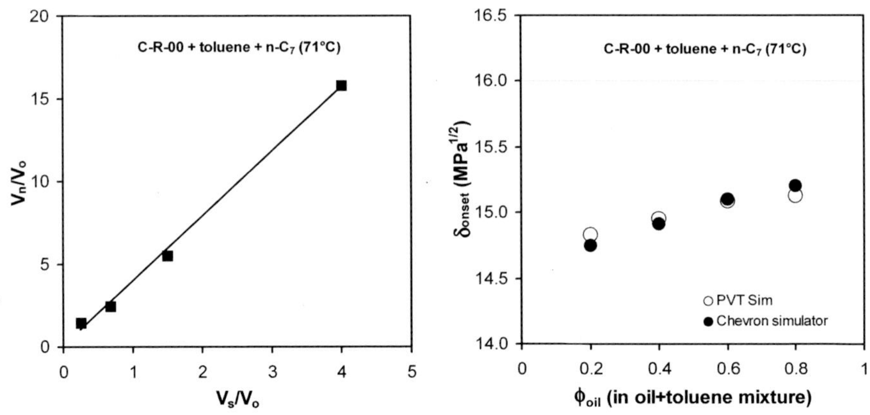

Figure 16.6. EOS calculations of onset solubility parameters (shown in the right-hand graph) from experimental onset observations (shown in the graph on the left) for mixtures of C-R-00 STO with toluene and *n*-heptane.

onset mixtures of C-R-00 STO with toluene and *n*-heptane at 71°C. The existing oil PVT description, tuned with an older sample from the same field, was matched to the STO density at 20°C. Two different simulations were performed. An in-house Chevron Peng Robinson-based EOS simulator was used to calculate solubility parameters from departure functions.[51] A commercial PVT simulator[52] does not have a routine to calculate solubility parameters, but the information necessary to do so can be generated (see Appendix III). Figure 16.6 shows on the left the onset volumes measured for mixtures of C-R-00 STO with varying amounts of toluene and *n*-heptane at a temperature of 71°C. Solubility parameters calculated for those compositions by the two simulators are shown in Figure 16.6 on the right. Both simulations confirm that δ_{onset} is not constant, but changes with dilution, in agreement with previously reported trends.[49]

1.7. Flory–Huggins Predictions: The Asphaltene Solubility Model (ASM)

Composition, solubility parameters, and molar volumes of the oil components can, in principle, be used to find conditions that minimize the free energy of mixing (Eq. (16.1)). In practice, however, there are significant problems in applying the Flory–Huggins model to asphaltenes. The simplest approach is to treat the crude oil as a mixture of two pseudo-components, lumping the asphaltic components in the asphaltene pseudo-component and including everything else (saturates, aromatics, and resins) in the oil pseudo-component. Asphaltenes, however, are not a well-defined chemical class; widely varying estimates of molecular weight have appeared in the literature.[19,53] Molar volume, solubility parameter, and molar concentration of the asphaltenes are not known and if they could be measured, would undoubtedly vary depending on the nonsolvent used in the separation process. The simplifying assumptions necessary to calculate liquid–liquid phase behavior are significant. It is surprising, therefore, that this approach to describing asphaltene stability works as well as it does.

The use of different simplifying assumptions produces somewhat different results. Hirschberg *et al.*[30] and de Boer *et al.*[54] assumed that the asphaltene-rich phase that forms at the onset of instability is "pure" asphaltene and that oil is saturated with asphaltene at reservoir conditions.[54] Cimino *et al.*[5] assumed that the asphaltene concentration in the solvent phase is zero and that the asphaltene-rich phase is 80% asphaltene and 20% oil. Wang and Buckley[31] made neither of these simplifying assumptions. Instead their Asphaltene Solubility Model (ASM) used measurements of onset conditions for a set of seven STOs to constrain the selection of the calculated free energy curves that best correspond to the experimentally observed onset condition.

Figure 16.7 shows the experimental onsets for Mars-Pink crude oil determined with nonsolvents from *n*-pentane to *n*-pentadecane and compares the best fit by each of these three implementations of the Flory–Huggins approach to these data. In this and all subsequent examples, solubility parameters were calculated from measured RI values. As shown previously,[31] ASM produces a family of solutions in terms of the coupled variables, δ_a and v_a. For purposes of comparison, the molar volume of the asphaltene phase, v_a, was set at 2500 cm^3/mol in all cases, a

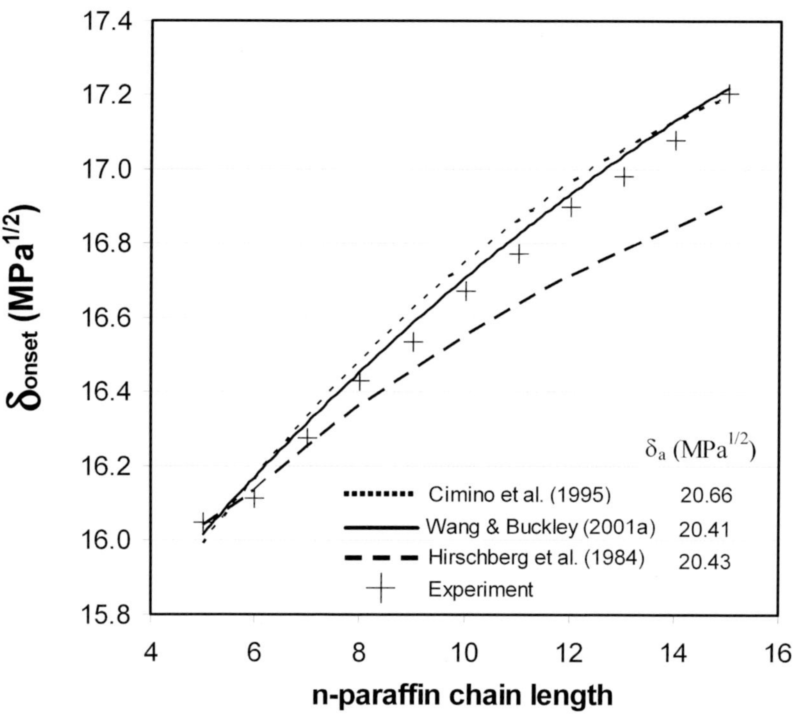

Figure 16.7. Comparison of three predictions for Mars-Pink (after Wang and Buckley[31]). For purposes of comparison, all predictions assume that the molar volume of the asphaltene phase is 2500 cm^3/mol. Details of the equations, assumptions, and algorithm used to produce these fits are given in Wang and Buckley.[31]

value that may represent an aggregate of three to five asphaltene molecules, based on recent measurements of asphaltene molecular weight.[53] The Hirschberg model cannot give the correct slope of the onset solubility parameter as a function of nonsolvent carbon chain length; the asphaltene solubility parameter, δ_a, can be changed to allow the predicted curve to match a different data point, but the slope remains too small to match more than a few measurements. The other two models give much better fits to the experimental data.

2. Asphaltene Instability Trends (ASIST)

2.1. ASIST Established by Titrations with *n*-Alkanes

The solubility conditions at the onset vary monotonically with the number of carbons in the *n*-paraffin chains of the nonsolvent used to destabilize the asphaltenes, as shown in Figure 16.7. An even simpler relationship has been observed between molar volume of the nonsolvent and the onset solubility conditions. Figure 16.8 shows the data for Mars-Pink, replotted as a function of $v_p^{1/2}$ where v_p is the molar volume of the *n*-paraffin precipitants. The experimental solubility parameters follow a linear trend with $v_p^{1/2}$, defining an extremely useful relationship, which we call the ASphaltene InStability Trend (ASIST). ASIST can be determined by titrations with as few as two *n*-paraffins, although it is safer to measure at least three points, spanning the range of the liquid *n*-paraffins, so that deviations from linearity that might result from experimental uncertainties can be detected. To date we have established linear ASIST relationships for more than 40 crude oil samples. The properties of these oils and ASIST parameters at various temperatures are given in Appendix IV.

ASIST is primarily a measure of the portion of asphaltene stability due to chemical properties of the asphaltenes themselves. Stability conferred by the solvent properties of the rest of the oil is related to the solubility parameter of that oil. Saturates have lower and aromatics higher RI for a given molecular weight; resins and asphaltenes are higher in both molecular weight and RI than either the saturate or aromatic fractions. Overall asphaltene stability depends on the difference between the solubility parameters of oil and oil-plus-precipitant at the onset conditions and on the molar volume of the nonsolvent.

Asphaltenes can be ranked by comparing values of δ_{onset} for a given value of $v_p^{1/2}$. Higher values of δ_{onset} correspond to less stable asphaltenes. The range of measured trends is illustrated in Figure 16.9 for the samples from the highest to the lowest *y*-intercepts at 20°C and 60°C (from the data in Table A-IV-2).

The oil solubility parameter contributes to asphaltene stability. SARA analysis can be used to estimate oil solubility parameter[55] because the relative contributions to RI, and hence solubility parameter, of each chemical sub-group is approximately constant. Overall, asphaltene stability can be assessed by comparing the onset condition to the solubility parameter of the parent oil. As shown in Figure 16.10 for 27 different STO samples, the most stable (on the left) have high δ_o and low δ_{onset} whereas the least stable (on the right) have low δ_o and high δ_{onset}.

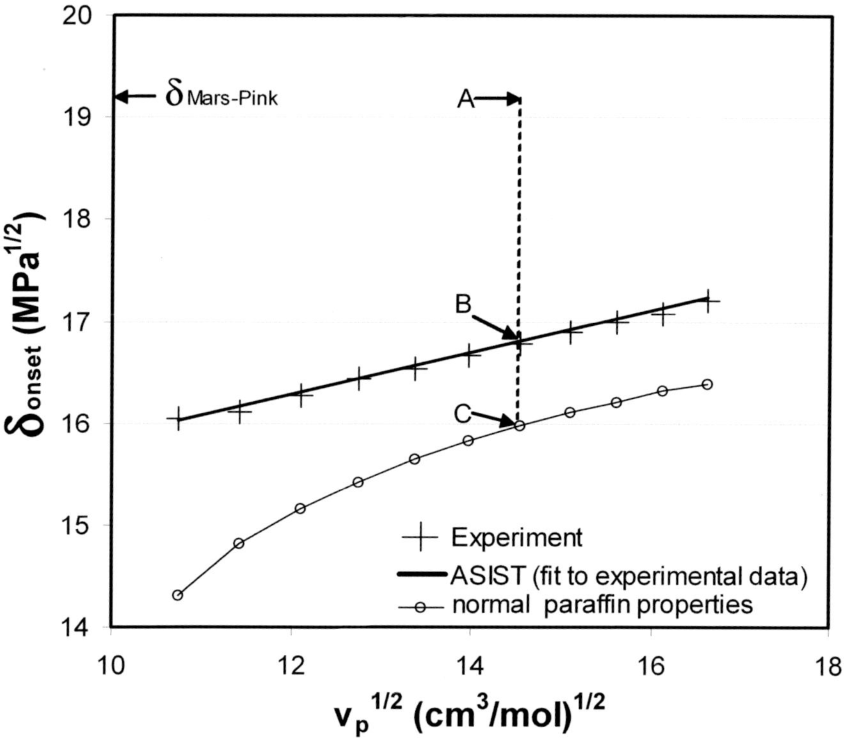

Figure 16.8. The linear ASphaltene InStability Trend (ASIST) defined by δ_{onset} vs. $v_p^{1/2}$ of n-alkane nonsolvents is illustrated with data for Mars-Pink STO and n-alkanes from n-pentane to n-pentadecane at 20°C. The experimental onset solubility parameters are a linear function of $v_p^{1/2}$ where v_p is the molar volume of n-alkane nonsolvents. Volumetric data can be extracted graphically from the onset plots using Eq. (16.3) and the known values of δ for the STO and n-alkane. For example, in a titration of Mars-Pink with normal undecane, the volume fraction of n-C_{11} added at the onset, ϕ_n, is proportional to the ratio of the lengths of the line segments AB/AC; the volume fraction of Mars-Pink STO in the onset mixture, ϕ_o, is given by BC/AC.

Between these two extremes, however, trends are less obvious. For example, an oil with $\delta_o = 17.6$ MPa$^{1/2}$ is more stable than one with $\delta_o = 18.6$ MPa$^{1/2}$ because the latter oil has asphaltenes that are inherently less stable, as shown by the values of the n-C_7 δ_{onset} (16 and 17.8 MPa$^{1/2}$, respectively).

In most cases, the solubility parameter of an STO is higher than the onset solubility condition for flocculation of n-C_7 asphaltenes as shown in Figure 16.10. However, asphaltene flocs are often observed in STO samples. The solubility parameter at which ASIST intersects the solubility parameter of a particular STO can help to explain this observation, as shown for two crude oils in Figure 16.11. Although δ_{onset} for n-C_7 asphaltenes is less than $\delta_{LagraveSTO}$, only slightly larger n-alkanes (with carbon chains of only 12 to 13 carbons) would destabilize asphaltenes and asphaltenes are, in fact, observed in Lagrave STO. The situation with Mars-Pink is different: $\delta_{\text{Mars-Pink STO}}$ is much greater than δ_{onset} for its n-C_7 asphaltenes.

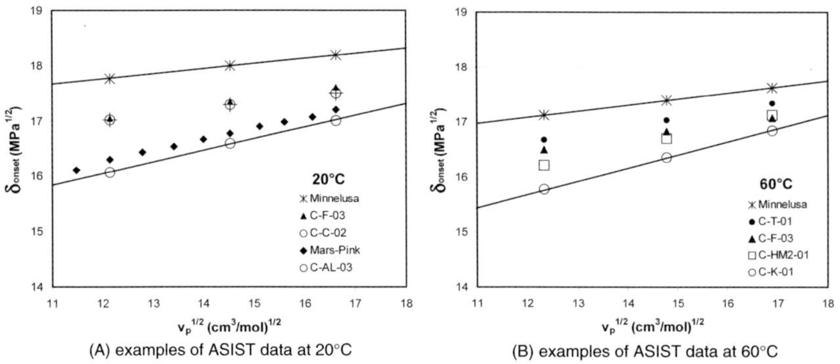

Figure 16.9. The range of ASIST relationships observed for different oils is illustrated by samples with y-intercepts representing the minimum and maximum measured to date, along with several intermediate examples at (A) 20°C and (B) 60°C. In general, samples with higher y-intercept also have lower slopes.

In fact, ASIST for Mars-Pink predicts that n-alkanes with more than 46 carbons (based on the value of v_p at intersection between δ_{oil} of Mars-Pink and ASIST) would be required to destabilize asphaltenes, species that are probably present only in very small concentrations in any crude oil. Wang and Buckley[49] reported that most oils for which δ_{STO} intersects ASIST at molar volumes corresponding to chain lengths greater than 28 carbons do not have asphaltenes in the STO whereas many samples with intersections of 27 and less carbons do have at least a small amount of flocculated asphaltene in the STO samples.

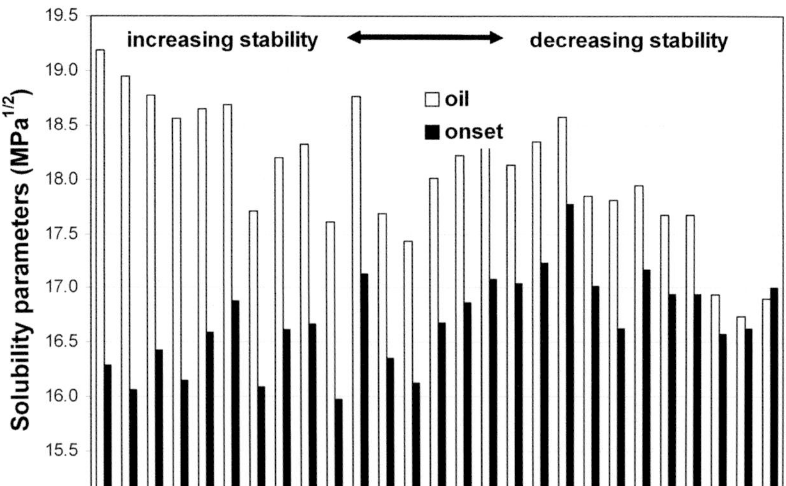

Figure 16.10. Solubility parameters of STOs and onset mixtures with n-C_7 at 20°C. Asphaltene stability is proportional to the difference between δ_o and δ_{onset}.

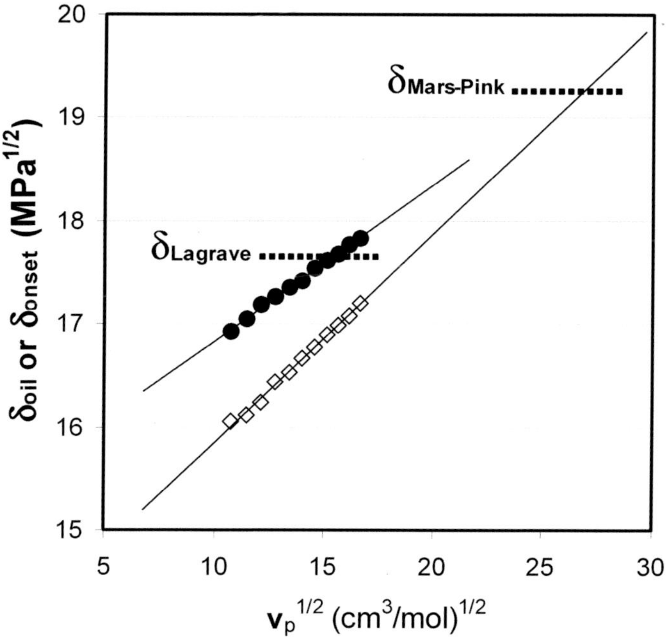

Figure 16.11. The intersections of ASIST and δ_{STO} give a rough estimate of the size of *n*-alkanes that would destabilize asphaltenes at STO conditions. Asphaltenes are stable in Mars-Pink STO, but there are flocculated asphaltenes in Lagrave STO.

ASIST can be used to predict *n*-alkane asphaltene onsets. Linear interpolation between onsets measured with any two *n*-paraffins gives the onsets for other, intermediate *n*-paraffins. Asphaltene stability in the presence of mixtures of two or more *n*-paraffins can also be estimated on the basis of the average molar volume of the *n*-paraffin mixture. A more interesting application is extrapolation to predict the effects of lower molecular weight linear paraffins that are more difficult to investigate because they are gases at ambient conditions. Since live oils can contain substantial amounts of these materials in their light ends, the ability to relate their onset conditions to those measured with liquid alkanes using ASIST is potentially very powerful.

2.2. Use of ASIST to Predict Onset Pressure

In a live oil, asphaltene precipitation is caused by incompatibility of asphaltenes with methane and other light, saturated alkane components. During depressurization, the solubility parameter of a live oil mixture decreases with decreasing density until the bubble point pressure is reached. In some cases, therefore, the live oil solubility parameter can fall below the asphaltene instability trend and asphaltene can begin to flocculate. Wang and Buckley[48] describe in detail how to extrapolate the results of liquid alkane titration experiments to estimate the onset pressure during depressurization of a live oil. A key assumption is that the

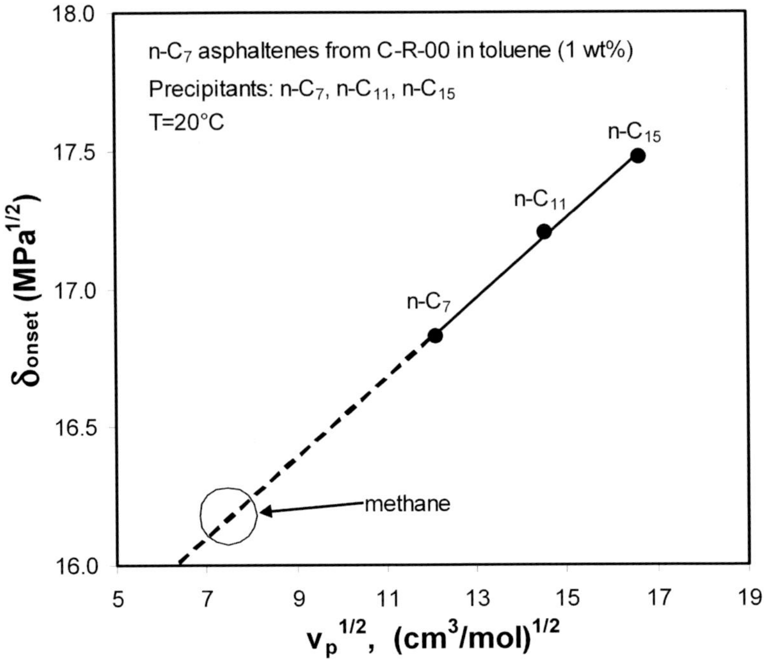

Figure 16.12. Onset characterization of a mixture of 1% n-C_7 asphaltenes from C-R-00 STO dissolved in toluene. The partial molar volume of methane is roughly in the range indicated and is a function of pressure, temperature, and composition.

extrapolation follows ASIST, as established by titrations with liquid alkanes. That assumption was tested using the depressurization onset results of Ting[56] with a model oil, composed of asphaltenes dissolved in toluene to which varying amounts of methane were added. Figure 16.12 shows the liquid titration results that were extrapolated to the partial molar volume of methane in the asphaltene solution. The experimental and predicted onset pressure results are shown in Figure 16.13. Onset pressures are extremely dependent on the amount of methane dissolved in the model oil. The matches show some deviation from measured onset pressures, but the trend of onset pressure with the amount of dissolved methane is accurately predicted. This result provides strong evidence in support of the hypothesis that methane precipitates asphaltenes by the same mechanism as heptane and other liquid alkanes.

As reservoir pressure decreases, fluid properties change in ways that affect asphaltene stability. As the oil is depressured from reservoir pressure to bubble point pressure, the mass and molar composition is constant, but the overall density of the oil is decreasing and the volumetric composition is changing. The volumes occupied by the C_{6-} components are increasing more rapidly than those of the C_{7+} fraction, which consists of the less compressible components of the oil. Whether asphaltenes become unstable depends on the stability properties of the asphaltenes and on the solvent properties of the oil as a function of pressure.

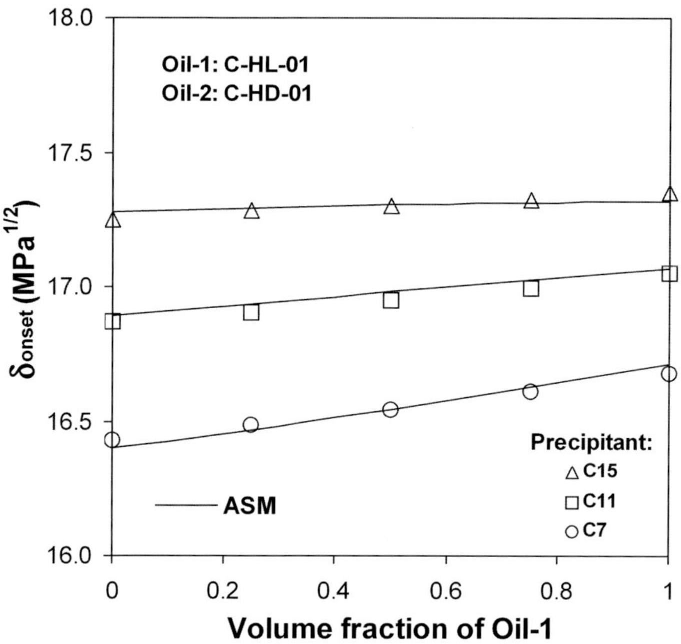

Figure 16.15. Mixtures of two oils from the same platform. The onset of asphaltene instability of mixtures is intermediate between the onsets for the individual oil samples.

n-alkanes. Instead, the asphaltene solubility model (ASM), which includes solubility parameters and molar volumes for both asphaltene and nonasphaltene components, can be used to calculate free energies of mixing and predict the appearance of a separate asphaltene phase in mixtures, from the properties of two different oils. Figure 16.15 shows experimental onset data for titrations of two oils from the same platform with n-C_7, n-C_{11}, and n-C_{15}. Three mixtures of these two oils have onset properties that are intermediate between the unmixed oil endpoints, lying almost on straight lines that join the endpoints. These results are consistent with calculations using ASM, as shown. Note that titrations of all mixtures with all three *n*-alkanes are well represented by ASM using only one adjustable parameter—the asphaltene solubility parameter—for each oil.

Specific interactions are not needed to explain these observations, but they cannot be eliminated on the basis of these results since the oils may have a common source. However, mixtures of two oils from completely different sources, when titrated with n-C_7, gave similar results, as shown in Figure 16.16. In this case, both the experimental results and the ASM-calculated onsets for the mixtures fall slightly below a straight line joining the endpoints. A second set of data is shown for n-C_7 asphaltenes, separated from the same two oils, dissolved in toluene to give 1% by weight solutions. Note that the stabilities of the n-C_7 asphaltenes are more similar to one another than the asphaltenes in the STO samples. No specific interactions are required to explain these results, even though the oils are completely unrelated.

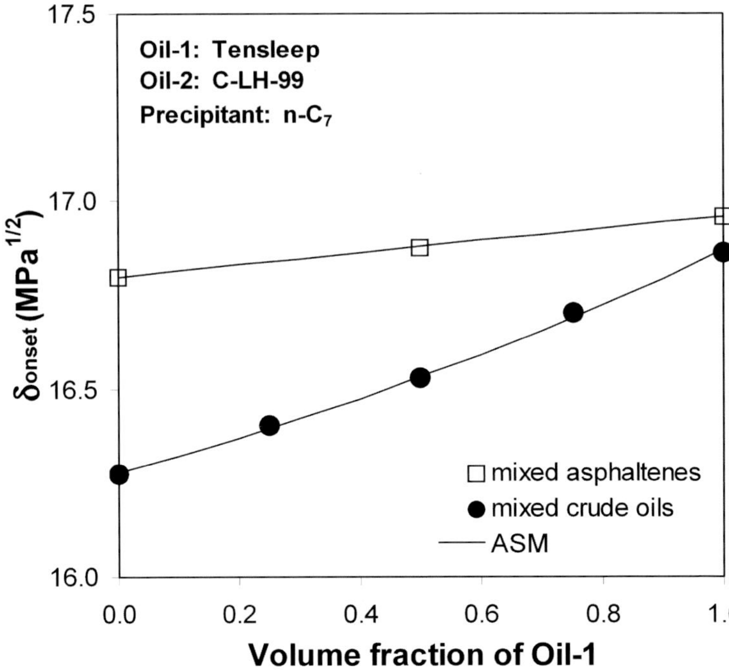

Figure 16.16. Mixtures of two unrelated oils and mixtures of toluene solutions (1 wt%) of their n-C_7 asphaltenes.

The mixtures in the above examples are not incompatible; in all cases non-solvents were added to destabilize the asphaltenes. If asphaltene stability were always intermediate between the stabilities of the parent oils, as in these examples, it is hard to imagine how unexpected incompatibilities might occur. These are, however, special cases of oils with similar amounts of asphaltenes (C-HD-01 has 2.7%, C-HL-01 has 2.3%, Tensleep has 3.2%, and C-LH-99 has 2.8% n-C_7 asphaltenes; the asphaltene solutions in toluene both have 1% asphaltenes). Mixtures of oils with very different asphaltene contents behave differently because the solubility parameter of the mixture varies linearly with volume fractions of the mixed oils (Eq. 16.3), but the asphaltene properties are dominated by the oil with the larger amount of asphaltenes. This is illustrated in another example (Figure 16.17). As in the previous case, the two oils (C-F-03 and E-1XO-00) are from different geographic locations. In this case, however, the oils have significantly different asphaltene contents (n-C_7 asphaltenes: 6.0% for C-F-03 and 0.8% for E-1XO-00). Both experimental measurements and ASM predictions (using one value of $\delta_{asphaltene}$ for each oil to match all of the experimental data with no additional fitting parameters) confirm a nonlinear interpolation of onset conditions for mixtures of these two oils.

In the above example, all the oil mixtures are compatible because the solubility parameters of oil mixtures ($\delta_{mixture}$) are always higher than onset conditions (δ_{onset}) and some precipitant must be added to reach the onset condition. Situations might arise, however, when two stable oils with significantly different asphaltene

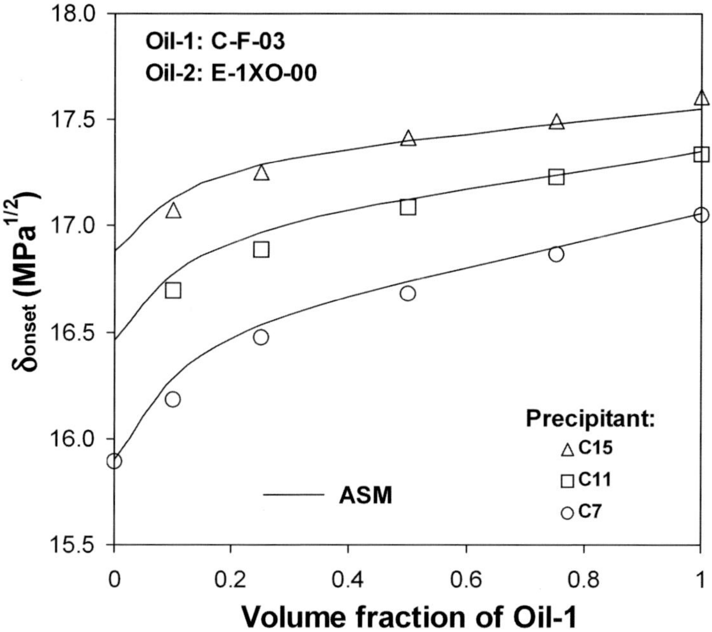

Figure 16.17. Mixtures of oils with different amounts of asphaltenes (n-C_7 asphaltene: 6.0% for C-F-03 and 0.8% for E-1XO-00).

contents become incompatible if solubility parameter of each of the oils is only moderately higher than its asphaltene onset condition. ASM can be used to explore how asphaltene stability would be affected in mixtures of such oils. In Figure 16.18, the asphaltene onsets are shown for three different cases. The straight line connecting the onset points for Oils 1 and 2 was calculated assuming equal amounts

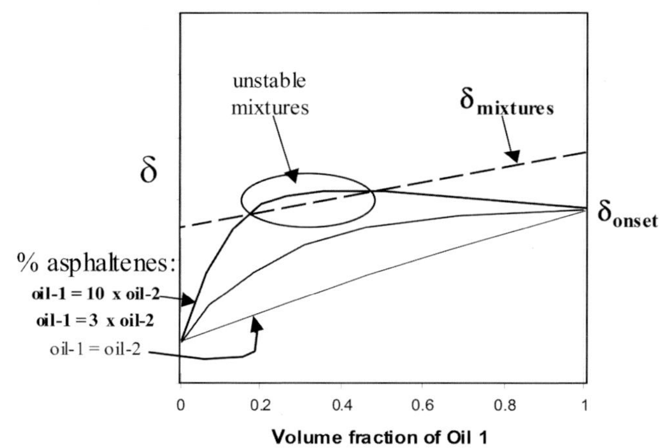

Figure 16.18. Hypothetical mixtures of oils with varying amounts of asphaltenes. Asymmetric mixtures, with much more asphaltene in one oil than the other, can be unstable.

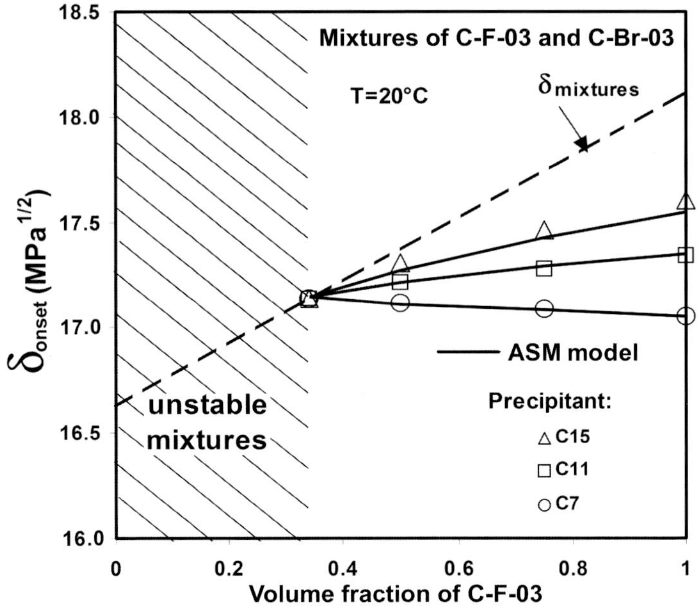

Figure 16.19. Mixtures of an asphaltic oil (C-F-03) with a condensate (C-Br-01) are unstable if the amount of condensate exceeds 65%. ASM predicts the observed onset conditions for stable mixtures.

of asphaltene in both oils. For the two curved lines, the amount of asphaltene in Oil-2 was reduced to one-third and one-tenth the amount in Oil-1, without any other changes in asphaltene or oil properties. As the amount of asphaltene in Oil-2 decreases, the mixture properties are increasingly dominated by the asphaltenes in Oil-1. When the amount of asphaltene in Oil-2 is one-tenth that in Oil-1, ASM predicts that the mixture onset line crosses the solubility parameter line of the mixed oils; asphaltenes in mixtures for which δ_{onset} is greater than $\delta_{mixture}$ are predicted to be unstable. Similar destabilizing effects can be predicted by varying the oil molar volumes.

Asphaltene stability in a case where an asphaltic oil (C-F-03) and a condensate (C-Br-01) form incompatible mixtures is shown in Figure 16.19. Mixtures with more than about 65% condensate are unstable without adding any nonsolvents. Mixtures with less than 65% condensate can be titrated to determine the mixture onset conditions. Asphaltenes flocculate in response to the mixtures of condensate and *n*-alkanes, so the trends with all three *n*-alkanes converge to the point where condensate alone can cause instability. As in the case of the compatible oils, the asphaltene stability of the mixtures can be predicted by ASM from the properties of the asphaltic oil and condensate.

4. Some Remaining Problems

The simple, two-component thermodynamic approach using solubility parameters and molar volumes to describe asphaltene and nonasphaltene portions of an oil is remarkably successful in reproducing the stability of the least stable

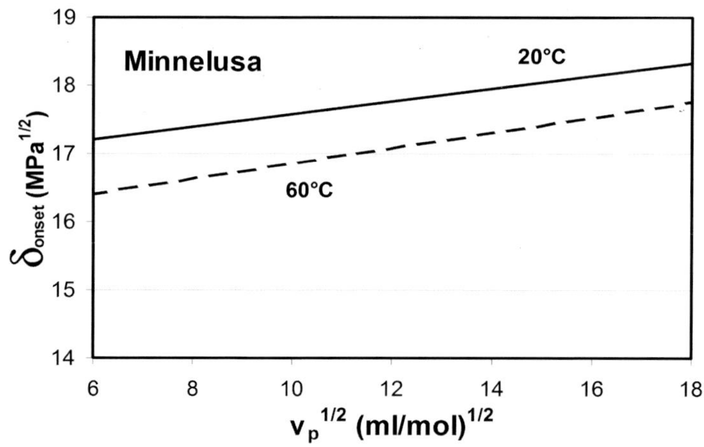

Figure 16.20. Increasing temperature decreases the *y*-intercept solubility parameter. The ASIST slope usually increases slightly with an increase in temperature.

asphaltenes. Nevertheless, there are aspects of asphaltene phase behavior that cannot be captured in such a simple model. Some of the most important outstanding questions include how temperature affects asphaltene phase behavior, how much asphaltene will precipitate as a function of solubility conditions, and what impact asphaltene stability will have in oil reservoirs, production facilities, pipelines, and refineries.

4.1. Effect of Temperature on ASIST

No theory (including Flory–Huggins) is currently able to predict the effect of changing temperature on polymer stability a priori. In the case of Flory–Huggins models, changing temperature can have different effects on molar volumes of solvent and solute, invalidating basic model assumptions about the occupation of lattice sites by solvent and solute that were made in the original derivation of the model. The effect of temperature can vary from one system to another and is usually described empirically. For weakly aggregated asphaltenes, increasing temperature probably alters the extent of aggregation, as suggested by the results of some SANS tests.[58]

The empirically observed effect of increasing temperature usually is to decrease the *y*-intercept and slightly increase the ASIST slope as shown for Minnelusa crude oil in Figure 16.20. Many other examples are given in Table A-IV-2. There is some variability in the shifts in ASIST curves with temperature and efforts are underway to predict the effects of temperature on asphaltene stability with an equation of state based on statistical association fluid theory (SAFT).[56,59]

4.2. Polydispersity and Amount of Asphaltene

The thermodynamic models of asphaltene phase behavior that employ the Flory–Huggins approach discussed in this work have been limited to two pseudo-components, a very severe simplification considering the actual complexity of crude oils and their asphaltene fractions. They are quite successful in describing the onset of asphaltene instability with a minimum of characterization data and fitting parameters, but are less able to predict the amount of asphaltene that will separate at any solubility condition since more information about the distribution of material properties is required. Models using more extensive characterization have been reported by Correra and Donaggio[60] and Alboudwarej *et al.*[61] SAFT is also being developed to handle polydisperse asphaltene systems.[56,59] The model developed should be able to improve predictions of the amount of asphaltene that separates as a function of solubility conditions beyond the onset of instability.

4.3. Wetting, Deposition, and Coprecipitation

Quantifying asphaltene stability is the first step toward systematic study of many asphaltene-related phenomena. For example, previous research has shown that the effects of asphaltenes on wettability alteration vary with their stability.[11] Similarly, deposition has been shown to occur in a capillary tube during the flocculation process.[6]

A great deal of effort has been devoted to "purifying" asphaltenes, removing coprecipitated or codeposited material that does not strictly meet the solubility criteria that define asphaltenes as insoluble in *n*-heptane and soluble in toluene. In so doing, we run the risk of ignoring important interactions among the asphaltenes, waxes, and other materials that tend to separate into an asphaltene-rich phase, possibly magnifying the impact of unstable asphaltenes. More work is needed to understand the overall phase behavior of asphaltenes in real oils, rather than that of asphaltenes separated from the remainder of the oil system.

4.4. Model Systems and Standards

There probably are questions that can best be addressed using model systems, but it is also possible to be misled if a model is qualitatively different from reality. In the case of asphaltenes in crude oils, reality consists of very complex mixtures of chemical species. Separated asphaltenes dissolved in mixtures of toluene and heptane retain some of the complexity of the asphaltene fraction, but they sacrifice all of the complexity of the nonasphaltene portions of the oil. Figure 16.16 clearly illustrates the difference between stability of asphaltenes in their parent oils and stability of asphaltenes removed from those oils by precipitation with *n*-heptane. The separated asphaltenes, redissolved in toluene, were less stable than the same asphaltenes in their STOs. Presumably, asphaltic and resinous materials that do not separate with the n-C_7 asphaltenes help to stabilize the n-C_7 asphaltenes in the STO.[49] In terms of stability, the separated asphaltenes from these two oils were much more similar to one another than were the asphaltenes in their parent oils.

Another approach, the use of individual model compounds, may be able to help resolve questions about interactions between asphaltene-like molecules that are not amenable to study in the complex environment of a crude oil, but only if the right models are selected and the right questions are asked.

5. Conclusions

- Although asphaltenes consist of an extremely complex mixture of species that are not well-defined chemically, some aspects of their phase behavior can be described by straight-forward thermodynamic models that take into account the size difference between solute and solvent species. Thermodynamic description of the appearance of the least soluble asphaltenes as a separate phase that can be detected by optical microscopy has been demonstrated.
- There are many reasons for differences between observations in different laboratories. Most important among these are real differences in samples and experimental details including detection sensitivity and duration of the experiments compared to the kinetics of asphaltene flocculation, which are slowest near the onset of asphaltene instability.
- It is not possible to determine the solubility parameters of the onset conditions and the original oil from experiments in dilute solutions; the solubility parameter at the onset condition is not a constant, but varies with dilution. An independent measure of the solubility parameter of the original oil is essential for interpretation of onset experimental data.
- Because crude oil is mostly nonpolar and the interactions between asphaltene particles at the onset of instability are primarily the nonpolar van der Waals interactions, refractive index can be used to estimate oil and onset solubility parameters.
- A linear relationship between the onset solubility parameters and the square-root of n-alkane nonsolvent molar volumes can be demonstrated experimentally and predicted thermodynamically. We call this linear relationship the asphaltene instability trend or ASIST and report values for more than forty oil samples.
- Use of ASIST to predict asphaltene onset pressures during depressurization of live fluids is demonstrated for a model fluid and a live crude oil.
- Asphaltene stability in mixtures can be predicted from asphaltene stability in unmixed oils. Even in cases where asphaltenes are stable in both parent oils, the instability is predictable. No special interactions with other crude oil fractions such as resins need to be invoked to explain asphaltene stability.

Acknowledgments

The authors thank NPTO (US DOE), Chevron, and the Deepstar Consortium for support and their collaborators at Rice University (George Hirasaki, Walter Chapman, David Ting, and Doris Gonzales) and Oilphase-DBR (a division of Schlumberger) for helpful discussions and access to data and equipment. The authors also thank Oliver Mullins for the opportunity to contribute to this volume.

References

[1] Oliensis, G.L. (1933). In: *ASTM, Proceedings. of the 36th Annual Meeting*, Chicago 33, Part II, p. 715.
[2] Mullins, O.C. (2004). In: *5th International Conference on Petroleum Phase Behaviour and Fouling*, Banff.
[3] Haskett, C.E. and M. Tartera (1965). *J. Petrol. Technol.* **17**, 387.
[4] Tuttle, R.N. (1983). *J. Petrol. Technol.* **35**, 1192.
[5] Cimino, R., S. Correra, A. Del Bianco, and T.P. Lockhart (1995). In: E.Y. Sheu and O.C. Mullins (eds.), *Asphaltenes: Fundamentals and Applications*. Plenum, New York, p. 97.
[6] Wang, J.X., J.S. Buckley, and J.L. Creek (2004). *J. Disp. Sci. Tech.* **25**, 287.
[7] Førdedal, H., Ø. Midttun, J. Sjöblom, O.M. Kvalheim, Y. Schildberg, and J.-L Volle (1996). *J. Coll. Interface Sci.* **182**, 117.
[8] McLean, J.D. and P.K. Kilpatrick (1997). *J. Colloid Interface Sci.* **196**, 23.
[9] Silverman, L.D., S. Winkler, J.A. Tiethof, and A. Witoshkin (1986). In: *NPRA Annual Meeting*, Los Angeles.
[10] Asomaning, S. and A.P. Watkinson (1999). In: T.R. Bott (ed.), *Understanding Heat Exchanger Fouling and its Mitigation*. Begell House, New York, p. 283.
[11] Al-Maamari, R.S.H. and J.S. Buckley (2003). *SPE REE* **6**(4), 210.
[12] Hotier, G. and M. Robin (1983). *Revue de l'IFP* **38**, 101.
[13] Fotland, P., H. Anfindsen, and F.H. Fadnes (1993). *Fluid Phase Equilibria* **82**, 157.
[14] Jamaluddin, A.K.M., N. Joshi, M.T. Joseph, D. D'Cruz, B. Ross, and J. Creek *et al.* (2000). Paper # 2000–68. In: *Canadian International Petroleum Conferences.*
[15] Escobedo, J. and G.A. Mansoori (1995). *SPE Prod Fac.* **10**(2), 115.
[16] Correra, S., F. Capuano, and N. Panariti (2004). In: *5th International Conference on Petroleum Phase Behaviour and Fouling*, Banff.
[17] Yudin, I.K., G.L. Nikolaenko, E.E. Gorodetsky, V.I. Kosov, V.R. Melikyan, and E.L. Markhashov *et al.* (1998). *J. Pet. Sci. Eng.* **20**, 297.
[18] Wiehe, I.A. and R.J. Kennedy (2000). *Energy Fuels* **14**, 56.
[19] Speight, J.G., D.L. Wernick, K.A. Gould, R.E. Overfield, B.M.L. Rao, and D.W. Savage (1985). *Revue de l'IFP* **40**, 51.
[20] Altgelt, K.H. and M.M. Boduszynski (1994). *Composition and Analysis of Heavy Petroleum Fractions*. Marcel Dekker, New York.
[21] Andersen, S.I., A. Keul, and E. Stenby (1997). *Petrol. Sci. Technol.* **16**, 611.
[22] Ravey, J.C., G. Decouret, and D. Espinat (1988). *Fuel* **67**, 1560.
[23] Tanaka, R., E. Sato, J.E. Hunt, R.E. Winans, S. Sato, and T. Takanohashi (2004). *Energy Fuels* **18**, 1118.
[24] Maham, Y., M. Chodakowski, X. Zhang, and J.M. Shaw (2004). In: *5th International Conference on Petroleum Phase Behaviour and Fouling*, Banff.
[25] Sirota, E.B. (2004). In: *5th International Conference on Petroleum Phase Behaviour and Fouling*, Banff.
[26] Nellensteyn, F.I. (1924). *J. Inst. Petrol. Technol.* **10**, 211.
[27] Nellensteyn, F.I. (1938). In: A.E. Dunstan (ed.) *The Science of Petroleum*, Vol. 4. Oxford University Press, London, p. 2760.
[28] Pfeiffer, J.P. and R.N. Saal (1940). *J. Phys. Chem.***44**, 139.
[29] Andersen, S.I. and K.S. Birdi (1991). *J. Colloid Interface Sci.* **142**, 497.
[30] Hirschberg, A., L.N.J. deJong, B.A. Schipper, and J.G. Meijer (1984). *SPEJ* **24**, 283.
[31] Wang, J.X. and J.S. Buckley (2001). *Energy Fuels* **15**, 1004.
[32] Hildebrand, J.H., J.M. Prausnitz, and R.L. Scott (1970). *Regular and Related Solutions—The Solubility of Gases, Liquids, and Solids*. Van Nostrand Reinhold, New York.
[33] Mitchell, D.L. and J.G. Speight (1973). *Fuel* **52**, 149.
[34] Burke, N.E., R.D.Hobbs, and S.F. Kashou, (1990). *J. Petrol. Tech.* 1440.
[35] Wiehe, I.A. (1996). Fuel Sci. Tech. Int. **14**, 289.
[36] Wiehe, I.A. (1995). *Ind. Eng. Chem. Res.* **34**, 661.

[37] Mertens, E.W. (Dec. 1960). Predicting weatherability of coating grade asphalts from asphaltene characteristics. *ASTM Bulletin* (TP 218–222), 40–44.
[38] Waxman, M.H., C.T. Deeds, and P.J. Closmann (1980). SPE 9510, ATCE, Dallas.
[39] Reichert, C., B.J. Fuhr, and L.L. Klein (Sept.–Oct. 1986). Measurement of asphaltene flocculation in bitumen solutions. *J. Cnd. Pet. Tech.*, 33–37.
[40] Andersen, S.I. (1999). *Energy Fuels* **13**, 315.
[41] Donaggio, F., S. Correra, and T.P. Lockhart (2001). *Petrol. Sci. Technol.* **19**, 129.
[42] Buckley, J.S., G.J. Hirasaki, Y. Liu, S. Von Drasek, J.X. Wang, and B.S.Gill, (1998). *Petrol. Sci. Technol.* **16**, 251.
[43] Israelachvili, J.N. (1991). *Intermolecular and Surface Forces*, 2nd edn. Academic Press, San Diego.
[44] Buckley, J.S. and J.X. Wang (2002). *J. Pet. Sci. Eng.* **33**, 195.
[45] Prausnitz, J.M. and F.H. Shair (1961). *Am. Inst. Chem. Eng. J.* **7**, 682.
[46] Barton, A.F.M. (1991). *CRC Handbook of Solubility Parameters and Other Cohesion Parameters*, 2nd edn. CRC Press, Boca Raton, FL.
[47] Taylor, S.D., J. Czarnecki, and J. Masliyah (2001). *Fuel* **80**, 2013.
[48] Wang, J.X. and J.S. Buckley (2001). In: *SPE 64994, Oilfield Chem. Symposium*, Houston.
[49] Wang, J.X. and J.S. Buckley (2003). *Energy Fuels* **17**, 1445.
[50] Wang, J.X., J.S. Buckley, and J.L. Creek (2004). In: *5th International Conference on Petroleum Phase Behaviour and Fouling*, Banff.
[51] Reid, R.C., J.M. Prausnitz, and T.K. Sherwood (1977). *The Properties of Gases and Liquids*, 3rd edn., McGraw Hill, New York.
[52] PVTSim (Ver. 11) (2002). *Calsep.*
[53] Groenzin, H. and O.C. Mullins (2000). *Energy Fuels* **14**, 677.
[54] de Boer, R.B., K. Leerlooyer, M.R.P. Eigner, and A.R.D. van Bergen (1995). *SPE Prod. Fac.* **10**(1), 55.
[55] Fan, T., J.X. Wang, and J.S. Buckley (2002). In: *SPE 75228, SPE/DOE IOR Symposium*, Tulsa.
[56] Ting, P.D. (2003). PhD Thesis, Rice University, Houston.
[57] Wang, J.X., J.S. Buckley, N.E. Burke, and J.L. Creek (2004). *SPE Prod Fac.* **19**, 152.
[58] Thiyagarajan, P., J.E. Hunt, R.E. Winans, K.B. Anderson, and T. Miller (1995). *Energy Fuels* **9**, 829.
[59] Gonzalez, D.L., D.P. Ting, G.J. Hirasaki, and W.G. Chapman (2004). In: *5th International Conference on Petroleum Phase Behaviour and Fouling*, Banff.
[60] Correra, S. and F. Donaggio (2000). In: *SPE 58724, Int. Symp. on Formation Damage*, Lafayette.
[61] Alboudwarej, H., K. Akbarzadeh, J. Beck, W.Y. Svrcek, and H.W. Yarranton (2003). *AIChE J.* **49**, 2948.
[62] van Kerkvoort, W.J., A.J.J. Nieuwstad, and M. van der Waarden, (1952). In: *IV Congr. Intern. Chauffage Ind.* (Preprint No. 220), Paris.
[63] Heithaus, J.J. (1962). *J. Inst. Petrol.* **48**(458), 45.
[64] Bichard, J.A. (1969). in: *19th Canadian Chem. Engr. Conf. and 3rd Symp. on Catalysts*, Edmonton, Alberta.
[65] Kesler, M.G. and B.I. Lee (1976). Hydrocarbon Processing, **55**, 153.

Appendix I: Asphaltene Onset Detection by Batch Titration

The procedure used to determine the onset of asphaltene flocculation in this work is outlined in Figure A-I-1.

Step 1. Microscopic examination. A drop of STO oil is observed microscopically at high magnification ($\sim 320X$). If the sample is free from any particles no special pretreatment is required. Proceed to Step 3.

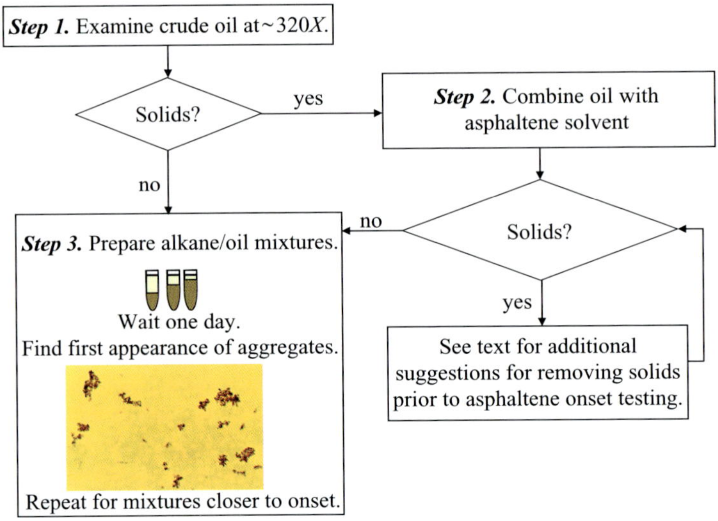

Figure A-I-1. Outline of the asphaltene onset determination scheme.

Often crude oils contain aggregated asphaltenes, wax crystals, emulsified water, and other unidentified solid particles. Asphaltenes can usually be distinguished from wax crystals by observing a drop of oil between polarizing filters oriented at 90° to one another. Crystalline wax particles transmit the polarized light and appear as bright spots. Amorphous asphaltenes do not transmit polarized light. Highly spherical objects are likely to be emulsified liquids. Existing aggregates or particles may obscure the onset observation. Strategies for removing sources of interference are outlined in Step 2.

Step 2. Treatment of sample to remove sources of interference. Preexisting asphaltene aggregates, waxes, and other particulates might be removed by filtration or centrifugation. To avoid changes in composition, however, a better strategy is to redisperse the asphaltene aggregates by addition of a good solvent such as toluene or 1-methylnaphthalene (1-MN). 1-MN is the stronger solvent, but it is more viscous than toluene so the dispersion time is longer. To redisperse asphaltene aggregates, the following steps are recommended:

1. Prepare a mixture of oil with a minimal amount of the chosen solvent. Sonicate the mixture for several hours, heat the mixture to 60–70°C for another several hours, and allow the mixture to equilibrate for 1 day at room temperature. Use sealed vials during these treatments to prevent evaporation.

2. Examine the mixture microscopically. If asphaltene aggregates remain, add more solvent to the mixture; repeat as above until the minimum volume fraction of solvent necessary to completely redisperse asphaltenes is found.

After the minimum solvent amount has been found, prepare at least one more oil/solvent mixture with more solvent than the minimum requirement and measure the volume ratios at onset conditions for each mixture. Use the linear trend between V_n/V_o and V_s/V_o (Eq. 16.4) to extrapolate the onset volume ratio of V_n/V_o for oil without solvent. Refractive index of mixture of oil + nonsolvent (without added solvent) at onset condition (P_{RI}) can then be calculated by:

$$P_{RI} = \frac{1}{1+x}(x \cdot RI_n + RI_o), \qquad \text{(A-I-1)}$$

where x is the extrapolated value of V_n/V_o without solvent, RI_n and RI_o represent refractive indices for nonsolvent and oil, respectively. Preparation of additional oil + solvent mixtures can improve the accuracy of the extrapolation.

If there are wax crystals visible in the oil, it is preferable to conduct the onset measurements at a temperature above the wax melting point. Dissolving wax back to the solution helps to reduce the interference during onset observation and restores the contributions from all paraffinic components, including the heavy ends, to asphaltene instability.

Emulsified water may require additional treatment by centrifugation and/or washing with distilled water, especially if asphaltenes are associated with the oil/water interfaces (indicated by brownish color).

Step 3. Onset determination. The onset condition can be determined in batch mode to a reasonable tolerance by examination of mixtures of oil (or oil/solvent mixture) and precipitant that vary by 1% (by volume) in composition. For example, if aggregates are observed in a mixture consisting of 50% oil and 50% heptane, but a mixture with 51% oil and 49% heptane is clear, then the 50:50 mixture is the onset mixture.

RI for the onset oil/precipitant mixture (P_{RI}) can be directly measured and used to approximate onset solubility parameter (δ_{onset}) from Eq. (16.9), if no solvent is involved. In the case where solvent is used, P_{RI} can be obtained indirectly as described in Step 2. RI is measured with a temperature-controlled automatic refractometer (e.g., Index Instruments, model GPR 11-37). Transfer mixture to the refractometer prism; seal immediately to prevent evaporation. Temperature equilibration requires a few seconds. It is important to record the initial stabilized RI to minimize time-dependent effects of adsorption and/or deposition of asphaltenes on the prism.

Always add nonsolvent (e.g., heptane) to oil, not oil to nonsolvent. A glass vial that can be sealed immediately with a PTFE-lined cap is a good mixing vessel. Shake vigorously to avoid high local concentrations of nonsolvent that might cause premature precipitation. Agitate the vials in an ultrasonic bath for a few minutes for added mixing. Allow the mixtures to rest at the desired temperature for about 24 hr—an arbitrary period that should accommodate the slow flocculation kinetics without unduly prolonging the analysis. Microscopic examination, as described above, can be repeated to find the onset of flocculation. Some of the components in the mixture are volatile so the examination should be made quickly. Differential evaporation rates can cause changes in solvent quality and asphaltene stability. Once the onset with n-heptane has been determined, that value can be used to estimate the onset conditions for other n-alkanes, reducing the number of mixtures required to pinpoint their onset values.

Appendix II: Historical Interpretations of *n*-Alkane Titration Data

A summary of historical interpretations of titrations to determine the onset of asphaltene instability in dilute solutions, all of which are based on either volume or mass of solvents and nonsolvents, is given in Table A-II-1.

Appendix III: Calculation of Solubility Parameters Using PVTsim

The Hildebrand solubility parameter for liquid, δ, is defined as the square root of the density of cohesive energy:

$$\delta = \sqrt{\frac{\Delta U}{v}}, \tag{A-III-1}$$

where ΔU is the energy of vaporization and v is the liquid molar volume. By definition, ΔU is the energy required to vaporize one mole of a given substance from its liquid state to ideal gas state at the same temperature, i.e.,

$$\Delta U = U^{\text{ideal gas}} - U^{\text{liquid}}, \tag{A-III-2}$$

where $U^{\text{ideal gas}}$ and U^{liquid} are internal energies at ideal gas state and liquid state, respectively. From thermodynamic relationships, we know that

$$U^{\text{ideal gas}} = H^{\text{ideal gas}} - RT \tag{A-III-3}$$

$$U^{\text{liquid}} = H^{\text{liquid}} - pv, \tag{A-III-4}$$

where $H^{\text{ideal gas}}$ and H^{liquid} are enthalpies at ideal gas state and liquid state, respectively, p is pressure, R is gas constant, and T is absolute temperature. Substituting (A-III-3) and (A-III-4) to (A-III-2), we get:

$$\Delta U = (H^{\text{ideal gas}} - H^{\text{liquid}}) - RT + pv. \tag{A-III-5}$$

We know that enthalpy is a physical property relative to a reference point. It is its change that is physically meaningful. If we choose the ideal gas state as the reference point, we can decompose H^{liquid} into two parts:

$$H^{\text{liquid}} = H^{\text{ideal gas}} + H^{\text{residual}} \tag{A-III-6}$$

and simplify (A-III-5) as:

$$\Delta U = -H^{\text{residual}} - RT + pv, \tag{A-III-7}$$

where H^{residual} is the residual enthalpy, which equals the enthalpy change from ideal gas to the liquid state (negative).

Equation (A-III-6) is employed in PVTsim to calculate the overall enthalpy. Unfortunately, only the overall enthalpy H^{liquid} is output, while its components

Table A-II-1. Summary of Interpretations of Titration Data Compared to Eq. (16.4)

Observation expressed as:	Interpretations:	$S =$	$I =$	Reference and notes
$FR = V_s / (V_s + V_n)$ $X = (V_s + V_n)/ V_o$ FR_{max} is the value of FR as $V_o \rightarrow 0$ X_{min} is the value of X when $V_s = 0$	Peptizability of asphaltenes: $p_a = 1 - FR_{max}$ Peptizing power of dispersing medium: $p_o = FR_{max}\ (X_{min} + 1)$ State of peptization: $P = p_o/(1 - p_a) = X_{min} + 1$	$\frac{1 - FR_{max}}{FR_{max}}$	X_{min}	van Kerkvoort *et al.*,[62] c.f. also Heithaus[63]
$V_n/M_o = S_1(V_s/M_o) + I_1$	Dispersability of the least soluble asphaltenes: $p_a = V_n/(V_n + V_s)$ as $V_o \rightarrow 0$ $p_a = (\delta_s - \delta_{Cr})/(\delta_s - \delta_n)$	$\frac{V_n - V_{n\,min}}{V_s}$	$\frac{V_{n\,min}}{V_o \rho_o}$	Mertens[37]
$V_s/M_o = S_2\ (V_n/M_o) + I_2$	Solvent quality: $\cot \phi = (V_n - V_{nmin})/ V_s$	$\frac{V_n - V_{n\,min}}{V_s}$	$\frac{V_{n\,min}}{V_o \rho_o}$	Bichard[64] cf also Waxman *et al.*[38]
$\sum \alpha_i\ V_i = 0$	By convention: $\alpha < 0$ for solvents, $\alpha > 0$ for precipitants	$-\alpha_s/\alpha_n$	$-\alpha_o/\alpha_n$	Hotier and Robin[12]
$M_n/M_o = c_o + c_1 M_s/M_o$		$c_1 \rho_s/\rho_n$	$c_0 \rho_o/\rho_n$	Cimino *et al.*[5]
$\sum \delta_i \phi_i = \delta_{Cr}$		$\frac{\delta_s - \delta_{Cr}}{\delta_{Cr} - \delta_n}$	$\frac{\delta_{o,Cr} - \delta_{Cr}}{\delta_{Cr} - \delta_n}$	Andersen[40]
$\sum \delta_i \phi_i = \delta_{Cr}$ $\frac{100\ V_s}{V_n + V_s} = I_N + \frac{100\ V_o}{V_n + V_s}\left[\frac{I_N - S_{BN}}{100}\right]$	$I_N = \frac{\delta_{Cr} - \delta_n}{\delta_s - \delta_n}$ $S_{BN} = \frac{\delta_{o,Cr} - \delta_n}{\delta_s - \delta_n}$	$\frac{100 - I_N}{I_N}$	$\frac{S_{BN} - I_N}{I_N}$	Wiehe and Kennedy[18] Note: (1) *S* and *I* reduce to the same solubility parameter ratios as in Andersen above; (2) *S* and *I* also reduce to the Heithaus[63] expressions.

V = volume (cm^3), M = mass (g), ρ = density (g/mL), ϕ = volume fraction (dimensionless), δ= solubility parameter (MPa)$^{1/2}$; subscripts: Cr = critical, o = oil, o,Cr = oil (calculated assuming δ_{Cr}), s = solvent, n = nonsolvent, n min = minimum amount of nonsolvent (when $V_s = 0$). S and I are defined by Eq. (16.4). Other slopes and intercepts are indicated by subscripts 1 and 2. All other symbols are as defined in columns 1 and 2 (see references for additional details).

$H^{\text{ideal gas}}$ and H^{residual} are not given. Let us first analyze how these two components are calculated.

The ideal gas enthalpy, $H^{\text{ideal gas}}$, is calculated as:

$$H^{\text{ideal gas}} = \int_{T_{\text{ref}}}^{T} C_{\text{p}}^{\text{ideal gas}}\, dT, \tag{A-III-8}$$

where T_{ref} is the reference temperature, T is the temperature of interest, and $C_{\text{p}}^{\text{ideal gas}}$ is the isobaric heat capacity which is calculated from an empirical equation:

$$C_{\text{p}}^{\text{ideal gas}} = C_1 + C_2T + C_3T^2 + C_4T^3. \tag{A-III-9}$$

Coefficients C_1, C_2, C_3, and C_4 can either be obtained from literature (Reid et al.,[51] for pure compounds) or estimated from normal boiling point, specific gravity and acentric factor (Kesler and Lee[65] for heavy hydrocarbons). T_{ref} is chosen as 0°C (273.15 K) in PVTsim, but can be set to any other temperature. In other words, *the enthalpy calculated from PVTsim is the total enthalpy relative to the enthalpy at 0°C.*

On the other hand, H^{residual} is calculated from the equation-of-state by:

$$H^{\text{residual}} = -RT^2 \frac{\partial \ln f}{\partial T}, \tag{A-III-10}$$

where f is fugacity coefficient. In other words, as long as the EOS is well-tuned, the residual enthalpy, H^{residual}, can be calculated from Eq. (A-III-10), either for a pure compound or a mixture of compounds.

The analyses above show that if we can set $H^{\text{ideal gas}}$ equal to zero, the output from PVTsim will be $H^{\text{liquid}} = H^{\text{residual}}$. Conceptually, this could be done by setting $T_{\text{ref}} = T$ in Eq. (A-III-8), but that is not an option since the user cannot change T_{ref}. Alternatively, we can set $C_{\text{p}}^{\text{ideal gas}} = 0$. Since the user has access to the coefficients C_1, C_2, C_3, and C_4, $C_{\text{p}}^{\text{ideal gas}}$ can be forced to be zero by setting all four coefficients in Eq. (A-III-9) equal to zero. It should be emphasized at this point that we should first tune the PVT model parameters to match the experimental measurements (densities, bubble points, etc.) before setting C_1, C_2, C_3, and C_4 equal to zero to obtain H^{residual}. Once the PVT dataset is tuned, setting C_1, C_2, C_3, and C_4 equal to zero for all the components and pseudo-components does not affect calculations of physical properties other than the ideal gas enthalpy. Once we have H^{residual}, the solubility parameter can be readily calculated from Eqs. (A-III-7) and (A-III-1).

Appendix IV: Oil and Asphaltene Properties

Oil properties are summarized in Table A-IV-1; Table A-IV-2 lists the ASIST parameters for fits to onset data.

Table A-IV-1. Summary of Crude Oil Properties

Oil ID	°API	δ_{oil} at 20°C ($MPa^{1/2}$)	n-C_7 asphaltene (wt%)	Acid#	Base#	Avg. MW (Da)	Saturates[a] (wt%)	Aromatics[a] (wt%)	Resins[a] (wt%)
				(mg KOH/g oil)					
A-93	25.5	18.65	5.4	0.14	2.42	255	56.0	18.2	18.1
A-95	25.2	18.54	8.7	0.24	2.20	236	51.0	20.5	19.7
B-BQ-02	29.7	18.09	3.7	0.01	1.73	240	65.6	20.5	10.1
Black Mtn-04	21.8	18.77	5.0	0.59	2.73	315	52.6	19.6	21.8
Blindberry	39.8	17.43	0.4	0.07	0.72	183			
C-AG1-02	43.6	16.90	0.4						
C-AG2-02	45.7	16.74	0.2	0.30	0.70	165			
C-AG3-02	43.5	16.95	0.2	<0.01	0.74	175			
C-AL-03	18.7	18.95	2.4	1.79	4.94	484	53.8	18.4	25.2
C-B2-01	36.3	17.51	1.3	0.50	2.67	196	68.5	17.5	12.7
C-Br-01	48.0	16.64	0.1	0.08	0.11	148	74.8	24.5	0.7
C-BrR-04	33.0	17.68	0.9	0.23	0.56	246	73.2	16.1	9.8
C-C-02	31.2	17.85	3.7	0.29	1.75	232	63.3	20.3	13.4
C-CM1-02	16.8	19.12	3.5	1.82	5.83	382			
C-CM2-02	17.2	19.04	3.0	2.18	5.46	386			
C-CS3-02	28.7	17.98	4.4	0.39	2.74	310			
C-F-03	29.5	18.14	6.0	0.16	1.52	282	61. 7	18.4	13.5
C-F2-03	27.9	18.20	2.0	0.70	1.32	335	63.2	17.9	16.2
C-GC-T1-03	31.1	17.88	4.6	0.03	1.69				
C-GC18A-02	26.5	18.41	11.2	0.12	2.20	266	62.0	15.1	13.3
C-GC21A-02	27.2	18.30	8.2	0.08	2.06	259	62.8	18. 6	11.6
C-GC21B-02	30.3	18.06	6.5	0.03	1.84	236	68.8	15.9	9.4
C-GC237-32-02	30.7	17.67	3.7	2.15	2.08	232	64.9	16.3	12.6
C-HD-01	20.7	18.78	2.7	0.91	2.58	346	64.1	14.5	17.9
C-HL-01	29.2	18.00	2.3	<0.01	1.80	259	67.3	14.9	15.1
C-HM2-01	23.0	18.69	8.8	0.13	3.59	306	49.8	18.5	21.2
C-K-01	18.9	18.95	3.5	2.44	5.19	319	52.8	19.0	24.8
C-L-01	34.0	17.63	1.4	0.21	2.30	216	66.3	16.7	13.2
C-LH-99	22.6	18.56	2.8	1.90	6.05	268	49.5	21.5	25.6
C-P-01	31.9	17.81	3.8	0.54	1.55	236	63.5	16.7	16.0
C-R-00	31.1	17.82	1.9	0.25	0.49	235	70.6	16.3	11.4
C-T-01	26.2	18.35	9.8	0.38	1.87	280	60.6	17.0	11.6
C-T-02	34.2	17.71	1.4	0.01	1.30	214	61.1	22.8	14.6
C-T1-00	31.6	17.87	2.7	0.10	3.60	210	63.4	16.5	17.4
C-T2-00	31.2	17.92	2.6	0.22	3.58	233	62.8	15.8	18.7
Cottonwood-03	26.4	18.32	2.5	0.04	1.87	262	57.9	22.7	16.5
E-1XO-00	21.9	18.57	0.8	3.42	2.57	264	57.8	22.1	19.4
Mars-Pink	16.5	19.19	4.8	3.92	2.30	309	38.4	29.8	25.8
Minnelusa	24.6	18.58	9.0	0.17	2.29	280	60.1	18.9	11.3
Minnelusa-02	24.3	18.57	8.8	0.01	2.01	264	58.0	20.2	13.9
S-A5-01	32.5	17.94	8.6	<0.01	1.17	239	61.1	21.2	9.0
S-A6-01	28.4	18.22	7.1	0.08	1.78	267	60.2	19.7	13.0
S-Ven-39	28.8	18.15	5.8	0.14	1.68	240	51.1	28.3	14.5
SQ-95	37.2	17.60	1.3	0.17	0.62	213	65.2	18.3	13.9
Tensleep	31.2	17.89	3.2	0.16	0.96	271	69.3	16.0	9.4

[a]Saturates, aromatics, and resins were measured after precipitation of the *n*-hexane asphaltenes.[55] The weight % of *n*-C_6 asphaltene is approximately equal to [100-(saturates + aromatics + resins)]

Table A-IV-2. Summary of Asphaltene Instability Trends (ASIST)

Oil ID	T (°C)	Slope	Intercept ($MPa^{1/2}$)	R^2	Oil ID	T (°C)	Slope	Intercept ($MPa^{1/2}$)	R^2
A-93	20.0	0.1440	14.85	0.915	C-GC237-32-02	20.0	0.1192	15.50	1.000
A-95	20.0	0.1966	14.70	0.966	C-GC237-32-02	60.0	0.1356	14.75	1.000
B-BQ-02	71.0	0.1311	15.44	0.998	C-HD-01	20.0	0.1839	14.20	1.000
Black Mtn-04	20.0	0.2054	14.64	0.956	C-HD-01	60.0	0.2052	13.37	1.000
Blindberry	20.0	0.1351	14.49	0.981	C-HL-01	20.0	0.1499	14.86	0.999
C-AG1-02	20.0	0.0779	16.06	0.998	C-HL-01	60.0	0.1719	13.97	1.000
C-AG1-02	60.0	0.0951	15.22	0.999	C-HM2-01	20.0	0.1446	15.13	0.999
C-AG2-02	20.0	0.1037	15.37	1.000	C-HM2-01	60.0	0.2041	13.69	1.000
C-AG3-02	20.0	0.1400	14.89	0.993	C-K-01	60.0	0.2403	12.80	1.000
C-AL-03	20.0	0.2110	13.51	0.999	C-L-01	60.0	0.1944	13.22	1.000
C-B2-01	60.0	0.1979	13.16	1.000	C-L-01	70.0	0.1981	13.02	1.000
C-BrR-04	20.0	0.1404	14.65	0.997	C-LH-99	20.0	0.2148	13.55	0.992
C-BrR-04	50.0	0.1386	14.45	0.998	C-P-01	20.0	0.1858	14.38	1.000
C-BrR-04	60.0	0.1443	14.23	0.999	C-R-00	71.1	0.1562	14.17	1.000
C-C-02	20.0	0.1060	15.74	0.998	C-T-01	20.0	0.1369	15.57	1.000
C-C-02	60.0	0.1238	14.96	0.999	C-T-01	60.0	0.1494	14.83	1.000
C-CM1-02	40.0	0.2275	12.63	1.000	C-T-02	20.0	0.1566	14.20	1.000
C-CM1-02	73.3	0.2507	11.80	1.000	C-T-02	60.0	0.2077	12.90	1.000
C-CM2-02	40.0	0.2269	12.64	1.000	C-T1-00	60.0	0.1668	13.96	0.999
C-CM2-02	73.3	0.2532	11.77	1.000	C-T1-00	70.0	0.1733	13.73	0.999
C-CS3-02	40.0	0.1798	13.61	1.000	C-T2-00	60.0	0.1659	14.01	1.000
C-CS3-02	73.3	0.1956	12.87	1.000	C-T2-00	70.0	0.1705	13.80	1.000
C-F-03	20.0	0.1240	15.54	1.000	Cottonwood-03	20.0	0.1787	14.51	0.979
C-F-03	60.0	0.1281	14.92	0.998	Mars-Pink	20.0	0.1999	13.87	0.997
C-F2-03	20.0	0.1600	14.68	0.998	Minnelusa	20.0	0.0932	16.64	1.000
C-GC-T1-03	60.0	0.1138	15.20	0.999	Minnelusa	60.0	0.1133	15.72	1.000
C-GC-T1-03	80.0	0.1140	14.99	0.999	Minnelusa-02	60.0	0.1104	15.76	1.000
C-GC18A-02	60.0	0.1280	15.25	0.999	S-A5-01	20.0	0.1300	15.59	0.999
C-GC18A-02	80.0	0.1271	14.94	0.996	S-A6-01	20.0	0.1720	14.79	1.000
C-GC21A-02	60.0	0.1366	15.03	1.000	S-Ven-39	60.0	0.1293	14.95	0.986
C-GC21A-02	80.0	0.1440	14.66	0.999	SQ-95	20.0	0.1729	13.89	0.992
C-GC21B-02	60.0	0.1121	15.42	0.999					
C-GC21B-02	80.0	0.1048	15.32	0.999					

Appendix V: Prediction of Live Oil Asphaltene Stability from ASIST

Calculation of live oil RI from PVT data: As shown previously,[42] RI of live oil can be calculated from PVT data and measured RI for STO using Eq. (A-V-1):

$$\left(\frac{n^2-1}{n^2+2}\right)_{\text{live oil}} = \frac{1}{B_o}\left(\frac{n^2-1}{n^2+2}\right)^{\text{o}}_{\text{STO}} + 7.52\times10^{-6}\frac{R_s}{B_o}\sum_i x_i R_i, \qquad \text{(A-V-1)}$$

where n, refractive index, dimensionless; B_o, formation volume factor, rb/stb; R_s, gas/oil ratio, scf/stb; x_i, molar fraction of component i in light end, dimensionless; R_i, molar refraction of component i in light end, cm^3/mol; STO, stock tank oil; superscript ° indicates standard conditions (1 atm, 60°F).

The parenthesis in the first item in the right-hand side of (A-V-1) indicates that RI of STO is measured at standard conditions.

Calculation of v_p for light ends: The thermal expansion per barrel of STO from 60°F to reservoir temperature at 1 atm is ($B_o(1)$-1), where $B_o(1)$ is B_o at 1 atm. We assume this volume change is independent of compression.

The volume change due to compression can also be estimated from PVT data, as shown in the following two steps.

For one barrel of STO, the volume at 60°F and a given pressure P, without considering the thermal expansion, is

$$V = \rho_o/\rho, \tag{A-V-2}$$

where ρ_o is density at standard condition (1 atm, 60°F), and ρ is density at 60°F and pressure P, which can be calculated from an EOS simulator such as PVTsim.

Now, assume that the volume of dissolved gas at reservoir temperature and pressure is V_{gas}; the total volume for one barrel STO at reservoir conditions (including gas) is given by:

$$B_o(P) = \rho_o/\rho + (B_o(1) - 1) + V_{gas} \tag{A-V-3}$$

which can be rearranged to give:

$$V_{gas} = B_o(P) - B_o(1) - \frac{\rho_o}{\rho} + 1. \tag{A-V-4}$$

Since one barrel is equal to 1.5899×10^5 cm^3, and one scf is equal to 1.1957 mole of ideal gas, the approximate partial molar volume of the dissolved gas at reservoir condition is

$$v(\mathrm{cm}^3/\mathrm{mol}) = \frac{V_{gas} \times 1.5899 \times 10^5}{R_s \times 1.1957}. \tag{A-V-5}$$

Dissolved gas is assumed to be the precipitant. Hence, the value of v_p, the partial molar volume of nonsolvent in live oil at reservoir conditions, is that given by Eq. (A-V-5).

Assess asphaltene stability in live oil: At least two ASIST trends are established from onset measurements at two temperatures. The ASIST at reservoir temperature can be extrapolated from these two measured ASIST trends. v_p is used to extrapolate the potential onset condition (δ_{onset}) from ASIST at reservoir temperature. Calculated RI for live oil from Eq. (A-V-1) is used to approximate $\delta_{\text{live oil}}$. Finally, $\delta_{\text{live oil}}$ is compared with δ_{onset} at each pressure. If $\delta_{\text{live oil}} > \delta_{onset}$, asphaltene is stable, otherwise precipitation will occur. The first crossover of $\delta_{\text{live oil}}$ line with δ_{onset} line along the depressurization path indicates the onset pressure.

17

Dynamic Light Scattering Monitoring of Asphaltene Aggregation in Crude Oils and Hydrocarbon Solutions

Igor K. Yudin and Mikhail A. Anisimov

1. Introduction

An overview of dynamic light scattering (DLS) studies of asphaltene aggregation in hydrocarbons and crude oils is presented. A special optical scheme, to make light scattering measurements in "nontransparent" light-absorbing colloids possible, has been designed and tested. The modified DLS is an effective technique for real-time monitoring of petroleum colloids. Combined with accurate viscosity measurements, DLS is a powerful tool for investigating the colloid nature of crude oils, heavy petroleum fractions, and asphaltene solutions. Various regimes of asphaltene aggregation have been investigated by DLS in crude oils and hydrocarbon mixtures. Crossover between reaction-limited aggregation and diffusion-limited aggregation has been observed in hydrocarbon solutions of asphaltenes. Asphaltene colloidal structures, originally persisting in some crude oils, have been detected. Using n-heptane as precipitant, we have studied stability of crude oils with respect to asphaltene aggregation. We have found that the oils with different aromatics/saturates ratios exhibit different aggregation kinetics. Formation of stable asphaltene aggregates, originating from ultrasonic agitation and surfactant addition, has been also studied. A DLS microrheology in a crude oil has been demonstrated.

Crude oils and heavy petroleum fractions are complex multicomponent fluids exhibiting properties of either molecular solutions or colloids depending on the origin and on the properties under consideration. Non-Newtonian rheology of petroleum systems (especially those with high-content asphaltenes and paraffins) indicates the presence of a supra-molecular mesoscopic structure. However, the physical nature of this structure is not yet well understood. It is commonly accepted

Igor K. Yudin • Oil and Gas Research Institute, Russian Academy of Sciences, Moscow 117971, Russia; and PhotoCor Instruments, Inc., College Park, Maryland 20740-822. **Mikhail A. Anisimov** • Department of Chemical and Biomolecular Engineering and Institute for Physical Science and Technology, University of Maryland, College Park, Maryland 20742.

that the mesoscopic structure is somehow associated with asphaltenes. Asphaltenes are customarily defined as colloid-like heavy fractions of crude oils that are completely miscible with aromatic hydrocarbons such as toluene or benzene but are insoluble in n-alkanes. The latter property is used for extraction of asphaltenes.[1–3] Although asphaltenes are poorly characterized and the above definition is in a certain sense sounds arbitrary, they often behave like distinct species during refining.[4] Petroleum asphaltenes are complex aromatic macro-cycle molecules surrounded and possibly linked by alkanes. The molecular structure of asphaltenes varies significantly depending on their origin, method of petroleum recovery, and history of extraction.[5] Attempts to find a correlation between extensive data on asphaltene structure and their colloidal properties have not been very successful.

Asphaltenes are the most intriguing fraction of petroleum fluids. Polarity and complex structure of asphaltenes determine the viscosity of crude oils as well as their tendency to flocculate and precipitate during the course of oil recovery and refinery. Adsorbed asphaltenes plug the porous media of reservoir rocks resulting in severe decrease of wellbore area permeability. Asphaltene colloids cause a dramatic reduction in oil flow as they change the rheological properties of crude oil and form deposits in wellbore tubing and auxiliary equipment. Coking and deactivation of catalysts affecting oil refining are also induced by asphaltenes. The asphaltene content directly affects the physical properties of crude oils, in particular, it determines their non-Newtonian rheology.[1,2]

The formation of colloidal asphaltene associates in several organic solvents and in crude oil was observed more than half a century ago.[6–8] In aromatic solvents, asphaltenes can form micelle-like aggregates above a certain concentration of asphaltenes, which reminds the critical micelle concentration (CMC) commonly observed in solutions of surfactants. As a result of these early observations one may conclude that (a) a crude oil is a colloidal system and (b) asphaltene and resin molecules are the two major components that form petroleum micelles.

The analysis of current experimental data indicates that formation of associates in crude oil and petroleum liquid fractions is more complicated and cannot be described by a simple mechanism.[9–12] It is still not clear whether one should consider asphaltenes in a good solvent below the CMC either as a true molecular solution or as stable molecular associates. Micellization ability of asphaltenes and resins has been discussed by Andersen and Speight.[13,14] They concluded that while the micellization above a certain concentration is a typical phenomenon for asphaltenes of different origin dissolved in various solvents, very little evidence is found for a similar behavior of other petroleum fractions.[15] The asphaltenes apparently are active species in micelle formation, while for resins this question is still open. CMC data for asphaltenes obtained by many authors differ widely, *e.g.*, the CMC in toluene was reported in the range from 0.1 to 3.24 g/L.[13–16] This can be explained by the diversity of the techniques applied and by the variety of asphaltenes (origin, extraction, impurities) studied. Most recently, Andreatta *et al.*,[17] using *high-Q* ultrasonic spectroscopy, have unambiguously detected the CMC in toluene for various asphaltenes at roughly 0.1 g/L. The authors name this threshold a "critical nanoaggregate concentration." It seems reasonable that the previously reported much higher CMCs in toluene probably correspond to higher-level aggregation of asphaltenes.

Asphaltene solutions in mixtures of toluene and n-heptane are suitable for investigating the stability and precipitation of asphaltenes and can be used as a model system to study aggregation phenomena in colloidal petroleum fluids. At a certain threshold value of the ratio n-heptane/toluene, asphaltene solutions become unstable and asphaltene particles start to aggregate. Although the aggregation kinetics of colloidal particles has been a subject of extensive experimental studies during the last decade, the mechanism of aggregation, the process of formation of fractal aggregates, and the crossover between different regimes of aggregation are still not fully understood, especially for such complex colloidal systems as asphaltene solutions. It has become evident that a new scientific approach that utilizes advanced physical techniques as well as a fundamental theory are required to clarify the colloidal properties of asphaltenes and the mechanism of their flocculation and precipitation.

An important task in such studies is to monitor in real time the kinetics of asphaltene aggregation and measure a changing size of the aggregates. The dynamic light scattering (DLS), also known as the photon correlation spectroscopy (PCS), is the most promising technique to perform such monitoring.[18] This technique directly measures the diffusion coefficient of Brownian particles suspended in a viscous medium and, if the kinetics is slow enough, enables one to monitor the particle growth. However, the application of light scattering to crude oils and asphaltene solutions is also a significant challenge since such systems are almost nontransparent for light.

2. Dynamic Light Scattering Technique

DLS (synonym of PCS) is commonly used to determine the size of nanoparticles in a suspension by measuring the dynamics of the light-scattering intensity fluctuations. Colloidal particles or macromolecules suspended in fluid undergo Brownian motion. This motion causes fluctuations in the local concentration of the particles resulting in local inhomogeneities of the refractive index. This in turn leads to the so-called Rayleigh light scattering. The time-dependent autocorrelation function $G(\tau)$ of the scattered light is a function of the delay time τ and for a single-exponential decay can be represented by

$$G(\tau) = b[1 + \varepsilon \exp(-\tau/\tau_c)], \tag{17.1}$$

where the baseline ("accidental") correlation level b is proportional to the total intensity I. The coefficient ε depends on the amount of stray light and on the aperture, but this coefficient is always less than unity.

There exist two techniques to measure the dynamic correlation function: heterodyning and homodyning. In heterodyne measurements, which are most suitable for small intensities, the scattered light is mixed coherently with a static light source at the incident wavelength, and the static field is added to the scattered fields at the photodetector. In homodyne measurements the photodetector receives scattered light only. Homodyning is most suitable for strong scattering (e.g., colloidal particles or critical opalescence). The diffusion coefficient D of the particles is directly related to the decay rate τ_c of the time-dependent correlation function. In the homodyne regime the relation between D and τ_c reads[18]:

$$D = \frac{1}{2\tau_c k^2}, \tag{17.2}$$

where k is the wave number of the scattered light, $k = \frac{2\pi n}{\lambda} \sin\frac{\theta}{2}$, n is the refractive index of the solvent, λ is the wavelength of the incident light in vacuum, and θ is the scattering angle. Equation (17.2) follows from a Fourier representation of the common diffusion equation that describes diffusion of a particle at the scale of $\frac{2\pi}{k}$.[18]

The mean hydrodynamic radius of the particles, R, then can be calculated by using the Stokes–Einstein equation:

$$D = \frac{k_B T}{6\pi\eta R}, \tag{17.3}$$

where k_B is Boltzmann's constant, T is the temperature, and η is the shear viscosity of the solvent. The size calculated from Eq. (17.3) is called the hydrodynamic radius. It may be larger than the radius of the bare particles because of possible layers of solvent, surfactant molecules, or (for charged particles) adsorbed ions. In most cases these layers add a negligible correction to the size except for the smallest sizes measurable. Equations (17.1) and (17.2) are valid for noninteracting spherical particles. If the particles are involved in an aggregation process, they certainly interact. However, these equations are still applicable to monitor the change of the apparent ("effective") particle size if the characteristic time of aggregation kinetics is much larger than the time of the measurements.

We have developed an original technique to study nontransparent disperse systems, including petroleum systems, when strong opacity makes particle-size measurements by light scattering to be a challenging problem.[19–21] The problem involves two interrelated obstacles: strong light absorption in a sample making the detection of the scattered light difficult and a complex refractive index making the interpretation of the measurements difficult.[22]

Two optical schemes to perform particle-size measurements in opaque fluids, in particular, to study aggregation phenomena of petroleum asphaltenes, are presented in Figure 17.1.

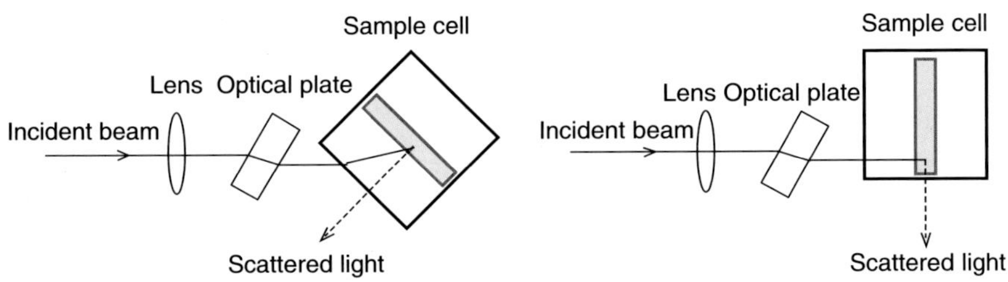

Figure 17.1. Two optical arrangements for dynamic light scattering in nontransparent fluids (upper view): "backscattering" (left) and "corner" (right).

In the "backscattering" geometry a rectangular sample cell is placed at an angle of 45° with respect to the incident laser beam emanating from a low-intensity He–Ne laser (~2 mW, 632.8 nm). The optical axis of the photon counting system is normal to the front plane of the sample cell. The scattered light is collected from the region where the laser beam enters the sample. To obtain optimal conditions for collection of the scattered light and for a suitable alignment of the incident laser beam, an optical plate unit is used. The optical plate unit is based on a precision parallel window with thickness of 10 mm. Adjustment (turning) of this window precisely shifts the incident laser beam. With use of the optical plate unit the position of the scattering volume can be adjusted so as to study light scattering near the wall (not less than 0.1 mm from the wall to exclude the influence of the surface and heterodyning from the stray light). Validity of the homodyne regime was verified by performing particle size measurements at different distances from the wall of the sample cell. Such a scheme enables one to measure a particle size in many apparently nontransparent liquids.

The "corner" geometry is suitable for semitransparent dispersions. Such a scheme enables one to minimize stray light. It is also important to note that for opaque systems there is no problem with interpretation of dynamic light-scattering data in terms of single-photon scattering since the strong light absorption eliminates any multiply scattered light. Standard square 4-mm-path-length optical-glass cells produced by Hellma Inc. were used for this study.

The principal scheme of the Photocor Complex DLS setup[23] is shown in Figure 17.2. It has features that are typical in multipurpose installations for studying both static (intensity) and dynamic (time-dependent correlation function) light scattering. The optical bench and precision goniometer are based on a massive base. A He–Ne laser and the focusing system are fixed on the bench. The thermostat and the cell holder are coaxial with the goniometer axis. The receiving system (photon counting system) is located on the rotating goniometer arm. The photon counting system consists of a set of pinholes for selection of aperture, a low-noise photomultiplier tube (PMT), and an amplifier-discriminator. The output signal produced by the photon counting system occurs as a sequence of standard pulses corresponding to escaped photoelectrons. The raw signal from the photon counting system is analyzed by a single-board correlator Photocor-FC,[23] which is plugged into a personal computer. The personal computer performs data analysis and instrument control. The correlator is a multichannel device that performs real-time data processing and accumulation of the correlation function of the light scattering fluctuations. The correlator Photocor-FC has two time-scale modes of operation, i.e., a linear mode with equidistant points of the measured correlation function and a so-called multiple-tau mode with the logarithmic order of the correlation points in time. Operation of the correlator in multiple-tau mode needs no tuning of time-scale and is suitable for measurements of multi-peak particle distributions. The linear scale of the correlator enables one to achieve maximum accuracy for measurements of a mono-modal narrow distribution.

Examples of the measured correlation function with Photocor-FC for both modes of operation are presented in Figure 17.3 (A and B). These correlation functions are accumulated for a transparent sample of the monodisperse polystyrene

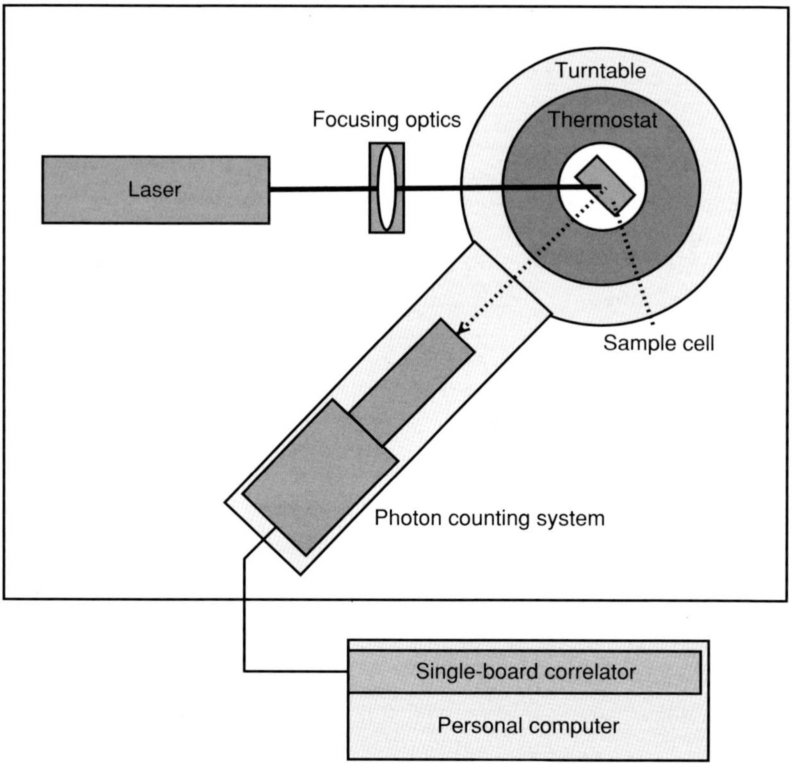

Figure 17.2. Schematic representative of the dynamic light-scattering setup for particle sizing in nontransparent fluids.

latex and demonstrate high accuracy of DLS measurements in the "ideal" case. The exponential function is precisely valid for this case. The results of fitting to Eq. (17.1) are also presented in Figure 17.3. For such quality of samples it is possible not only to obtain the mean correlation time and, correspondingly, the mean size of particles but also the particle-size distribution for a polydisperse sample. Unfortunately, the petroleum opaque systems are not simple in any sense for light-scattering studies. This is why only the mean radius of the particles (aggregates) can be reliably obtained. Two examples of the DLS correlation functions for aggregation process in crude oil are presented in Figure 17.4 (A and B). Accumulation time of the both correlation functions is the same. At the initial stage of aggregation the particle size and concentration of particles are relatively small. As a result, the accuracy of the correlation function is relatively poor (Figure 17.4A). For the developed aggregation process characterized by larger particle size and concentration the accuracy of the correlation function is essentially better (Figure 17.4B), nevertheless, not sufficiently to obtain a polydispersity distribution of aggregates. Furthermore, as the particle size continuously rises, it is impossible to increase accumulation time to statistically improve the quality of measurements.

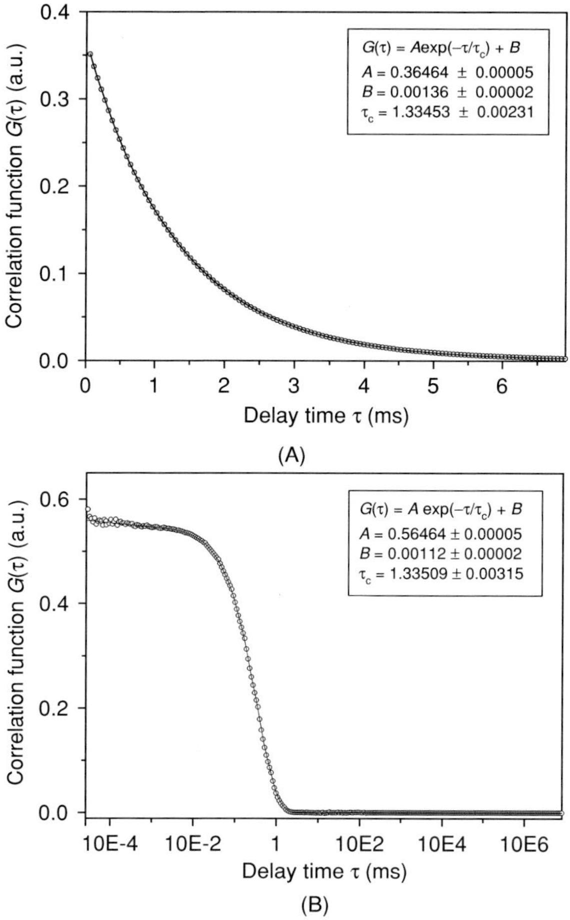

Figure 17.3. Examples of DLS autocorrelation functions for latex particles with 100 nm radius: (A) linear time-scale correlator; (B) quasi-logarithmic time-scale correlator.

Strong light absorption in opaque petroleum samples may cause a thermal-lens effect due to local overheating of the sample. As a result of the temperature gradients, a directed motion of fluid takes place. In this case the measured correlation function is not exponential and looks like a Lorenz curve that allows an easy identification of the appearance of the thermal-lens effect. To avoid this effect, the laser beam power should be attenuated by neutral-glass filters.

Though dynamic light scattering is an absolute method of particle sizing that needs no preliminary calibration, a number of test experiments have been carried out to make sure that this is a correct operation for opaque dispersions. Two opaque systems with different nature of light absorption were investigated: (1) with light-absorbing disperse particles, and (2) with light-absorbing disperse medium. In Figure 17.5, the results of size measurements of light-absorbing particles (dispersed

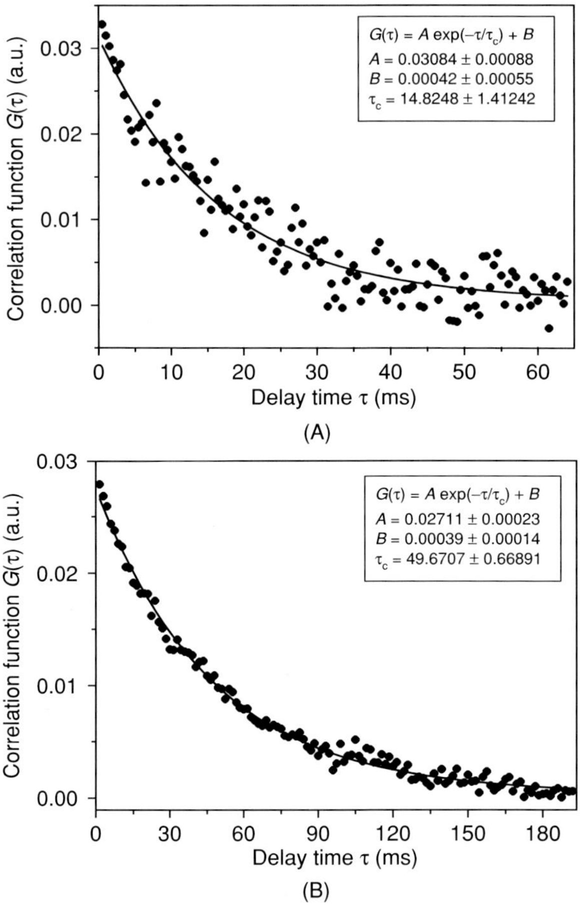

Figure 17.4. Examples of DLS autocorrelation functions for asphaltene aggregation in Unocal Swanson River crude oil with addition of 5% vol. *n*-heptane: (A) after 13 min of aggregation, the radius of aggregates is 0.44 μm; (B) after 3,450 min of aggregation, the radius of aggregates is 1.49 μm.

black aniline dye) dissolved in water are shown. A commercial aniline dye for fabric was used. The measured particle size, 537 ± 16 nm, turned out to be independent on the dye concentration. This fact proves the validity and availability of the method suggested.

In Figure 17.6, we show the measurement results of calibrated latex particles in water with the addition of black ink. In this case there are nonabsorbing colloidal particles of latex dissolved in an absorbing liquid. The certified radius of particles $R = 100 \pm 3$ nm. The measured radius in the opaque solvent is 101.0 ± 3.6 nm.

The dye and ink concentrations were changed from that corresponding to perfectly transparent to completely opaque solutions. In the latter case the light penetration into the sample did not exceed 0.1 mm. As can be seen from Figures 17.5 and 17.6, for all concentrations of dye and ink, the measured value of the radius

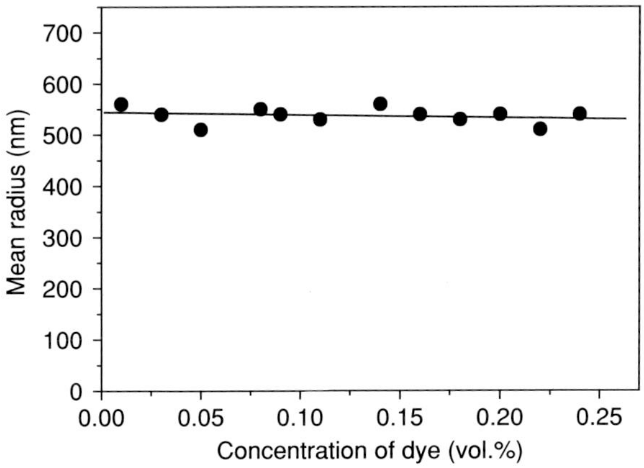

Figure 17.5. Size of light absorbing colloidal particles (black aniline dye) in water for various dye concentrations.

is stable within ±5%. The measurements were carried out with "backscattering" 135° geometry. The conventional 90° geometry in a cylindrical cell has also been used for one concentration corresponding to the transparent solution.

A possible influence of multiple light scattering for high concentrated opaque dispersions has also been studied. For multiple scattering there is problem of correct interpretation of the measured correlation functions as the actual wave vector of the received scattered light does not correspond to the angle under which the photoreceiver is set up. For many random scattering acts it is impossible to know the actual wave vectors of the multiply scattered photons. Use of the single-scattering equations to describe multiple scattering data will result in the calculated particle size less than the real one. The latex sample with very high particle's concentration of about 5%, which exhibits multiple light scattering only, has been prepared. The

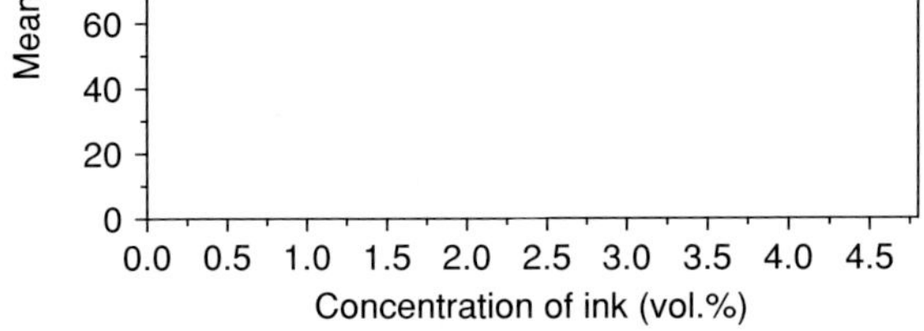

Figure 17.6. Size of calibrated latex particles in water for various concentrations of light-absorbing opaque liquid (black ink).

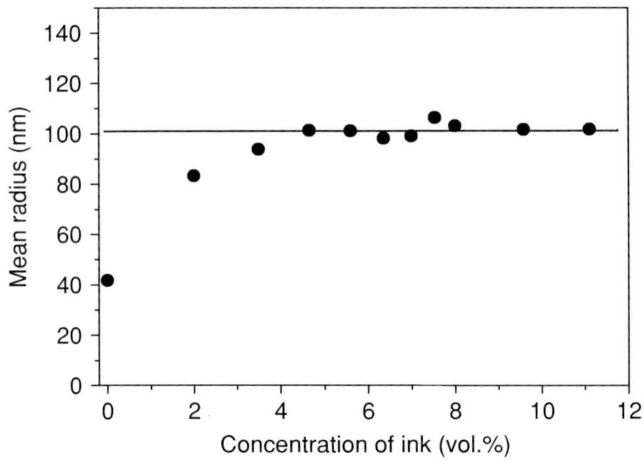

Figure 17.7. Apparent size of calibrated latex particles in water for various concentrations of light-absorbing opaque liquid (black ink). Initial concentration of latex in water is 5%. Such a sample exhibits only the multiple scattering of light.

results of particle-size measurements for various opacities of solution (various concentrations of ink) are presented in Figure 17.7.

For the low absorption of the solution the measured (apparent) particle size, as expected, is far from the certified radius ($R = 100 \pm 3$ nm). However, for the opaque systems with concentration of ink above 4%, where multiple scattering is negligible, the measured size turns out to be equal to the certified radius. These results prove the validity and accuracy of the modified dynamic light scattering method for investigations of light-absorbing colloids including various petroleum systems.

3. Aggregation of Asphaltenes in Toluene–Heptane Mixtures

According to recent investigations, the structure and the shape of asphaltene micelle-like associates appear to be much more complicated than those of classic surfactant micelles.[24–26] These studies indicate that asphaltene micelle-like associates are formed as a result of multiple interactions: aggregates with size of about 4.5 nm are present either in a molecular form or formed by strong intermolecular attraction forces, while the larger particles are self-associates, formed from these small aggregates by a weaker attraction potential. One should also note that asphaltene micelle-like associates, formed in toluene above the CMC, are characterized by significant polydispersity.[27] In fact, the problem of small asphaltene associates in toluene is not well investigated and smaller than 4.5 nm particles (undetectable by DLS) can exist (O.C. Mullins, Private communication).

Molecular thermodynamics approach[28–30] enables one to consider phase equilibrium of a crude oil in the presence of asphaltene precipitation. The thermodynamics calculations describe experimental results for a few crude oils reasonably well. Nevertheless, a universal prediction of asphaltene precipitation behavior is still not possible and experimental data for particular crude oils are highly desired.

The complexity of the asphaltene solutions imposes a number of limitations on the conditions of the light-scattering experiment. This is why we made several precautions to improve the quality of the measurements.[31] Petroleum samples were dehydrated and micro-pore filtered. Filtered *n*-heptane was used as a precipitant. To avoid the thermal-lens effect a neutral-glass optical filter was used. A fresh sample was prepared for each experiment. The particles sizing was started immediately after addition of *n*-heptane. To obtain a correlation function of a reasonably good quality, the accumulation time for each measurement lasted from 1 to 15 min depending on the rate of the aggregation process. The total duration of each experimental run varied from 10 min to several days. All these experiments were performed at ambient conditions.

The experimental procedure included both the light-scattering intensity measurements and accumulation of the dynamic autocorrelation function in the homodyne regime. The resulting correlation function was fitted to a single exponential decay with one cumulant; then the effective diffusion coefficient was deduced and the mean radius of particles was calculated with Eq. (17.3).[23]

The asphaltenes used for the measurements were extracted from crude oils of various regions: Arabian and Venezuelan were provided by *Institut France du Petrole*; Alaskan was supplied by Shell, and Siberian and Kazakhstan were supplied by Moscow Technical University of Oil and Gas. As a result, we were able to compare colloidal properties and aggregation phenomena for asphaltenes of various origins.

The results presented below have been obtained for the asphaltenes extracted from Karazhanbas crude oil (Kazakhstan).[33,34] An asphaltene content of these crude oils is 6.2%. Three initial solutions of asphaltenes in toluene with the asphaltene concentration of 1 g/L, 5 g/L, and 10 g/L were prepared. All these concentrations are well above the most recent value of the CMC., however, a next level of asphaltene self-association is possible at about 3–4 g/L.[13,16] One can expect that asphaltene colloidal particles ("micelles") initially presented in toluene may affect the aggregation phenomena after adding a precipitant.

For each initial concentration of asphaltenes, we determined the threshold concentration of *n*−heptane in the following way: heptane was gradually added to the asphaltene solutions until aggregation of asphaltene started. We observed the growth of asphaltene aggregates for solutions with the threshold concentration of *n*-heptane, as well as for solutions with a number of other concentrations of *n*-heptane above the threshold concentration. As a whole, the aggregation kinetics for three concentrations of *n*-heptane (initial concentration of asphaltene 1 g/L) and seven concentrations of *n*-heptane (initial concentration of asphaltene 10 g/L) have been studied.[34,35]

Generally, an aggregation process may involve two characteristic times: a diffusion time τ_D and a reaction time τ_R. If $\tau_D \gg \tau_R$, the aggregation kinetics is controlled by diffusion and referred to as diffusion-limited aggregation (DLA). In the case, when $\tau_R > \tau_D$, the aggregation kinetics is determined by the rate of the "reaction" and referred to as reaction-limited aggregation (RLA).[36–38] Following

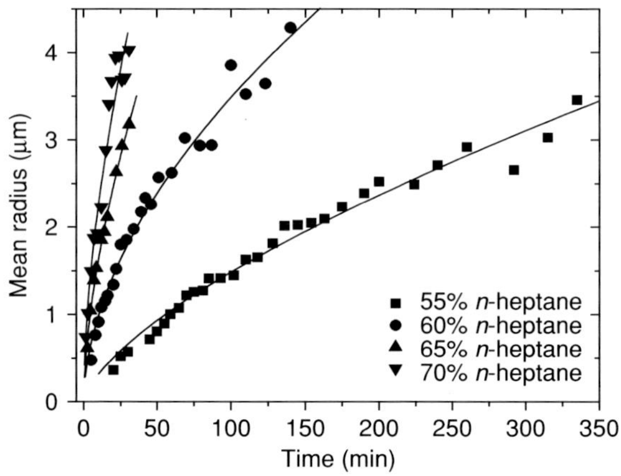

Figure 17.8. Mean radius of aggregates as a function of time in an initial solution of 1 g/L of asphaltenes in toluene upon adding 55, 60, 65, and 70% vol. of *n*-heptane. The solid curves represent the power law (Eq. (17.4)) representing DLA.

from a solution of the Smoluchowski's equation,[39] DLA and RLA kinetics processes are described by the following simple formulae:

$$\text{DLA}: \qquad N = 1 + t/\tau_D \qquad R = R_0(1 + t/\tau_D)^{1/d_f} \tag{17.4}$$

$$\text{RLA}: \qquad N = \exp(t/\tau_R) \qquad R = R_0 \exp(t/\tau_R d_f), \tag{17.5}$$

where N is the average number of particles in an aggregate, R is the mean radius of an aggregate, R_0 is the initial particle size and t is time. In Eqs. (17.4) and (17.5) d_f is the fractal dimensionality according to the definition $N = (R/R_0)^{d_f}$. It is important to note that d_f and R_0 can be different for DLA and RLA.

In Figure 17.8, we show the mean radius of the aggregates as a function of time, formed when four different amounts of *n*-heptane were added in the toluene solution with an initial concentration of 1 g/L of asphaltenes. The solid curves represent fits to the power law (17.4) for DLA. The characteristic time τ_D of aggregation depends on the concentration of *n*-heptane. The coefficients R_0 and d_f in Eq. (17.4) were treated as adjustable parameters in the fit (Table 17.1). The fractal dimension d_f is almost the same for all the fits, and its average value is $d_f \approx 1.7 \pm 0.2$. This agrees well with the well-known result for classic colloidal systems in which $d_f \approx 1.75 \pm 0.05$.[37,39] Upon changing the concentration of *n*-heptane from the near-threshold value of 55% vol. to 70% vol., the rate of aggregation increases by a factor of 50. The *n*-heptane concentration of 70% vol. corresponds to the highest aggregation rate that could be measured under the given experimental conditions (especially, accounting for the considerable opacity of the sample).

The nature of the aggregation process in the solution with an initial concentration of 10 g/L of asphaltenes, as shown in Figure 17.9, is obviously different. In this case, the mean radius of the aggregates increases exponentially with time and

Table 17.1. Parameters of the Asphaltene Aggregation Kinetics in Toluene–Heptane Mixtures

	Asphaltene (g/L)	n-heptane (% vol.)	R_0 (μm)	τ_{char} (min)	d_f
	1.0	55[a]	0.1	1.8	1.49
DLA	1.0	60	(fixed for all	0.15	1.83
	1.0	65	concentrations)	0.09	1.68
	1.0	70		0.04	1.77
	10.0	52[a]	0.11	58.3	2
	10.0	53	0.2	28.9	(fixed for all
	10.0	54	0.35	11.9	concentrations)
RLA	10.0	55	0.29	5.17	
	10.0	56	0.21	3.93	
	10.0	57	0.41	3.74	
	10.0	58	0.24	1.07	
Crossover	5.0	55[a]	0.14	$\tau_D = 1.04$, $\tau_R = 18.9$	2 (fixed)

[a] Near-threshold concentration.

can be fitted to Eq. (17.5) for RLA. It should be noted that the rate of aggregation strongly depends on the amount of n-heptane added to the asphaltene–toluene solution and increases by a factor of 50 with an increase of the concentration from the near-threshold value of 52% vol. to 58% vol.

The experimental results for three initial concentrations of asphaltenes, low, high, and intermediate, are represented in Figure 17.10. The intermediate concentration demonstrates neither pure DLA nor RLA kinetic behavior. Instead, crossover from RLA to DLA dominates the experimental range of the monitoring time.

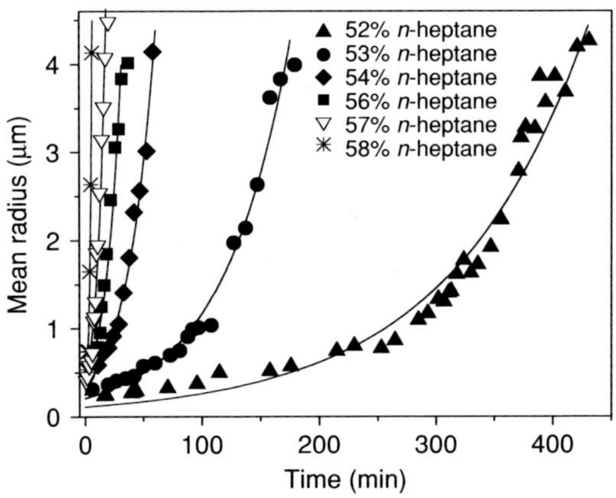

Figure 17.9. Mean radius of aggregation as a function of time in an initial solution of 10 g/L of asphaltenes in toluene upon adding 52, 53, 54, 56, 57, and 58% vol. of n-heptane. The solid curves represent the exponential growth (Eq. (17.5)) representing RLA.

To describe the experimental data, we proposed a simple equation to incorporate a crossover behavior from RLA to DLA[40]:

$$\frac{dN}{d\tau} = \frac{N}{\tau_D \gamma N + \tau_R}, \tag{17.6}$$

where $\gamma = (R/R_{0(D)})^{d_f}$ is a constant. For $\tau_D \gamma N$ much smaller than τ_R, this equation implies RLA. As N grows, $\tau_D \gamma N$ becomes larger than τ_R. Ultimately, when $\tau_D \gamma N \gg \tau_R$, DLA prevails. Integration of Eq. (17.6) gives the following crossover equation for the aggregation number N:

$$(N-1)\left(\frac{\tau_D \gamma}{\tau_R}\right) + \ln N = \frac{\tau}{\tau_R} \tag{17.7}$$

and for the mean radius R:

$$\tau_D \left(\frac{R}{R_{0(D)}}\right)^{d_f} + \tau d_f \ln\left(\frac{R}{R_{0(R)}}\right) = \tau. \tag{17.8}$$

The solid curves in Figure 17.10 represent fits according to Eq. (17.4) for DLA, Eq. (17.5) for RLA, and to Eq. (17.8) for the crossover behavior.

The two limiting regimes of aggregation and crossover between them for 5 g/L asphaltene solution exhibit a universal behavior as shown in Figure 17.11. If one uses scaled variables R/R_0 and $\tau^* = \tau/\tau_{char}$ with the characteristic time $\tau_{char} = \tau_R$ for the RLA and the crossover curves, and with $\tau_{char} = \tau_D$ for the DLA curves, the four DLA curves and the six RLA curves collapse into two master curves, and the crossover between the two master curves is clearly seen (the solid circles). The scaling factor, $\tau_D \gamma/\tau_R = 1/43$, is the ratio of the two characteristic times and it is found from the fit of Eq. (17.8) to the data exhibiting crossover behavior. For DLA the aggregation number N corresponds to the average number

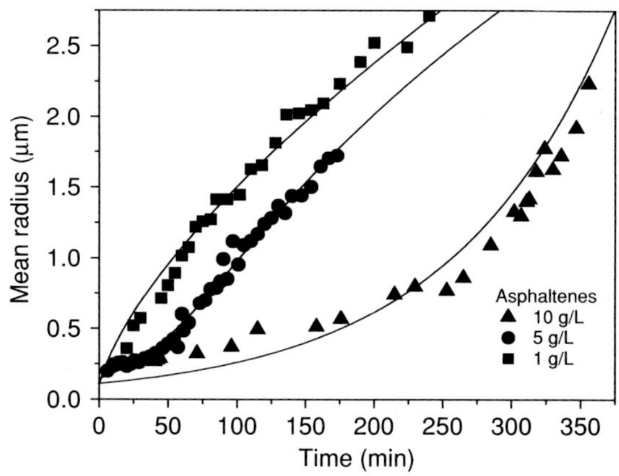

Figure 17.10. Mean radius of aggregation as a function of time for initial solutions of 1, 10, and 5 g/L of asphaltenes in toluene upon adding a near-threshold concentration of *n*-heptane. The solid curves represent fits according to Eq. (17.4) for DLA, Eq. (17.5) for RLA, and to Eq. (17.8) for the crossover behavior.

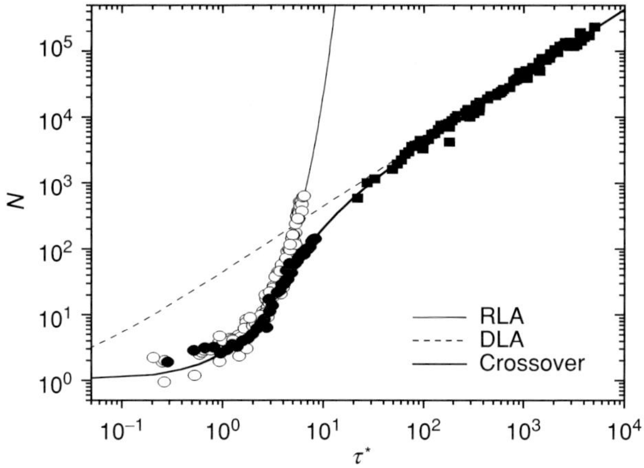

Figure 17.11. The aggregation number N as a function of the scaled time t^*. Open circles represent data for 10 g/L asphaltenes/toluene solution exhibiting RLA; solid squares represent data for 1 g/L asphaltenes/toluene solution exhibiting DLA; and solid circles represent data for 5 g/L asphaltenes/toluene solution exhibiting crossover aggregation kinetics. The thick solid curve is the crossover behavior described by Eq. (17.7).

of asphaltene "primary particles" in an aggregate, while for RLA and for the crossover aggregation the aggregation number N has the meaning of the average number of higher-level associates in an aggregate. These two quantities are related by the ratio γ.

While the crossover equation is a purely phenomenological interpolation of two limits, DLA and RLA, there is a natural physical picture behind it. As the concentration of asphaltenes in toluene goes up, the size of primary associates, R_0, grow up either. On the other hand, the aggregation number of nonequilibrium aggregates (at the presence of precipitant) grows up with time. Simultaneously, the structure of the aggregates also may change with concentration and with time. Moreover, the addition of n-heptane to toluene asphaltene solution can change the size of primary associates and their structure. We note that the characteristic reaction time is much larger than the characteristic diffusion time. In a plausible scenario one can image the RLA "barrier" as internal primary-aggregate diffusion that serves as the rate limiting step until R becomes much larger than R_0. This can explain why crossover from DLA to RLA occurs with increase of concentration of asphaltenes in toluene (growing R_0) while in the limit of large aggregation number (growing R) the DLA prevails._

The results of the fits are presented in Table 17.1.[34] At the lower concentration of asphaltenes in toluene, 1 g/L (but already above CMC, ~0.1 g/L),[17] the nonequilibrium aggregates are formed from elementary asphaltene nanosize associates and the character of the aggregation is solely determined by the diffusion of the aggregates. Above 5 g/L, the nonequilibrium aggregates are probably formed from larger equilibrium associates. As the asphaltene concentration increases, the associates become larger and the potential barrier becomes higher. This is why the RLA rate slows down with increasing asphaltene concentration.

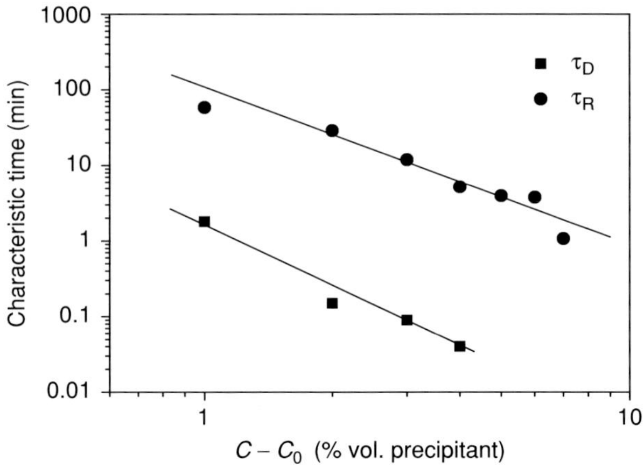

Figure 17.12. Characteristic times τ_D for DLA and τ_R for RLA as functions of the concentration of precipitant (*n*-heptane).

The curves depicting the dependence of the DLA and RLA characteristic times on the concentration, c, of n-heptane are shown in Figure 17.12. These dependencies can be roughly described by a power law, $\tau_{D,R} \propto (c - c_0)^{-x}$ where c_0 is the threshold concentration of n-heptane, with $x = 1$, $c_0 = 54.5\%$ vol. for DLA and $x = 2.3$, $c_0 = 51\%$ vol. for RLA.

It is remarkable that the aggregation kinetics of asphaltenes in toluene–heptane mixtures is well described by the classic theory of colloid aggregation. One can state that asphaltene solutions are typical colloid systems. Similar aggregation behavior has been observed for asphaltenes extracted from other crude oils, such as Alaskan and Arabian crude oils.[26,34,41]

4. Aggregation of Asphaltenes in Crude Oils

Three different samples of crude oils (courtesy by Shell) were investigated.[41,42] The SARA analysis for the crude oils is presented in Table 17.2. According to the information from Shell these oils had asphaltene instability problems. In particular, the first oil causes troubles during oil recovery and transportation. With respect to aromatics/saturates ratio, this oil differs from the others, therefore, we could expect different aggregation behavior.

The effective Brownian diffusion coefficient detected in Swanson River oil was found to change with monitoring time. Initially the diffusion coefficient was

Table 17.2. Representative SARA Values for the Crude Oil Samples

	Crude oil	Saturates (%)	Aromatics (%)	Resins (%)	Asphaltenes (%)
1	Swanson River	53.1	35.2	7.2	4.53
2	Europe SkyBlue	27.13	56.72	9.4	6.75
3	Mars A12	24.3	55.6	14.9	5.15

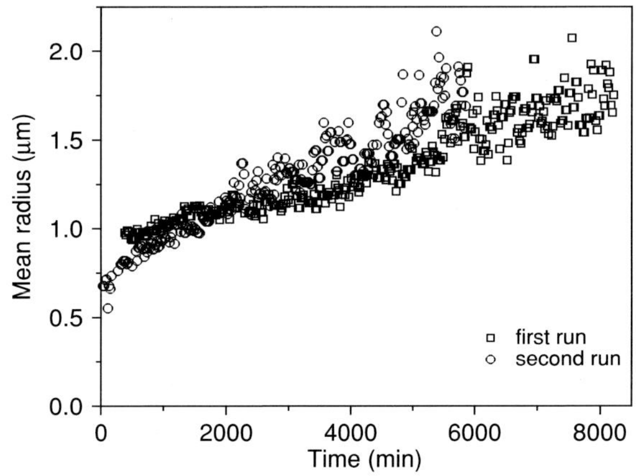

Figure 17.13. Size of aggregates as a function of monitoring time in Unocal Swanson River crude oil.

about $5 \cdot 10^{-10}$ cm^2/s that corresponds to a particle size of about 1 μm, but within 5 days the particle size increased by a factor of two (Figure 17.13). We were able to obtain the average particle size using Eq. (17.3) as the shear viscosity of the samples was measured independently (see Section 6). During the experiment, the laser beam constantly entered the sample, and the growth of the particle size might be somehow affected by laser radiation. Therefore, this effect requires further investigation. To exclude any thermal lens effect, we reduced the laser power with a neutral filter with about 12 times attenuation.

Spectacular aggregation kinetics was observed in the mixtures of this oil with n-heptane added. Typical curves for oil/precipitant mixtures with different n-heptane concentration that represent a slow increase of particle size in the double logarithmic scale are shown on Figure 17.14.

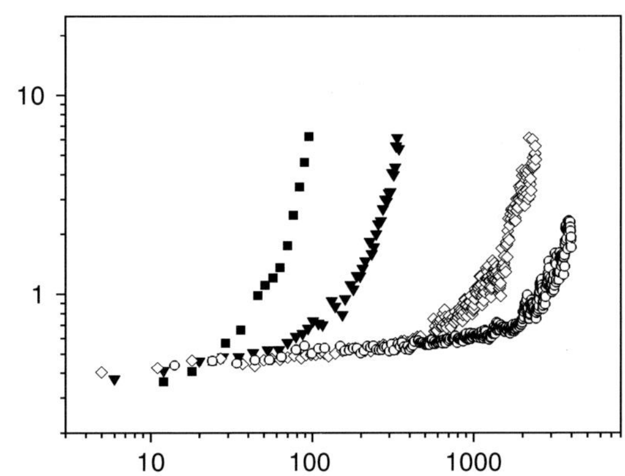

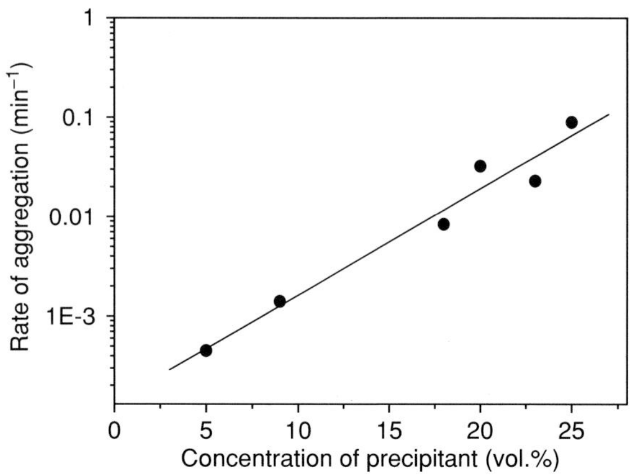

Figure 17.15. Exponential growth of the aggregation rate as a function of precipitant volume concentration for Unocal Swanson River oil/precipitant mixtures.

The scattered intensity was found to increase during the experiments. The process is considered a colloidal particle aggregation process and it represents exponential growth of particle size. The rate of aggregation strongly depends on the precipitant concentration (Figure 17.15). An exponential fit agrees reasonably well with the experimental data and gives us a "characteristic threshold concentration" for this oil sample, equal to about 3% of *n*-heptane. The addition of the precipitant reduces the energy barrier of reaction of aggregation, and a grater amount of the precipitant results in higher aggregation rate. Such behavior corresponds to the reaction-limited aggregation observed in toluene/heptane solutions of asphaltenes.

The particle size growth in mixtures with 5, 9, and 18% of *n*-heptane involves two different stages. The initial part of the experimental data obeys a power law with an exponent close to 1/3. This extremely slow process is similar to the process observed in the pure oil. The second part of the experimental data represents exponential growth with a characteristic time of 2,300 min in the 5% mixture (Figure 17.16). However, the initial stage of this exponential growth process is concealed by the power law process as larger particles dominate. This overlapping area prevents us from distinguishing these two processes with dynamic light scattering. The point of intercept of these two processes is shifted to lower times with increase of the precipitant concentration (Figure 17.14).

This crude oil was found close to aggregation instability. There is no well-defined threshold concentration of the precipitant for aggregation onset. Asphaltene particle growth was observed in a solution with concentrations of *n*-heptane as low as 3%.

Unocal Swanson River oil clearly reveals an intrinsic colloidal structure that appears kinetically unstable. The mixtures of this oil with *n*-heptane show spectacular aggregation phenomena with the rate depending on the oil/precipitant ratio.[41] According to the SARA analysis, the content of this oil is quite different from the other oils (large fraction of saturated hydrocarbons (see Table 17.2).

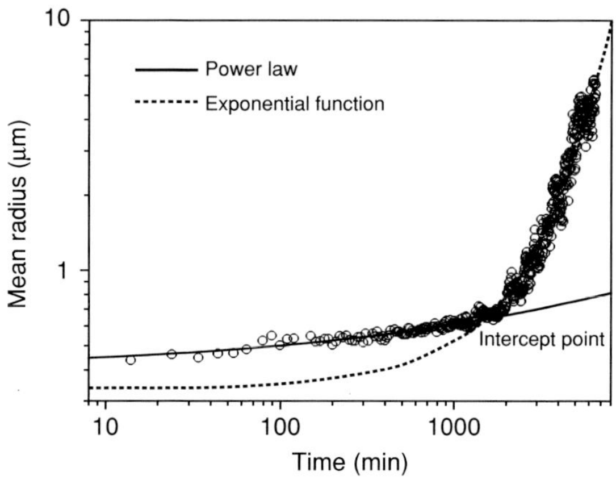

Figure 17.16. Mean radius versus the monitoring time for Unocal Swanson River oil with 5% vol. of the precipitant. Two dynamic processes, a power law growth and an exponential growth, are clearly seen in the double logarithmic scale.

So, the analysis gives us a good indication of crude oil stability with respect to asphaltene aggregation.

An effective diffusion coefficient, about $2{\cdot}10^{-9}$ cm^2/s, that was measured in Europe SkyBlue crude oil does not change with monitoring time within at least 24 hr. This diffusion coefficient corresponds to particles of about 20 nm in diameter. In a mixture with 20% of the precipitant added, the diffusion coefficient increased up to 10^{-8} cm^2/s but the particle size remained the same. The experiment with the mixture lasted for 24 hr and showed no significant change in the size of the particles and the scattered intensity during the experiment (Figure 17.17).

In solutions with concentrations of n-heptane 25% and greater, two effective particle sizes were detected by analyzing the dynamic correlation function in

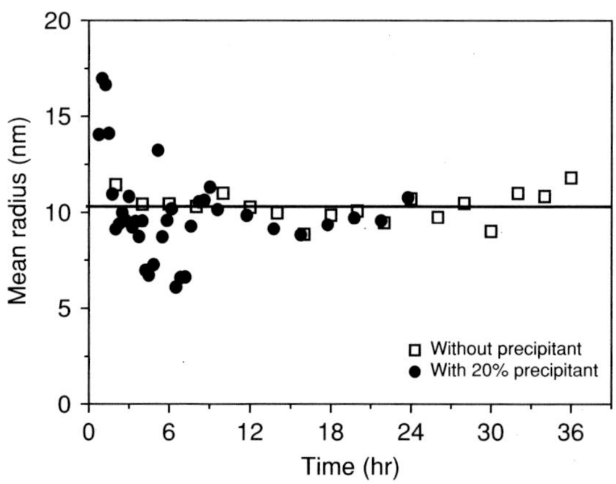

Figure 17.17. Stable small particles in SkyBlue crude oil without precipitant and with 20% of n-heptane as a function of the monitoring time.

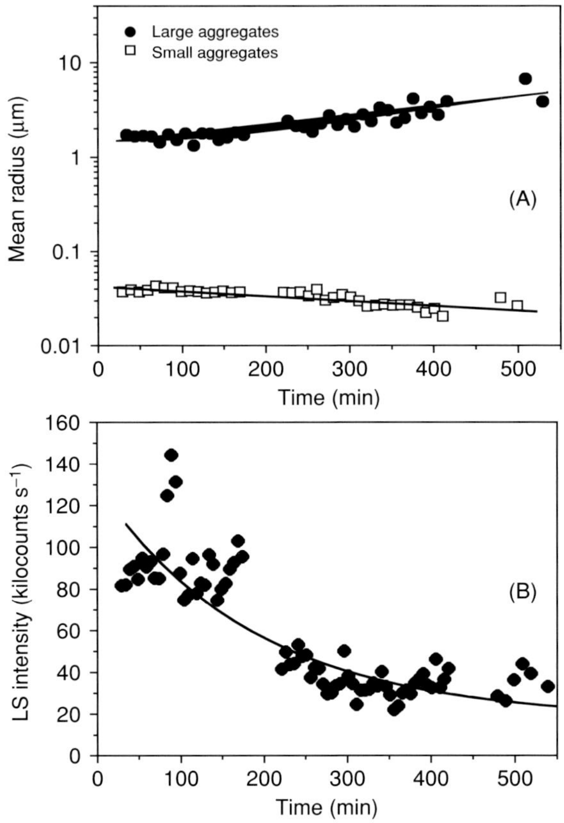

Figure 17.18. Mean radius (A) and light-scattering intensity (B) for SkyBlue oil with 44% vol. of the precipitant as a function of the monitoring time.

different correlation time scales. While the smaller particles remained in the system, larger aggregates formed. While the intensity of the scattered light decreased with monitoring time, the diffusion coefficients did not change by more than a factor of two. This change can be attributed to worse experimental conditions, namely, a low scattered intensity. An example of the process for a mixture with 44% of n-heptane is shown in Figure 17.18 (A and B). The size of the larger aggregates depends on the amount of the precipitant added and grows up to several microns (Figure 17.19).

Depending on the concentration of the precipitant, we were able to measure the particle size over a time period of more than 20 hr for a solution with 30% concentration of n-heptane, over less then 10 hr for a solution with 44% n-heptane, over less then 6 hr for a solution with 49% n-heptane, and over 2 hr for a solution with 53% n-heptane. After the aforementioned periods, the scattered intensity became so low that we could not make measurements any longer. The effective particle sizes were restored after the sample was shaken and disappeared within 1.5 hr. No slow change of particle size was observed in this crude oil sample.

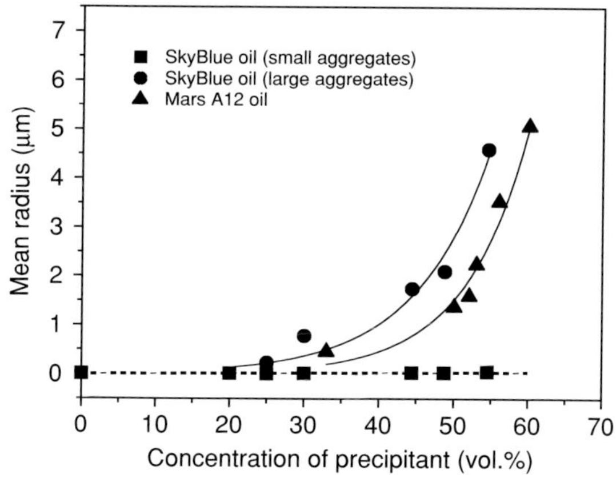

Figure 17.19. Average size of asphaltene particles for various concentrations of the precipitant.

From the above results we can conclude that Europe SkyBlue oil has an intrinsic colloidal structure. The structure remains stable both in pure oil and in solution with up to 20% of the precipitant. The size of asphaltene particles suspended in oil media is about 20 nm. Addition of the precipitant at concentrations higher than the threshold concentration initiates fast aggregation. Since the aggregates are growing too fast, their size change with time cannot be measured. Addition of *n*-heptane causes formation of large (several microns) aggregates; however, their size remains stable. Eventual sedimentation follows the aggregation and the number of particles in the scattering volume decreases with time. The sedimentation rate depends on the concentration of the precipitant. Higher precipitant concentration leads to higher sedimentation rate and vice-versa. The threshold concentration for this oil sample is about 25% vol. of *n*-heptane.

No Brownian diffusion coefficient was detectable in both the Mars A12 crude oil sample and in a solution of Mars A12 with 20% of *n*-heptane. A very low diffusion coefficient, that indicates very large particles, was observed in solutions of 30–60% concentration of *n*-heptane. It did not change significantly with monitoring time. The scattered intensity remained unchangeable as well. The diffusion coefficient was about $6 \cdot 10^{-10} cm^2/s$ for mixtures with different oil/heptane ratios. However, the size of the particles grew from about 1 μm for 33% vol. concentration of *n*-heptane to 10 μm for 60%. A greater than 60% amount of the precipitant causes formation of very large clusters which cannot be measured. The dependence of the size of aggregates on the concentration of the precipitant for Sky-Blue and Mars A12 oils is shown in Figure 17.19. Both SkyBlue and Mars A12 exhibit similar aggregation behavior. The threshold concentration of aggregation is clearly observed.

We have concluded that Mars A12 oil has no detectable colloidal structure. Adding *n*-heptane induces the formation of stable colloidal particles. The threshold concentration for its formation is about 20% of *n*-heptane. A considerable amount of the precipitant added (more than 60%) breaks the stability of the system and causes flocculation of the particles into huge clusters.

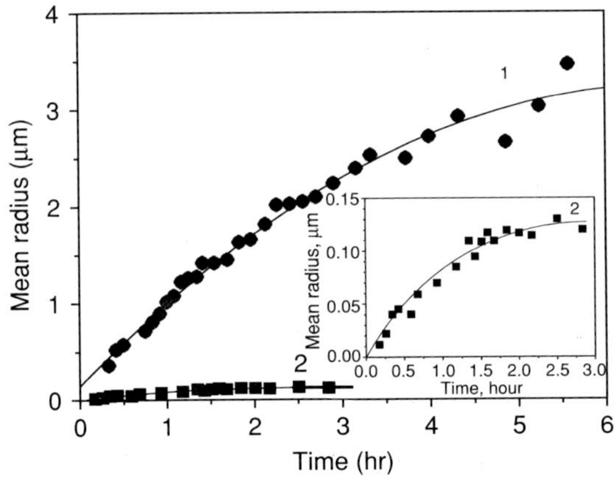

Figure 17.20. Asphaltene aggregation in toluene–heptane mixture (35/65). 1—Unrestricted aggregation. 2—Aggregation with surfactant (DDAB) addition (shown also as insert). Asphaltene aggregates are stabilized at the size of 0.12 μm.

5. Stabilization of Asphaltene Colloids

Asphaltene colloids in many cases are kinetically unstable. However, they can be stabilized even in solvents as poor as water by surfactants or/and by ultrasonic agitation. Stabilization of asphaltene colloids is an important problem for petroleum industry and attracts attention of many investigators.[43–45] The addition of surfactant and ultrasonic agitation affect the stability of asphaltene solutions at different stages of aggregation. The experimental results of such studies presented in this section have not been previously published.

Figure 17.20 shows the results of DLS measurements of asphaltene particle-size growth in the toluene (35%)–heptane (65%) mixture. Initial concentration of asphaltenes extracted from Arabian Light (courtesy by *Institut France du Petrole*) in toluene was 1 g/L. The curve 1 represents the unrestricted aggregation resulting in the ultimate deposition of asphaltene flocks. The curve 2 represents the particle growth in the same solution with addition of 0.4 g/L didodecyldimethylammonium bromide (DDAB). This addition results in the ultimate stabilization of the particle size at 0.12 μm.

Figure 17.21 demonstrates effects of mechanical perturbations on asphaltene aggregation. Initial concentration of asphaltenes in toluene was 1 g/L. After addition of 65% of *n*-heptane and initial shaking of the mixture during 1 min, the measurements of particle size started. At about 400 min after the start of the aggregation process, an intense 1-min shaking reduced the aggregates size more than twice. After the shaking stopped, aggregation resumed with a higher rate. At about 540 min of the process an ultrasonic agitation was applied for 2 min. The agitation reduced the aggregate size down to 0.2 μm. After the agitation stopped, the aggregation resumed with the initial rate. This qualitative experiment was performed to estimate structural stability of the growing aggregates under

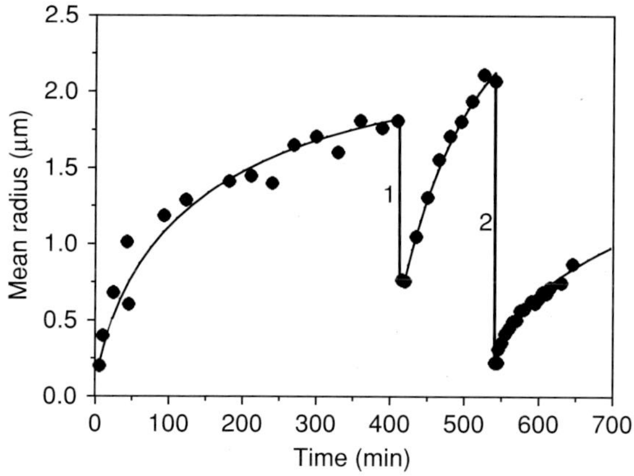

Figure 17.21. Effects of mechanical perturbations on asphaltene aggregation in the mixture of toluene (35%) with n-heptane (65%) and with 1 g/L of asphaltenes in toluene. 1—Destruction of asphaltene aggregates by shaking (1 min) with following re-aggregation. 2—Destruction of asphaltene aggregates by ultrasonic agitation (2 min) with following re-aggregation.

different level of mechanical disturbance and the results should be regarded as preliminary.

While in the good-solvent/precipitant mixture the ultrasonic agitation prevents the asphaltene aggregation only during its action, in poor solvents an intense (44 kHz, 1–10 W) 2-min ultrasonic agitation can produce kinetically stable colloid asphaltene solution.

Figure 17.22 demonstrates the behavior of asphaltene aggregates in various solvents. The curve 1 shows the unlimited asphaltene aggregation in the mixture of

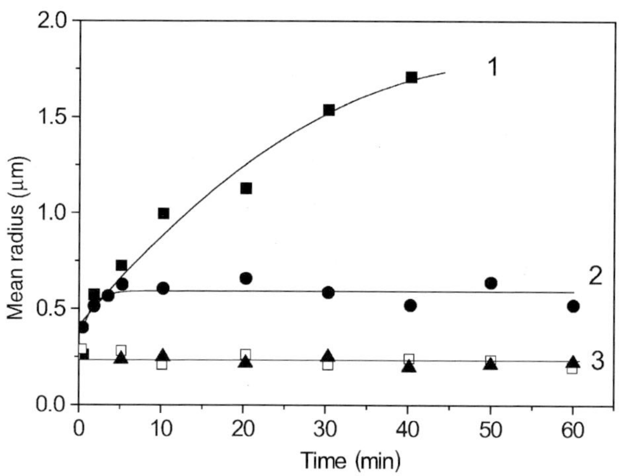

Figure 17.22. Asphaltene aggregates in various solvents at ambient conditions. 1—Unlimited asphaltene aggregation in the toluene (35%)–n-heptane (65%) mixture. 2—Formation of stable asphaltene particles in n-heptane as a result of ultrasonic agitation. 3—Formation of stable asphaltene particles in water as a result of ultrasonic agitation (triangles—2 min agitation, squares—5 min agitation).

toluene (35%) and n-heptane (65%). Initial concentration of asphaltenes in toluene was 1g/L. The curvc 2 averages the size of stable asphaltene particles in n-heptane as a result of ultrasonic agitation. The curve 3 averages the size of stable asphaltene particles in water as a result of intense ultrasonic agitation. An increase in the time of agitation from 2 to 5 min does not considerably change the size of aggregates.

It is also remarkable that the stable asphaltene colloids in poor solvents are almost as optically transparent as the pure solvents. The mechanism of stabilization of asphaltene aggregates in poor solvents by intense ultrasonic agitation is not yet understood. It is possible that the ultrasonic agitation in poor ("asphaltene-phobic") solvents can lead to restructuring of the aggregates and formation of an amphiphilic surfactant-like layer on the aggregate surface.

6. Viscosity and Microrheology of Petroleum Systems

To obtain a particle size from dynamic light-scattering experimental data one needs to know shear viscosity of the medium. Classic viscosity measurements are widely used for investigation of asphaltene aggregation phenomena.[46–48] In our studies the shear viscosity of the crude oils and oil/precipitant mixtures was measured using Rheometrics Fluids Spectrometer (RFS II). The RFS II is a laboratory instrument for study of the viscous and viscoelastic properties of both structured and unstructured fluids. The instrument applies a steady rotational or a dynamic oscillatory shear, or either form of shear sequentially, to the test material. By measuring the torque transmitted through the material, the material's storage (elastic) modulus, the loss (viscous) modulus, and the shear viscosity are derived.

As most of the light scattering experiments in crude oils were carried out at room temperature, viscosity was measured at two different temperatures close to room temperature. Then the viscosity was extrapolated to a required temperature. The procedure of viscosity measurements was as follows. After the temperature of the cap and the bob was stabilized, crude oil was transported into the cap and mixed for some time. After that the crude oil viscosity value was taken. Then n-heptane was gradually added and viscosity of mixtures at different oil/precipitant ratios was obtained. The shear viscosity of oil/precipitant mixtures for Swanson River and SkyBlue oils is presented in Figure 17.23.

Heavy crude oils are usually viscoelastic non-Newtonian fluids. This nature of crude oils often causes difficulties in interpretation of viscosity measurements. In Figure 17.24, the apparent viscosity of Mars A12 is presented as a function of the measurement time. The measurements were performed in the rotational regime. As seen in the figure, the viscosity steadily increases with time. The increase continues after the rotation resumes. A change in the rotation rate (up to 8 times) and in the direction of rotation did not cause any significant change in the viscosity. One can speculate that the steady shear rate results in the formation of a supramolecular structure that causes the increase of viscosity. Such interpretation is supported by a comparison of the viscosity measurements in the rotational and oscillating regimes (Figure 17.25). As the oscillation possibly does not result

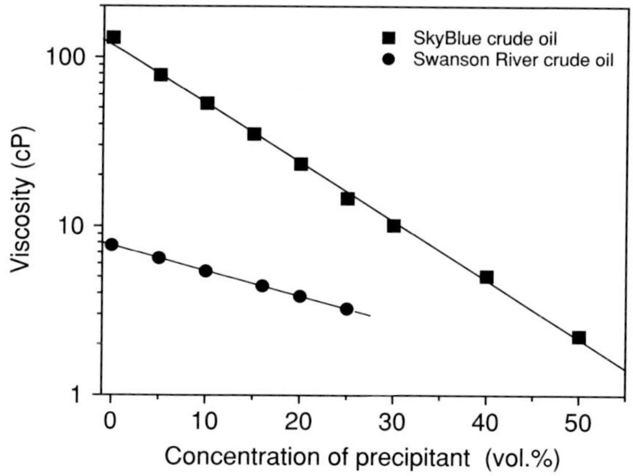

Figure 17.23. Viscosity of crude oil–*n*-heptane mixtures at ambient temperature.

in the formation of a supra-molecular structure, the viscosities measured in this regime are systematically lower.

During the recent years investigations of rheological properties of soft-matter materials at the mesoscale by measuring the shear viscosity with dynamic light scattering from calibrated Brownian particles have become popular. This method is known as "microrheology."[49,50] Diffusion coefficient measurements of known-size particles dispersed in a liquid enables one to calculate viscosity with use of Stokes–Einstein formula (17.3). It is important to note that the particles must be spherical and noninteracting. Application of microrheology to crude oils and heavy petroleum fractions is restricted as it is difficult to suspend calibrated Brownian particles in such systems.[51] We have found the way to solve this problem by using a crude oil (Utkinskaya, Russia) containing traces of water. This oil is a natural dilute microemulsion of water. The DLS sizing of this oil has shown the existence

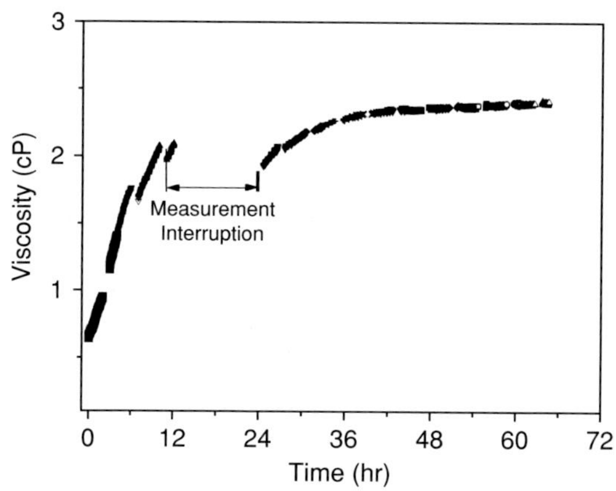

Figure 17.24. Viscosity dynamics induced by shear rate at ambient temperature.

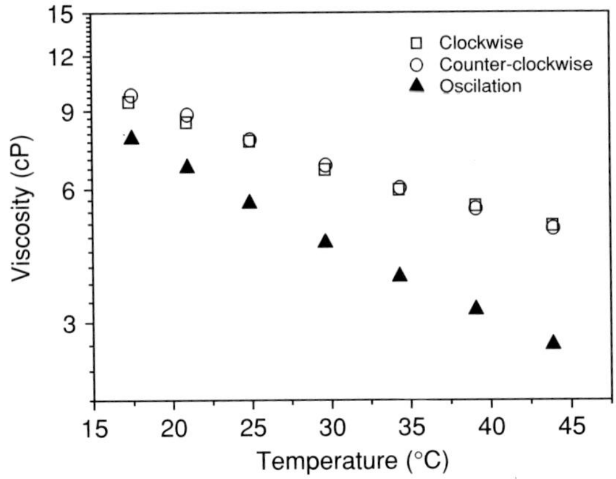

Figure 17.25. Comparison of the viscosity measurements in rotational (circles and squares) and oscillating (triangles) regimes.

of stable water droplets of 70 nm. Heating and ultrasonic agitations do not destroy the droplets. However, the droplets disappear after careful dehydration of the oil sample. Emulsion microdroplets are spherical. Since the trace concentration of the water is very low one can regard the emulsion droplets as noninteracting particles. It means also that this trace concentration of emulsion does not affect the crude oil viscosity. We measured the viscosity of the dehydrated oil by a rotational viscometer and use this viscosity to obtain the droplet radius with Eq. (17.3). Then, assuming the size to be independent of temperature, we obtained the viscosity, defined by Eq. (17.3), as a function of temperature. A comparison between the "macroviscosity" (measured by viscometer) and "microviscosity" (measured by DLS) is presented in Figure 17.26. This comparison shows reasonable agreement between the DLS microrheology and the classic macrorheology.

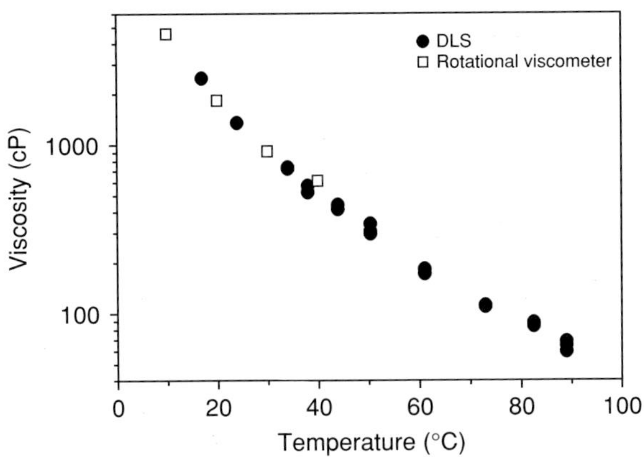

Figure 17.26. Viscosity of crude oil–solvent mixtures measured by DLS microrheology (close circles). Open squares–results of viscosity measurements with a rotational viscometer.

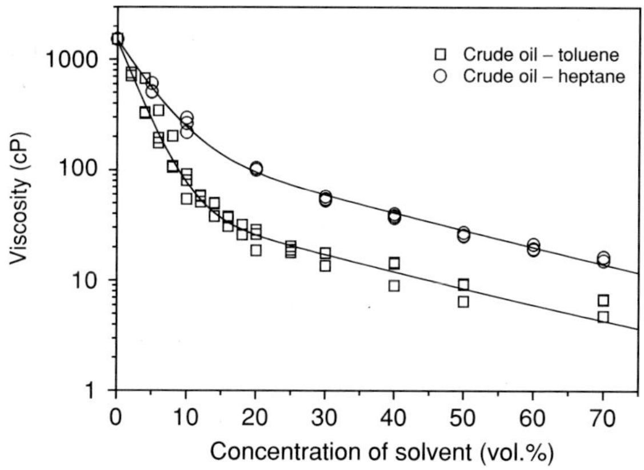

Figure 17.27. Viscosity of crude oil–*n*-heptane and crude oil–toluene mixtures at ambient temperature measured by DLS microrheology.

In Figure 17.27, the viscosity of Utkinskaya crude oil mixed with two different solvents, toluene and *n*-heptane, is presented as a function of solvent concentration. The viscosity was obtained by the DLS microrheology. It is interesting that the viscosity of the oil/toluene mixtures is systematically lower than that of the oil/*n*-heptane mixtures. This fact can be explained by the destruction of supramolecular asphaltene structures in this very viscous crude oil by the good solvent, toluene.

7. Conclusions

We have demonstrated that dynamic light scattering is an effective technique for real-time monitoring of stability of petroleum colloids. Combined with accurate viscosity measurements, it is a powerful tool for investigating the colloid nature of crude oils and solutions of heavy petroleum fractions such as asphaltenes. To make light-scattering measurements in poorly transparent (strongly light-absorbing) colloids possible, a special optical scheme has been designed and tested.

We have found that the aggregation of asphaltene particles in hydrocarbon solutions obeys two different kinetics laws, namely, diffusion-limited aggregation and reaction-limited aggregation. For the asphaltene solutions with a concentration well above the nanosize association threshold and with a concentration of *n*-heptane near the precipitation threshold, crossover between these two regimes has been observed. Similar phenomena are observed in classical colloids. In terms of scaled variables, our data clearly demonstrate a universal behavior that, as we expect, may be present in other colloidal systems.

We have also investigated colloidal properties and kinetics of asphaltene aggregation in crude oil samples with a dynamic light-scattering method adapted to opaque fluids. For two oil samples there exists an intrinsic colloidal-like structure with particles of about 0.02 and 1 μm size. Our experiments show that for one sample this structure may somehow be affected by laser radiation, however, the effects of local asphaltene precipitation as a result of photochemistry including possible resonance light absorption or local overheating in crude oils require further investigation. In the future experiments it will be desirable to use lasers with various visible and infrared wavelengths to study effects of light radiation on stability and precipitation of asphaltenes.

We assume that the particles observed are asphaltene/resin colloidal particles. Using n-heptane as precipitant, we have studied the stability of crude oil systems with respect to asphaltene aggregation. One of the oils exhibits an extremely slow aggregation process. The greater the amount of the precipitant added the faster the rate of aggregation. There is no well-defined threshold concentration of the precipitant for this oil. This oil is intrinsically close to instability. Two other oils exhibit different behavior. The second sample exhibits fast aggregation, which starts at the threshold concentration of the precipitant and lasts a short period of time. The process leads to the formation of a colloidal structure with stable asphaltene aggregates. The third sample has no initial colloidal structure but it does occur in a solution with 20% of n-heptane and remains stable. Addition of more than 60% vol. of precipitant completely breaks the stability of the structure and causes flocculation of particles into huge clusters, which cannot be measured by DLS. Our results show that the oils with high aromatics/saturates ratio have more stable structure and exhibit no slow aggregation kinetics, while the oil with much lower ratio is on the edge of stability and shows very slow aggregation kinetics.

Regarding the future studies of asphaltene, aggregation by dynamic light scattering, measurements under high pressures and in a broad range of temperatures are highly desirable. The elucidation of the nature of stabilization of asphalted aggregates in poor solvents by ultrasonic agitation would be very important for industrial applications. Application of microrheology (measuring the shear viscosity at the mesoscale by dynamic light scattering from calibrated Brownian particles[51]) to crude oils and petroleum systems would be also very promising.

Acknowledgments

It is our pleasure to thank V.A. Agayan, Y. Briolant, Y.G. Burya, A.I. Chernoutsan, V.A. Dechabo, I.A. Dmitrieva, D. Frot, E.E. Gorodetskii, V.I. Kosov, E.L. Markhashov, V.R. Melikyan, V.V. Nikitin, G.L. Nikolaenko, J.V. Sengers, and H. Toulhoat for collaborating in these studies and O.C. Mullins for valuable discussions. We are grateful to Shell International E&P—Westhollow Technology Center, *Institut France du Petrole*, and Moscow Oil and Gas Technical University for providing us with various crude oil and asphaltene samples.

References

[1] Sheu, E.Y. and D.A. Storm (1995). Colloidal properties of asphaltenes in organic solvent. In: E.Y. Sheu and O.C. Mullins (eds.). *Asphaltenes: Fundamentals and Applications*. Plenum, New York, p. 1.

[2] Yen, T.F. (1998). Asphaltenes: Types and sources. In: O.C. Mullins and E.Y. Sheu (eds.). *Structure and Dynamic of Asphaltenes*. Plenum, New York, p. 1.
[3] Long, R.B. (1981). The concept of asphaltenes. In: J.W. Bunger and N.C. Li (eds.). *Chemistry of Asphaltenes*. Advances in Chemistry, Series 195. American Chemical Society, Washington. DC.
[4] Storm, D.A. and E.Y. Sheu (1993). *Fuel* **72**, 233.
[5] Groenzin, H. and O.C. Mullins (2000). *Energy Fuels* **14**, 667.
[6] Mack, C. (1932). *Phys. Chem.* **36**, 2901.
[7] Nellensteyn, F.J. (1933). *Proc. World Petroleum Congress*. **2**, 616.
[8] Pfeiffer, J.P. and R.N.J. Saal (1940). *J. Phys. Chem.* **44**, 139.
[9] Buckley, J.S., G.J. Hirasaki, Y. Liu, S. Von Dracek, J-X. Wang, and B.S. Gill (1998). *Petroleum Sci. Technol.* **16** (34).
[10] Wang, J. (2000). *Predicting Asphaltene Flocculation in Crude Oils*. PHD thesis, New Mexico Institute of Mining & Technology, Socorro, New Mexico, 87801.
[11] Mason, T.G. and M.Y. Lin (2003). *Phys. Rev. E* **67**, 050401(R).
[12] Mason, T.G. and M.Y. Lin (2003). *J. Chem. Phys.* **119**, 565.
[13] Andersen, S.I. and J.G. Speight (1993). *Fuel* **72**, 1343.
[14] Andersen, S.I. and S.D. Christensen (2000). *Energy Fuels* **14**, 38.
[15] Sheu, E.Y., M.M. De Tar, D.A. Storm, and S.J. DeCanio (1992). *Fuel* **71**, 299.
[16] Andersen, S.I. and K.S. Birdi (1991). *J. Colloid. Interface Sci.* **142**, 497.
[17] Andreatta, G., N. Bostrom, and O.C. Mullins (2005). *Langmuir* **21**, 2728.
[18] Oliver, C.J. (1974). Correlation techniques. H.Z. Cummins and E.R. Pike (eds.). *Photon Correlation and Light Beating Spectroscopy*. Plenum, New York, p. 151.
[19] Anisimov, M.A., I.A. Dmitrieva, and I.K. Yudin (1988). *Optika i Spectroskopiya* (in Russian) **49**, 144.
[20] Yudin, I.K., G.L. Nikolaenko, E.E. Gorodetskii, and V.I. Kosov (1996). Colloid Aggregation Kinetics in Opaque Petroleum Mixture. In: *Photon Correlation and Scattering*. OSA Technical Digest Series. OSA, Washington DC, **14**, p. 128.
[21] Yudin, I.K., G.L. Nikolaenko, V.I. Kosov, V.A. Agayan, M.A. Anisimov, and J.V. Sengers (1997). *Int. J. Thermophys.* **18**, 1237.
[22] Yudin, I.K. and G.L. Nikolaenko (1997). Photon correlation spectroscopy of opaque fluids. In: E.R. Pike and J.B. Abbiss (eds.), *Light Scattering and Photon Correlation Spectroscopy*. Kluwer Academic Publishers, The Netherlands, p. 341.
[23] http://www.photocor.com. Opaque-dispersions.htm
[24] Herzog, P., D. Tchoubar, and D. Espinat (1988). *Fuel* **67**, 245.
[25] Ravey, J.S., G. Ducouret, and D. Espinat (1988). *Fuel* **67**, 1560.
[26] Anisimov, M.A., I.K. Yudin, V.V. Nikitin, G.L. Nikolaenko, A.I. Chernoutsan, H. Toulhoat *et al.* (1995). *J. Phys. Chem.* **99**, 9576.
[27] Sheu, E.Y., K.S. Liang, S.K. Sinha, and R.E. Overfield (1992). *J. Colloid Interface Sci.* **153**(2), 399.
[28] Wu, J., J.M. Prausnitz, and A. Firoozabadi (1998). *AIChE J.* **44**, 1188.
[29] Wu, J., J.M. Prausnitz, and A. Firoozabadi (2000). *AIChE J.* **46**, 197.
[30] Firoozabaddi, J.A. (1999). *Thermodynamics of Hydrocarbon Reservoirs*. McGraw-Hill, New York.
[31] Burya, E.G., I.K. Yudin, V.A. Dechabo, V.I. Kosov, M.A. Anisimov (2001). *Appl. Opt.* **40**, 4028.
[32] Yudin, I.K., G.L. Nikolaenko, E.E. Gorodeskii, V.I. Kosov, V.R. Melikyan, E.L. Markhashov *et al.* (1998). *J. Petrol. Sci. Eng.* **20**, 297.
[33] Nadirov, N.K., L.H. Batrakov, S.M. Burkitbaev, and S.B. Nurzhanov (1987). *Dokl. Akad. Nauk* (in Russian, available in English), **295**, 1177.
[34] Yudin, I.K., G.L. Nikolaenko, E.E. Gorodetskii, E.L. Markhashov, D. Frot, Y. Briolant *et al.* (1998). *J. Petrol. Sci. Technol.* **16**, 395.
[35] Yudin, I.K., G.L. Nikolaenko, E.E. Gorodeskii, E.L. Markhashov, V.A. Agayan, M.A. Anisimov *et al.* (1998). *Physica A* **251**, 235.
[36] Weitz, D.A., J.S. Huang, M.Y. Lin, and J. Sung (1985). *Phys. Rev. Lett.* **54**, 1416.
[37] Weitz, D.A. and M.Y. Lin (1986). *Phys. Rev. Lett.* **57**, 2037.

[38] Lin, M.Y., H.M. Lindsay, D.A. Weitz, R.C. Ball, R. Klein, and P. Meakin (1990). *Phys. Rev. A.* **41**, 2005.
[39] Von Smoluchowski, M. (1916). *Phys. Z.* **17**, 593.
[40] Yudin, I.K., G.L. Nikolaenko, E.E. Gorodetskii, E.L. Markhashov, V.A. Agayan, and M.A. Anisimov (2000). Crossover from reaction-limited aggregation to diffusion-limited aggregation of asphaltenes in hydrocarbon solutions. In: *Porous Media: Physics, Models, Simulation.* World Scientific, Singapore, p. 75.
[41] Burya, E.G., I.K. Yudin, V.A. Dechabo, and M.A. Anisimov (2001). *Int. J. Termophys.* **22**, 1397.
[42] Yudin, I.K., V.I. Kosov, V.A. Deshabo, E.E. Gorodetskii, Y.G. Burya, and M.A. Anisimov (2000). Aggregation phenomena in petroleum colloids studied by dynamic light scattering. In: *Photon Correlation & Scattering.* OSA Technical Digest Series. OSA, Washington, DC, p. 30.
[43] Clark, J P.F. and B.B Pruden (1997). *Fuel* **76**, 607.
[44] McLean, J.D. and P.K. Kilpatrick (1997). *J. Coll. Interface Sci.* **189**, 242.
[45] McLean, J.D. and P.K. Kilpatrick (1997). *J. Coll. Interface Sci.* **196**, 23.
[46] Escobedo, J., and G.A. Mansoori (1995). *SPE Production Facilities* **May**, 115.
[47] Escobedo, J., and G.A. Mansoori (1997). *SPE Production Facilities* **May**, 116.
[48] Fenistein, D., L. Barre, D. Broseta, D. Espinat, A. Livet, J-N. Roux *et al.* (1998). *Langmuir* **14**, 1013.
[49] Mason, T.G. and D.A. Weitz (1995). *Phys. Rev. Lett.* **74**, 1250.
[50] Crocker, J.C., M.T. Valentine, E.R. Weeks, T. Gisler, P.D. Kaplan, A.G. Yodh *et al.* (2000). *Phys. Rev. Lett.* **85**, 888.
[51] Williams, D.F. and C.H. Byers (1986). *J. Phys. Chem.* **90**, 2534.

18

Near Infrared Spectroscopy to Study Asphaltene Aggregation in Solvents

Kyeongseok Oh and Milind D. Deo

1. Introduction

Petroleum asphaltenes are extremely complex and are difficult to characterize. Asphaltenes are a solubility class, defined as that portion of the oil (or organic material) that is soluble in toluene and insoluble in normal heptane. (Sometimes, other alkane solvents are used to define this solubility class, the most common other solvent being, normal pentane.) Asphaltenes present numerous problems during production, transportation, and processing of crude oils because they are on the higher polarity and molecular weight end of the crude oil compositional spectrum. As a result, this solubility class has been widely studied. There are over a thousand papers on various aspects of asphaltenes, including their chemistry, molecular weight, solubility, phase behavior, reactivity, etc.[1–4]

Crude oils are mixtures of thousands of polar and nonpolar compounds. Asphaltenes are the most polar of crude oil compounds, and are often compared to large amphiphilic molecules. Self-aggregation properties of amphiphilic molecules are well established. In that context, one question has often been posed—do asphaltene molecules self-aggregate? If they do, at what concentrations do they exhibit this behavior? What are sizes and other properties of these aggregates? Based on surface tension measurements, Sheu and coworkers[5] answered the first question in the affirmative. They observed distinct break points in surface tensions of asphaltene–solvent mixtures as function of concentration. They were working at concentration ranges of the order of 0.1–10 g/L. Since then, it is generally believed that asphaltenes aggregate; however, different terms have been used to describe the aggregates and the aggregation process. In keeping with the amphiphilic molecule comparison, the term "micelle" has been commonly used to describe an asphaltene aggregate. The term micelle has been used to describe both the so-called asphaltene–resin complexes and the self-associated asphaltene molecules. Because of the polydispersed nature of asphaltenes, several researchers[6–8] have argued that micelle is not an appropriate term for these aggregates. Some research

---

Kyeongseok Oh and Milind D. Deo • Department of Chemical Engineering, University of Utah, Salt Lake City, UT.

groups[9] have argued that asphaltenes do not self-associate. Recently, Groenzin and Mullins[10] have observed asphaltene aggregation at much lower concentrations, of the order of 10–60 mg/L. It was argued that the nature of this aggregation is different and could be related to the formation of dimmer–trimer–tetramer type structures. There is also the question of differences in asphaltene behavior when dissolved in oils or in organic solvents.

Thus, there may be different aggregation/association regimes for asphaltenes in solvents and in oils. This study focuses on asphaltene aggregation behavior in organic solvents, at gram/liter concentration levels, which sometimes is referred to as *mesoaggregation*.[11] The vast amount of literature accumulated in experimental studies related to the aggregation of asphaltenes pointed to the existence of a critical micelle concentration (CMC) in the concentration range, 0.1–10 g/L.[9,12–15] (*The term CMC is used here only for historical reference, recognizing that it may not be appropriate for asphaltenes*). These studies recognized the fact that asphaltenes are amphiphilic molecules, and even though they are polydispersed, they exhibit a property common to surfactants. Behavior of asphaltenes in solvents is a complex phenomenon related to the solubility of asphaltenes in these solvents along with the aggregation behavior. In the literature where asphaltene aggregation measurement is the target, solubility aspects are often not examined. For example, surface tension values of asphaltenes in toluene or pyridine show distinct break points with respect to asphaltene concentration. These break points were interpreted to be indicative of the presence of CMC-type phenomena in asphaltene mixtures.[1] This type of abrupt property change was also observed in viscosity measurements.[1,15] Recent studies have questioned the concentration range over which this behavior occurred and argued that the CMC-like behavior is observed at much lower concentrations (of the order of 1–60 mg/L).[6–8,10] However, property transitions are positively identified (particularly with respect to surface tension) in the higher concentration range.

This chapter is a discussion of near-infrared (NIR) spectroscopic methods to study the aggregation behavior of asphaltenes. The technique, apart from being fast, accurate, and nondestructive, provides additional information about asphaltene solubility. The asphaltene precipitation onset measurement with respect to the asphaltene concentrations in solvent mixtures provides the means of establishing a transition point. The critical concentration measured is termed critical aggregation concentration (CAC) to distinguish it from the lower concentration phenomenon. The measured CAC values are compared to the CMC values (historical definition) in the literature obtained by surface tension measurements or through calorimetric studies.[1,9,12–15] This analysis is extended further to asphaltene subfractions.

2. Literature

The reported molecular weights of asphaltenes have ranged from 500 to 1,000,000 g/mol.[16] Some of these differences can be attributed to the fact that asphaltenes were obtained from different sources. However, the range of measured variation can only be explained by considering some degree of aggregation in asphaltene–solvent mixtures.

Since 1920s, it has been hypothesized that asphaltenes may exist in the form of a colloidal phase.[17] The asphaltenes were thought to be high molecular weight molecules peptized by the less heavy fractions, such as resins and aromatic compounds. Pfeiffer and Saal[18] suggested that the peptized asphaltenes form the asphaltene micelles. Dickie and Yen[3] proposed four asphaltene entities; (1) a unit fused-aromatic sheet (molecule), (2) a particle, (3) a micelle, and (4) aggregates. In this representation, the molecular association of asphaltenes forms stacked aromatic sheets in solution held together by π–π association and hydrogen bonds. There may be occasional charge interactions. The associations of asphaltene molecules with resins and other aromatic compounds present in the oil were said to form asphaltene micelles. Finally, aggregates of asphaltene micelles are formed giving a precipitate. Thus, asphaltene precipitation can be postulated to occur in the following steps:

(1) molecular self-association into dimer or tetramer like molecules;
(2) pseudo-micellization with resins and aromatic compounds;
(3) aggregation of 'micelles' into precipitates.

The literature is replete with measurements of the critical micelle concentration for asphaltenes in various organic solvents. The values of CMC are obtained by measuring surface tension and interfacial tension[1,12–15] or heat effects in microcalorimeter.[9,19] Ultrasonic measurements[11] have also been reported recently. Most CMC values have been reported in the asphaltene concentration range of 0.1–10 g/L. Rogel and coworkers[14] reported CMC values up to asphaltene concentrations of 30 g/L. However, the concept of asphaltene micelle formation is not completely accepted. Using vapor pressure osmometry (VPO), Yarranton and coworkers[15] found that 2–6 asphaltene molecules aggregate at concentrations of 10–20 g/L, and that there is no evidence of micellization from the interfacial tension data. This may be due to the fact that CMC analyses were carried out with fractionated asphaltenes (e.g., heptane insolubles), which do not have enough resinous materials to form the micelles. In any case, asphaltene self-association is fundamentally different from traditional micellization because of the nonuniformity of the postulated asphaltene micelles. However, there are a number of other papers in the literature that provide evidence of molecular-level association of asphaltenes at very low concentrations.[6–8,10]

Groenzin and Mullins[10] measured asphaltene molecular weights using fluorescence depolarization measurements and found values between 500–1000 g/mol. They suggested that asphaltenes self-associate (dimer formation) at 60 mg/L in toluene, based on an observed signal shift at this concentration. Acevedo and coworker[8] observed a minimum value of thermo-optical diffusivity at asphaltene concentrations of 50 mg/L. They attributed this minimum as evidence of the asphaltene molecular association. Evdokimov and coworkers[6,7] proposed stepwise asphaltene aggregation from asphaltene dimers to higher aggregates at increasing asphaltene concentrations by measuring NMR relaxation times. They suggested that molecular, association-free asphaltene solutions are possible only below 1–2 mg/L, and that the asphaltene molecular nanoclusters (tetramers or dimer pairs) were formed at around 100 mg/L.

There has been some confusion in attempts to explain the discrepancy between the low-concentration association (concentrations in the order of about 1–60 mg/L) and higher apparent CMC values (0.1–10 g/L). The asphaltene's self-association that takes place at very low concentrations does not represent even the apparent CMCs for asphaltenes. Two mechanistic transition points have been measured by a few researchers. Priyanto and coworkers[20] presented two different asphaltene aggregation steps in a concentration range of 0.1–10 g/L in 1-methyl naphthalene. They attributed the lower concentration to the CMC and the higher concentration aggregation to the micelle coacervation point (MCP). Their first transition point may be related to a molecular association phenomenon, while the second may indicate CMC-like aggregation. Pietraru and Cramb[21] found fluorescence spectral shifts at certain asphaltene concentrations. The shift was observed at concentration values that were relatively higher than values in an earlier study.[10] They interpreted the shift to occur at a critical aggregation concentration (CAC) (range 0.2–1.5 g/L) and suggested that the CAC was caused by the formation of the larger asphaltene molecular aggregates from the smaller asphaltene molecular aggregates. The CAC values reported in this study were close to the MPC values measured by Priyanto and coworkers.[20] Both MCP and CAC, nevertheless, are close to CMC values reported in the literature. VPO measurements[15] and small angle neutron scattering[5] have shown that the asphaltene aggregates have a limited size at higher concentrations.

Yudin and coworkers[22] reported that the heptane-induced asphaltene aggregations were also affected by the initial asphaltene concentration. They prepared low (1 g/L) and high concentration (10 g/L) asphaltene–toluene mixtures. They measured asphaltene particles sizes in the range of 0.2–4.5 μm using photon correlation spectroscopy. Rapid particle size growth was observed when they added larger amounts of heptane above the onset concentration. Diffusion-limited particle aggregation kinetics was observed for low concentration (1 g/L) asphaltene–toluene mixtures (below the CMC value that they proposed), and rate-limited particle aggregation was reported for high-concentration (10 g/L) asphaltene–toluene mixtures (above the CMC).

In this study, Rangely field asphaltenes were used and precipitation onsets were measured using NIR spectroscopy and microscopic observations. Onset is defined as the minimum amount of heptane necessary to induce asphaltene precipitation. These onset data provide the solubility of asphaltene in various organic solvents. Some surface tension measurements are also provided.

3. Experimental

Three types of asphaltenes were used in this study; two different Rangely field-asphaltenes (A1 and A2) and heptane insolubles asphaltene (C7-asphaltenes) from the Rangely crude oil. Rangely oil is a medium gravity, low asphaltenic crude oil being produced from the Rangely field in northwestern Colorado. A carbon dioxide enhanced oil recovery flood is underway in Rangely. A1 and A2 in this study were the solids that were deposited on subsurface tubing and pumps, and on surface equipment. C7-asphaltenes were prepared from the oil using a procedure described by the modified IP 143/57.[23,24] The Rangely oil was mixed with *n*-heptane at volume ratio of 1:30 and stirred for 16 hr at room temperature. C7-asphaltenes were filtered with a-2.7 μm pore size filter, washed with *n*-heptane, and were dried. The asphaltenes obtained were stored in the laboratory at normal conditions. The original amount of asphaltenes in the crude oil was 1.7 wt%. Four organic solvents were used in the study (toluene, trichlroethylene (TCE), tetrahydrofuran (THF), and pyridine). The solvents used were over 99% pure and used as received.

A fixed amount of A1, with an accuracy of ±0.0001 g, was dissolved in the solvents separately. Each asphaltene–solvent mixture was stirred for over 24 hr. The concentration range was 1–15 g/L. Mixtures were also prepared by dilution of a primary 15 g/L concentration (20 g/L for pyridine) mixture to ensure that asphaltenes with identical characteristics were present in all the solutions. Experiments were performed with both types of solutions for comparison. A2 and C7-asphaltenes in toluene mixtures were also prepared in the concentration ranges of 1–15 g/L and 1–10 g/L, respectively.

The amount of precipitant, such as heptane or pentane, needed to initiate asphaltene precipitation at a given concentration is measured by NIR. Asphaltene precipitation onsets were measured by heptane titration method using NIR spectrometer (Quantum-1200 plus from LT industries, Inc.). The path length was 6 mm. Titrations were carried out with 40 mL of asphaltene–solvent mixtures. The flow rate was controlled by a peristaltic pump. Onset measurements were recorded using NIR spectra with the probe placed directly inside the mixture. The optical densities (OD) at the wavelength of 1600 nm were collected every minute. The onset was determined to be the heptane amount corresponding to the minimum OD. After measuring the heptane volumes corresponding to the minimum OD, the OD were recorded at every 30 s at near onset data. The onset measurement was carried out once more to verify the onset values. The error in each onset data point was ±0.0125 mL-heptane/mL-mixture.

Apparent molecular weights (vapor pressure osmometry, Knauer, K-7000) were measured at 60°C in pyridine as a solvent. The error range (reproducibility) in molecular weight measurements was <1% for standards (benzyl, polyethylene glycol 1000) and in the range of 2–3% for asphaltenes.

Whilhelmy plate (Krüss K10T) was used to measure the surface tension of asphaltene solutions at room temperature. Carbon, hydrogen, nitrogen, and sulfur concentrations of the asphaltenes were analyzed using the CHNS-932 (Leco Corp.). Thermal gravimetric analysis was carried out in a nitrogen and oxygen atmosphere using TGA Q500 (TA instruments) with small amounts (about 20 mg) of asphaltenes.

4. Results and Discussion

4.1. Asphaltene Aggregation or Self-Association

When a certain physicochemical property of a solution changes abruptly over a narrow concentration range, the solution is said to be characterized by critical micelle concentration (CMC). This property change has been attributed to

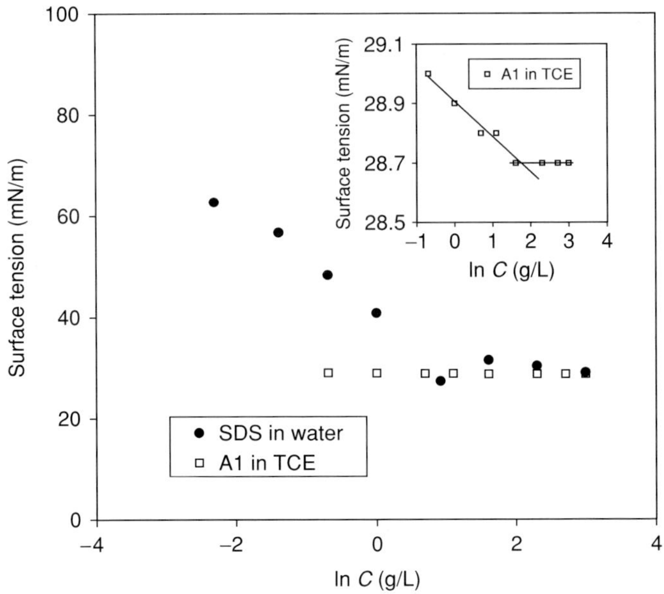

Figure 18.1. Surface tension data for sodium dodecyl sulfate in water and A1 asphaltene in trichloroethylene (TCE).

the formation of aggregates. Surface active materials like sodium dodecyl sulfate (SDS) in aqueous solutions show these property shifts at certain concentrations. There is a discontinuous point in the surface tension data as a function of SDS concentrations in water. It can be easily observed that the surface tension decreases rapidly before CMC followed by a gradual decrease above CMC. It was reported that further SDS addition into the aqueous phase produces micelles rather than the further growth of existing micelles.[25]

The surface tension data for the sodium dodecyl sulfate–water system and for field-asphaltene (A1) in trichloroethylene (TCE) system are presented in Figure 18.1. SDS forms a spherical micelle from approximately 60 combined molecules in an aqueous phase.[26] The surface tension values were in the concentration range of 62.7–27.3 mN/m in the SDS–water system, and the minimum value was observed at a concentration of 2.5 g/L (0.92 on the logarithm axis, 0.0087 mol/L). The reported CMC value of SDS is 0.008 M (mol/L). This difference may be due to an impurity in SDS as postulated in Miles and Shedlovsky.[26] In A1-TCE mixtures, the surface tension data were observed in the range of 29–28.7 mN/m. This variation in surface tension values is small, as a result of which it is difficult to document the surface tension break point. The range of variation in surface tension values for SDS–water mixtures, for example, is of the order of 30 mN/m. The surface tension break point for the A1-TCE mixture was at a concentration of 5.3 g/L. The CMC values of various asphaltene–solvent mixtures have been reported in the concentration range of 0.1–10 g/L.[9,12–15]

Measurements of surface tension versus concentration of field asphaltene A1 in pyridine are shown in Figure 18.2A. There is a break point in the data

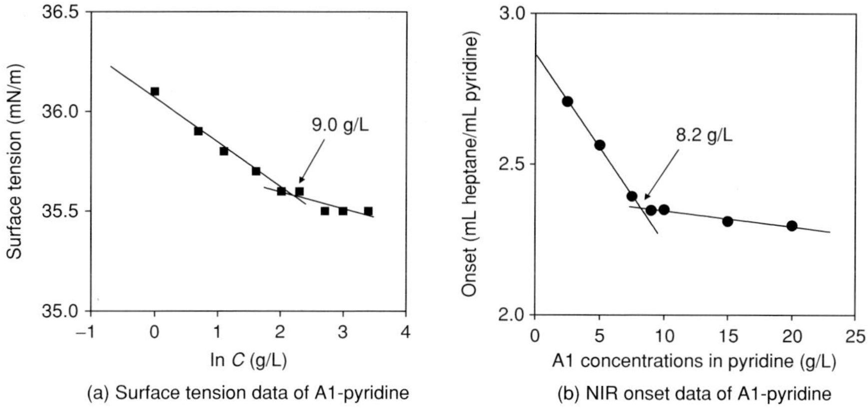

Figure 18.2. Surface tension data and NIR onset data of A1-pyridine mixtures. Onsets were measured using heptane flow rate of 1 mL/min.

at 9.0 g/L recorded as a change in the slope of the surface tension data. The break point in the surface tension is defined as the CMC in the previous literature and is interpreted in this study as the concentration at which critical numbers of asphaltene aggregates (to facilitate precipitation) begin to form. NIR precipitation onset measurements were also obtained for the same asphaltene–pyridine mixtures (for which the surface tension data are reported). The onset data as a function of concentration are plotted in Figure 18.2B. A break point similar to the one observed with surface tension measurements is seen on this plot at a concentration of 8.2 g/L. The NIR data is a direct measurement of the onset of asphaltene precipitation at different asphaltene concentrations.

It is clear from Figure 18.2B that as the asphaltene concentrations increase, the precipitant to solvent ratio (PSR) required at onset decreases. But at the break point this dependence changes—the slope of the plot flattens. After the break point, increase in concentration does not lead to as much decrease in the PSR as before the break point. This indicates that a critical aggregation point may have been reached. Aggregation is now limited, to a certain extent, by factors other than concentration. This may be the transition to rate-limited aggregation as described by Yudin et al.[22] In this study, henceforth, the break point is called the critical aggregation concentration (CAC), the same terminology employed by Pietaru and Cramb.[21] The surface tension decreases observed were relatively small in the range of 36.1–35.5 mN/m. For some solvent mixtures the surface tension differences are very low, much lower than detections limits of the instrument. The NIR-onset technique adopted in this paper offers a clear and robust alternative of detecting CAC.

4.2. Onset of Asphaltene Precipitation

Figure 18.3 shows the NIR optical densities in the wavelength range of 1200–2200 nm after various amounts of heptane are added to 40 mL of 10 g/L A1–toluene mixture. Spectra are shown after additions of 31–100 mL heptane,

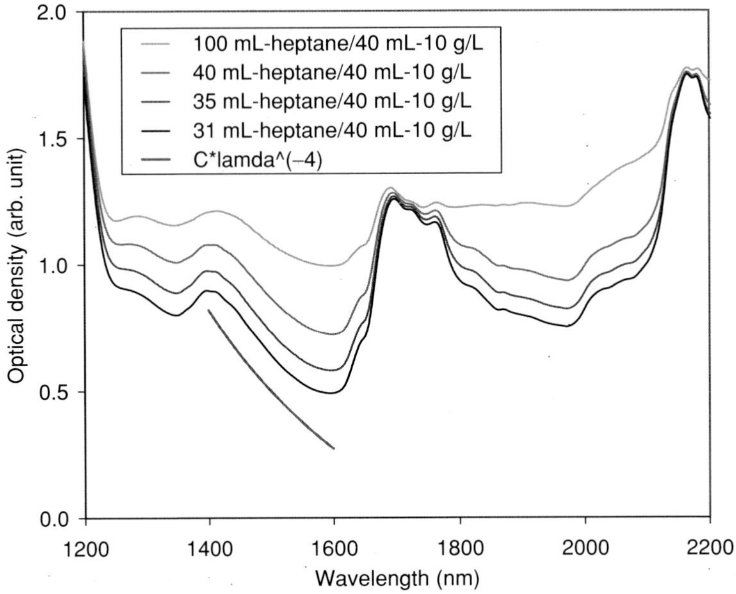

Figure 18.3. NIR optical densities at wavelengths between 1200 and 2200 nm when various amounts of heptane (indicated in the legend) are added to a-40 mL mixture of A1 asphaltene and toluene at a concentration of 10 g/L.

with the precipitation onset being at 31 mL. The optical densities were observed to decrease due to the dilution of the mixture with heptane until the onset point was reached, and then, the optical densities increased due to the generation of asphaltene particles. Addition of heptane is not the same as true dilution, however. This is because addition of heptane, even prior to the onset of precipitation, is moving the system toward particle generation.

Peak correspondence to the presence of different groups has been examined in detail in earlier papers.[27,28] Joshi and coworkers[28] reported a wavelength-dependent and a wavelength-independent particle growth in NIR spectra. The wavelength-dependent particle growth is observed during the growth of a smaller particle. In this case, NIR optical densities have larger increases in the shorter wavelength region (at about 1400–1600 nm) than in the longer wavelength region (at about 1800–2100 nm). The wavelength-independent growth is observed in the formation of bigger particle sizes, where the OD changes are the same over the entire wavelength range. The data in Figure 18.3 show that the wavelength-dependent particle size growth was obtained near the onset. After that, wavelength-independent particle growth is observed. The data in Figure 18.3 were for a 10 g/L mixture. When lower concentrations of asphaltene solutions are titrated with heptane, the difference between wavelength dependent and independent regions is not very clear.

It is important to note that the OD is the sum of the effects of light absorption and light-scattering contributions. After the precipitation of asphaltene particles, the OD increases mainly by light scattering of these particles. Rayleigh scattering of the particles is a reasonable approximation when all the particles are much smaller than the wavelength employed. When the particles are formed by nucleation, they should all be of similar size and small compared to the wavelength of NIR light. Rayleigh scattering gives an OD that varies with the minus four power of wavelength.[29] As a result, a curve showing $C^*\lambda^{-4}$ in the wavelength range of 1400–1600 nm was added alongside the minimum OD plot—the point where particles are nucleated (the value of C was chosen to approximate the slope of the onset plot and was 5.07×10^{12}). This curve is very similar in slope and curvature to the OD data, suggesting that particles are responsible for the NIR data in this wavelength range. As the particles get larger, their scattering can be followed by the evolution of OD as a function of wavelength. A flattening of the curve as the amount of heptane is increased indicates that the particles are growing (while continuing to nucleate) and broadening their size distribution while aggregating together.

Minus four power dependence on λ is usually applied to the Rayleigh region. Joshi and co-workers[28] used power values of 2 and 0.2 to describe particle sizes of approximately 0.85 and 2.3 μm, respectively. The particle sizes in their case were in the range of wavelengths used (1000–2500 nm). The applicability of the power law fit starts to break down when the particles are much bigger than the wavelengths employed. This is believed to be the case in the post-onset spectra in Figure 18.3. Yudin and coworkers[22] studied asphaltene particle growth using photon correlation spectroscopy. The maximum particle size that could be measured through this technique was 4.5 μm. They found that the particle growth was very rapid, post-onset. They also postulated two different mechanisms of aggregation; diffusion limited at lower concentrations and rate limited at higher concentrations. Figure 18.3 provides clues about particle sizes at onset and post-onset, but it is not possible to extract particle sizes from this data. Since a power of 4 describes the onset curve, it can be postulated that the particles are less than 100 nm in size, which is where the Rayleigh scattering really applies. Qualitatively, a power of less than 4 describes post-onset curves. But here, it is speculated that the particles are of sizes ranging from 10 μm and thus exceed the wavelength range (1200–2400 nm) employed.

The amount of asphaltenes precipitated as a function of time, when a fixed amount of heptane (minimum amount, as determined from onset studies), was added is plotted in Figure 18.4. Figure 18.4A shows the data for a 1 g/L solution, while the data for the 10 g/L is shown in Figure 18.4B. The OD at 1600 nm at these times is recorded on the right-hand y-axis. It is seen from these figures that the OD tracks the amount of asphaltenes precipitated until a certain time (3 hr for 1 g/L solution and 2 hr for 10 g/L solution), and then the two curves separate. The amount of asphaltenes flattens out, while OD shows a mild increase. This data clearly demonstrate that the particles are generated for a certain time (the time until which the two curves track), but the particle growth continues beyond this point. In the initial period of particle generation, the spectrum is affected by particle generation and growth. Hence it is difficult to deduce only the particle growth information from the NIR spectrum.

Figure 18.5 shows the OD changes when five different mixtures were titrated with n-heptane. Optical densities at a wavelength of 1600 nm are shown. The single wavelength analyses at 1600 nm have been reported previously.[30–32] The OD

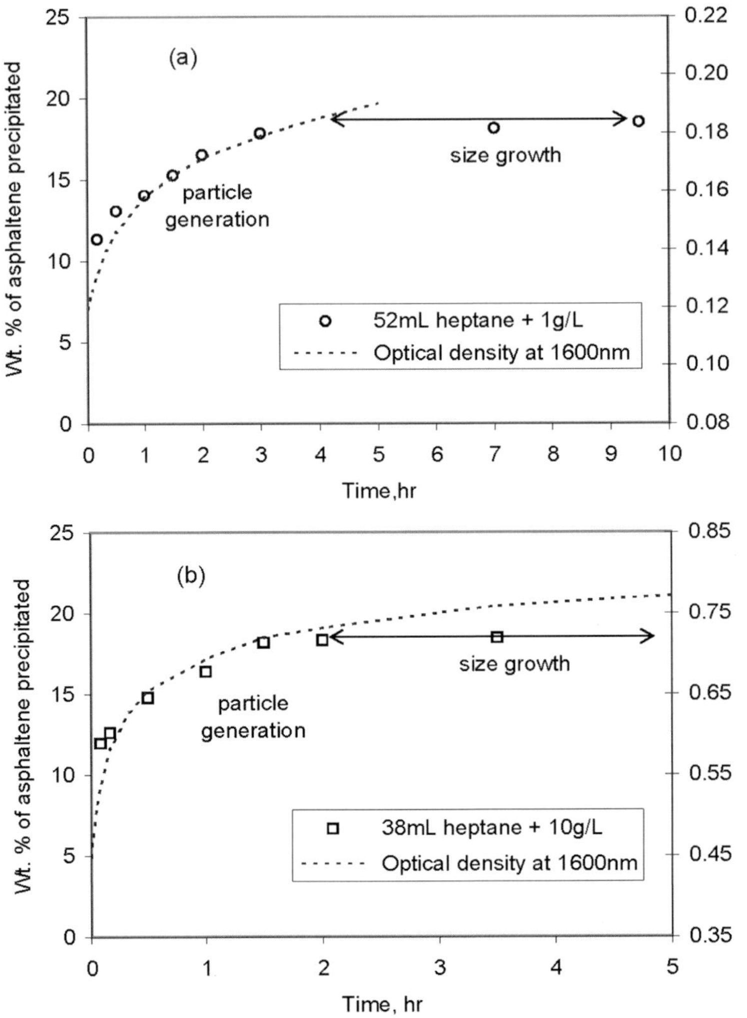

Figure 18.4. The amount of asphaltenes precipitated as a function of time, when a fixed amount of heptane (minimum amount, as determined from onset studies), was added. The optical density values at 1600 nm are also shown. (A) data for 1 g/L; (B) data for 10 g/L.

values were relatively higher as the asphaltene concentrations increased. The OD values decrease initially and reach a minimum point followed by a signal increase. The initial decrease is due to heptane addition. The NIR signal is affected by a number of factors, and it does not decrease linearly as solvent is added. The absorbing molecules in the asphaltene–solvent mixtures undergo changes that affect the NIR absorbance. The minimum in OD corresponds to the onset of asphaltene aggregation.[30–32] The experiments were repeated for reproducibility.

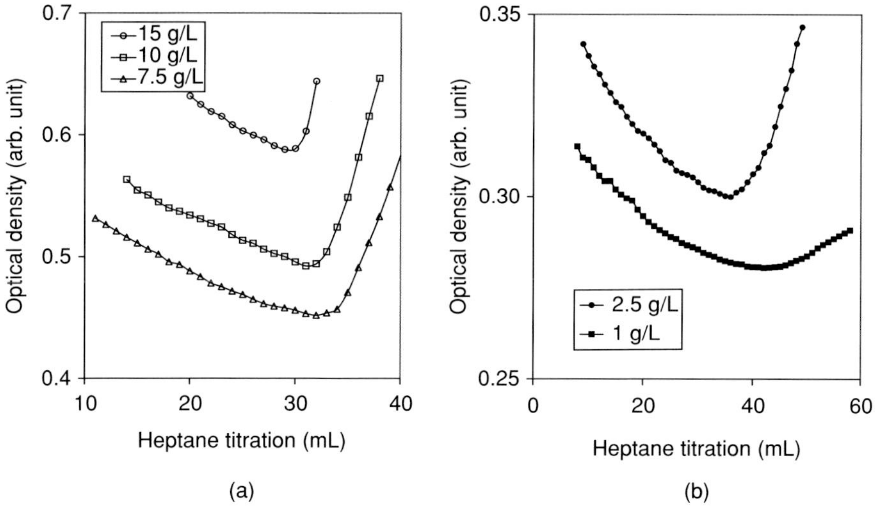

Figure 18.5. NIR optical density curves for various concentrations of A1 asphaltenes–toluene mixtures at a wavelength of 1600 nm as a function of heptane added. The heptane flow rate used was 1 mL/min.

4.3. Effect of the Solvent

Figure 18.6 shows the onset points plotted as solvent (L)/asphaltenes (g) versus heptane (L)/asphaltenes (g). The linear trends reported by earlier researchers[33,34] are also observed. The slopes are distinctly different for toluene, TCE, THF, and pyridine, and increase as the solvent solubility parameters increase. Buckley et al.[35] found that as different solvents are added to the crude oil, different amounts of precipitants are necessary. However, the refractive indices of the mixtures at onset were approximately the same. Buckley et al.[35] primarily studied the so-called polarizability of compounds in nonpolar mixtures, such as crude oils. They stated that the assumption of the dominance of London dispersion forces would not be valid if significant amounts of polar materials were added to mixtures. There are two significant differences between the present study and the results published by Buckley et al.[35] Buckley et al. worked with oils and with predominantly hydrocarbon (nonploar) additives and solvents. Data in the current paper are measured using asphaltenes dissolved in polar and nonpolar solvents. One of the major findings of Buckley et al. was that precipitant to solvent ratios are specific to crude oils and to the additives used. They found, for example, that addition of α-methylnaphthalene (a nonpolar compound) to the oil increases the amount of precipitant necessary to induce precipitation. Results in Figure 18.5 are in a sense complimentary to findings of Buckley et al.[35] It is seen that solvent polarity is also very important in determining precipitant to solvent ratios. Finding with respect to refractive index at onset was not verified in this work, since refractive indices were not measured. Precipitant to solvent ratio, identified by Buckley et al. as important, was further examined as function of asphaltene concentration.[33]

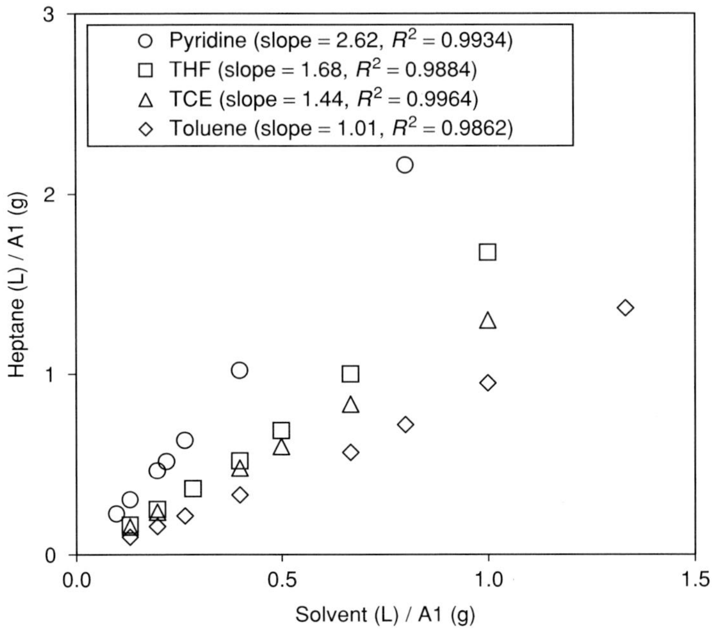

Figure 18.6. Onset of asphaltene aggregation data for the different organic solvents. Plotted in this manner, linear trends are observed. Onsets were measured using heptane flow rate of 1 mL/min.

The data in Figure 18.6 were replotted in Figure 18.7 with the ratio of heptane (mL) to solvent (mL) on the y-axis and A1 concentration (range 0.5–15 g/L) in the solvent on the x-axis. In Figure 18.6, there are two slopes in each plot separated by a critical point, the critical aggregation concentration, CAC. In the part of the plot at lower A1 concentrations, the onsets decrease linearly with a steep slope. In the second part of the plot, at higher A1 concentrations (after the CAC) the slope also decreases linearly, but less steeply than in the first part. Identical trends are

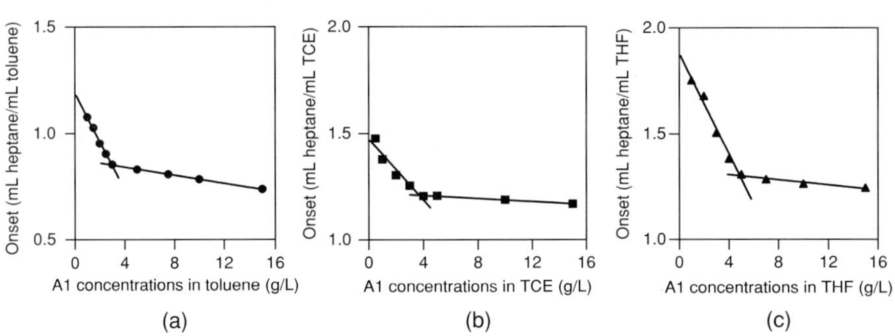

Figure 18.7. Onset data plotted versus A1 asphaltene concentration, arranged according to the organic solvent used. (A) Toluene; (B) TCE; (C) THF. Onsets were measured using heptane flow rate of 1 mL/min.

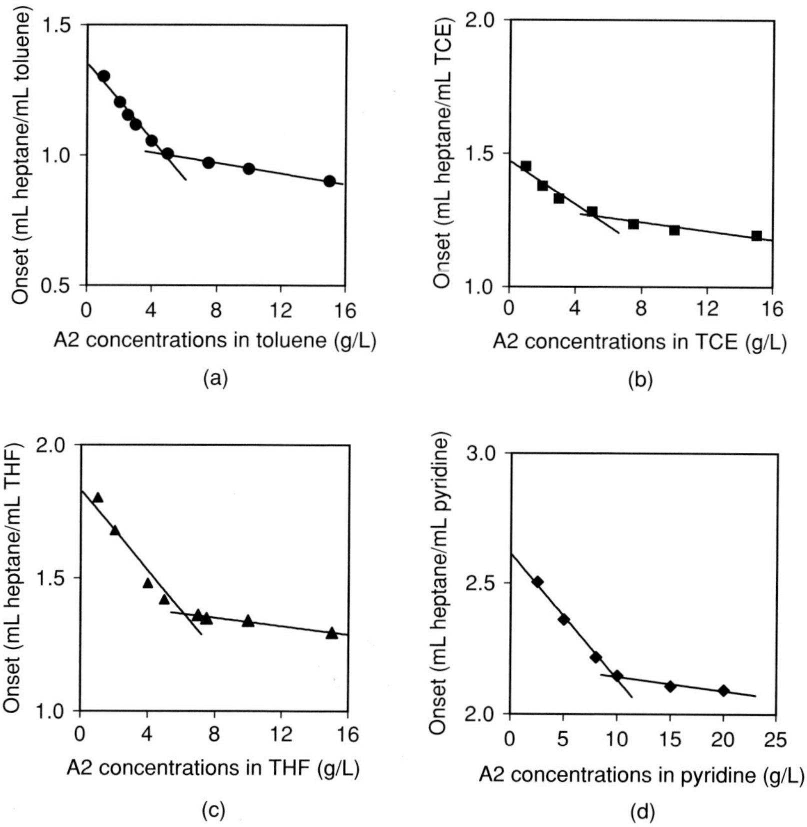

Figure 18.8. Onset data plotted versus A2 asphaltene concentration, arranged according to the organic solvent used. (A) Toluene; (B) TCE; (C) THF; (D) pyridine. Onsets were measured using heptane flow rate of 1 mL/min.

observed for each of the solvents with CACs occurring at different locations. The CACs were at asphaltene concentrations of 3.0, 3.7, 5.0, and 8.2 for toluene, TCE, THF, and pyridine (Figure 18.2B) respectively.

Figure 18.8 shows the onset data of A2–solvent mixtures by heptane titration. Similar CACs were observed at the concentration range of 1–10 g/L. These results are compared with Figures 18.2B and 18.7. All CAC values were plotted with respect to the solvent solubility parameters in Figure 18.9. More or less linear increase of CAC values was observed.

The CACs in Figures 18.2B, 18.7, and 18.8 can be interpreted to signify a mechanism shift, since a clear transition is observed in the onset data with respect to asphaltene concentration. At lower concentration, a relatively larger amount of precipitant is required to initiate aggregation since the asphaltenes are more effectively dissolved in the solvent. As the concentration increases, a threshold value (CAC) is reached. The CAC value was interpreted as the CMC in a number of previous papers. The CAC measured here does not represent a threshold

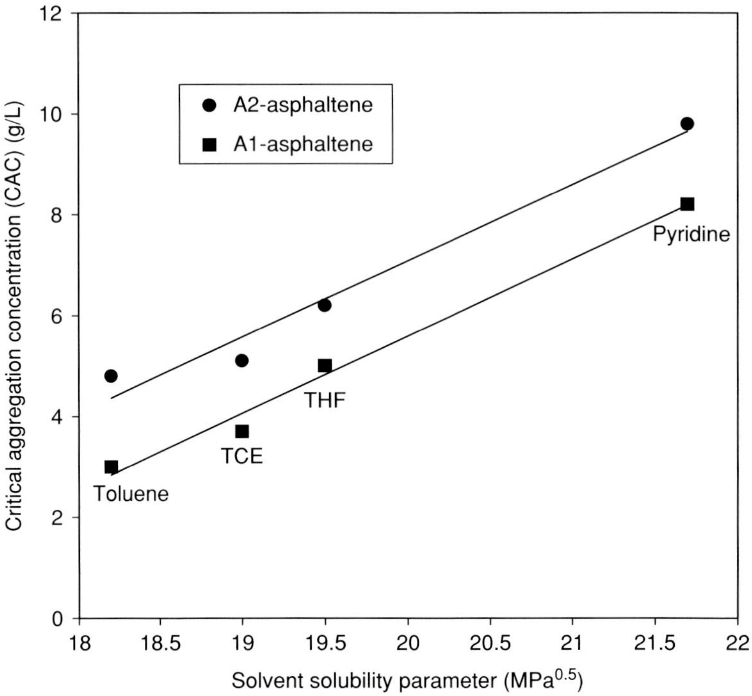

Figure 18.9. Linear onset trends with respect to the solvent solubility parameters.

concentration of asphaltene molecular self-association. It has been reported that the asphaltene aggregates do not grow with increasing concentration.[5,15] Beyond the CAC, the asphaltenes exhibit a limited degree of particle growth, as seen by the OD following the wavelength to the minus four power dependence for several dilutions higher than that of the onset in Figure 18.3, and a great deal of particle aggregation. This aggregation takes place because relatively smaller amounts of precipitants are adequate in inducing flocculation.

The CAC values interpreted in this work and the CMC values reported in previous studies are provided for comparison purposes in Table 18.1. The solvent solubility parameters are also tabulated. The data reported by Andersen and Birdi[9] and by Rogel and coworkers[14] did not show consistent trends (either increasing or decreasing) for CMC values when they used different solvents. This could be related to the measurement methods and/or to the nature of the asphaltenes. Meanwhile, Mohamed and coworkers[13] measured CMC using surface and interfacial tensions for C5I (pentane insoluble asphaltene) and C7I (heptane insoluble asphaltene). They reported that the CMC values of C5I were higher than the CMC values of C7I. The amount of C5I is generally higher than C7I with a wide asphaltene molecular weight distribution. They also showed that the CMC values decreased with increasing the solvent solubility parameters. It is observed in Figure 18.9 that the CAC values increased more or less linearly with the solvent solubility parameter. This means that as the solvent solubility parameter increases, it is relatively more difficult for the induction of flocculation. It should be noted that the onset

Table 18.1. Asphaltene CMC (g/L) Values from the Literature in Comparison to CAC Values Measured in This Study

Reference	Asphaltene type[a]	CMC (g/L)	Solvent	S. P.[b] (MPa$^{1/2}$)	Method[c]
	C7I	1.0–5.9	Cyclohexane	16.7	ST
	C7I	5.2–18.6	Carbon tetrachloride	17.6	ST
[14]	C7I	10.5–22.9	Toluene	18.2	ST
	C7I	1.7–10.8	THF	19.5	ST
	C7I	0.9–30.7	Nitrobenzene	21.8	ST
	C5I	14.8–15.4	Toluene	18.2	ST, IFT
	C5I	12.1	Pyridine	21.7	ST
[13]	C5I	8.2–9.0	Nitrobenzene	21.8	ST, IFT
	C7I	6. 5	Toluene	18.2	ST
	C7I	5.6	Pyridine	21.7	ST
	C7I	6.1	Nitrobenzene	21.8	ST
	C7I	2.93	*n*-Decylbenzene	17.3	CAL
	C7I	0.60	Carbon tetrachloride	17.6	CAL
[9]	C7I	3.24	Toluene	18.2	CAL
	C7I	4.49	Quinoline	22.0	CAL
	C7I	3.91	2,4-Dimethylpyridine	22.2	CAL
	A1	3.0	Toluene	18.2	Onset
	A1	3.7	TCE	19.0	Onset
	A1	5.3	TCE	19.0	ST
	A1	5.0	THF	19.5	Onset
This study	A1	8.2	Pyridine	21.7	Onset
	A1	9.0	Pyridine	21.7	ST
	A2	4.8	Toluene	18.2	Onset
	A2	5.1	TCE	19.0	Onset
	A2	6.2	THF	19.5	Onset
	A2	9.8	Pyridine	21.7	Onset

[a] C5I: Pentane insoluble asphaltenes, C7I: heptane insoluble asphaltenes, and A1 and A2: field-asp; field-asphaltenes deposited in the oil production wells.
[b] S. P.: solubility parameter.
[c] ST: surface tension, IFT: interfacial tension, and CAL: calorimeter.

values over the entire concentration range are shifted to higher values as the solvent solubility parameter increases (Figures 18.7, 18.8, and 18.2B).

In Figure 18.10, aggregation onset points for the C7-asphaltenes are plotted. Toluene was the solvent in these experiments. All of the onset points for C7-asphaltenes have higher *y*-axis values in comparison to the onsets for A1 or A2 in toluene. This means that the C7-asphaltenes are more stable than field-asphaltenes (A1 and A2) in toluene. It is possible that C7-asphaltenes are lighter (with higher concentrations of relatively lighter compounds) compared to the field asphaltenes. This hypothesis was tested using thermal weight loss analysis. Figure 18.11 shows the thermal weight loss of A1, A2 and C7-asphaltenes in inert (nitrogen, up to 700 °C) and oxygen (above 700 °C) atmosphere. The weight loss curve for the field-asphaltenes is consistently above that of C7-asphaltenes, indicating an overall lighter constitution. The CAC measurements for C7-asphaltenes in toluene mixtures were observed at a concentration of 3.1 g/L.

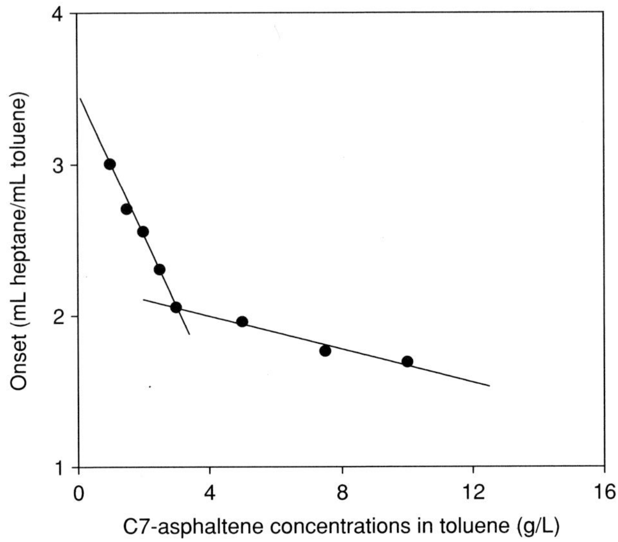

Figure 18.10. Onset data for C7-asphaltene in toluene plotted versus asphaltene concentrations. Onsets were measured using heptane flow rate of 1 mL/min.

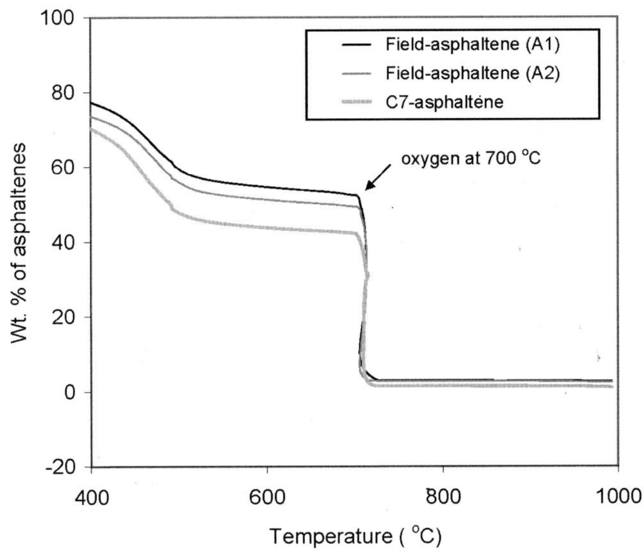

Figure 18.11. Thermal gravimetric plot of A1, A2, and C7-asphaltene in the temperature range of 400–1000°C.

4.4. Asphaltene Subfractions

The result from Figures 18.7A and 18.8A indicates that A2 asphaltenes are relatively more stable in toluene than A1. The onset break points were observed at 3.0 g/L for A1 and 4.8 g/L for A2. The average molecular weights of the two asphaltenes were measured using vapor pressure osmometer (Knauer, K-7000). The average molecular weights were 1149 for A1 and 930 for A2.

To obtain asphaltene subfractions, dichloromethane (DCM) was used as the solvent and pentane as the precipitant. The volumetric ratio of DCM to pentane was 1:2.5. A fixed amount of A2 (10 g) was dissolved in DCM (200 mL). The solution was mixed (with a magnetic stirring device) for about 12 hr, following which 500 mL of pentane was added to the mixture. The resultant system was mixed for six more hours. Then, the mixture was filtered under vacuum with a 11-μm filter paper. The filtered asphaltenes were termed A2-F11. The remaining mixture was dried under vacuum conditions evaporating the solvent. The fraction left behind was named A2-F12. The weight of A2 was more or less equally partitioned between A2-F11 and A2-F12. When a different volumetric ratio (1:2) of DCM to pentane was used for A2 fractionation, about 42 wt% of the asphaltenes was precipitated in the first fraction (A2-F21). Hydrogen-to-carbon ratios of A1 and A2 are more or less identical at 1.02 and 1.01, respectively. Heavier fractions, A2-F11 and A2-F21, have higher contents of nitrogen and sulfur. Heteroatoms in the most polar fraction play a major role in asphaltenes dissolution.[36,37] A2-F11 was more aromatic than A2-F12 based on the fractionation sequence, and this was confirmed by the elemental analysis with H/C ratios of 0.88 and 1.15, respectively, for the two fractions. It was also expected that A2-F11 was heavier than A2-F12, which was confirmed by VPO molecular weight measurements.

Onsets of the four asphaltenes/asphaltenes fractions (A2, A2-F11, A2-F12, and A2-F21) are plotted in Figure 18.12. The onset break points are observed at 3.1, 3.6, 4.8, and 5.9 g/L for A2-F21, A2-F11, A2, and A2-F12, respectively. The CAC values of asphaltenes vary with the solvent used for aggregation and origin of asphaltenes. A2-F11, the more aromatic and heavier fraction had the lowest onset break point, and A2-F12 the highest. The onset break point for A2 was between these two values.

Attention should also be focused on the actual onset values. The onset values for A2-F11, for example, range from about 0.75 mL/mL to about 0.5 mL/mL, signifying its lower solubility in comparison to the parent fraction. The onset values for A2-F12 are uniformly higher, as expected. The onset break point of A2-F21 occurred at 3.1 g/L. Thus, the CAC values of A1 and A2-F21 are almost identical at 3 g/L. It should be noted that these two solids have very different characteristics (Table 18.2). Andersen and Christensen[38] reported CMC values (interpreted as CAC in this study) with asphaltene subfractions. Lower molecular weight subfractions obtained by the solubility based separations had higher CMC values and higher solubility, and higher molecular weight subfraction had lower CMC values and lower solubility. Hence, asphaltenes from the different sources or origin could have similar CAC values, which support the idea that the CAC values

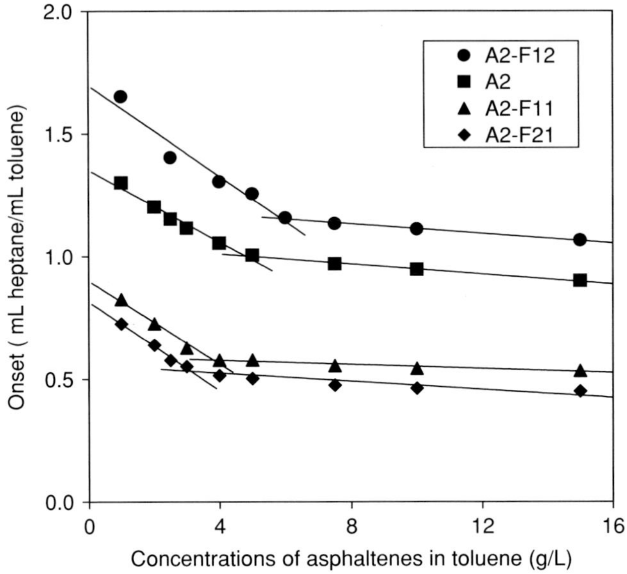

Figure 18.12. Onset data of A2-F12, A2, A2-F11, and A2-F21 with respect to their concentrations in toluene.

may not be the only factor to evaluate asphaltene solubility. The onset values for A1 range from about 1.25 mL/mL to about 0.75 mL/mL, while the onset values for A2-F21 are from about 0.8 mL/mL to about 0.5 mL/mL. This reinforces the idea that precipitant to oil (or solvent in case of asphaltene–solvent mixtures) ratios are characteristic of the type of oil or asphaltenes.[35] A1 is much more soluble in toluene in comparison to A2-F21. A1 has a wider molecular distribution in comparison to A2-F21, which is heavier with a narrow molecular weight distribution.

Table 18.2. Elemental Analysis and Molecular Weights of Asphaltenes

	C (%)	H (%)	N (%)	S (%)	H/C	VPO-MW[a]
A1	83.34	7.11	0.83	3.53	1.02	1149
A2	87.13	7.33	0.79	2.81	1.01	930
A2-F11	87.33	6.39	0.91	3.50	0.88	3601
A2-F12	82.90	7.92	0.62	2.27	1.15	523
A2-F21	87.40	6.33	0.94	3.64	0.87	3756
C7-asphaltene	86.45	8.25	0.71	3.18	1.15	975

[a] VPO-MW: number average molecular weight as measured by VPO .

5. Conclusions

Asphaltenes have been known to aggregate. Different aggregation regimes have been the subjects of study over a number of years. This study deals with aggregation of asphaltenes dissolved in predominantly polar solvents in concentrations of 1–10 g/L. Aggregation onset points of asphaltenes were determined using NIR spectroscopy. There were break points in the data when the onsets were plotted with respect to the asphaltene concentrations. This transition was defined as the critical aggregation concentration, indicating that the transition represented formation of larger clusters from smaller aggregates and not the molecular-level phenomenon hypothesized at lower concentrations. CAC determination based on the onset of asphaltene aggregation provides data on asphaltene solubility as well. Asphaltene solubility increased when the solvent solubility parameters increased. These CAC values were more or less linear with respect to the solvent solubility parameters. Earlier studies have shown that polarizability is important in keeping asphaltenes in solution. This study has shown that when asphaltenes are in mixtures' polar solvents, solvent polarity is also important in establishing asphaltene solubility and asphaltene aggregation characteristics, when aggregation is induced by adding a precipitant. CAC was also observed for heptane insoluble asphaltenes in toluene. Compared to the CMC values measured by surface tension measurements, the CAC values observed by onset measurement were at slightly lower values.

Aggregation studies were also conducted with asphaltene subfractions. Lower molecular weight (VPO data) fractions from the parent asphaltenes had higher solubility and higher CAC. Heavier molecular weight (measured by VPO) fractions had lower solubility and lower CAC values. Examination of the onset data for carefully fractionated asphaltenes established that two asphaltenes with different characteristics could have the same CAC values. For instance, a heavier fraction with a narrow molecular weight distribution had the same CAC value as a different parent fraction, which was lighter with a broader molecular weight distribution. This work shows that CAC data need careful evaluation since they are not the only factor that determines asphaltenes precipitation and aggregation.

Acknowledgments

Financial support from the U.S. Department of Energy for the contract DE-AC26-98BC15107 is gratefully acknowledged. The authors would like to acknowledge the Chevron-Texaco Rangely Office for providing oil and asphaltene samples. The authors also appreciate to the support of the Utah State Government through funding for the Petroleum Research Center (PERC, www.perc.utah.edu) in the University of Utah.

References

[1] Sheu, E.Y. and D.A. Storm (1995). In : E.Y. Sheu and O.C. Mullins (ed.), *Asphaltenes: Fundamentals and Applications*. Plenum Press, New York, pp. 1–52.

[2] Speight, J.G. (1999). *The Chemistry and Technology of Petroleum*, 2nd (edn.). Marcel Dekker, New York.

[3] Dickie, J.P. and T.F. Yen (1967). *Anal. Chem.* **39**, 1847.

[4] Ancheyta, J., G. Centeno, F. Frejo, G. Marroquín, J.A. García, E. Tenorio, and A. Torres (2002). *Energy Fuels* **16**, 1121.
[5] Sheu, E.Y., D.A. Storm, and M.M. De Tar (1991). *J. Non-Crys. Solids* **131–133**, 347; Sheu, E.Y., M.M. De Tar, D.A. Storm, and S.J. DeCanio (1992). *Fuel* **71**, 299.
[6] Evdokimov, I.N., N. Yu. Eliseev, and B.R. Akhmetov (2003). *J. Pet. Sci. Eng.* **37**, 135.
[7] Evdokimov, I.N., N.Y. Eliseev, and B.R. Akhmetov (2003). *Fuel* **82**, 817.
[8] Acevedo, S., M.A. Ranaudo, J.C. Pereira, J. Castillo, A. Fernàndez, P. Pérez, and M. Caetano (1999). *Fuel* **78**, 997.
[9] Andersen, S.I. and K.S. Birdi (1991). *J. Colloid Interface Sci.* **142**, 497.
[10] Groenzin, H. and O.C. Mullins (1999). *J. Phys. Chem.* A **103**, 11237; Groenzin, H. and O.C. Mullins (2000). *Energy Fuels* **14**, 677.
[11] Mullins, O.C. Personal Communication.
[12] Rogacheva, O.V., D.K. Rimaev, V.Z. Gubaidullin, and D.K. Khakimov (1980). *Colloid J. USSR* **490**.
[13] Mohamed, R.S., A.C.S. Ramos and W. Loh (1999). *Energy Fuels* **13**, 323.
[14] Rogel, E., O. León, G. Torres, and J. Espidel (2000). *Fuel* **79**, 1389.
[15] Yarranton, H.W., H. Alboudwarej, and R. Jakher (2000). *Ind. Eng. Chem. Res.* **39**, 2916.
[16] Speight, J.G., D.L. Wernick, K.A. Gould, R.E. Overfield, B.M.L. Rao, and D.W. Savage (1985). *Rev De L'Inst Français Du Pétr.* **40**, 51.
[17] Nellensteyn, F.J. (1924). *J. Inst. Pet. Tech.* **10**, 311.
[18] Pfeiffer, J.P. and R.N.J. Saal (1940). *J. Phys. Chem.* **44**, 139.
[19] Andersen, S.I. and S.D. Christensen (2000). *Energy Fuels* **14**, 38.
[20] Priyanto, S., G.A. Mansoori, and A. Suwono (2001). *Chem. Eng. Sci.* **56**, 6933.
[21] Pietraru, G.-M. and D.T. Cramb (2003). *Langmuir* **19**(4), 1026.
[22] Yudin, I.K., G.L. Nikolaenko, E.E. Gorodetskii, E.L. Markhashov, V.A. Agayan, M.A. Anisimov, and J.V. Sengers (1998). *Physica* A **251**, 235.
[23] Satndards for Petroleum and Its Products. Standard no. IP143/57. Institute of Petroleum, London.
[24] Speight, J.G., R.B. Long, and T.D. Trowbridge (1984). *Fuel* **63**, 616.
[25] Evans, D.F. and H. Wennerström (1994). *The Coloidal Domain—Where Physics, Chemistry, Biology, and Technology Meet*. VCH Weinheim.
[26] Miles, G.D. and L. Shedlovsky (1944). *J. Phys. Chem.* **48**, 57.
[27] Siesler, H.W., Y. Ozaki, S. Kawano, and H.M. Heise (2002). *Near-Infrared Spectroscopy: Principles, Instruments, Applications*. Wiley-VCH, Weinheim.
[28] Joshi, N.B., O.C. Mullins, A. Jamaluddin, J. Creek, and J. McFadden (2001). *Energy Fuels* **15**, 979.
[29] Kerker, M. (1969). *The Scattering of Light and Other Electromagnetic Radiation*. Academic Press, New York.
[30] Leontaritis, K.J. (1997). *World Oil* Nov., 101.
[31] Fuhr, B.J., C. Cathrea, L. Coates, H. Kalra, and A.I. Majeed (1991). *Fuel* **70**, 1293.
[32] Oh, K. and M.D. Deo (2002). *Energy Fuels* **16**, 694.
[33] Cimino, R., S. Correra, and P.A. Sacomani (1995). *SPE International Symposium on Oilfield Chemistry*, Society of Petroleum Engineers, San Antonio, TX, USA, Feb. 1995, Paper SPE28993.
[34] Reichert, C., B.J. Fuhr, and L.L. Klein (1986). *J. Can. Pet. Tech.* Sept.–Oct. 33.
[35] Buckely, J.S., G.J. Hirasaki, Y. Liu, S. Von Drasek, J-X. Wang, and B.S. Gill (1998). *Pet. Sci. Technol.* **16** (3&4), 251.
[36] Nalwaya, V., V. Tangtayakom, P. Piumsomboon, and S. Fogler (1999). *Ind. Eng. Chem. Res.* **38**, 964.
[37] Kaminski, T.J., H.S. Fogler, N. Wolf, P. Wattana, and A. Mairal (2000). *Energy Fuels* **14**, 25.
[38] Andersen, S.I. and S.D. Christensen (2000). *Energy Fuels* **14**, 38.

19

Phase Behavior of Heavy Oils

John M. Shaw and Xiangyang Zou

1. Introduction

The phase behavior and physical properties of heavy oil are complex and present significant experimental and modeling challenges. Only through coordinated use of diverse experimental techniques from X-ray transmission videography to small angle X-ray scattering (SAXS) to calorimetry can the behaviors of these materials be elucidated at the diverse length scales required for their characterization and processing. Available data sets tend to be fragmentary and interpretation can be ambiguous. However, heavy crudes and crude fractions, such as Maya crude and Athabasca vacuum bottoms, appear to comprise a minimum of two phases over a broad range of conditions, and when combined with light hydrocarbons as many as four fluid phases are observed. The behavior of asphaltenes within these phases remains a subject of ongoing inquiry as does hydrocarbon speciation more generally, and as a consequence, phase behavior and physical property models remain correlative in nature.

Large heavy oil deposits, comprising dense viscous crudes with high asphaltene and heteroatom contents and found principally in Canada and Venezuela, with smaller deposits in Mexico and at sites in Asia and elsewhere, are becoming an increasingly important resource worldwide. Recoverable deposits in Canada alone, where both production and refining rates are increasing sharply with time, exceed 300 billion barrels. The composition, phase behavior, and physical properties of these mixtures in situ, during production and refining, are key inputs for the adaptation and development of technologies for their production and refining. For example, while some reservoirs are amenable to pumping, production technologies predicated on the injection of sweep gases, solvents and steam (steam-assisted gravity drainage—SAGD, vapor extraction—VAPEX, etc.), as well as mining (near surface bitumen deposits) have found application. On the refining end, deasphalting technologies are being adapted. Flow assurance issues associated with offshore heavy oil production are particularly complex, and the stakes are high.

The phase behavior of heavy oil can be viewed as part of the continuum for hydrocarbon mixtures of industrial interest. Early work on physical properties and

---

John M. Shaw • Department of Chemical and Materials Engineering, University of Alberta, Edmonton, Canada. **X. Zou** • Oilphase-DBR, Schlumberger, Edmonton, Canada.

phase behavior of a number of heavy oils is summarized in a series of Alberta Oil-sands Technology and Research Authority (AOSTRA) sponsored publications.[1,2] Such mixtures comprise innumerable components with diverse physiochemical characteristics,[3,4] and frequently exhibit multiphase behavior. Reservoir fluids have been shown to exhibit multiphase equilibria in and of themselves[5] or on addition of other fluids.[6,7] In particular, the addition of carbon dioxide, nitrogen, and light hydrocarbon gases to reservoir fluids as envisaged in a number of production and carbon dioxide sequestration process schemes can induce or enhance complex phase behavior over the ranges of pressure and temperature encountered in hydrocarbon reservoirs. For example, Shelton and Yarborough[6] found that the addition of carbon dioxide or an equimolar mixture of methane, ethane, and propane to a condensate rich reservoir fluid yielded a four organic phase zone (solid–liquid–liquid–vapor) at intermediate pressures as well as large three phase (solid–liquid–liquid and solid–liquid–vapor) zones. Solids drop out, solid–liquid, or solid–vapor phase behavior, can also occur during normal reservoir depletion.[8] Multiphase behavior with solids has also been observed with heavy oils or bitumen + diluents,[9–11] and with model reservoir systems.[12] Further, asphaltenes, which comprise a significant fraction of heavy oils, are poorly defined. The nature and amount of asphaltenes separated from petroleum mixtures change with the type and amount of precipitant, temperature, pressure, and the pore size of the filters employed.[13] Asphaltenes have a tendency to self-associate to form aggregates. The mean molar mass, shape and structure at both the molecular and aggregate levels, and the reversibility of phase transitions involving asphaltenes[14] remain unresolved in the literature. These uncertainties, as well as co-precipitation with wax and mineral matter,[15] and experimental limitations, described below, convolute the phase behavior of heavy oils, and hamper phase behavior measurement, interpretation, and prediction. Multiple techniques are needed to define the phase behavior of heavy oils with any degree of certainty.

2. Origin of Multiphase Behavior in Hydrocarbon Mixtures

The maximum possible number of phases present at equilibrium for a mixture with a given number of components is governed by the phase rule. This rule, first derived in its simplest form by Gibbs over a century ago, has been augmented over time to allow for magnetic and other field effects that increase the number of degrees of freedom (where relevant) and critical points, where more than one phase possesses identical compositions and physical properties, that reduce the number of degrees of freedom. In petroleum science and engineering, constraints imposed by critical phenomena are important and the relevant phase rule has become:

$$F = N - \pi + 2 \qquad \text{for } k = 0 \tag{19.1a}$$

$$F = N - \pi + 2 - (k - 1) \qquad \text{for } k = 2, 3 \ldots \tag{19.2b}$$

where F is the number of degrees of freedom, N is the number of components, π is the number of phases, and k is the number of phases that are critically identical.

For a binary mixture, two noncritical phases may coexist over a range of compositions and pressures at fixed temperature, or over a range of temperatures and compositions at fixed pressure. By contrast, a binary mixture may only exhibit two-critical phases, such as liquid–liquid or liquid–vapor, along a curve in composition–pressure–temperature space as there is only one degree of freedom, that is, once the composition is fixed the pressure and temperature are fixed as well. This curve is called a critical locus when projected into the pressure–temperature plane. Finally, binary mixtures may exhibit up to four noncritical phases in equilibrium or up to three phases in equilibrium when two of the phases are critically identical. Simple mixtures exhibiting multiphase behavior include methane + hexane which exhibits liquid–liquid–vapor phase behavior[16] and phenol–water–n-hexane which exhibits five coexisting phases (one vapor phase + four liquid phases); the constituent binaries of phenol–water–n-hexane mixture exhibit a maximum of three phases in equilibrium.[17]

At temperatures below the melting temperature of one or both of the components in binary mixtures, solid phases may appear. Compound solids comprising fixed mole ratios of components can even be stable at temperatures above the melting points of the individual components. Solidification has a dramatic impact on the appearance of phase projections and diagrams.[18–21] Figures 19.1A and B illustrate the impact of solidification on the appearance of a type I binary projection[22] where only simple solids occur. In Figure 19.1A, solid–liquid–vapor curves ($S_A LV$, $S_B LV$) emerge from the triple points, open triangles, of pure components A and B, respectively, and converge at a four phase eutectic point, solid square, where $S_A S_B LV$ phase behavior arises at a specific composition, temperature, and pressure. $S_A S_B L$ and $S_A S_B V$ curves also arise within the projection. If the triple point of component A occurs at a temperature well above the critical temperature of component B, the $S_A LV$ curve intersects the critical locus, $L = V$, and the projection shown in Figure 19.1B results. So, solid–solid–liquid–vapor phase behavior is a common four-phase behavior even for binary mixtures. In addition to four-phase eutectics ($S_A S_B LV$), Q-points ($S_A L_1 L_2 V$), and other multiphase phenomena arise in binary mixtures.[12] It is also worth noting that LLV phase behavior, which arises in type II through type VI binaries, can be suppressed in whole or in part by solidification if the triple point temperature of component A is sufficiently high.[20]

Multicomponent mixtures possess additional degrees of freedom, and diverse multiphase behaviors can arise over broad ranges of temperature, pressure, and composition. Phase diagrams for such mixtures are inherently multidimensional. In practice, construction and interpretation of phase diagrams for multicomponent systems is similar to and based on those of binary systems.[23–26] Pressure–temperature diagrams at fixed composition and triangular phase diagrams or prisms are frequently prepared. However, due to the large numbers of degrees of freedom for such mixtures, sharp transitions in the number and/or nature of phases present may arise with small changes in pressure, temperature, or composition.[27] This is particularly true near the bubble pressure of such mixtures. For example, Vengu[28] shows a transition from L to SLLV phase behavior for a reservoir fluid + sweep gas mixture.

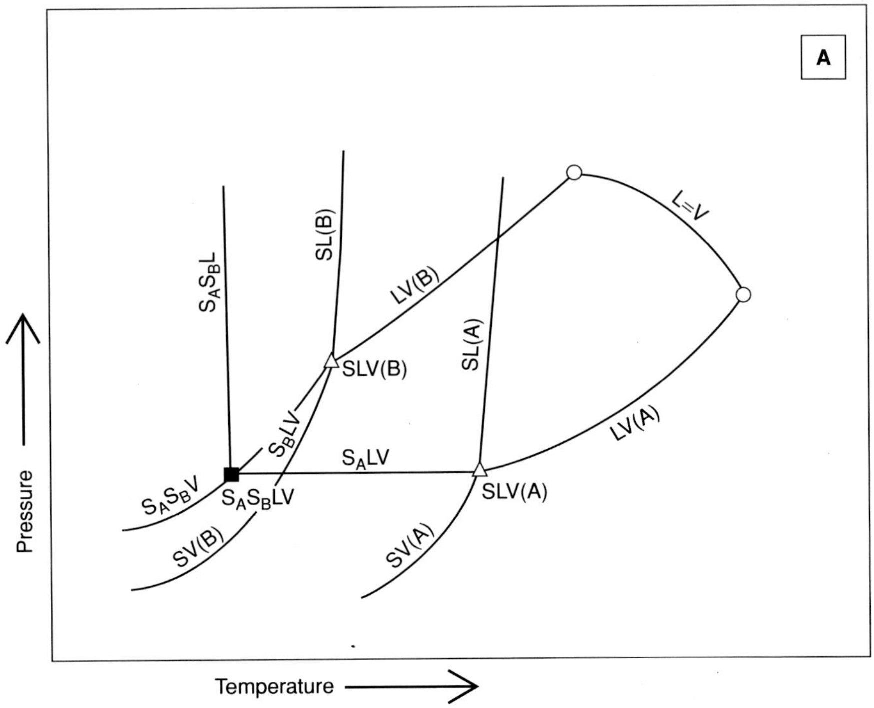

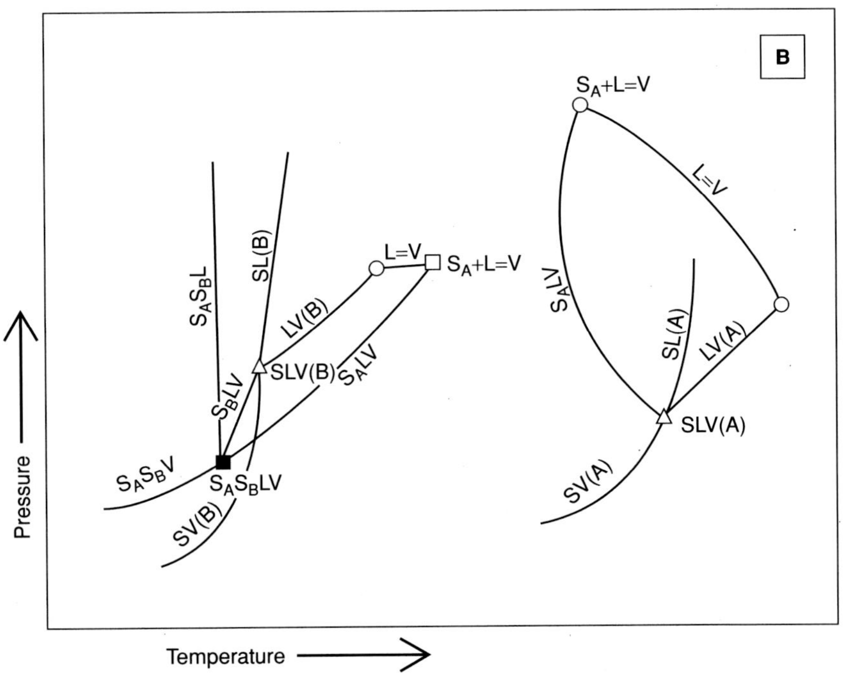

Figure 19.1. Type I phase projections for binary mixtures including solids: (A) the melting points of the two components are similar; (B) the melting temperature of component A is much greater than the melting temperature of component B. The liquid–vapor critical locus is labeled L = V in both images. The curves labeled (A) and (B) refer to phase behaviors of the pure components A and B respectively.

3. Phase Behavior Prediction

3.1. Bulk Phase Behavior Prediction for Hydrocarbon Mixtures

Phase equilibrium is established when the Gibbs free energy of a mixture (G_{mix}) as a whole is minimized and where the chemical potentials of individual components (μ_i) in each of the phases present are the same. At fixed temperature and pressure, Gibbs free energy for both gases and liquids of arbitrary composition can be computed from equations of state to create a Gibbs free energy surface in the composition variables. For a single-phase case, the chemical potentials of each component are related to the Gibbs free energy of the mixture and to the derivatives of Gibbs free energy with respect to composition. For a two-component single-phase mixture, the chemical potential of component 1 is

$$\mu_1 \equiv G_{\text{mix}} + \left.\frac{\partial G}{\partial x_1}\right|_{x1,\text{mix}} (1 - x_{1,\text{mix}}) \tag{19.1}$$

and the chemical potential of component 2 is

$$\mu_2 \equiv G_{\text{mix}} - \left.\frac{\partial G}{\partial x_1}\right|_{x1,\text{mux}} (x_{1,\text{mix}}) \tag{19.2}$$

If more than one phase is present, one replaces the composition of the mixture with the compositions for each phase into equations (19.1) and (19.2) and evaluates the chemical potentials of both components in each phase separately. As the Gibbs free energy of a multiphase mixture is a linear combination of the Gibbs free energy of the phases present, weighted by the mole or mass fraction of each phase present, Gibbs free energy for a mixture as a whole is minimized if each of the phases shares the same tangent line (binary case) and where the tangent line does not intersect the Gibbs surface. For a multicomponent mixture, this tangent line becomes a tangent hyperplane to the Gibbs surface with respect to the composition variables. Thus, a multicomponent mixture is single phase if, at the global composition of the mixture, the tangent hyperplane is tangent to the surface at only one point and does not intersect the Gibbs surface at other compositions. If this is not the case, then Gibbs free energy for the mixture as a whole can be reduced if the mixture splits into two or more phases with differing compositions/densities but sharing the same tangent hyperplane.

Phase behaviors for hydrocarbon mixtures are commonly predicted using cubic equations of state in conjunction with the Gibbs free energy and the tangent plane criterion described above.[29,30] Computational routines based on equations of state are now built into commercial process simulators. While phase behavior and phase composition computations are frequently reliable for well-defined mixtures at low pressures (vapor–liquid equilibrium) where interaction parameters and acentric factors are identified, one must be wary of the results obtained, in general, because errors can be substantial. This is particularly so at higher pressures where the number, nature and composition of phases present at fixed composition can be predicted incorrectly,[10,31] or where the wrong phase behavior type can be predicted for a mixture as a whole.[32] Some equations of state also predict phase behaviors that do not arise in nature. Custom regression of interaction parameters based on experimental data and careful attention to the numerical methods employed mitigate but do not eliminate these difficulties even for model mixtures because identification of minima in multivariable computations where differences among possible phase states is small remains a challenge. Baker *et al.*[31] show examples where chemical potentials are equal but where Gibbs free energy is not minimized, that is, where metastable or unstable LL or LV phase behaviors are computed instead of LLV phase behavior. Additional complexities arise when equation of state methods are applied to heavy oils where ambiguities in the critical constants and acentric factors for pseudocomponents employed increase the number of adjustable parameters in the phase behavior models significantly.

Pure solid–fluid phase behavior can be modeled in principle if the triple point of the pure solid and the enthalpy of fusion are known, the physical state of the solid does not change with temperature and pressure, and a fugacity model, with known coefficients, for solid constituents is available. These conditions are rarely met even for simple mixtures, and it is difficult to generalize multiphase behavior prediction results involving even well-defined solids.[12] The presence of polymorphs, solid–solid transitions, and compound solids provide additional modeling challenges. For example, ice, gas hydrates, and solid hydrocarbons all have multiple forms.

Due to the number of components involved, constituents of mixtures of industrial interest are typically lumped into pseudocomponents for computation. The reservoir fluid and heavy oil literatures diverge on how the lumping is performed.[6] For reservoir fluids, the typical scheme is to treat low molar mass species individually and intermediate hydrocarbons as pseudocomponents based on carbon number. For example, an ill-defined maltene fraction may comprise one or more pseudocomponents and asphaltenes one or two others. For bitumen and heavy oil mixtures, the lightest components (hydrogen, hydrogen sulfide, methane) are frequently ignored and one encounters pseudobinary and more typically ternary diagrams where the composition variables are asphaltenes, and one or two hydrocarbon fractions. Thus the impact of the variability of the phase behavior of the high molar mass constituents on the phase behavior of reservoir fluids is largely ignored in one literature and the impact of low molar mass constituents on the phase behavior of heavy feedstocks is largely ignored in the other literature. The phase behavior of hydrocarbon mixtures, as noted above, is too sensitive for such an approach to be successful from a general phase behavior modeling perspective. As a consequence, the phase behavior and phase diagrams for individual mixtures are obtained primarily from experiments.

3.2. Asphaltene Precipitation and Deposition Models

Asphaltene precipitation and deposition are both linked computationally to multiphase behavior. Commercially available correlative tools based on equations of state WinProp (CMG), PVTsim (Calsep), or DBR Solids (Schlumberger-Oilphase) can be used to interpolate data sets. De Boer plots[33] that correlate the likelihood of precipitation and deposition problems based on reservoir fluid density, reservoir pressure minus fluid bubble pressure, and the maximum saturation of the reservoir fluid at the bubble pressure are not definitive. There are no models available for predicting, a priori, either precipitation or deposit formation. Experimentally, deposition appears to arise within portions of some multiphase zones. Zones such as the L1L2V, and L1L2 are prone to the production of adherent asphaltene-rich deposits, whereas other zones appear less susceptible.[34] The physical state of the deposits formed appears variable (viscous liquids, solids, slurries).[35,36] Pressure, temperature, and composition boundaries for phase behaviors prone to precipitation provide conservative estimates for zones at risk for asphaltene deposition. Frequently, phase behaviors likely to lead to adherent deposit formation comprise small zones within phase diagrams that are both difficult to observe and compute.[12,37] Direct computational approaches for asphaltene precipitation based on the SAFT equation of state,[38] which regrettably is yet to be included in commercial simulators, and molecular thermodynamics[39] offer some promise. As precipitation does not always lead to the formation of adherent deposits, in the laboratory or in the field, further refinement of zones at risk to deposition will require inclusion of relevant hydrodynamic, kinetic, and surface phenomena. Work on these latter topics is ongoing.[40–42]

4. Experimental Methods and Limitations

Bulk phase behaviors of hydrocarbon mixtures are typically observed using variable-volume view cells employing visible and infrared light. Sophisticated phase behavior measurement apparatus and services based on these techniques are available from suppliers worldwide.[43] The amount of precipitated phases present is determined directly by filtration or indirectly by heat transfer, electrical conductivity, and viscosity measurements. Incipient asphaltene and wax precipitation can be discriminated and dispersed phase formation detected using light scattering,[44] microscopic visualization,[45] surface tension measurements,[46] liesegang ring tests,[47] and refractive index. Nonintrusive techniques on this list typically fail to identify all bulk phases present with asphaltene mass fractions exceeding 5 wt% due to the opacity of the mixtures.

Recently, a variable volume view-cell technique based on X-ray transmission video that does not suffer from this composition limitation has become available. This latter technique has a broad operating range (from room temperature to 720 K, and from subatmospheric pressure to 30 MPa) and provides rapid online measures of phase volumes and densities for fluids with high asphaltene contents. Dispersed solid phases are also detected. The apparatus comprises a polychromatic tungsten X-ray source that emits K, L, and M band radiation, an X-ray transparent beryllium view cell, an X-ray detector, and an image processing system. The method, analogous to medical radiography, is described in detail elsewhere,[48] and only key features are described here. The view cell comprises a 0.15 m long, 0.17 m outside diameter/ 0.05 m inside diameter beryllium tube that is bolted to stainless steel base plates. A stainless steel bellows is attached to the top plate so that the volume can be varied. A beryllium insert attached to the base plate facilitates internal

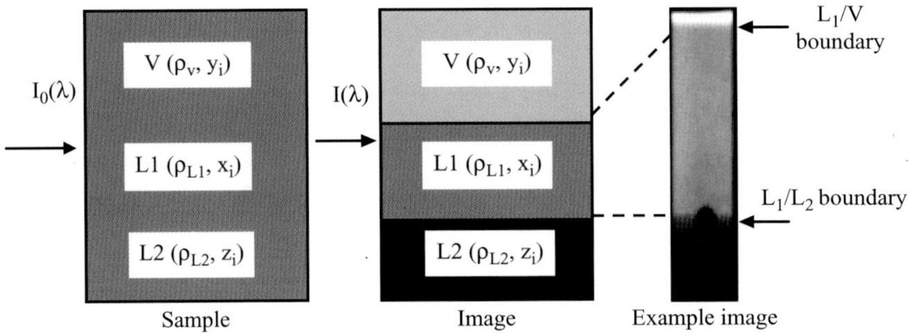

Figure 19.2. An illustration of the response of the X-ray view cell technique to a three-phase mixture.

calibrations for phase density and permits detection of the small volumes of dense liquid phases which frequently arise in heavy oil or heavy oil + diluent mixtures.[49] A magnetically driven mixer sits on top of the insert.

As with visible light, specific wavelengths of X-rays are absorbed differentially by various elements (carbon, hydrogen, sulfur, etc.). Absorbance is also proportional to mass density. So, if polychromatic X-rays are employed, and the relative intensities of specific wavelengths are varied, then both the density and elemental composition of a sample can be determined as long as the path length is fixed and solids, which scatter X-rays, are not present. If a fixed intensity is employed, and the composition range of the fluids is restricted, then phase densities can be calibrated against standards, with the same restrictions noted above. For the example shown in Figure 19.2, which is opaque to visible light, a vapor, V, and two liquids, L1 and L2, are readily discriminated using X-rays. The vapor phase absorbs little energy and appears light. The phase boundary between the vapor and low-density liquid (L1) is well defined, as is the boundary between the low- and high-density liquid (L2) phases. This latter boundary intersects the mixer (black circle). The insert, which restricts the view cell volume at the base of the cell, is also clearly visible in the video still. As the volume of the view cell is calibrated with respect to elevation and the response of the detector is calibrated with respect to phase density for hydrocarbon liquids, differences between the actual density of a mixture (mass inserted divided by the volume occupied) and the apparent density from X-ray transmission indicates the presence of dispersed solids[34] or highly ordered structures which scatter X-rays. An example is shown in Figure 19.3. The apparent density of saturated pentane liquid and the actual density agree to within experimental error (0.005 g/cm^3) at 50°C and 159 kPa at all elevations within the liquid phase, Figure 19.3A, whereas the actual and apparent densities of a 50 wt% Athabasca vacuum bottoms (ABVB, 525°C + cut comprising 32 wt% asphaltenes) + decane mixture at 130°C and 35 kPa differ substantially at all elevations, indicating that a liquid phase and a dispersed solid phase or ordered structures are present—Figure 19.3B. Without density calibration, the dispersed material would not have been detected. This latter mixture is opaque to visible light, and the dispersed material would normally be detected only using destructive techniques such as filtration.

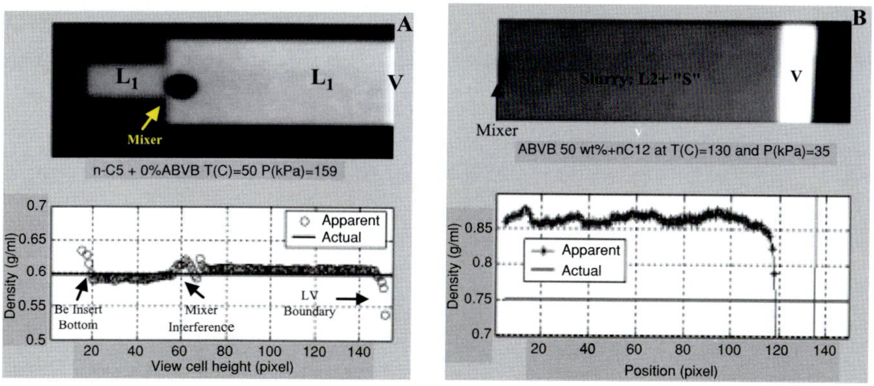

Figure 19.3. Apparent vs. actual density values for condensed phase cases: (A) pure pentane at 50°C and 159 KPa (a liquid), and (B) 50 wt% ABVB + dodecane 130°C and 35 kPa (solid + liquid).

Calorimetric and high-resolution microscopy measurements can also play a key role in determining the phase behavior of heavy oils, particularly at low temperatures where parts of the mixtures may undergo glass transitions or where thermal cycling reveals polymorphic behavior.[50–52]

The physical state and the state of aggregation of asphaltenes present in heavy oils can be obtained directly by neutron and X-ray scattering[46] or indirectly using viscometry,[53] but scattering experiments are costly and a number of restrictions apply.[46,54] The asphaltene concentration and aggregate size range that can be explored with any given technique are limited. Data related to asphaltene structure are available for near atmospheric pressure in dilute systems. However, there is a growing recognition that asphaltenes coexist in a variety of forms in hydrocarbon mixtures, so one cannot extrapolate current structural results, obtained under ambient conditions, to relevant reservoir or refining conditions with any degree of certainty. Currently there are no techniques available that provide simultaneous bulk phase behavior measurements in conjunction with in situ measures of asphaltene state or state of aggregation, by phase, for asphaltene rich mixtures. While there is interest in creating such an apparatus, useful results are likely to emerge first from coordinated synchrotron and conventional or X-ray view-cell experiments where the structures of asphaltenes in specific phases are assessed individually.

5. Phase Behavior Observations and Issues

5.1. Heavy Oil

Key physical property and thermodynamic data such as density, vapor pressure and viscosity, and the solubility of light gases are presented elsewhere in detail.[1,2,55–57] The focus here is the phase behavior of heavy oils themselves over a broad range of length scales and temperatures. As these mixtures are opaque to visible and infrared light, one must rely on other techniques for assessing the number and the nature of phases present. Heavy oils, such as Maya crude, and heavy oil fractions, such as Athabasca vacuum bottoms (ABVB), which contain 15 wt% and 32 wt% pentane asphaltenes, respectively, scatter X-rays consistently and reversibly over a broad range of conditions. Small angle X-ray scattering (SAXS) measurements are reported in Figures 19.4 for these two fluids. For the sets of histograms shown in Figures 19.4A and B, the samples were heated from room temperature to 310°C and then cooled. Changes in the histograms and in the mass of scatterers present, Figure 19.4C, with temperature are small and reversible. The leading dimensions of the scatterers, determined by fitting uniform histograms of spheres to the SAXS emission data, indicate radii which span the detection range for SAXS measurements (from 0.5 to 50 nm). The largest dimension for each category is reported as a parameter in Figure 19.4. From the shape of the size distribution of the scatterers present in ABVB, structures larger than 50 nm are clearly indicated. Wide angle X-ray scattering (WAXS) or simultaneous SAXS and WAXS measurements are clearly needed to capture the full range of structures present in heavy oils.

The question as to whether these scatterers constitute a separate phase is an open one in the literature. Interpretation of differences between apparent and actual density measurements for ABVB + alkane mixtures suggests that a portion of the scatterers comprise a separate solid phase. Calorimetric data showing evidence of polymorphism,[50,51] and the presence of endotherms and exotherms in the apparent heat capacity of ABVB and Maya crude,[52] Figure 19.5, are also consistent with the presence of two or more phases over broad ranges of temperature. In the absence of phase change, heat capacities of both amorphous and crystalline organic solids are simple, single-valued, and well-defined functions of temperature. Their heat capacity values can be computed on the basis of vibration spectra.[58,59] *n*-Alkanes are the most flexible organic molecules and possess the greatest heat capacity values; polynuclear aromatic molecules are the least flexible and possess the lowest values in both the solid and liquid state. Asymptotic values for the heat capacity of long *n*-alkane chains and polynuclear aromatic sheets provide extrema for pure solid and liquid hydrocarbons and are included in Figure 19.5 for comparison purposes.[52] Apparent heat capacities for both Maya crude and ABVB fall outside these bounds and are not single-valued functions of temperature over broad ranges of temperature. They are clearly subject to distributed transitions beginning at ~200 K and extending to well above room temperature. The nature of the transition(s), possibly a plastic transition or solid to liquid transition or both for a fraction of these mixtures, is not yet clear. The interpretation of the calorimetric data is further complicated because asphaltenes release energy when dissolved in "good solvents".[60] Numerous interpretations are possible. However, a minimum of two condensed phases appears to be present over a broad range of conditions. Further, in the case of ABVB, the transitions observed calorimetrically appear to be unrelated to scatterers present as both their mass and size distribution are largely invariant over the temperature range where the transitions occur. By contrast, the viscosity of ABVB decreases several orders of magnitude over the same temperature interval.

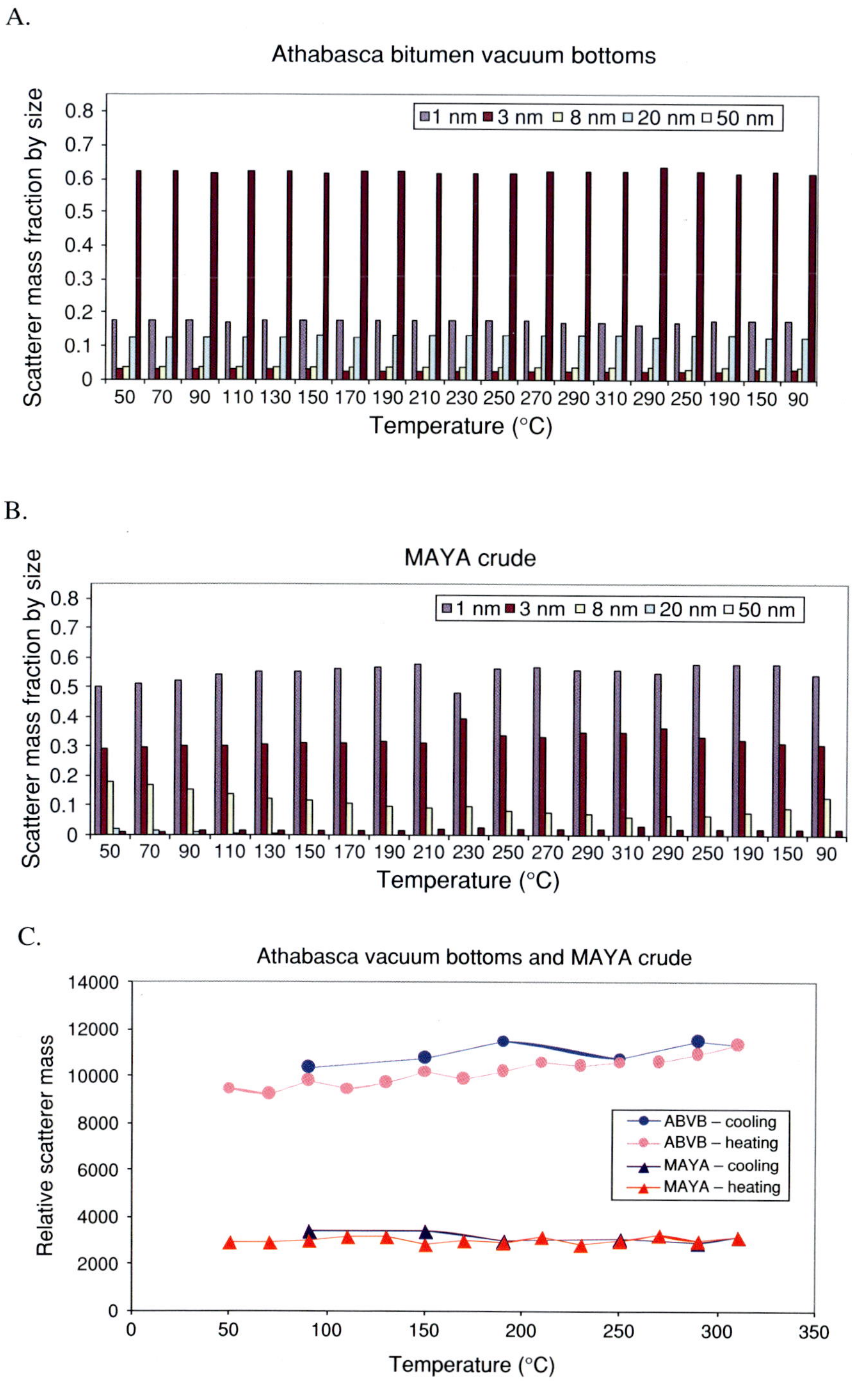

Figure 19.4. Leading dimensions for asphaltene aggregates in the size range from 0 to 50 nm on heating and subsequent cooling for: (A) Athabasca vacuum bottoms, (B) Maya crude, and (C) relative mass of scatterers present.[58]

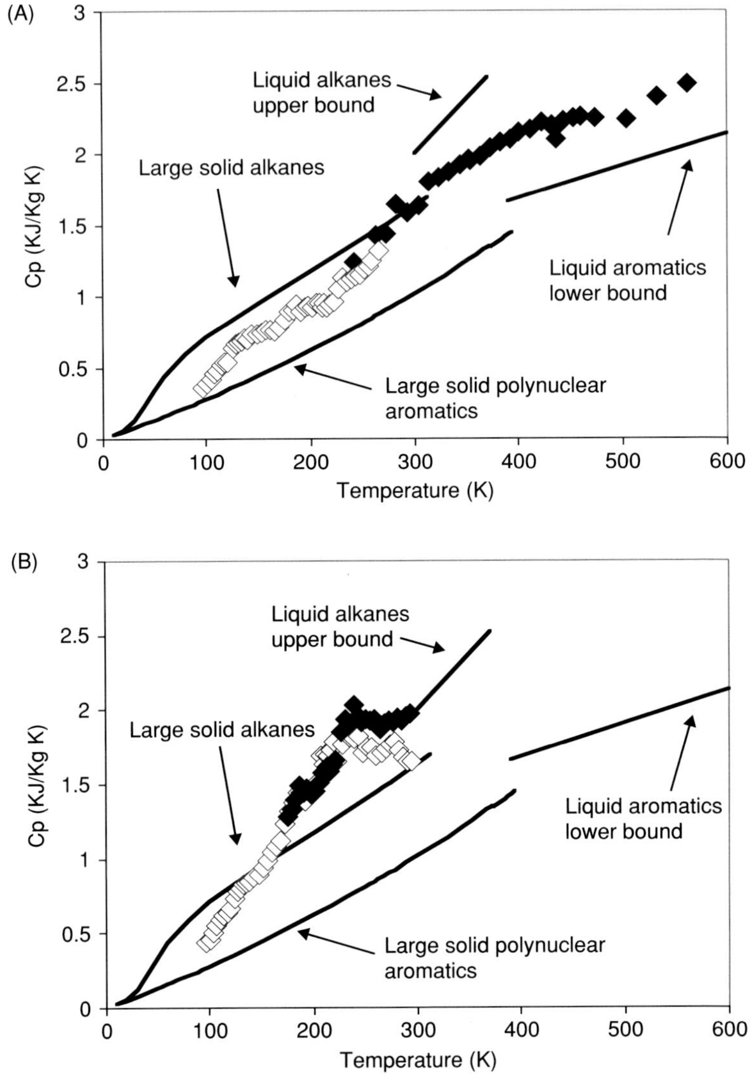

Figure 19.5. Apparent heat capacity of (A) Athabasca vacuum bottoms and (B) Maya crude. Open symbols are for samples cooled from room temperature to 100 K prior to apparent heat capacity measurements. Closed symbols are for samples heated from room temperature to over 473 K prior to cooling to 100 K and subsequent apparent heat capacity measurement.

5.2. Heavy Oil + Solvent Mixtures

Given the range of constituents present, heavy oil + solvent mixtures are expected to exhibit multiphase behavior. High-pressure viscosity data for Athabasca and Peace river bitumen + ethane mixtures are consistent with LL phase behavior.[2] Further, phase diagrams for Athabasca vacuum bottoms (ABVB) plus pentane have been prepared.[11] A representative isotherm at 160°C, illustrated in Figure 19.6A as a series of video stills taken at the bubble pressure of the respective mixtures from pure pentane at the left to pure "ABVB" at the right, and shown schematically in Figure 19.6B as a pressure composition diagram, illustrate the complexity of the

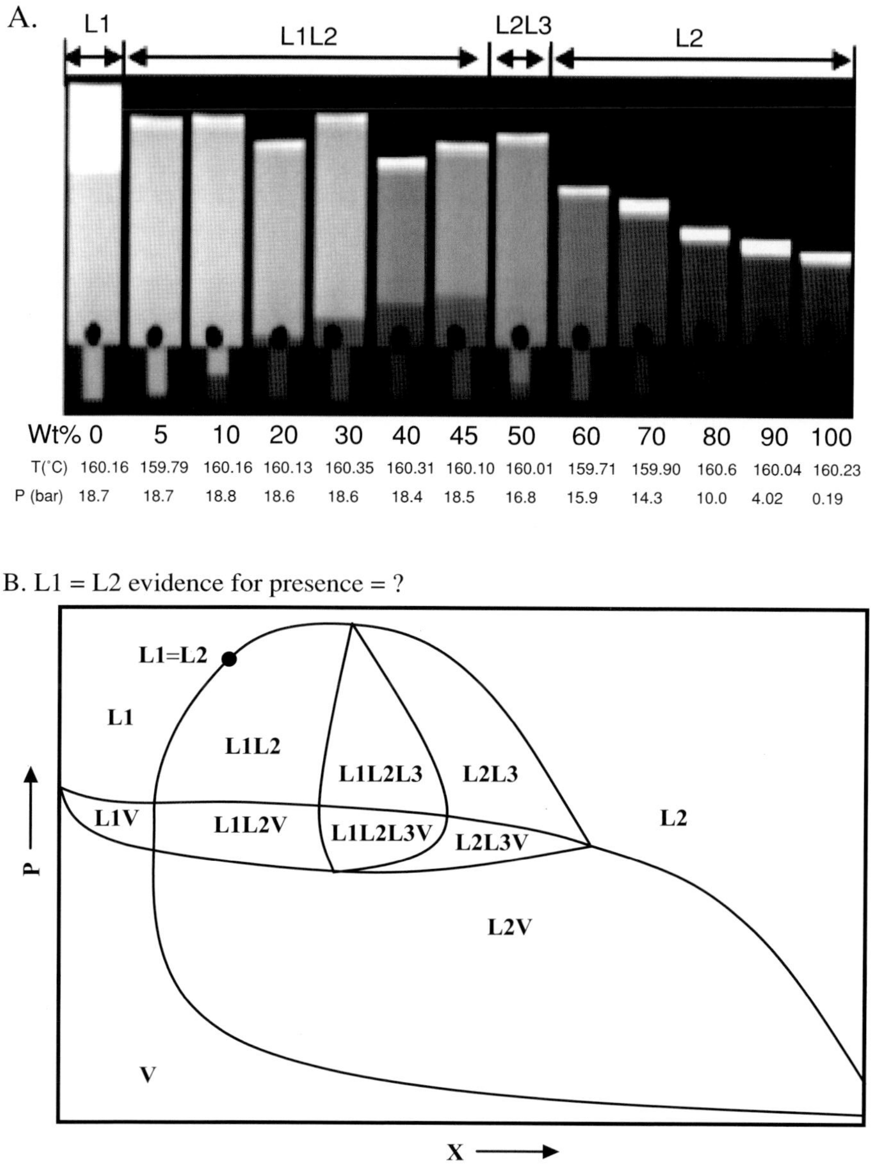

Figure 19.6. Example phase behavior data for the mixture pentane + ABVB. (A) Video stills at 160°C , just below the bubble pressure of the mixtures. (B) Pressure–composition sketch for ABVB + pentane at 160°C. Only fluid phases are shown. (C) L1L2L3V phase behavior at 120°C for 45 wt% ABVB + pentane.

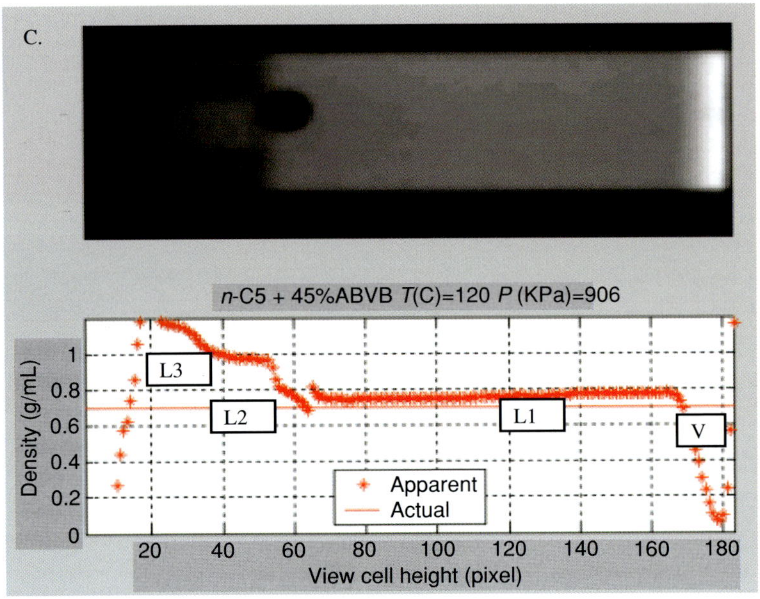

Figure 19.6. *(Continued)*

phase behavior of such mixtures. Figure 19.6A shows that at the bubble pressures of the individual mixtures, liquid–liquid–vapor phase behavior is exhibited from less than 5 wt% ABVB to more than 50 wt% ABVB. Observations at other pressures but at the same temperature and at other temperatures and pressures are not shown. The phase behavior assignments, L1, L2, and L3, are not arbitrary. They are dictated by these observations. For example, pentane + 45 wt% ABVB exhibits L1L2L3V phase behavior over a narrow range of pressure as illustrated in Figure 19.6C. Thus three liquid phases are anticipated and must be reflected in the phase diagram. By tracking the volumes of the phases as a function of pressure, from 5 wt% ABVB to 40 wt% ABVB at 160°C, we observed that the low-density liquid phase grew at the expense of the vapor phase as pressure was increased and vice versa yielding L1L2 phase behavior at high pressure and L2V phase behavior at low pressure. The volume of the L2 phase changed little with pressure. At 50 wt% ABVB, the constituents in L3 became miscible with those in L2 as the pressure was reduced, that is, as the mass fraction of pentane in L2 decreased. At low-pressure L2V phase behavior results and at high pressure L2L3 phase behavior results. As it is not clear whether the X-ray scatterers comprise a separate phase, solids are not included in the phase diagram shown in Figure 19.6B. From the X-ray transmission measurements, the density of L2 at 60 wt% ABVB is much less than the density of the L3 phase at 50 wt% ABVB. So, bulk phase transitions L1V $\rightarrow$ L1L2V $\rightarrow \cdots \rightarrow$ L2L3V $\rightarrow$ L2V are observed with increasing ABVB mass fraction at 160°C. The transition to and then from L1L2L3V phase behavior as the ABVB mass fraction is increased is not observed but must be present and is observed at other temperatures.

The small mass and volume fractions of phases present pose challenges for their detection and phase nature determination and were the motivation for placing the insert at the base of the view cell. Even so, not all phase behaviors and phase behavior transitions are observed. Incomplete phase behavior data sets are the norm for such complex mixtures. For example, the video stills for 60 wt% ABVB and 70 wt% ABVB, Figure 19.6A, are similar visually whereas X-ray scattering obtained by the difference between the apparent and actual density of the two fluids, Figure 19.7A, suggests a significant difference. The scattered energy at 60 wt% ABVB in pentane (18 wt% asphaltenes) is an order of magnitude smaller than the energy scattered at 70 wt% ABVB in pentane (22 wt% asphaltenes), reflecting a large increase in the number and/or size of the scatterers present over a narrow composition range. A similar step change over the same composition range

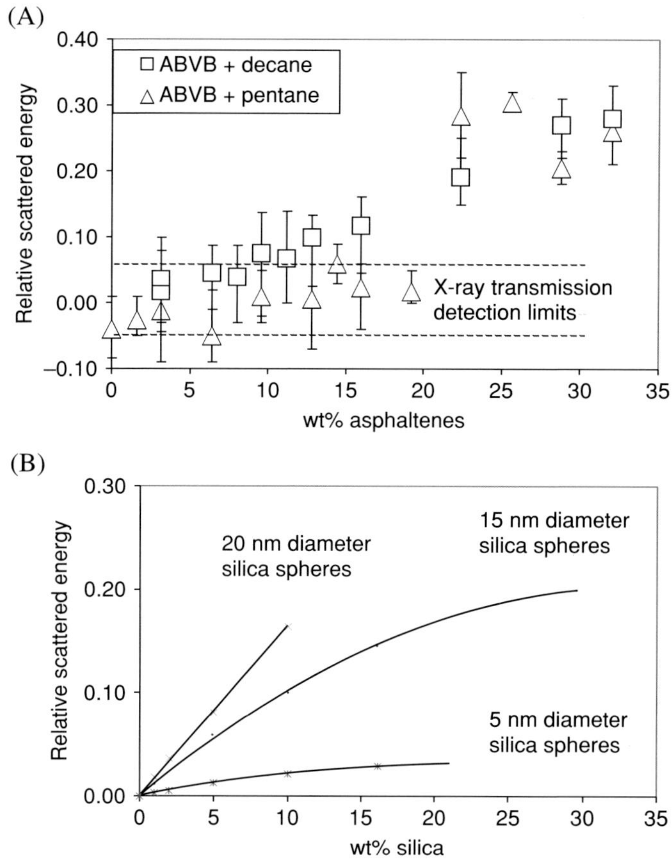

Figure 19.7. (A) Relative scattered energy for Athabasca vacuum bottoms (32 wt% asphaltenes) + alkane mixtures for the temperature range from 300 to 580 K based on X-ray transmission data—a minimum of 33 measurements per point with the average and range indicated. The detection limit for this difference measurement is also indicated. (B) Total SAXS emission, a direct measure of scattered energy for 5, 15, and 20 nm diameter monodispersed silica spheres in water.

is exhibited by ABVB + decane mixtures.[61] The step change is consistent with the appearance of a phase boundary, for example, such a step change was used to evaluate an LV/SLV boundary in pentane + anthracene mixtures,[34] but is equally consistent with significant growth of existing aggregates over a narrow composition range. This ambiguity is readily illustrated with a control experiment. Total scattered energy from SAXS data by 5, 15, and 20 nm diameter monodispersed silica particles in water at room temperature, Figure 19.7B, illustrate the impact of particle size and concentration on the relative amount of energy scattered for over the same particle size and concentration range.[62] For example, at 10 wt% silica in water, 20 nm diameter particles scatter ten times as much energy as 5 nm diameter particles.

5.3. Phase Behavior Reversibility

Phase behavior reversibility is a key topic because it is an explicit assumption in the development of predictive phase behavior models. The selection of remedial schemes for resolving asphaltene deposition problems, for example, depends on whether or not phase behavior is reversible and on the time scale of phase transitions. The bulk phase behavior of heavy oils + solvents is sensitive to small variations in pressure, temperature and global composition as exemplified in Figure 19.6. The state of aggregation of asphaltenes also varies with phase composition (Figure 19.4A, ABVB, vs. Figure 19.8A, 2 wt% ABVB in decane, and Figure 19.8B, 2 wt% ABVB in 1-methyl naphthalene). The total mass of scatterers in these two latter cases is comparable, Figure 19.8C, but the distributions differ markedly. In decane, the distribution is dominated by small scatters while in 1-methyl naphthalene, the distribution is dominated by larger scatterers. Even though scatterers with radii less than 0.5 nm or larger than 50 nm are not detected using SAXS, one expects asphaltenes in some phases to pass through a filter more readily than those in others due to differences in mean aggregate size. Consequently, variation in the relative amounts of phases present can yield apparent irreversibility with respect to asphaltene phase behavior—if asphaltene mass fraction is based on filtration. Such variations arise within view cells as well as in sampling systems where pressure control is difficult. Variations in separation conditions can explain, in part, the apparent irreversibility or partial reversibility of asphaltene behavior.[63] Variation in phase separation kinetics with operating conditions has also been advanced speculatively as a partial explanation. Nonintrusive methods suggest that asphaltene flocculation pressures at fixed temperature can be identified reversibly.[65] It is not clear whether these thresholds for "change" reflect transitions in phase behavior per se or merely aggregate growth to the point where they become observable with a particular experimental technique. The SAXS measurements reported in Figure 19.8B and C for the 2 wt% ABVB + 1-methyl naphthalene mixture suggest that the disaggregation occurring at high temperature is reversible on cooling but that the kinetics are slow. It is only once the sample is recooled to near room temperature that the total scatterer mass and the scatterer size distribution both return to the low-temperature values.

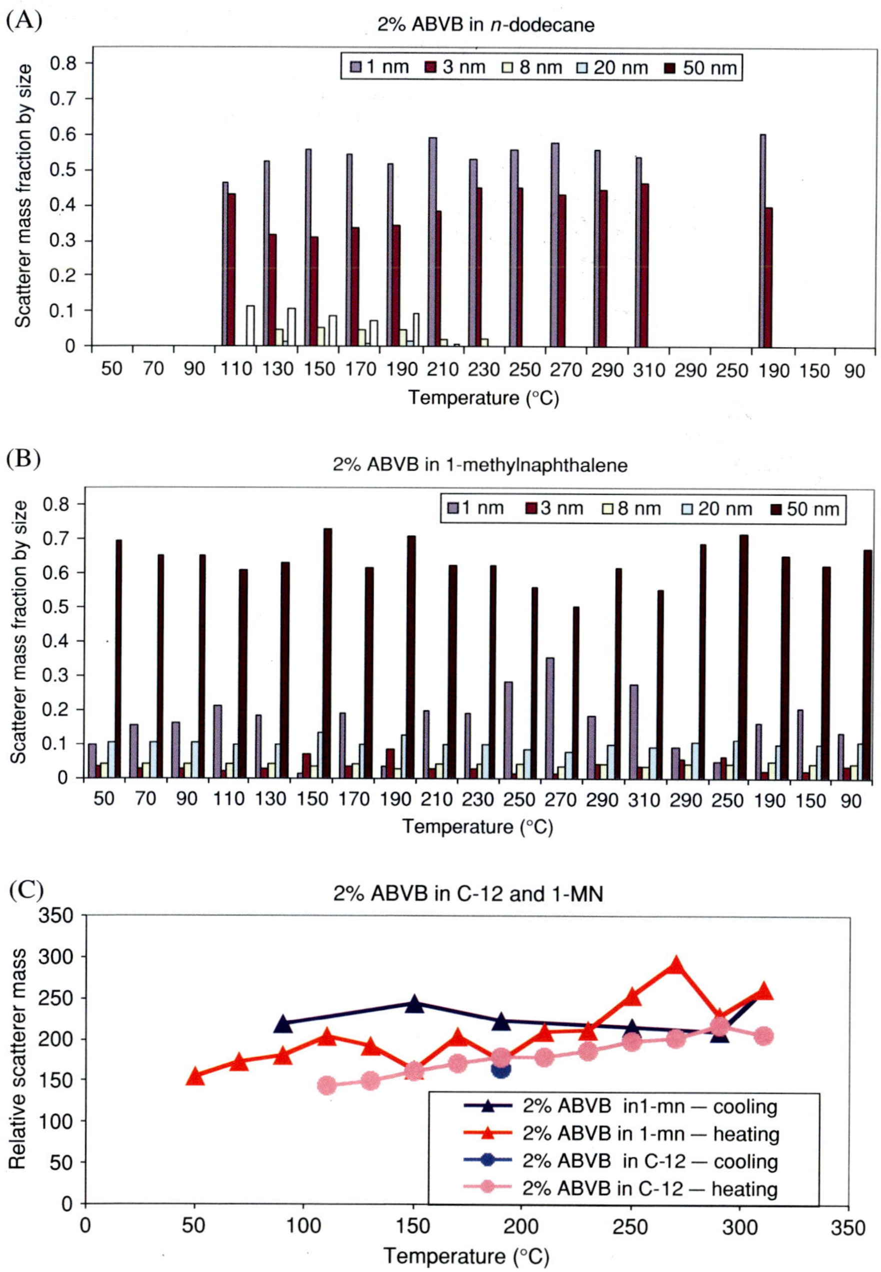

Figure 19.8. Leading dimensions for asphaltenes aggregates in the size range from 0 to 50 nm on heating and then cooling for: (A) 2 wt% ABVB in dodecane, (B) 2 wt% ABVB in 1-methyl naphthalene, and (C) relative scatterer mass for the cases cited in (A) and (B).[58]

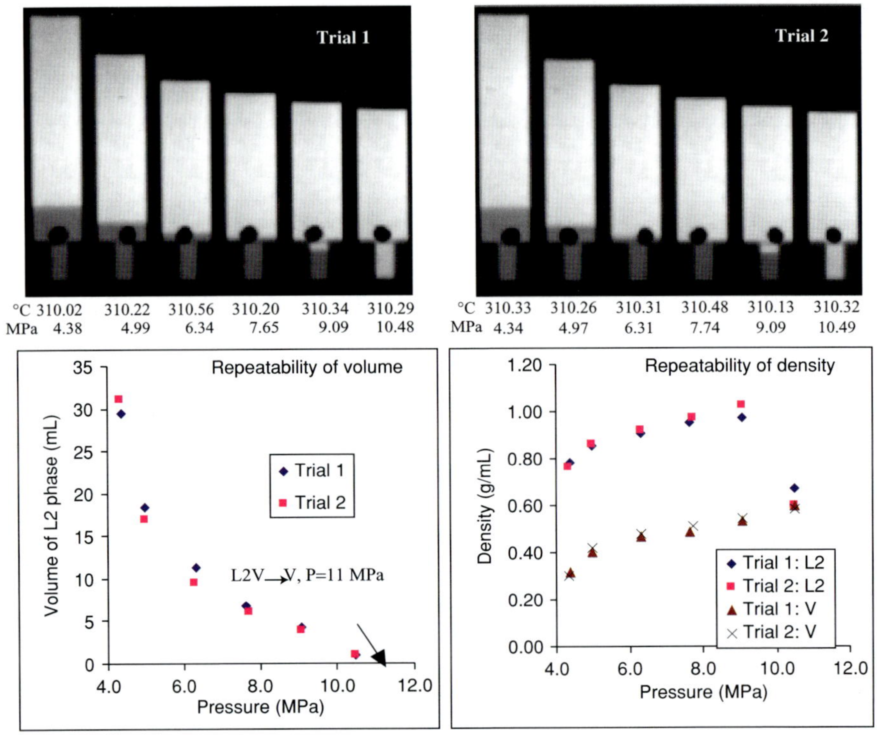

Figure 19.9. Phase behavior measurement repeatability for the mixture heptane + ABVB (35 wt%) at 310°C. Phase densities, phase volumes, and phase behavior boundaries are reported. The samples were cooled to room temperature between measurements.

By contrast, bulk phase behaviors, including phase volume, phase density, and phase boundary data, are repeatable to within fine tolerances as shown in Figure 19.9 for a 25 wt% ABVB + heptane mixture at 310°C. The two trials were performed 3 months apart on the same sample. The sample was stored in the view cell at room temperature between measurements. It is worth noting that this mixture, comprising about 9 wt% asphaltenes, exists as a dense vapor at elevated pressures at 310°C.

6. Conclusions

The detection and definition of phases present in heavy oil and heavy oil mixtures is challenging as diverse and novel approaches must be applied. Calorimetric data clearly show that heavy oils comprise two or more phases over a broad range of temperatures. These phases, possible combinations of glass, liquid, and solid, appear to be independent of nanostructures that scatter X-rays—normally interpreted as asphaltenes or a fraction of asphaltenes present. The phase behavior of heavy oil + *n*-alkane mixtures is equally complex with behaviors ranging from dense vapors at elevated temperatures and pressures to liquid–liquid–liquid–vapor phase behavior at less extreme conditions. Asphaltenes present in these mixtures exhibit varying degrees of association depending on the temperature, pressure, and composition of the mixtures. However, interpretation of data obtained for asphaltenes is ambiguous because changes in the mean size of asphaltene aggregates present do not constitute a change of state. The appropriateness of the treatment of the aggregates as associative structures within a phase or as a separate phase remains unresolved in the literature but is an active area of study.

The success of phase behavior prediction models for heavy oils and heavy oil + solvent/antisolvent mixtures has been limited by the complexity of the phase behaviors (the sharp phase behavior transitions associated with them, the incomplete nature of the phase behavior data typically available), as well as by shortcomings related to hydrocarbon speciation and equation of state based models.[10,63] Issues unresolved in the literature, such as the physical state and state of aggregation of asphaltenes which appear to span the range of both SAXS and WAXS measurements, the state and nature of precipitated phases, etc. also warrant further study. Commercially available correlative tools can be used to interpolate phase behavior within existing data sets but there is no model for a priori prediction of either phase behavior or adherent deposit formation for asphaltene-bearing hydrocarbon mixtures. Future success will also hinge upon the availability of detailed compositional analysis of the maltene fraction now possible[4] as constituents reporting to the maltene fraction have a tremendous impact on the nature and complexity of the phase behavior of hydrocarbon mixtures.[27]

Acknowledgments

Use of the Advanced Photon Source was supported by the U.S. Department of Energy, Office of Science, Office of Basic Energy Sciences, under contract no. W-31-109-ENG-38. We also acknowledge support from the Alberta Synchrotron Institute and our host at APS-BESSRC-CAT 12.

Sponsors for the NSERC Industrial Research Chair in Petroleum Thermodynamics include: Alberta Energy Research Institute, Albian Sands Energy Inc., Computer Modelling Group Ltd., ConocoPhillips, Imperial Oil Resources, NEXEN Inc., Natural Resources Canada, Petroleum Society of the CIMM, Oilphase-DBR, Oilphase—a Schlumberger Company, Schlumberger, Syncrude Canada Ltd., NSERC.

References

[1] Robinson, D.B. (ed.). (1984). *The Thermodynamic and Transport Properties of Bitumens and Heavy Oils*. AOSTRA, Edmonton, Canada. p. 229.

[2] Hepler, L.G. and C. Hsi (ed.). (1989). *Technical Handbook on Oilsands, Bitumens and Heavy Oils*. AOSTRA, Edmonton, Canada. p. 376.

[3] Speight, J.G. (1999). In: J.G. Speight (ed.), *Petroleum Analysis and Evaluation, Petroleum Chemistry and Refining*. Taylor & Francis Press, Washington, DC.

[4] Henry, C.M. (2003). Fine look at crude oil. *Chem. Eng. News* **81**(13), 39.

[5] Robinson, D.B. (1989). The interface between theory and experiment. *Fluid Phase Equilib.* **52**, 1–14.

[6] Shelton, J.L. and L. Yarborough (1976). Multiple phase behavior in porous media during CO_2 or rich-gas flooding. Presented at *SPE-AIME Conference*, Tulsa, OK, March 22–24, 1976. Paper # SPE 5827.

[7] Turek, E.A., R.E. Metcalfe, and R.E. Fishback (1988). Phase behaviour of several CO_2/West Texas reservoir oil systems. *SPERE SPE 13117*, May 1988, 505–516.

[8] Burke, N.E., R.E. Hobbs, and S.F. Kashou (1990). Measurement and modeling of asphaltene precipitation. *JPT* November 1990, 1440–1446.

[9] Abedi, S.J., S. Seyfaie, and J.M. Shaw (1998). Unusual retrograde condensation and asphaltene precipitation in a model heavy oil system. *Pet. Sci. Technol.* **16**(3,4), 209–226.

[10] Cartlidge, C.R., L. Dukhedin-Lalla, P. Rahimi, and J.M. Shaw (1996). Preliminary phase diagrams for ABVB + n-dodecane + hydrogen. *Fluid Phase Equilib.* **117**, 257–264.

[11] Zou, X-Y. (2003). Selective removal of fine inorganic solids, heavy metals and sulfur from heavy oils/bitumen. PhD Thesis, University of Toronto, Toronto.

[12] Shaw, J.M. and E. Béhar (2003). SLLV phase behavior and phase diagram transitions in asymmetric hydrocarbon fluids. *Fluid Phase Equilib.* **209**, 185–206.

[13] Hu, Y–F. and T.M. Guo (2001). Effects of temperature and molecular weight of n-alkane precipitants on asphaltene precipitation. *Fluid Phase Equilibr.* **192**, 13–25.

[14] Cimino, R., S. Correra, A. Del Bianco, and T.P. Lockhart (1995). In: E.Y. Sheu and O.C. Mullins (eds.), *Asphaltenes: Fundamentals and Applications*. Plenum Press, New York, p. 126.

[15] Kotlyar, L.S., B.D. Sparks, and J.R. Woods (1999). Solids associated with asphaltenes fraction of oil sands bitumen. *Energy & Fuel* **13**, 346–350.

[16] Chen, R.J.J., P.S. Chappelear, and R. Kobayashi (1976). Dew-point loci for methane-n-hexane and methane-n-heptane binary systems. *J. Chem. Eng. Data* **21**(2), 213–219.

[17] Heidemann, R.A. and R.M. Abdel-Ghani (2001). A ternary system with five equilibrium phases. *Chem. Eng. Sci.* **56**(24), 6873–6881.

[18] Peters, C.J. (1994). In: E. Kiran and J.M.H. Levelt Sengers (eds.), *Multiphase Equilibria in Near-Critical Solvent. Supercritical Fluids: Fundamentals for Application*. Kluwer Academic Press, Dordrecht. pp. 117–145.

[19] Miller, M.M. and K.D. Luks (1989). Observations on the multiphase equilibria behavior of CO_2 rich and ethane rich mixtures. *Fluid Phase Equilib.* **44**, 295–304.

[20] Wisniak, J., A. Apelblat, and H. Segura (1998). Prediction of gas-solid equilibrium using equations of state. *Fluid Phase Equilib.* **147**, 45–64.

[21] Poot, W., K-M. Krüger, and T.W. de Loos (2003). High-pressure phase behavior of the binary systems (butane + adamantane) and (butane + diamantane). *J. Chem. Thermodyn.* **35**(4), 591–604.

[22] van Konynenburgh, P.H. and R.L. Scott (1980). Critical lines and phase equilibria in binary van der Waals mixtures. *Philos. Trans. R. Soc. London* **298**, 495–540.

[23] Gregorowicz, J., Th. W de Loos, and J.A. de Swaan (1992). Unusual retrograde condensation in ternary hydrocarbon systems. *Fluid Phase Equilib.* **73**, 109–115.

[24] Gregororwicz, J., Th.W. de Loos, and J.A. de Swaan (1993). Liquid–liquid–vapor phase equilibria in the system ethane+propane+eicosane: Retrograde behavior of the heavy liquid phase. *Fluid Phase Equilib.* **84**, 225–250.

[25] Shaw, J.M., Th.W. de Loos, and A.J. de Swaan (1993). Prediction of unusual retrograde condensation in model reservoir fluids. *Fluid Phase Equilib.* **84**, 251–266.

[26] Shaw, J.M., Th. W. de Loos, and A.J. de Swaan (1997). An explanation for solid–liquid–liquid–vapor phase behavior in reservoir fluids. *Pet. Sci. Technol.* **15**(5/6), 503–521.

[27] Shaw, J.M. (2002). Towards common generalized phase diagrams for asphaltene containing hydrocarbon fluids. *ACS Pet. Chem. Div. Prepr.* **47**(4), 338–342.

[28] Vengu, T. (1983). Propriétés thermodynamiques d'un pétrole brut en presence de gaz carbonique. Application à la récupération assistée. Ph.D. thesis, Université Pierre et Marie Curie, Paris VI.

[29] Michelsen, M.L. (1982). The isothermal flash problem. Part I: Stability. *Fluid Phase Equilib.* **9**, 1–19.

[30] Michelsen, M.L. (1982). The isothermal flash problem. Part II: Phase-split calculation, *Fluid Phase Equilib.* **9**, 21–40.
[31] Baker, L.E., A.C. Pierce, and C.D. Luks (1982). Gibbs energy analysis of phase equilibria. *SPE J.* **22**, 731–742.
[32] Minicucci, D., X-Y. J. Zou, and M. Shaw (2002). The impact of liquid–liquid–vapor phase behavior on coke formation from model coke precursors. *Fluid Phase Equilib.* **194–197**, 353–360.
[33] De Boer, R.B., K. Leerlooyer, M.R.P. Eigner, and A.R.D. van Bergen (1995). Screening of crude oils for asphalt precipitation: Theory, practice and the selection of inhibitors. *SPE Prod. Facil.* February 1995, 55–61.
[34] Zou, X-Y. and J.M. Shaw (2004). Dispersed phases and dispersed phase deposition issues arising in asphaltene rich hydrocarbon fluids. *J. Pet. Sci. Technol.* **22**(7/8), 759–771.
[35] Kokal, S.L., J. Najman, S.G. Sayegh, and A.E. George (1992). Measurement and correlation of asphaltene precipitation from heavy oils by gas injection. *J. Can. Pet. Technol.* **31**(4), 24.
[36] Yarranton, H.W. and J.H. Masliyah (1996). Molar mass distribution and solubility modeling of Asphaltenes. *AIChE J.* **42**, 3533–3543.
[37] Zou, X.Y. and J.M. Shaw (2004). Challenges inherent in the development of predictive deposition tools for asphaltene containing hydrocarbon fluids. *J. Pet. Sci. Technol.* **22**(7/8), 773–786.
[38] Ting, P.D., G.J. Hirasaki, and W.G. Chapman (2003). Modeling of asphaltene phase behavior with the SAFT equation of state. *J. Pet. Sci. Technol.* **21**(3/4), 647–661.
[39] Wu, J., J.M. Prausnitz, and A. Fasabooti (2000). *AIChE J.* **46**, 197–209.
[40] Alboudwarej, H. (2003). Asphaltene deposition in flowing systems. Ph.D. thesis, University of Calgary.
[41] Branco, V.A.M., G.A. Mansoori, L.C. De Almeida Xavier, S. J. Park, and H. Manafi (2001). Asphaltene flocculation and collapse from petroleum fluids. *J. Pet. Sci. Eng.* **32** (2–4), 217–230.
[42] Mansoori, G.A. (1997). Modeling of asphaltene and other heavy organic depositions. *J. Pet. Sci. Eng.* **17**(1/2), 101–111.
[43] Fujisawa, G., O.C. Mullins, C. Dong, A. Carnegie, S.S. Betancourt, T. Terabayashi, S. Yoshida, A.R. Jaramillo, and M. Haggag (2003). Analyzing reservoir fluid composition in situ in real time: Case study in a carbonate reservoir. In: *SPE Meeting*, Denver CO, October 5–8, 2003. SPE 84092.
[44] Ferworn, K.A. and W.Y. Svrcek (1998). In: O.C. Mullins and E.Y. Sheu (eds.), *Structure and Dynamics of Asphaltenes*. Plenum Press, New York. p. 227.
[45] Keinitz, W. and S.I. Anderson (1997). Paper presented at the 8th *Symposium on Oil Field Chemistry*, Geilo, Norway, March.
[46] Sheu, E.Y. (1998). In: O.C. Mullins and E.Y. Sheu (eds.), *Structure and Dynamics of Asphaltenes*, Plenum Press, New York. p. 115.
[47] Sheu, E.Y. (2002). Asphaltene precipitation and Liesegang ring. In: *International Conference on Heavy Oil Deposition*, Puerto Vallarta, Mexico, November 2002.
[48] Abedi, S.J., H-Y. Cai, S. Seyfaie, and J.M. Shaw (1999). Simultaneous phase behavior, elemental composition and density measurement using X-rays imaging. *Fluid Phase Equilib.* **158–160**, 775–781.
[49] Zou, X.Y., L. Dukhedin-Lalla, X. Zhang, and J.M. Shaw (2004). Selective rejection of inorganic fine solids, heavy metals, and sulfur from heavy oils/bitumen using alkane solvents. *Ind. Eng. Chem.* **43**, 7103–7112.
[50] Masson, J.-F. and G.M. Polomark (2001). Bitumen microstructure by modulated differential scanning calorimetry, *Thermochim. Acta* **374**, 105–114.
[51] Masson, J.-F. and G.M. Polomark (2004). Erratum to Bitumen microstructure by modulated differential scanning calorimetry, *Thermochimica Acta* **374** (2001) 105–114. *Thermochim. Acta* **413**, 273.
[52] Maham, Y., M.G. Chodakowski, X. Zhang, and J.M. Shaw (2005). Asphaltene phase behavior: Prediction at a crossroads. *Fluid Phase Equilib.* **227**, 177–182.
[53] Escobedo, J. and G.A. Mansoori (1998). Viscometric principles of the onset of colloidal asphaltene flocculation in paraffinic oils and asphaltene micellization in aromatics. *SPE Prod. Facil.* May 1998, 116–122.

[54] Scotti, R. and L. Montanari (1998). In: O.C. Mullins and E.Y. Sheu, (eds.), *Structure and Dynamics of Asphaltenes*. Plenum Press, New York. p. 79.

[55] Schwarcz, B.J. and J.M. Prausnitz (1987). Solubilities of methane, ethane and carbon dioxide in heavy fossil-fuel fractions. *Ind. Eng. Chem. Res.* **26**, 2360–2366.

[56] Schwarcz, B.J., J.A. Wilhelm, and J.M. Prausnitz (1987). Vapour pressure and saturated liquid densities of heavy fossil fuel fractions. *Ind. Eng. Chem. Res.* **26**, 2353–2360.

[57] Cai, H-Y., J.M. Shaw, and K.H. Chung (2001). Hydrogen solubility measurements in heavy oil and bitumen cuts. *Fuel* **80**, 1055–1063.

[58] Pyda, M., M. Bartkowiak, and B. Wunderlich (1998). Computation of heat capacities of solids using a general tarasov equation. *J. Thermal Anal.* **52**, 631–656.

[59] Wunderlich, B. (2003). In: S.Z.D. Cheng (ed.), *Heat Capacity of Polymers, Handbook of Thermal Analysis and Calorimetry, Volume 3: Applications to Polymers and Plastics*, (1), pp. 1–95.

[60] Zhang, Y., T. Takanohashi, S. Sato, T. Kondo, and I. Saito (2003). *Energy Fuels* **17**, 101–106

[61] Zhang, X. (2005). *The Impact of Multiphase Behaviour on Coke Deposition in Heavy Oil Hydroprocessing Catalysts*, Ph.D. thesis, University of Alberta.

[62] Chodakowski, M. (2005). Ph.D. thesis, University of Alberta (in progress).

[63] Szewczyk, V. and E. Béhar (1999). Compositional model for predicting asphaltenes flocculation. *Fluid Phase Equilib.* **158–160**, 459–469.

[64] Anderson, S.I. and J.G. Speight (1999). Thermodynamic models for asphaltene solubility and precipitation. *J. Pet. Sci. Eng.* **22**, 53–66.

[65] Joshi, N.B., O.C. Mullins, A. Jamaluddin, J. Creek, and J. McFadden (2001). *Energy Fuels* **15**, 979–986.

20

Selective Solvent Deasphalting for Heavy Oil Emulsion Treatment

Yicheng Long, Tadeusz Dabros, and Hassan Hamza

1. Introduction

Solvents are often used to treat heavy oil emulsions such as bitumen froth produced in the recovery of bitumen from oil sands. It has been found that the aromaticity of the solvent and the solvent-to-bitumen dilution ratio (S/B, by wt) have profound effects on the stability of water-in-bitumen emulsions. An aliphatic solvent at a certain S/B not only partially precipitates asphaltenes from the solvent-diluted bitumen solution, but also, at the same time, promotes aggregation of the emulsified water droplets (WD), dispersed solids (DS), and precipitated asphaltenes (PA). The WD/DS/PA aggregates exhibit zone settling in solvent-diluted bitumen, and various distinct zones and corresponding interfaces develop during settling. Typically, the top zone is a clean oil phase containing less than 0.1 wt% water–plus–solids, the middle is an emulsion zone containing the WD/DS/PA aggregates, and the bottom is a compaction zone of the settled WD/DS/PA aggregates. The position of the sharp interface that develops between the oil phase and the emulsion zone can be tracked by visual observation with the help of proper illumination. An in-line transflectance probe coupled with a spectrophotometer via a fiber-optic cable is also developed for monitoring the settling of the WD/DS/PA aggregates. The settling rate depends on many factors and conditions including solvent type, S/B, temperature, and mixing conditions. Treatment of bitumen emulsions with aliphatic solvents not only produces clean bitumen but also leads to significant reductions in asphaltene content, coking propensity, and the concentrations of metals, sulfur, and nitrogen. The viscosity of the bitumen is also reduced, which, in turn, significantly reduces the amount of light diluent that must be blended with the bitumen in order to meet the pipelining specification for viscosity. Therefore, proper selection of solvent, S/B, and other process conditions make it possible to produce bitumen products to satisfy a variety of specifications.

The term *heavy oils* usually refers to crudes with API gravity of less than 20°. Heavy oils with API gravity less than 10° are also called *extra heavy oils*

Yicheng Long, Tadeusz Dabros, and Hassan Hamza • CANMET Energy Technology Centre–Devon, Natural Resources Canada, Devon, Alberta, T9G 1A8, Canada.

(for viscosity less than 10 Pa s at reservoir conditions) or *bitumen* (for viscosity higher than 10 Pa s at reservoir conditions).[1] As conventional oil reserves decline, production of these heavy oils becomes increasingly important. The reserves of extra heavy oil and bitumen worldwide are comparable to those of conventional oils. There are about 1.2 trillion barrels of extra heavy oil deposited in the Orinoco Oil Belt in eastern Venezuela. The proven amount of bitumen in place as oil sands in Canada is estimated to be about 1.7 trillion barrels, of which about 300 billion barrels of bitumen can be recovered based on present technology.[2] The major oil sands deposits are the Athabasca, Cold Lake, and Peace River deposits in northern Alberta, Canada.[3,4]

Due to the warm reservoir temperature of Venezuela extra heavy oil, cold production similar to the production of conventional oils is feasible although a variety of other techniques can be used to improve recovery. Canadian bitumens have very high viscosities at reservoir temperature (e.g., $>$1000 Pa s at 11°C). For shallow oil sands deposits up to about 75 m deep, bitumen is recovered via open-pit mining. For deposits more than 400 m deep, bitumen is presently produced through in situ methods such as cyclic steam injection and steam-assisted gravity drainage (SAGD).[2,5] Bitumen production from Canadian oil sands currently accounts for about 1.1 million barrels per day of synthetic crude (750 000 daily barrels) and thermally produced bitumen (350 000 daily barrels) in diluted products. The industry is now well on its way to producing 2 million barrels daily in aggregate by 2012.[5] It is anticipated that production of bitumen from Canada's oil sands will reach 5 million barrels per day by 2030.[5] The cost of producing bitumen from oil sands via open-pit mining is currently about US $8 per barrel.[6]

Demulsification is one of the major challenges during extra heavy oil and bitumen processing due to the serious problems arising from the emulsified water that may contain significant amounts of corrosive chloride[6] and dispersed solids (detrimental to downstream bitumen upgrading processes) in the oil. In this chapter, bitumen chemistry and the stability of water-in-bitumen emulsions are reviewed. Partial rejection of asphaltenes from bitumen, the demulsification of water-in-bitumen emulsions, and the relationship between these two are discussed.

2. Bitumen Chemistry

There is a large volume of literature characterizing bitumen.[3,7–9] Table 20.1 lists the elemental composition and properties of a typical Athabasca bitumen.[7,9] Bitumen differs from conventional oils in many regards but most notably it has high viscosity and high molecular weight and contains detrimental amounts of heteroatoms such as sulfur, nitrogen, vanadium, and nickel. The hydrogen-to-carbon atomic ratio (H/C) of bitumen is lower than that of conventional oils.

Figure 20.1 shows the boiling point distributions (obtained with GC simulated distillation) of a few samples of Athabasca bitumen. On average, Athabasca bitumen contains only about 3 wt% light ends with normal boiling point lower than 250°C and more than 45 wt% heavy ends with normal boiling point higher than 550°C. Obviously, bitumen is a very heavy crude.

Table 20.1. Elemental Composition and Properties of Typical Athabasca Bitumen

Elemental composition (wt%)			
Carbon	81–84	Sulfur	4.6–5.6
Hydrogen	10.0–11.0	Vanadium	160–300 ppm
Nitrogen	0.3–0.6	Nickel	60–100 ppm
Oxygen	0.8–1.6	H/C[a]	1.46–1.50
Properties			
API gravity	8.05°	Viscosity at 24°C	323 Pa s
Ramsbottom carbon	10–13.7 wt%	Average molecular weight[b]	490–620 g/mol

[a] Hydrogen-to-carbon atomic ratio.
[b] Determined using VPO.

Bitumen is often characterized chemically by its hydrocarbon class compositions such as saturates–aromatics–resins–asphaltenes (SARA) contents.[9] Asphaltenes are a solubility class defined as insolubles in light aliphatic solvent (e.g., *n*-pentane or *n*-heptane) and solubles in aromatic solvent (e.g., toluene). In this work, asphaltene content is determined using *n*-pentane as precipitant by following a standard procedure outlined in various works.[9–11] The deasphalted oil (DAO) is usually referred to as maltenes. Resins are the polar components in maltenes. A representative Athabasca bitumen contains about 17 wt% saturates, 40 wt% aromatics, 26 wt% resins, and 17 wt% asphaltenes.[9] Obviously, the concentration

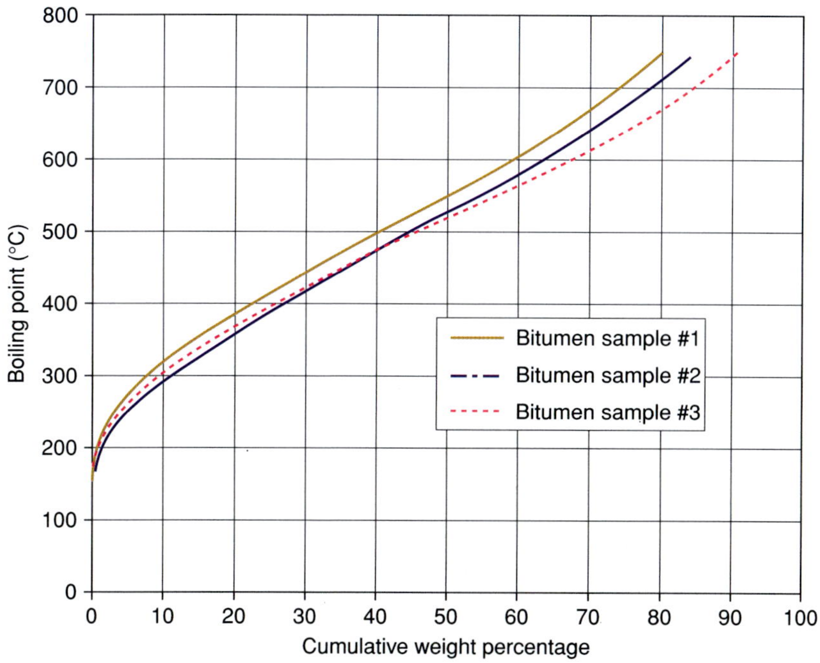

Figure 20.1. Boiling point distributions of Athabasca bitumen samples.

of asphaltenes in bitumen is relatively high and they contribute to a variety of problems such as high stability of water-in-bitumen emulsion, high viscosity of bitumen for pipelining, as well as high-coking propensity and catalyst poisoning during bitumen upgrading and refining.

Asphaltenes are not a chemical class by definition, but there is a large body of literature on asphaltene structure.[12–15] Although the chemical and physical properties of asphaltenes have been investigated for decades using a wide range of techniques such as VPO, SEC, FTIR, NIR, NMR, MS, fluorescence depolarization, spectral fluorescence microscopy, rheology, interfacial tension, refractive index, STM, SAXS, and SANS, their molecular structure and thermodynamic state in various systems are still debated. Pfeiffer and Saal[16] pictured an asphaltene solution as a resin-stabilized colloidal suspension. Sheu et al.[15,17–19] then described a solvation model for asphaltene solutions. According to the solvation model, asphaltenes can be solvated by hydrocarbon media in general.

The molecular weight of asphaltenes is the subject of a long-standing controversy. The reported average molar mass of asphaltenes ranges from 400 to 10,000 g/mol depending on the technique used to determine molecular weight.[12,13,20,21] According to the most recent studies, the average molecular weight of asphaltenes is around 750.[12] Groenzin and Mullins[12] gave the idealized molecular structures for asphaltenes as depicted in Figure 20.2. According to this presentation, the asphaltene molecules possess, on average, about seven fused aromatic rings and the diameters of asphaltene molecules are 10 to 20 Å.[22] Due to the relatively large

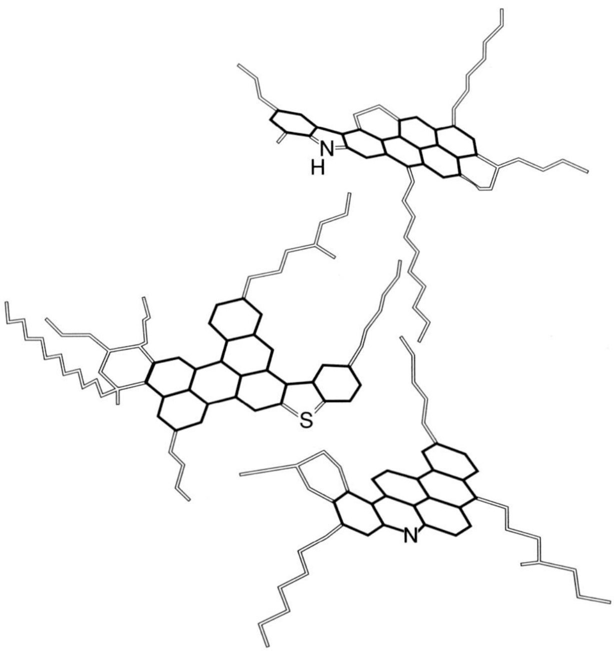

Figure 20.2. Idealized molecular structures of asphaltenes depicted by Groenzin and Mullins.[12]

Table 20.2. Compositions of a Few Typical In Situ Bitumen Emulsions

In situ bitumen emulsion sample	Bitumen (wt%)	Water (wt%)	Solids (wt%)
In situ bitumen emulsion #1	82.5	17.3	0.2
In situ bitumen emulsion #2	66.3	33.3	0.4
In situ bitumen emulsion #3	69.1	30.3	0.6
In situ bitumen emulsion #4	90.7	8.3	1.0

conjugated π system in an asphaltene molecule, the $\pi - \pi$ interaction between asphaltene molecules can lead to the formation of asphaltene aggregates and even precipitation of asphaltenes if the solvent does not have sufficient solvation power.

3. Stability of Water-in-Bitumen Emulsions

3.1. In situ Bitumen Emulsion and Bitumen Froth

The SAGD method for in situ bitumen production utilizes pairs of superposed parallel horizontal wells drilled into an oil sands formation. Steam is injected continuously into the upper well to heat the oil sands formation. Consequently, the bitumen softens and drains into the lower production well. The produced fluid in SAGD is a complex mixture of bitumen, water, and sand. After sand removal and free water knockout, the oil-continuous phase is a water-in-bitumen emulsion (referred to as in situ bitumen emulsion in this work). Table 20.2 presents the compositions of a few typical in situ bitumen emulsion samples from the Cold Lake area. The in situ bitumen emulsion contains a considerable amount of water, but its solids content is usually lower than 1 wt%.

In the processing of mined oil sands, bitumen is extracted by means of a hot water process pioneered by Karl Clark et al.[23] An intermediate material called bitumen froth is produced via an air flotation process. The air in the aerated bitumen froth is removed before further treatment, but the material is still referred to as bitumen froth after deareation. Table 20.3 gives compositions of a few typical bitumen froth samples. On average, bitumen froth contains about 60 wt% bitumen, 30 wt% water, and 10 wt% solids. In this work, we refer collectively to the in situ bitumen emulsion and the bitumen froth as bitumen emulsion.

The contaminants such as water (carrying corrosive chloride) and solids in either the in situ bitumen emulsion or the bitumen froth have to be removed before

Table 20.3. Compositions of a Few Typical Bitumen Froths

Bitumen froth sample	Bitumen (wt%)	Water (wt%)	Solids (wt%)
Bitumen froth #1	47.7	39.2	13.1
Bitumen froth #2	58.8	30.9	10.2
Bitumen froth #3	44.2	36.5	19.3
Bitumen froth #4	68.3	20.2	11.5

the bitumen can be further upgraded or refined. For example, the total water and solids content (BS&W, bottom solids and water analyzed by centrifugation) has to be less than 0.5 vol% in order to meet the specification for pipelining. Furthermore, the bitumen must contain only minimal amounts of impurities (e.g., less than a few hundred ppm of ash) if it is to be used as feedstock for high-liquid conversion hydrocracking processes.[24]

Bitumen typically has a density of 1.01 g/mL[9,25] and a viscosity of the order of 100 Pa s [26] at ambient conditions. Therefore, bitumen emulsion treatment frequently involves dilution with organic solvent to facilitate separation of water and solids by reducing the viscosity of the oil phase and, at the same time, increasing the density difference between the oil and water phases. Sometimes, bitumen emulsions are treated at high temperatures (e.g., higher than 120°C) to enhance demulsification. The current commercial oil sands operations at Syncrude Canada Ltd. and Suncor Energy Inc. use naphtha for bitumen froth treatment at a solvent-to-bitumen mass ratio (S/B) of 0.6–0.7.[25] After multistage centrifugation (with centrifugal force up to 2500 *g*), together with demulsifier addition, the solvent-diluted bitumen product still contains 2–5 wt% water and 0.3–1 wt% mineral solids (determined as ash content)[6,25], which indicates that a significant amount of water and solids in bitumen froth is difficult to remove. This portion of the water is held in a very stable water-in-diluted bitumen emulsion, and the remaining solids are fine particles that are well dispersed in the oil phase.

3.2. Size Distributions of Emulsified Water Droplets and Dispersed Solids

Residual water content in diluted bitumen during settling of solvent-diluted bitumen emulsion is a function of water droplet size, the densities of both the dispersed phase and the medium, the viscosity of the continuous phase, the population of water droplets, and acceleration due to external forces. We have developed a method to determine the size distribution of water droplets based on experimental data from a settling test. Detailed discussion of the method is beyond the scope of this chapter; nevertheless, we have found that there are two main types of water droplet size distribution curves among a variety of solvent-diluted bitumen froth samples. Figures 20.3 and 20.4 give examples of the two types as represented by froth samples A and B. We thus classify the water in solvent-diluted bitumen emulsions into three categories: (1) emulsified water with water droplets less than 10 μm in diameter, (2) dispersed water with droplet diameters between 10 and 60 μm, and (3) free water with droplet/pocket sizes larger than 60 μm. The total water content in bitumen froth A diluted with naphtha at S/B = 0.7 is 17.6 wt%, of which 3.2 wt% is emulsified water as shown in Figure 20.3 and the rest is free water (14.4 wt%). There is no significant amount of dispersed water in this type of bitumen emulsion. On the other hand, bitumen froth B diluted with naphtha at S/B = 0.7 contains 28.5 wt% water, of which 4.0 wt% is emulsified water as shown in Figure 20.4. There are more than 11 wt% of dispersed water and about 13.5 wt% free water in this sample.

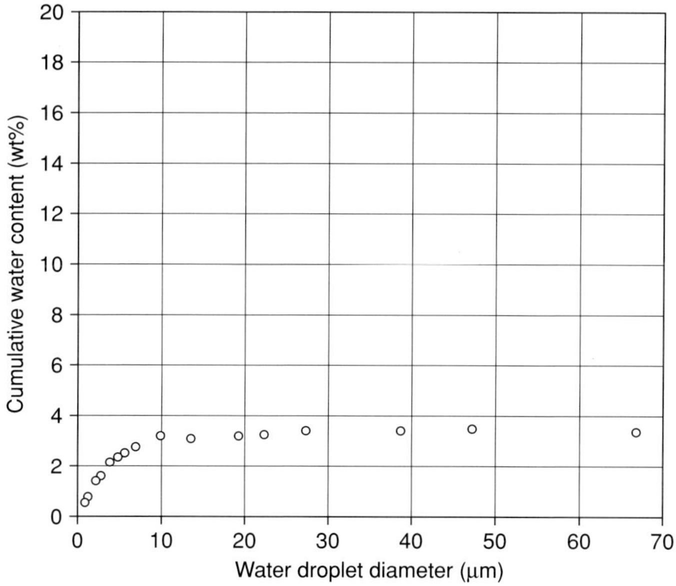

Figure 20.3. Size distribution of water droplets in solvent-diluted bitumen emulsion, bitumen froth A.

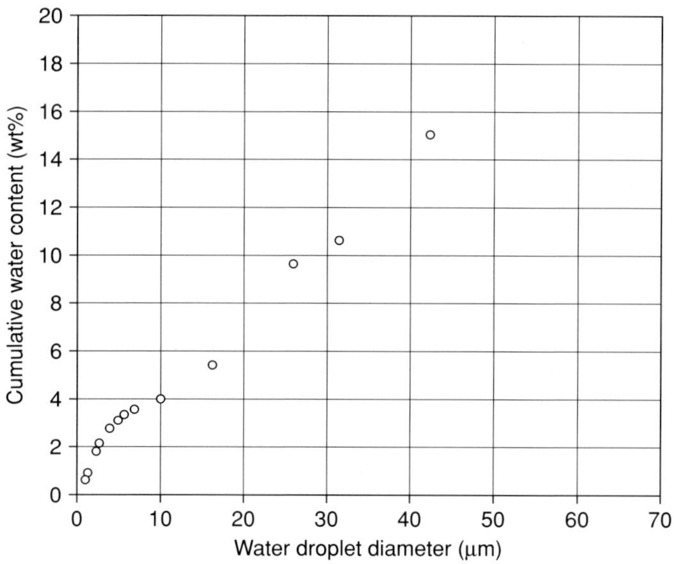

Figure 20.4. Size distribution of water droplets in solvent-diluted bitumen emulsion, bitumen froth B.

For micrometer-sized water droplets, there is little or no separation of water by settling in diluted bitumen. Therefore, the emulsified water cannot be resolved from the oil phase unless larger droplets are achieved through coalescence or aggregation of water droplets. If there is no efficient aggregation and/or coalescence among the small water droplets, the level of emulsified water in the system ultimately dictates the water content of the solvent-diluted bitumen product after bitumen emulsion treatment. The dispersed water in the oil phase settles faster than the emulsified water, but sufficient time and/or additional external acceleration forces (e.g., centrifugation) are required. Excessive amounts of dispersed water result in complications in bitumen emulsion treatment. The free water, on the other hand, settles almost instantly upon dilution of the bitumen emulsion with solvent, and has no significant negative effect on demulsification.

There are indications that the emulsified water and the dispersed water may be formed via different mechanisms. Dabros et al.[27] described a new emulsification mechanism—the budding mechanism. They found that micrometer-sized water droplets detach from a larger water droplet (e.g., >10 μm) when it contracts in toluene-diluted bitumen ($>10\%$ bitumen). In industrial operation, water droplets undergo deformations even under gentle agitation. New emulsion droplets can thus be formed through budding at much lower shear rates than those required for a conventional break-up mechanism.[6] In other words, the existence of micrometer-sized water droplets in bitumen emulsion is probably the nature of the system.

Bitumen froth contains considerable quantities of mineral solids. The original solids in bitumen froth have a particle size distribution from submicrometer to 500 μm in diameter.[25] The addition of solvent to bitumen froth results in the separation of most of the solids from the solvent-diluted bitumen, but fine solids still remain. Fine solids were separated from the oil phase by centrifugation at 67 000g-force followed by washing with toluene. Figure 20.5 shows the particle size distribution of the fine solids analyzed by XDC (X-ray disk centrifuge). The majority of the fine solids are seen to be of submicrometer diameter. It should be noted that Brownian motion is significant for these small particles while dispersed in oil phase. It has been reported that the fine solids are mainly clays with some small particles of quartz, metal oxides, sulfides, and carbonates.[25,28,29]

3.3. Stabilization Mechanism of Bitumen Emulsions

Bitumen emulsions are very complex systems. There are a variety of natural surface-active components (such as asphaltenes, resins, and fine solids) that can contribute to the stabilization of water-in-bitumen emulsions.[30] Alberta bitumen typically contains 15–21 wt% asphaltenes, and many researchers have studied their role in the stabilization of water-in-oil emulsions. For example, Gu et al.[31], through water extraction and multispeed centrifugation tests, found that asphaltenes are the key stabilizers of water-in-toluene-diluted bitumen emulsion and that the asphaltenes associated with the water/oil interface have lower values of H/C and higher values of oxygen-to-carbon atomic ratio (O/C) than the asphaltenes in

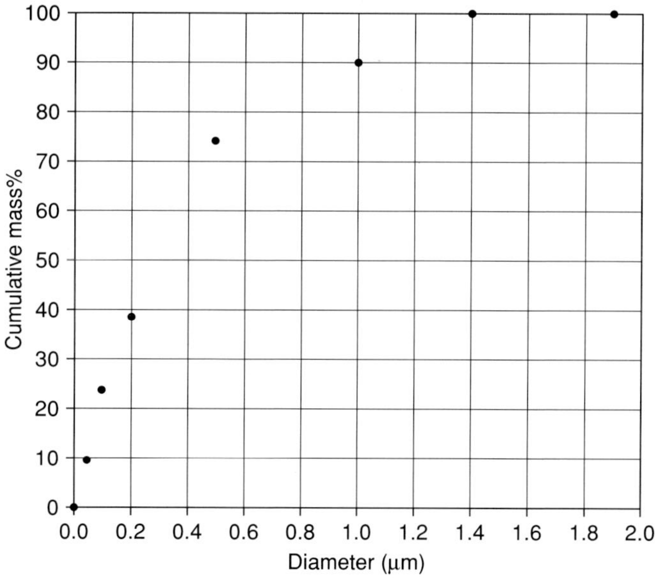

Figure 20.5. Particle size distribution of fine solids in solvent-diluted bitumen emulsion.

the bulk phase. Kloet et al.[32] studied the coalescence of water droplets using a micropipette technique and found that asphaltenes play a dominant role in stabilizing water droplets in solvent-diluted bitumen. They also found that fine solids enhance the stability of water droplets but do not necessarily provide an efficient barrier to droplet coalescence. Yan et al.[33] concluded that asphaltenes and fine solids are the important stabilizers of water-in-diluted-bitumen emulsion, and the emulsion is most stable when both asphaltenes and fine solids are present. Sjoblom et al.[30] pointed out that interplay between asphaltenes and resins can explain most of the stability.

Yeung et al.[34–36] studied the interfacial properties of water droplets in toluene-diluted bitumen using a micropipette technique. They found that there is a rigid skin surrounding the water droplet at certain conditions as a result of adsorbed layers at the water/oil interface. The adsorbed layers are responsible for a steric barrier to droplet coalescence as suggested by Czarnecki.[6]

4. Effect of Solvent on Bitumen Emulsion Stability

For bitumen emulsion treatment, solvent was originally used to create driving forces for demulsification of water-in-bitumen emulsions by increasing the density difference between the oil and water phases and, at the same time, reducing oil phase viscosity. Since asphaltenes have been identified as closely involved in the stabilization of bitumen emulsions, solvent aromaticity should play an important

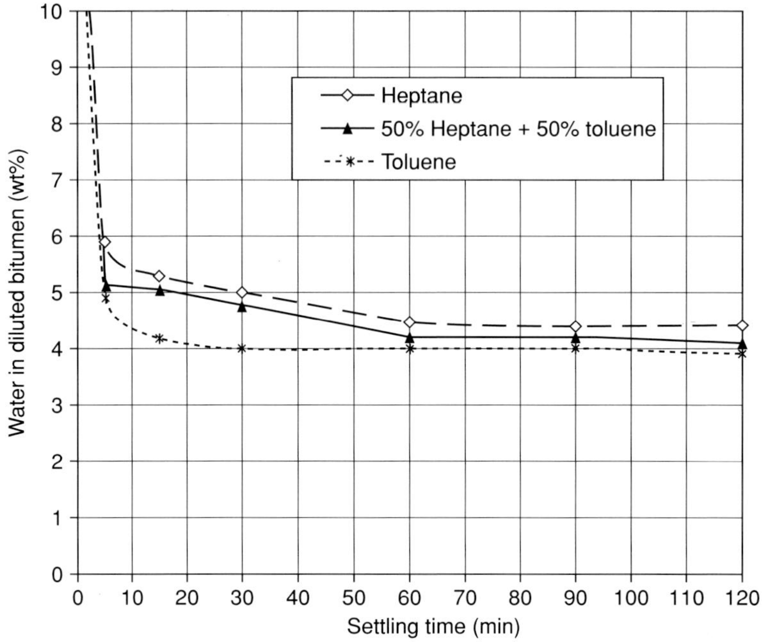

Figure 20.6. Separation of water from bitumen froth diluted with heptane, toluene, and heptane/toluene mixture at 80°C and S/B = 0.70.

role in the demulsification of bitumen emulsions, considering that the phase behavior of asphaltenes depends on solvent aromaticity.

Tipman and Long[37] performed a series of settling tests with a bitumen froth and heptane/toluene solvents at 80°C and S/B = 0.70. Figure 20.6 shows the water content in diluted bitumen as a function of settling time. The solvent with higher aromaticity (toluene) speeds up water resolution from oil phase slightly more than the solvent with lower aromaticity (heptane), but the residual water content in the diluted bitumen product remains as high as about 4 wt% for all cases. These results indicate that water-in-bitumen emulsions cannot be efficiently treated simply by increasing solvent aromaticity, although asphaltenes do have higher solubility in an aromatic solvent than in an aliphatic solvent.

Dabros[38] studied the interactions of water droplets in heptane- and toluene-diluted bitumen (10 wt% bitumen). In his experiments, a stationary water droplet 40–50 μm in size was placed in a microscopic cell. Solvent-diluted bitumen containing micrometer-sized emulsified water droplets was then pumped through the microscopic cell. Interactions of the micrometer-sized water droplets in the diluted bitumen emulsion with the larger stationary water droplet were investigated. He found that the mobile micrometer-sized water droplets were not intercepted by nor did they coalesce with the stationary water droplet when the solvent was toluene, indicating that the water-in-toluene-diluted bitumen emulsion is stable. When the system was switched to heptane solvent, he found that the stationary water droplet

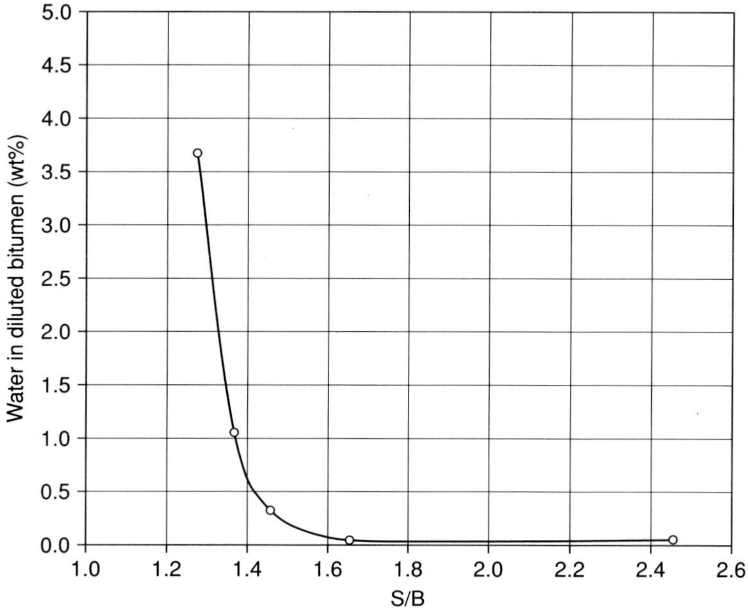

Figure 20.7. Water content in diluted bitumen product after treatment of a bitumen froth sample with *n*-heptane at 80°C, 30 min settling time.

could intercept the micrometer-sized water droplets in the heptane-diluted bitumen emulsion and form aggregates of water droplets, indicating that aliphatic solvent promoted the aggregation of water droplets.

Tipman and Long[37] used a variety of aliphatic solvents to treat bitumen froth over a wide range of S/B. They found that complete demulsification of bitumen emulsions could be achieved with an aliphatic solvent at an S/B higher than a certain value (described as S/F, the solvent-to-froth ratio in Tipman and Long[37]). Figure 20.7 is an example showing water content in diluted bitumen product after treatment of a bitumen froth with *n*-heptane at 80°C. Using this process, a product containing less than 0.1 wt% water was produced when S/B was higher than 1.6 at the tested conditions. The bitumen product contained less than 0.1 wt% solids (analyzed as ash content). The chloride content was reduced to less than 1 ppm.[37]

Through the continued research work at CANMET Energy Technology Centre–Devon[10,11,39–41,43,47] and collaborative technology development efforts with the oil sands industry, the fundamental understanding of aliphatic solvent treatment for bitumen emulsions has been greatly advanced and the associated technology is now an effective new alternative froth treatment process for the oil sands industry.[5] The main outcomes of this process are (1) complete demulsification of the emulsified water together with complete solids removal and (2) partial upgrading of the bitumen through selected rejection of asphaltenes. As a result, the concentrations of metals, sulfur, and nitrogen in bitumen are reduced significantly.[10,11,39–41] Moreover, the bitumen upgrading/refining units can be located remotely from the mining/production site as the bitumen product is pipelinable.

Yeung et al.[34] used a micropipette technique to study interfacial properties of micrometer-sized water droplets in crude oil. They noted that, at less than 0.1 vol% bitumen in a 50/50 by mass solvent mixture of toluene and heptane, the bitumen/water interface exhibited surface rigidity. At about 1 vol% bitumen concentration the interface was flexible. These results indicate that the toluene-diluted bitumen at certain bitumen concentrations changes the properties of the bitumen/water interface.

It has been reported that a naphtha solvent (a mixture of paraffinic, naphthenic, and aromatic components) can also achieve complete demulsification of bitumen emulsions at much higher S/B (e.g., >4) than that required when using aliphatic solvent.[42] Therefore, a wide range of solvents can be used to break bitumen emulsions efficiently. In general, lighter solvents (e.g., n-pentane) without aromatic and naphthenic constituents require lower S/B while heavier solvents (e.g., naphtha) with aromatic and naphthenic components require higher S/B.

5. Treatment of Bitumen Emulsions with Aliphatic Solvents

5.1. Behavior of Bitumen Emulsion upon Dilution

A bitumen froth sample was mixed with toluene or n-heptane at 25°C at various S/B values. After the free water and coarse solids had settled, the solvent-diluted bitumen emulsions were examined using light microscopy. Figure 20.8 shows the micrograph, taken under bright field illumination, of the bitumen emulsion diluted with n-heptane at S/B = 1.0. Toluene (aromatic) and n-heptane (aliphatic) solvents have similar effects on the bitumen emulsion at this S/B; a significant of amount of water (about 4.0 wt%) remains in the oil phase as mobile water droplets less than 5 μm in diameter. Coalescence or aggregation of the water droplets was never observed, even when water droplets collided with each other during prolonged observation under a light microscope. Figure 20.9 is the dark field micrograph of the bitumen emulsion diluted with n-heptane at S/B = 1.0. Solid particles reflect light and show up as bright objects on a darker background under dark field illumination from above. At S/B = 1.0 with n-heptane solvent, the solid particles (mainly submicrometer- to micrometer-sized clays) are well dispersed and appear to move around freely in the oil phase.

Similar experiments to those above were performed at S/B = 2.0. It was found that the results for toluene-diluted bitumen emulsion at this S/B were very similar to those for toluene or n-heptane solvents at S/B = 1.0. Figure 20.10 shows the bright field micrograph of the bitumen emulsion diluted with n-heptane at S/B = 2.0. A portion of asphaltenes is precipitated at this S/B. All water droplets (WD), dispersed solids (DS), and precipitated asphaltenes (PA) form aggregates, indicating that the aliphatic solvent at this S/B causes drastic changes in the properties of the emulsion system. The WD/DS/PA aggregates are much larger than the original water droplets and solids. Settling of these aggregates from diluted bitumen eventually leads to

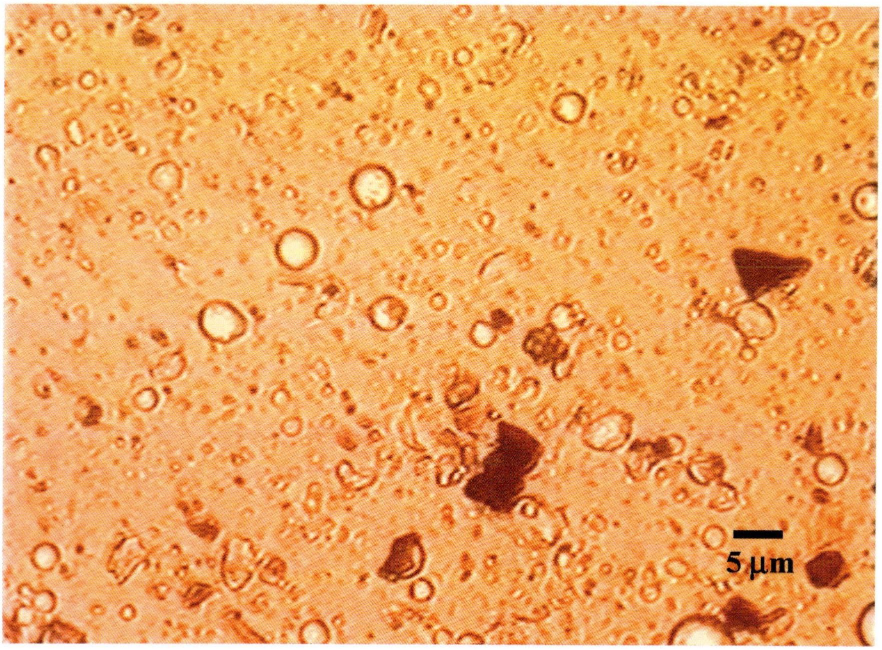

Figure 20.8. Bright field micrograph of bitumen emulsion diluted with n-heptane at S/B = 1.0.

Figure 20.9. Dark field micrograph of bitumen emulsion diluted with n-heptane at S/B = 1.0.

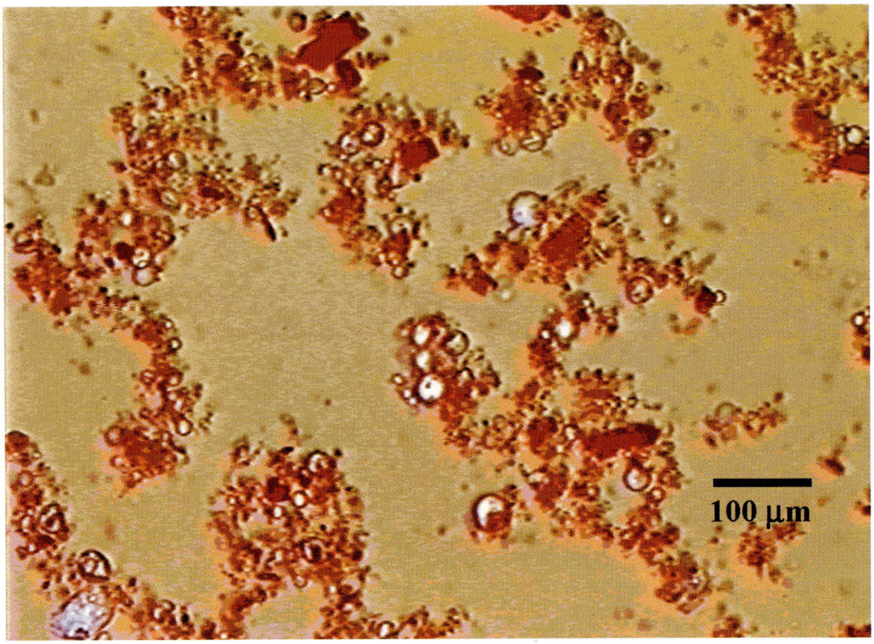

Figure 20.10. Bright field micrograph of bitumen emulsion diluted with n-heptane at S/B = 2.0.

a diluted bitumen product free of water and solids during solvent treatment of bitumen emulsions.

5.2. Settling Characteristics of Bitumen Emulsions Diluted with Aliphatic Solvent

It is known that particles settle at velocities determined by their sizes and densities when the suspension is dilute. A concentration gradient (i.e., segregation) develops as larger particles accumulate in the lower part of the settler, while the smaller ones lag behind in the upper part. As the concentration of the suspension increases, for a given sufficiently narrow size distribution of particles, the particles settle in bulk independently of individual sizes. This behavior is called zone-settling. The zone-settling mode has important practical implications since the supernatant from zone settling is free of particles.

Figure 20.11 shows an example of the water and ash content profiles during settling of the bitumen froth diluted with n-heptane at S/B = 3.0 and 25°C. Three distinct zones develop during settling. The upper part is a clean oil phase containing little water and solids (<0.1 wt%). The middle layer is an emulsion zone containing the WD/DS/PA aggregates (about 3 wt% water and 1.4 wt% solids at the tested condition). One of the characteristics of the emulsion zone is that the water and ash contents remain almost constant from the top to the bottom of the zone during settling. There are abrupt changes in the concentrations of water and solids at the

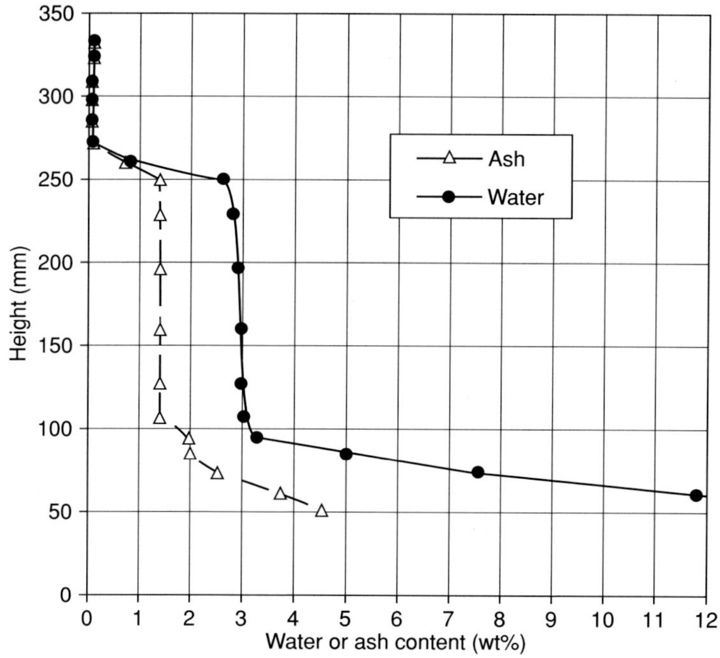

Figure 20.11. Profiles of water and ash contents in bitumen froth diluted with n-heptane at S/B = 3.0, 25°C, settling time = 20 min.

interface between the upper supernatant and the middle emulsion zone. The bottom part is a compaction zone containing the settled WD/DS/PA aggregates.

A camcorder or a microscope was used to take images of the various zones developed during settling of the WD/DS/PA aggregates in bitumen froth diluted with aliphatic solvent. Figure 20.12 depicts the experimental setup. Fiber-optic

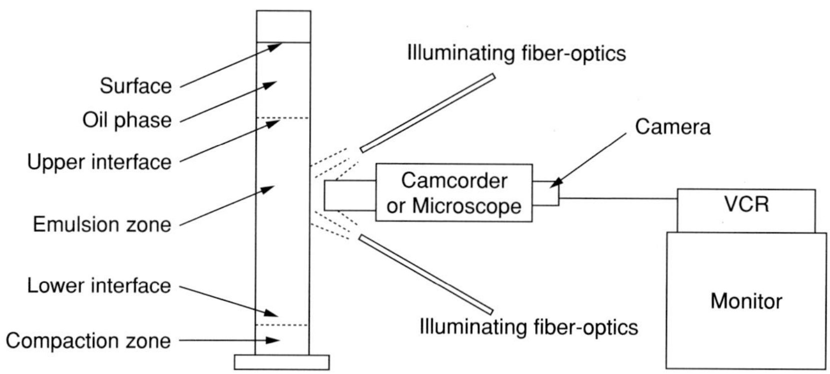

Figure 20.12. Experimental setup for investigating settling of bitumen emulsions diluted with aliphatic solvent.

Figure 20.13. Micrograph of the emulsion zone and the upper interface between the oil phase and the emulsion zone. Image taken using an Infinity microscope.

cables were used to illuminate the settler. Figure 20.13 presents a micrograph of the emulsion zone and the upper interface between the supernatant and the emulsion zone during settling of a bitumen froth diluted with n-heptane at 25°C and S/B = 3.0. The image was taken using an Infinity™ microscope. The existence of a sharp interface between the upper oil phase and the middle emulsion zone is clearly evident. The individual WD/DS/PA aggregates can be identified in this micrograph.

A dark field illumination micrograph of a representative WD/DS/PA aggregate is shown in Figure 20.14.[10] The image was taken using a light microscope with the same setup shown in Figure 20.12. The aggregates are composed of primary particles (WD, DS, PA) and are also highly porous. Significant amounts of solvent-diluted bitumen are trapped in the porous structures.

5.3. Settling Curve and Settling Rate of WD/DS/PA Aggregates

Zone settling of WD/DS/PA aggregates in bitumen diluted with aliphatic solvent develops characteristic interfaces. Using the experimental setup shown in Figure 20.12, the interface position during settling can be tracked by visual observation. Figure 20.15 shows representative settling curves for bitumen froth diluted

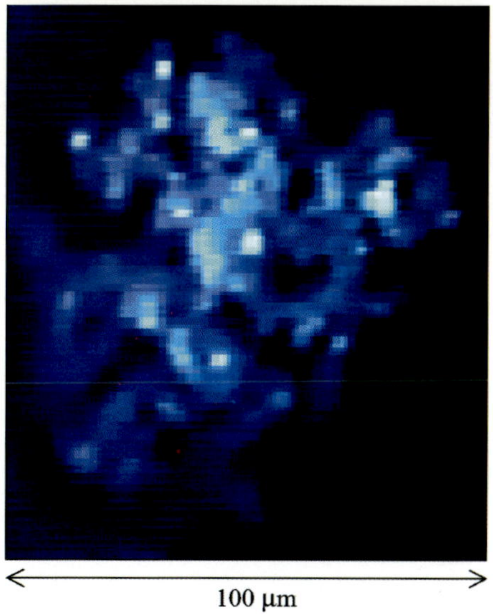

Figure 20.14. Micrograph of a representative WD/DS/PA aggregate. Image taken using a light microscope.

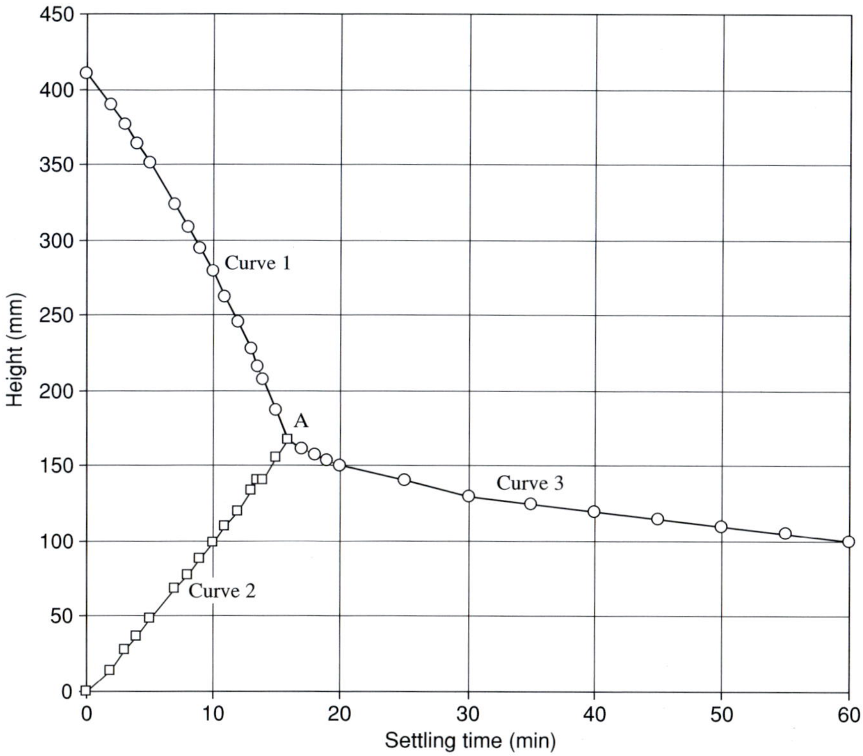

Figure 20.15. Settling curves of bitumen froth diluted with n-pentane/n-hexane solvent (50/50 by weight mixture) at S/B = 2.8, 22°C.

with n-pentane/n-hexane solvent (50/50 by weight mixture) at 22°C and S/B = 2.8. Curves 1 and 2 are the positions of the downward-moving oil phase/emulsion zone interface and the upward-propagating emulsion zone/compaction zone interface, respectively. These two interfaces merge at point A, at which compaction starts. Accordingly, curve 3 represents the compaction of the settled WD/DS/PA aggregates.

The settling rate of the WD/DS/PA aggregates is one of the most important parameters for bitumen emulsion treatment with aliphatic solvent. It dictates the size of settler required to achieve a certain throughput. From the experimental settling curve shown in Figure 20.15, we calculate the average settling rate of the aggregates as the slope of curve 1 by fitting a linear equation to the experimental data points. The settling rate of the WD/DS/PA aggregates in bitumen diluted with aliphatic solvent depends on many factors and variables including solvent type, S/B, temperature, and mixing conditions.

Figure 20.16 presents the experimental settling rates for bitumen froths diluted with aliphatic solvent as a function of solvent carbon number. n-Alkanes (n-pentane, 50/50 mixture by weight of n-pentane and n-hexane, n-hexane, and n-heptane) were used as solvents at S/B = 2.8 for these tests, and the temperature was set at 22°C. A light alkane gives a much greater settling rate than a heavier alkane at the same S/B. Obviously, the density and viscosity of the oil phase (medium) changes when the solvent is varied. Assuming all other parameters (i.e., WD/DS/PA aggregate structure, the concentration of the aggregates in the medium, and the interactions among aggregates) do not change when the solvent used to treat the bitumen froth is varied from n-pentane to n-heptane, the variation

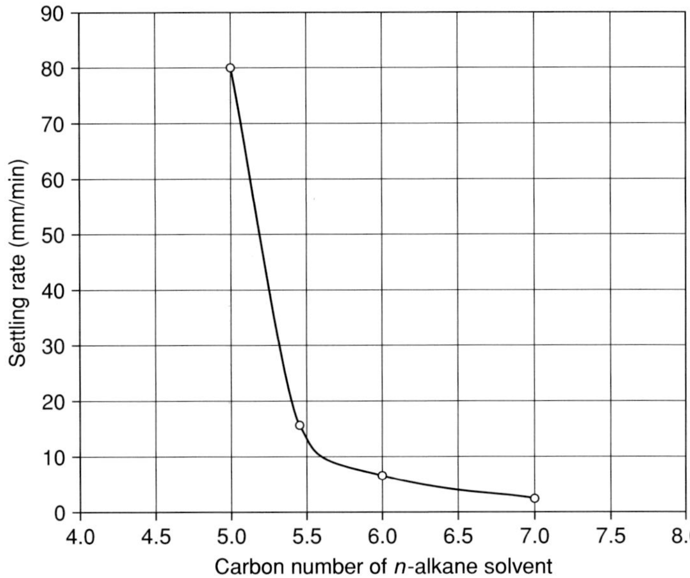

Figure 20.16. Settling rate of bitumen froth diluted with n-alkane at S/B = 2.8, 22°C.

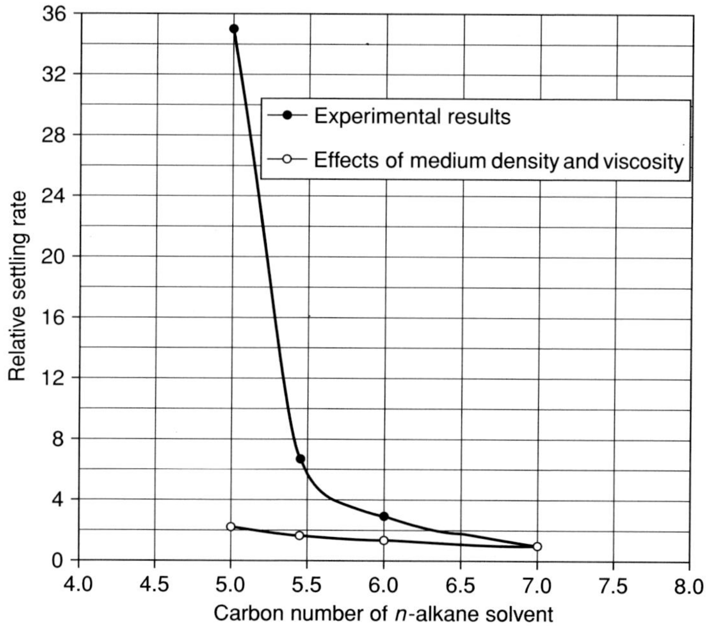

Figure 20.17. Relative settling rates of bitumen froth diluted with n-alkanes at S/B = 2.8, 22°C.

in the driving force for settling could be calculated according to the Stokes law (equation (20.2)) for a creeping flow regime (Reynolds number: $Re \ll 1$). Figure 20.17 compares the experimental relative settling rate with the relative settling rate calculated by accounting for the contributions of only density and viscosity changes of the oil phase. The results indicate that density and viscosity variations of the oil phase do not explain the settling rate differences. Therefore, structural differences of the WD/DS/PA aggregates must be the dominating factor.

Figure 20.18 shows the effect of S/B on the settling rate of bitumen froth diluted with n-pentane/n-hexane solvent (50/50 by weight mixture) at 22°C. Higher S/B results in much faster settling of the WD/DS/PA aggregates for a given solvent. Similar analysis to that above reveals that differences in the WD/DS/PA aggregate structure have to be taken into account in order to explain the settling rate increase with S/B.

The authors[43] studied the effect of mixing temperature on settling rate for bitumen froth diluted with n-alkane solvents. In these tests, the settling temperature was fixed at 30°C and the mixing temperature was varied from 30 to 100°C. A 50/50 by weight mixture of n-pentane and n-hexane at S/B = 2.0 was used as solvent. The results are shown in Figure 20.19. The settling rate of the WD/DS/PA aggregates is 24 mm/min for a 30°C mixing temperature. The settling rate increases to 98 mm/min when the mixing temperature is 100°C. The results indicate that the structure of the aggregates strongly depends on mixing conditions for froth and solvent.

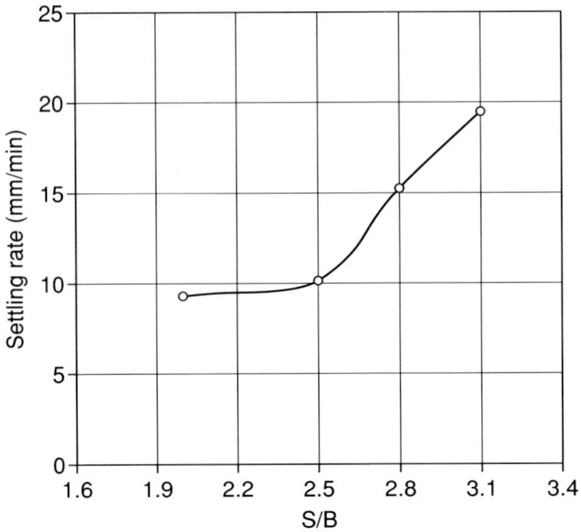

Figure 20.18. Settling rate vs. S/B for bitumen froth diluted with n-alkane solvent (50/50 by weight n-pentane/n-hexane) at 22°C.

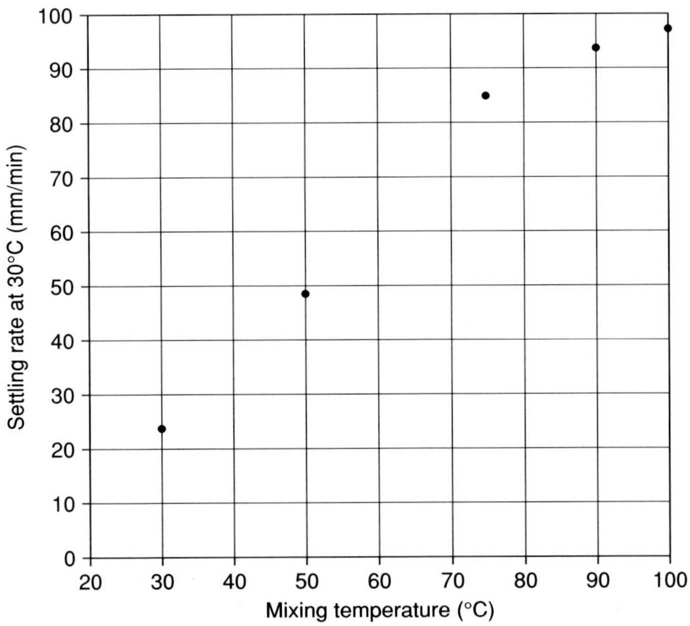

Figure 20.19. Effect of mixing temperature on settling rate of bitumen emulsions diluted with n-alkanes; settling temperature = 30°C; 50/50 by weight n-pentane/n-hexane solvent at S/B = 2.0.

5.4. Structural Parameters of WD/DS/PA Aggregates

As discussed above, the settling rate of the WD/DS/PA aggregates in solvent-diluted bitumen depends on many factors and conditions, but the key to achieving a high settling rate under a certain set of conditions is to obtain aggregates that are larger and more compact. Therefore, it is important to quantitatively describe the structure of the WD/DS/PA aggregates at various conditions. The authors[10,43] have modeled the hindered settling rate of the WD/DS/PA aggregates in solvent-diluted bitumen, analyzed the effect of the porous structure on settling rate, and developed a series of procedures to determine the structural parameters of the WD/DS/PA aggregates.

The settling rate of the WD/SD/PA aggregates (before reaching the point of compaction) can be modeled well by Kynch's theory of kinematic sedimentation.[44] The formula for the hindered settling rate, u, can be derived from the Richardson–Zaki approximation[45]:

$$u = u_0(1 - \phi)^n \tag{20.1}$$

where ϕ is the volume fraction of the aggregates, n is a coefficient that can vary from 4 to 22, and u_0 is the free settling rate of the aggregates. For an impermeable sphere in a Newtonian creeping flow regime, u_0 can be described by the Stokes law:

$$u_0 = \frac{g \Delta\rho D^2}{18\eta} \tag{20.2}$$

where g is the gravitational constant, $\Delta\rho$ is the difference between the aggregate effective density and the continuous medium density, D is the diameter of the aggregates, and η is the viscosity of the continuous medium. For a porous aggregate with porosity of ξ:

$$\Delta\rho = (\rho_E - \rho_m) = (1 - \xi)(\rho_P - \rho_m) \tag{20.3}$$

where ρ_E is the effective density of the aggregates, ρ_P is the density of the primary particles composing of the aggregates, and ρ_m is the density of the medium. Substituting equation (20.3) to equation (20.2) and considering equation (20.1), the hindered settling rate of impermeable aggregates, u, is obtained as

$$u = \frac{g(1 - \xi)(\rho_P - \rho_m)D^2}{18\eta}(1 - \phi)^n. \tag{20.4}$$

Experimentation has shown that the WD/DS/PA aggregates are highly porous, so the effect of the permeability of the aggregates on settling rate must be addressed. For permeable aggregates, the settling rate, u_p, can be described as

$$u_p = u/\Omega \tag{20.5}$$

where Ω accounts for the reduction of the drag force due to the porous structure and is defined as the hydrodynamic resistance experienced by a permeable sphere relative to the resistance experienced by an impermeable sphere with the same diameter. The Brinkman's extension of Darcy's law has been recognized as one of the most appropriate models for Ω[46]:

$$\Omega = \frac{2\beta^2\left(1 - \dfrac{\tanh\beta}{\beta}\right)}{2\beta^2 + 3\left(1 - \dfrac{\tanh\beta}{\beta}\right)} \tag{20.6}$$

where

$$\beta = \frac{D}{2\sqrt{\kappa}}$$

and κ is the permeability. The Brinkman model is appropriate for low- to high-porosity flocs and reads

$$\kappa = \frac{d_P^2}{72}\left[3 + \frac{4}{1-\xi} - 3\sqrt{\frac{8}{1-\xi} - 3}\right] \tag{20.7}$$

where d_P is the diameter of primary particles. Figure 20.20 shows the calculated Ω as a function of aggregate porosity for a 100-μm aggregate composed of 5-μm primary particles according to the Brinkman model.[43] The value of Ω is always larger than 0.97 for porosity up to 0.80. The effect of permeability of the aggregates on settling rate is significant only when their porosities are larger than about 0.80. Our results presented below will show that typical WD/DS/PA aggregates have porosities less than 0.75. Equation (20.4) is therefore valid for modeling the settling rate of the WD/DS/PA aggregates.

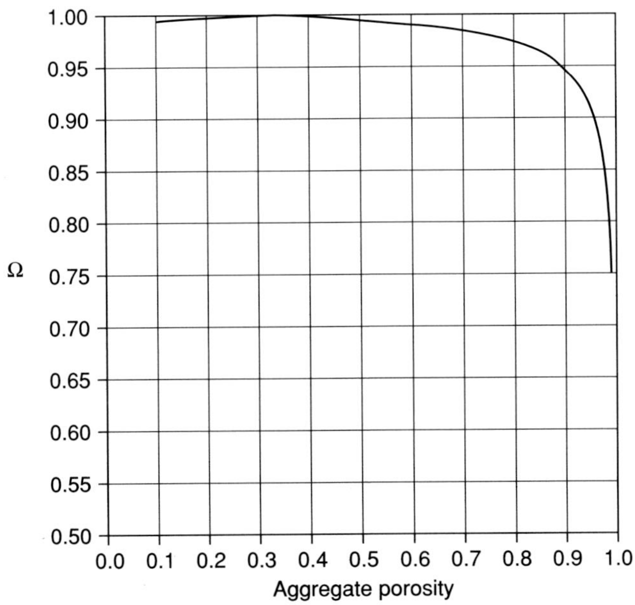

Figure 20.20. Ω as a function of aggregate porosity as modeled by the Brinkman equations. $D = 100$ μm; $d_P = 5$ μm.

According to equation (20.4), the determining parameters for settling rate are the structural parameters of the aggregates (ϕ, n, ρ_P, ξ, and D) and properties of the oil phase (η and ρ_m). The viscosity and density of the medium can be measured directly. In order to determine the structural parameters of WD/DS/PA aggregates, a suspension containing WD/DS/PA aggregates is prepared via settling and then combined with diluted bitumen at various ratios to obtain a series of diluted suspensions. The diluted suspensions are mixed by inverting the container ten times and then allowed to settle. The interface is tracked and the settling rate is calculated. Assuming there is no significant alteration of aggregate structure while preparing the diluted suspensions (or the aggregate structure recovers after suspension dilution), the settling rate of the diluted suspension relative to that of the original suspension, u_α, is

$$u_\alpha = \left(\frac{1 - \alpha\phi}{1 - \phi}\right)^n \tag{20.8}$$

where ϕ is the volume fraction of the WD/DS/PA aggregates in the original suspension and α is the volume fraction of the original suspension in the diluted suspension. Therefore, a series of experimental data of u_α vs. α are sufficient for determining the values of n and ϕ that are needed to model the settling rate. Figure 20.21 shows examples of the experimental data of u_α vs. α. Table 20.4 lists examples of determined ϕ and n values.[43] The lighter solvent (n-pentane/n-hexane

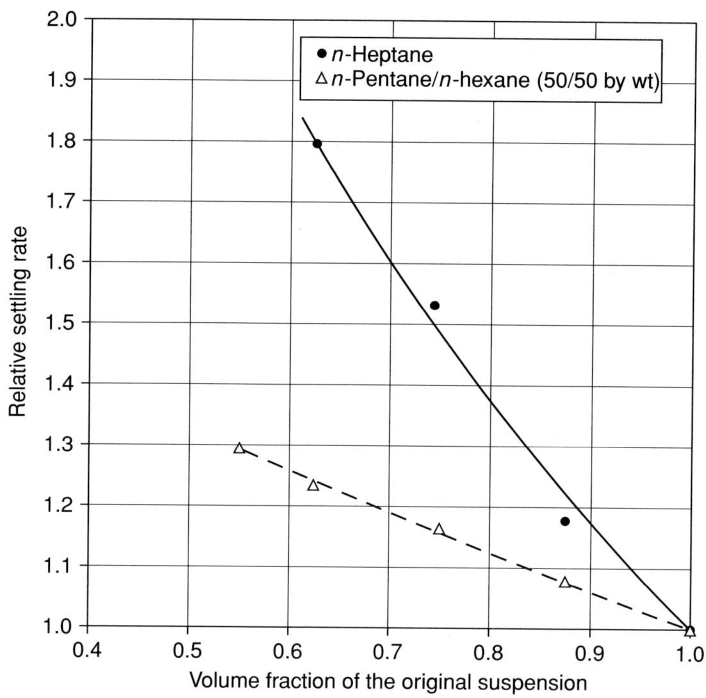

Figure 20.21. Relative settling rate, u_α, as function of volume fraction of the original suspension, α; temperature = 30°C.

Table 20.4. Values of ϕ and n for WD/DS/PA Aggregates

Solvent	Mixing temperature (°C)	Settling temperature (°C)	ϕ	n
n-Heptane	70	30	0.127	7.02
n-Heptane	30	30	0.123	11.43
n-Pentane/n-hexane (50/50 by wt)	30	30	0.121	4.27

mixture) gives a smaller n value than the heavier solvent (n-heptane), which is an indication that lighter solvent leads to more compact aggregate structures.

The value of ρ_P for the WD/DS/PA aggregates can be determined by analyzing the composition of the original suspension knowing densities of water droplets, solids, and precipitated asphaltenes. The value of ξ can be determined by analyzing the composition of the compacted WD/DS/PA aggregates, assuming that the network of aggregates collapses to a structure with overall porosity similar to the original individual aggregates after long compaction time since the WD/DS/PA aggregates have structures that are relatively soft and porous. The aggregate diameter D can therefore be determined using equation (20.4). Table 20.5 shows a few examples of aggregate diameter and aggregate porosity determined in this way. The results clearly indicate that higher mixing temperature leads to larger aggregates and a lighter solvent results in much more compact aggregates. Therefore, light aliphatic solvent and high temperature are the preferred conditions for bitumen emulsion treatment with aliphatic solvent with respect to settling rate.

5.5. Measuring Settling Rate of WD/DS/PA Aggregates Using In-Line Fiber-Optic Probe

Although the settling rate of the WD/DS/PA aggregates can be determined accurately by visual observation using the experimental setup shown in Figure 20.12, the method is difficult to apply to tests at elevated temperature and/or under pressure. It is also not an appropriate method for monitoring the settling of the aggregates in a settler in a plant setting. Long and Dabros[47] successfully used an in-line transflectance probe to monitor the settling of the WD/DS/PA aggregates in solvent-diluted bitumen. The probe is coupled with a near infrared (NIR) spectrophotometer via a fiber-optic cable. Figure 20.22 illustrates the experimental setup. A specially designed diffuse transflectance probe is inserted horizontally through a side opening in the settler and is, therefore, immersed in the solvent-diluted bitumen emulsion, thus providing direct in-line monitoring of the settling

Table 20.5. Aggregate Size and Porosity

Solvent	n-Pentane/n-Hexane (50/50 by wt)		n-Heptane	
Mixing temperature (°C)	30	75	30	70
Settling temperature (°C)	30	30	30	30
Aggregate diameter (μm)	62	114	56	90
Aggregate porosity	0.404	0.437	0.716	0.748

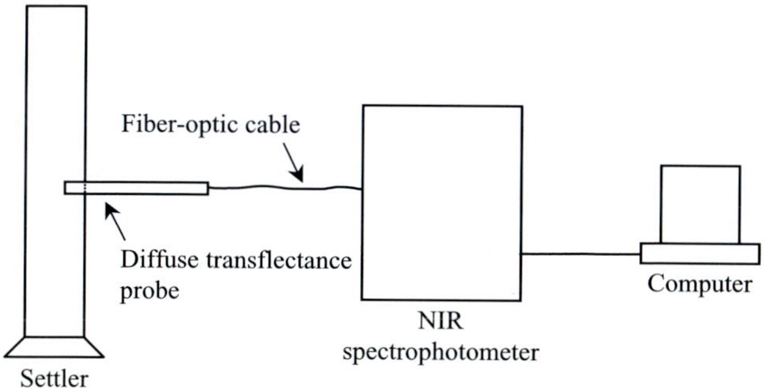

Figure 20.22. Experimental setup for monitoring settling of WD/DS/PA aggregates in solvent-diluted bitumen.

process. Incident monochromatic light is partially scattered and absorbed by the solution and the WD/DS/PA aggregates. The scattered light is detected by the in-line probe. The signal returning to the spectrophotometer thus contains information that reflects the instantaneous local composition of the solvent-diluted bitumen emulsion.

Figure 20.23 presents a series of NIR spectra acquired during settling of a bitumen froth diluted with *n*-pentane/*n*-hexane solvent (50/50 by weight mixture)

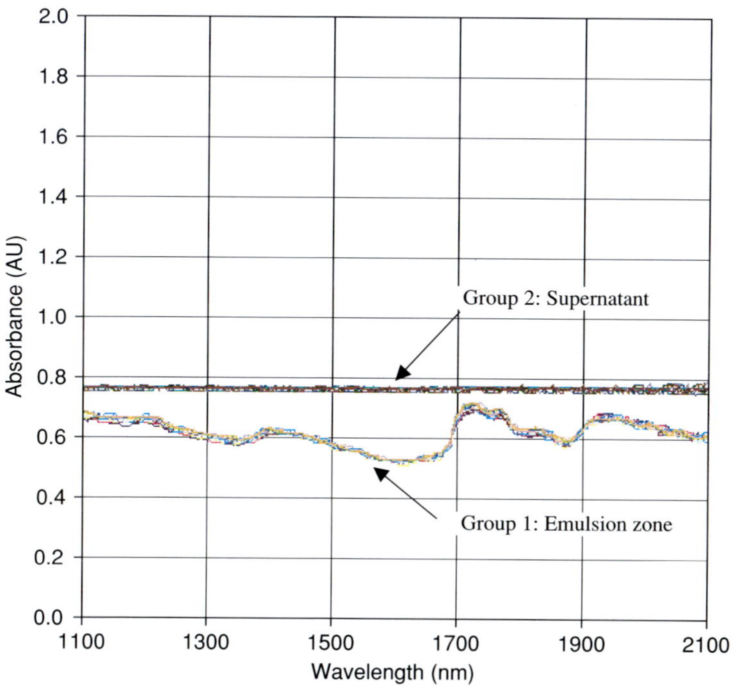

Figure 20.23. NIR spectra obtained using the in-line probe during settling of a bitumen froth diluted with *n*-pentane/*n*-hexane solvent (50/50 by weight mixture) at S/B = 3.1 and 22°C.

at S/B = 3.1 and 22°C. The apparent absorbance at wavelength i (A_i) is defined as

$$A_i = \log(1/R_i) = \log(I_i^0/I_i) \quad (20.9)$$

where R_i is the transflectance at wavelength i; I_i^0 and I_i are, respectively, the incident light intensity and the detected light intensity. The results in Figure 20.23 show that there are two groups of NIR spectroscopic signals. The probe gives group 1 spectra when it is in the emulsion zone and group 2 spectra when it is in clean solvent-diluted bitumen (i.e., the interface has passed below the probe as a result of settling of the WD/DS/PA aggregates). It is noted that all spectra in each group overlap, which indicates, characteristic of zone settling, that the compositions of both the emulsion zone and the clean oil phase are constant during settling. The NIR spectral features of the emulsion zone acquired through the diffuse transflectance probe are a combination of many factors. Nevertheless, the second-order C–H overtone band at 1200 nm, the first-order C–H overtone peak at 1750 nm, and the combination band at around 1400 nm can be identified.

Figure 20.24 compares the visually observed settling curve with the probe reading at 1650 nm. The probe is positioned at 210 mm. The probe reading agrees well with the visual observation and has a jump at 11 min, demonstrating that the in-line probe can effectively detect the time when the interface between the

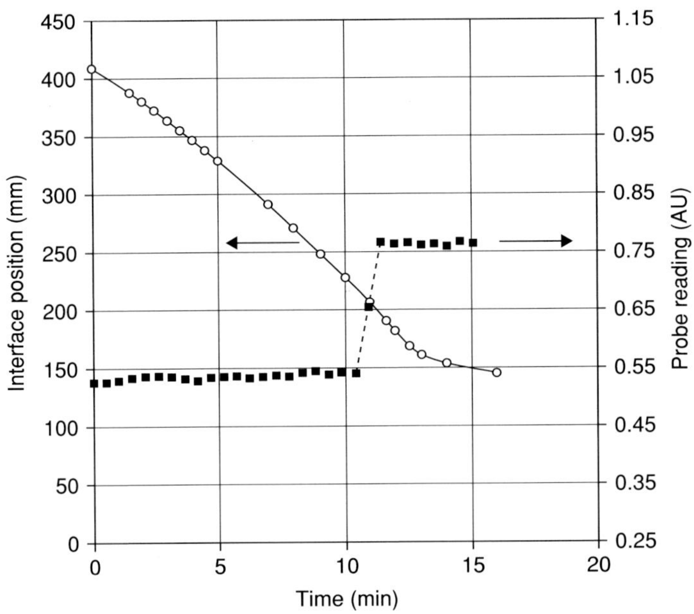

Figure 20.24. In-line NIR probe reading at 1650 nm as a function of settling time (filled squares) and the settling curve obtained by visually tracking the interface during settling (open circles); n-pentane/n-hexane solvent (50/50 by weight mixture) at S/B = 3.1 and 22°C. The in-line probe was located at 210 mm.

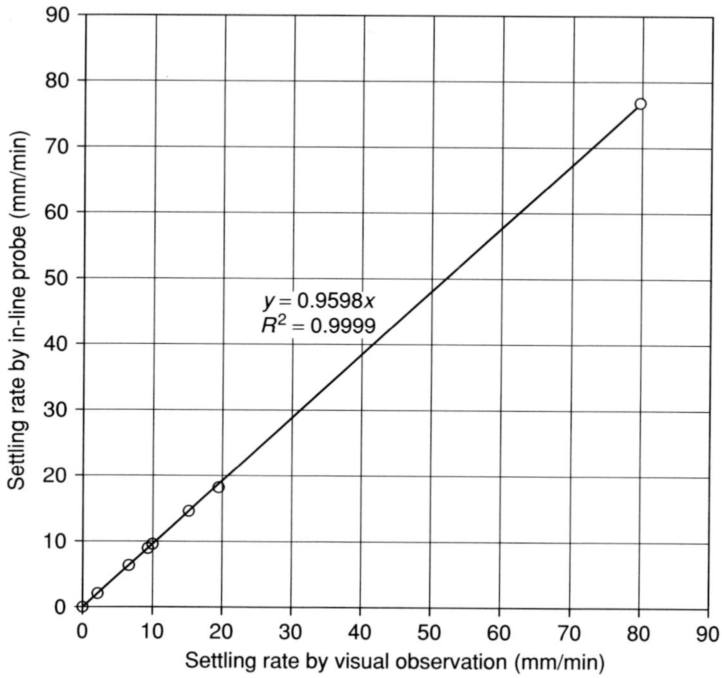

Figure 20.25. Settling rate determined by in-line NIR probe vs. settling rate measured by visually tracking the interface.

supernatant and the emulsion zone passes the probe and thus can be used to monitor settling of the WD/DS/PA aggregates.

Knowing the position of the in-line NIR probe in the settler and the time when the interface passes the probe, the average settling rate of the WD/DS/PA aggregates can be easily calculated. Figure 20.25 shows the settling rate determined by the in-line NIR probe vs. the settling rate determined from the settling curve obtained by visually tracking the interface. Excellent agreement is found between the results from the two different methods.

5.6. Asphaltene Rejection

During treatment of bitumen emulsions with aliphatic solvents, a portion of asphaltenes is rejected from the original bitumen, at the same time, when complete demulsification is achieved. The level of asphaltene rejection depends on many factors including solvent type, S/B, and temperature. Figure 20.26 shows asphaltene content in bitumen after treatment as a function of S/B with n-pentane, n-hexane, and n-heptane solvents at 25°C. The lighter n-pentane rejects more asphaltenes than the heavier n-heptane at the same S/B, and a higher S/B leads to more asphaltene rejection for the same solvent. The threshold S/B value for the onset of asphaltene precipitation is 1.0 for n-pentane, 1.35 for n-hexane, and 1.6 for n-heptane at 25°C.

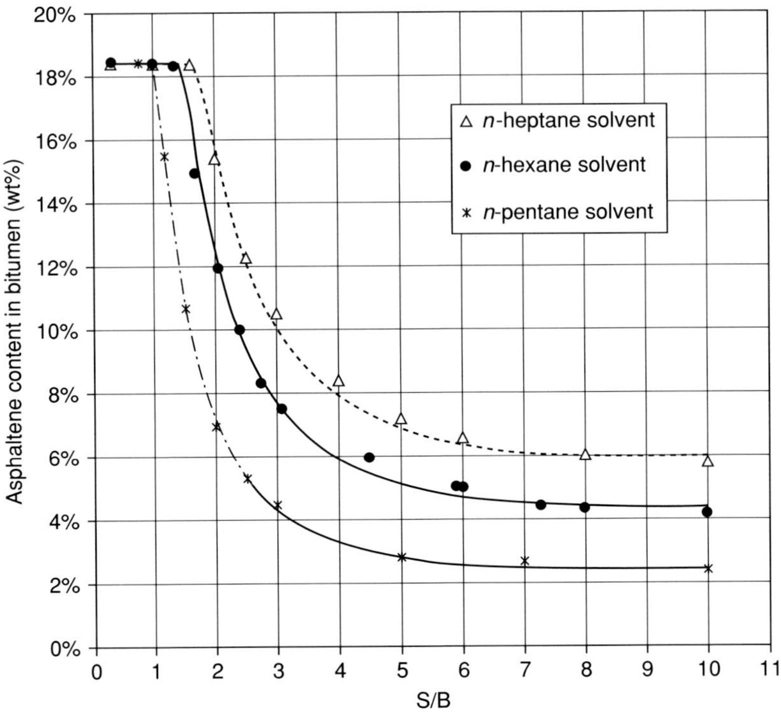

Figure 20.26. Asphaltene content in bitumen after treatment with aliphatic solvents at 25°C.

Figure 20.27 presents an example of the effect of temperature on asphaltene content in bitumen after treatment with n-pentane at S/B = 1.4 and 400 psi. Asphaltene solubility increases from 30 to 100°C but decreases at temperature above 100°C. A detailed fundamental treatment of asphaltene solubility as a function of temperature is beyond the scope of this work. Nonetheless, it is considered likely that the decrease in asphaltene solubility in pentane-diluted bitumen at high temperatures may result from an excessive reduction in solvating power of the solvent due to its reduced density at elevated temperatures.

5.7. Product Quality—Water and Solids Contents

The primary criteria to assess product quality after treatment of bitumen emulsions are water content in the solvent-diluted bitumen and solids content in the bitumen. Water content is usually analyzed using Karl Fischer titration and solids content is often analyzed as ash content. Figure 20.28 shows water content in diluted bitumen product after treatment of a bitumen froth with various n-alkane solvents at 25°C. The water content reaches about 0.05 wt% when the S/B is higher than a certain value for each solvent. Comparing the results in Figures 20.26 and 20.28, the minimum S/B that is required to produce water-free diluted bitumen always coincides with the onset S/B for asphaltene precipitation for each solvent. It is not absolutely clear yet whether asphaltene precipitation is a precondition for aggregation of water droplets and solids or whether the asphaltenes and the

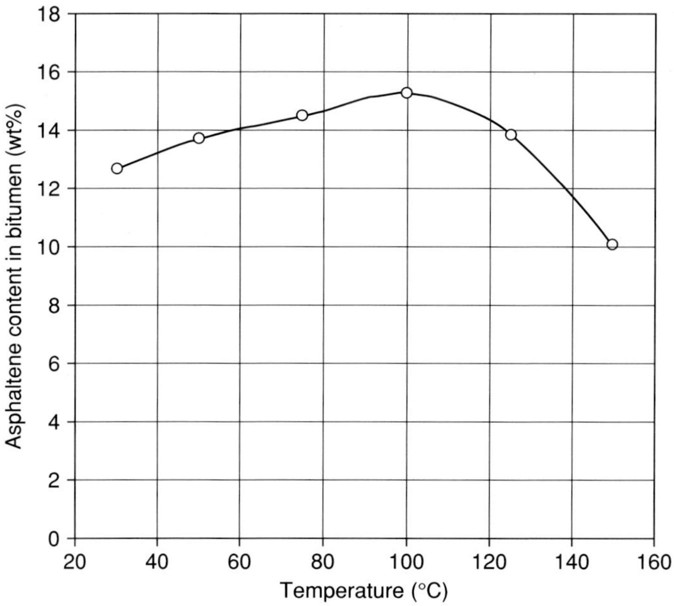

Figure 20.27. Asphaltene content in bitumen after treatment with n-pentane solvent as a function of temperature at S/B = 1.4.

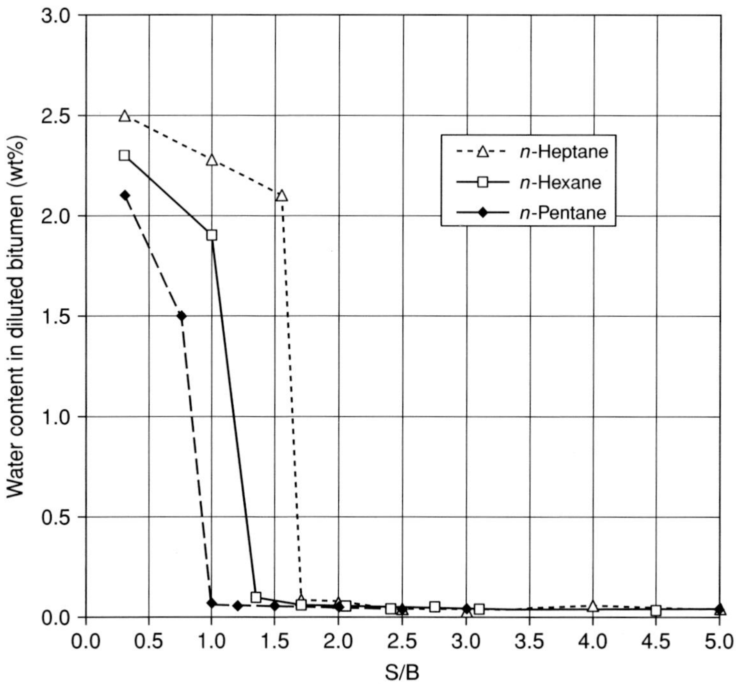

Figure 20.28. Water content in diluted bitumen product after solvent treatment of bitumen froth vs. S/B.

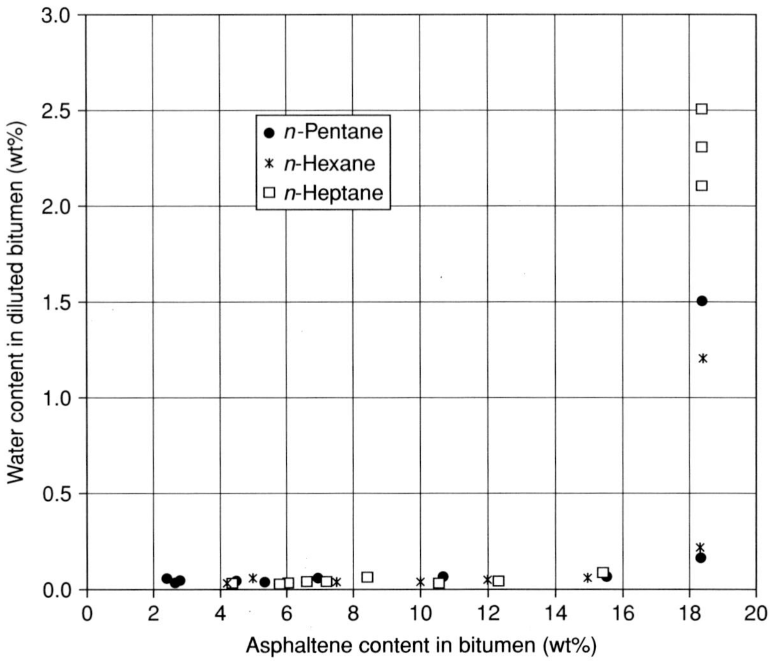

Figure 20.29. Water content in diluted bitumen product after solvent treatment of bitumen froth vs. asphaltene content in bitumen product.

water-in-bitumen emulsion just happen to become destabilized at about the same S/B. Nevertheless, experimental observations have proved that asphaltenes always form aggregates together with water droplets and dispersed solids as long as the onset conditions for asphaltene precipitation are satisfied. Figure 20.29 shows water content in diluted bitumen product vs. asphaltene content in bitumen product. Water-free product is always guaranteed as long as there is partial asphaltene rejection during the treatment with aliphatic solvents.

It has been found that the ash content in bitumen product follows the same trend as the water content shown in Figure 20.29. When the S/B is higher than the onset value for asphaltene precipitation, ash content in bitumen product is always minimized. The ash content ranges from 0.2 to 2 wt% when the S/B is lower than the onset value for asphaltene precipitation.

Figure 20.30 shows a bright field micrograph of the diluted bitumen after treatment of a bitumen froth with n-hexane at S/B = 1.5 and 25°C. There are no remaining water droplets after the treatment.

5.8. Product Quality—Micro-Carbon Residue (MCR)

Asphaltenes have a high-coking propensity. Partial rejection of asphaltenes thus offers the advantage of producing bitumen that has reduced coking propensity. Figure 20.31 shows experimentally determined micro-carbon residue (MCR) as a

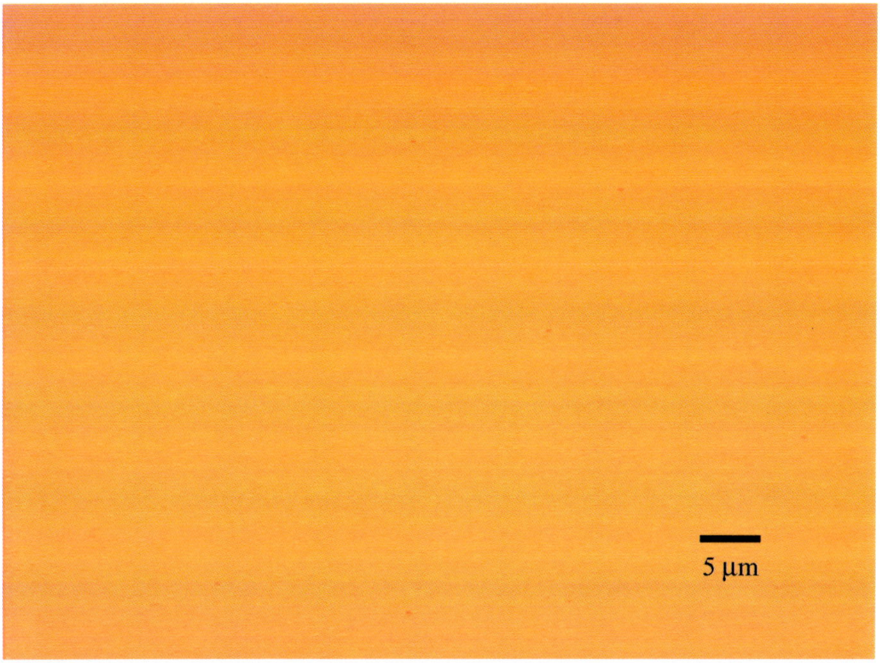

Figure 20.30. Bright field micrograph of clean diluted bitumen product after treatment of bitumen emulsion with aliphatic solvent.

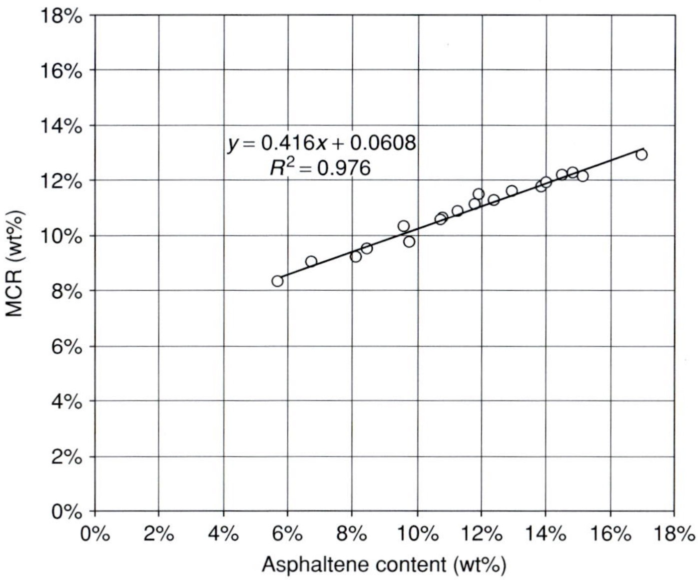

Figure 20.31. Micro-carbon residue as a function of asphaltene content in bitumen.

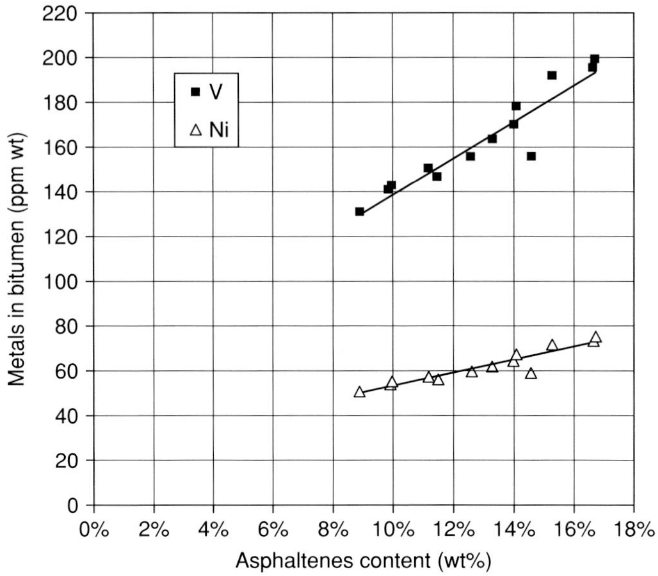

Figure 20.32. Nickel and vanadium contents in bitumen as functions of asphaltene content.

function of asphaltene content in bitumen.[39,41] The MCR in the original bitumen is 13.5 wt%, and it is reduced to about 9.7 wt% when half of the asphaltenes (equivalent to about 9 wt% asphaltene content in bitumen product) are rejected during solvent treatment.

5.9. Product Quality—Metals Contents

Nickel and vanadium are part of the porphyrin component in bitumen. Figure 20.32 shows nickel and vanadium contents in bitumen vs. asphaltene content.[39,41] Partial rejection of asphaltenes after solvent treatment results in significant reductions in nickel and vanadium contents in the bitumen product. The original bitumen contains 205 ppm vanadium and 75 ppm nickel. Rejection of half of the asphaltenes reduces vanadium and nickel contents to 130 and 50 ppm, respectively, which is equivalent to about a one-third reduction.

5.10. Product Quality—Sulfur and Nitrogen Contents

Asphaltenes contain relatively high concentrations of sulfur and nitrogen. Therefore, partial rejection of asphaltenes reduces the concentrations of sulfur and nitrogen in the bitumen. Figure 20.33 shows sulfur and nitrogen contents in bitumen vs. asphaltene content.[39,41] The original bitumen contains 5.05 wt% sulfur and 0.38 wt% nitrogen. Sulfur and nitrogen are reduced to 4.67 and 0.30 wt%, respectively, when half of the asphaltenes are removed from the original bitumen.

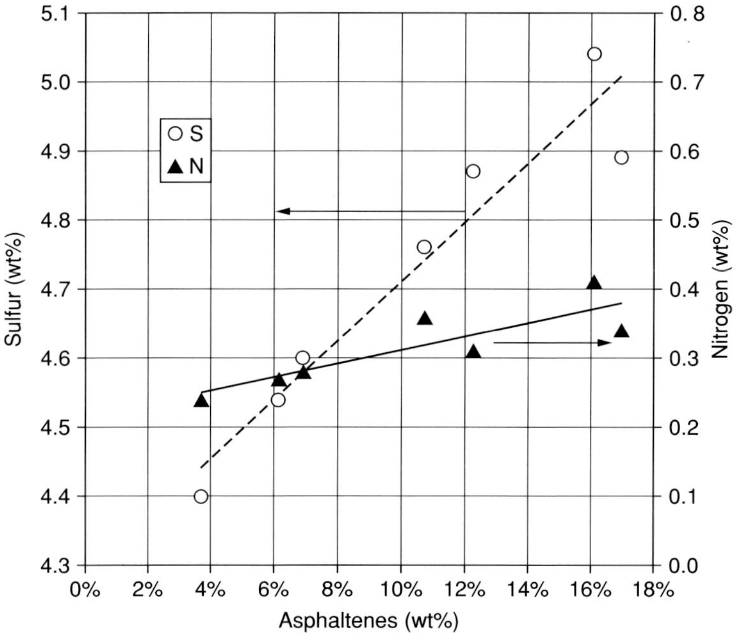

Figure 20.33. Sulfur and nitrogen contents in bitumen as a function of asphaltene content.

5.11. Viscosity of Bitumen

Asphaltenes are the heaviest components in bitumen, and it is known that they greatly increase crude viscosity.[48] Due to the relatively high concentration of asphaltenes in bitumen (15–21 wt%), bitumen viscosity is very high. In order to meet the pipelining specification (<350 cSt at pipelining temperature), 20–30 vol% light diluents (e.g., natural gas condensate) are typically blended in.[5] As discussed above, the treatment of bitumen emulsions with aliphatic solvent not only produces clean product (<0.1 wt% water plus solids) but also rejects a portion of the asphaltenes. Figure 20.34 shows an example of the effect of asphaltene content on bitumen viscosity. The viscosity of the bitumen is 2800 mPa s at 50°C, and it drops to about 200 mPa s when the asphaltenes are reduced to 6 wt%. Accordingly, significantly less solvent (e.g., <15 vol% hexanes) is required for blending with this bitumen to achieve the viscosity specification for pipelining.

6. Conclusion

Solvents are often used to treat heavy oil emulsions such as bitumen froth produced as an intermediate material in the recovery of bitumen from oil sands. It has been found that the aromaticity of the solvent and the solvent-to-bitumen dilution ratio (S/B, by wt) have profound effects on the stability of water-in-bitumen emulsions. An aliphatic solvent at a certain S/B not only partially precipitates

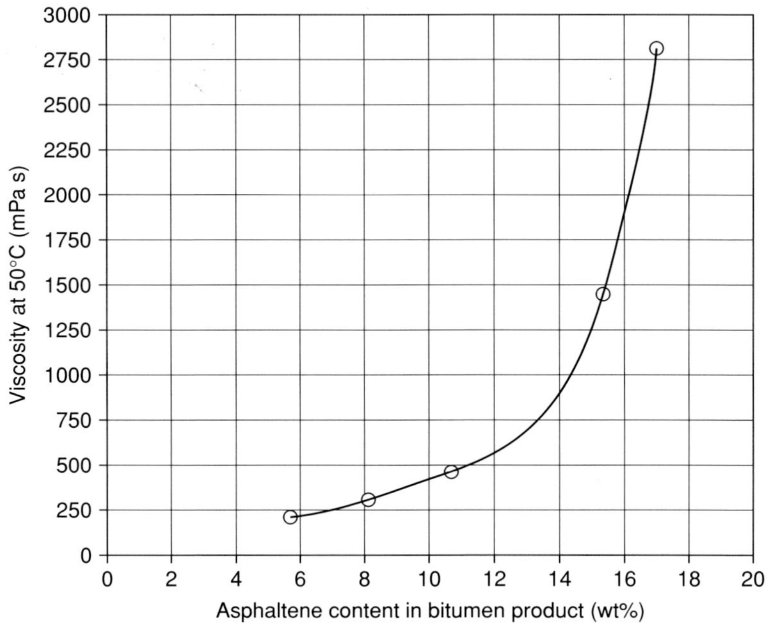

Figure 20.34. Viscosity of bitumen as a function of asphaltene content.

asphaltenes from the solvent-diluted bitumen solution but also, at the same time, promotes aggregation of the emulsified water droplets (WD), dispersed solids (DS), and precipitated asphaltenes (PA). The minimum S/B for achieving aggregation of water droplets and solids in oil phase always coincides with that required for the onset of asphaltene precipitation. A lighter aliphatic solvent has a lower minimum S/B (e.g., 1.0 for n-pentane) for the onset of asphaltene precipitation and formation of WD/DS/PA aggregates than a heavier aliphatic solvent (e.g., 1.6 for n-heptane). A solvent containing aromatic and/or naphthenic components can also promote aggregation, but much higher S/B (e.g., >4 for naphtha) is required. The WD/DS/PA aggregates exhibit zone settling in solvent-diluted bitumen, and various distinct zones and corresponding interfaces develop during settling. The top zone is a clean oil phase containing less than 0.1 wt% water-plus-solids, the middle is an emulsion zone containing the WD/DS/PA aggregates, and the bottom is a compaction zone of the settled WD/DS/PA aggregates. The position of the characteristic sharp interface that develops between the oil phase and the emulsion zone during settling can be monitored by visual observation with the help of proper illumination. An in-line diffuse transflectance probe coupled with a spectrophotometer via a fiber-optic cable has also been developed for monitoring of the settling of the WD/DS/PA aggregates. Settling rates at various conditions were measured. It was found that the settling rate depends on many factors and conditions including solvent type, S/B, temperature, and mixing conditions. A lighter aliphatic solvent gives lower-porosity (e.g., about 0.4 with n-pentane compared to about 0.7 with n-heptane) WD/DS/PA aggregates and higher temperature leads to larger aggregates. Treatment of bitumen emulsions with aliphatic solvents not only produces clean bitumen but also leads to significant reductions in asphaltene content, coking propensity, and concentrations of metals, sulfur, and nitrogen. The viscosity of the bitumen after partial rejection of asphaltenes is also reduced significantly, which, in turn, greatly reduces the amount of light diluent that must be blended with the bitumen in order to meet the pipelining specification for viscosity. Therefore, proper selection of solvent, S/B, and other process conditions makes it possible to produce bitumen products to satisfy a variety of specifications. The resulting new method allows bitumen upgrading/refining units to be remotely located from the mining/production site since the bitumen product is pipelinable.

Acknowledgments

The authors thank Anita Jahn, Derek Chao, James Mason, John Ioannides, and Tom Kratchmer for their help in performing some of the laboratory experiments, and Conrad Gietz for his assistance in preparing this chapter.

References

[1] Mandil, C. (2002). Environmental and technological issues associated with non-conventional oil. In: *Proceedings, IEA Conference on Non-Conventional Oil: Prospects for Increases in Production*, Calgary, Alberta, Canada, November 25–26, 2002.

[2] Petroleum Communication Foundation (2000). *Canada's Oil Sands and Heavy Oil.* PCF, Calgary, Alberta, Canada.

[3] Speight, J.G. (1999). *The Chemistry and Technology of Petroleum*, 3rd edition, Marcel Dekker, Inc., New York.

[4] Berkowitz, N. and J.G. Speight (1975). The oil sands of Alberta. *Fuel* **54**, 138–149.

[5] Alberta Chamber of Resources (2004). *Oil Sands Technology Roadmap: Unlocking the Potential.* Alberta Chamber of Resources, Edmonton, Alberta, Canada.

[6] Czarnecki, J. (2001). Water-in-oil emulsions in recovery of hydrocarbons from oil sands. In: J. Sjoblom (ed.), *Encyclopedic Handbook of Emulsion Technology.* Marcel Dekker Inc, New York.

[7] Strausz, O.P. (1989). Bitumen and heavy oil chemistry. In: L.G. Hepler and C. Hsi (eds.), *AOSTRA Technical Handbook on Oil Sands, Bitumens, and Heavy Oils.* Alberta Oil Sands Technology and Research Authority, Edmonton, Alberta, Canada.

[8] Angle, C.W. (2001). Chemical demulsification of stable crude oil and bitumen emulsions in petroleum recovery—A review. In: J. Sjoblom (ed.), *Encyclopedic Handbook of Emulsion Technology.* Marcel Dekker Inc., New York.

[9] Peramanu, S., B.B. Pruden, and P. Rahimi (1999). Molecular weight and specific gravity distributions for Athabasca and Cold Lake bitumens and their saturate, aromatic, resin, and asphaltene fractions. *Ind. Eng. Chem. Res.* **38**, 3121–3130.

[10] Long, Y., T. Dabros, and H. Hamza (2002). Stability and settling characteristics of solvent-diluted bitumen emulsions. *Fuel* **81**, 1945–1952.

[11] Long, Y., T. Dabros, and H. Hamza (2004). Analysis of solvent-diluted bitumen from oil sands froth treatment using NIR spectroscopy. *Can. J. Chem. Eng.* **82**, 776–781.

[12] Groenzin, H. and O.C. Mullins (2000). Molecular size and structure of asphaltenes from various sources. *Energy Fuels* **14**, 677–684.

[13] Sheu, E.Y. (2002). Petroleum asphaltene—Properties, characterization, and issues. *Energy Fuels* **16**, 74–82.

[14] Dabir, B., M. Nematy, A.R. Mehrabi, H. Rassamdana, and M. Sahimi (1996). Asphalt flocculation and deposition III. The molecular weight distribution. *Fuel* **75**, 1633–1645.

[15] Storm, D.A. and E.Y. Sheu (1995). Characterization of colloidal asphaltenic particles in heavy oil. *Fuel* **74**, 1140–1145.

[16] Pfeiffer, J.P. and R.N.J. Saal (1940). Asphaltic bitumen as colloid system. *Phys. Colloid Chem.* **44**, 139–149.

[17] Sheu, E.Y., M.M. De Tar, and D.A. Storm (1992). Structure, interaction and phase transition of vacuum residue in apolar solvents. *Fuel Sci. Technol. Int.* **10**, 607–647.

[18] Storm, D.A. and E.Y. Sheu (1993). Rheological studies of Ratawi vacuum residue at 366 K. *Fuel* **72**, 233–237.

[19] Sheu, E.Y., M.M. De Tar, D.A. Storm, and S.J. DeCanio (1992). Aggregation and kinetics of asphaltenes in organic solvents. *Fuel* **71**, 299–302, and *Fuel* **73**, 1368, (1994).

[20] Andersen, S.I. (1994). Effect of precipitation temperature on the composition of *n*-heptane asphaltenes. *Fuel Sci. Technol. Int* **12**, 51–74.

[21] Miller, J.T., R.B. Fisher, P. Thiyagarajan, R.E. Winans, and J.E. Hunt (1998). Subfractionation and characterization of Mayan asphaltenes. *Energy Fuels* **12**, 1290–1298.

[22] Groenzin, H. and O.C. Mullins (1999). Asphaltene molecular size and structure. *J. Phys. Chem. A* **103**, 11237–11245.

[23] Clark, K.A. and D.S. Pasternak (1932). Hot water separation of bitumen from Alberta bituminous sand. *Ind. Eng. Chem.* **24**, 1410–1416.

[24] Baldassari, M.C. (2003). Influence of partially deasphalted feed on LC-FINNING unit performance. In: *Proceedings of the 3rd NCUT Meeting on Upgrading and Refining of Heavy Oil, Bitumen, and Synthetic Crude Oil*, Edmonton, Alberta, Canada, September 22–23.

[25] Tipman, R.N. and R.C. Shaw (1993). Recent advances in the treatment of oil sands froth. In: *Proceedings of Oil Sands Our Petroleum Future Conference*, Edmonton, Alberta, Canada, April 4–7.

[26] Schramm, L.L. and J.C.T. Kwak (1998). The rheological properties of an Athabasca bitumen and some bituminous mixtures and dispersions. *J. Can. Petrol. Technol.* January–February, 26–35.

[27] Dabros, T., A. Yeung, J. Masliyah, and J. Czarnecki (1999). Emulsification through area contraction. *J. Colloid Interface Sci.* **210**, 222–224.

[28] Mikula, R.J., V.A. Munoz, and W.W. Lam (1989). Microscopic characterization of oil sands processing emulsions. *Fuel Sci. Technol. Int.* **7**, 727–749.

[29] Axelson, D.E., R.J. Mikula, and Z.M. Potoczny (1989). Characterization of oil sands mineral components and clay-organic complexes. *Fuel Sci. Technol. Int.* **7**, 659–673.

[30] Sjoblom, J. *et al.* (2003). Our current understanding of water-in-crude oil emulsions. Recent characterization techniques and high pressure performance. *Adv Colloid Interface Sci.* 100–102, 399–473.

[31] Gu, G., Z. Xu, K. Nandakumar, and J.H. Masliyah (2002). Influence of water-soluble and water-insoluble natural surface active components on the stability of water-in-toluene-diluted bitumen emulsion. *Fuel* **81**, 1859–1869.

[32] Kloet, J.V., L.L. Schramm, and B. Shelfantook (2001). The influence of bituminous froth components on water-in-oil emulsion stability as determined by the micropipette technique. *Colloid Surfaces A: Physicochem. Eng. Aspects* **192**, 15–24.

[33] Yan, Z., J.A.W. Elliott, and J.H. Masliyah (1999). Roles of various bitumen components in the stability of water-in-diluted-bitumen emulsions. *J. Colloid Interface Sci.* **220**, 329–337.

[34] Yeung, A., T. Dabros, J. Czarnecki, and J. Masliyah (1999). On the interfacial properties of micrometer-sized water droplets in crude oil. *Proc. R. Soc. Lond.* A **455**, 3709–3723.

[35] Yeung, A., T. Dabros, and J. Masliyah (1998). Does equilibrium interface tension depend on method of measurement. *J. Colloid Interface Sci.* **208**, 241–247.

[36] Yeung, A., T. Dabros, J. Masliyah, and J. Czarnecki (2000). Micropipette: a new technique in emulsion research. *Colloids Surfaces* A **174**, 169–181.

[37] Tipman, R.N. and Y. Long (1999). Solvent process for bitumen separation from oil sands froth. Canadian Patent 2,149,737. Also US Patent 5,876,592, (1999).

[38] Dabros, T. Private technical communication.

[39] Hamza, H.A. (2003). Effect of solvent selection on froth treatment for mining operations. In: *Proceedings of the 3rd NCUT Meeting on Upgrading and Refining of Heavy Oil*, Bitumen and Synthetic Crude Oil, Edmonton, Alberta, Canada, September 22–23, 2003.
[40] Long, Y. Modeling asphaltene phase equilibria in solvent-diluted bitumen. *Fluid Phase Equil.* submitted.
[41] Long, Y., T. Dabros, and H. Hamza (2001). Asphaltene rejection from bitumen/heavy oil upon dilution with aliphatic solvents. presented to the *2001 CSChE Conference*, Halifax, Nova Scotia, Canada.
[42] Yang, X. and J. Czarnecki (2002). The effect of naphtha to bitumen ratio on properties of water in diluted bitumen suspensions. *Colloid Surface A: Physicochem Eng. Aspect* **211**, 213–222.
[43] Long, Y., T. Dabros, and H. Hamza (2004). Structure of water/solids/asphaltenes aggregates and effect of mixing temperature on the settling rate in solvent-diluted bitumen. *Fuel* **83**, 823–832.
[44] Kynch, G.J. (1952). A theory of sedimentation. *Trans. Farad. Soc.* **48**, 166–176.
[45] Richardson, J.F. and W.N. Zaki (1954). Sedimentation and fluidization I. *Trans. Inst. Chem. Engs. (London)* **32**, 35–53.
[46] Tang, P. and J.A. Raper (2002). Modeling the settling behaviour of fractal aggregates–A review. *Powder Technol.* **123**, 114–125.
[47] Long, Y. and T. Dabros (2005). Monitoring settling of water-solids-asphaltenes aggregates using in-line probe coupled with near infrared spectrophotometer. *Energy Fuels* **19**, 1542–1547.
[48] Sheu, E.Y., M.M. De Tar, and D.A. Storm (1991). Rheological properties of vacuum residue fractions in organic solvents. *Fuel* **70**, 1151–1156.

21

The Role of Asphaltenes in Stabilizing Water-in-Crude Oil Emulsions

Johan Sjöblom, Pål V. Hemmingsen, and Harald Kallevik

1. Introduction

Stable water-in-oil emulsions may form during the production of crude oil, as coproduced water is mixed with the oil from reservoir to separation facilities. Such emulsions introduce technical challenges, as they must be resolved to provide the specified product quality. Asphaltenes and resins indigenous to the oil are acknowledged as the most important components in respect to stabilization of the interface against coalescence. Fine solids may also contribute to the stabilization, as may the presence of naphthenic acids. The combination of these components creates a complex picture of several contributing mechanisms to the stability of water-in-oil emulsions. It is also established that the pressure conditions will influence the behavior of active components and the properties of the interface. In order to successfully mitigate the problems of stable emulsions, a thorough knowledge of component properties, behavior, interactions, and effect on water/oil interfacial properties must be developed for pressures ranging from ambient to high. This chapter seeks to bring to light recent findings related to these topics.

The theoretical and experimental impacts to create a better understanding of association phenomena and molecular interaction in oil-based systems are of recent date. One can say that these needs significantly emerge from the crude oil industry where a lot of problems of practical nature are closely related to colloid chemistry. In this chapter, we will concentrate on a description of water-in-crude oil (or model oil) emulsions and topics related to these.

The coproduction of water and crude oil in the form of an emulsion is highly undesirable from a process and product quality point of view. When the crude oil is processed from the well head to the manifold, there is usually a substantial pressure reduction with a pressure gradient over chokes and valves where the

Johan Sjöblom, Pål V. Hemmingsen • Ugelstad Laboratory, Department of Chemical Engineering, Norwegian University of Science and Technology, Trondheim, Norway and **Harald Kallevik** • Statoil R&D Center, Trondheim, Norway.

mixing of oil and water can be intense. After this the well-stream is entering the separator train where most of the water is separated from the crude. The final treatment may include an electrocoalescer after which the level of water should be below 0.5%. The processing of oil and water offers several possibilities to vigorously mix the phases and create an emulsion. However, this does not explain the manifold of stability levels found for emulsions based on different well-streams but similar process conditions. Obviously, indigenous components in the crude oil can cause higher or lower levels of emulsion stability. Our present view is that an interplay between asphaltenes and resins can explain most of the differences in stabilities. In addition fine solids (both organic and inorganic) will contribute to the stabilization.[1–3]

The chemistry of asphaltenes is dependent on pressure, solvency, and the occurrence of organic dispersants. On the way from the reservoir to the well head, the fluid system experiences a rather large pressure reduction and a slight temperature reduction. As the pressure decreases toward the bubble point, the light components (e.g., methane, ethane, and propane) will gradually occupy more and more volume relative to the rest of the components (see Figure 21.1). This will of course change the solubility parameter of the fluid. At some point this might lead to asphaltene flocculation and/or precipitation. Below the bubble point line, light components will form a gas phase. Upon a further pressure reduction, the influence of the light components on the solubility will be reduced. Thus, at some point the asphaltene aggregates might become solvable again. In the absence of efficient dispersants a large scale precipitation can occur. The dispersants will however suspend the small asphaltene particles in the crude oil. With the addition of water and the formation of a fresh water/oil (w/o) interface, a change in the interaction pattern between the components will take place. Since the dispersant normally has the highest affinity toward the fresh water/oil interface, an initial monolayer might be formed. Without the protecting dispersant layer, the asphaltene particles will

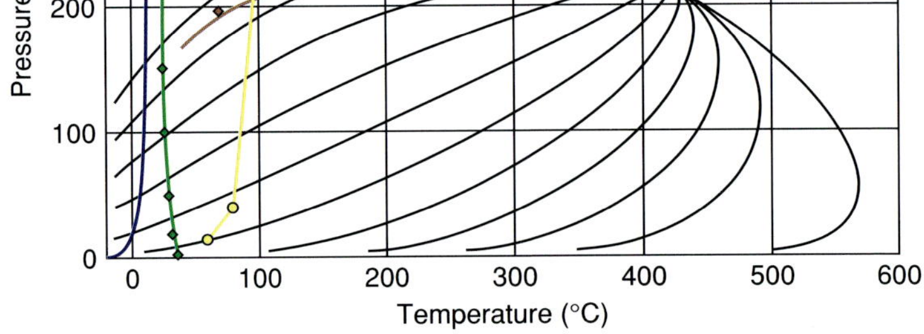

Figure 21.1. Pressure–temperature phase diagram for a crude oil system.

precipitate and will be accumulated at the surface of the water droplets. This is the basis for the formation of a mixed interfacial layer. In order to give high stability against coalescence, the interfacial rheological properties must be correct. The layer should be more elastic than viscous. In this way the interfacial relaxation will be much faster. This approach explains the stability level of w/o emulsions based on heavy crude oil components.[4–8] Both steric and particles stabilization normally account for the final stability level. Acidic crude oils can in addition to these mechanisms have a multilayer stabilization around the water droplets due to the special association of partially free and partially undissociated naphthenic acids close to their pKa.[9–14]

In order to understand the complex nature behind water-in-crude oil emulsion stability under real process condition, a thorough knowledge at both ambient and high pressures should involve the properties of the crude oil components (mainly the heavy ones), their association tendencies and accumulation at w/o interfaces, their solubilities and sensitivity toward changes in pressure and temperature, the special features of acidic components and their association structures in water/oil systems, and above all how these components by themselves or by an intricate interplay form highly elastic interfaces protecting the droplets against coalescence. This chapter attempts to elucidate the most recent findings covering these topics.

2. Chemistry of Crude Oils and Asphaltenes

2.1. Analytical Separation of Crude Oil Components

Crude oils have a complex nature, and it is not possible to determine the individual molecular constituents. Compositional studies are usually done by fractionation into predefined chemical families. The SARA group type fractionation separates the crude oil into the following classes: saturates, aromatics, resins, and asphaltenes. The definition of the SARA fractions is well accepted and will not be presented in detail here. Speight[15] gives a thorough introduction to the chemistry and structure of crude oil. Asphaltenes are defined as the solubility class of crude oils that precipitate in the presence of aliphatic solvents, while the resin fraction is defined as the fraction soluble in light alkanes but insoluble in liquid propane.[15] Asphaltenes and resins are known as large, polar, polynuclear molecules consisting of condensed aromatic ring systems and heteroatoms like sulfur, nitrogen, and oxygen. Asphaltenes and resins have gained increased interest with the knowledge of the large effect of these constituents in particular on the overall performance of heavy crude oils.

Although the SARA fractionation method is just a rough sorting of the crude oil constituents, it provides an important classification of crude oils. As will be seen from other examples of applications in this chapter, SARA characteristics are frequently used when discussing different crude oil properties. The SARA fractionation method has found great utility in combination with high-performance liquid chromatography (HPLC).[16–20] In a normal-phase liquid chromatographic process, a nonpolar mobile phase permeates through a porous solid stationary phase

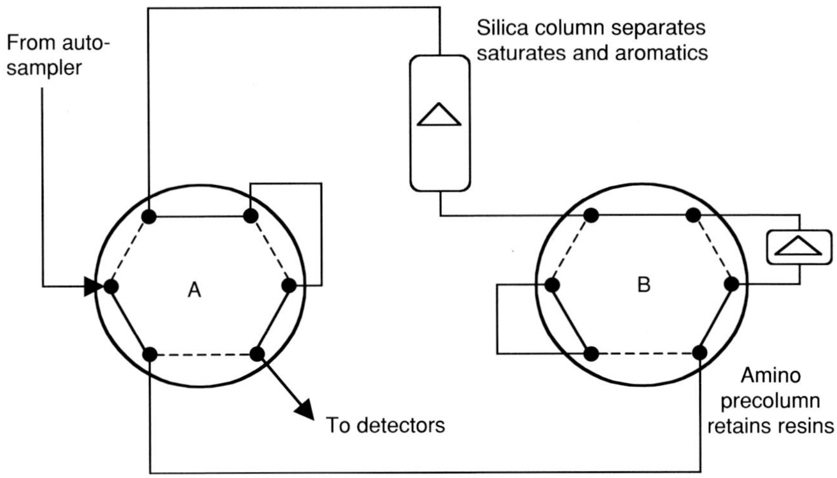

Figure 21.2. Schematic representation of HPLC system with main modules.

(polar) usually in the form of small uniform particles, packed into a cylindrical column. The sample is injected into the mobile phase, travels through the column, and is retained by the stationary phase mainly depending on polarity. Adsorption interactions between sample components, the mobile phase, and the stationary phase can be manipulated by the choice of mobile and stationary phases and flow conditions.

As an example of the SARA fractionation method we will discuss the procedure from a study by Hannisdal and coworkers.[21]

In this study, 20 crude oil samples at ambient temperature and pressure were received from exploration sites on the Norwegian Continental Shelf and sites located in Brazil, France, the South China Sea, the Atlantic Ocean, and the Gulf of Mexico. These oils were quantitatively fractionated into saturates, aromatics, resins, and asphaltenes (SARA) by asphaltene precipitation in *n*-hexane and preparative HPLC.

The HPLC system used in this study was built up from the modules shown schematically in Figure 21.2. Two columns were used: one 21.2 × 250 mm column packed with unbonded silica 15 μm and one 21.2 × 50 mm amino (10 μm) column. Dichloromethane (99.8%) and *n*-hexane (95%) were used as mobile phases. Generally, preparative separations of organic samples favor the use of unmodified silica as column packing material.[22] However, experiments on the preparative silica column showed that dichloromethane did not have the necessary solvent strength to elute the most polar components in heavy crude oils. These resins were irreversibly adsorbed onto the silica packing material. Instead an amino column, weaker in column strength, was used for this study in a multidimensional column setup. The two selection valves shown allow an individual reloading of the two columns. Valve A controls the flow direction through columns (normal/backflush), while valve B includes or excludes the precolumn. The new setup reduced the time of analysis from 70 to 26 min.

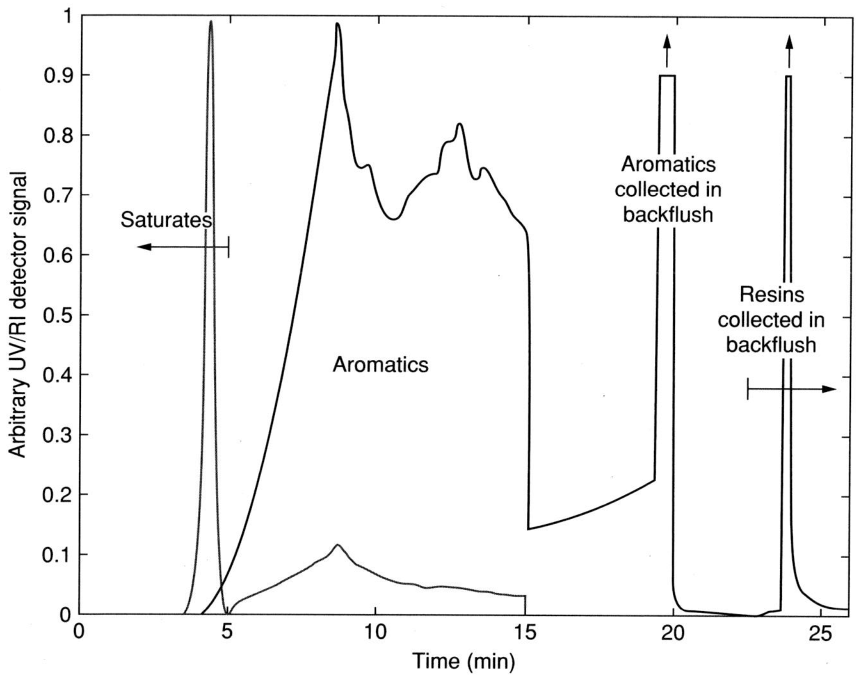

Figure 21.3. Typical chromatogram from preparative HPLC analysis. The solid line represents the RI signal, while the broken line represents the UV chromatogram at 254 nm. Peaks are cut for better visualization.

A sample of 5 mL asphaltene-free crude oil in *n*-hexane (corresponding to 0.6 g crude oil) was injected into the 20 mL/min flow (isocratic) of filtrated and degassed *n*-hexane. Crude oil components travel through the columns and are retained mainly depending on polarity. Saturates, having no retention on the columns, are collected from RI signals (3–5 min). After 15 min, all aromatics have left the amino precolumn but some condensed polycyclic aromatic ring systems are not yet eluted from the silica column. These are collected from the preparative column by dichloromethane backflush (<23 min), while the resin fraction is desorbed from the amino precolumn by dichloromethane backflush, 23–26 min after sample injection. This procedure provides the chromatogram in Figure 21.3 where the three fractions are well resolved. The SAR weight fractions are determined gravimetrically after controlled evaporation of solvents in N_2-atmosphere.

This example shows how SARA values are determined using a procedure for asphaltene preprecipitation followed by automated preparative high performance liquid chromatography. The discussed HPLC method had a sample capacity corresponding to 0.6 g of heavy crude oil. Since the determination of the SARA components is done gravimetrically, larger sample capacity will improve the accuracy of the procedure significantly. Moreover, the large sample capacity allows further characterization of the individual SARA fractions.

Saturates and aromatics: The saturates (or aliphatics) are the nonpolar compounds containing no double bonds and include both the alkanes and the cycloalkanes. Wax is a subclass of the saturates. The aromatics consist of all compounds with one or more benzene rings. These ring systems may be linked up with naphtene rings and/or aliphatic side chains.

Resins: This fraction is composed of polar molecules often containing heteroatoms such as nitrogen, oxygen, or sulfur. This fraction is operationally defined, and one common definition of resins is as the fraction soluble in light alkanes such as pentane and heptane, but insoluble in liquid propane.[15,23,24] Naphthenic acids are a part of this fraction.

Asphaltenes: Asphaltenes are polar molecules that can be regarded as similar to the resins, but with higher molecular weight, typically 500–1500 g/mol. The asphaltene fraction, like the resins, is defined as a solubility class, namely the fraction of the crude oil precipitating in light alkanes like pentane, hexane, or heptane. The precipitate is soluble in aromatic solvents like toluene and benzene. The asphaltene fraction contains the largest percentage of heteroatoms (O, S, N) and organometallic constituents (Ni, V, Fe) in the crude oil. The structure of asphaltene molecules is believed to consist of polycyclic aromatic clusters, substituted with varying alkyl side chains.[15,23]

2.2. Solubility and Aggregation of Asphaltenes

Asphaltenes tend to self-associate, a fact that has led to much controversy regarding asphaltene molecular weights. Molecular weights from a few hundred to several million have been reported.[15,25–27] Geographic origin is believed to influence structure, heteroatom content, and molecular weight. The self-association of asphaltenes is affected both by solvent conditions, pressure and temperature, and has been studied in detail by several authors.[5,6,27–36] Generally, increasing amount of aliphatic solvents and pressure reductions. toward the bubble point are the main contributors to increasing asphaltene aggregate size.

Asphaltenes are believed to be suspended as colloids in the crude oil consisting of nanometer-sized particles.[23] Each particle consists of one or more aromatic sheets of asphaltene monomers, with adsorbed resins acting as surfactants to stabilize the colloidal suspension. Under unfavorable solvent conditions, resins desorb from the asphaltenes, leading to an increase in asphaltene aggregate size and eventually precipitation of large asphaltene aggregates. This asphaltene aggregation model is the so-called steric stabilization model developed by Leontaritis and Mansoori[37–39] based on the earlier work by Nellensteyn and others.[40–42]

An alternative view on asphaltene stabilization exists, referred to as the thermodynamic model, and first reported by Hirschberg *et al.*[43] In this approach, the resins are not considered explicitly, but are treated as an undifferentiated part of the solvent medium. This view implies that the asphaltene monomers and aggregates are in thermodynamic equilibrium, and solvated by the surrounding medium. Thus, the critical distinction between the thermodynamic and steric stabilization models lies in the issue whether the asphaltene colloids are solvated or suspended in hydrocarbon media.[23] Asphaltene aggregation models have been reviewed by Andersen and Speight.[44]

The reversibility of the asphaltene aggregation is also a subject of some controversy in the literature. Hirschberg *et al.*[43] assumed that the aggregation was reversible, but probably very slow. Joshi *et al.*[45] found the precipitation from a live crude oil to be reversible in the matter of minutes, except for a subtle irreversibility observed for the first depressurization of the crude oil. They also discussed the different behavior of asphaltenes precipitated from crude oils with excess *n*-alkanes and by asphaltenes contained in the original crude oil. Hammami *et al.*[46] also found that the aggregation is generally reversible, but that the kinetics of the redissolution varies significantly depending on the physical state of the system. Peramanu *et al.*[47] reported differences in the reversibility of solvent- and temperature-induced aggregation.

Due to the mentioned uncertainty in asphaltene aggregation behavior, the prediction of asphaltene aggregation is a difficult task for the oil industry. Aggregation problems may be encountered both during mixing of crude oils and pressure depletion from reservoir conditions.[45,46] The aggregation and precipitation of asphaltenes may cause a large variety of problems, from formation damage, equipment plugging, catalyst deactivation, etc. In addition, asphaltene aggregation has a great influence on emulsion stability of crude oil–water emulsions.[48]

2.3. Characterization of Crude Oils by Near Infrared Spectroscopy

Over the last 30 years, near infrared spectroscopy has been increasingly used as an analytical tool, particularly by the food and agricultural industries, but also by the textile and polymer industries in addition to the petroleum industry.[49] The increasing popularity is due to the four principal advantages of the method, namely low time consumption, simplicity, multiplicity of analysis from a single spectrum, and the nonconsumption of the samples. The main disadvantages of NIR spectroscopy have always been the insensitivity to minor constituents and the broad absorption bands.[50] The latter drawback, however, is compensated by the rapid development of advanced and user-friendly software for multivariate analysis. Near infrared spectroscopy is very central in the crude oil characterization (see also sections in this chapter). The main advantage of NIR for our purposes is the possibility of obtaining both chemical and physical information on the crude oils. In the following a brief introduction to near infrared spectroscopy is given.

One of the main advantages of near infrared spectroscopy when working with colloidal systems like crude oils is the ability to gain information on the state of the system. In addition to absorption, near infrared spectra will display a baseline elevation due to light scattering by aggregates or particles in solution. For slightly lossy dielectric particles within the Rayleigh limit, the scattering and absorption processes contribute separately to the extinction coefficient[51]:

$$\sigma_{\mathrm{tot}} = \sigma_{\mathrm{abs}} + \sigma_{\mathrm{sc}} \tag{21.1}$$

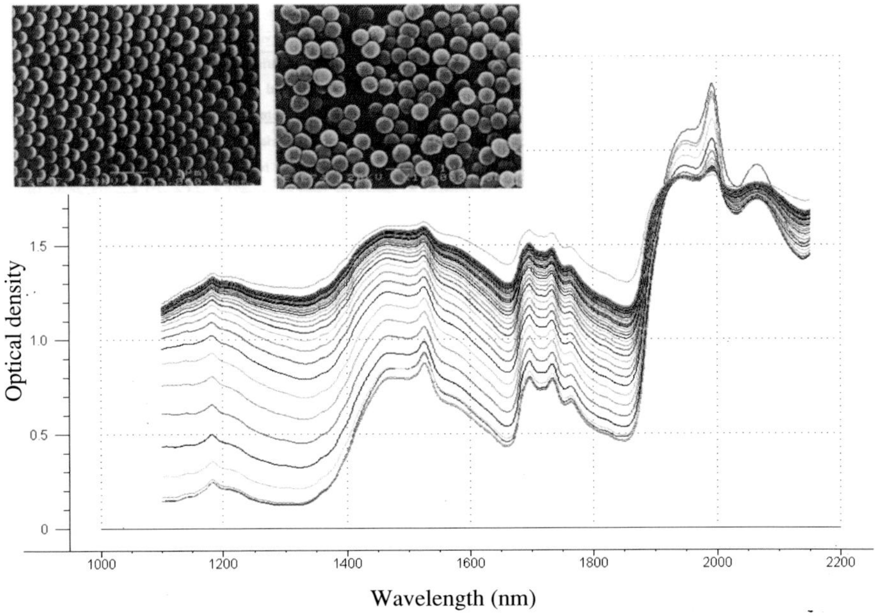

Figure 21.4. NIR spectra of silica particles formed by sol–gel process. Monodisperse silica particles with $d = 330$ and $d = 630$ nm shown.

where σ_{tot}, σ_{abs}, and σ_{sc} are the total, absorption, and scattering cross-sections, respectively. Also, the ratio of scattering to absorption scales with r^3, indicating the importance of particle size on the total light extinction. Thus the influence of scattering increases extensively with increasing diameter of the particles. Figure 21.4 demonstrates the effect of particle size on NIR spectra. During a sol–gel preparation of silica particles, NIR spectra were taken at intervals and the increasing baseline level due to increasing particle size is clearly seen.

Several authors have used NIR spectra for particle size determinations.[52–55] NIR spectroscopy has also found widespread use in the determination of both physical and chemical properties of crude oils and related materials.[49,56–61]

2.4. Asphaltene Aggregation Studied by High-Pressure NIR Spectroscopy

The aggregation behavior of asphaltenes has recently been studied by use of a high-pressure NIR system shown in Figure 21.5.[62] The system combines the accurate pressure and temperature control offered by a high-pressure rig in which a high-pressure transmittance NIR cell is connected. Specially designed software enables direct principal component analysis (PCA) of the spectra in order to easily detect asphaltene aggregation. By plotting principal components in the so-called score plots, samples are easily distinguished based on subtle changes in their respective NIR spectra. Changes caused by, for instance, compressibility or particle

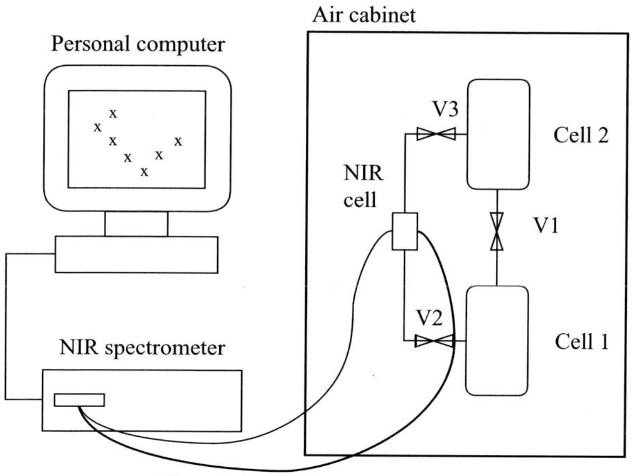

Figure 21.5. High-pressure near infrared spectroscopy setup.

aggregation in the samples can also be identified by the principal component analysis.

Both a recombined crude oil and model systems of pentane, toluene, and asphaltene were studied. The model systems were designed to be close to asphaltene precipitation, and two systems of 1.2 wt% asphaltene in 35 and 40 wt% pentane-in-toluene solvents were made. The crude oil contained 0.8 wt% asphaltene, had a reservoir bubble pressure of 155 bar, and was known to be susceptible to asphaltene precipitation by pressure depletion. The systems were pressurized to 100 or 300 bar and charged to the high-pressure rig at temperatures from 100°C to 150°C. The systems were then depressurized in steps, and the resulting NIR spectra at each pressure level were analyzed by the PCA routine. About 20 min of equilibration was allowed at each pressure level. It was found that at the asphaltene aggregation onset pressure a distinct shift in the PCA score plots could be seen. Reducing the pressure even further, the bubble point was detected by NIR spectra displaying low absorbance due to gas evolution. In Figure 21.6, the procedure of asphaltene onset detection for the crude oil from the NIR spectra via the PCA analysis to the impact on the phase envelope is demonstrated. At high pressures, there is a decrease in optical density with pressure depletion due to the compressibility of the crude oil. A soon as aggregates start to grow, the optical density increases due to increased light scattering. Although this is difficult to see directly from the spectra it is clearly seen from the score plot. In the phase envelope, the pressure depletion path is indicated along with the detected asphaltene aggregation onset pressure of 180 bar.

Performing the same type of experiments on the model systems (containing only asphaltenes in addition to solvents) produced the same type of results, thus confirming that the shift in the score plot indeed is caused by asphaltene aggregation. Depressurization of a model system without asphaltenes showed no shift in the score plot. Table 21.1 summarizes the experimental conditions and detected

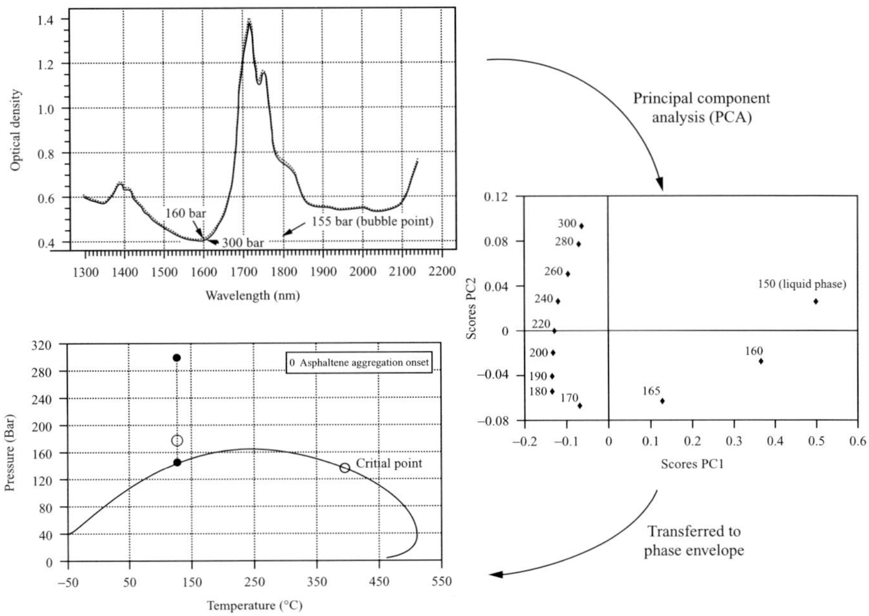

Figure 21.6. Detection of asphaltene aggregation onset pressure from high-pressure NIR spectroscopy and PCA.

onset pressures and bubble points for the systems studied. The bubble point pressures for the low-temperature model systems were too low to be detected. As expected, the onset pressure for the 40 wt% pentane solvent is higher than for the 35 wt% pentane system. In addition, the onset pressure at 150°C was found to be higher than that at 100°C.

The reversibility of the asphaltene aggregation was also studied for the crude oil and the 35 wt% pentane system at 150°C. This was done by repressurizing the systems stepwise from the bubble point to 100 bar and 300 bar, respectively, for the model system and the crude oil. When the original pressure was reached, the system was left to equilibrate and NIR spectra were recorded in intervals until no further spectral change was observed. In Figure 21.7, the optical density at 1600 nm for both the depressurization and the repressurization for the two systems is shown.

Table 21.1. Bubblepoint and Asphaltene Aggregation Onset Pressures

System	*Temperature (°C)*	*Bubblepoint (bar)*	*Onset pressure (bar)*
Crude oil	125	155	180
35 wt% pentane	100	<2.5	20
35 wt% pentane	150	10	30
40 wt% pentane	100	<2.5	40
40 wt% pentane, no asphaltene	150	14	–

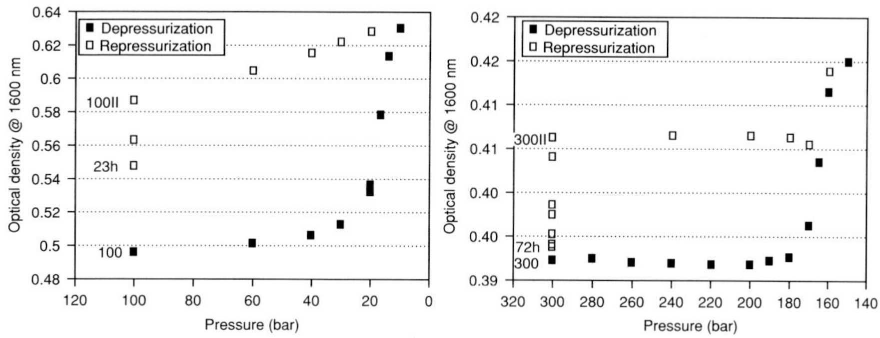

Figure 21.7. Asphaltene aggregation and reversibility of model system and crude oil.

For the crude oil, at the left-hand side in the figure, considerable redissolution of the aggregates was seen when increasing the pressure from 150 to 170 bar, as seen by the decrease in optical density at 1600 nm. However, no significant redissolution was then observed until the crude oil was left at the original pressure of 300 bar (300II). After 72 hr, the aggregate size had practically returned to its original state. For the model system on the other hand, the aggregates seemed to redissolve steadily during repressurizing, but the redissolution came to a stop after approximately 23 hr at the original pressure of 100 bar. Thus, while the asphaltene aggregation of the crude oil was more or less completely reversible, the model system asphaltene aggregation was only partial reversible. The asphaltenes of the model system are stripped for the resin fraction when they were prepared. And this can explain the different behavior of the crude oil and the model system.

Using NIR spectroscopy in combination with multivariate analytical techniques like principal component analysis has been shown to be a very efficient tool in studying asphaltene aggregation behavior. Very small spectral changes are detected in a score plot, enabling an accurate determination of the onset pressure. In addition bubble points are easily detected from the spectral features.

2.5. Disintegration of Asphaltenes Studied by NIR Spectroscopy

It is currently accepted that asphaltenes consist of aromatic compounds with π–π interactions, which undergo acid–base interactions and self-associate through hydrogen bonding.[63,64] The main strategy for the disintegration of asphaltene aggregates should hence be to break these bonds and stabilize the smaller asphaltenes aggregates from precipitation. Resins are usually thought to function as a dispersant of asphaltenes in crude oil. Auflem and coworkers[65] have introduced near infrared (NIR) spectroscopy as a potential tool for studying the effect of chemicals as dispersants of asphaltene aggregates. The structures and molecular weights of the various chemicals are summarized in Table 21.2.

Table 21.2. A Summary of the Amphiphiles and Naphthenic Acids Used as Additives in This Study

Name	Molecular weight	Chemical structure	Source	Purity (wt%)
5-β(H)-cholanoic acid (CHOL)	361	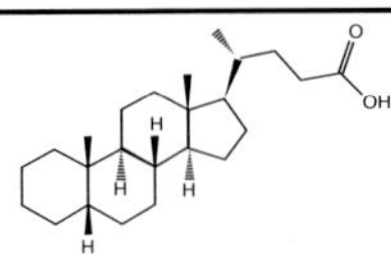	Chiron AS	>95
1-Naphthalenepentanoic acid, decahydro (2C4)	238	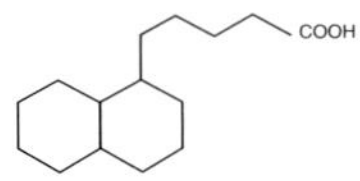	Chiron AS	>90
1-Naphthalenoic acid, decahydro-2-buthyl (C42)	238	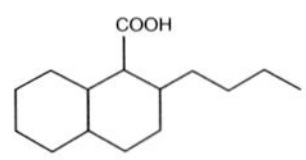	Chiron AS	>90
Crude naphthenic acid (CNA)	250	Mixture	Corn van der Locke	–
Fluka naphthenic acid (Fluka)	240	Mixture	Fluka	–
Naphthenic acid from North Sea (North Sea)	400	Mixture	Extracted	–
2-Ethyl-1-hexanol	130.23	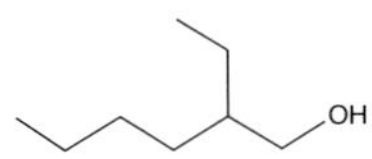	Merck	>99
1-Octanol	130.23	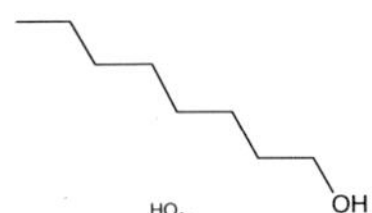	Merck	>99
Benzyl alcohol	108.14	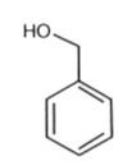	Fluka	>98
Hexylamine	101.94	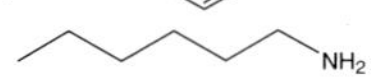	Merck	>98
n-Alkylbenzenesulfonic acid, ($n = C_{10} - C_{13}$) (ABSA)	385.5	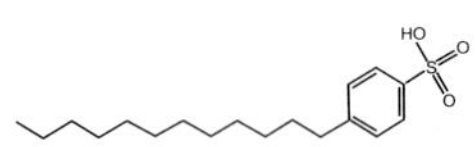	Alfa Aesar	96 +
Inhibitor A	–	Mixture	Tros / Dyno	–
n-Heptane	100.21		Merck	>99
Toluene	92.14		Merck	>99

[a] Up to 30 wt% unsaturated bonds in cyclic part of molecular structure.
[b] Up to 10 wt% unsaturated bonds in cyclic part of molecular structure.

Naphthenic acids: In crude oils the naphthenic acids are normally incorporated in the group of resin molecules. It is generally believed that these acids, which are complex mixtures of condensed ring structures where the number of saturated and unsaturated rings and alkyl moieties, arrangement of the COOH groups and hydrophile–lipophile balance will vary, may interact with the asphaltenes. The interactions are thought to occur between the acid groups and basic components in the asphaltene molecule. Thus, the dispersing power is determined by the molecular structure, and naphthenic

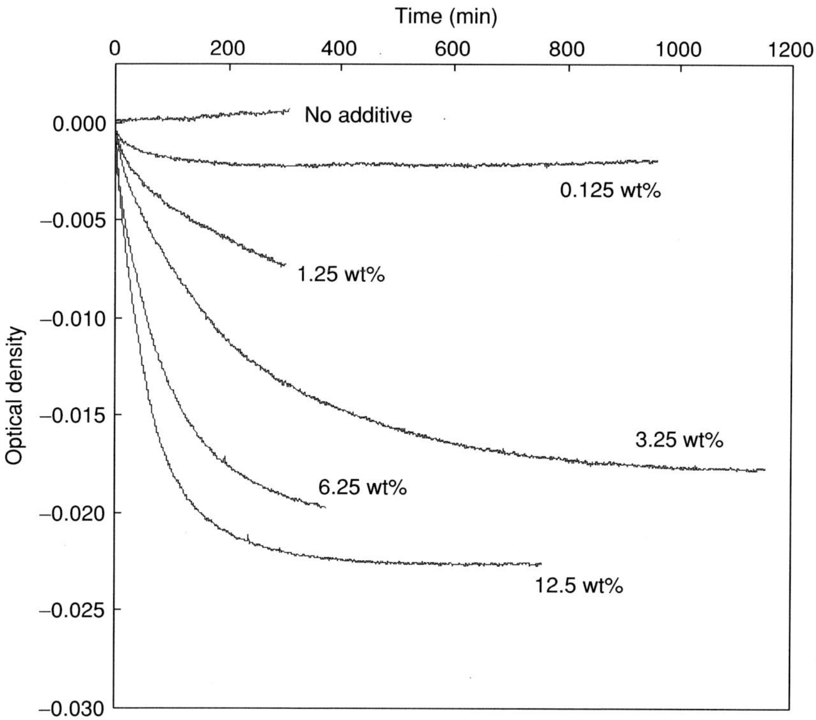

Figure 21.8. NIR scattering measurements at 1600 nm for 0.125 wt% asphaltenes in a 70/30 *n*-heptane/toluene mixture with crude naphthenic acid (CNA) added in various concentrations.

acid with different structures will possess different dispersing powers. In this study, we investigated the following monodisperse and polydisperse naphthenic acids: 5-β(*H*)-cholanoic acid (CHOL), 1-napthalenepentanoic acid, decahydro (2C4), 1-naphthalenoic acid, decahydro-2-butyl (C42), crude naphthenic acid (CNA), Fluka naphthenic acid, and North Sea naphthenic acid.

Inhibitor A. This is a commercial blend consisting of fatty amines and acids in polar solvents.

The experiments were performed by continuously measuring the change in scattering represented as optical density (OD) at 1600 nm wavelength, upon addition of chemicals to a solution of asphaltenes in heptane/toluene (70/30 vol%/vol%). At that aromatic/paraffinic ratio, the asphaltenes were expected to form rather large aggregates, and any effect on the size should be easy detectable.

Figure 21.8 shows the relative OD at 1600 nm versus time for toluene/*n*-heptane/asphaltene mixtures with different amounts of CNA is shown in. The influence of additive upon aggregate size is depicted as the decrease of scattering as a function of time. The relative OD versus time for toluene/*n*-heptane/asphaltene mixtures with various naphthenic acids is plotted in Figure 21.9. The commercial Fluka naphthenic acid and the naphthenic acids extracted from a North Sea crude

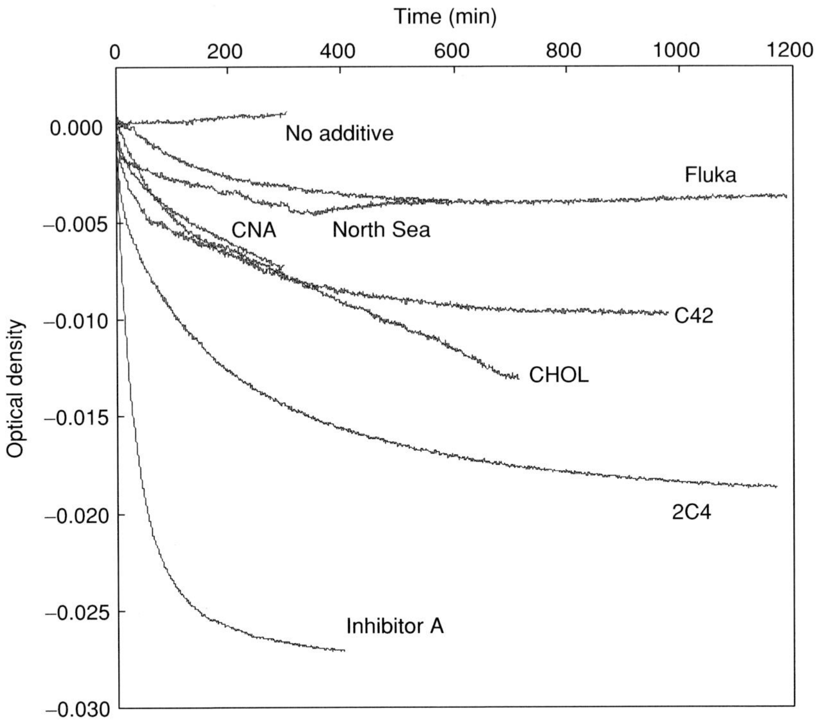

Figure 21.9. NIR scattering measurements at 1600 nm for 0.125 wt% asphaltenes in a 70/30 *n*-heptane/toluene mixture with 1.25 wt% of various naphthenic acids added.

seem to affect the state of the asphaltenes only to a minor extent. CNA is most efficient of these polydisperse naphthenic acids.

A comparison between 2C4, C42, and CHOL shows these species to be somewhat more efficient than the previous group. Especially the 2C4 molecule has a very efficient breakdown to start with and also attains a low final value. It is interesting to see that the molecular structure affects the results to this extent.

It was shown that NIR spectroscopy is a powerful method to follow the disintegration of asphaltene aggregates upon addition of chemicals. The method is based on the scattering from preferentially large aggregates. The NIR technique, which is very fast and accurate, is a good choice for the initial screening of large numbers of chemicals for asphaltene inhibition. The results show that additives, which are efficient in replacing hydrogen bonds, possess dispersive power and can serve as inhibitors. Commercial blends of active molecules gave the best results. It was also interesting to observe that the NIR technique could differentiate between the efficiency of the different naphthenic acids.

It should be mentioned that the disintegration of asphaltene aggregates had a profound influence on the corresponding emulsion stability. Prior to the addition of chemicals, the emulsions exhibited high stability, while the aggregate disintegration resulted in complete destabilization.

2.6. Asphaltene Aggregation Studied by NMR

In an investigation by Östlund and coworkers,[66] PFG-SE NMR (pulsed field gradient-spin echo nuclear magnetic resonance) measurements were combined with NIR (near infrared) spectroscopy to evaluate potential interactions between asphaltenes and naphthenic acids. The NIR experiments were performed upon systems where the asphaltenes were slightly above the precipitation point, as opposed to the PFG-SE NMR experiments where the systems were below this point. The experiments were run with to types of asphaltenes, one extracted from an acidic crude (asphaltene 1) and one from a neutral crude (asphaltene 2). The naphthenic acids employed in the experiments were synthetic monodisperse acids (CHOL and 2C4). A concentration series with asphaltenes in pure toluene was also prepared and studied, in order to obtain information about self-association of the asphaltene molecules.

PFG-SE NMR measurements of the concentration series of asphaltenes 1 dissolved in toluene-d_8 are presented in Figure 21.10. As can be seen, the median diffusion coefficient of the asphaltenes decreased as a function of increased asphaltene concentration. Östlund *et al.*[67] have shown that the obstruction effect in asphaltenic systems is large, due to the asphaltenes having a disk-like shaped structure. However, the decrease observed in this system was significantly larger than previously reported. It was thus likely that the asphaltenes investigated were not only subjected to obstruction, but also to self-association with an onset of flocculation at 0.1 wt% asphaltenes.

Further, it was interesting to note from the results shown in Figure 21.11 that the diffusion of asphaltene 2 decreased in all cases independently of which naphthenic acid that had been added. This indicated that both CHOL and 2C4 interacted with asphaltene 2. The diffusion coefficient of asphaltene 1, on the other hand, did not change and thus, it appeared as if there were no or only weak interactions between the naphthenic acids and asphaltenes of type 1.

2.7. Adsorption of Asphaltenes and Resins Studied by Dissipative Quartz Crystal Microbalance (QCM-D™)

The adsorption of asphaltenes and resins onto a hydrophilic gold surface as a function of bulk concentration was investigated in Ekholm *et al.*[68] The measurements were performed by a quartz crystal microbalance with dissipation measurements (QCM-D™), which allows for simultaneous measurements of frequency, f, and energy dissipation factor, D. The change in frequency is related to the mass adsorbed onto the surface of the sensor crystal, and from the change in dissipation factor, information about the interfacial processes can be resolved.

The results showed that the resins in pure heptane adsorb onto a gold surface, and pack into a compact monolayer (see Figure 21.12). However, there was no tendency of resin aggregation on the surface. With increasing amount of aromaticity in the solvent, the adsorbed quantity decreased, and was practically zero in pure toluene. This was related to an increased solvency of the resins which lower

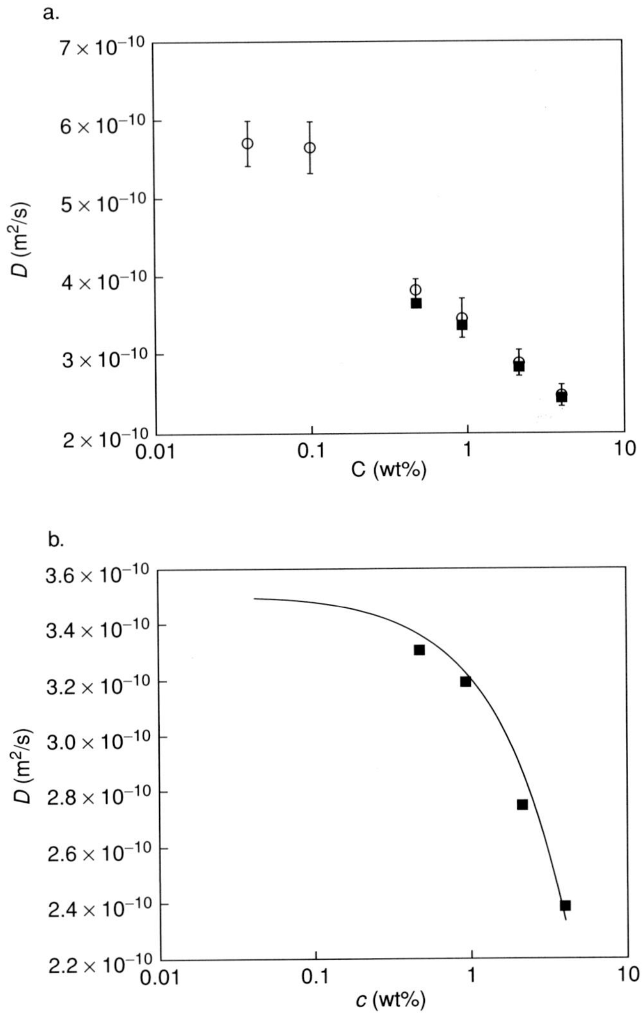

Figure 21.10. The median diffusion coefficients are displayed as a function of the asphaltene concentration (○). Also included are the calculated diffusion coefficients of the asphaltene micelles (■).

the affinity toward the surface. The asphaltenes in heptane/toluene mixtures, or pure toluene, adsorbed to a larger extent (see Figure 21.13). The adsorption was higher than observed for typical nonassociating polymers indicating aggregate adsorption. At lower concentrations they formed a rigid layer. When higher concentrations were injected it was possible to obtain further adsorption, which was related to the strong tendency of aggregation of asphaltenes in bulk solution. Desorption studies showed that resins were not able to desorb preadsorbed asphaltenes from the surface. Neither did they adsorb onto the asphaltene-coated surface. On the other hand, resins and asphaltenes associated in bulk liquid, and the adsorption from mixtures containing both resins and asphaltenes, was markedly different to

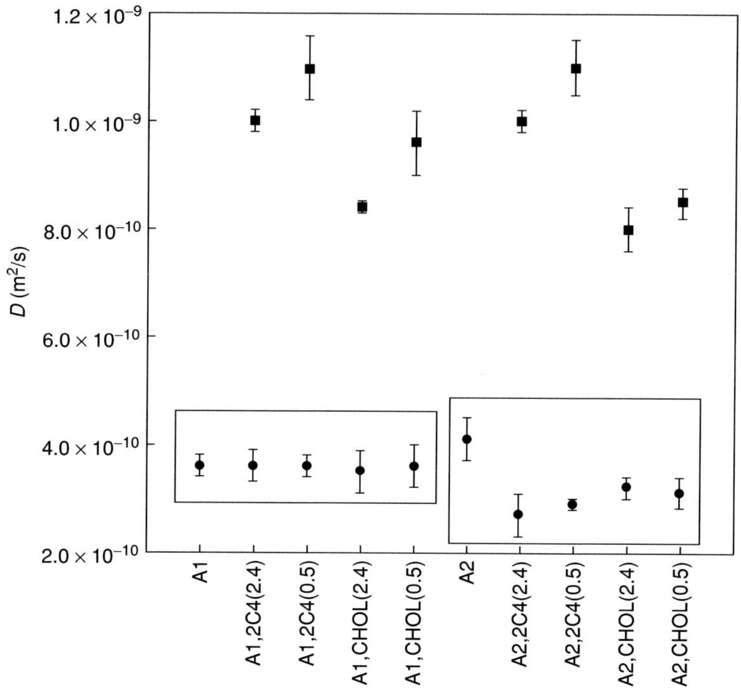

Figure 21.11. A presentation of results from samples containing naphthenic acid and asphaltenes. ■Corresponds to the diffusion of the naphthenic acid (0.5 or 2.4 wt% of CHOL alternatively 2C4), while ●corresponds to the diffusion of asphaltene 1 (A1) or asphaltene 2 (A2). The diffusion of only asphaltene 1 or 2 in toluene-d_8 (reference samples) has been included. Frames have been put around the diffusion coefficients from asphaltenes of the same kind (A1 or A2).

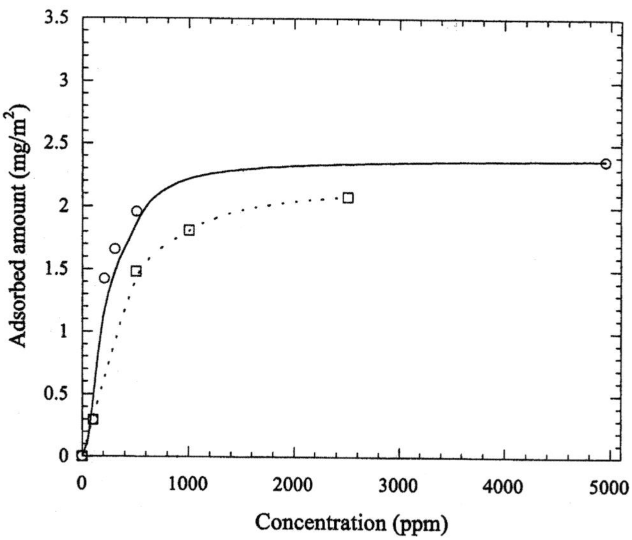

Figure 21.12. Frequency and dissipation shift as a function of time for the adsorption of resins onto a hydrophilic gold surface in a n-heptane solution.

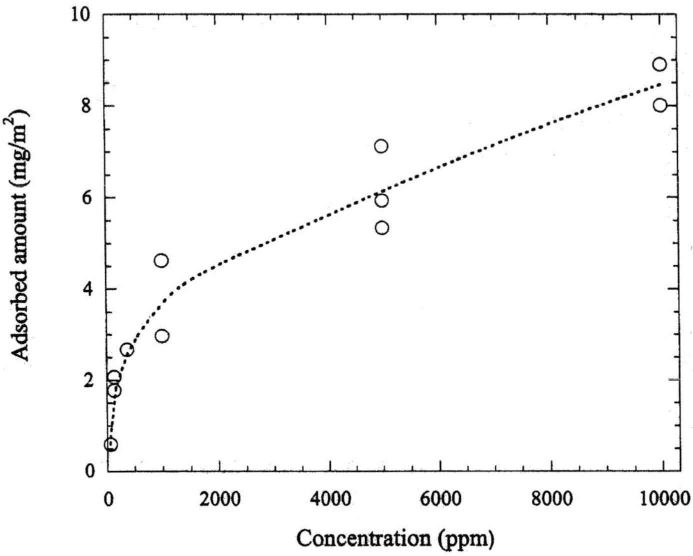

Figure 21.13. Adsorption isotherm for asphaltenes onto a hydrophilic gold surface, as a function of asphaltene concentration in pure toluene.

that observed for the pure components. Hence, it was concluded that preformed resin-aggregates adsorb to the surface.

The irreversibly adsorbed amount for a crude oil solution was smaller than for the asphaltene and resin mixture but quite similar to that of the separate fractions. The resin and asphaltene molecules, arranged in a different way in the adsorbed layer, when the effects from other constituents like paraffin and wax were absent. These constituents could be incorporated in the adsorbed layer, or affect the interaction forces in the bulk of the crude.

The QCM technique can obviously give some insight in the adsorption behavior and affinity of different components in the crude oil. It should be mentioned that while a gold and a water surface/interface are different in nature, the driving forces for adsorption should be quite similar.

2.8. Interfacial Behavior and Elasticity of Asphaltenes

It has been shown that different components in the crude oil have varying affinities for a fresh w/o interface. In order to understand the stability of water-in-crude oil emulsions, it is imperative to understand the formation of different kinds of interfacial structures. From interfacial tension measurements (as exemplified in Figure 21.14), one can draw the conclusion that it is the resin fraction that has the highest affinity toward the fresh interface. The resins can reduce the IFT to values close to 15 m/Nm. The asphaltene remained at 25 m/Nm as the limiting value. The value for the entire crude oil at 30 m/Nm reveals that there are other indigenous components influencing the IFT than resins and asphaltenes. There might be

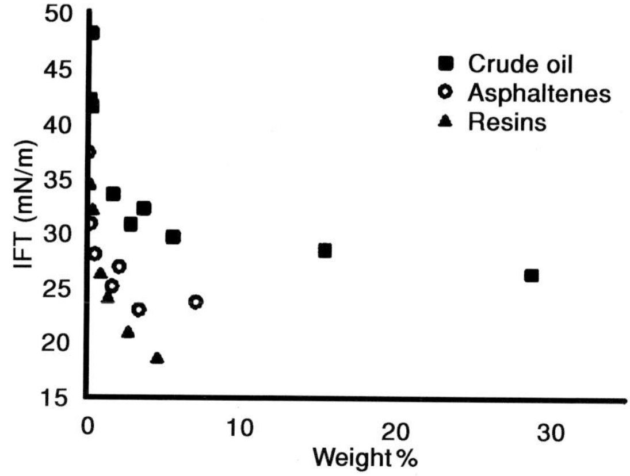

Figure 21.14. Measurements of IFT at the interface between distilled water and a crude oil, extracted asphaltenes and resins dissolved in cyclohexane (redrawn from Ref.[123]).

obstruction phenomena at the w/o interface resulting in different molecular structures in real systems and in model systems.

Generally, when an interface with adsorbed interfacially active molecules is stretched, interfacial tension gradients is generated. The tension gradients will oppose the stretching and try to restore the uniform interfacial tension state, that is the interface will behave elastically. This is the so-called Gibbs–Marangoni effect. The main function of interfacially active molecules is not the interfacial tension lowering they produce, but that their presence can lead to such gradients in interfacial tension able to resist tangential stresses. In practice, emulsion droplets being stretched can resist coalescence due to the elastic membrane, providing the droplet interfaces with a self-healing mechanism.[69]

Interfacial rheological properties are measured through the so-called interfacial dilatational modulus. This property gives a measure of resistance to the creation of interfacial tension gradients, and the rate at which such gradients disappear after the deformation. The interfacial dilatational modulus, ε, is defined as the increase in interfacial tension for a unit of relative increase in surface area[70]:

$$\varepsilon = \frac{d\gamma}{d \ln A} \tag{21.2}$$

where γ is the interfacial tension and A is the interfacial area. For a small deformation of the area, the change in interfacial tension can be written as a sum of an elastic and a viscous contribution

$$\Delta\gamma = \text{elastic} + \text{viscous} = \varepsilon_{\mathrm{d}} \Delta \ln A + \eta_{\mathrm{d}} \frac{d \ln A}{dt} \tag{21.3}$$

where ε_{d} and η_{d} is the interfacial dilatational elasticity and viscosity, respectively. The two contributions can be measured separately by subjecting the interface to small, periodic oscillations at a given frequency. In an oscillatory experiment, the interfacial area is varied with time, t, according to

$$\Delta \ln A \propto \exp(i\omega t) \tag{21.4}$$

where ω is the frequency of the oscillations. Based on the above equations, the interfacial dilatational modulus can be written as a complex number

$$\varepsilon = \varepsilon_d + i\omega\eta_d \tag{21.5}$$

where the first term will be equal to the elastic contribution and the second term will be proportional to the viscous contribution. Only the elastic contribution will be discussed in the following.

In order to measure the interfacial dilatational modulus an oscillating pendant drop apparatus was utilized. It operates by oscillating an oil drop in an aqueous phase at constant frequency. Photographs are continuously taken of the drop profile, and by knowing the densities of the two phases the interfacial tension is calculated by use of the Laplace equation.[71] As both the changes in interfacial area and interfacial tension are known, the interfacial elasticity may be calculated.

For emulsion systems, the interfacial concentration of surfactants will be low due to the large interfacial area created by the small droplets formed. Therefore, in practice, the oil systems under study is diluted with some solvent in order to better simulate the interfacial concentrations encountered in real systems. The left-hand side of Figure 21.15 shows the measured elasticity as a function of crude oil concentration and solvent composition for a diluted crude oil.[4] The sample was diluted in heptane/toluene solutions containing 0, 50, 95, and 100 vol% heptane (heptol(0)-heptol(100)) at concentrations of 0.002, 0.01, and 0.02 mL oil/mL solvent. In addition all the solutions were characterized by NIR spectroscopy, and on the right-hand side of the figure are shown the corresponding optical density at 1600 nm from the NIR spectra of the crude oil solutions. This value indicates the level of asphaltene aggregation in the solutions.

In poor solvents (high-heptane content), the interfacial elasticity decreases with increasing oil concentration. In good solvents, the interfacial elasticity

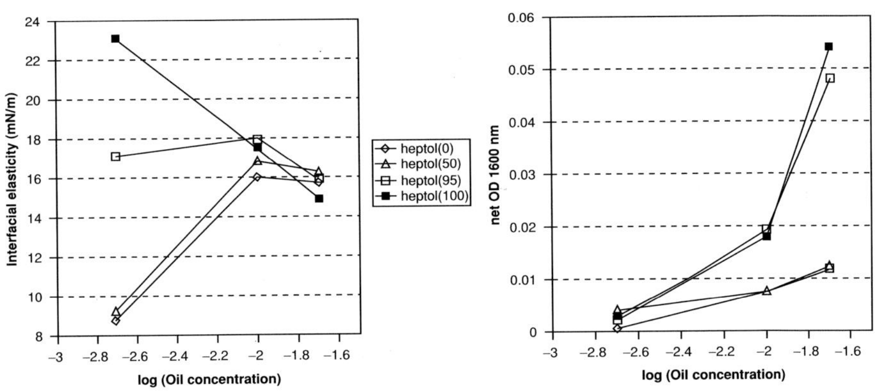

Figure 21.15. Interfacial elasticity and optical density for a crude oil diluted with various ratios of heptane and toluene.

generally increases with oil concentration. From the optical density results, the formation of larger aggregates is seen to be much more dominating in poorer solvents. The lowering of the interfacial elasticity thus appears to be caused by the formation of large, weakly interfacially active asphaltene aggregates. At the lowest oil concentration, the highest elasticity values are obtained in the poorest crude oil surfactant solvents. In these solvents, the interfacial activity of resins and asphaltenes is high, and more or less no aggregation is taking place as seen from optical density values. At higher oil concentrations, the effect of interfacial activity seems to be opposed by the formation of asphaltene aggregates.

Similar results as shown above have been found for other crude oils, and demonstrate that measuring interfacial rheological properties of crude oil systems is highly dependent both on the oil concentration and what solvent is used for dilution.[4] The effect of high-interfacial activity in aliphatic solvents is quickly opposed by asphaltene aggregation when increasing the oil concentration. When comparing the interfacial rheological properties of different crude oils, both these effects have to be taken into consideration. The findings are consistent with the asphaltene aggregation model proposed by Kilpatrick and coworkers.[5,6,72] They found that precipitated asphaltenes had a lower ability to form elastic interfaces. This was attributed to lower interfacial activity of such precipitated aggregates and possibly high amounts of defects in films of precipitated material.

The emulsion stability in different crude oil-based systems is highly dependent on the rheological properties of the w/o interface. Generally, a high elasticity will increase the level of stability.

3. Chemistry of Naphthenic Acids

The interest in the chemical properties of naphthenic acid derivatives originates from an ever-increasing coproduction of crude oils as acidic crudes.[73,74] It is well known that these naphthenic acids show a polydispersity in size and structure.[75] The smallest molecules are readily dissolved in the aqueous phase at pHs around 5, while the larger molecules are preferably oil-soluble. Most of these homologues are dissolved in an aqueous phase at pHs around 10–11.[76–78]

There are some immense crude oil production problems related to the occurrence of naphthenic acids. They cause corrosion problems at the refineries when being processed. The mechanism of this process is not clarified in detail, since it takes place in an oil environment.[79] Since the naphthenic acids and their soaps are surface\interfacially active they will accumulate at w/o interfaces and stabilize colloidal structures.[75,78,80,81] The worst scenario from an operational point of view is the stabilization of water-in-oil emulsions, which will cause problems in topside separators.[82] If Ca- or Mg-soaps are formed, the precipitation of corresponding particulate soap with low-water solubility will be a consequence. The metal soaps tend to accumulate at water/oil interfaces. The agglomeration of these particles to voluminous precipitates will cause operational problems in the separation process with shutdown periods, during which thorough cleaning must take place. The formation of naphthenate mixed scale particles in deep-offshore fields has recently been studied by Rousseau *et al.*[74]

In contrast the low-molecular weight and water-soluble derivatives will cause quality problems for the wastewater.

With this background it is evident that there is a need to gain a better understanding of the chemistry of naphthenic acid and their derivatives. Although the corresponding paraffinic fatty acids are very well characterized with regard to association and micellization in water, phase equilibrium, formation of lyotropic liquid crystals and microemulsions, monomolecular film properties, etc.[83–91], little information is available for naphthenic acids.

3.1. Origin and Structure

Naphthenic acids are classified as monobasic carboxylic acids of the general formula RCOOH, where R represents a cycloaliphatic structure. The term "naphthenic acid" is often used to account for all carboxylic acids present in crude oil, including acyclic and aromatic acids. Nearly all crude oils contain some naphthenic acids. Heavy crudes from geologically young formations have the highest acid content while paraffinic crudes usually have low-acid content.[92] Naphthenic acids are known to be produced during the in-reservoir biodegradation of petroleum hydrocarbons,[93] and they are considered to be a class of biological markers, closely linked to the maturity and the biodegradation level of the fields.[74]

The naphthenic acids are extremely complicated mixtures. Many different methods and analytical techniques have been used for analyzing these acids in the past.[75,81,94–97] An overview shows them to be C_{10}–C_{50} compounds with 0–6 fused saturated rings and with the carboxylic acid group apparently attached to a ring with a short side chain.[98] The distribution of carbon number and ring content varies with crude oil source and distillate fraction. Naphthenic acids with similar total acid number (TAN) and average molecular weight can exhibit significantly different profiles.[99]

3.2. Phase Equilibria

Professor Stig E. Friberg was a pioneer in linking the importance of phase equilibria in water/oil/stabilizer systems with emulsion stability. He documented very clearly the importance of the existence of a lamellar liquid crystal (a so-called D-phase) and a corresponding increase in emulsion stability.[9–12]

The basic idea is to cover the emulsion droplets with a multiple layer of surfactant/water to enhance the rigidity of the interfacial w/o-layer. In this way a barrier against coalescence is build up. Per today, no detailed and complete phase equilibria of a naphthenic acid (different pH)/water/oil or sodium naphthenate/naphthenic acid/water/oil has been published. Horváth-Szabó has, in cooperation with groups in Edmonton, published qualitative phase equilibria of sodium naphthenates in aqueous solution[100] and sodium naphthenates/toluene/water system[101] and found evidence for the existence of lamellar liquid crystals.

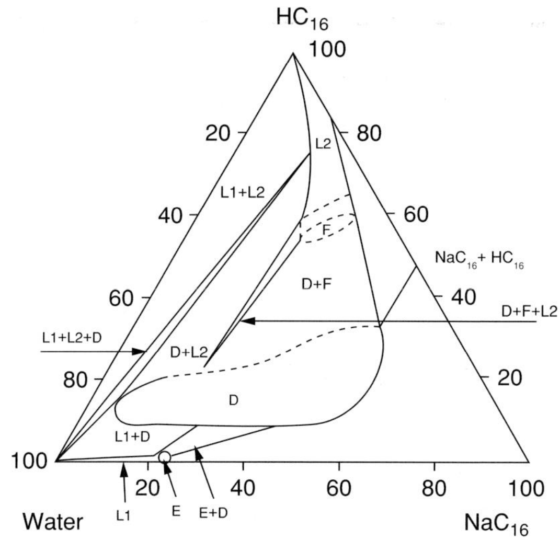

Figure 21.16. Phase equilibria in the system water/hexadecanoic acid (HC16)/sodium hexadecanoate (NaC16) at 70°C.[83] The diagram is expressed in weight percent.

It is our opinion that the phase equilibria in the system water/sodium hexadecanoate (NaC16)/hexadecanoic acid (HC16) at 70°C is also representative for naphthenic acid based systems. The phase diagram of this system has been published by Skurtveit *et al.*[83] and is given in Figure 21.16.

In this system, existence of four or five isotropic phases can be observed. These are a micellar L_1-phase (with ordinary micelles), a hexagonal liquid crystalline E-phase (with long rods in an aqueous environment), a lamellar liquid crystalline D-phase (with a bilayer structure), and a reversed solution phase L_2 (with reversed micelles). It is questionable if a F-phase with a reverse hexagonal structure exists. In this diagram, the importance of the stabilization of water/oil emulsions, due to the D-phase has been documented.[83]

For practical reasons, it is of importance to map the existence of the lamellar D-phase. In Figure 21.16, it is noticeable that an upper limit of phase equilibria including the D-phase, seems to be HC16/NaC16 $\approx$ 85/15, and a lower limit is HC16/NaC16 $\approx$ 15/85. Hence it can be claimed that in an equivalent naphthenic acid-based system, a D-phase can exist for ratios of naphthenic acid to sodium naphthenate ~0.1–6. It is sufficient to have just 10% of a naphthenate salt to obtain the D-phase.

In order to view the competition between D-phase stabilization and asphaltene-particle stabilization of water/oil emulsions, the following experiment has been performed.[14] A D-phase was prepared based on the equilibria in Figure 21.16. A dichloromethane solution, containing dissolved asphaltenes, was mixed with the D-phase and the solvent was evaporated. After 24 hr, there was a clear evidence of asphaltene particles present in the D-phase. This D-phase, modified with different amount of asphaltenes, was used to stabilize w/o-emulsions. In order

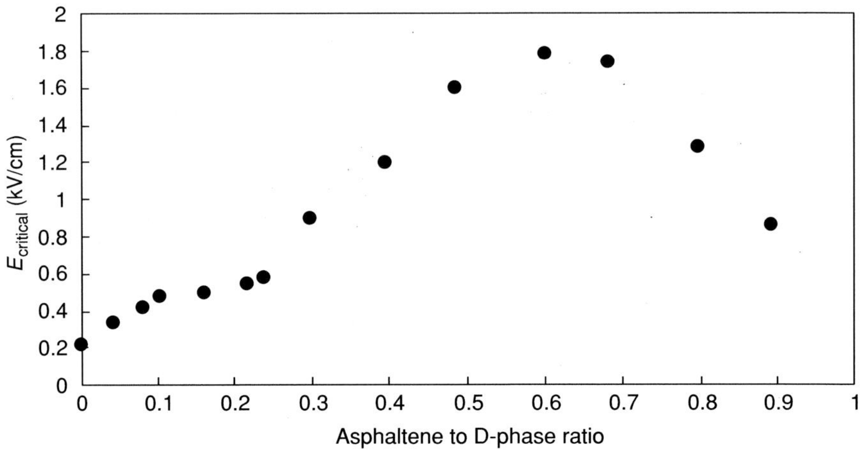

Figure 21.17. $E_{critical}$ as a function of asphaltene to D-phase ratio for water in *n*-decane emulsions.[14]

to measure the emulsion stability, the $E_{critical}$ was determined. The result is shown in Figure 21.17.

Obviously the addition of asphaltene particles can enhance the w/o-emulsion stability from the case with just a D-phase present. But a too high amount of asphaltenes particle will give a destabilization. A ratio of 0.6, that is 60 wt% particles and 40 wt% D-phase seems to be optimal. The experiment in Figure 21.17 is very important in visualizing different mechanisms behind emulsion stability in crude oils with high amounts of asphaltenes and naphthenic acids. Representatives to these are heavy crudes, and bitumens.

A special relevance of the experiments in Figure 21.17 is for ageing systems, where the lamellar D-phase can form gradually at the w/o interface as a function for a combined effect if multilayers and asphaltene particles in such systems.

4. Water-in-Crude Oil Emulsions

4.1. Stability Mechanisms

An emulsifying agent must be present to form stable water-in-crude oil emulsions. Such agents include scale and clay particles, added chemicals or indigenous crude oil components like asphaltenes, resins, waxes, and naphthenic acids.[8] Asphaltenes are believed to be the main contributor to emulsion stability, and McLean and Kilpatrick[5,6] have postulated that the dominating mechanism whereby crude oil emulsions are stabilized is through the formation of a viscoelastic, physically cross-linked network of asphaltenic aggregates at the oil-water interface. The ability of asphaltenes and resins to form elastic crude oil-water interfaces has been emphasized by several other authors as an important factor with regard to emulsion stability.[102–105] The aggregation state of the asphaltenes in the crude oil is very decisive with regard to the emulsion stability properties. Several authors have pointed out that emulsions are to a small extent stabilized by individual asphaltene molecules as compared to emulsions stabilized by colloidal asphaltene aggregates.[5,106–109]

4.2. Characterization by Critical Electric Fields

Demulsification by electrocoalescence is common in the oil industry. Water is separated from the emulsion by applying a high-electric field of 1–10 kV/cm to cause flocculation and coalescence of water droplets in the oil phase.[110] The principle behind the electrically induced coalescence is that the electric field acts to enlarge the droplets of the dispersed phase by inducing coalescence. After this initial step, the water phase settles out under gravity at an increased rate. There is a variety of factors influencing the electrically induced coalescence, such as the dielectric properties of the dispersed and the continuous phase, the volume fraction of the dispersed phase, conductivity, size distribution of the dispersed droplets, electrode geometry, electric field intensity, the nature of the electric field (AC or DC), etc.[111]

The electric field strength needed to cause such coalescence can be used as a measure of the emulsion stability. In low-electric fields, water droplets surrounded by a rigid interfacial film will attain a chain-like configuration. Increasing the field strength the droplets will bridge the gap between the electrodes. Ultimately, an irreversible rupture of the interfacial films between the droplets will increase the conductivity through the emulsion sample.[107,110] This may then be defined as a measure of the emulsion stability, the so-called $E_{critical}$. Figure 21.18 demonstrates the effect of increasing the electric field strength over an emulsion.

An electric field cell has recently been developed for the determination of emulsion stability.[112] The method is similar to the one employed by Kallevik *et al.*[113] It consists of a Teflon plate with a-10 mm diameter hole in the center, with a brass plate on each side. The distance between the plates is 0.5 mm, and

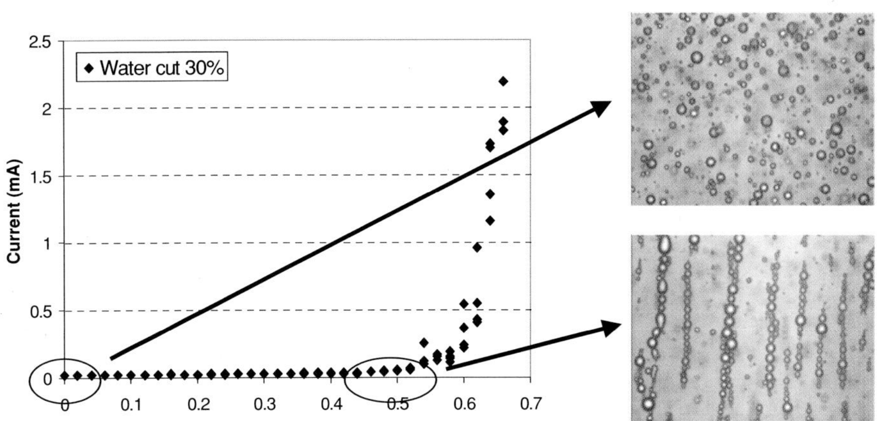

Figure 21.18. Emulsion droplets in an increasing electric field.

the upper brass plate has two small holes for sample injection. The brass plates are connected to a power supply, which can increase the applied voltage in user-defined steps. The system is held together with isolating plexiglass plates. The power supply delivers a maximum of 100 V DC, corresponding to a maximum electric field of 2.0 kV/cm. Emulsions are injected into the cell volume, and the electric field over the emulsion sample is increased until E_{critical} is reached, as seen by increased conductivity.

4.3. Multivariate Analysis and Emulsion Stability

Principal component analysis (PCA) is a projection method that helps to visualize the most important information contained in a data set. PCA finds combinations of variables that describe major trends in the data set. Mathematically, PCA is based on an eigenvector decomposition of the covariance matrix of the variables in a data set. Given a data matrix $\mathbf{X}$ with m rows of samples and n columns of variables, the covariance matrix of $\mathbf{X}$ is defined as

$$\text{cov}(\mathbf{X}) = \frac{\mathbf{X}^{\text{T}}\mathbf{X}}{m-1}. \tag{21.6}$$

The result of the PCA procedure is a decomposition of the data matrix $\mathbf{X}$ into principal components called score and loading vectors

$$\mathbf{X}_{n\times m} = \mathbf{t}_1\mathbf{p}_1^{\text{T}} + \mathbf{t}_2\mathbf{p}_2^{\text{T}} + \mathbf{t}_i\mathbf{p}_i^{\text{T}} + \cdots + \mathbf{t}_k\mathbf{p}_k^{\text{T}} + \mathbf{E}_{n\times m}. \tag{21.7}$$

Here $\mathbf{t}_i$ is the score vector, $\mathbf{p}_i$ is the loading vector, and $\mathbf{E}$ is the residual matrix. The score and loading vectors contain information on how the samples and variables, respectively, relate to each other. The direction of the first principal component $(\mathbf{t}_{1,}\mathbf{p}_1)$ is the line in the variable space that best describes the variation in the data matrix $\mathbf{X}$. The direction of the second principal component is given by the straight line that best describes the variation not described by the first principal component and so on. Thus, the original data set can be adequately described using a few orthogonal principal components instead of the original variables, with no significant loss of information. When plotting principal components against each other, relations between samples are easily detected.[114]

Regression is used to fit a model to observed data in order to quantify the relationship between two groups of variables. The regression model may then be used to predict properties of new samples. Regression between the $\mathbf{X}$ data and the response $\mathbf{y}$ is often impossible by ordinary least-squares methods due to strongly correlated and redundant information in the data set. This problem is especially prominent for spectral data sets. One solution to this is to decompose the $\mathbf{X}$ matrix as shown in equation (21.7), and perform the regression between the resulting score vectors and the response. PCA decomposition followed by regression is called principal component regression (PCR). The partial least-squares (PLS) technique is another alternative. In PCA the scores and loadings are the vectors that best describes the variance of the $\mathbf{X}$ matrix. In PLS, the scores and loadings (called latent variables) are the vectors that have the highest covariance with the response vector

y. The decomposition is followed by a regression between the latent variables and the response.

Due to the risk of overfitting the regression model, the optimum number of latent variables to be used is determined. One way of doing this is the cross-validation technique. Cross-validation checks a model by repeatedly taking out different subsets of calibration samples from the model estimation, and instead using them as temporary, local sets of secret test samples. If the model parameter estimates are stable against these repeated perturbations, this indicates that the model is reliable.[115] In the simplest case, each subset contains only one sample, which is called full cross-validation, and is the technique employed in the regression modeling in this study.

Data may be transformed prior to multivariate analysis in order to make them more suitable for a powerful analysis. *xy*-Transformations, thus including logarithmic transformations, are especially useful to make the distribution of skewed variables more symmetrical.

The critical electric field cell above was developed to measure the emulsion stability of a set of characterized crude oils and condensates. By multivariate analysis, the emulsion stability was then correlated to the physico-chemical properties of the samples. In addition to SARA-data and interfacial elasticity, properties like density, interfacial tension, molecular weight, total acid number, and viscosity of the samples were known. The aim of this study was to gain insight into which parameters govern the emulsion stability properties of the crude oils. In addition NIR spectra of the crude oils were also used as input data. Table 21.3 shows the data matrix, including E_{critical} values for water cuts of 20% and 30%. t1 and t2 are the first and second score vectors from a principal component analysis of the NIR spectra. These two variables illustrate, in a condensed form, large parts of the variation between the samples found in the NIR spectra.

A PLS regression model was built using the average E_{critical} values at water cuts of 20% and 30% as response. Due to a skewed distribution the E_{critical} values was logarithmically transformed before modeling. This excluded the condensates from the model since their E_{critical} values were zero. Figure 21.19 shows the regression coefficients and the predicted vs. measured plot for the resulting

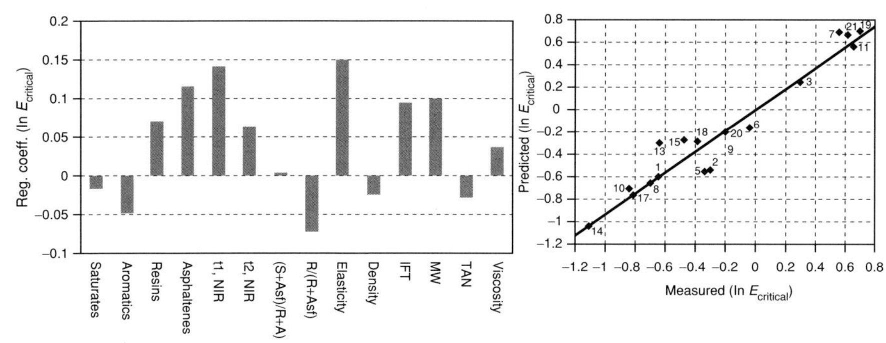

Figure 21.19. Regression coefficients and predicted vs. measured plot for the emulsion stability model (from Ref.[112]).

Table 21.3. Data Matrix on Crude Oils and Condensates

	Origin	WC20 [KV/cm]	WC30 [Kv/cm]	*S* (wt%)	*A* (wt%)	*R* (wt%)	Asph. (wt%)	(S+Asph)/(R+A)	R/(R+Asph)
1	West Africa	0.58	0.47	47.9	36.5	15.2	0.4	0.93	0.97
2	North Sea	0.87	0.61	48.0	37.5	14.2	0.3	0.93	0.98
3	West Africa	2.00	0.68	41.2	36.4	20.4	2.1	0.76	0.91
4	North Sea	0.00	0.00	82.7	13.4	3.9	0.0	4.78	1.00
5	North Sea	1.00	0.64	62.7	23.6	12.2	1.5	1.79	0.89
6	North Sea	0.91	1.03	45.5	37.1	16.0	1.4	0.88	0.92
7	North Sea	2.00	1.50	35.3	36.8	24.5	3.5	0.63	0.88
8	North Sea	0.55	0.45	56.0	29.6	14.1	0.3	1.29	0.98
9	North Sea	0.84	0.59	41.8	38.8	18.7	0.6	0.74	0.97
10	North Sea	0.53	0.33	50.9	34.6	14.0	0.5	1.06	0.97
11	West Africa	2.00	1.85	40.6	32.1	20.6	6.6	0.90	0.76
12	North Sea	0.00	0.00	79.8	16.5	3.6	0.1	3.98	0.97
13	West Africa	0.61	0.45	57.3	27.9	13.5	1.3	1.42	0.91
14	North Sea	0.59	0.08	60.6	30.0	9.2	0.2	1.55	0.98
15	West Africa	0.85	0.40	42.4	36.1	20.5	1.0	0.77	0.95
16	North Sea	0.00	0.00	65.0	30.7	4.3	0.0	1.86	1.00
17	North Sea	0.47	0.42	44.1	41.6	13.8	0.5	0.81	0.97
18	North Sea	0.93	0.43	50.3	31.4	17.5	0.7	1.04	0.96
19	North Sea	2.00	2.00	54.5	28.8	14.9	1.8	1.29	0.89
20	West Africa	0.91	0.72	55.4	28.3	12.9	3.4	1.43	0.79
21	France	2.00	1.70	24.4	43.4	19.9	12.4	0.58	0.62

	Origin	t1	t2	OD, 1600 nm	Elasticity (mN/m)	Density (g/cm^3)	IFT (mN/m)	Mw (g/mol)	Visc., 25°C (kg/ms)	TAN
1	West Africa	−5.45	0.11	0.168	11.5	0.914	20.5	234	18.7	1.10
2	North Sea	−3.16	0.26	0.240	7.3	0.916	24.8	279	57.4	3.10
3	West Africa	8.91	0.16	0.659	7.8	0.916	26.4	310	143.0	1.50
4	North Sea	−7.42	0.17	0.107	−1.1	0.839	37.1	166	1.8	0.69
5	North Sea	−3.62	−0.43	0.236	16.5	0.844	12.8	201	5.4	0.18
6	North Sea	−0.39	0.52	0.332	3.7	0.862	31.9	244	14.5	0.02
7	North Sea	5.99	0.81	0.544	11.7	0.945	27.4	333	386.6	2.30
8	North Sea	−5.02	−0.10	0.185	6.7	0.850	24.7	216	6.6	0.17
9	North Sea	−3.53	0.28	0.229	8.4	0.914	11.8	284	51.0	3.10
10	North Sea	−4.54	0.12	0.197	5.0	0.885	22.8	234	11.6	2.70
11	West Africa	17.50	−0.49	0.991	10.6	0.888	29.0	260	27.8	0.49
12	North Sea	−5.84	−0.60	0.165	−1.4	0.796	34.2	157	1.7	0.01
13	West Africa	−2.87	−0.05	0.257	–	0.873	24.6	235	17.7	0.50
14	North Sea	−5.11	−0.39	0.186	0.2	0.857	22.9	227	10.5	0.04
15	West Africa	1.33	−0.02	0.395	9.2	0.921	16.2	295	105.8	3.60
16	North Sea	−7.36	0.18	0.107	−0.6	0.796	27.6	142	1.2	0.02
17	North Sea	−4.94	−0.37	0.192	8.1	0.847	19.9	223	11.7	0.15
18	North Sea	−4.24	0.14	0.206	14.8	0.898	19.5	249	19.1	1.20
19	North Sea	9.54	0.11	0.684	24.9	0.840	27.4	298	63.1	0.36
20	West Africa	4.91	−0.41	0.525	5.6	0.873	19.0	248	15.3	0.44
21	France	15.31	0.00	0.893	12.9	0.939	13.4	303	278.9	0.20

model. The R^2 value of the model is 0.93. From the regression coefficients, the amount of asphaltenes and interfacial elasticity is seen to be main contributors to emulsion stability. In addition the NIR spectra (through the scores t1) seem to contain much information with regard to emulsion stability properties. On the opposite, high values of aromatics and the relation resins/(resins + asphaltenes) seem to be the two main variables decreasing emulsion stability.

Based on the full model, a reduced model, containing only a few of the most important variables, was built. Five parameters from the full model were preserved, the four SARA-parameters and the interfacial elasticity. All four SARA-parameters were kept since they are determined in the same experiment. This model produced a R^2 value of 0.82. This means that data from only two experimental procedures, the SARA-separation and the interfacial elasticity determination, are able to give a reasonable estimate of the emulsion stability properties of a crude oil. Earlier attempts to correlate the emulsion stability to SARA-data and some of the other variables in the data matrix were not successful, underlining the importance of knowing the interfacial properties of crude oils when discussing emulsion stabilization/destabilization. Crude oil 19 may serve as an example by forming very stable emulsions despite its relatively low asphaltene content (1.8 wt%). However, from the data table this crude oil is seen to produce by far the highest values of interfacial elasticity.

The NIR spectra appeared to contain information on the emulsion-stabilizing properties of the crude oils due to the high value of the t1 regression coefficient. Based on this, a third PLS model using NIR spectra of the crude oils as X-data was constructed. Wavelengths from 1300 to 2200 nm were used to predict the same E_{critical} values as in the other two models. Figure 21.20 shows NIR spectra and the predicted vs. measured plot of the resulting model. The model produced a R^2 value as high as 0.95, confirming that near infrared spectroscopy is a very good source of information for crude oil properties. We have earlier demonstrated that SARA-parameters could be predicted from NIR spectra,[116] and in addition NIR spectra have been shown to contain information on the aggregation state of the

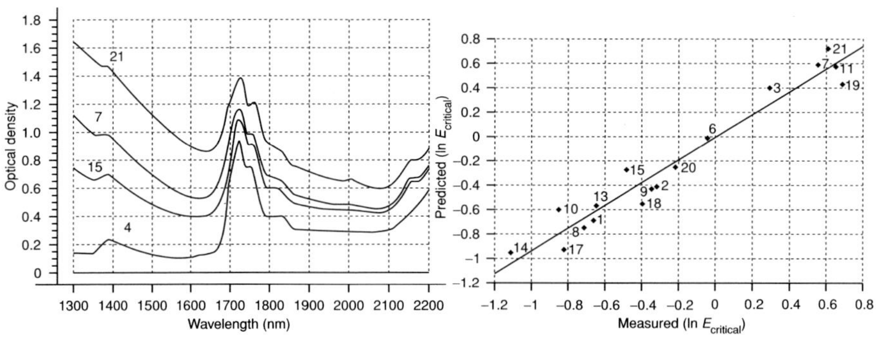

Figure 21.20. Example NIR spectra and predicted vs. measured plot for the NIR model (from Ref.[112]).

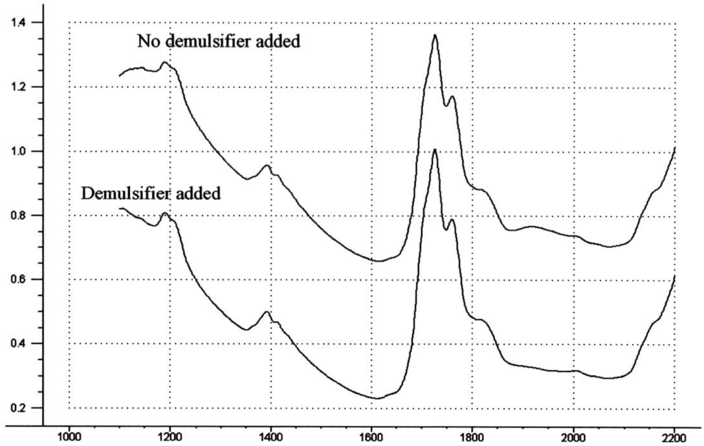

Figure 21.21. Effect on NIR spectra of crude oil with addition of demulsifier.

asphaltenes in the crude oils.[62] These are all factors contributing to emulsion stability, and explain some of the good predictive power of near infrared spectroscopy.

One of the problems associated with the prediction of emulsion stability from single parameters like SARA, elasticity, viscosity etc. is that detecting the effect of constituents of very low concentration in the crude oil may be difficult. However, such constituents may have a dramatic effect on the emulsion stability properties. One example is chemical demulsifiers added during crude oil production at ppm-levels. A SARA-determination will not be able to distinguish between a crude oil sample with and without such additives. An indirect analytical procedure like near infrared spectroscopy, on the other hand, can be able to detect such additives by measuring the effect they have on the crude oil, that is, their functionality. Figure 21.21 shows two NIR spectra of the same North Sea crude oil, the only difference being the addition of 10–15 ppm of demulsifier. It is seen that this has an adverse effect on the NIR spectra, a difference that a SARA-analysis would not have revealed. The emulsion stability of the crude oil with no added demulsifier (as measured by E_{critical}) is twice as high as for the sample with addition.

4.4. High-Pressure Performance of W/O Emulsions

4.4.1. High-Pressure Rig for Emulsion Stability Studies

In order to study separation of emulsions under realistic conditions, a high-pressure separation rig has been constructed at Statoil's R&D Center. The rig can be used to prepare emulsions and monitor the separation of oil and water, as well as any stable foam formation, in a horizontal batch separation cell. The separator cell is made from sapphire assuring full visibility of the separation processes, it has a volume of 0.5 L and can stand pressures up to 200 bar. A schematic drawing of the rig is presented in Figure 21.22. The Rig includes four 600 cc high pressure sample cylinders. With the aid of four motor-driven high-capacity piston pumps, water and

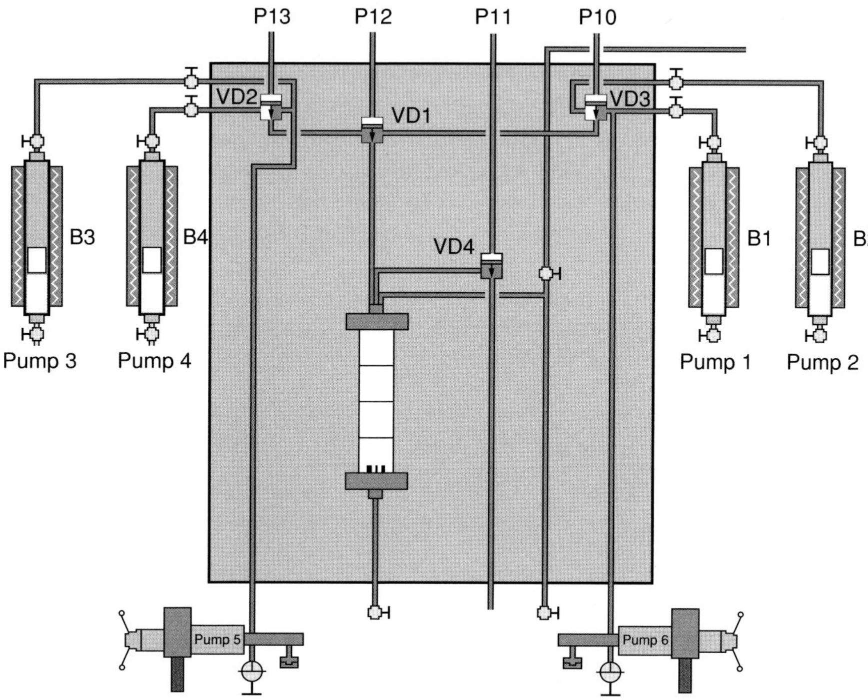

Figure 21.22. Schematic drawing of the high-pressure separation rig.

oil are pumped from the sample cylinders, through the choke valves and into the separator. The four pumps can be controlled independently, however, the total flow rate is usually kept constant. Pressure drops through the choke valves are back-pressure controlled. In order to control the separation pressure, the separation cell is pressurized with gas, inert or natural gas, and the pressure is regulated by another back-pressure controlled valve. To ensure temperature control of the system, a thermostatted cabinet encloses the separation cell and provides temperatures in the range of −7°C to 175°C.

The principle of the rig is that two flows of pressurized fluids meet and stream through a choke valve (VD1) and into the batch separator. As the fluid mixture passes through the choke valve it undergoes a pressure drop, which provides the shear force necessary to create more interface between the oil and water phases, and water droplets are formed and dispersed into the oil. If the pressure drops below the bubble point pressure, a gaseous phase appears. The gas evolved may form a foam layer, as well as influence the sedimentation and coalescence processes, which the water droplets go through. The quantity of the different phases, foam, oil, emulsion/dispersion and water, can thereafter be recorded as a function of time. To aid in the monitoring process, video cameras has been connect to the rig. After a given separation time, oil, water, and emulsion layers are sampled and analyzed. Upstream of the choke valve VD1 there is two other choke valves, one on each separate line (VD2 and VD3). Through these choke valves the same processes take place as described for VD1, and dispersions of oil and water can be prepared. To study the effect of emulsion and foam inhibitors or demulsifiers, the rig is equipped with two independent high-precision pumps 5 and 6, which delivers volumes down to 0.03 cc/hr. In this way, low concentration chemicals can be injected in to any of the flow lines. Injections can also be made in the bottom of the cell, where a stirrer can be used to distribute the chemicals.

The high pressure separation rig is a batch separator, and the results will therefore not apply exactly for a field separation process where the separation is continuous. Nevertheless, the results will show trends for temperature, pressure, pressure drop, and mixing with other oils, etc. It will also indicate whether there is a need for chemical treatments (demulsifier, foam inhibitor, etc).

4.4.2. Video Microscopy Measurement of the Droplet Size Distribution (DSD) in Emulsions at High Pressures

When crude oil is produced from offshore oil fields water and oil can be cotransported from reservoir to downstream separation facilities on the production platforms. The reservoir pressures vary greatly, and as production at very deep waters and from very deep reservoirs is initiated, emulsions form at pressures far from what ordinary lab facilities are designed to handle. It is a fact that system pressure does affect the asphaltene (the most important crude oil components in relation to emulsion stabilization) state, for example, aggregate size. Asphaltene aggregates and particles are coadsorbed into the interfacial film, giving rise to high-droplet stability against coalescence. Too large particles will not coadsorb, and very small aggregates and monomers give weaker, less stabilizing films. By the development of the high pressure separation facilities (see section "High-Pressure Rig for Emulsion Stability Studies"), it is now possible to sample emulsions at high pressure that have been created under realistic process conditions and measure the DSD at the same pressure, the advantage of which is that one avoids possible re-agitation during pressure release (and subsequent changes in the DSD). The video microscopy has been equipped with a flow-cell which can withstand pressures up to 600 bar which enables image capture from the flowing emulsion. The image analysis is of course the same as for ambient pressure studies.

The high pressure video microscopy system has been used for studying the effect of system pressure changes on DSD related to changes in asphaltene aggregate formation (see also section "Asphaltene Aggregation Studied by High-Pressure NIR Spectroscopy"). Studies have been performed on separation behavior as function of mixing of oils from different fields, in order to predict possible separation problems in coprocessed streams. DSD measurements have also been used for determining changes in droplet size arising from mixing at valves (and series of valves) due to variations in the pressure drop across the valves. Further, the effects of variations in pipe length and flow distance after mixing have been examined. All under genuine processing conditions.

Figure 21.23 shows how the distribution of a pressure drop across two consecutive valves shifts the DSD towards smaller droplet sizes than is the case when

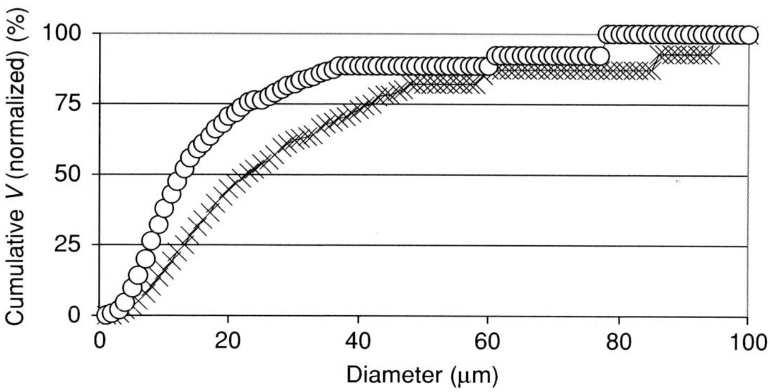

Figure 21.23. When the pressure drop ($\Delta p_{total} = 20$ bar) is distributed equally across two consecutive valves (open circles) the droplet size is smaller than when the entire pressure drop is located at one valve only (%). The emulsion is 30% water-in-crude oil, and the initial system pressure is 31 bar.

the entire pressure drop is located at one valve only. It is natural to assume that this is due to the fact that at the second valve what is mixed is an emulsion created at the first valve, rather than separate phases.

The video microscopy set above is the first attempt to measure droplet sizes under true pressures. The experimental setup has shown clear evidence of how sampling, emulsion processing, and pressure release will influence the DSD. These factors are very important when designing large-scale process facilities.

4.4.3. High-Pressure Experiments on W/O Emulsions

The objective of this chapter is to present studies of the effects of separation pressure, pressure drop, and solvency as well as release of gas bubbles on the stability of crude oil emulsions. In the next section ("New Destabilization Mechanisms"), the conclusions are summarized.

Auflem *et al.*[117] give experiments where a North Sea crude oil was recombined with dry natural gas to a separator pressure of 11 bar. The sample was thereafter mechanically pressurized further to 100 bar, or in some cases 182 bar, by a piston pump connected to the sample cylinder. The experiments were performed in a high-pressure separation rig, which is described in section "High-Pressure Rig for Emulsion Stability Studies." In the rig, water and oil can be mixed (choke valve mixing) under conditions similar to manifold/upstream separator conditions, and then pumped across another choke valve into a vertical batch separator cell.

In the experiments, several effects on the separation were observed: (1) an increased separation with increasing pressure drop below the bubble pressure, (2) a increased pressure drop gave more stable emulsions for separation above the bubble pressure, and (3) toluene dilution of the crude oil resulted in less stable emulsions.

The results in (1) and (2) were accounted for by a flotation effect from gas bubble on the stabilizing material. As the oil phase is depressurized, the solubility of light end molecules decreases, and a gas phase evolve. The gas phase will then rise through the solution in the form of small bubbles, which rip off surface-active material from the water–oil interface.

A higher energy input, due to a larger pressure drop, resulted in smaller water droplets and consequently a slower separation process, at separation pressures above the bubble point. The relation between energy input and droplet size has been shown before by several authors[118–122] and came thus as no surprise. Also a destabilizing effect from diluting the crude with toluene was expected. McLean and Kilpatrick[5] showed in 1996 (and also Førdedal[108]) that as the aromaticity of the oil phase increased, the asphaltene aggregates were dissolved, and the stability of the emulsions was reduced. Further, the foamability was also affected by the toluene addition. For increasing content of toluene in the oil phase, the capacity of the system to form foam decreased as a result of dissolution of stabilizing material. More interesting was a comparison of experiments performed on (1) a recombined oil phase, and (2) a recombined oil phase that had been degassed and repressurized mechanically. That is, the recombined oil phase was depressurized to atmospheric pressure, while allowing the gas phase to evaporate. The degassed sample was thereafter repressurized mechanically by use of a piston pump to the original pressure (100 bar). Thereafter, the two types of oil samples were put through an identical emulsification procedure as follows. The oil was mixed with the pressurized formation water, 35 vol%, by pressure drops through two succeeding choke valves: From 100 to 11 bar and from 11 to the separation pressure in the separator of 7 or 1 bar. For the recombined samples, there were a significant foam formation and relatively fast separation for both separations at 7 and 1 bar. Interesting to notice was that the experiment with the largest pressure drop over the second choke valve, separated fastest. For the degassed samples there were no foam formation, and both the separation at 7 and 1 bar were equally poor, as shown in Figure 21.24.

4.4.4. New Destabilization Mechanisms

Proposed mechanisms for breaking of oil-continuous emulsions (Figure 21.25) are

The droplet rupturing effect: The polar gas dissolved into the aqueous phase (the droplets) will rapidly coalesce and form small bubbles upon a pressure reduction to below the bubble point. Due to gravity reasons these bubbles will propagate through the emulsified system. When the bubbles leave the water droplet, they have to pass the interfacial film built up by indigenous polar components (asphaltenes, resins, and waxes). As a consequence the interface will be ruptured. If the CO_2 (g) bubbles carry with them surface-active material from the interface (flotation effect), the time for the interface to reform will, most likely, be much longer than the coalescence time. Hence the system will break and water and oil phases should appear. Application pressures could be about 60 bar depending on the chemical system and the whole process design.

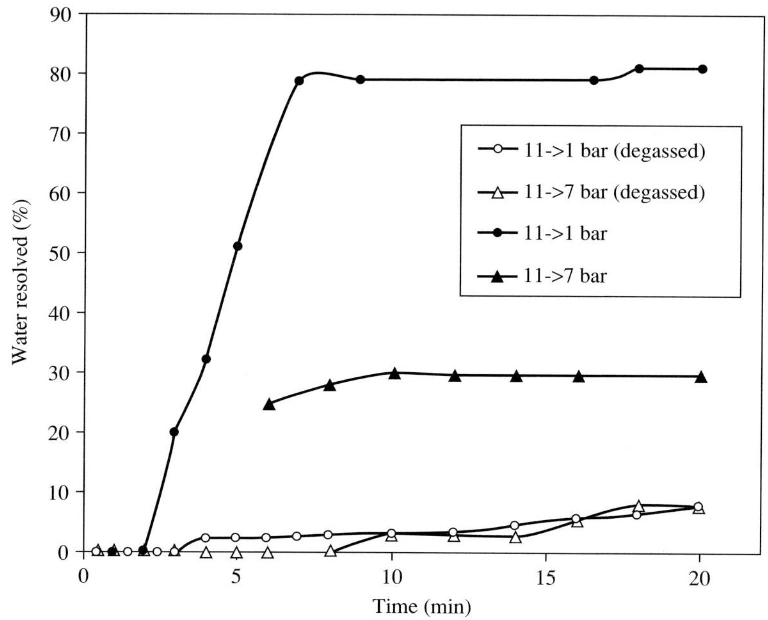

Figure 21.24. Resolution of water from water-in-oil emulsions made from recombined samples and degassed recombined samples at 7 and 1 bar separation pressure.

The film drainage effect: The gas dissolved in the oil phase (the continuous phase) will also rapidly coalesce and form bubbles upon a pressure reduction below the bubble point. The buoyancy forces will cause the bubbles to propagate through the emulsified system. In doing so they will tear off surface-active material from the o/w interface described as a flotation effect. This effect should be common for all oil soluble gases below the bubble point.

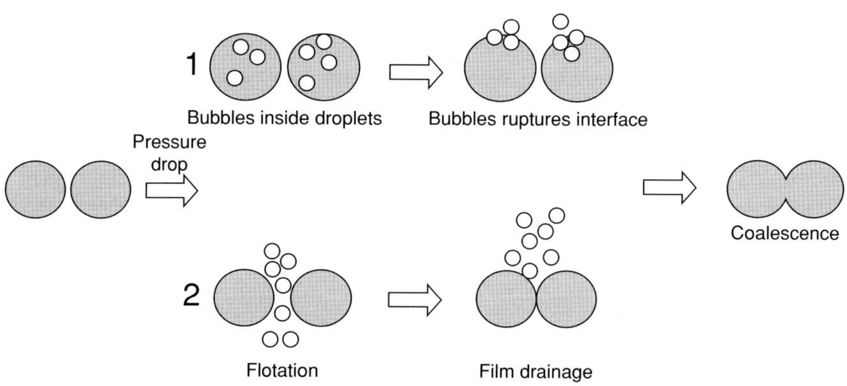

Figure 21.25. Illustration of the proposed effects from CO_2 gas bubbles on water droplets in an oil-continuous phase.

Acknowledgments

I would like to acknowledge my former PhD students Narve Aske, Inge Auflem, Trond Havre and Øystein Sæther for their impact. We would also like to acknowledge the collaboration with Chalmers University in the field of NMR, YKI/KTH in the field of QCM, and Norsk Hydro R&D Center in the field of interfacial elasticity.

References

[1] Sjöblom, J. (ed.). (1996). *Emulsions and Emulsion Stability*. Marcel Dekker, New York.
[2] Sjöblom, J. (ed.). (2001). *Encyclopedic Handbook of Emulsion Technology*. Marcel Dekker, New York.
[3] Sjöblom, J., N. Aske, I.H. Auflem, Ø. Brandal, T.E. Havre, Ø. Sæther, A. Westvik, E.E. Johnsen, and H. Kallevik (2003). *Adv. Colloid Interface Sci.* **100**, 399–473.
[4] Aske, N., R. Orr, J. Sjöblom, H. Kallevik, and G. Oye (2004). Interfacial properties of water-crude oil systems using the oscillating pendant drop. Correlations to asphaltene solubility by near infrared spectroscopy. *J. Dispersion Sci. Technol.* **25**(3), 263–275.
[5] McLean, J.D. and P.K. Kilpatrick (1997). *J. Colloid Interface Sci.* **189**, 242.
[6] McLean, J.D. and P.K. Kilpatrick (1997). *J. Colloid Interface Sci.* **196**, 23.
[7] Sjöblom, J., Ø. Sæther, Ø. Midttun, M.-H. Ese, O. Urdahl, and H. Førdedal (1998). In: E. Sheu and O.C. Mullins (eds.), *Structures and Dynamics of Asphaltenes*. Plenum Press, New York, Ch. 11.
[8] Sjöblom, J., E.E. Johnsen, A. Westvik, *et al.* (2001). In: J. Sjöblom (ed.), *Encyclopedic Handbook of Emulsion Technology*. Marcel Dekker, New York, p. 595.
[9] Friberg, S., L. Mandell, and M. Larsson (1969). *J. Colloid Interface Sci.* **29**, 155.
[10] Friberg, S. (1971). *J. Colloid Interface Sci.* **37**, 291.
[11] Friberg, S., P.O. Jansson, and E. Cederberg (1976). *J. Colloid Interface Sci.* **55**, 614.
[12] Friberg, S. and C. Solans (1986). *Langmuir* **2**, 121.
[13] Sjöblom, J., O. Urdahl, K.G.N. Børve, L. Mingyuan, J.O. Sæten, A.A. Christy, and T. Gu (1992). *Adv. Colloid Interface Sci.* **41**, 241.
[14] Havre, T. and J. Sjöblom (2003). Combined D-phase and particle stabilization of water-in-crude oil emulsions, *Colloid Surf.* A **228**, 131.
[15] Speight, J.G. (1998). *The Chemistry and Technology of Petroleum*, 3rd edn., Marcel Dekker, Inc., New York.
[16] Ali, M.A. and W.A. Nofal (1994). Application of high performance liquid chromatography for hydrocarbon group type analysis of crude oils. *Fuel Sci. Technol. Int.* **12**(1), 21–33.
[17] Fan, T.G. and J.S. Buckley (2002). Rapid and accurate SARA analysis of medium gravity crude oils. *Energy Fuels* **16**(6), 1571–1575.
[18] Hammami. A., *et al.* (1998). Asphaltenic crude oil characterization: An experimental investigation of the effect of resins on the stability of asphaltenes. *Pet. Sci. Technol.* **16**(3,4), 227–249.
[19] Radke, M., H. Willlsch, and D.H. Welte (1980). Preparative hydrocarbon group type determination by automated medium pressure liquid chromatography. *Anal. Chem.* **52**(3), 406–411.
[20] Suatoni, J.C. and R.E. Swab (1976). Preparative hydrocarbon compound type analysis by high-performance liquid chromatography. *J. Chromatogr. Sci.* **14**(11), 535–537.
[21] Hannisdal, A., P.V. Hemmingsen, and J. Sjöblom (2005). Group-type analysis of heavy crude oils using vibrational spectroscopy in combination with multivariate analysis. *Ind. Eng. Chem. Res.* **44**(5), 1349–1357.
[22] Snyder, L.R., J.J. Kirkland, and J.L. Glajch (1988). *Practical HPLC Method Development*, 2nd edn. John Wiley and Sons, Inc., New York, vol xvi, p. 260.
[23] Sheu, E.Y., and O.C. Mullins (1995). *Asphaltenes: Fundamentals and Applications*. Plenum Press, New York.

[24] Andersen, S.I. and J.G. Speight (2001). *Pet. Sci. Technol.* a**19**, 1.
[25] Dabir, B., M. Nematy, A.R. Mehrabi, H. Rassamdana, and M. Sahimi (1996). *Fuel* **75**, 1633.
[26] Peramanu, S., B.B. Pruden, and P. Rahimi (1999). *Ind. Eng. Chem. Res.* **38**, 3121.
[27] Groenzin, H. and O.C. Mullins (2000). *Energy Fuels* **14**, 677.
[28] Rogel, E., O. Leon, G. Torres, and J. Espidel (2000). *Fuel* **79**, 1389.
[29] Loh, W., R.S. Mohamed, and A.C.S. Ramos (1999). *Pet. Sci. Technol.* **17**, 147.
[30] Leon, O., E. Rogel, J. Espidel, and G. Torres (2000). *Energy Fuels* **14**, 6.
[31] Andersen, S.I. and K.S. Birdi (1991). *J. Colloid Interface Sci.* **142**, 497.
[32] Andersen, S.I. and E.H. Stenby (1996). *Fuel Sci. Technol. Intl.* **14**, 261.
[33] Park, S.J. and G.A. Mansoori (1988). *Energy Sources*, **10**, 109.
[34] Yudin, I.K., G.L. Nikolaenko, E.E. Gorodetskii, *et al.* (1998). *J. Pet. Sci. Eng.* **20**, 297.
[35] Anisimov, M.A., I.K. Yudin, V. Nikitin, *et al.* (1995). *J. Phys. Chem.* **99**, 9576.
[36] Castillo, J., A. Fernandez, M.A. Ranaudo, and S. Acevedo (2001). *Pet. Sci. Eng.* **19**, 75.
[37] Leontaritis, K.J. and G.A. Mansoori (1987). In: *SPE International Symposium on Oilfield Chemistry*, Richardson, TX.
[38] Leontaritis, K.J. (1989). In: *SPE Production Operations Symposium*, Oklahoma City, OK.
[39] Mansoori, G.A. (1997). *J. Pet. Sci. Eng.* **17**, 101.
[40] Nellensteyn, F.J. (1939). *Chem. Weekblad* **36**, 362.
[41] Swanson, J. (1942). *J. Phys. Chem.* **46**, 141.
[42] Pfeiffer, J.P. and R.N. Saal (1940). *Phys. Chem.* **44**, 139.
[43] Hirschberg, A., L.N.J. DeJong, B.A. Schipper, and J.G. Meijer (1984). Influence of Temperature and Pressure on Asphaltene Flocculation. *Soc. Pet. Eng. J.* **24**(3), 283–293.
[44] Andersen, S.I. and J.G. Speight (1999). *J. Pet. Sci. Eng.* **22**, 53.
[45] Joshi, N.B., O.C. Mullins, A. Jamaluddin, J. Creek, and J. McFadden (2001). *Energy Fuels* **15**, 979.
[46] Hammami, A., C.H. Phelps, T. Monger-McClure, and T.M. Little (2000). *Energy Fuels* **14**, 14.
[47] Peramanu, S., C. Singh, M. Agrawala, and H.W. Yarranton (2001). *Energy Fuels* **15**, 910.
[48] Spiecker, P.M. and P.K. Kilpatrick (2002). In: *3rd International Conference on Petroleum Phase Behavior & Fouling*, New Orleans, LA.
[49] Blanco, M., S. Maspoch, I. Villarroya, X. Peralta, J.M. Gonzalez, and J. Torres (2001). *Anal. Chim. Acta*, **434**, 133.
[50] McClure, W.F. (1994). *Anal. Chem.* A **66**, 43.
[51] Mullins, O.C. (1990). *Anal. Chem.* **62**, 508.
[52] Gossen, P.D., J.F. MacGregor, and R.H. Pelton (1993). *Appl. Spectrosc.* **47**, 1852.
[53] Frake, P., I. Gill, C.N. Luscombe, D.R. Rudd, J. Waterhousea, and U.A. Jayasorriya (1998). *Analyst* **123**, 2043.
[54] Pasikatan, M.C., J.L. Steele, C.K. Spillman, and E. Haque (2001). *J. Near Infrared Spectrosc.* **9**, 153.
[55] Santos, A.F., E.L. Lima, and J.C. Pinto (1998). *J. Appl. Polym. Sci.* **70**, 1737.
[56] Parisi, A.F., L. Nogueiras, and H. Prieto (1990). *Anal. Chim. Acta* **238**, 95.
[57] Kelly, J.J. and J.B. Callis (1990). Nondestructive analytical procedure for simultaneous estimation of the major classes of hydrocarbon constituents of finished gasolines. *Am. Chem. Soci.* **62**(14), 1444–1451.
[58] Hidajat, K. and S.M. Chong (2000). *J. Near Infrared Spectrosc.* **8**, 53.
[59] Chung, H. and M.-S. Ku (2000). *Appl. Spectrosc.* **54**, 239.
[60] Ku, M.-S. and H. Chung (1999). *Appl. Spectrosc.* **53**, 557.
[61] Kim, M., Y.-H. Lee, and C. Han (2000). *Comput. Chem. Eng.* **24**, 513.
[62] Aske, N., H. Kallevik, E.E. Johnsen, and J. Sjöblom (2002). Asphaltene aggregation from crude oils and model systems studied by high pressure NIR spectroscopy. *Energy Fuels* **16**(5), 1287–1295.
[63] Chang, C.-L. and S.H. Fogler (1994). *Langmuir* **10**, 1758.
[64] Chang, C.-L. and S.H. Fogler (1994). *Langmuir* **10**, 1749.
[65] Auflem, I.H., T.E. Havre, and J. Sjöblom (2002). Near infrared study on the dispersive effects of amphiphiles and naphthenic acids on asphaltenes in model heptane–toluene mixtures. *Colloid Polym. Sci.* **280**, 695–700.

[66] Ostlund, J.-A., M. Nyden, I.H. Auflem, and J. Sjöblom (2003). Interactions between asphaltenes and naphthenic acids. *Energy Fuel* **17** (1), 113–119.

[67] Ostlund, J.-A., S.-I. Andersson, and M. Nyden (2001). *Fuel* **80**, 1529.

[68] Ekholm, P., E. Blomberg, P. Claesson, I.H. Auflem, J. Sjöblom, and A. Kornfeldt (2002). *J. Colloid Interface Sci.* **247**, 342.

[69] Lucassen-Reynders, E.H. (1996). In: P. Becher (ed.), *Encyclopedia of Emulsion Technology*, Vol. 4. Marcel Dekker, New York, p. 63.

[70] Lucassen-Reynders, E.H. (1993). *Food Struct.* **12**, 1.

[71] Benjamins, J., A. Cagna, and E.H. Lucassen-Reynders (1996). *Colloids Surf.* **114**, 245.

[72] Kilpatrick, P.K. and P.M. Spiecker (2001). In: J. Sjöblom (ed.), *Encyclopedic Handbook of Emulsion Technology.* Marcel Dekker, New York, p. 707.

[73] Goldszal, A., C. Hurtevent, and G. Rousseau (2002). Scale and naphthenate inhibition in deep-offshore fields. In: *SPE Oilfield Scale Symposium*, Aberdeen, UK.

[74] Rousseau, G., H. Zhou, and C. Hurtevent (2001). Calcium carbonate and naphthenate mixed scale in deep offshore fields. In: *SPE Oilfield Scale Symposium*, Aberdeen, UK.

[75] Hsu, C.S., G.J. Dechert, W.K. Robbins, and E.K. Fukuda (2000). *Energy Fuels* **14**, 217.

[76] Rudin, J. and D.T. Wasan (1992). *Colloids Surf.* **68**, 67.

[77] Rudin, J. and D.T. Wasan (1992). *Colloids Surf.* **68**, 81.

[78] Sjöblom, J., E.E. Johnsen, A. Westvik, L. Bergflødt, I.H. Auflem, T.E. Havre, and H. Kallevik (2000). Colloid chemistry in sub sea petroleum and gas processing. In: *Second International Conference on Petroleum and Gas Phase Behaviour and Fouling*, Copenhagen, Denmark.

[79] Slavcheva, E., B. Shone, and A. Turnbull (1999). *Br. Corros. J.* **34**(2), 125.

[80] Marquez, M.L.H. (1999). Interfacial activity of native acids in heavy crude oil. In: *AICHE Spring National Meeting Session T6005*, Houston, TX.

[81] Acevedo, S., G. Escobar, M.A. Ranaudo, *et al.* (1999). *Energy Fuels* **13**(2), 333.

[82] Pathak, A.K. and T. Kumar (1995). Study of indigenous crude oil emulsions and their stability. In: *Proceedings of PETROTECH-95, Technology Trends in Oil Industry*, New Delhi.

[83] Skurtveit, R., J. Sjöblom, and H. Høiland (1989). *J. Colloid Interface Sci.* **133**(2), 395.

[84] Sjöblom, J., R. Lindberg, and S.E. Friberg (1996). *Adv. Colloid Interface Sci.* **95**, 125.

[85] Gillberg, G., H. Lehtinen, and S.E. Friberg (1970). *J. Colloid Interface Sci.* **33**(1), 40.

[86] Friberg, S.E., L. Mandell, and P. Ekwall (1969). *Kolloid-Z. Z. Polym.* **233**(1,2), 955.

[87] Lindman, B. and H. Wennerstroem (1980). *Top. Curr. Chem.* **87**, 1.

[88] Wennerstrom, H. and B. Lindman (1979). *Phys. Rev.* **52**, 1.

[89] Ekwall, P., I. Danielsson, and P. Stenius (1972). In: M. Kenken (ed.), *MTP Rev. Sci. Phys. Chem. Ser. 1*, Vol. 7. Butterworths, London, p. 97.

[90] Ekwall, P. (1975). In: G.H. Brown (ed.), *Advances in Liquid Crystals*, Vol. 1. Academic Press, New York, p. 1.

[91] Ekwall, P., L. Mandell, and K. Fontell (1969). *Mol. Cryst. Liq. Cryst.* **8**, 157.

[92] Brient, J.A., P.J. Wessner, and M.N. Doyle (1995). In: K. Othmer (ed.), *Encyclopedia of Chemical Technology*, John Wiley & Sons, New York. p. 1017.

[93] Meredith, W., S.-J. Kelland, and D.M. Jones (2000). *Org. Geochem.* **31**(11), 1059.

[94] Fan, T.-P. (1991). *Energy Fuels* **5**(3), 371.

[95] Koike, L., L.M.C. Reboucas, F.D.A. Reis, A.J. Marsaioli, H.H. Ichnow, and W. Michaelis (1992). *Org. Geochem.* **18**(6), 851.

[96] Tomczyk, N.A., R.E. Winans, J.H. Shinn, and R.C. Robinson (2001). *Energy Fuels* **15**(6), 1498.

[97] Qian, K., W.K. Robbins, C.A. Hughey, H.J. Cooper, R.P. Rodgers, and A.G. Marshall (2001). *Energy Fuels* **15**(6), 1505.

[98] Robbins, W.K. (1998). *Prepr-Am. Chem. Soc. Div. Pet. Chem.* **43**(1), 137.

[99] Brient, J.A. (1998). *Prepr. Am. Chem. Soc. Div. Pet. Chem.* **43**(1), 131.

[100] Horvath-Szabo, G., J. Czarnecki, and J. Masliyah (2001). *J. Colloid Interface Sci.* **236**, 233.

[101] Horvath-Szabo, G., J. Masliyah, and J. Czarnecki (2001). *J. Colloid Interface Sci.* **242**, 247.

[102] Strassner, J.E. (1968). *J. Pet. Technol.* **20**(3), 303.

[103] Eley, D.D., M.J. Hey, and M.A. Lee (1987). *Colloids Surf.* **24**, 173.

[104] Mohammed, R.A., A.I. Bailey, P.F. Luckham, and S.E. Taylor (1993). *Colloids Surf.* **80**, 223.

[105] Mohammed, R.A., A.I. Bailey, P.F. Luckham, and S.E. Taylor (1993). *Colloids Surf.* **80**, 237.

[106] Førdedal, H., E. Nodland, J. Sjöblom, and O.M. Kvalheim (1995). *J. Colloid Interface Sci.* **173**, 396.

[107] Førdedal, H., Y. Schildberg, J. Sjöblom, and J.-L. Volle (1996). *Colloids Surf.* **106**, 33.

[108] Førdedal, H., Ø. Midttun, J. Sjöblom, O.M. Kvalheim, Y. Schildberg, and J.-L. Volle (1996). *J. Colloid Interface Sci.* **182**, 117.

[109] Djuve, J., X. Yang, I.J. Fjellanger, J. Sjöblom, and E. Pelizzetti (2001). *Colloid Polym. Sci.* **279**, 232.

[110] Chen, T.Y., R.A. Mohammed, A.I. Bailey, P.F. Luckham, and S.E. Taylor (1994). *Colloids Surf.* **83**, 273.

[111] Førdedal, H. (1995). WyO emulsions in high electric fields as studied by means of time domain dielectric spectroscopy. Ph.D. thesis, University of Bergen, Norway.

[112] Aske, N., H. Kallevik, and J. Sjöblom (2002). Water-in-crude oil emulsion stability studied by critical electric field measurements. Correlation to physico-chemical parameters and near-infrared spectroscopy. *J. Pet. Sci. Eng.* **36**(1-2), 1–17.

[113] Kallevik, H., O.M. Kvalheim, and J. Sjöblom (2000). *J. Colloid Interface Sci.* **225**, 494.

[114] Wise, B.M. and N.B. Gallagher (1996). *J. Proc. Cont.* **6**, 329.

[115] Martens, H., and M. Martens (2001). *Multivariate Analysis of Quality. An Introduction.* John Wiley & Sons, Chichester.

[116] Aske, N., H. Kallevik, and J. Sjöblom (2001). *Energy Fuels* **15**, 1304.

[117] Auflem, I.H., H. Kallevik, A. Westvik, and J. Sjöblom (2001). *J. Pet. Sci. Eng.* **31**(1), 1.

[118] Hinze, J.O. (1955). *AIChE J.* **1**(3), 289.

[119] Karabelas, A.J. (1978). *AIChE J.* **24**(2), 170.

[120] Davies, J.T. (1985). *Chem. Eng. Sci.* **40**(5), 839.

[121] Sleicher, C.A.J. (1962). *AIChE J.* **8**(4), 471.

[122] Meijs, F.H. and R.W. Mitchell (1974). *J. Pet. Technol.* May, 563.

[123] Eiken, M.B. (1999). M.Sc. thesis, Norwegian University of Science and Technology, Trondheim.

22

Live Oil Sample Acquisition and Downhole Fluid Analysis

Go Fujisawa and Oliver C. Mullins

1. Introduction

All aspects of crude oil production depend on the proper acquisition and analysis of petroleum samples from subsurface formations. The analyses of hydrocarbon samples that are acquired in exploration wells guide the subsequent production strategies and facility designs. Nevertheless, hydrocarbon sample acquisition and analysis have been subjects of considerable uncertainty, contributing to a frequent inability to predict properly fluid (and volumetric) production parameters. This mismatch can be inordinately costly in important production arenas such as deepwater oil development, thus new solutions are mandated. The recently developed technology Downhole Fluid Analysis (DFA) is significantly improving several problematic processes involving fluid samples. The technology is sufficiently cost-effective to be routinely utilized in many economic settings. DFA can identify unwanted phase transitions and determine unacceptable contamination levels, thus is indispensable for sample acquisition. In addition, DFA can readily identify fluid differences at various points in the reservoir (those that have been penetrated by a well), thus DFA can uncover fluid complexities that are now appreciated to be common. It has long been recognized that fluid analysis can identify different reservoir production units known as compartments. However, the geoscientist has been greatly impeded in this quest due to the high cost of essentially blind sample acquisition. DFA is the "missing link" in this chain, enabling identification of the samples that are needed for characterization of reservoir architecture. Finally, as in forensic science, proof of sample provenance and validity is a quintessential objective, but which has been largely absent in the oil field. DFA has enabled implementation of the "chain of custody" for reservoir hydrocarbon samples. The chain of custody provides quantitative proof that even delicate hydrocarbon samples that travel thousands of miles and to other continents for analysis in fact reflect virgin reservoir fluid properties. This chapter delineates aspects of the setting and the technology for sample acquisition. DFA is introduced in concept and in specific

---

Go Fujisawa • Schlumberger K.K., Sagamihara, Kanagawa, Japan
Oliver C. Mullins • Schlumberger-Doll Research, Ridgefield, CT.

technology in accord with its sundry applications. Field examples are provided, documenting the major advances made by DFA. This fertile new ground provides new opportunities for petroleum scientists to establish utility in this important, cost-conscious industry.

Fundamental questions that must be addressed in the production of hydrocarbon fluids include: "what is the size, structure, and character of each hydrocarbon-bearing zone?", "what is the nature of the contained hydrocarbon fluid to be produced?", and "what is the interaction between rock and fluid?" These formidable objectives include characterization of all chemistry and physics of solid and fluid matter and their interaction in the subsurface. Efficient production of gas and oil requires correct answers to these questions; the proper analyses of reservoir fluids play a vital role in this quest. In some circumstances, such as inexpensive land development of oil fields, errors from improper prediction of hydrocarbon production can still be corrected after production is initiated. For instance, in this environment, if actual oil production is smaller than expected due to reservoir compartmentalization, one can modify and scale production plans accordingly. If asphaltene deposits are formed unexpectedly, it may still be economically feasible to perform periodic remediation to resolve the issue. In certain economic settings, it may be preferable to remediate the subset of wells that exhibit problems rather than to pay for proper analysis prior to production for each well. However, in certain settings such as deepwater oil development, rather complete forward modeling is required due to the inordinate costs of unanticipated intervention. Development costs for individual fields in deep water can exceed one billion dollars. Post-production retroactive changes in subsea facilities design or production strategy may destroy profitability. Thus, the accuracy of production forward modeling becomes of paramount importance. Fluids properties including phase transitions, gas–oil ratios, and viscosities must be known before a development plan is put into place. In addition, the risk associated with uncertain characterization of reservoir architecture must be reduced. For both objectives, proper acquisition and analysis of fluid samples is required. Downhole Fluid Analysis is indispensable for both sample acquisition and analysis.

In order to study properties of subsurface hydrocarbon fluids, nothing can substitute for obtaining live, representative fluid samples. Live, representative samples exist as they are in the formations; thus, they have dissolved gases, dissolved waxes, and stably suspended asphaltenes. However, acquisition of representative hydrocarbon fluid is difficult and requires sophisticated methodologies. Improper sampling can result from the addition of impurities (e.g., drilling fluid), from the removal of original components (e.g., H_2S or precipitated asphaltenes), or from incorrect component proportions caused by multiphase breakout with differential flow. There are two methods to acquire samples from newly drilled wells. One method is to utilize an appropriate tool package suspended from a cable or wireline to collect samples downhole. Wireline acquisition of samples has the benefits of taking a relatively short period of time and requiring relatively low costs. The other method to acquire the fluid sample utilizes the drillstring—termed drillstem testing (DST). Packers are placed to isolate the targeted permeable zone, and by controlling the pressure inside the drillstring stem, a large volume of formation fluid can be produced to surface. In certain settings such as deepwater oil development, the substantial cost and environmental concerns associated with DST sampling prohibit its routine deployment. The cost of a DST can approach the cost of a new well, and the new well would be far more valuable. Wireline sample acquisition is often much less than 10% of the DST cost, particularly in deep water. As a result, DST sampling often is replaced by wireline fluid sampling.[1,2]

A typical wireline job commences immediately after drilling has been completed. At this stage the well is "openhole," that is, not yet cased with steel pipe (at least at the well depth of interest). The well is filled with drilling mud. The mud maintains borehole pressure, conveys rock cuttings to surface during drilling, and lubricates drilling among other functions. The drilling mud is weighted to maintain a higher borehole pressure than formation fluid pressure to prevent blowouts. Clay and other solids are added to the mud in order to limit fluid loss into permeable zones. At permeable zones, the drilling mud initially flows into the formation. The mud is designed to form an impermeable mud cake on the borehole wall to limit fluid loss. Nevertheless, some mud filtrate seeps into the formation. Drilling muds are formulated either as water-based mud (WBM) or oil-based mud (OBM). OBMs provide better lubrication, stabilize reactive shale, and build better quality mud cakes, thus are generally preferred for deepwater drilling. Sample acquisition and analysis must be achievable in both WBM and OBM settings.

This chapter treats new innovations in wireline fluid sampling. In particular, DFA in the form of optical spectroscopy is shown to be invaluable in the sample acquisition process. In addition, new and surprising applications of DFA are treated, addressing the most important concerns in the production of crude oil. Fluids are often distributed in a heterogeneous manner in the reservoir. DFA provides a cost-effective means to identify important spatial variations of fluids. In addition, flow connectivity, or its inverse, compartmentalization, is often difficult to identify. The heterogeneous distribution of hydrocarbons can be helpful in identifying flow units because different compartments are often filled with different fluids. DFA has also enabled the forensic concept of chain of custody to be invoked for hydrocarbon samples. The DFA measurement fingerprint provides a quantitative means to prove sample validity. DFA has significantly increased the value of fluid sample acquisition and analysis via both improved efficiencies and new fluid applications.

2. Wireline Fluid Sampling Tools

The archetypical wireline fluid sampling tool is found in the Schlumberger MDT* (modular formation dynamics tester). Figure 22.1 shows the probe module of the MDT which is used to establish hydraulic or flow communication with the formation. We use the MDT tool as the example to explain the process of wireline fluid sampling. A typical MDT tool configuration for DFA and sample acquisition, along with pressure measurement, is shown schematically in Figure 22.2. From

---

*Trademark of Schlumberger.

Figure 22.1. A photograph of the MDT tool, probe module section. The probe that contacts the borehole wall, a thick protruding metal tube, is seen on the extreme left of the photograph. Two standoffs seen on the right side of the picture are used to press against the opposing borehole face balancing the forces. With high-applied forces, the interior of the probe is hydraulically isolated from the interior of the borehole. The rubber packer surrounding the probe pushes against the formation helping to seal.

bottom to top, the tool string consists of two single-probe modules, LFA* (live fluid analyzer), pumpout module, CFA* (composition fluid analyzer), sample-chamber modules, and a power and telemetry cartridge. The modules have a flowline for fluid flow, and when they are assembled and connected, the flowline forms the backbone of the MDT tool. The power cartridge converts AC power sent from surface through the wireline to DC power necessary to power all modules attached in the toolstring. The single-probe module contains the probe assembly that hydraulically connects the formation fluid with the interior of the MDT flowline so that formation fluid can flow into the toolstring. The sampling port consists essentially of a thick-walled tube that is pushed against the formation with great force. This tube or port is surrounded by a rubber packer, which hydraulically isolates the sample point from the surrounding borehole environment. The hydraulic module (not depicted

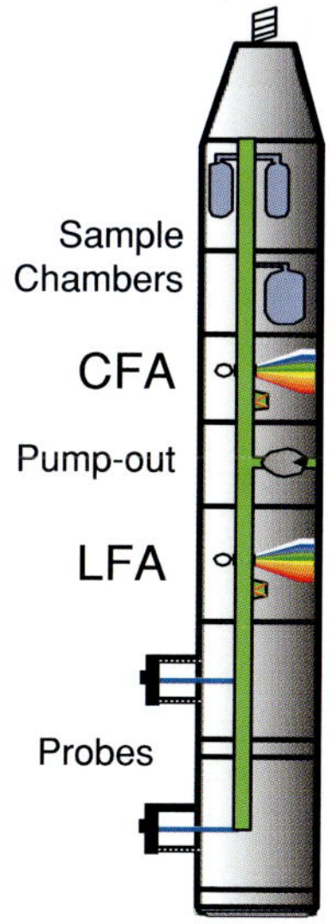

Figure 22.2. An example of a wireline fluid sampling tool configuration is shown. Downhole fluid analysis tools are the LFA and CFA.

in Figure 22.2) provides the necessary hydraulic power for setting, retracting the probe and operating the pump out module. Figure 22.2 shows incorporation of two single-probe modules in the toolstring, enabling some advanced applications and providing some redundancy. The LFA module is one of the DFA modules and is equipped with an optical spectrometer and an optical refractometer for gas detection.[3] The CFA module has a second, different optical spectrometer and a fluorescence cell.[4] The LFA and CFA analyzers monitor formation fluid as it flows into the MDT toolstring. The pumpout module creates the pressure drop, called drawdown, in the flowline to cause fluids to flow. Fluids can be either discharged into the borehole or directed into a sample bottle. There are various types of sample chambers arrayed in different sample modules of the MDT suite. An advanced sample module is the single-phase multisample chamber (SPMC) that consists of six independent sample chambers. Each SPMC bottle contains a piston assembly and nitrogen gas charge, allowing the sample to be kept at downhole pressures indefinitely. This is necessary to prevent asphaltene flocculation and other unfavorable phase segregations. It is very difficult to reconstitute precipitated asphaltenes especially if mixed with inorganic solids, and water that often reside on the bottom of the sample bottles.

3. Downhole Fluid Analysis with Wireline Tools

3.1. Measurement Physics

Visible (VIS) and/or near-infrared (NIR) absorption spectroscopy are utilized in many industrial settings for a wide variety of purposes. For example,

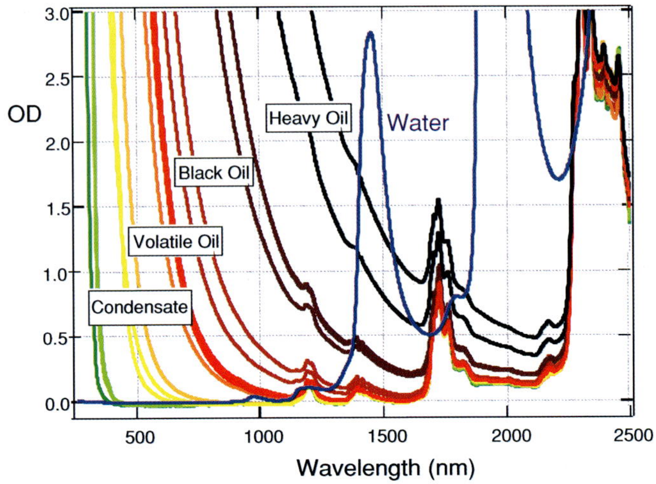

Figure 22.3. VIS/NIR Spectra of crude oil and water. Both vibrational mode absorption peaks and electronic excitation absorption profiles are seen in this figure. Spectral differentiation of oil and water is facile. Crude oils have similar NIR peaks but very distinct electronic absorption profiles.

constituent bulk chemical content in food substances,[5] BTU content of natural gas,[6] and the gas–oil ratio (GOR) of crude oils[7] are all suited to NIR spectroscopy and analysis. The NIR spectra record vibrational overtones of hydrogen-containing chemical bonds, which are ubiquitous in organic matter. The NIR is a natural choice for implementing spectroscopic measurements in typical oil reservoir conditions since it utilizes conveniently long sample pathlength, and the appropriate light source, moreover, detectors are readily available and are robust. For materials such as crude oils, where color conveys meaningful compositional information, the NIR spectral measurement can be extended to include the visible spectral range with essentially the same instrumentation. For crude oil, light absorption in the VIS/NIR range is caused by two different molecular mechanisms: vibrational mode excitation and electronic excitation. Examples of crude oil spectra are shown in Figure 22.3. A brief description of the relevant molecular science and spectral principles is given here, and more detailed descriptions are found in many standard textbooks.[8]

Optical spectroscopy records the optical transmission through a sample as a function of wavelength. The intensity of the incident light I_0 is reduced by absorption A and scattering S yielding exiting light intensity I. Light transmission is defined as $T = I/I_0$. The optical density (OD) of a sample is given by

$$\mathrm{OD} \equiv -\log\left(\frac{I}{I_0}\right). \tag{22.1}$$

Often, the OD $= A + S$. For homogeneous samples with no scattering ($S = 0$), Beer's law applies

$$A = \sum_i \varepsilon_i c_i l, \tag{22.2}$$

where ε_i is the extinction coefficient of component i, c_i is the concentration of component i, and l is the optical pathlength through the sample. Both A and ε_i are a function of wavelength. Optical spectra of materials are characterized by the extinction coefficient and concentration of their constituents. Light scattering is an additional complication that can be strong when sample heterogeneity is comparable to or larger than the wavelength of light. Emulsions, foams, mists, sands, mud fines, and asphaltene flocculation can and do produce light scattering downhole.

Molecular vibrations can be considered as mechanical oscillators; the atoms are masses and the bonds are springs. The corresponding resonant frequency ω is given by

$$\omega = \sqrt{\frac{k}{\mu}}, \tag{22.3}$$

where k is the bond force constant and μ is the reduced mass of the mode. Modes involving hydrogen atoms are high frequency due to their small reduced mass and reasonable bond strength. The resonant frequencies of these fundamental vibrational modes for organic molecules are found in the mid-IR region.

Figure 22.4 shows the molecular potential and energies of the vibrational states plotted vs. bond length parameter ξ in both the harmonic approximation and

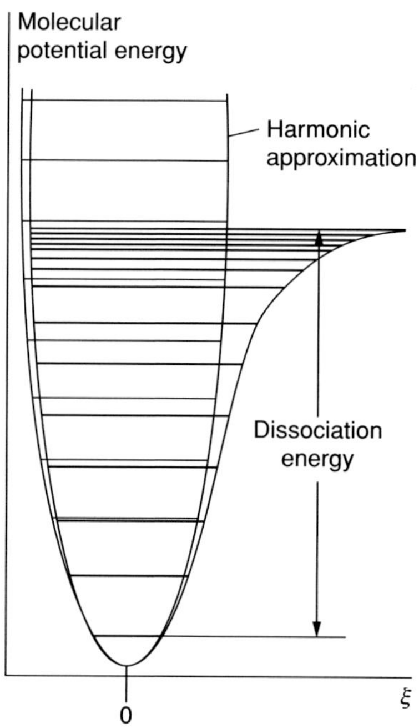

Figure 22.4. The potential energy for an harmonic oscillator and the actual molecular potential energy plotted vs. bond length parameter ξ. The quantized vibrational levels are also shown for the two different potentials. The actual potential curve differs from the harmonic model; thereby producing optically excited overtone modes.

actual potential. At low energies of excitation, the vibrational potential approximates the harmonic oscillator.

The vibrational states are quantized, appearing at discrete energies only. These energies occur where the corresponding vibrational wavefunction ψ_n for state n can satisfy the potential-well boundary conditions. At low energies, the vibrational energy states are equally spaced (cf. Figure 22.4). While the lowest energy transitions, called fundamental modes, carry rich information about molecular structures and have been extensively studied, they are not particularly suited for DFA applications for a number of reasons. Transitions between states adjacent in energy, which occur in the mid-IR, are allowed. This very strong absorption of fundamental modes necessitates very short pathlength (the order of 10 μm) for hydrocarbon liquids; this short pathlength is not convenient for bulk flow analysis. Transitions between nonadjacent molecular vibrational levels are formally forbidden in the harmonic approximation. In reality, the anharmonicity in the molecular potential (Figure 22.4) modifies the vibrational wavefunction creating a nonzero transition cross-section for optical excitation between nonadjacent vibrational levels. In addition, nonlinear variations of the bond electric dipole moment vs. bond length parameter ξ also cause overtone amplitude. At the lowest energies in the potential well, the deviations of vibrational energies and wavefunctions from the harmonic approximation are only slight (cf. Figure 22.4), thus the first overtone, the optical excitation of the ground vibrational state to the third vibrational state, is quite small. An even smaller cross-section occurs for the second overtone, the optical excitation of connecting the ground state to the fourth vibrational state. The first and second overtones of the CH stretch mode of the alkanes, which are all forbidden, are seen in Figure 22.3 at wavelengths ~1.7 micron (μm), and ~1.2 μm. (The combination band CH "two-stretch + bend" can also be seen at ~1.4 μm.) Note the decreasing peak height (cross-section) with increasing overtone. These forbidden transitions of the NIR are roughly 100 times weaker than the fundamental vibrational bands of the mid-IR. Correspondingly, the NIR bands require much longer pathlength (~1 mm) for significant light absorption which is quite convenient for DFA purposes.

Figure 22.5 shows the first overtone of the CH stretch mode. The spectrum for methane is quite distinct from that of n-heptane. The n-heptane bands that result from the -CH_3 (methyl) group appear as two shoulder peaks on the high energy side of the 1725 nm peak. The largest peak at 1725 nm results primarily from the -CH2- (methylene) group. This peak is very useful for DFA purposes.

The fundamental vibrational transitions are well described by normal mode analysis; many atoms are involved in concerted vibrational motions. Vibrational transitions from the ground to highly excited states are well described by local mode analysis; all of the excitation energy is essentially dumped into a single CH vibrator involving two atoms. Among other shortcomings, the high overtones are too weak to be of use for DFA. The low overtones involving CH vibrations are neither normal nor local modes. This makes their analysis somewhat complicated because individual resonance peaks are often not cleanly associated with specific functional groups. In spite of this complication, the NIR is highly useful for chemical analysis.

Light sources and detectors for the VIS/NIR are more effective than those of the mid-IR (MIR), especially at elevated temperatures of operation. The

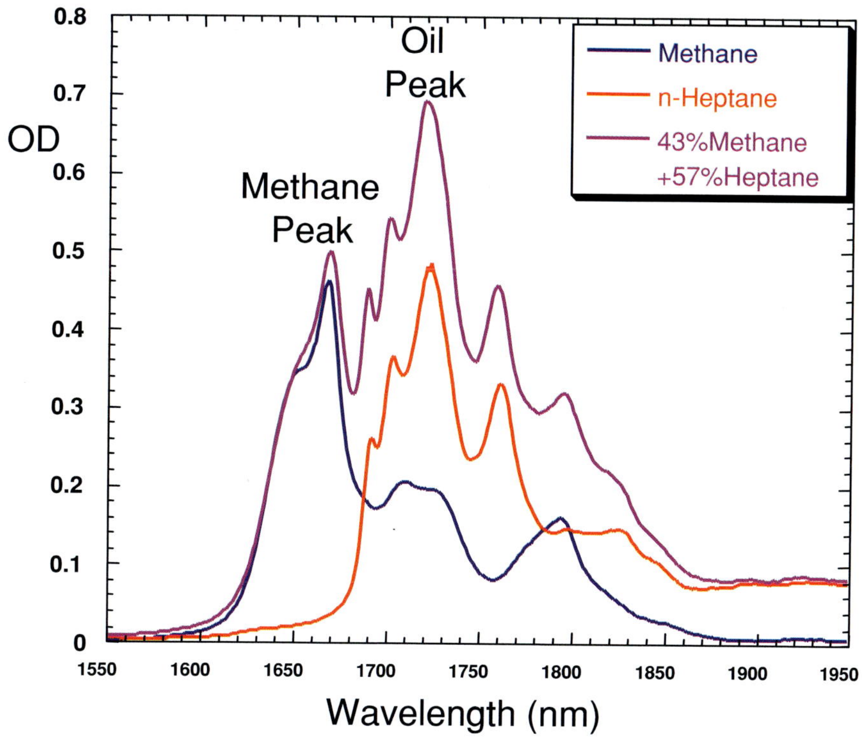

Figure 22.5. The NIR CH two-stretch overtone. The methane peak is distinguished from alkane peaks. Detailed analysis of this CH two-stretch overtone yields compositional information of crude oils.

Stefan–Boltzmann law states that the emissive power E (radiated energy per unit time per unit area) of a blackbody radiator increases with the fourth power of temperature T, $E \propto T^4$. Wien's displacement law, $T\lambda_M = 0.2898\,\text{cm} \cdot \text{K}$ shows that for a blackbody radiator, the wavelength of maximum emission λ_M varies inversely with temperature T. For tungsten–halogen lamps ($\sim$3000 K radiator), λ_M corresponds to about 1 μm with the source spectrum covering the VIS/NIR range. The equivalent radiation source for peak emission in the MIR is of much lower emissive power due to lower source temperature. Detector technology is also better developed for the VIS/NIR region over MIR, because silicon (Si) and indium gallium arsenide (InGaAs) detectors are available for the VIS/NIR. Si detectors have a cutoff wavelength of detection at $\sim$1.1 μm, while InGaAs detectors have a cutoff wavelength at $\sim$ 1.7 μm, which can be extended somewhat. Thus, these photodetectors have reasonably large bandgaps limiting thermal noise. In addition, silica fiber-optics for the VIS/NIR light transport are far simpler than for the MIR equivalent.

A series of studies[7,9–11] on live hydrocarbon fluid spectra at high pressures and high temperatures has shown that the alkane spectra are essentially linear in the concentrations of constituent molecules, that is, Beer's law is applicable. The primary effect of both pressure (P) and temperature (T) on these bands is via fluid density effects, for P and T of interest in oil reservoirs. Nontrivial P and T

effects on line width can be seen for methane in gas samples below 4000 psi. Large pressure-dependent changes occur in the methane spectrum below 2000 psi. The overtone vibration modes are useful for measuring approximate fluid compositions because different chemical function groups have different resonance wavelengths. The CH_4, $-CH_3$, and $-CH_2-$ groups have somewhat similar but still distinctive absorption spectra (cf. Figure 22.5), and their concentrations can be directly determined by spectroscopic measurement. These chemical groups in turn map into chemical composition. The CH_4 group maps uniquely to methane, the $-CH_3$ functional group maps mainly to nonmethane gaseous hydrocarbon (ethane, propane, etc.) but is found in liquid hydrocarbons as well, and the $-CH_2-$ functional group mainly maps to liquid hydrocarbons although it is found in some hydrocarbon gas, too.

The other molecular absorption mechanism that occurs in the VIS/NIR for crude oils is electronic excitation. In large polycyclic aromatic hydrocarbons (PAHs), π electrons are delocalized, thereby lowering their transition energies. This is in accord with the familiar quantum particle in a box; generally, the bigger the box (or aromatic ring system) the lower the transition energies. For example, benzene, a small aromatic compound is colorless (its lowest π electronic transition is at 250 nm), while graphite, a giant aromatic compound, is black. Graphite's electronic absorption bands extend far below the visible. This effect is well known in classical physics as well; notes played on a guitar are lowered as the nodes are moved further apart. Ring geometry in addition to ring number also influences the transition energies. A linear arrangement of aromatic rings, a catacondensed configuration, produces low energy electronic transitions. A circular arrangement of aromatic rings, essentially pericondensation, produces higher energy electronic transitions (cf. Chapter 4). The aromatic ring systems of asphaltenes are predominantly sextet aromatic carbon which can be pericondensed or catacondensed (cf. Chapter 5).

The spectrally structureless increase in optical absorption with shorter wavelength produces the characteristic yellow, tan, brown, or black coloration of crude oil.[12] This same absorption mechanism is responsible for the coloration of resins and asphaltenes.[12] Crude-oil color extends beyond the visible into the NIR for heavy crude oil.[12, 13] Figure 22.6 shows the VIS/NIR spectra of many crude oils; crude oil coloration is very systematic and is seen to extend into the NIR.

It has been shown that the characteristic coloration of crude oils occurs because of overlapping spectra from many constituent polycyclic aromatic hydrocarbons (PAHs) in these polydispersed systems.[12] However, the crude-oil coloration is not at all random but very systematic. Crude-oil coloration exhibits the "Urbach tail," first noted in semiconductors.[12–14] In semiconductors, the Urbach tail in absorption spectra corresponds to the Boltzmann distribution of thermally activated electronic absorber sites. Thermally excited electronic absorber sites have correspondingly lower electronic transition energies by an amount equal to the thermal excitation. Crude-oil coloration does not result from thermally excited benzene. Instead, the thermal production of large PAHs from small PAHs coupled with alkylated PAH solubility dictate the population distribution of the largest ring systems of any given crude oil. This process is similar to

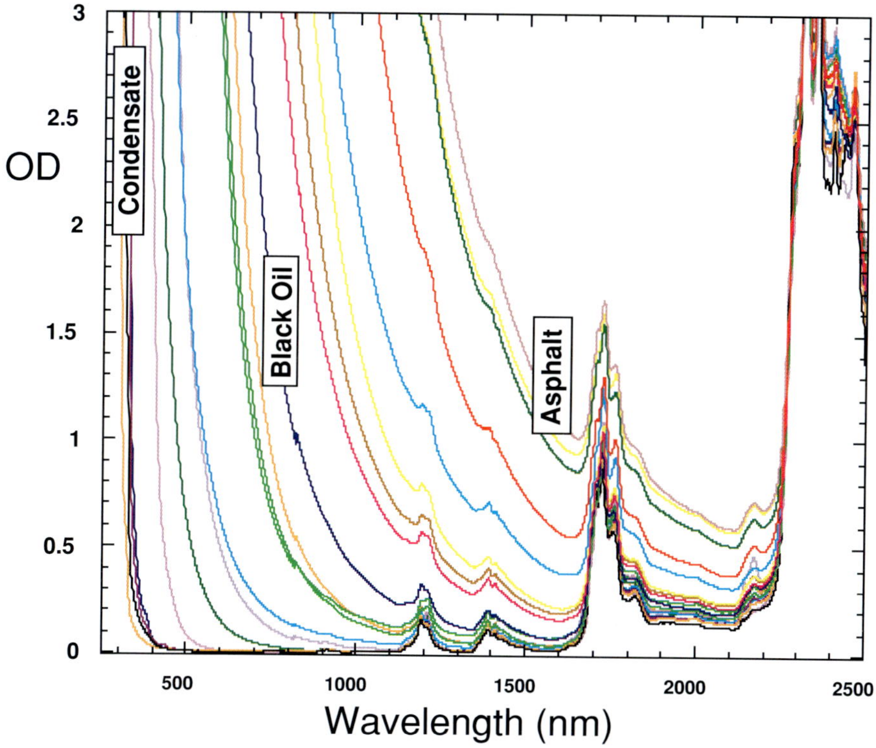

Figure 22.6. The VIS/NIR spectra of many crude oil. The 'color' absorption for different crude oils is very different but is very systematic.

toasting white bread. Initial heating produces yellow or tan coloration. Continued toasting produces dark brown to black coloration in accordance with producing large PAH ring systems from small PAHs. Crude oils and asphaltenes do have an additional complication: this cooking process involves a disproportionation reaction of immature kerogen into fluid hydrocarbons plus mature kerogen or of an oil into a lighter oil and an immovable pyrobitumen.[15] A further complication is that larger PAHs have a decreasing solubility in crude oil. Nevertheless, crude-oil electronic spectra exhibit the Urbach tail as shown in Figure 22.7. That is, the electronic absorption edge is linear when the optical density is plotted on a log scale against photon energy (1/wavelength). This absorption spectrum directly reflects the population distribution of PAHs in crude oils and corresponds to an exponentially decreasing population of molecules containing increasingly large PAHs.[12] This systematic coloration of all crude oils and asphaltenes is useful for achieving sampling objectives. In addition, larger aromatics imparting color to the fluid are hydrogen deficient and therefore are of higher density. Thus, there is an approximate correlation between crude-oil density and color, that is, the darker the fluid color, the greater the fluid density as shown in Figure 22.8.

Finally, we note that some crude oils have a green appearance. Green coloration, a well-known spectrum, is never seen in any crude-oil absorption spectrum.

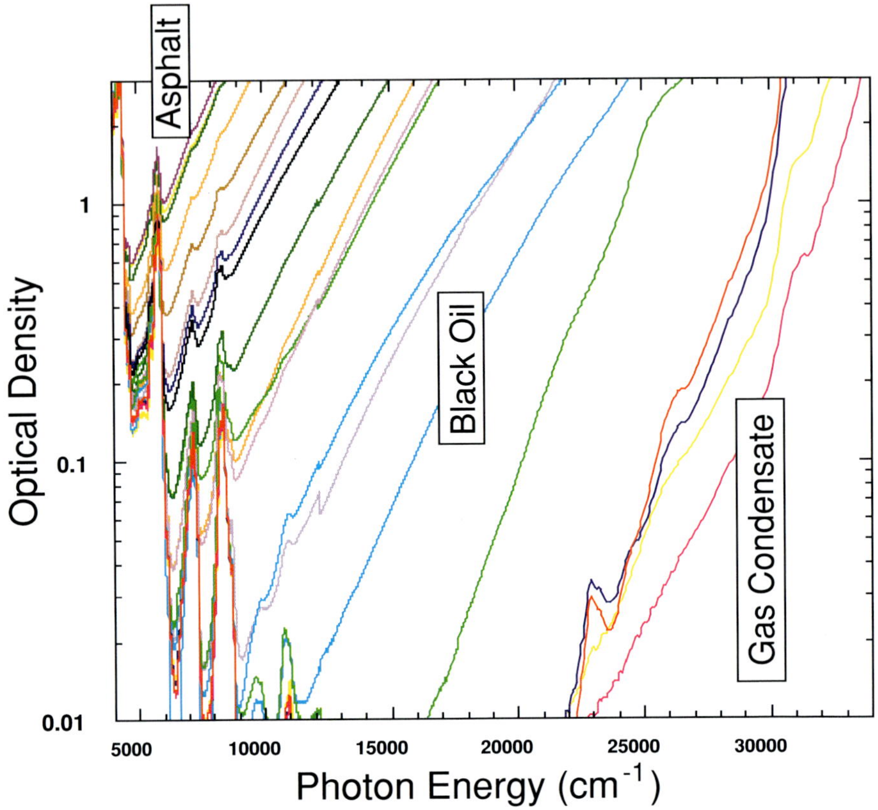

Figure 22.7. Crude oil spectra exhibit the Urbach tail, the slope of the electronic absorption edge for all crude oils is the same—but the spectral locations of the edge differ.

The green appearance of some crude oils is most likely a result of some fluorescence emission induced by blue/UV content of the source illumination coupled with the optical absorption of the crude oil. As such, these crude oils will have very different appearances when illuminated with incandescent light (3000 K radiator), sunlight (6000 K radiator), or fluorescent light (strong line source contribution with significant blue content).

The DFA suite of measurements includes fluorescence. Fluorescence is the radiation process that can follow photoexcitation. Photoabsorption is a necessary first step in the fluorescence process; materials that do not absorb the excitation light will not fluoresce. For example, for visible excitation, colorless materials will not fluoresce. Fluorescence is a dynamic process which is strongly influenced by many factors.[16] Essentially, fluorescence is a slow process taking on the order of nanoseconds or more. In crude oils, other electronic de-excitation processes compete with fluorescence, such as collisional energy transfer[17] and quenching,[18] intersystem crossing (with subsequent quenching), thermally induced quenching,[19] and quenching from complex formation. For example, heavy crude oils, which have a much higher concentration of fluorophores than light crude oils, have a much

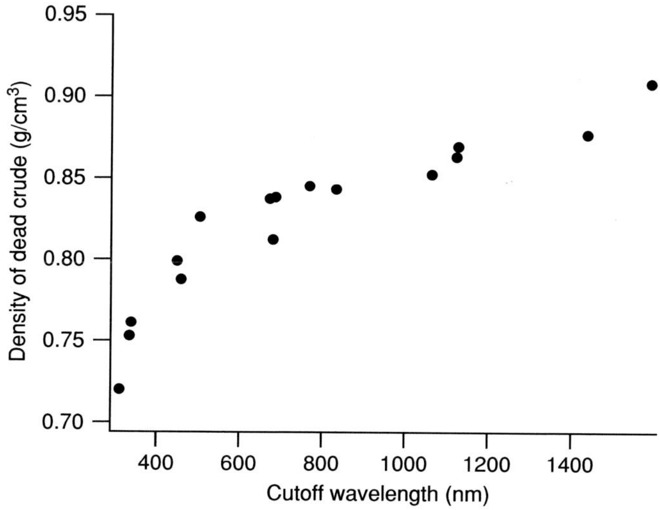

Figure 22.8. A correlation between cutoff wavelength and mass density is shown. Here, the cutoff wavelength is defined as the wavelength at which electronic absorption equals OD = 1 for a-2 mm pathlength. The more dense the crude oil, the longer the cutoff wavelength. More dense crude oil contains a larger fraction of big aromatic compounds that absorb longer wavelength light.

lower fluorescence quantum yield due to various quenching mechanisms.[18] These mechanisms are also responsible for the dull red color of fluorescence from heavy oils as opposed to the brighter orange, yellow, or even blue fluorescence color of increasingly lighter crude oils, respectively, when excited by UV/blue light. The primary governing principles of crude-oil fluorescence have been worked out and are presented elsewhere.[12] For example, crude oils are one of the few systems that have been shown to obey the well-known Exponential Energy Gap Law,[18] which governs the fluorescence intensity and color from crude oils. Crude oils and asphaltenes exhibit broad systematics in their fluorescence properties, which have been used for decades in the oil field, and which are also quite useful for DFA objectives.

3.2. DFA Implementation in Wireline Tools

Both the LFA and CFA tools employ VIS/NIR spectroscopy to achieve downhole fluid analysis objectives. Performing spectroscopic analysis on live fluids inside a wireline tool presents many challenges, including high temperature, high-fluid pressure, mechanical shock and vibration, and limited available space. Wireline spectrometer tools need to be designed to perform under such hostile conditions. Both the LFA and CFA tools share some similarities in their hardware designs. They use a tungsten-halogen lamp as the light source. They couple randomized, bifurcated optical fiber bundles with optical filters to detect transmitted light at chosen wavelengths. This approach yields a robust spectrometer, but it limits the number of wavelength channels and spectral resolution, particularly if high-light flux, thus high-dynamic range, is desired for each channel.

The LFA tool measures fluid optical density at wavelengths chosen to measure fluid color (electronic excitation), water quantity, methane concentration, and (liquid) oil concentration. Discriminating water from hydrocarbon using NIR is the most basic fluid measurement and has been utilized for a decade in oil wells to identify fluid type (water or hydrocarbon).[3] Measuring methane and liquid oil concentrations in a single-phase, live crude oil enables determination of the gas–oil ratio (GOR) of the oil.[7] GOR is the first DFA answer product in the petroleum industry.[20] The GOR of a crude oil is the volumetric ratio of the gas phase to the liquid phase of a crude oil (defined) at conditions of one atmosphere and 60° F. The exact GOR value depends on the depressurization and equilibration process, that is, GOR is not a state function for a crude oil. The NIR spectrum can predict the GOR (largely single-stage flash) of the live oil[20] using the following equation,[21]

$$\mathrm{GOR} = 8930 \frac{M_{\mathrm{m}}}{M_{\mathrm{o}} - 0.193 M_{\mathrm{m}}} \mathrm{scf/bbl} \tag{22.4}$$

where M_m and M_o are the mass concentration of methane and oil, respectively, determined from the NIR spectra.[20, 21] This equation accounts for dissolved methane and other hydrocarbon gases, and for the vapor pressure of other alkanes. Figure 22.9 shows that the spectral GOR matches the laboratory-flash GOR reasonably well over a range of 0–2,000 scf/bbl (standard cubic feet of gas per barrel of liquid). To obtain GOR in m^3/m^3, divide by ~5.6.

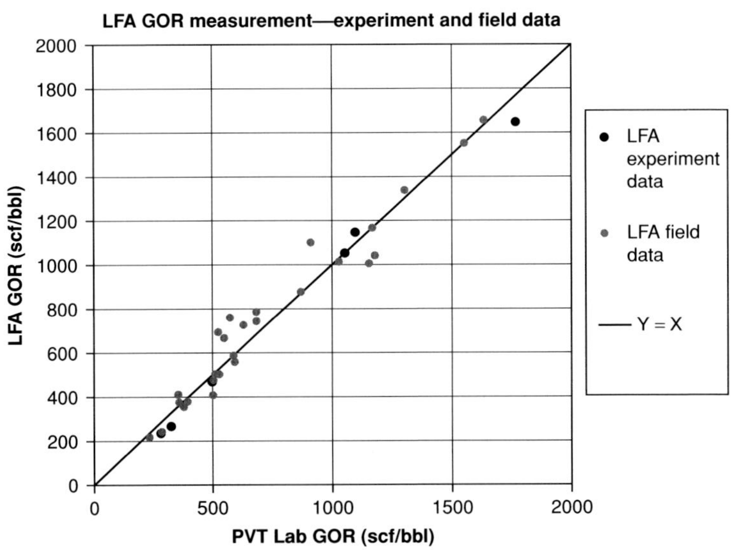

Figure 22.9. Laboratory GOR and LFA GOR compare well. LFA GOR data are both from shop testing and downhole field testing.

The CFA tool measures the NIR spectrum with emphasis on the CH two-stretch overtone peak.[4] Since the targeted problem of compositional analysis is known to behave approximately as a linear system,[7,9–11] it is expected that a chemometrics technique based on linear algebra should give a good result. Here, principal components analysis (PCA) and principal components regression (PCR) are employed for the data analysis and model building, however, other techniques such as partial least squares (PLS) should give similar results for this type of problem. These analytical techniques are well explained in many standard chemometrics textbooks and related literature.[22] The hydrocarbon fluid is analyzed in terms of concentrations for four components or psuedocomponents, methane (C1), non-methane hydrocarbon gases (C2–C5), liquids (C6+), and carbon dioxide (CO_2).[11]

The actual CFA analysis algorithm was constructed in the following manner. First, NIR spectra for 20 different, known hydrocarbon samples were measured at pressures ranging from 4,000 psi to 15,000 psi at temperatures ranging from 20°C to 150°C. This allows construction of the calibration data matrix $\boldsymbol{D}_{t\times m}$, where t represents the number of channels and m represents the number of the calibration spectra.[11] We denote a matrix with a bold capital letter followed by two suffix numbers showing the size of the matrix. A vector is expressed with a bold lowercase letter, and a scalar is expressed with a normal lowercase letter. For the sake of consistency, we also add suffix numbers to vectors and scalars. Following the standard PCA procedure, the $\boldsymbol{D}_{t\times m}$ matrix is decomposed into two orthogonal matrices, $\boldsymbol{R}_{t\times t}$ (loadings) and $\boldsymbol{C}_{t\times m}$ (scores).

$$\boldsymbol{D}_{t\times m} = \boldsymbol{R}_{t\times t}\boldsymbol{C}_{t\times m}. \tag{22.5}$$

We have more samples than channels ($t < m$), which is unusual for spectrometers in laboratories but is the case for wireline tool spectrometers due to limited channel numbers. Among t principal components, only f of them are useful for predicting concentrations of composition groups. So we can reduce the dimensions of $\boldsymbol{R}_{t\times t}$ and $\boldsymbol{C}_{t\times m}$ without losing meaningful information. Now the data matrix $\boldsymbol{D}_{t\times m}$ is approximated by $\underline{\boldsymbol{D}}_{t\times m}$, so that

$$\boldsymbol{D}_{t\times m} \sim \underline{\boldsymbol{D}}_{t\times m} = \underline{\boldsymbol{R}}_{t\times f}\underline{\boldsymbol{C}}_{f\times m}, \tag{22.6}$$

where $\underline{\boldsymbol{R}}_{t\times f}$ and $\underline{\boldsymbol{C}}_{f\times m}$ represent reduced loading matrix and reduced score matrix, respectively. In the PCR model, the concentration of each chemical component y is independently related to the score of spectra by a regression vector $\boldsymbol{b}$. Using m calibration dataset, the relation is given as follows:

$$\boldsymbol{y}_{1\times m} = \boldsymbol{b}_{1\times f}\underline{\boldsymbol{C}}_{f\times m} \tag{22.7}$$

Since both $\boldsymbol{y}_{1\times m}$ and $\underline{\boldsymbol{C}}_{f\times m}$ are known quantities, the regression vector $\boldsymbol{b}_{1\times f}$ can be determined using a least-squares approach. Once the regression vector $\boldsymbol{b}_{1\times f}$ is determined from the calibration dataset, the concentration of each composition group in an unknown reservoir fluid is predicted by the following model equation:

$$y_{1\times 1} = \boldsymbol{b}_{1\times f}\left(\underline{\boldsymbol{R}}^T_{t\times f}\underline{\boldsymbol{R}}_{t\times f}\right)^{-1}\underline{\boldsymbol{R}}^T_{t\times f}\boldsymbol{m}_{t\times 1} \tag{22.8}$$

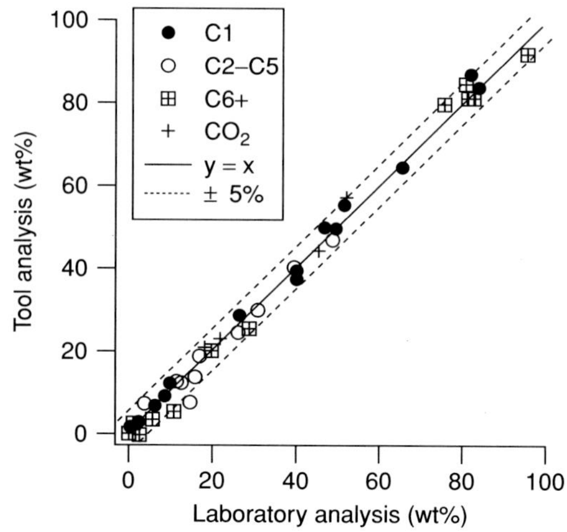

Figure 22.10. CFA composition measurements compare well with laboratory results.

where $\boldsymbol{m}_{t\times 1}$ is a spectrum of reservoir fluid measured with a t-channel wireline spectrometer tool.[11]

The CFA analyses based on this model agree well with laboratory analysis as shown in Figure 22.10. The CFA gives more accurate fluid GOR for high-GOR fluid such as retrograde gas. For relatively low-GOR fluid, both the LFA and CFA tools naturally give similar analysis results since the influence of C2–C5 group concentration to overall fluid GOR is very small for such fluid. For very low GOR, the LFA is expected to be more accurate due to its ability to account for oil coloration at the hydrocarbon peaks used for GOR analysis.

4. Live Oil Sampling Process

4.1. Contamination

The first order of business for wireline fluid sampling is to obtain a large number of pressure points (50 is not uncommon) as a function of depth. The toolstring depicted in Figure 22.2 is set against the borehole wall, pressure communication with the formation is established, and the formation pressure is measured. By plotting pressure versus true vertical depth, one obtains a pressure gradient, thus formation fluid density, and this can be used to identify oil, water, and gas within the fluid column provided there is vertical pressure communication. With this knowledge, a sampling program can be established.

Next, the MDT tool is again set against the wall, but this time for the purpose of extracting formation fluid. The pumpout module reduces the flowline pressure below formation pressure thereby initiating flow; this pressure difference is called drawdown pressure. Flowline fluids are either discharged to the borehole or are acquired in sample bottles. Since the formation near the borehole is invaded by drilling-fluid filtrate, the first fluid flowing into the tool is a mixture of the original reservoir oil plus filtrate (also known as contamination) and has no value as a fluid sample. The OBM filtrate is particularly problematic since it is miscible with crude oil. As pumping time continues, contamination is reduced. The time required to clean up the formation fluid depends on many factors (e.g., invasion depth, flow rate, formation permeability) and is difficult to predict; the cleanup often takes more than 1 hr. Thus, monitoring the flowline contents using DFA is mandated. The basic fluid type (oil, gas, and water) is readily distinguishable by optical spectroscopy measurement.[3] Miscible contamination is monitored using the oil coloration[23] (OBM filtrates are typically lighter in color compared to crude oil) and dissolved methane content[20] (OBM filtrates contain no gas). Employing an approximate equation for the clean up process presuming a point sink, one can calculate the contamination level quantitatively.[20] This oil-based mud contamination monitoring (OCM) algorithm facilitates efficient acquisition of samples. Figure 22.11 shows an example of the clean up process; flowline fluid coloration and dissolved methane content increase with pumping time. The real-time contamination prediction based on the OCM algorithm agrees with the subsequent laboratory analysis results.

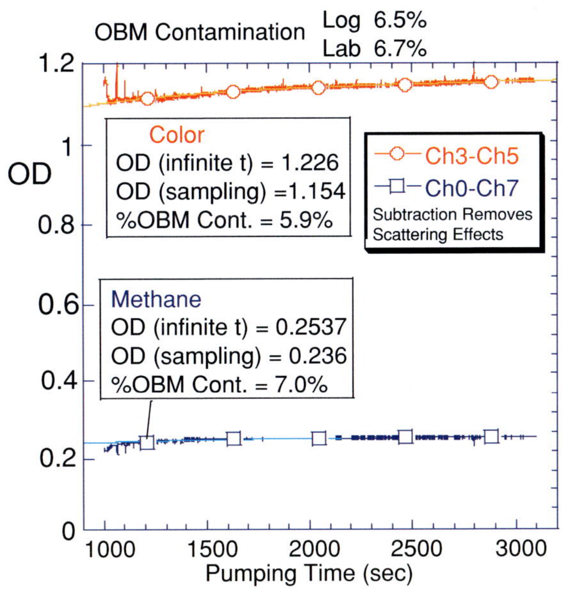

Figure 22.11. A color channel "Ch3" is selected for real-time OBM contamination monitoring (OCM algorithm) in this example. Influence of wavelength-independent light scattering is eliminated first by channel subtraction, here Ch3–Ch5. The asymptote of the scattering-free buildup curve named "Ch3–Ch5" indicates the OBM-contamination-free OD (or coloration), which is OD = 1.226 for this fluid. The current level of OBM contamination is estimated from an OD value at the time. In this case, OD = 1.154 translates as 5.9% contamination. Similar analysis for the methane channel yields 7%. The post-job lab measurement is in close agreement.

4.2. Phase Transition

Another important sampling parameter is the proper drawdown pressure. A drawdown that is too small unnecessarily reduces flow rate thereby increasing the clean up time. If the drawdown pressure is too large, the sample can break into two phases. If this happens, it is likely that the high-mobility phase flows preferentially, yielding a nonrepresentative sample. Gas evolution is detected by an index-of-refraction measurement of the flowline fluid.[24] Retrograde dew can be detected using fluorescence, because the liquid portion is highly enriched in PAHs. The liquid film that covers the window after dew formation increases the fluorescence signal.[25] Phase segregation in the pumpout module greatly enhances the ability to detect phase transitions.[25] Asphaltene drop out is detected by light scattering, the preferred laboratory method,[26] but downhole, this signature is not unique. Sample acquisition is best performed when no light scattering is evident. DFA is very important to establish acquisition of valid single-phase samples.

The reservoir fluid is captured in a sample chamber once it satisfies the aforementioned two criteria; acceptably low filtrate contamination and single-phase flow. Typically, fluid is captured in more than one sample bottle at each sampling depth for redundancy and quality-control purposes. The exact depths and numbers of stations for fluid sampling are planned prior to a job but can be optimized in real time with DFA. This optimization often becomes necessary to confirm spatial variations of fluid observed by DFA in real time. For example, prior to the actual job, it is impossible to know the existence of a fluid gradient in composition or GOR and to plan a sampling job to quantify it. If the existence of such a fluid gradient is suspected from DFA during the job, it is necessary to change the plan to confirm the observation. The tool is tripped back to surface after all sampling objectives are met. Fluid temperatures cool, and the sample may break into two phases. Sample bottles are detached from the MDT, and the samples are transferred after being reconditioned to reverse possible phase transitions for well-site analysis and for shipment (if different bottles are required). Best practices call for comparisons of DFA, well-site, and subsequent laboratory analyses. For laboratory analysis, the fluids are reconditioned with heat and agitation before sample transfer. If all of the above processes are done properly, the sample is ready for analysis.

Often, there are problems in the sample acquisition, transfer, and analysis process. Asphaltenes may precipitate out of fluid and irreversibly mix with solids and water in the bottle. A part of the multiphase fluid may be lost in dead volumes in sample chambers or during the process of fluid transfer. Bottles can leak, and if the corresponding sample is two phase, then bottle leakage invalidates the sample. In the past, quality control has been performed via replicate runs on different bottles of ostensibly the same fluid. Sometimes different labs are used. Discrepancies have occurred, but this process does not identify and resolve the failure. In a more general sense, there are often large discrepancies between oil properties predicted from wireline samples and those obtained from (ultimately) produced samples. These discrepancies likely involve many varied factors; nevertheless, reducing uncertainty in the acquisition and analysis of wireline samples can only help.

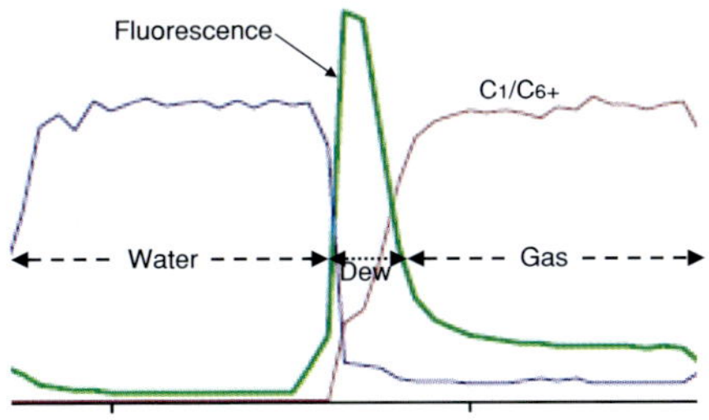

Figure 22.12. The log shows how multiphase fluid flow changes with time as measured by a ratio of C_1 concentration/C_{6+} concentration, fluorescence intensity, and water concentration. Water was flowing initially in this log section. Then, the retrograde dew was seen as indicated by intense fluorescence signal and low C_1/C_{6+} composition ratio. Finally, gas phase arrived as seen by low fluorescence signal and high C_1/C_{6+} composition ratio.

Phase transitions can be used to identify the fluid type. Figure 22.12 shows a section of the DFA log for a retrograde condensate. The existence of retrograde dew was confirmed by dropping pressure below the suspected dew point and observing the phase segregation. The corresponding tool configuration is shown in Figure 22.2. The CFA unit, where this log was taken, was placed immediately downstream of the pumpout module. The multiple phases and fluids segregate due to density contrasts while resident inside the pumpout module. The fluids flow sequentially according to density, exiting the pumpout module, and flowing into the CFA module. The exact fluid scenario is a bit complicated because the pumpout module has two chambers and some dead volumes in the fluid transfer lines. Detailed analysis of the timing and volume of the different phases is described elsewhere.[25] In this section of the log example, water was seen first, followed by small amount of retrograde liquid (high fluorescence, low C_1/C_{6+} composition ratio), followed by large amount of gas (low fluorescence, high C_1/C_{6+} composition ratio).

4.3. Chain of Custody

DFA provides initial analysis of the samples and provides a new method for quality control, substantially reducing required sample redundancy and fluid sampling operation cost. One way to eliminate uncertainty in the sample acquisition and analysis process is to "fingerprint" the fluid immediately as it enters the sampling tool.[27] This fingerprint can then be checked repeatedly at any subsequent time. If the sample integrity has not been compromised, then the fingerprint will remain unchanged. The DFA measurement suite provides this fingerprint. These best-practice ideas are embodied in the "chain of custody" methodology, which is currently in development for oilfield samples.[27] Chain of custody is mandated in other arenas such as forensic science.

Figure 22.13 shows the comparison of the DFA spectrum acquired with the LFA tool and the corresponding equivalent spectrum acquired in the laboratory months later. The laboratory spectrum was acquired with a high-resolution Cary 5 UV-VIS-NIR spectrometer. The corresponding spectrum was subject to a

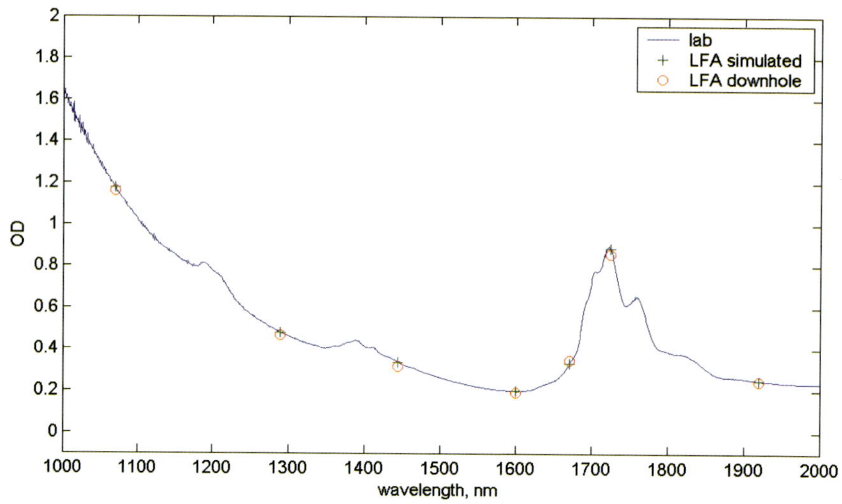

Figure 22.13. A fluid spectrum measured by the LFA module in downhole agrees well with a fluid spectrum measured in the laboratory, indicating fluid was handled and transferred properly.

proprietary filtering algorithm that mimics the downhole spectrometer of the LFA. The excellent agreement between the downhole and laboratory NIR fingerprint validates that sample transfer and storage were done properly.

5. "What Is the Nature of the Hydrocarbon Fluid?"

Hydrocarbon fluids can be disbursed in the formation in a very heterogeneous fashion. The type of hydrocarbon that evolves from a single source rock varies with time; higher maturity is associated with higher GOR.[28] Often, different source rocks of markedly different origin can contribute hydrocarbon charges to individual fields. Hydrocarbon fluids can vary vertically and laterally due to gravity, thermal gradients, biodegradation, real-time charging, water washing, leaky seals, which are often pressure dependent, high-temperature reaction with inorganics and hydrocarbon phase changes (e.g., tar mat formation due to a second hydrocarbon charge). Some of these factors enforce fluid disequilibrium. Both diffusion and convection promote equilibrium, but with much different time constants. Thus, the hydrocarbon fluids in a reservoir are often in dynamic flux. For instance, biodegradation can create large vertical compositional variations.[29] The microbes live in the water zone and preferentially consume the alkanes at the oil-water interface thereby concentrating the heavy ends at the bottom of the oil column. Diffusion acts to vertically mix the column, but often the rate of biodegradation exceeds the rate of vertical mixing, giving rise to a large gradient.[29] Because of the large number of complex factors that determine fluid variations within a reservoir, accurate fluid descriptions must be data driven. Herein lies a problem. While the geoscientist would caution that fluid complexities in the formation may exist, repeated fluid sampling from different depth stations is simply too expensive without strong a priori knowledge

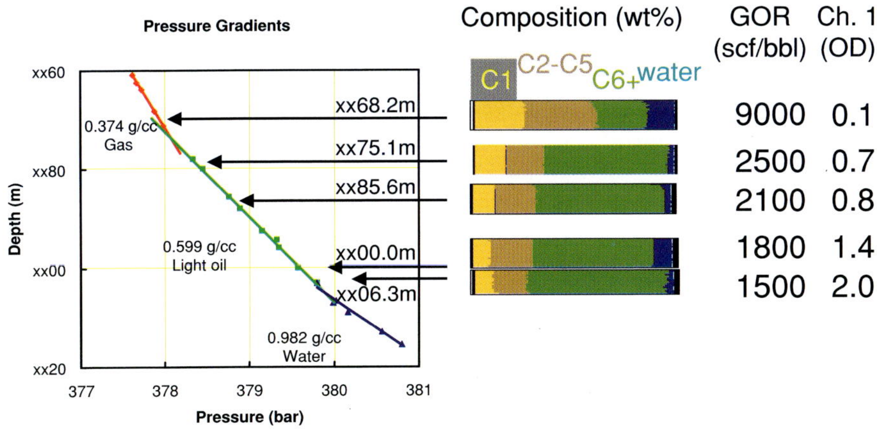

Figure 22.14. Large fluid gradient, missed by pressure measurements, was detected by DFA. The trend in the color measurement confirms the trend in the GOR.

that fluid complexities will be found. Consequently, operational considerations have resulted in the simplistic presumption that the reservoir oil is homogenous.

The solution to this state of affairs is to identify in real time during sample acquisition the hydrocarbons that are meaningfully different. DFA enables exactly this objective. DFA provides real-time analysis of fluids as they are being acquired, enabling the acquisition of high-quality samples that are distinct and worthy of detailed analyses. For instance, if an oil column has a single hydrocarbon fluid, then only one sample is needed. While if two or three distinct samples are observed in an oil column, each fluid should be sampled and analyzed.

Figure 22.14 shows an example from an appraisal well in which a large fluid gradient was detected by DFA.[30] Prior to the wireline fluid sampling job, conventional wireline logs were run to assess formation properties and a hydrocarbon-bearing zone was identified. The zone seemed free from an obvious impermeable layer, and there was no indication of any fluid gradient at this stage. Then, wireline pressure testing was performed to measure the formation pressure at multiple points in this interval. Pressure measurements revealed three distinctive pressure gradient slopes corresponding to gas, oil, and water from top to bottom (Figure 22.14, left section); however, there was still no indication that the oil column in the middle section had a large fluid compositional variation. DFA was used to investigate the reservoir fluids from this interval, it quickly became obvious that the oil column section was not homogeneous, as it might look from pressure measurements, but it had a significant variation in terms of composition and GOR (Figure 22.14, middle section).[30] The fluid sampling program was quickly changed to confirm this observation. The fluid gradient discovered by DFA required a curved pressure gradient. This caused the calculated gas-oil contact and the oil-water contact to separate further, resulting in more reserves being booked. The lowest DFA station in Figure 22.14 confirmed the lower oil-water contract. DFA enables real-time optimization of the sampling and analysis program as highlighted in this case study.

6. "What Is the Size and Structure of the Hydrocarbon-Bearing Zone?"

This is probably the most fundamental and important question for upstream operations in oil companies, as it is directly related to the quantity of recoverable hydrocarbon. Seismic surveys provide information about the overall size of the geologic units that potentially contain hydrocarbons. Wireline logging provides near-wellbore measurements to identify properties of the formations and the oil column. One of the most difficult questions to answer involves the size of the individual flow units, or compartments, in the reservoir. The reservoir might contain many small compartments or a few big compartments. Obviously, a few big compartments (or even better, a single giant compartment) are much easier to drain than many small compartments. The best technical approach to discern the extent of compartmentalization is to perform an extended flow test of the well–a drill stem test (DST). However, in the highest cost settings, the cost of a DST can approach the cost of the well and does not deliver the value of a new well, so is not routinely done. A seismic survey may reveal major faults, but in general it cannot discern whether these faults are sealing. Nor can seismics discern small impermeable barriers critical to assess compartment size. Near-wellbore measurements of the formation cannot confidently predict sealing barriers, since strata can change away from the borehole. The petroleum industry has settled on the presumption that pressure communication equals flow communication. If different layers follow a single pressure gradient, then the assumption has been made that the layers are in flow communication. The problem with this assumption is that pressure communication can occur over geologic time ($\sim 10^7$ years) while flow communication must occur over production time (~ 10 years). Furthermore, flow communication requires higher permeability than pressure communication. Predicting production based on pressure communication is inaccurate by up to nine orders of magnitude and naturally leads to overly optimistic production scenarios.

Very costly problems have arisen due to compartmentalization. A typical scenario is that due to pressure communication, a reservoir is presumed to consist of one or a few very large compartments. Large reserve estimates are announced, expensive production facilities are built, large contracts for guaranteed markets are signed, and then after a brief initial spurt—production falls dramatically, often in months. The culprit is often compartmentalization. The question remains how to address compartmentalization in a cost-effective manner. Fluid analysis and DFA in particular afford this opportunity. Because fluids are often heterogeneously dispersed,[28, 29] different compartments are likely to be filled with different fluids. It is widely known that laboratory geochemical analyses can reveal different fluids, thus allowing detection of different compartments. The problem has been to identify which fluids in the formation should be sampled and analyzed in the laboratory. DFA is the "missing link" in this chain, identifying which fluids merit laboratory analysis.[31]

DFA can identify compartments in a very simple fashion—finding fluid density inversions. Figure 22.15 shows an example of determination of a flow

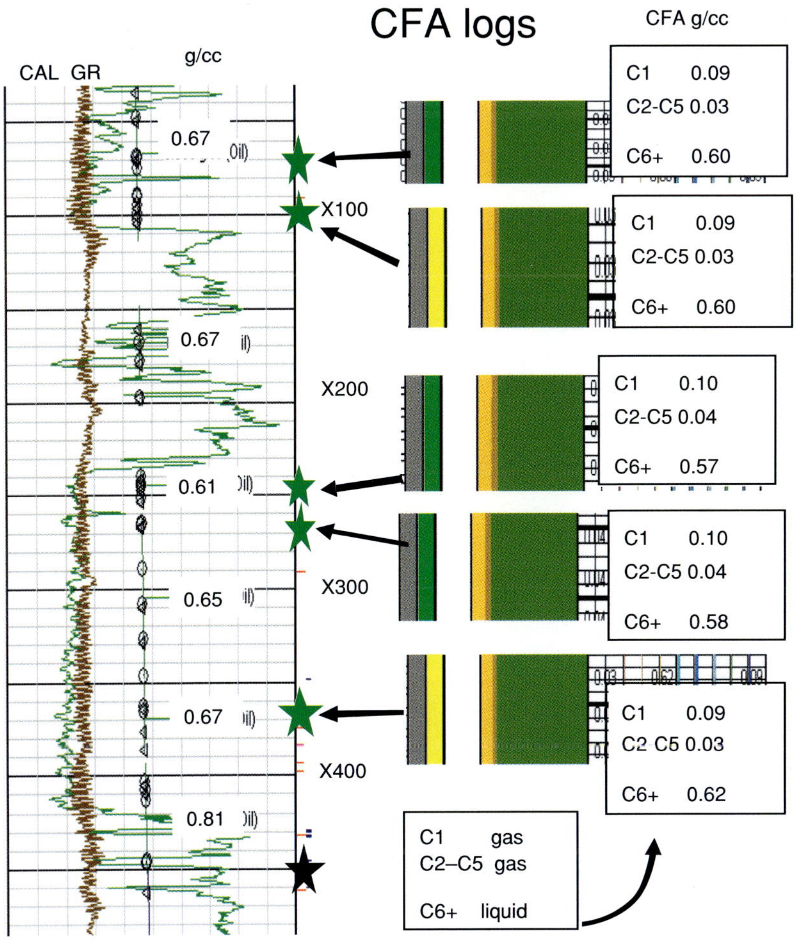

Figure 22.15. DFA shows compartmentalization by fluid density inversion. The fluid at ×100′ depth is more dense than the fluid at ×250′ depth. This is discerned from the liquid vs. gas (sum of C1 and C2–C5) partial densities shown in the figure. Thus a flow barrier exists between these depths—in accord with the observed shale. There is a subtle compositional gradient observed between ×250′ and ×400′ depth.

barrier by a fluid density inversion. That is, there is a lower density fluid lower in the oil column. The CFA determines the mass density of methane and of the C2–C5 fraction in the single-phase live fluid. The CFA also determines the mass density of the liquid fraction in a single-phase live fluid. Higher density fluids have a higher liquid partial density and a lower gas partial density contributing to the overall fluid density. In Figure 22.15, one observes a density inversion when comparing fluids at ×100 feet depth and ×250 feet depth. The strong indication is that the intervening shale is a sealing barrier.[32]

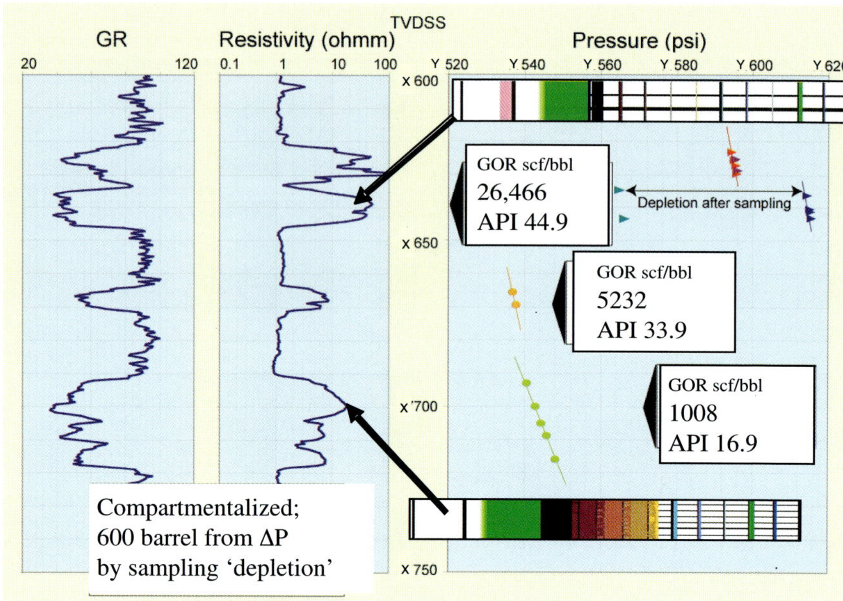

Figure 22.16. DFA coupled with pressure measurement proves a very small compartment volume. DFA shows three different types of hydrocarbons in three sands (two DFA logs are shown). The very high pressure dictates that these separate phases are not in equilibrium. The lack of pressure communication in these sands is also observed. The gas containing sand suffered a 50 psi pressure drop upon removing less than one barrel of fluid. By estimating the compressibility of this light hydrocarbon one obtains that the compartment volume is only 600 barrel—a shockingly small number. This deepwater sand is not economic by orders of magnitude.

In a different example of compartmentalization, DFA coupled with pressure measurements proves a shockingly small volume of a compartment. Figure 22.16 shows several sand bodies separated by shales in a column. Three hydrocarbons are shown utilizing DFA (two DFA logs are shown). At the ambient pressures ($\sim$10,000 psi) these hydrocarbons cannot be in equilibrium implying there are different compartments. That is, 10,000 psi effectively precludes a gas cap above an oil. There is also no pressure communication proving the lack of flow communication. In the gas sand, the formation pressure was reduced by 50 psi upon sampling (and removing) less than one barrel from the formation.[33] By estimating the compressibility of this gaseous hydrocarbon, one obtains a compartment volume of 600 barrels. This is not economic by orders of magnitude.

DFA in a single well is performed with the same tool, same operator, same time, same temperature, and same calibration. Consequently, the measured difference between two fluids at two sampling stations in the formation is very accurate. The absolute fluid measurements have a larger error bar, but the relative differences are measured accurately. Consequently, subtle density differences can be determined. If a high-density fluid is above a low-density fluid in the reservoir, the most likely explanation is compartmentalization. The gas–oil ratio serves as a proxy for

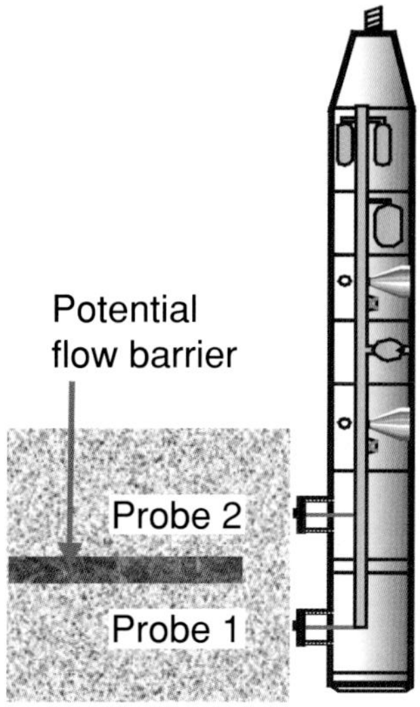

Figure 22.17. The preferred tool configuration for testing fluid communication between two closely spaced, potentially compartmentalized zones. The tool straddles the potential flow barrier. Fluid samples are taken from both above and below the potential flow barrier for direct comparison by switching the active probe without moving the tool string.

fluid density; gas-rich fluids are of lower density. GOR is around zero for a heavy oil and is infinity for a dry gas. This infinite dynamic range makes GOR ideally suited to detect density inversions. Nature produces density inversions (in different compartments) in the natural course of events. Minimal catagenesis of kerogen, obtained at minimal depth, yields heavy hydrocarbons.[15] High-temperature catagenesis of kerogen causes extensive cracking, producing light hydrocarbons. Often the deepest hydrocarbon targets are dry gas, while shallow reservoirs such as the Athabasca tar sands contain the heaviest hydrocarbons. DFA—with its accurate differential measurement can be employed to find density inversions.

Vertical compartmentalization can be tested with a single tool set, provided the barrier is sufficiently thin. For this purpose, two single probes are employed, as shown in Figure 22.17. In this particular case, a thin (<6 feet) barrier was suspected.[34] The MDT was set with the two single probes straddling the potential permeability barrier. First, flow was initiated through the lower probe, below the potential barrier. After a long cleanup period, a sample was acquired. Then flow was initiated through the probe set above the potential barrier. The log shown in Figure 22.18 was obtained. The first part of the log, up to 5400 s, corresponds to the fluid lower in the column. After 5400 s, the fluids entered the tool through the higher probe, thus were higher in the column. After a short period of time, the assessment was made that a GOR inversion was obtained and the barrier is in fact a flow barrier.[34] The relative comparison of the two fluids at the same point in time with the same tool enabled sensitive fluid comparison. This job shows the power of DFA. The operating company was going to acquire a sample in any event. For a small cost increase of a little extra pumping time, the operating company

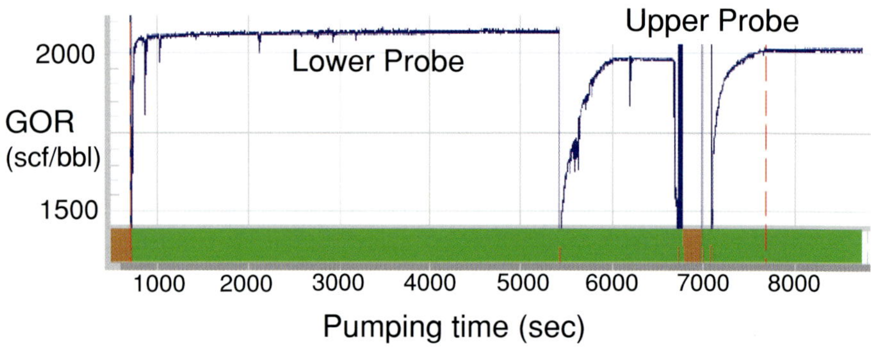

Figure 22.18. The first fluid (up to 5400 s) was sampled from the probe below the barrier and has a higher GOR than the second fluid sampled ($t > 5400$ s) above the barrier. The different and higher density fluid higher in the column indicates that the barrier is in fact sealing against flow, direct evidence of compartmentalization.

determined that the fluids indicated vertical compartmentalization. There are many examples of large increases in GOR lower in the column. Some of these cases also exhibit a lack of pressure communication between the high-GOR fluid lower in the column and shallow, low-GOR fluid. Clearly, vertical compartmentalization is implicated. Other cases exhibit a single pressure gradient but the density inversion of the fluids indicates a lack of flow communication. The integration of DFA with pressure and other measurement is now recognized as essential in reservoir evaluation.[35]

7. Conclusions

The petroleum industry has long suffered from the inability to reliably and accurately characterize reservoirs and contained fluids prior to actual production. The cost of these technical shortcomings is exacerbated by the growing importance of technically challenging arenas such as the deepwater production of oil. An integral part of proper fluid and reservoir evaluation is the analysis of live formation fluid samples acquired in openhole wireline logging. Recently introduced Downhole Fluid Analysis technologies are improving many aspects of reservoir assessment in a variety of ways. DFA has been shown to improve acquisition and handling of wireline fluid samples by assuring low filtrate contamination, single-phase fluid sampling, and the forensic concept of chain of custody. In addition, DFA is key to identifying fluid compositional variation. Sample acquisition programs can be optimized in situ based on DFA results. DFA can also be used to identify reservoir compartmentalization, a subject of major concern in the petroleum industry. DFA is the missing link connecting efficient sample acquisition with sophisticated laboratory studies which can address many important production issues, such as flow assurance, fluid compositional grading, and compartmentalization. DFA is rapidly becoming an indispensable component in the characterization of reservoirs.

References

[1] Witt, C.J., A. Crombie, and S. Vaziri (1999). A Comparison of Wireline and Drillstem Test Fluid Samples from a Deepwater Gas-Condensate Exploration Well, *SPE Annual Technical Conference and Exhibition*, SPE 56714.

[2] Whittle, T.M., J. Lee, and A.C. Gringarten (2003). Will Wireline Formation Tests Replace Well Tests, *SPE Annual Technical Conference and Exhibition*, SPE 84086.

[3] Smits, A.R., D.V. Fincher, K. Nishida, O.C. Mullins, R.J. Schroeder, and T. Yamate (1993). In-Situ Optical Fluid Analysis is an Aid to Wireline Formation Sampling, *SPE Annual Technical Conference and Exhibition*, SPE 26496.

[4] Fujisawa, G., O.C. Mullins, C. Dong, A. Carnegie, S.S. Betancourt, T. Terabayashi, S. Yoshida, A.R. Jaramillo, and M. Haggag (2003). Analyzing Reservoir Fluid Composition In-Situ in Real Time: Case Study in a Carbonate Reservoir, *SPE Annual Technical Conference and Exhibition*, SPE 84092.

[5] Williams, P. and K. Norris (2001). *Near-Infrared Technology: In the Agricultural and Food Industries*, 2nd ed. American Association of Cereal Chemists, St. Paul, MN.

[6] Brown, C.W. and S.-C. Lo (1993). Feasibility of on-line monitoring of the BTU content of natural gas with a near-infrared fiber optic system, *Appl. Spectrosc.* **47**, 812.

[7] Mullins, O.C., T. Daigle, C. Crowell, H. Groenzin, and N.B. Joshi (2001). Gas-oil ratio of live crude oils determined by near-infrared spectroscopy, *Appl. Spectrosc.*, **55**, 197.

[8] Atkins, P.W. (1983). *Molecular Quantum Mechanics*, 2nd ed. Oxford University Press, New York.

[9] Mullins, O.C., N.B. Joshi, H. Groenzin, T. Daigle, C. Crowell, M.T. Joseph, and A. Jamaluddin (2000). Linearity of alkane near-infrared spectra, *Appl. Spectrosc.* **54**, 624.

[10] van Agthoven, M.A., G. Fujisawa, P. Rabbito, and O.C. Mullins (2002). Near-infrared spectral analysis of gas mixtures, *Appl. Spectrosc.* **56**, 593.

[11] Fujisawa, G., M.A. van Agthoven, F. Jenet, P.A. Rabbito, and O.C. Mullins (2002). Near-infrared compositional analysis of gas and condensate reservoir fluids at elevated pressures and temperatures, *Appl. Spectrosc.* **56**, 1615.

[12] Mullins, O.C. (1998). Optical interrogation of aromatic moieties in crude oils and asphaltenes. In: O.C. Mullins and E. Y. Sheu (eds.), *Structures and Dynamics of Asphaltenes.* Plenum Press, New York, Ch. 2.

[13] Mullins, O.C., S. Mitra-Kirtley, and Y. Zhu (1992). Electronic absorption edge of petroleum. *Appl. Spectrosc.* **46**, 1405.

[14] Mullins, O.C. and Y. Zhu (1992). First observation of the Urbach tail in a multicomponent organic system, *Appl. Spectrosc.* **46**, 354.

[15] Tissot, B.P. and D.H. Welte (1984). *Petroleum Formation and Occurrence*, 2nd rev. edu. Springer-Verlag, New York.

[16] Turro, N.J. (1978). *Modern Molecular Photochemistry*. Benjamin/Cummings Publishing, Menlo Park, CA.

[17] Downare, T.D., O.C. Mullins (1995). Visible and Near-Infrared Fluorescence of Crude Oils. *Appl. Spectrosc.* **49**, 754.

[18] Ralston, C.Y., X. Wu, and O.C. Mullins (1996). Quantum yields of crude oils, *Appl. Spectrosc.* **50**, 1563.

[19] Zhu, Y. and O.C. Mullins (1992). Temperature dependence of fluorescence in petroleum, energy and fuels, *Energy & Fuels* **6**, 545.

[20] Mullins, O.C., G.F. Beck, M.E. Cribbs, T. Terabayashi, and K. Kegasawa (2001). Downhole Determination of GOR on Single Phase Fluids by Optical Spectroscopy, *SPWLA 42nd Annual Symposium,* Houston, TX, paper M.

[21] Dong, C., P.S. Hegeman, H. Elshahawi, O.C. Mullins, G. Fujisawa, and A. Kurkjian (2003). In-Situ Contamination Monitoring and GOR Measurement of Formation Samples, *SPWLA 44th Annual Logging Symposium*, paper FF.

[22] Malinowski, E.R. (1991). *Factor Analysis in Chemistry*, 2nd edn. Wiley-Interscience, New York.

[23] Mullins, O.C., J. Schroer, and G.F. Beck (2000). Real-time Quantification of OBM Filtrate Contamination in the MDT using OFA data, *SPWLA 41st Annual Logging Symposium*, Houston, TX, paper SS.

[24] Mullins, O.C., R.J. Schroeder, and P. Rabbito (1994). Gas detector response to high pressure gases, Applied Optics, *Appl. Opt.* **33**, 7963.
[25] Betancourt, S.S., G. Fujisawa, O.C. Mullins, K.O. Eriksen, C. Dong, J. Pop, and A. Carnegie (2004). Exploration Applications of Downhole Measurement of Crude Oil Composition and Fluorescence, *SPE Asia Pacific Conference*, SPE 87011.
[26] Hammami, A., C.H. Phelps, T. Monger-McClure, and T.M. Little (2000). Asphaltene precipitation from live oils: An experimental investigation of onset conditions and reversibility, *Energy & Fuels* **14**, 14–18; N.B. Joshi, O.C. Mullins, A. Jamaluddin, J. Creek, J. McFadden, Asphaltene Precipitation from Live Crude Oils, Energy & Fuels 15, 979, (2001)
[27] Betancourt, S.S., J. Bracey, O.C. Mullins, G. Gustavson, G. Syriac. Chain of Custody via Spectroscopy for Oil Field Samples, Submitted to Applied Spectroscopy.
[28] England, W.A. (1990). The organic geochemistry of petroleum reservoirs, *Org. Geochem.* **16**, 415–425.
[29] Koopmans, M.P., S.R. Larter, C. Zang, B. Mei, T. Wu, and Y. Chen (1999). Biodegradation and Mixing of Crude Oils in the Liaohe Basin, In: *Proceedings of the 19th International Meeting on Organic Geochemistry*, Istanbul, extended abstract 63–64.
[30] Fujisawa, G., S.S. Betancourt, O.C. Mullins, T. Torgersen, M. O'Keefe, T. Terabayashi, C. Dong, and K.O. Eriksen (2004). Large Hydrocarbon Compositional Gradient Revealed by In-Situ Optical Spectroscopy, *SPE Annual Technical Conference and Exhibition*, SPE 89704.
[31] Elshahawi, H., M.N. Hashem, O.C. Mullins, G. Fujisawa (2005). The missing link—identification of reservoir compartmentalization by downhole fluid analysis, SPE 94709.
[32] Mullins, O.C., H. El-Shahawi, M. Hashem, and G. Fujisawa (2005). Identification of Vertical Compartmentalization and Compositional Grading by Downhole Fluid Analysis: Towards a Continuous Downhole Fluid Log, *SPWLA 46th Annual Logging Symposium*, paper K.
[33] Mullins, O.C., G. Fujisawa, M.N. Hashem, and H. Elshahawi (2005). Identification of Vertical Compartmentalization and Compositional Grading by Downhole Fluid Analysis: Towards a Continuous Downhole Fluid log, *Int. Petrol. Technol. Conf.* Nov. 2005, Qatar, paper 10036.
[34] Mullins, O.C., M. Hashem, H. Elshahawi, G. Fujisawa, C. Dong, S.S. Betancourt, and T. Terabayashi (2004). Hydrocarbon Compositional Analysis In-Situ in Openhole Wireline Logging, *SPWLA 45th Annual Logging Symposium* paper FFF. June 6–9, The Netherlands.
[35] Hashem M.N., H. Elshahawi, G. Ugueto (2004). A Decade of Formation Testing—Dos and Don't and Tricks of the Trade, *SPWLA 45th Ann. Log. Symp.* June 6–9, The Netherlands.

23

Precipitation and Deposition of Asphaltenes in Production Systems: A Flow Assurance Overview

Ahmed Hammami and John Ratulowski

1. Introduction

The movement of production systems to deepwater and subsea environments in recent years has increased the importance of fluid property related flow assurance issues. Asphaltene precipitation and deposition is one of these potential problems. While not as common as wax or scale, the impact of asphaltene is often catastrophic. Asphaltene can cause reservoir impairment, plugging of wells and flowlines through deposition, separation difficulties, and fouling in facilities. Offshore, the cost of remediating an unexpected asphaltene problem is excessive. It is imperative that the behavior of asphaltenes in an offshore production system be understood in the design stage of the project. Proper control and remediation strategies must be built into the system from the beginning. In this chapter, we will review the current process of sampling and analysis that provide the initial assessment of asphaltene stability in the reservoir fluid. We will relate these measurements in the laboratory to expected behavior in the field and address uncertainties. We will then discuss how improved characterization and deposition measurements will decrease uncertainty and allow less conservative design and operation strategies. Finally, we will briefly review the types of asphaltene precipitation models and discuss their respective underlying assumptions.

Land-based sources of conventional crude oil and gas have been decreasing for the last couple of decades. As a result, the petroleum industry has turned to explore, develop, and produce hydrocarbon fluids from more challenging fields including deep and ultra deep offshore areas such as the North Sea, Gulf of Mexico, West Africa, and East coast Canada.[1] Productions in these regions are under extreme conditions where temperature can be near freezing point, and pressure drops from reservoir to production facility are quite large. These conditions often lead to

Ahmed Hammami • Schlumberger Oilfield Services, Edmonton, Alberta, Canada, and **John Ratulowski** • Schlumberger Well Completion and Productivity Subsea-Flow Assurance, Rosharon, Texas, USA

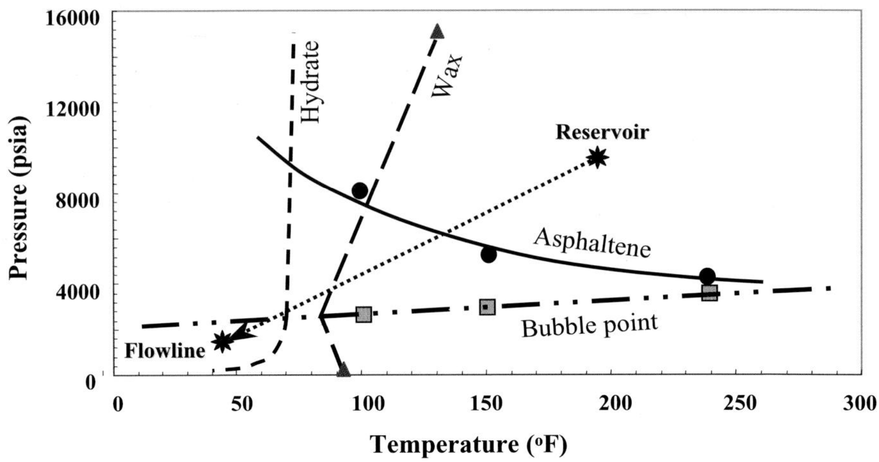

Figure 23.1. Gulf of Mexico black oil phase diagram.[2]

the precipitation and deposition of common organic solids (i.e., waxes, asphaltenes, and hydrates) and inorganic scales. Some less common solids include diamondoids, elemental sulfur, and naphtenates.[2]

Depending on the design and operation of the production system, some or all of the solid phase boundaries may be crossed as the fluid moves from the reservoir through the flow line to arrival conditions at the host facility. Figure 23.1 shows a phase diagram for typical Gulf of Mexico deepwater black oils.[2] Precipitation and, more importantly, deposition of these solids can have detrimental effects on the profitability of production systems, especially in offshore operations. These solids may deposit on surfaces, collect in low-energy regions or increase the effective viscosity of the flowing fluid. The net effect is an increase in pressure drop for a given rate through reduction in flowing area or viscosity change. In some cases the system may be completely blocked. It is, therefore, imperative that the potential for and severity of organic solids deposition problems be assessed early in the design process.[3–5] If deposition is likely, provisions for control and remediation must be incorporated into both the system design and operating strategy at an early stage. The risk and cost of these measures can influence the decision to proceed with the development of a prospect. Therefore, this decision must be based on sound laboratory data obtained from representative samples.[3]

Flow assurance is a self-descriptive term. The discipline of flow assurance addresses a variety of fluid property related issues that impact the flow of oil, gas, and water through production systems. The goal of the flow assurance engineer is to assure that fluids flow through the system as designed. For all aspects of petroleum production, transportation, and processing, it is necessary to understand the properties and the behavior of petroleum fluids under consideration. Due to the diverse heterogeneous nature of petroleum products, composition, and operating conditions of temperature and pressure, it is essential to comprehend the behavior of complex mixtures of components with wide ranges of physical and chemical properties. The ability to measure a minimum number of these properties in the laboratory and subsequently predict other properties based on the measured data is of vital economic importance to the petroleum industry.

Prior to addressing the details of petroleum fluid phase behavior and related flow assurance studies, it is useful to provide an overview of petroleum fluid chemistry and the current understanding of the key parameters affecting the stability of organic solids in general and asphaltenes in particular.

2. Chemistry of Petroleum Fluids

Petroleum is a mixture of mainly hydrocarbons plus organic compounds of sulfur, nitrogen, and oxygen as well as compounds containing metallic elements including vanadium, nickel, and iron. Hydrocarbon contents may range from as high as 97% in lighter paraffinic oils or as low as 50% in heavy, asphaltenic crudes. Despite this bulk characterization, the nonhydrocarbon molecules in the crude are usually in the form of large hydrocarbon structures with only one or two substituted heteroatoms. Therefore, heavy crudes with as little as 50% hydrocarbon components are still assumed to retain most of the essential characteristics of hydrocarbons.[6] Of the data available in the literature, it appears that the proportions of elements in petroleum sources vary only slightly despite the highly differing nature of their sources and overall characteristics.[7,8] This consistency is shown in Table 23.1.

The isolation of pure hydrocarbon constituents from crude oils began in earnest in the 1930s with the separation of light paraffinic molecules.[9] It soon became obvious, however, that the resolution of heavier, individual paraffin molecules from crudes was impossible due to the increasing numbers of isomers that exist. For example, while there is only one possible structure for methane, ethane, and propane, two structures may exist for butane: straight-chain normal butane, and branched-chain isobutane. As is shown in Table 23.2, the number of possible structures increases phenomenally with increasing carbon number.

Two other difficulties in the characterization and property prediction arise from the immense numbers of isomers of petroleum molecules. First, molecules

Table 23.1. Ultimate Elemental Composition Ranges of Crude Oils

Carbon	83.0–87.0%
Hydrogen	10.0–14.0%
Nitrogen	0.1–2.0%
Oxygen	0.05–1.5%
Sulfur	0.05–6.0%

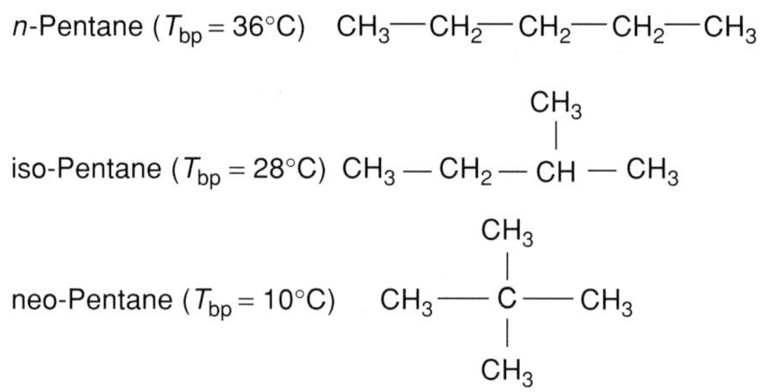

Figure 23.2. Boiling points of isomers of pentane.

of similar molecular weights but varying structures may have very different properties. For example, there are marked differences in normal boiling point between three isomers of paraffinic pentane (Figure 23.2).

Second, the number of isomers shown in Table 23.2 is restricted to only those structures possible for paraffins. If naphthenic and aromatic isomers are also considered, along with the possible substituted nonhydrocarbon components, it is obvious that even if every individual component could be experimentally distinguished, the number of components and data requirements would overwhelm thermodynamic models. It is also significant to note that a fraction given in a typical true boiling point distillation or simulated distillation analysis represents the concentration of all structures (paraffins, napthenes, and aromatics) that boil within the range of the given paraffinic compound. For example, in a typical crude oil breakdown, the C10 fraction contains all hydrocarbons that boil between the normal boiling points of *n*-C9 and *n*-C10, irrespective of molecular weight or structure. More specifically, a 345–425°C (653–797°F) cut of a typical crude oil may contain *n*-paraffins of molecular weights 282–380 g/mol and *n*-alkylnaphthalenes of considerably lower molecular weight (240–296 g/mol). As a result, it appears that the most comprehensive characterization scheme should include the measurement of molecular weights (or molecular weight distributions) in conjunction with a structural analysis.

Although there are no absolute rules for describing the change in crude characteristics toward higher boiling ranges, Speight[9] has shown that as boiling

Table 23.2. Possible Isomers of Paraffin Hydrocarbons

Carbon atoms paraffin molecule	Possible isomers
6	5
7	9
9	35
12	355
18	60,523
40	6.24×10^{13}

ranges (or proportionally molecular weights) increase in a petroleum sample, the concentration of paraffinic molecules decreases. This reduction is paralleled by an increase in polynuclear aromatics and polycycloparaffins (polynaphthenes).

In subsequent sections, a very general classification scheme is used to define the various components typically found in petroleum fluids. The scheme is specifically designed to address special phase behavior and solid deposition issues. In general, petroleum constituents are classified under two major groups, namely the well-defined and volatile C6- fraction and the poorly defined and relatively nonvolatile C6+ fraction.

The C6- fraction consists of all pure hydrocarbon components (and nonhydrocarbons) with carbon numbers up to C5. These include all isomers in each carbon number range; the physical properties of each of the pure component species are well understood and recorded in the literature. The C6+ fraction, on the other hand, is far more complex due to the multiple isomer combinations available to hydrocarbons with increasing chain length.[8] This group of components is classified as paraffins (P), naphthenes (N), aromatics (A), resins (R), and asphaltenes (A). The combined fraction of paraffins and naphthenes is also termed the saturate (S) fraction.

2.1. Saturates

Saturates are nonpolar and consist of normal alkanes (*n*-paraffins), branched alkanes (iso-paraffins) and cyclo-alkanes (also known as naphthenes). Examples of each of these classes of chemicals along with the aromatics, resins, and asphaltenes have been reported elsewhere.[10–12] Saturates are the largest single source of hydrocarbon or petroleum waxes, which are generally classified as paraffin wax, microcrystalline wax, and/or petrolatum.[13] Of these, the paraffin wax is the major constituent of most solid deposits from crude oils.

2.2. Aromatics

Aromatics are hydrocarbons, which are chemically and physically very different from the paraffins and naphthenes. They contain one or more ring structures similar to benzene. The atoms are connected by aromatic double bonds.

2.3. Resins

Resins are thought to be molecular precursors of the asphaltenes. The polar heads of the resins surround the asphaltenes, while the aliphatic tails extend into the oil. Resins may act to stabilize the dispersion of asphaltene particles and can be converted to asphaltenes by oxidation. Unlike asphaltenes, however, resins are assumed soluble in the petroleum fluid. Pure resins are heavy liquids or sticky (amorphous) solids and are as volatile as the hydrocarbons of the same size. Petroleum fluids with high-resin content are relatively stable.

Figure 23.3. Variation of asphaltene texture and character with destabilization method and/or conditions.

2.4. Asphaltenes

Asphaltenes are arbitrarily defined as a solubility class of petroleum that is insoluble in light alkanes but soluble in toluene or dichloromethane.[1,5,10] They are composed of aromatic polycyclic clusters variably substituted with alkyl group and contain heteroatoms (N, S, O) and trace metals (e.g., Ni, V, Fe). The actual chemical structure of asphaltenes is difficult to define using existing analytical tools, and it remains the subject of ongoing research and contentious debates.[15–21] Creek[8] has recently argued that structure is important provided it is related to function and reactivity. He suggests an ensemble approach be considered, rather than a stylized "molecule" for asphaltenes.

It is often assumed that asphaltenes do not dissolve in petroleum but are dispersed/suspended in the fluid as colloids (evidence of this is mixed). The amount, chemical constituency, and physical structure of asphaltenes precipitated vary with precipitant type, pressure and temperature (Figure 23.3). Pure asphaltenes are black dry powders and are nonvolatile; they tend to crack before boiling.

3. Petroleum Precipitates and Deposits

Solid materials precipitating from hydrocarbon systems in the field will rarely be composed of components derived from one of the above general component groups. To the contrary, hydrocarbon precipitates are most often mixtures of components from the different fractions. As a result, the following subclassifications have been developed to address the solids precipitation issues.

3.1. Petroleum Waxes

Petroleum waxes refined from crude oils are generally classified as paraffin wax, microcrystalline wax, and petrolatum (wax and oil). The commonly reported crystal habits of petroleum waxes are needles, plates, and malcrystalline masses.

Such differing crystal structures and compositions provide waxes with a wide range of properties.

In field operations, asphaltenes and residual oil components can coprecipitate with the waxes and result in varying appearance (color) and texture to the precipitated solids. As would be expected, waxes from condensates and wet gases tend to be cleaner and more pure than those from heavier crudes. The predominately waxy character of the solid can only be defined based on analysis of the solid and the remedial techniques that can be used to redissolve such a solid. In general, only small amounts of aromatic components coprecipitate with waxes and the solid material usually melts by applying heat.[22]

3.2. Asphaltene Deposits

Our experience indicates that field asphaltene deposits consist of a wide range of components from all the groups outlined previously. The definition of a predominately asphaltenic deposit must be based on the analysis of the solid and the remedial techniques required to redissolve the material. In general, significant quantities of wax, aromatics, and resins can coprecipitate with asphaltenes, however, the majority of the solid does not melt upon the application of heat.

3.3. Diamondoids

Diamondoids are very stable solid compounds, which usually precipitate directly from the gas phase. Diamondoid solids often precipitate from gases that have undergone thermochemical sulfate reduction. These gases tend to be sour and very dry and are associated with high temperatures and carbonate structures containing anhydrite. They contain principally saturated, cyclic hydrocarbon compounds with a diamond structure; hence the name diamondoids. The molecular weights of diamondoids range from 136 to more than 270 g/mol. The carbon skeleton chemical structures of the three major solid-forming compounds, namely adamantane, diamantane, and triamantane are shown in Figure 23.4.

3.4. Gas Hydrates

Hydrates are a common problem in hydrocarbon production and transportation systems where free water is present and in contact with the C6 -components of the hydrocarbon fluid under certain conditions of temperature and pressure. Under the proper conditions, the C6- hydrocarbon molecules can occupy the geometric lattices of the water molecules in the aqueous phase. When these lattices are filled with the light hydrocarbon compounds, a semistable solid compound similar to ice forms at temperatures far in excess of 0°C (32°F). Hydrates can be identified in the field by their behavior. Without sufficient pressure to maintain the hydrocarbon molecules in the water lattices, the hydrate structures will decompose into water and gas. This behavior can be easily identified in practice. Examples of typical gas hydrate structure and textures are provided in Figure 23.5.

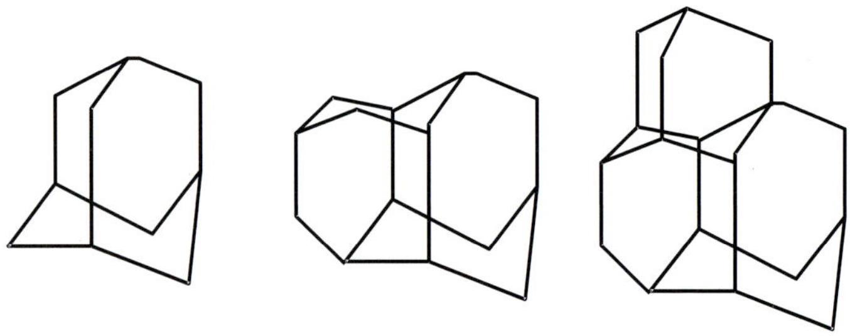

	Adamantane	Diamantane	Triamantane
Melting Point (°F)	514	464	430
Specific Gravity	1.07	1.21	-
Molecular Weight	136	188	240

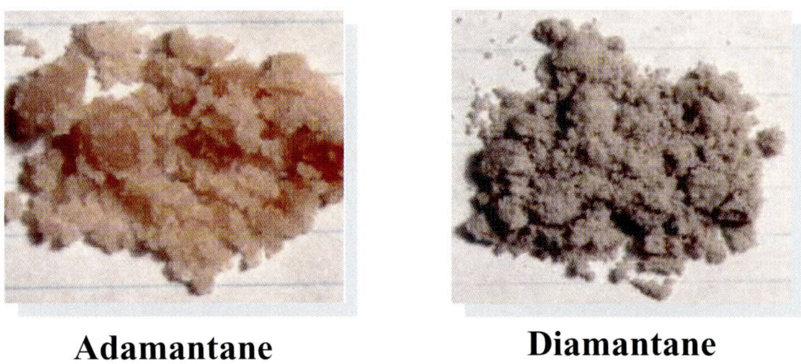

Figure 23.4. Principal diamondoid compound structures, properties, and sample textures.

4. Terminology: Precipitation vs. Deposition

The term deposition has often been used to describe the precipitation process. It is important to clarify the difference between the two.[22] While the precipitation may be defined as the formation of a solid phase out of a liquid phase, deposition can be described as the formation and growth of a layer of the precipitated solid on a surface. Further, a necessary but not a sufficient condition for deposition is the precipitation of a solid phase from liquid solution. That is, precipitation although a precursor to deposition, does not necessarily ensure deposition. Moreover, whereas the precipitation is mainly a function of thermodynamic variables such as composition, pressure, and temperature, the deposition is also dependent on the flow hydrodynamics, heat and mass transfer, and solid–solid and surface–solid

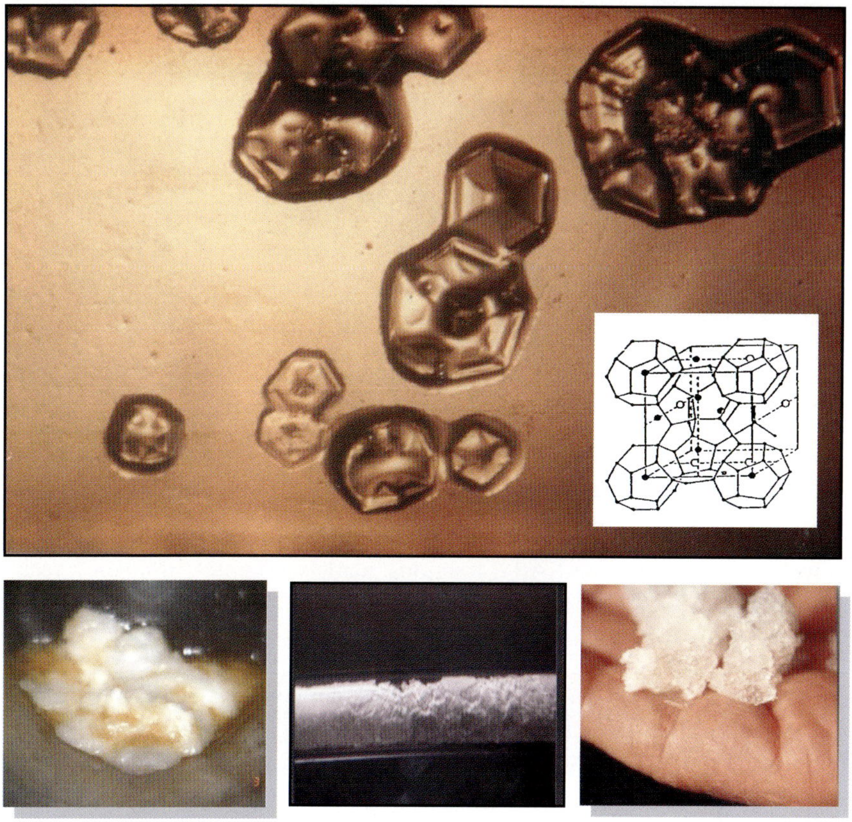

In Live Oil | In Gas Pipe | At Ambient

Figure 23.5. Typical hydrate structure and textures.

interactions. The issue of deposition is beyond the scope of this chapter; precipitation is our main focus unless otherwise specified.

5. Mechanisms of Asphaltene Precipitation: What We think We Know and Why?

The available laboratory and field data indicate that asphaltenes separated from crude oils consist of various particles with molecular weights ranging from ~700 to several thousands. Such an extensive range of asphaltene size distribution suggests that asphaltene may be partly dissolved and partly suspended/peptized in the crude oil.[23] While the first scenario is a relatively well-understood reversible thermodynamic process, the latter is a more complex colloidal irreversible mechanism. The discussion that follows focuses on the colloidal aspect of asphaltene deposition and is augmented by referring to the schematic shown as Figures 23.6–23.10.[1,24–26]

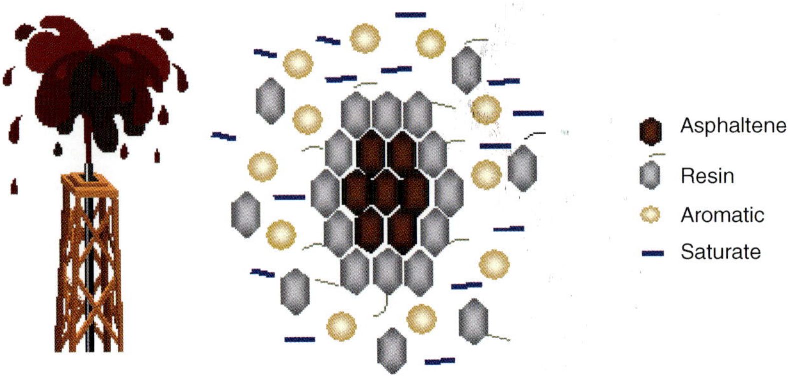

Figure 23.6. Simplified view of petroleum.

5.1. Colloidal Model

Figure 23.6 is a simplistic schematic representation of crude oils in terms of the well-known four major SARA fractions, namely saturates, aromatics, resins, and asphaltenes. Asphaltene molecules are believed to be surrounded by resins that act as peptizing agents; that is, the resins maintain the asphaltenes in a colloidal dispersion (as opposed to a solution) within the crude oil. The resins are typically composed of a highly polar end group, which often contains heteroatoms such as oxygen, sulfur, and nitrogen, as well as long, nonpolar paraffinic groups. The resins are attracted to the asphaltene micelles through their end group. This attraction is a result of both hydrogen bonding through the heteroatoms and dipole–dipole interactions arising from the high polarities of the resin and asphaltene. The paraffinic component of the resin molecule acts as a tail making the transition to the relatively nonpolar bulk of the oil where individual molecules also exist in true solution. The aromatics (such as toluene) are relatively good solvents for both wax and asphaltenes.

Field experience[5,27–29] and experimental observations[1,25,26,30–33] indicate that asphaltene stability is dependent on various factors including (but not limited to) composition, pressure, and temperature of the oil. The general consensus is that the effect of composition and, in turn, pressure on asphaltene precipitation is stronger than the effect of temperature. However, there still exists some disagreement in the literature regarding the effect of temperature on asphaltene precipitation.[16,31,33]

5.2. Effect of Compositional Change

Asphaltene precipitation can occur in situ during mixing of incompatible hydrocarbon fluids, miscible flooding, CO_2 flooding, gas lift operation using rich gases and/or acidizing jobs.[1,5] The addition of compounds with molecules that differ greatly from resins and asphaltenes in terms of size and structure, and therefore, solubility parameter, shifts the equilibrium that exists in the nonpolar portion of

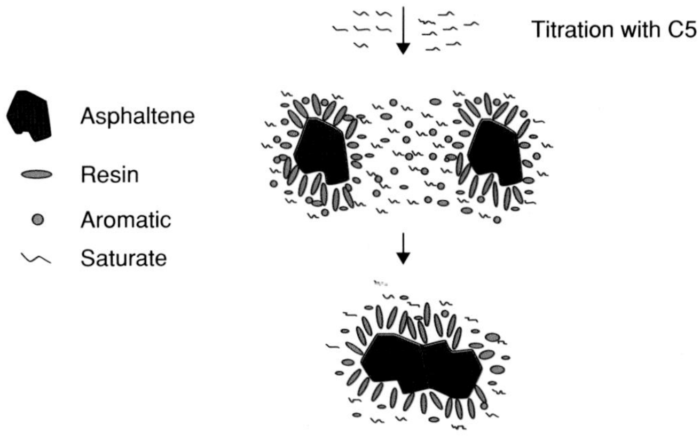

Figure 23.7. Schematic of asphaltene flocculation mechanism due to titration with *n*-C5.[24,26]

the crude oil. For example, normal alkane liquids (such as pentane, hexane, or heptane) are often added to crude oils in an attempt to reduce heavy oil viscosities. The result of this introduction is an alteration in the overall characteristics of the crude oil making it lighter. In response, resin molecules desorb from the surface of the asphaltenes in an attempt to re-establish the thermodynamic equilibrium that existed in the oil.[24,26] This desorption of peptizing resins forces the asphaltene micelles to agglomerate in order to reduce their overall surface free energy (see Figure 23.7). If sufficient quantities of the particular titrant are added to the oil, the asphaltene molecules aggregate to such an extent that the particles overcome the Brownian forces of suspension and begin to precipitate.[24,26]

The above description suggests that the quantity and type of solvent added to the crude oil may be crucial to the amount and characteristics of the asphaltenes precipitated. As illustrated in Figure 23.8, *n*-alkanes induced total "asphaltenes" from the same dead oil (using 40:1 *n*-alkane:oil ratio) decrease with increasing titrant carbon number.[25] Visual inspection of the corresponding precipitates reveals variation in "asphaltene" textures and characters with titrant carbon number. While short *n*-alkanes yield tacky and sticky "asphaltenes," longer *n*-paraffins produce powdery and dry "asphaltenes." The precipitated "asphaltenes" so obtained are clearly different both qualitatively and quantitatively. It is widely recognized that the short/light *n*-alkanes tend to coprecipitate resins along with the asphaltenes.[25] In addition, it has been established that wax and occasionally resins would likely coprecipitate with asphaltenes when the titration is conducted at room temperature.

Figure 23.9(a) shows the experimental approach used to investigate the effects of native resins separated from the parent North Sea oil on the stability of the corresponding asphaltenes.[26] Briefly, a sufficient volume of dead oil was fractionated using high performance liquid chromatography (HPLC) into the four major SARA fractions. As can be seen in Figure 23.9(b), the critical *n*-C5 concentration required to induce asphaltene precipitation increases essentially linearly with increasing native resin concentration at 25°C and 690 kPa.[26] Similarly, the

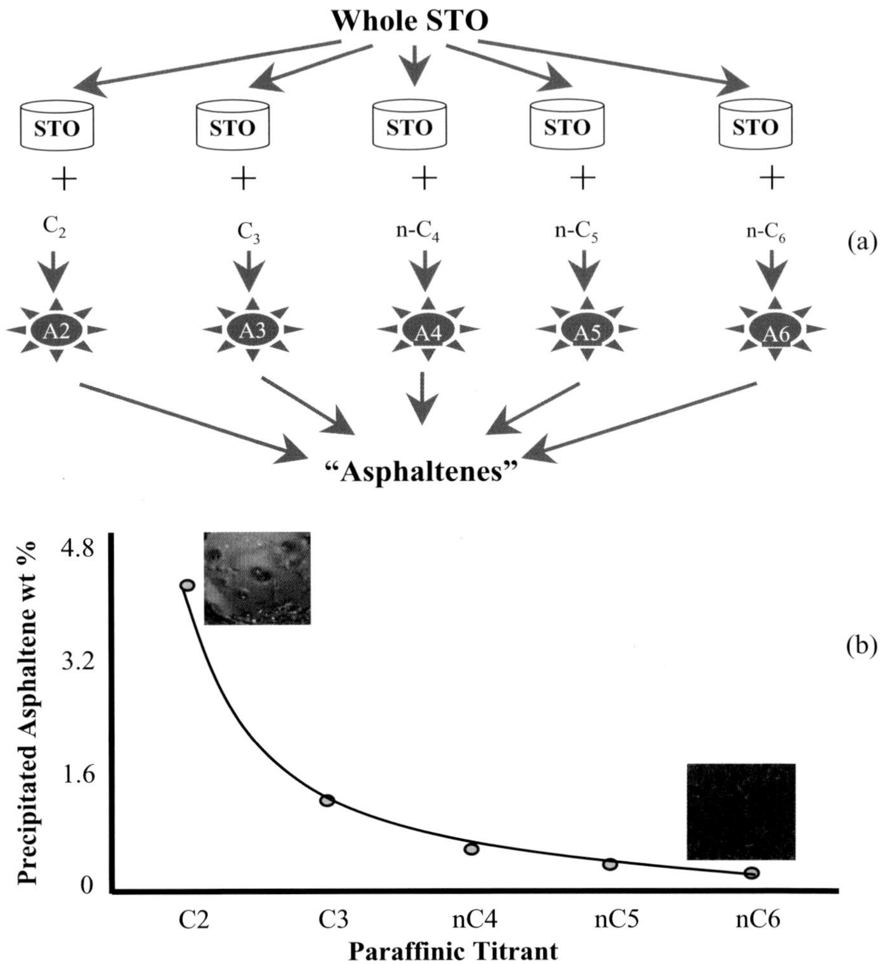

Figure 23.8. Effect of paraffin (titrant) carbon number on "asphaltenes"[25]: (a) experimental methodology (titrant to oil ratio = 40:1) and (b) amounts and textures of "asphaltenes."

concept of using deasphalted oil to dissolve precipitated asphaltenes has been introduced and validated in another independent study.[34]

5.3. Effect of Pressure Change

Asphaltene precipitation and deposition can also occur in oil-well tubing below the depth at which the oil becomes saturated.[1–5,28,29,32,35–37] This phenomenon is largely ascribed to the different extents of compressibility of the light ends and the heavy components (e.g., resins and asphaltenes) of the undersaturated crude. In fact, the relative volume fraction of the light ends within the liquid phase increases as the pressure of the undersaturated reservoir fluid approaches its bubble point. Such an effect is similar to adding a low molar mass hydrocarbon (precipitant) to a crude oil causing asphaltene depeptization. This concept

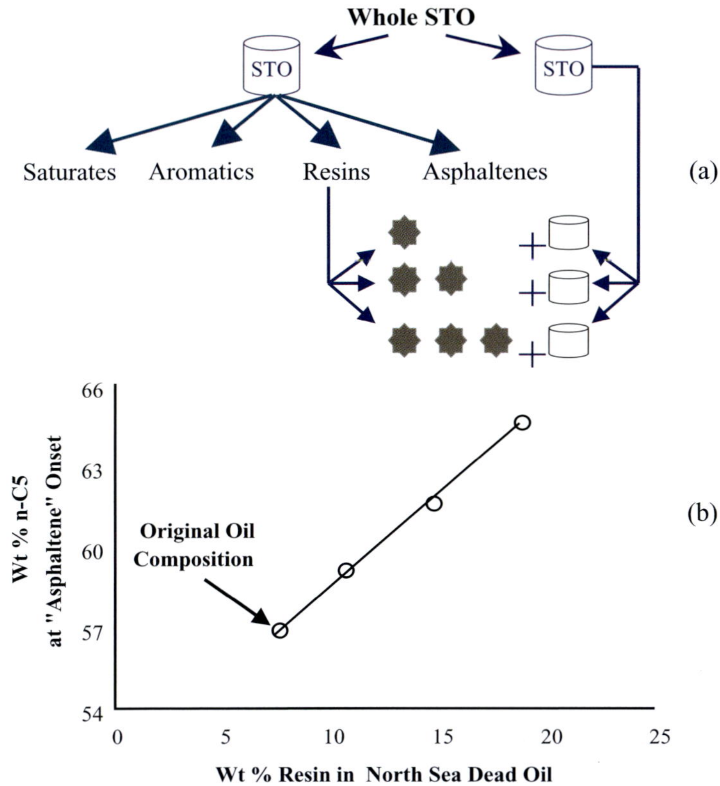

Figure 23.9. Effect of resins on asphaltene stability at 25°C and 690 kPa.[26] (a) Experimental methodology and (b) Titration results.

is illustrated schematically in Figure 23.10(a) and augmented by referring to the Peng–Robinson EOS model simulation results of Figure 23.10(b).[38] Clearly, the partial molar volume of readily compressible molecules (e.g., methane) changes noticeably compared to that of pentane as the system pressure is decreased isothermally. Such a change is even less pronounced for the case of C7–C12 fraction. It is, therefore, conceivable that the much heavier petroleum fractions (i.e., resins and asphaltenes) would exhibit relatively minimal partial molar volume change with decreasing pressure at the same temperature.

It has been established that the bulk of asphaltene precipitation occurs in the proximity of the bubble point pressures of undersaturated crude oils.[1,28,35] Below the bubble point, the volatile hydrocarbons evaporate from the liquid as a gas phase causing an increase in the density of the liquid phase (i.e., change of liquid composition). Recognizing that light ends as well as asphaltenes compete for solvency in the crude oil, the loss of light ends implies better solubility of the asphaltenes in the crude oil.[1,24,28,29,35] This behavior was indeed observed/confirmed in the field of Hassi Messaoud as early as 1965.[27] The operator encountered major asphaltene production problems (deposition in production tubing) for such a relatively light

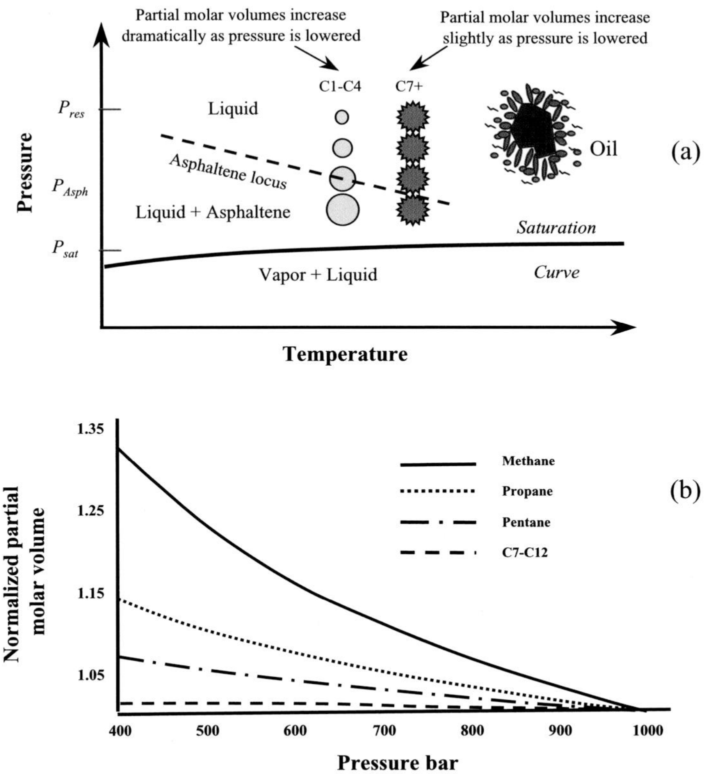

Figure 23.10. Effect of pressure on the partial molar volume of petroleum components.[38] (a) Schematic representation and (b) Peng–Robinson EOS simulations.

crude oil ($\sim$40 API with 0.01 wt% asphaltenes) at pressures above the bubble point. Below the saturation pressure, the plugging problems "magically" disappeared.[27]

5.4. The de Boer Plot

de Boer *et al.*[28] reported and compared the properties of some crude oils from the North Sea and Kuwait in which asphaltene problems were encountered during production to those that were operated problem-free. In general, problems were encountered with light crude oils that are high in C_1–C_3 ($>$37 mol%), relatively low in C_{7+} fraction ($<$46 mol%), low in asphaltene content ($<$1 wt%), significantly undersaturated (i.e., difference between reservoir and bubble point pressures $>$30 MPa) and of relatively high compressibility. While Hirshberg's data set indicated only low asphaltene content oils exhibited instability,[41] more complete data sets show little correlation between stability and asphaltene content. Live oil compressibility correlates to in situ density of the crude; therefore, a generalized plot (hereafter denoted by *de Boer plot*) of the degree of under-saturation versus in situ density with defined regions of potentially problem-free to very problematic crude oils was proposed.[28] Accordingly, the lighter the prospect crude oil is, the greater its tendency to precipitate asphaltenes during production will be.[1] The *de Boer plot* is generally considered a conservative screen that is best used to identify nonproblematic oils.

5.5. Reversibility of Asphaltene Precipitation

Many researchers believed that asphaltene precipitation is not reversible mainly due to experimental observation of the colloidal behavior of asphaltene suspensions.[39] Hotier and Robin[40] have argued that the results from titration experiments show that asphaltene precipitated by the addition of a precipitant can be redissolved by the addition of a solvent. This is, however, no firm indication for reversibility because the addition of a solvent is not the reverse process of the addition of a precipitant. Hirschberg *et al.*[41] have assumed that asphaltene precipitation is reversible but is likely very slow. Fotland[32] and Wang *et al.*[33] have speculated that asphaltene precipitation is less likely to be reversible for crude oils subjected to conditions well beyond those of the onset.

It is noteworthy to point out that most of the experimental techniques employed to study asphaltenes before 1995 were limited in pressure rating; hence, conclusions were generally overextrapolated to live oil conditions based on dead oil analyses. It was difficult to test for asphaltene reversibility based on dead oil experiments whereby the sample composition appeared to be altered irreversibly.

Following the development of high-pressure high-temperature laser-based light transmittance technique in 1995, Hammami *et al.*[1] were the first to establish and report a strong tendency for pressure-induced apshaltenes to redissolve in undersaturated Gulf of Mexico reservoir fluids upon increasing the system pressure above the saturation pressures at corresponding reservoir temperatures. These tests were performed over typical production conditions of composition, pressure, and temperature. Evidence of asphaltene precipitation above and redissolution below the saturation pressure was also observed. That is, a pressure range for asphaltene instability was measured at constant temperatures. Such an observation is indeed in agreement with the reported Hassi Messaoud field experience first reported in mid 1960s.[27]

6. Sampling

Flow assurance measurements have led to a new awareness of the need to have representative samples. Changes in pressure and temperature can cause phase changes that lead to sample alteration. The test fluid in any laboratory investigation must be reasonably representative of the reservoir fluid if the data are to have any practical application. Therefore, prior to initiating any laboratory fluid property measurement, it is imperative that the appropriate time, effort, and money be invested wisely so as to insure that the utmost appropriate fluid sampling and fluid preparation procedures are followed for a given petroleum fluid.

Bottom hole sampling from hydrocarbon reservoirs is essential to understand the fluids to be produced. Fluid samples are also acquired at later stages in the life of the reservoir to evaluate the conditions of the fluids at a certain point of time or after some production activities; however, the point of comparison is always the original fluid properties. Thus, it is important to acquire quality, representative fluid samples and to manage this data.

Introduction of contaminates during the sample acquisition process can also alter the fluid composition. The most common source of contamination is from drilling fluids. Oil-based mud (OBM) is commonly used in drilling operations. OBM filtrates often invade the formation due to overbalance pressure in the mud column and, in turn, can contaminate adjacent formation fluids. The extent of contamination depends on the OBM type as well as the penetration depth (layer) in the formation. This leads to problems and challenges in sampling representative reservoir fluids. In that, the presence of even small concentration of OBM filtrate in the collected sample could significantly alter the phase behavior and fluid properties of the formation fluid.[42]

The perfect sample would be collected contamination-free from the reservoir at constant temperature and pressure and transported intact to the laboratory maintaining both temperature and pressure. In this way alteration associated with phase changes, transfers or contamination is eliminated. In practice, this is not possible today. A more realistic goal is to reduce the potential for phase changes through pressure and potentially temperature compensation. Figure 23.11 shows the effect of pressure compensation for a sample taken with a formation-testing tool.[2,43] Unfortunately, many of the solids that cause flow assurance problems come out of solution during the sampling process just as they do in production systems. If phase change does occur, it is important to minimize the time the sample remains in the multiphase condition and avoid sample transfers. If a sample must be moved

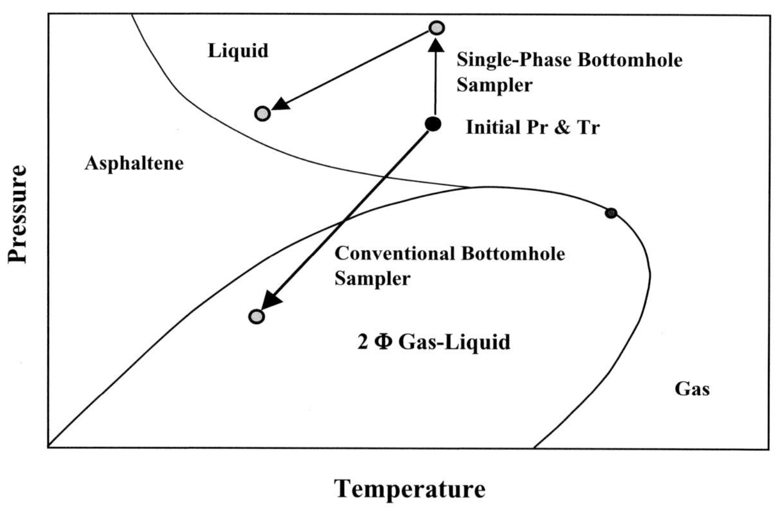

Figure 23.11. Pressure compensated sampling.[2,43]

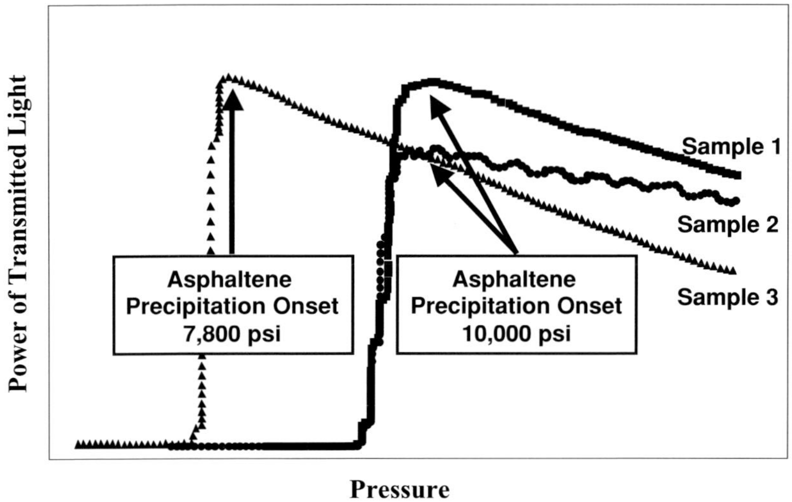

Figure 23.12. Effect of sample handling on asphaltene onset pressure.[38]

between vessels, it must be carefully equilibrated at reservoir conditions prior to transfer. Following each sample transfer, rinse the corresponding emptied sample cylinder thoroughly with toluene and collect the wash solution in properly labeled transport containers. Take multiple samples and perform consistency checks at all stages. A well thought out sample management program will include tracking the sample conditions from the point of acquisition through to the laboratory analysis. Pressures, temperatures, validation data, transfers, and volumes need to be documented.

Asphaltene measurements place the most stringent demand on hydrocarbon sample quality. Asphaltene is easily precipitated and lost during pressure and temperature reductions encountered during sampling. The stability of asphaltenes are also greatly effected by the presence of contaminates. Figure 23.12 shows the asphaltene precipitation pressure as measured by light scattering for three samples taken from the same sample point.[44] Sample 1 was pressure compensated and carefully equilibrated and transferred on site. Sample 2 was shipped in the sampling bottle to the laboratory directly and equilibrated for an extended period of time prior to transfer. Sample 3 was conditioned at a low temperature (below Tres) for a short period of time and transferred onsite. It is obvious that asphaltene was precipitated and lost from sample 3 giving a much lower precipitation pressure.

Figure 23.13 shows the effect of oil-based mud contamination on asphaltene precipitation pressure.[38,45] The actual reservoir fluid is very unstable with an onset significantly above the saturation pressure. As can be seen, the mud stabilizes the system so that with 19.4 wt% contamination the precipitation point disappears entirely. Data collected on these samples and fed into design or surveillance tools would give erroneous results that could lead to mistakes in both design and operating procedures. The effectiveness of both the design and surveillance efforts

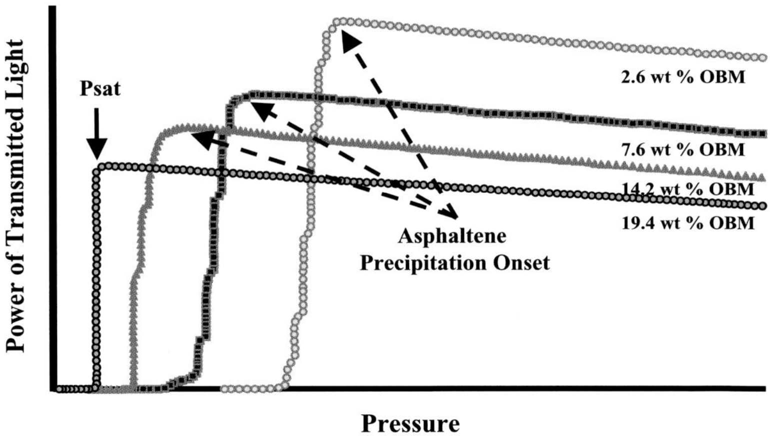

Figure 23.13. Effect of oil based mud contamination on asphaltene onset pressure.[38]

depends on the initial quality of the samples used to generate the flow assurance data set.

Fluid properties can vary both aerially and vertically even in well-connected reservoirs.[46] Samples taken with openhole formation testers represent a point value in the reservoir. Samples from a well test represent an average over the producing interval. Whenever possible, samples from multiple depths or aerial positions should be considered to identify and quantify variations. Understanding the magnitude and nature of compositional variation is important for system design.

7. Laboratory Sample Handling and Analyses

The analysis stage involves the measurement of relevant flow assurance related fluid properties of samples discussed above. The list of relevant fluid properties for asphaltenic fluids will depend on the expected system operating conditions. Typically, a phased approach to the design of an analysis program is taken.[1–3] Sample conditioning and validation is done first as summarized below.

7.1. Sample Handling and Transfer

Upon receiving appropriately collected samples from the reservoir in question, the first step is to properly prepare these samples for subsequent testing. The procedures followed vary slightly depending on the type of samples collected; however, the most important initial objective is to establish the compositional consistency of *multiple samples* from the same well and/or zone.

Initially, the transport sample cylinders would be checked for their opening pressures at room temperature, placed in an air bath oven, and heated to the reservoir temperature while maintaining constant pressure using a back pressure regulator.

Following a *minimum of 5 days* stabilization period under continuous rocking, an isobaric and isothermal transfer to a suitable high-pressure high-temperature cylinder would take place. This transfer involves the following steps.

First, a clean pistoned lab cylinder would be evacuated and filled with displacement fluid below the piston. The lab cylinder would then be connected to the shipping cylinder with a sight glass and high-pressure tubing. To begin the transfer, one pump would be used to displace the oil out of the shipping cylinder into the lab cylinder, while a second pump would be used to withdraw the displacement fluid from the lab cylinder. In the process, the transfer pressure is maintained well above saturation and likely asphaltene onset pressures. It is strongly recommended that the differential pressure between the two cylinders does not exceed 200 psi during the sample transfer. Following this, each emptied cylinder would be thoroughly rinsed with a sufficient amount of toluene to dissolve any organic deposit if any, and clean the transport cylinder. The corresponding toluene rinses are recovered in suitable pre-labeled containers.

Finally, each of the toluene rinses of the sampling chambers is subjected to compositional analyses up to C30+ (for waxy crudes) and/or rotoevaporation followed by asphaltene content of the resulting residues (for asphaltenic crudes). In the case of waxy crudes, the C30+ composition of the rinses is then compared to those of the corresponding dead oils on a toluene-free basis. The quality assurance/quality control (QA/QC) of onsite sample conditioning and transfer will be deemed acceptable provided the C30+ compositions of the rinses on a toluene-free basis and/or the asphaltene contents of the residues are confirmed to be reasonably similar to those of the parent oils.

7.2. Compositional Analyses

Subsamples of the reconditioned fluids are analyzed for composition (tyically up to C30+) by a flash procedure.[47] In this technique, an accurately measured volume of fluid is isobarically and isothermally displaced into a preweighed empty pycnometer where its density and mass are evaluated. The pycnometer is then connected to a gas–oil ratio (GOR) single stage flash apparatus where the fluid is flashed to ambient pressure and temperature conditions. Subsequently, the evolved gas phase is circulated through the residual liquid for a period of time to achieve equilibrium between phases. Following circulation, the volume of equilibrium vapor and the mass of liquid remaining in the pycnometer are measured. The vapor phase is resolved to C5 by natural gas chromatograph (GC), while the vapor C5+ fraction and the residual liquids are analyzed to C30+ using liquid GC. From the measured composition and total mass of each phase, the composition of the original live fluid may be calculated by mass balance.

7.3. Oil-Based Mud (OBM) Contamination Quantification

The most common technique used to quantify OBM concentrations in reservoir fluids is gas chromatography.[42] Gas chromatography provides a

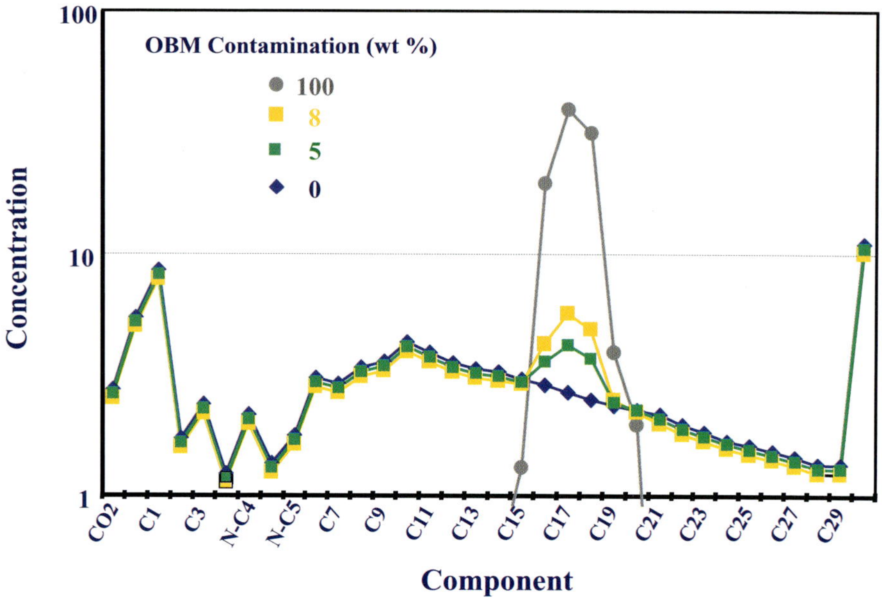

Figure 23.14. OBM contamination analysis by gas chromatography.

convenient visual presentation of OBM contamination level. Figure 23.14 shows an example composition distribution graph for a Gulf of Mexico black oil contaminated with a synthetic oil based mud (OBM). A chromatograph of the OBM filtrate sample is also superimposed on the same figure for ease of comparison. As can be seen, the log decay of the component concentration is essentially linear within the carbon range C11–C29 for uncontaminated oil sample. The observed peaks within the carbon range C15–C20 manifest the presence of differing OBM levels in the oil. The higher the OBM concentration is, the more pronounced the corresponding peak appears. The OBM concentration is commonly quantified using either a skimming technique (i.e., linear interpolation of component concentrations over the OBM carbon range based on the log decay slope) or a subtraction method (i.e., reduction of contaminated component concentrations by an optimized OBM fraction "constant multiplier" which yields the maximum R^2 for the log decay slope of the adjusted composition).[42] One word of caution, however, the subtraction and skimming techniques rely on the assumption of a log linear decay. Biodegraded oils and hybrid oils, oils consisting of mixtures of different oil/condensate types, may not exhibit a log linear decay and significant error may result from automated application of these standard techniques. A direct correlation between optics (i.e., fluid color) and OBM concentration in formation fluids has been developed;[48] it has since become the industry standard for monitoring the sample purity/quality in real time during sampling.

The level of contamination defines the sample quality. It is often difficult to correct for OBM increase with GOR. Contamination is mainly a concern for gas reservoirs and near critical fluids. In the case of black oils, less than 10 wt% OBM is usually considered acceptable for PVT but not for flow assurance studies.[2,37,38,49]

7.4. Dead Oil Characterization

Once duplicate bottom hole samples are verified to have similar C30+ composition, density, OBM concentration and GOR, a subsample (~100 cm^3) of one of the fluids (randomly selected) is flashed to ambient conditions. The resulting dead oil is then subjected to a suite of screening tests using different analytical techniques as detailed below.

Analytical methods have been available for some time for the separation and characterization of asphaltenes in heavy crudes and bitumens, and of waxes in waxy crudes.[13] Both of these component types can give rise to production, transportation, and processing problems related to the crudes or their products. Good data characterizing the asphaltenes in terms of their content and stability, and waxes in terms of their content and carbon number distribution, are usually required to help solve these problems. Increasingly both asphaltenes and waxes are being found in the same crudes, and the established methods for their separation are often not applicable. Improved separation techniques have been described, and their applicability to different crude types discussed.[13,50] Examples are given below for the types of characterization, which may be carried out on the separated asphaltenes, maltenes, waxes, and oils.

7.4.1. SARA Analysis

SARA analysis refers to the determination of the amounts of saturates, aromatics, resins, and asphaltenes by a combination of solvent induced precipitation and column chromatography.[5,13,50] Fuhr *et al.*[50] and Lira and Hammami.[13] published various procedures for the separation of crude oils into the SARA fractions depending on the nature and/or character of the oil being considered. Since the SARA determination can vary between various analytical laboratories, it is critical to follow the same procedure for all oils for relative comparison. An example SARA fractionation scheme for whole dead oil is shown as Figure 23.15.

Ideally, the crude oil is initially fractionated into two cuts by distillation (e.g., at 270°C/518°F which corresponds to the boiling point of C15). This step is necessary to separate the volatiles and, in turn, establish good mass balance in subsequent sample fractionation process. The light fraction (<270°C) is analyzed for its saturate (i.e., paraffins, iso-paraffins, and naphthenes) and aromatic split by supercritical fluid chromatography-mass spectrometry (SFC-MS) with N_2O chemical ionization (CI) and/or gas chromatography-mass spectroscopy (GC-MS); whereas, the residue (>270°C) fraction is subjected to SARA fractionation as detailed below. By combining the saturate and aromatic of the two fractions, the whole oil SARA is determined. Alternatively, the oil sample may be initially topped to constant weight at 60°C with flowing nitrogen to produce a volatile-free "stable" oil. The topped oil is then subjected to SARA fractionation as detailed below. This latter approach requires all oils be compared on a topped-oil basis. Assumptions

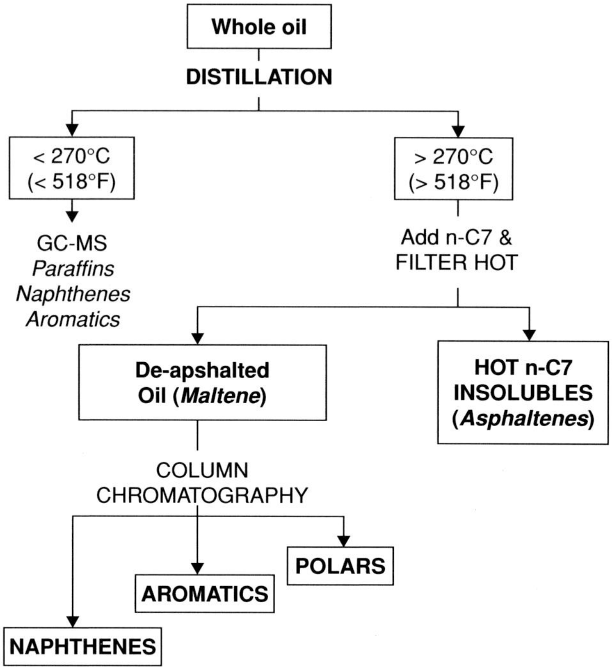

Figure 23.15. SARA fractionation protocol for waxy crudes.

are often made based on experience on the relative saturate to aromatic split for the lost (evaporated) light end fractions to express the SARA on whole oil basis.

7.4.1.1 Asphaltene fraction. Several methods for asphaltene content determination and variations thereof based on the long-standing analytical procedures, namely ASTM D2007, Universal Oil Products (UOP) 614-80, and/or IP 143, have been reported.[5,10,13,25,26,50,51] Typically, a subsample (~1–2 g) of the residue or stable dead oil is dissolved in 40 mL of paraffinic solvent (commonly n-heptane or n-pentane). It is also common to initially dissolve the 1 g of residue in 1 mL of toluene prior to adding 40 mL of paraffinic solvent. The resulting solution is then kept in the dark for several hours and under a nitrogen blanket at room temperature. Subsequently, the solution is heated to near its boiling temperature (to melt coprecipitated waxes and/or resins if any) and filtered hot using 0.45 μm filter under vacuum. Next, the filter is folded to confine the precipitate "i.e., asphaltene + residual oil," loaded into a Soxhlet apparatus (shown in Figure 23.16), and thoroughly washed using hot n-heptane (~75°C) until the solvent in the upper section of the Soxhlet becomes clear. Following this, the boiling flask containing n-heptane is replaced with another flask containing toluene and the extraction process is continued in the Soxhlet apparatus until all "asphaltenes" have been dissolved from the filter. The solution is then concentrated by evaporating the toluene

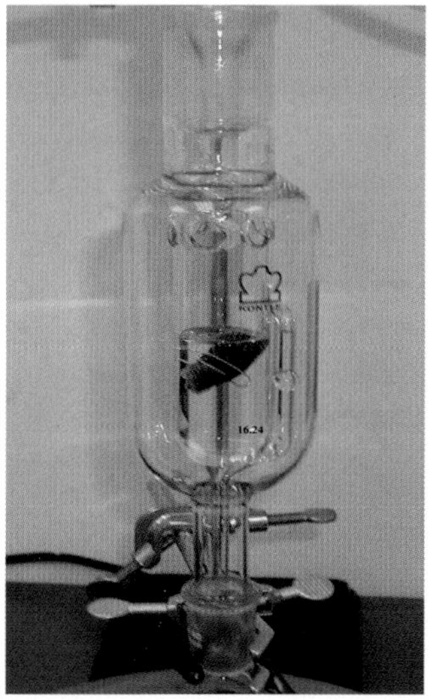

Figure 23.16. a Soxhlet apparatus.

and eventually transferred to a tared small vial for complete drying under nitrogen at 60°C. Finally, the vial is weighed and the "asphaltene" fraction is quantified.

7.4.1.2 Saturate, aromatics and resins fractions. The de-asphalted residue (i.e., filtrate) dissolved in paraffinic solvent obtained as described above is mixed thoroughly; an aliquot (in the order of 40 mg) is then subjected to liquid chromatographic fractionation using silica-packed column as illustrated in the schematic of Figure 23.17. Briefly, the sample diluted in paraffinic solvent is loaded onto the silica-packed column as a feed. While the aromatic and polar compounds adsorb onto the silica beads, the saturate (S) fraction elutes through the column along with the solvent (n-heptane or n-pentane) into a prelabeled glass jar. Subsequently, a

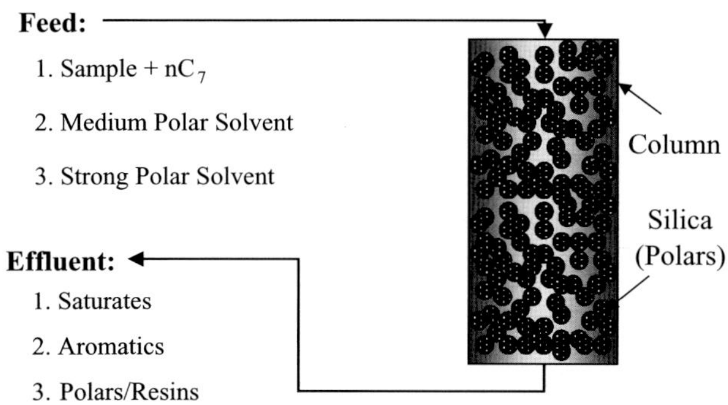

Figure 23.17. Schematic of liquid chromatographic separation of maltene (deasphalted oil) using silica-packed column.

medium polar solvent and a strong polar solvent are loaded onto the same packed column (one at a time) to recover/elute the aromatic (A) and resin (R) fractions, respectively. The corresponding effluents are collected in separate glass jars. Each solution is then subjected to roto-evaporation under nitrogen atmosphere to remove the respective solvents. The saturate fraction so obtained is observed to be white and opaque just like candle wax; whereas, the aromatic and polar fractions appear to be brownish and very dark, respectively.

7.5. Dead Oil Asphaltene Stability Tests

Various techniques and approaches are used to assess asphaltene stability of dead crude oils. In general, these include SARA screens and titration tests using *n*-pentane, *n*-hexane, or *n*-heptane.

7.5.1. SARA Screens

The reader is reminded that there are many variations of SARA fractionation method and procedures of crude oils. Hence, the SARA screens described herein must be used with caution. In other words, one must ensure the SARA fractions are obtained by strictly following the same experimental procedures and analytical techniques described by the developer of the SARA screen method in question.

One SARA screen is the colloidal instability index (CII) proposed by Asomaning and Watkinson.[52] CII is defined as the ratio of the unfavorable fractions (asphaltenes + saturates) to the favorable fractions (resins + aromatics) of the oil. Ratios greater than 1 mean the amount of unfavorable components exceed that of the favorable components in the system; thereby, asphaltenes are likely to be unstable. Another SARA screen is the cross-plot of two ratios—saturates/aromatics vs. asphaltenes/resins proposed by Stankiewicz *et al.*[5] The saturates/aromatics ratio is an indirect measure of the solvating power of the sample for asphaltenes (a high ratio implies poor solvating tendency). The asphaltene/resin ratio, on the other hand, relates to the measure of colloidal stability of the asphaltenes (a low ratio of asphaltene/resin implies good colloidal stabilization). A cross-plot of these ratios populated with data from over 30 asphaltene problem fields and more than 200 other fluids worldwide is reproduced here as Figure 23.18. The curve separating stable and unstable oils has been traced based on the authors' field observations and operational experience. As can be seen, two broad classifications of crude oils are readily observed. Some fit along the asymptotic horizontal curve for which asphaltene stability is predominately aromaticity controlled. Others are arranged along the asymptotic vertical curve for which asphaltene stability is predominately resin controlled.

7.5.2. Titration Tests

There are various aliphatic hydrocarbon titration tests (e.g., *P*-value,[5] refractive index measurements,[33,53] and asphaltene dispersant test (ADT) which measure the tendency of asphaltene solids to remain dispersed in hexane) that are

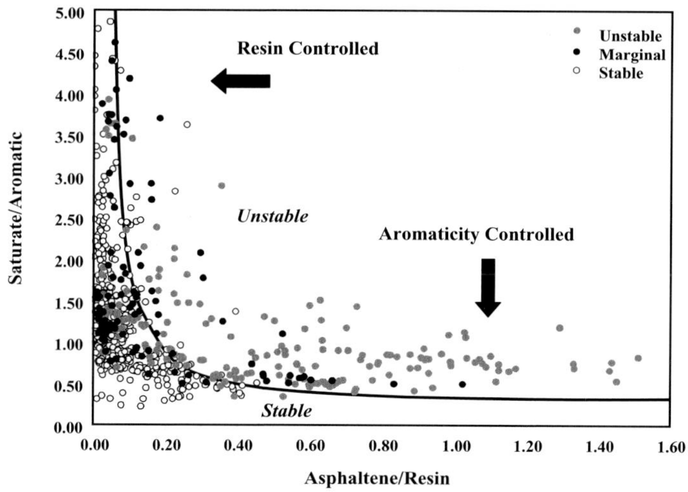

Figure 23.18. Shell's SARA cross plot stability screen.[5]

used to assess the stability of asphaltenes in dead crude oils. The general procedure involves either continuous (fast) or discrete (slow intermittent) addition of aliphatic titrant to the crude oil sample.

7.5.2.1 Continuous Titrations. In the case of continuous titration, the sample mixture is vigorously stirred and simultaneously monitored for asphaltene flocculation onset using typically near-infrared laser transmittance probes. The experimental technique used for this test is commonly referred to as the flocculation point apparatus (FPA). This system is analogous to the laser-based light transmittance solid detection system (SDS) described by Hammami and Raines.[47] As the titrant (typically heptane or pentane) concentration increases in the oil sample, the resulting mixture optical density decreases and, in turn, the light transmittance through the sample increases. At certain critical titrant concentration, however, asphaltenes flocs begin to appear; the corresponding light transmittance signal exhibits a maximum and begins to drop dramatically. Typically, the titrant concentration that corresponds to the maximum in the transmittance trace is deemed the FPA number, which is defined as the ratio of the volume of titrant to the initial mass of crude oil.[4,5] Figure 23.19 shows an example titration curve obtained using the infrared light transmittance probe.

The true asphaltene onset concentration, however, is almost always lower than the value, which corresponds to the observed maximum in light transmittance curve. This is because of the two simultaneous competing (counter) effects of the decrease in optical density vs. size and amount of precipitated asphaltene flocs with increasing titrant concentration. While the decrease in optical density leads to increase in light transmittance, the precipitation of asphaltenes results in a decrease of light transmittance. The asphaltene flocs must therefore reach a critical size and/or concentration before the light transmittance attains a maximum

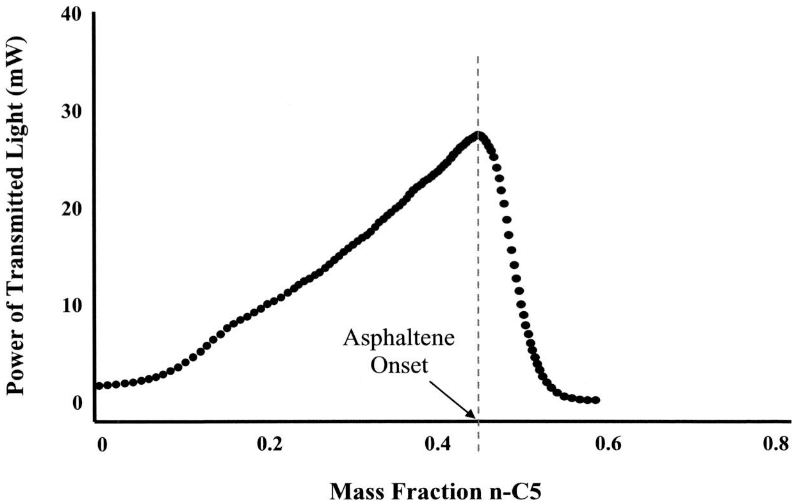

Figure 23.19. Typical light transmittance titration curve at constant temperature and pressure.

as a net effect of the two competing factors. Beyond the observed maximum in light transmittance, the effects of asphaltenes become stronger than those of the titrant dilution; thereby, light transmittance continues to drop with increasing titrant concentration.

7.5.2.2 Discrete Titrations. Discrete titrations account for slow kinetics of asphaltene instability and, thus, provide for better approximation of equilibrium measurements. Often the kinetics of titrant-induced asphaltene precipitation have been reported to be quite fast (on the order of few seconds) at relatively high-dilution rates.[24,54] At low-dilution rates, however, Wang[55] observed that the time required for a reasonable approximation to equilibrium for asphaltene precipitation with aliphatic titrant at ambient temperature is on the order of 24 hr. Anderson[56] reported solution formation to be time dependent, which is indicative of very slow kinetics in particle formation and reorganization. Long and coworkers[57] reported the precipitation of asphaltene from bitumen, induced by addition of aliphatic solvents, occurs in two distinct stages. The first stage entails fast and massive asphaltene flocculation after bitumen is contacted with the aliphatic solvent. The second stage is the continued slow precipitation of asphaltene with no further addition of aliphatic solvent. The asphaltene precipitation during the second stage was reported to undergo three phases, namely induction (slow nucleation), rapid growth, and fractal formation.

Wang and Buckley[57] and Buckely *et al.*[58,59] used refractive index measurements along with microscopic observations to evaluate various crude oils diluted with incremental proportions of aliphatic solvents at ambient temperatures. According to their refractive index screening approach, crude oils with refractive indices well above 1.45 at ambient temperature are generally quite stable with respect to asphaltene precipitation.

8. Live Oil Asphaltene Stability Techniques

The area of flow assurance measurements is still a developing field with new technologies becoming available regularly. This has both positive and negative consequences. On the positive side, our ability to measure and interpret changes in fluid behavior is continually improving. This leads to a better design that both optimizes performance and reduces flow assurance risks. However, the dynamic nature of the measurement technology has led to a lack of standardization and inconsistencies between measurements and modeling. Below we will discuss three established new technologies for flow assurance measurements that have been introduced in the past 10 years.

8.1. Light Transmittance (Optical) Techniques

The first is the laser-based light transmittance technique (also commonly referred to as solids detection system "SDS"). The SDS has gained wide acceptance in the late 1990s and became the industry standard for screening asphaltene stability in live formation fluids.[1–5,35–38,47,60,61] Details about this technique and corresponding measurement principles have been described elsewhere.[1,47] To this end, it is useful to summarize some of the main factors known to influence light transmittance (LT)[1]:

- LT is inversely proportional to density; hence, if the density decreases then LT increases. Density is proportional to pressure (above bubble point); hence, if the pressure decreases then LT increases.
- LT is inversely proportional to particle size (PS); hence, if the PS increases then LT decreases.
- LT is inversely proportional to the nucleation density of solids (NDS); hence, if the NDS increases then the LT decreases.

The suitability of this technique for measuring upper asphaltene precipitation locus and corresponding saturation curve is demonstrated through example Figure 23.20, which shows isothermal light transmittance traces measured for South American live oil. The corresponding sampling conditions and formation fluid properties are listed in Tables 23.3 and 23.4, respectively.[61] The South American crude is barely undersaturated at reservoir conditions. It is deemed to be black oil as characterized by the C_{7+} fraction (~44 mol%) and single stage GOR (~ 540 scf/STB). This fluid is not likely to precipitate asphaltenes during pressure depletion at reservoir temperature (i.e., 218°F). It is, however, prone to precipitate asphaltenes during isobaric temperature sweeps (Figure 23.21) and/or isothermal pressure depletions at lower temperatures. The onset pressure of asphaltene precipitation is observed to increase with decreasing temperature. The corresponding SDS depressurization traces are qualitatively similar to Gulf of Mexico reservoir fluids with minor asphaltene precipitation problems. The corresponding PT diagram is provided as Figure 23.22.[1,61] Note the isothermal asphaltene onset pressure at

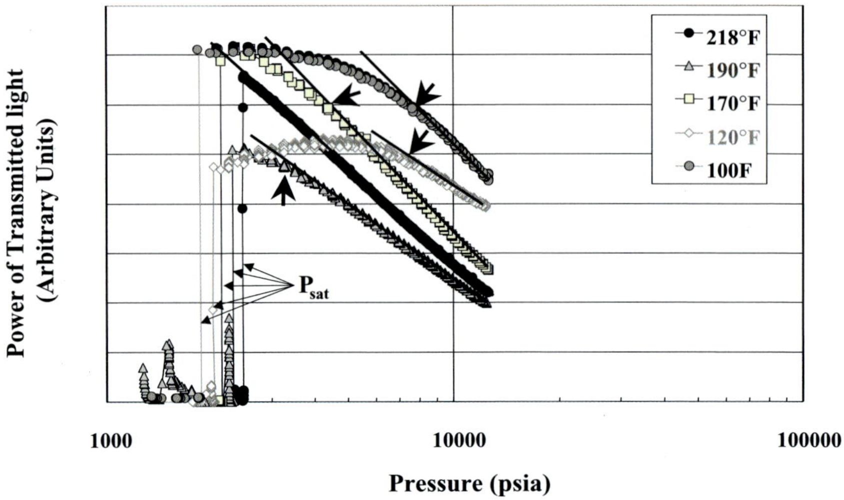

Figure 23.20. Isothermal depressurization of South American Crude at different temperatures.[61]

170°F corresponds to the isobaric asphaltene onset temperature at 4000 psi within experimental error as expected. Hence, the upper asphaltene locus is a true thermodynamic boundary. Additionally, some of the measured properties (such as SARA distribution, in situ density, and degree of undersaturation) compare well with those reported for fluids from the North Sea and Kuwait reservoirs that have no or minor production problems [28].

Figure 23.23 depicts an example of complete isothermal SDS pressure depletion trace of unstable Gulf of Mexico fluid at corresponding reservoir temperature (i.e., above and below the formation fluid saturation pressure). The semilog plot of LT power versus pressure clearly shows three major transitions. The highest transition pressure corresponds to the asphaltene onset pressure. The middle transition pressure is caused by appearance of vapor phase (i.e., bubble point pressure), and the lowest transition pressure is suspected to correspond to asphaltene redissolution pressure. It is speculated that below the saturation pressure, two competing effects emerge as solution gas is progressively liberated. LT tends to increase as precipitated asphaltenes commence to redissolve in the increasingly denser liquid phase. Simultaneously, LT also tends to decrease as the optical

Table 23.3. Reported Field Conditions[61]

	Conditions	
Sampling point	Pressure (psia)	Temperature (°F)
Reservoir (down hole)	2673	218
Well Head	1114	100
Separator	59	95

Table 23.4. Composition and Measured Properties of Bottom Hole Sample[61]

Component	Mol%
CO_2	6.383
N_2	0.113
C1	30.754
C2	2.638
C3	3.491
I-C4	1.646
N-C4	2.598
I-C5	1.823
N-C5	1.529
C6	4.45
C7+	44.58
Molecular weight (g/mol)	122.7
Density (g/cm^3) at 218°F and 7014 psia	0.771
Bubble point pressure at 218°F (psia)	2560
STO SARA fractions	
Saturates content (wt%)	56.4
Aromatics content (wt%)	26.6
Resins content (wt%)	10.9
Asphaltene (*n*-C5 insolubles) content (wt%)	6.1

density of the liquid phase continues to increase. Depending on which effect is stronger, the net result can either be an increase, plateau, or decrease in LT signal. Based on Figure 23.23, it appears that the effects of asphaltene redissolution are initially stronger than those of increasing optical density of liquid phase; hence,

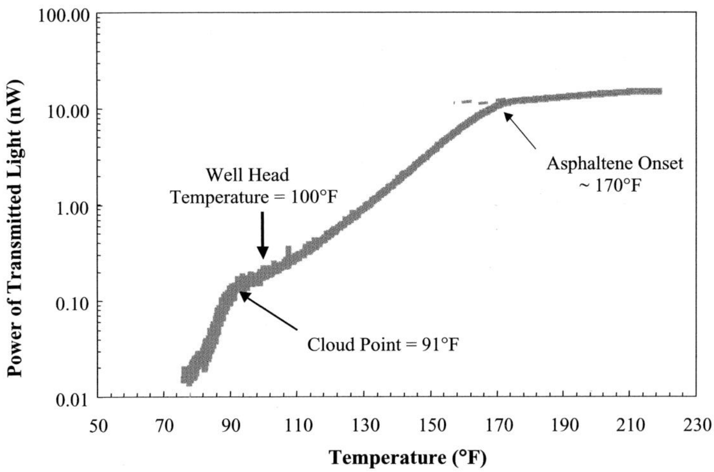

Figure 23.21. Isobaric temperature sweep of South American crude at 4000 psi.[61]

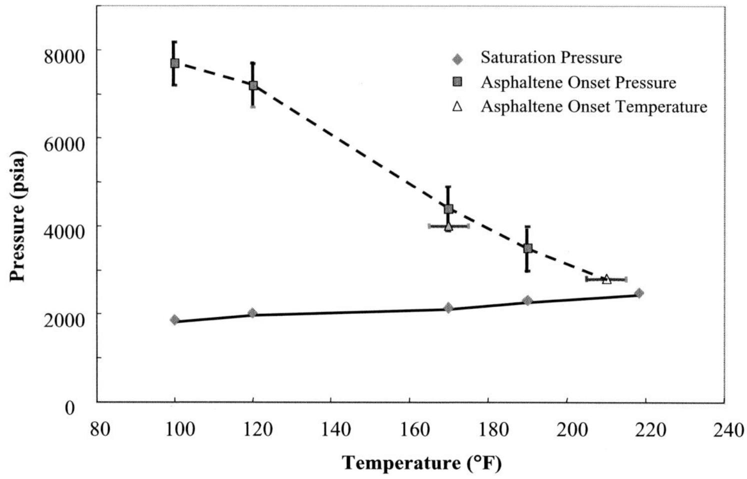

Figure 23.22. PT diagram of South American crude oil.[61]

the observed increase in LT. When the lowest transition occurs, the effects of increasing optical density of the liquid phase become stronger, thereby, the observed linear decrease in LT signal.

If the above speculations are correct, one must note that the lower asphaltene boundary/locus is not as well defined as the upper asphaltene precipitation locus. In fact, experience shows that asphaltene redissolution below the bubble point is kinetically slow. Often, depressurized samples of crude oils (i.e., dead oils) received in the laboratory contain asphaltene aggregates that are readily identified by microscopic examination.[53] They surely dissolve upon addition of toluene. Because of these limitations and/or uncertainties, a schematic phase diagram of unstable reservoir fluid is constructed as illustrated in Figure 23.24. In that, the lower asphaltene boundary is represented as a broad region instead of a clear locus.

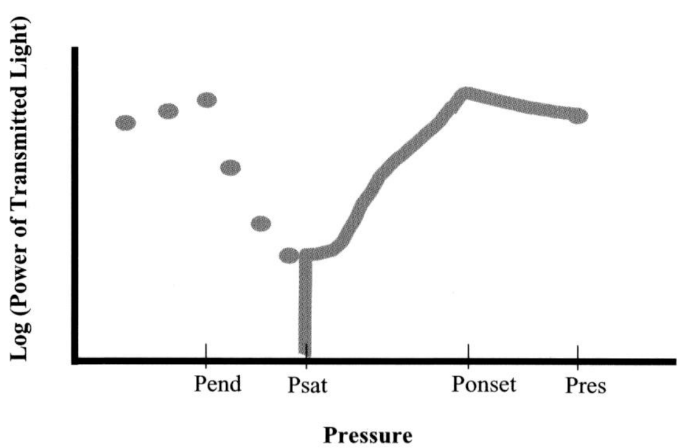

Figure 23.23. Isothermal discrete laser light transmittance pressure depletion trace for unstable GoM reservoir fluid at Tres.

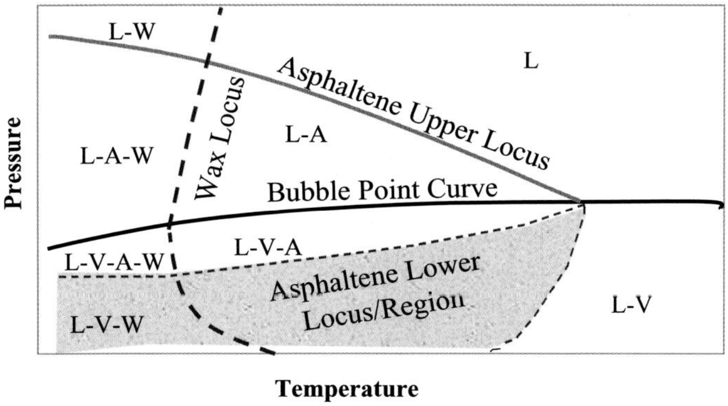

Figure 23.24. Schematic PT phase diagram for unstable reservoir fluids.

Nevertheless, the observed overall trend is quite consistent with reported Hassi Messaoud field observations,[27] namely severe asphaltene deposition problems in the producing wells were initially experienced while reservoir fluid was produced above its saturation pressure. The problem eventually disappeared when the production pressure was dropped below the fluid saturation pressure.

8.2. High Pressure Microscope (HPM)

Prior to visual capability, phase changes were measured with a variety of indirect methods. In addition to light scattering discussed above, other techniques include measuring changes in acoustic properties[62,63] and changes in viscosity or pressure drop across plugging filters.[64] The problem with these types of measurements is risk of subjectivity for the type and behavior of the solid phases could not be identified.

The first direct visual identification and confirmation of phase changes was made possible with the development and introduction of high-pressure microscope.[35,60] A schematic diagram of the HPM setup is shown in Figure 23.25. The HPM is rated to 200°C and 20,000 psi (138 MPa) and mounted in series with a mercury-free PVT cell equipped with a magnetic mixer (1400 rpm). The HPM consists of two sapphire windows cell, a long focal length camera of high resolution (~2 μm), positive displacement pumps, a receiving cylinder, and an air-bath oven. The HPM has a low-dead volume (<1 cm^3) and a small adjustable gap (100–400 μm) between the windows. The thin gap allows visualization in dark crude oils; it may also hinder/restrict the flow of large-sized particles and/or water droplets (i.e., >400 μm). The cell is connected to the bottom of the PVT cell inside the same air–bath oven. The HPM is also equipped with cross-polars which enable the simultaneous detection of both wax and asphaltene particles as they form/nucleate and subsequently grow. By rotating the polarizers on and off during an isobaric cooling experiment, one can easily observe the formation of wax and/or asphaltene as they form and/or coprecipitate. Waxes are crystalline in nature and, hence,

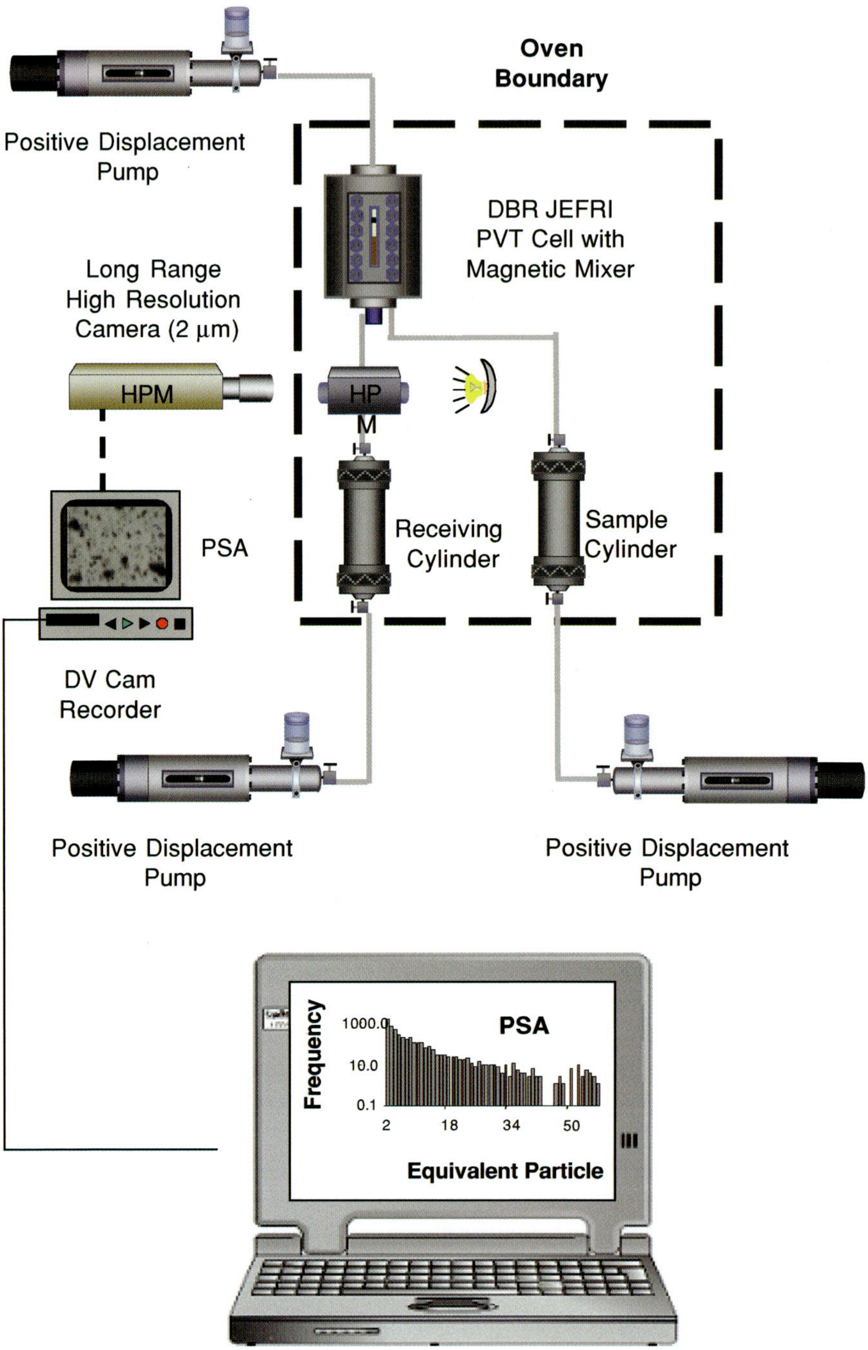

Figure 23.25. High pressure microscopy (HPM) + particle size analyzer (PSA).

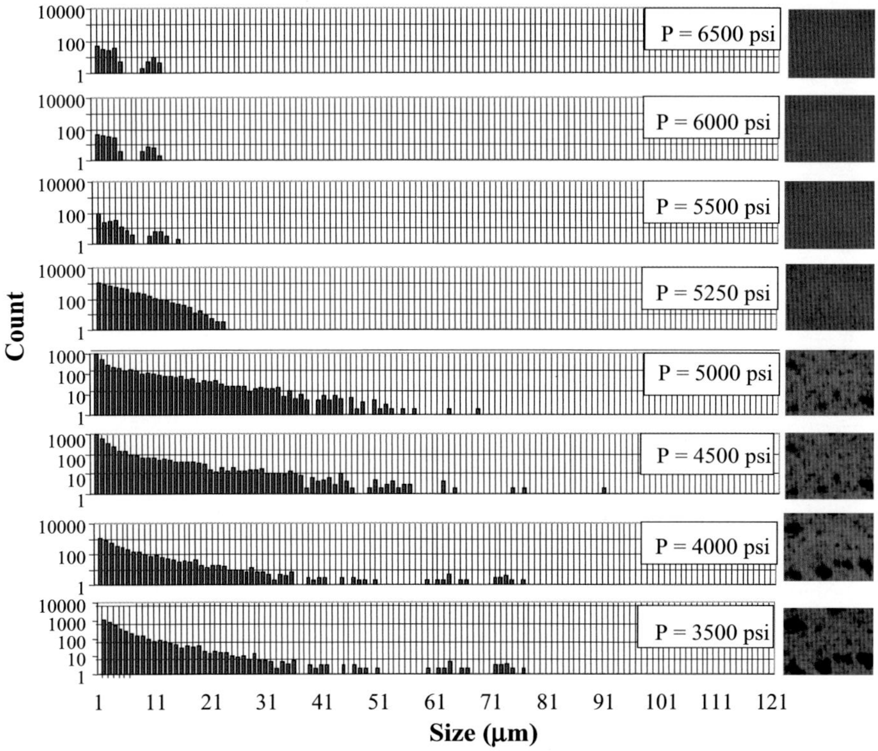

Figure 23.26. Sample PSA histograms for a discrete depressurization experiment.

are clearly visible under polarized light; whereas, asphaltene are predominantly amorphous and visible under normal light mode.

The HPM is complemented with a particle size analyzer (PSA), which uses proprietary image analysis software to determine the size and distribution of wax, asphaltene and /or water droplets as they form, nucleate, or grow. The PSA operates as a synchronous feature of the HPM system. The system has been calibrated using particle standards of varying concentrations and size (1–50 μm) as well as microscale objectives. The PSA software scans digital HPM photomicrograph images in real time at a high-sampling rate and simultaneously discretizes such images within a restricted field of view, on a pixel-by-pixel basis. From these data, histograms of particle size distribution are produced continuously and frequently over the residence time of an experimental sample contained by the HPM cell, thus providing information regarding relative abundance, morphology changes and onset conditions. Figure 23.26 depicts a sample PSA report generated during discrete depressurization of unstable live oil at reservoir temperature. The reported particle diameter is the diameter of a disk with an area equivalent to the projected area of an asphaltene particle. The number counts are cumulative value obtained from analysis of at least 20 images at a given pressure.

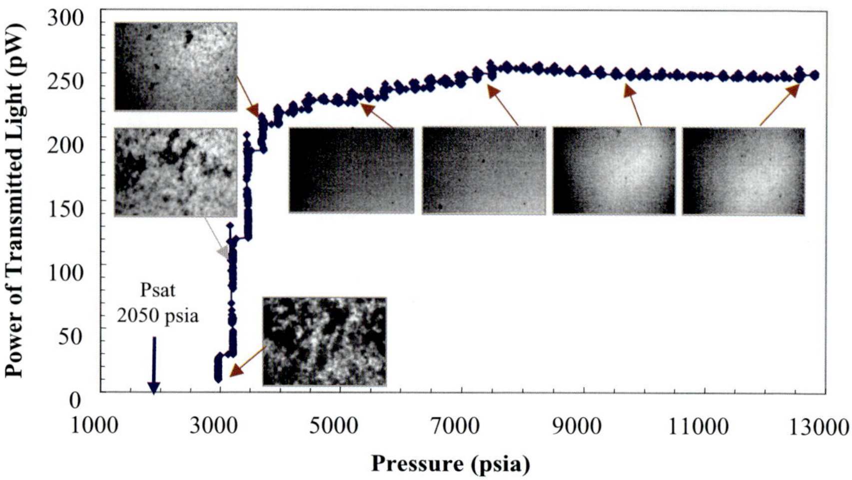

Figure 23.27. Isothermal pressure depletion light transmittance trace with superimposed high pressure microscope photomicrographs for a South American reservoir fluid.[2]

Figure 23.27 shows a transmitted light trace along with photomicrographs of a system that precipitates asphaltene.[3] At the onset of precipitation, asphaltene is present as small dispersed particles that flow easily through the system. At lower pressure another significant change takes place. Particles now stick to each other forming larger flocs and start to stick to the walls of the cell. One would expect the risk of deposition to be much greater at these conditions.[2,3]

Figure 23.28 shows photomicrographs of the complete isothermal pressure depletion experiment (i.e., above and below the saturation pressure) for the same unstable oil depicted in Figure 23.27. It is evident that for this oil sample, asphaltene instability occurs over a well-defined pressure range; three observations are readily made: (1) asphaltene start to precipitate well above the bubble point pressure at reservoir temperature, (2) the maximum amount of asphaltene precipitation occurs in the vicinity of the bubble point pressure (i.e., pressure-induced asphaltenes are least soluble at P_{sat}), and (3) precipitated asphaltenes begin to redissolve below the bubble point and completely disappear below 950 psi. This finding is clearly consistent with the SDS results as well as the reported Hassi Messaoud field experience previously discussed.

Finally, Figure 23.29 shows a partially cross-polarized picture of wax and asphaltene coprecipitating at 15,000 psi. At 75.5°C, we see only dark asphaltene particles. Below the wax appearance temperature (WAT) at 66.7°C, we see both dark asphaltene and white wax particles out of solution. Visual identification is the only unambiguous way to determine the wax phase boundary in this case. Unambiguous phase determination and the behavior and distribution of the phases are critical input to the flow assurance design and surveillance processes.[2]

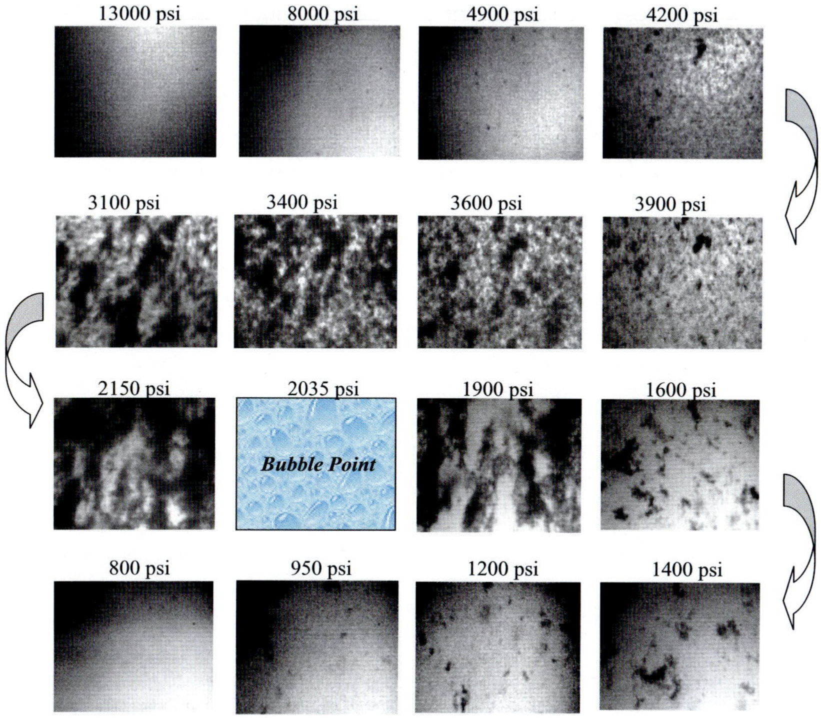

Figure 23.28. High-pressure microscope photomicrographs of South American reservoir fluid depressurized at reservoir temperature above and below its saturation pressure.

8.3. Deposition Measurements

While precipitation of solids can often lead to adverse rheological effects and/or fouling of process equipment, deposition or the growth of solid deposits on surfaces is the major mechanism for blockages in flowlines and wellbores associated with wax, asphaltene, and scales. The deposition process is not only a function of the thermodynamic and chemical properties of the fluid but also of

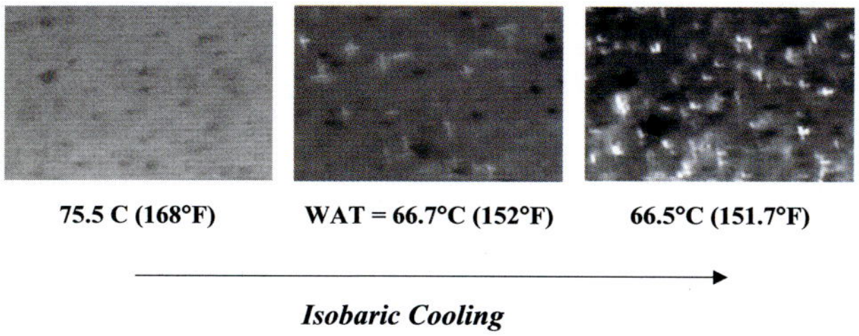

Figure 23.29. Visual identification of coprecipitation of wax and asphaltene at 15000 psia using varying polarization.[2]

the hydrodynamics, geometry, and surface characteristics of the system.[35] Current practice is to measure deposition tendency in the laboratory under conditions substantially different from those encountered in the production system and to then use empirical methods to scale the data. This is currently done for wax deposition. Low-shear dead oil deposition measurements are scaled to producing conditions. The empirical scaling tends to be conservative. Because asphaltene deposition occurs at elevated temperatures and pressures, deposition measurements are difficult. Therefore, no calibrated asphaltene deposition models are generally available. A very conservative approach is taken that assumes asphaltene will deposit if precipitation is observed. Qualitative information from visual observation of asphaltene behavior or tendency of asphaltene to adhere to laboratory equipment surfaces may also be considered.[61]

One new technology that addresses the deposition question is a shear deposition cell shown schematically in Figure 23.30.[65–68] Fully turbulent flow and wall shear conditions that span those found in flowlines can be obtained within the cell. Wall temperature and system pressure can also be independently controlled. Deposition surface type and roughness can be changed by insertion of a sleeve.[65] Figure 23.31 is a picture a shear cell wax deposit. Figure 23.32 shows a comparison of the shear cell deposit character and rate to data obtained from a 2-inch flow loop and a low shear cold finger experiment.[65–67] The shear cell was able to reproduce very similar results to the flow loop for both the deposit oil content and the normal paraffin deposition rate with only a fraction of the oil required for the loop. Uncertainty in empirical scaling procedures results in conservative predictions for wax deposition. These shear cell deposition data require less scaling to model flowline conditions; and therefore, result in less conservatism.

Figure 23.33 shows an asphaltene deposit formed under wellbore conditions.[68] This is, to our knowledge, the first direct formation of an asphaltene deposit at flowline conditions in the laboratory. Measurements like this access the true tendency of a fluid to deposit asphaltene and will hopefully form the basis of a predictive asphaltene deposition model.[65]

In the above three technologies, we see that important improvements have been made in our analytical capabilities and understanding of flow assurance data. However, there is a negative impact due to the dynamically changing nature of this field. Standardization and model development lags the development of measurement technology. Without a clear understanding of the procedure, it is difficult to interpret the data or compare it to similar data from other sources. Understanding the meaning of the measurement and incorporating it properly into the design and surveillance models are critical.

9. Asphaltene Precipitation Models

Because of the complex and variable nature of asphaltenes, thermodynamic models for asphaltene phase behavior have generally lagged those developed for wax. Even the mechanism for stabilizing or solubilizing asphaltenes has been the topic of many heated debates over the years. While nobody denies that asphaltenes

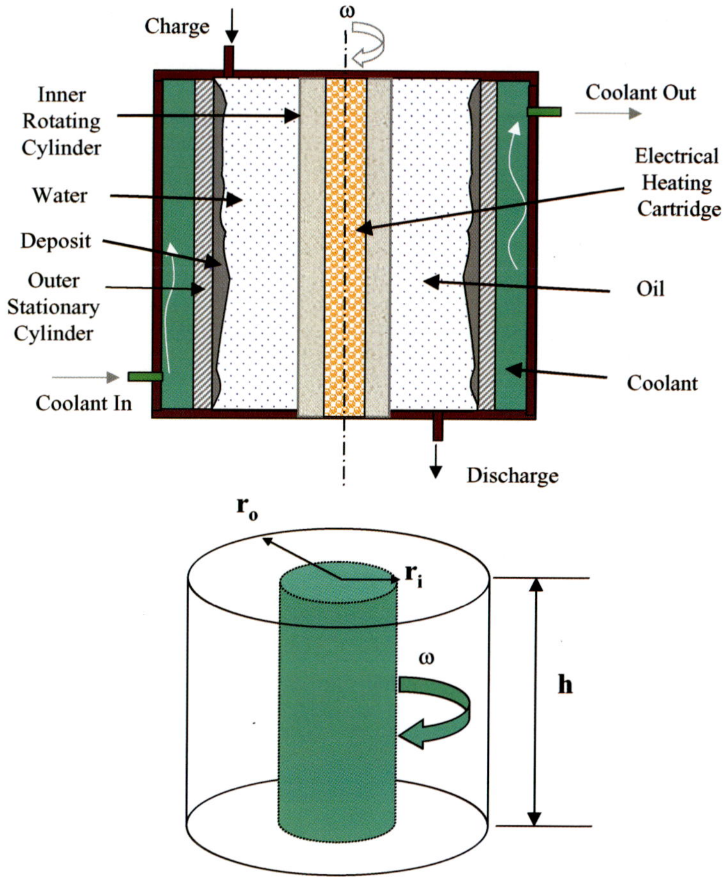

Figure 23.30. Schematic representation of the shear deposition cell based on the Couette–Tylor flow.[65]

are large and polar or polarizable molecules that tend to self-associate, arguments persist to the state of asphaltenes in solution: resin-stabilized colloids or solution-solubilized aggregates. The complexity of the system and disagreement on the stabilized mechanism has led to a large variety of models. The following is a brief overview of most of the major classes of asphaltene models. It should not be considered an exhaustive survey. Many more researchers have contributed to this area than are mentioned below.

Asphaltene precipitation/aggregation models can be broadly classified into two types: reversible and irreversible. Reversible models are based on equilibrium thermodynamics and constitute the largest class of models. Irreversible models are typically based on the resin-stabilized colloidal stabilization mechanism. All models can qualitatively describe the destabilization of asphaltenes through depressurization or compositional changes. However, only models that contain some element of reversibility can capture the redissolution of asphaltenes observed

Figure 23.31. Shear deposition cell wax deposit.[67]

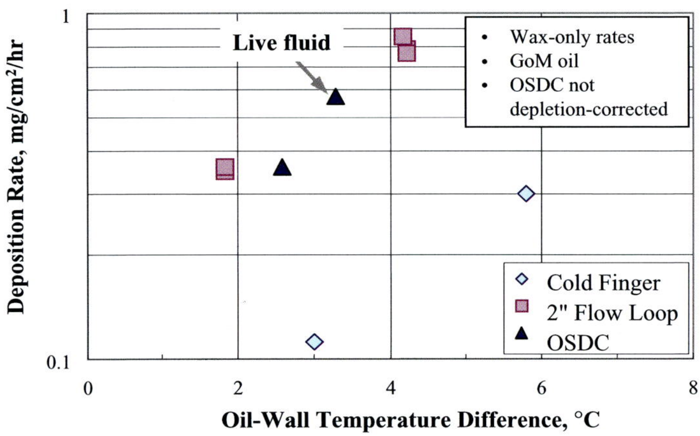

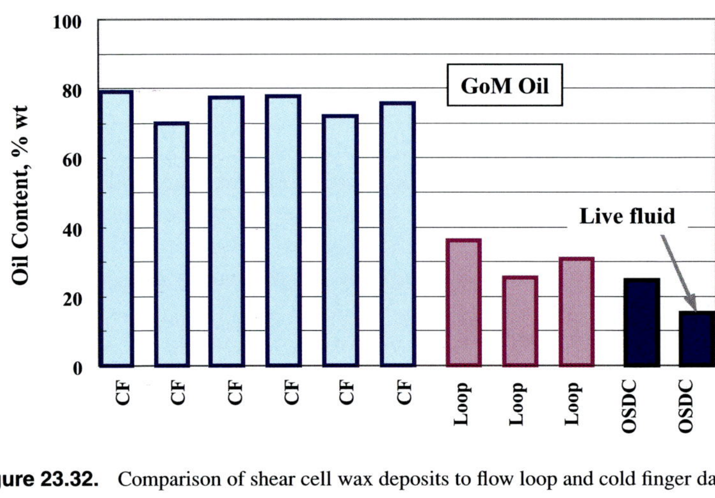

Figure 23.32. Comparison of shear cell wax deposits to flow loop and cold finger data.[67]

Figure 23.33. Shear deposition cell asphaltene deposit.[68]

in many systems. Models based strictly on equilibrium thermodynamics will always show complete redissolution if the system is returned to its initial single phase conditions (pressure, temperature, and composition). Experimentally, this is not always observed as slow dissolution kinetics may prevent reaching that condition in laboratory experiments. Because of the difficulty in characterizing asphaltene fraction, the predictive capability of most models is poor. As a result, the models are tuned to dead and live oil titration data and/or live oil depressurization data.

The simplest class of reversible models is based on regular solution theory. These models are strictly solution models in which nonpolar dispersion forces dominate. They do not address resin stabilization. Buckley[69] extended the original work by Hirshberg *et al.*[41] In Buckley's model, solubility parameters are related to indices of refraction. The use of critical solvent index of refraction to predict asphaltene stability in this model has been reasonably successful.

The second class of reversible models is based on cubic equations of state (EOS). In the first and simplest cubic EOS model, asphaltenes are modeled as either single or multiple heavy pseudocomponents. The critical properties and binary interaction parameters of these components are adjusted to cause a liquid–liquid phase split at the observed asphaltene precipitation point.[70] The resulting lower liquid or asphaltene phase contains all the oil components as well as a substantial amount of gas components. While some high gas content oils can exhibit this type of liquid–liquid phase split, these results are not consistent with typical asphaltene precipitation. To overcome this problem, solid phase fugacity models have been used in conjunction with cubic equations of state. Crystalline solid models similar to those used for waxes have been applied to asphaltene precipitation;[71,72] however, it is counterintuitive to define melting temperatures and heats of fusion for "amorphous" asphaltene solids.

More complex equations of state have been proposed to incorporate the effect of polar interactions and asphaltene self-association. In cubic plus association (CPA) models, an association term that accounts for polar interactions is added to the standard cubic formulation.[73] The statistical association fluid theory (SAFT) model has also been used to incorporate the effect of asphaltene self-association and polar interaction.[74,75] These models provide more flexibility in matching complex temperature-dependent behavior but do require additional parameters to characterize association energies. Firoozabadi and Pan[76] proposed a micellization model based on free energy minimization. This model explicitly addresses a resin-stabilized colloidal structure as in the case of association models.

Irreversible models assume asphaltenes exist as particles or colloidal suspensions stabilized by resins on the asphaltene surface. As the chemical potential of the resin in solution changes, the fractional coverage of the asphaltene surface changes. Beyond a certain critical value, there is an insufficient amount of resins on the asphaltenes surface to prevent agglomeration of the particles. The agglomeration process is assumed to be irreversible.[77,78] Mansoori and Leontaritis[78,79] also have proposed an irreversible fractal aggregation model. These models are, however, fairly outdated and less popular for they have been developed well before the recent advances made in solid phase behavior measurement techniques. These have produced overwhelming experimental evidence of the reversibility of asphaltene precipitation; this finding is quite consistent with field observations as discussed in the previous sections.

Acknowledgment

The authors would like to thank the engineers and technologists at Schlumberger Reservoir Fluids Center (formerly known as DBR) and Oilphase-DBR Product Centers & Laboratories for their work on fluids sampling, analyses and modeling. The authors also acknowledge their peers and colleagues at Shell Global Solutions (US) Inc. and Chevron for their valuable insights on the shear deposition cell section.

References

[1] Hammami, A., C. H. Phelps, T. Monger-McClure, and T. M. Little (2000). *Energy Fuels* **14**(1), 14–18.

[2] Ratulowski, J., A. Amin, A. Hammami, M. Muhammad, and M. Riding (2004). Flow assurance and sub-sea productivity: Closing the loop with connectivity and measurements. In: *2004 SPE Annual Technical Conference and Exhibition*, Houston, TX, Sept 26–29, 2004, SPE 90244.

[3] Ratulowski, J. and A. Hammami (1999). Planning for organic solids deposition in offshore systems. In: *3rd International Symposium on Colloid Chemistry in Oil Production. Asphaltene and Wax Deposition (ISCOP)*, Huatelco, Mexico, Nov. 14–17, 1999.

[4] Ellison, B.T., C.T. Gallagher, L.M. Frostman, and S.E. Lorimer (2000). The physical chemistry of wax, hydrates and asphaltene. In: *Offshore Technology Conference*, Houston, TX, May 1–4, 2000, OTC 11963.

[5] Stankiewicz, A.B., M.D. Flannery, N.A. Fuex, G. Broze, J.L. Couch, S.T. Dubey, S.D. Iyer, J. Ratulowski, and J.T. Westerich (2002). Prediction of asphaltene deposition risk in E&P operations. In: *Proceeding of 3rd International Symposium on Mechanisms and Mitigation of Fouling in Petroleum and Natural Gas Production, AIChE 2002 Spring National Meeting*, New Orleans, USA, March 10–14, paper 47C, pp. 410–416.

[6] Gruse, W.A. and D.R. Stevens (1960). *The Chemical Technology of Petroleum.* McGraw-Hill, New York.

[7] Draper, R.G., E. Kowalchuk, and G. Noel (1977). Analyses and characteristics of crude oil samples performed between 1969 and 1976. Report EPR/ERL 77-59, Energy, Mines, and Resources, Canada.

[8] Creek, J.L. (2005). Freedom of action in the state of asphaltenes: Escape from conventional wisdom. *Energy Fuels* **19**(4), 1212–1224.

[9] Hobson, G.D. and W. Pohl (1973). *Modern Petroleum Technology*, 4th edn. Halsted Press/John Wiley & Sons, New York.

[10] Speight, J.G. (1991). *The Chemistry and Technology of Petroleum*, 2nd edn. Marcel Dekkar Inc., New York.

[11] McCain, W.D. Jr. (1990). *The Properties of Petroleum Fluids*, 2nd edn. PennWell Publishing Co., Oklahoma.

[12] Cruse, W.A. and D.R. Stevens (1960). In: *Chemical Technology of Petroleum.* McGraw-Hill Book Publishers Inc., New York City, Chap. XXII.

[13] Lira-Galeana, C. and A. Hammami (2000). In: T.F. Yen and G. Chilingarian (eds.), *Asphaltenes and Asphalts.* Elsevier Science Publishers, Amsterdam., Chap. 21.

[14] Warth, A.H. (1956). In: *The Chemistry and Technology of Waxes*, 2nd edn. Reinhold, New York.

[15] Starusz, O.P., T.W. Mojelski, and E.M. Lown (1992). *Fuel* **71**, 1355–1364.

[16] Espinat, D. (1993). In: *SPE International Symposium on Oilfield Chemistry*, New Orleans, LA, March 2–5, 1993, SPE 25187.

[17] Speight, J.G. and R.G. Long (1996). *Fuel Sci. Technol. Int.* **14**(1/2), 1–12.

[18] Groenzin, H. and O.C. Mullins (1999). Asphaltene molecular size and structure. *J. Phys. Chem.* A **103**, 11237.

[19] Groenzin, H., O.C. Mullins, S. Eser, J. Mathews, M.-G. Yang, and D. Jones (2003). Asphaltene molecular size for solubility subfractions obtained by fluorescence depolarization. *Energy Fuel* **17**, 498.

[20] Murgich, J. (2005). Molecular mechanics study of the selectivity in the interaction between some typical resins and asphaltenes. In: *6th International Conference on Petroleum Phase Behavior and Fouling*, Amsterdam, the Netherlands, June 19–23, 2005.

[21] Rakotondradany, F., H. Fenniri, P. Rahimi, K.L. Gawrys, P.K. Kilpatrick, and M. R. Gray (2005). Synthesis and properties of model compounds for bitumen residue. In: *6th International Conference on Petroleum Phase Behavior and Fouling*, Amsterdam, the Netherlands, June 19–23, 2005.

[22] Hammami, A., J. Ratulowski, and J.A.P. Coutinho (2003). Cloud points: Can we measure or model them? *Pet. Sci. Technol.* **21**, 345–358.

[23] Leontaritis, K.J., and G.A. Mansoori (1987). Asphaltene flocculation during oil production and processing: A thermodynamic-colloidal model. In: *SPE Symposium on Oilfield Chemistry*, San Antonio TX, Feb. 4–6, 1987, paper SPE 16258.

[24] Ferworn, K.A. (1995). Thermodynamic and kinetic modelling of asphaltene precipitation from heavy oils and bitumens. Ph.D. dissertation, University of Calgary.

[25] Hammami, A., D. Chang-Yen, J.A. Nighswander, and E. Stange (1995). An experimental study of the effect of paraffinic solvents on the onset and bulk precipitation of asphaltenes. *Fuel Sci. Technol. Int.* **13**(9), 1167–1184.

[26] Hammami, A., K.A. Ferworn, J.A. Nighswander, S. Overå, and E. Stange (1998). Asphaltenic crude oil characterization: An experimental investigation of the effect of resins on the stability of asphaltenes. *Pet. Sci. Technol.*, **16**(3&4), 227–249.

[27] Haskett, C.E. and M. Tartera (1965). A practical solution to the problem of asphaltene deposits—Hassi Messaoud Field, Algeria. *J. Pet. Technol.*, April 387–391.
[28] de Boer, R.B., K. Leerlooyer, M.R.P. Eigner, and A.R.D. van Bergen (1995). Screening of crude oils for asphalt precipitation: Theory, practice and the selection of inhibitors. *SPE Prod. Facil.*, Feb., 55–61.
[29] Kokal, S.L. and S.G. Sayegh (1995). Asphaltenes: The cholesterol of petroleum. In: *SPE Middle East Oil Show*, Bahrain, March 11–14, 1995, paper SPE 29787.
[30] Fuhr, B.J., C. Cathrea, L. Coates, H. Kalra, and A.I. Majeed (1991). *Fuel* **70**, 1293–1297.
[31] Andersen, S.I. (1994). *Fuel Sci. Technol. Int.* **12**(1), 51–74.
[32] Fotland, P. (1996). *Fuel Sci. Technol. Int.* **14**(1/2), 313–325.
[33] Wang, J.X., K.R. Brower, and J.S. Buckley (1999). In: *SPE International Symposium on Oilfield Chemistry*, Houston, TX, Feb. 16–19, 1999, SPE 50745.
[34] Jamaluddin, A.K.M., T.W. Nazarko, S. Sills, and B.J. Fuhr (1996). Deasphalted oil: A natural asphaltene solvent. *SPE J. Prod. Facil.* **11**(3), 161–165.
[35] Karan, K., A. Hammami, M. Flannery, and A. Stankiewicz (2003). Systematic evaluation of asphaltene instability and control during production of live oils: A flow assurance study. *Pet. Sci. Technol.* **21**, 629–645.
[36] Joshi, N.B., O.C. Mullins, A. Jamaluddin, J. Creek, and J. McFadden (2001). Asphaltene precipitation from live crude oils. *Energy Fuels* **15**, 979.
[37] Joshi, N.B., M. Muhammad, J. Creek, and J. McFadden (2003). Flow assurance: A challenging path to well completions and productivity. In: *Offshore Technology Conference*, Houston, TX, May 5–8, 2003, OTC 15185.
[38] Ratulowski, J., A. Hammami, M. Muhammad, and J. Creek (1999). Considerations for determination of asphaltene stability and control—Effect of oil bases drilling fluid on the asphaltene precipitation onset pressure of live reservoir fluids. In. *IBC Forum Controlling Hydrates Waxes and Asphaltenes*, Aberdeen, UK, Oct. 28–29, 1999.
[39] Pfieffer, J.P. and R.N.J. Saal (1940). *J. Phys. Chem.* **44**, 139–149.
[40] Hotier, G. and M. Robin (1983). *Rev. L'institut Fr. Pét.* **38**, 101–120.
[41] Hirshberg, A., L.N.J. deJong, B.A. Schipper, and J.G. Meijer (1984). Influence of temperature and pressure on asphaltene flocculation. *SPE J.*, June, 283–293.
[42] Gozalpour, F., A. Danesh, D.-H. Tehrani, A.C. Todd, and B. Tohidi (1999). Predicting reservoir fluid phase and volumetric behaviour from samples contaminated with oil-based mud. *Society of Petroleum Engineers* # 56747, Houston, TX.
[43] Jamaluddin, A.K.M., B. Ross, D. Calder, J. Brown, and M. Hashem (2002). Single-phase bottomhole sampling technology. *JCPT* **41**(7), 25–30.
[44] Muhammad, M., M. McFadden, and J. Creek (2003). Asphaltene precipitation from reservoir fluids: Asphaltene solubility and particle size vs. pressure. *SPE International Symposium on Oilfield Chemistry*, Houston, TX, Feb. 5–8, 2003, SPE 80263.
[45] Muhammad, M. *et al.* (2004). Effect of oil based mud on live fluid asphaltene precipitation pressure. In: *5th International Conference on Petroleum Phase Behavior & Fouling*, Banff, Alberta, Canada, June 13–17, 2004.
[46] Ratulowski, F.A., J.T. Westrich, and J.J. Sieler (2003). Theoretical and experimental investigation of isothermal compositional grading. In: *SPEREE*, June, 2003, SPE 84777.
[47] Hammami, A. and M.A. Raines (1999). Paraffin deposition from crude oils: Comparison of laboratory results to field data. *SPE J.* **4**, 1–10.
[48] Andrews, R.J., G. Beck, K. Castelijns, A. Chen, M.E. Cribbs, F.H. Fadness, J. Irvine-Fortescue, S. Williams, M. Hashem, A. Jamaluddin, A. Kurkjian, B. Sass, O.C. Mullins, E. Rylander, and A. Van Dusen (2001). Quantifying contamination using color of crude and condensate. *Oilfield Rev.* Autumn, 24–43.
[49] Hashem, M.N., E.C. Thomas, R.I. McNeil, and O.C. Mullins (1997). Determination of producible hydrocarbon type and oil quality in wells drilled with synthetic oil-based muds. Society of Petroleum Engineers #39093, San Antonio, TX.
[50] Fuhr, B.J., L.R. Holloway, and A. Hammami (1999). Analytical considerations related to asphaltenes and waxes in the same crudes. *Energy Fuels* **13**, 336–339.

[51] Kaminski, T., H.S. Fogler, N. Wolf, P. Wattana, and A. Mairal (2004). Classification of asphlatenes via fractionation and the effect of heteroatom content on dissolution kinetics. *Energy Fuel* **14**, 25–30.

[52] Asomaning, S. and A.P. Watkinson (1998). Petroleum stability and heteroatom species effects in fouling of heat exchangers by asphaltenes. In: *AIChE Spring National Meeting* 1998, Preprint.

[53] Buckley, J.S. (1996). Microscopic investigation of the onset of asphaltene precipitation. *Fuel Sci. Technol. Int.* **14**, 55–74.

[54] Ferworn, K.A., A.K. Mehrotra, and W.Y. Svrcek (1993). Measurement of asphaltene agglomeration from Cold Lake bitumen diluted with *n*-alkanes. *Can. J. Chem. Eng.* **71**, 699–703.

[55] Wang, J. (2000). Predicting asphaltene flocculation in crude oils. Ph.D. dissertation. New Mexico Institute of Mining and Technology, Socorro, NM.

[56] Anderson, S.I. (2003). The state of asphaltenes in the region we called CMC! In: *4th International Conference on Petroleum Phase Behavior and Fouling*, Trondheim, Norway, June 2003.

[57] Long, Y., T. Dabros, H. Hamza, and W.J. Power (2001). Continued slow precipitation of asphaltene from aliphatic solvent-diluted bitumen solutions. In: *Symposium on Stability and Compatibility during Production, Transportation and Refining of Petroleum*, Calgary, Alberta, Canada, Sept. 17–18, 2001.

[58] Wang, J. and J.S. Buckley (2003). Asphaltene stability in crude oil and aromatic solvent—The influence of oil composition. *Energy Fuels* **17**(6), 1445–1451.

[59] Buckley, J.S., G.H. Hirasaki, Y. Liu, V. Drasek, J. Wang, and B.S. Gill (1998). Asphaltene precipitation and solvent properties of crude oils. *Pet. Sci. Technol.* **16**(3&4), 251–285.

[60] Karan, K., J. Ratulowski, and P. German (2000). Measurement of waxy crude properties using novel laboratory techniques. In: *Annual Technical Conference of Society of Petroleum Engineers*, Dallas, TX, Oct. 1–4, 2000, SPE paper 62945.

[61] Allenson, S.J., A. Hammami, H. Maeda, and K. Ohno (1999). Control of asphaltene deposition: Laboratory screening and evaluation of asphaltene inhibitors. In: *3rd International Symposium on Colloid Chemistry in Oil Production. Asphaltene & Wax Deposition* (ISCOP'99), Huatelco, Mexico, Nov. 14–17, 1999.

[62] Sivaraman, A., Y.F. Hu, F.B. Thomas, D.B. Bennion, A.K.M. Jamaluddin, and Y.F. Hu (1997). Acoustic resonance: An emerging technology to identify wax and asphaltene precipitation onset conditions in reservoir fluids. In: *48th Annual Technical Meeting of the Petroleum Society of CIM*, Calgary, Alberta, Canada, June 8–11, 1997.

[63] Kabir, C.S. and A.K.M. Jamaluddin (2002). Asphaltene characterization and mitigation in south Kuwait's Marrat reservoir. *SPE J.* **17**(4), 251–258.

[64] Aquino, O., B.E. Gonzales, S.I. Andersen, and C.L. Gaeana (2001). Investigation of inhibition of asphaltene pressure using bottomhole samples. *Energy Fuels* **15**, 236–240.

[65] Zougari, M., S. Jacobs, A. Hammami, G. Broze, M. Flannery, J. Ratulowski, and A. Stankiewicz (2006). Novel organic solid deposition and control device for live oils: Design and application. Accepted for publication in *Energy & Fuels*.

[66] Zougari, M., S. Jacobs, A. Hammami, G. Broze, and N. Fuex (2005). Live oils novel organic solid deposition and control device: Wax deposition validation. In: *14th SPIE Middle East Oil & Gas Show and Conference*, Bahrain, March 12–15, 2005, SPE 93558.

[67] Broze, G., P. Cornelisse, M. Zougari, S. Jacobs, A. Hammami, and J. Ratulowski (2003). A comparison of wax deposition under live and stock-tank conditions. In: *4th International Conference on Petroleum Phase Behavior and Fouling*, Trondheim, Norway, June 23–26, 2003.

[68] Flannery, M., P. Cornelisse, M. Zougari, and A. Hammami (2003). Asphaltene precipitation and deposition under live oil conditions in a novel experimental device. In: *4th International Conference on Petroleum Phase Behavior and Fouling*, Trondheim, Norway, June 23–26, 2003.

[69] Buckley, J.S. (1999). Predicting the onset of asphaltene precipitation from refractive index measurements. *Energy Fuels* **13**, 328–332.

[70] Rydhal, A.K., K.S. Pedersen, and H.P. Hjermstad (1997). In: *AICHE Spring National Meeting*, Houston, TX, March 1997.

[71] MacMillan, D.J., J.E. Tackett Jr., M.A. Jessee, and T.G. Monger-McClure (1995). A unified approach to asphaltene precipitation: Laboratory measurement and modeling. In: *SPE International Symposium on Oilfield Chemistry*, San Antonio, TX, Feb. 14–17, 1995, SPE paper 28990.
[72] Chung, T-H. Thermodynamic modeling for organic solid precipitation. In: *SPE Annual Technical Conference and Exhibition*, Washington, D.C. Oct. 4–7, 1992, SPE 24851.
[73] Edmonds, B., R.A.S. Moorwood, R. Szczepanski, X. Zhang, M. Heyward, and R. Hurle (1999). Measurement and prediction of asphaltene precipitation from live oils. In: *3rd International Symposium on Colloid Chemistry in Oil Production, Asphaltenes and Waxes Deposition* (ISCOP'99), Mexico, Nov. 1999.
[74] Wu, J., J.M. Prausnitz, and A. Firoozabadi (1998). *AIChE J.* **44**, 1188.
[75] Ting P.D., G.J. Hirasaki, and W.G. Chapman (2003). Modeling of asphaltene phase behavior with the SAFT equation of state. *Pet. Sci. Technol.* **21**, 647–661.
[76] Pan, H. and A. Firoozabadi. Thermodynamic micellization model for asphaltene precipitation from reservoir crudes at high pressures and temperatures. In: *SPE Annual Technical Conference and Exhibition*, Oct. 5–8, 1997, San Antonio, TX, SPE 38857.
[77] Takhar, S., P.D. Ravenscroft, and D.C.A. Nicholl (1995). Prediction of asphaltene deposition during production—Model description and experimental details. In: *European Formation Damage Conference*, The Hague, Netherlands, May 15–16, 1995, SPE 30108.
[78] Mansoori, G.A. (1994). The occurrance of asphaltene throughout production cycle: A comprehensive description of mechanism and factors influencing heavy organic deposition. In: *6th International Petroleum Exhibition and Conference*, Abu Dhabi, Oct. 16–19, 1994, paper ADSPE # 2. pp. 282–292.
[79] Leontaritis, K.J. and G.A. Mansoori (1987). Asphaltene flocculation during Oil production and processing: A thermodynamic-colloidal model. In: *SPE International Symposium on Oilfield Chemistry*, San Antonio, TX, Feb. 4–6, 1987, paper SPE 16258.

Index

Printed in Singapore